PHYSICAL SCIENCE

Judith River country, Montana, near Missouri River confluence.
(Courtesy David Muench)

PHYSICAL SCIENCE

Jerry S. Faughn, Ph.D.
Department of Physics
and Astronomy,
Eastern Kentucky University

Jonathan Turk, Ph.D.
Darby, Montana

Amos Turk, Ph.D.
Department of Chemistry,
City College of the
City University of New York

SAUNDERS GOLDEN SUNBURST SERIES

Saunders College Publishing
Harcourt Brace Jovanovich College Publishers
Fort Worth   Philadelphia   San Diego
New York   Orlando   Austin   San Antonio
Toronto   Montreal   London   Sydney   Tokyo

Text typeface: Baskerville
Compositor: York Graphic Services
Acquisitions Editor: John Vondeling
Associate Editor: Kate Pachuta
Managing Editor: Carol Field
Project Editor: Marc Sherman
Manager of Art and Design: Carol Bleistine
Art and Design Coordinator: Doris Bruey
Text Designer and Layout Artist: Ed Butler
Cover Designer: Lawrence R. Didona
Director of EDP: Tim Frelick
Production Manager: Bob Butler
Marketing Manager: Marjorie Waldron

Cover Credit: © 1991 THE IMAGE BANK/DaVinci Man and Grid in Space by Francesco Reginato

PHYSICAL SCIENCE

0-03-035353-X

Library of Congress Catalog Card Number: 90-053085

Printed in the United States of America

23   032   9 8 7 6 5 4 3

PREFACE

This text is an introduction to the elements of physics, chemistry, geology, meteorology, and astronomy for the nonscience major. Obviously these subjects cannot be treated in a comprehensive manner in a single textbook or in a single course. Therefore, authors and instructors alike are faced with various pedagogic choices about the strategy of the course—what to include, what to emphasize, or what to touch on lightly. We believe that the overall approach to a study of physical science should involve a central strategy based on *principles*.

The requirement for a foundation in basics imposes some necessary sequences. Motion cannot be taught without introducing Newton's laws; chemistry necessarily depends on an understanding of atomic structure; and any section on geology must include a discussion of plate tectonics. The integration between basics and examples is the fundamental pedagogic strategy of this book. The following features are incorporated into the text to aid in the integration of theory and practice.

FEATURES

● *Color* is used to add substance to the presentations. The following overall scheme for color renderings should be noted:

Axes of graphs and displacement vectors

Velocity vectors

Force vectors

Rotation and revolution motion indicators

Electric forces and fields

Magnetic forces and fields

Positive charges

Negative charges

Resistors, batteries, and switches

Ground symbol

Light rays

Lenses

Mirrors

Objects

Images

Most graphs are plotted with the curves in red and the axes in black.

In addition to the use of colors in the figures, various sections and features of the book are color coded for ease of identification and location:

Important equations

$$v = (9.8 \text{ m/s}^2) \text{ time}$$

Examples

Climbing higher and higher

Important statements

The orbit of a planet is an ellipse with the Sun at one focus. The other focus is empty.

- *Environmental and practical applications* are integrated throughout the text to broaden the students' understanding of theory, to retain their interest, and to relate their learning to the world we live in.

- *Flexibility* is built into the text to accommodate the instructor's needs in course length and rigor. Some courses are completed in one semester, others in two. Naturally, the abilities and backgrounds of different groups of students differ as well.

- *The use of mathematics is held to a minimum.* It is our belief that a beginning student must understand the basic concepts of science, which some simple calculations can clarify. Where mathematics is used, it is in the form of simple equations such as Ohm's law. Students need to know only basic arithmetic and simple algebra to do any of the computations.

- *Example problems* are incorporated into the body of the text, where mathematical manipulations are required, to show the student how to approach similar problems. For ease of location, these examples are set off from the body of the text with a colored bar, and most examples are given titles to describe their content.

- *Special focus sections* expose the student to various practical and interesting applications of physical principles. The focus sections are supple-

mental—in the sense that studying them is not required in order to continue through the book.

- *Writing style* is considered an aid for rapid comprehension. We have taken great care to ensure that the concepts of physical science are written not only correctly but in such a manner that the student can read the explanations with understanding. We hope to have developed a text in which concepts sometimes considered formidable—such as relativity, or chemical bonding—no longer seem so difficult, and in which learning science becomes an enjoyable undertaking.

- *Interviews* opening each section of the text—physics, chemistry, geology and astronomy—are intended to inspire and pique student interest. They provide an opportunity for students to learn how leading researchers started in their chosen fields, and enable those who have made great contributions to science to communicate what they find exciting in their areas of study.

ANCILLARIES

The following ancillaries are available with this text to assist the instructor teaching the course and to help the student learn the material covered in the course.

Instructor's Manual/Test Bank contains teaching suggestions and answers to selected problems in the text. Tests supply multiple-choice questions from the software disk. It is provided as another source of test questions and is helpful for the instructor who does not have access to a computer.

Computerized Test Bank is available for the IBM PC® and Macintosh computers and enables the instructor to create many unique tests. The system also permits the editing of questions as well as addition of new questions.

Overhead Transparencies are 100 full color figures selected from the text and printed on acetates for viewing in the classroom.

Laboratory Manual for Physical Science offers 30 experiments designed to supplement the learning of basic principles of the course. The laboratory manual features pre-lab assignments to increase student preparation. Every effort has been made to keep laboratory equipment costs to a minimum.

Instructor's Manual for Laboratory Manual contains a discussion of the experiment and teaching hints for the instructor.

Student's Study Guide contains chapter objectives and summaries, focus reviews of concepts stressed in text, and self-tests for each chapter.

ACKNOWLEDGMENTS

The reviewers selected for this text were both careful and incisive with their comments. We thank them all most sincerely. They are:

Robert Backes, Pittsburg State University
Basil Curnette, Kansas State University
Stewart Farrar, Eastern Kentucky University
Fred Gamble, Pensacola Junior College
Joe Greever, Delta State University
Ted Morishige, Central State University
Douglas Magnus, St. Cloud State University
Donald E. Rickard, Arkansas Tech University
Oswald Schuette, University of South Carolina
Harry Shipman, University of Delaware
Ralph Thompson, Eastern Kentucky University
Aaron Todd, Middle Tennessee State University
Maurice Witten, Fort Hays State University

Laura Faughn did an excellent job in typing much of the manuscript. Finally, we owe a debt of gratitude to our friends at Saunders College Publishing who worked diligently in our behalf: John Vondeling, Associate Publisher; Kate Pachuta, Associate Editor; Marc Sherman, Project Editor; and Carol Bleistine, Art Director.

▶ TO THE STUDENT

Many valuable facts about this textbook are presented in the preceding section. Pay particular attention to the key on page v for using color in the illustrations, since the use of color in this text has many functional purposes.

STUDY HINTS

One of the most commonly asked questions of any instructor is, "How do I study for a science class?" The answer is basically the same way that you do for any other class: Keep up on reading and doing assignments on a day-to-day basis rather than trying to cram the night before an exam. Set a study schedule for yourself and keep to it. There is plenty of time in your college or university life for outside activities and just plain fun. But, remember you should leave college knowing much more about the world in which you live than you did when you entered. The information discussed in science classes forms an important part of our understanding of how the world works. If approached with a positive attitude, you will find the task of understanding science can even be fun.

It is, of course, vital that you attend class regularly and pay attention. Ask questions, both inside and outside of class, whenever you do not understand a topic being covered. Take concise notes in class and review them regularly, filling in with any additional material necessary to make the information easier to comprehend and retain. Science, perhaps more than most other courses that you will take, requires that you understand *concepts*. Memorizing every detail of Newton's laws of motion will not help a great deal if you cannot see the "big picture" of their meanings and applications. To this end, we have presented many worked-out examples to aid you in your understanding.

Consider the homework problems as a way to test your new skills. Do not lose patience if you cannot work every problem assigned or if you cannot derive a complete answer for every question asked. The successful student will *try* to answer every question. If you have already made a diligent effort to work these homework problems, you will get much more out of an instructor's review of these problems in class.

The level of mathematics used in the text has been held to simple manipulation of a few algebraic equations. If you do not know how to do these already, some useful refreshers on math techniques can be found in the appendices of this text. If you perceive yourself as having a weakness in mathematics, there is no better time than the present to correct it. Your college career is not just for you to excel in what you can already do; it is also a time for you to correct deficiencies.

THE LABORATORY

Many of you will have a laboratory associated with this course, because science is based largely on experimental observations. You will find the lab helpful to you in placing the concepts studied in the textbook on a more concrete foundation, As you approach your lab, remember that the approach taken toward the development of a scientific theory is often called the *scientific method*.

THE SCIENTIFIC METHOD

The scientific method is common to all the sciences, and you should look for it in your experimental investigations. The five steps of the scientific method are often stated as:

1. Recognize the problem.
2. Hypothesize an answer.
3. Predict a result based on your hypothesis.
4. Devise and perform an experiment to check the hypothesis.
5. Develop a theory that links the hypothesis to previously existing knowledge.

These steps are the framework of every experiment you do in the lab, even though they may not be spelled out in this step-by-step order. Put yourself in the position of a great scientist of the past or, after you read the interviews opening each section of this text, a great scientist of the present and see show your investigations in the laboratory can improve your understanding of the day-to-day world.

CONCLUSION

Throughout your life you will make many decisions that will rely on your understanding of science. This textbook provides the fundamental principles which can aid you in making informed decisions. If you do not prepare yourself for such decisions, you will have to rely on the goodwill of others. An understanding of science is valuable to you and to the world. We have made every effort to make science understandable and interesting for you. Now you must do your part and become an active participant in the learning process.

J. S. Faughn
J. Turk
A. Turk

August 1990

CONTENTS OVERVIEW

The circular motion of the atmosphere is evident in this cyclonic storm north of Hawaii. (Courtesy NASA)

CONTENTS

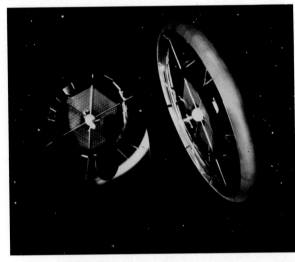

PART

PHYSICS

ONE

ANTHONY P. FRENCH

• INTERVIEW •

Anthony French graduated with a degree in physics from Cambridge University in 1942. In 1944 he went to the United States as a member of the British Mission at Los Alamos, New Mexico, to work on the atomic bomb. He returned to England in 1946 and continued his research in low-energy nuclear physics. In 1955 he emigrated to the United States to become a professor of physics at the University of South Carolina, and soon after became head of the physics department there. During this time his interest in basic physics education grew. He was invited to accept a visiting appointment at MIT to work on a physics curriculum development program. In 1964 he became a tenured professor on the MIT faculty and has worked there ever since.

Some of the many honors he has received for services to physics education include the University Medal of the Charles University of Prague, and the Bragg Medal of the Institute of Physics in London.

When did you first become interested in physics?

I always enjoyed science and spent lots and lots of time in the local library checking out many books, especially books dealing with the physical sciences, astronomy, and the like. When I was 14 or 15, the English educational system required me to take an examination which was overseen by the major British universities. On the strength of that, I had to choose whether I was going to go into science or into humanities. So, from age 15 and on, I was studying nothing but physics, chemistry, and mathematics.

What did you think you were going to do?

Actually I had no thought of going to college, but I had a wonderful mathematics teacher who told me that I ought to try for a science scholarship at Cambridge University. I was lucky enough to win a couple of scholarships that paid for my whole education there.

Tell us where your work in physics at Cambridge led you.

I went to Cambridge in 1939 as an undergraduate just after the war had started in Europe. In fact, I never expected to go there, having got these scholarships, and then the war started and I thought "Well, that's it, we'll all be called up [to go to war]." Then, during my second year, C.P. Snow (he was a Cambridge fellow) came up to Cambridge for several days and interviewed all of the science undergraduates to decide what would happen to them. (He later became famous for his book

Dr. French in the lab at the University of South Carolina, 1955.

meant that anyone could work on it—it was just a matter of who got there first.

But when I got to Los Alamos, I found myself assigned to a small group working on thermonuclear reactions for a bomb—before the fission bomb itself had ever been proved to be possible. Our problem was to measure the properties of the reactions of deuterium with deuterium, and deuterium with tritium, as the obvious candidates for making a thermonuclear explosion. For this purpose, we had to do our measurements with the first cubic centimeter of tritium gas that was ever produced. Edward Teller, as a matter of fact, was the chief theorist connected with the group and he was the one who had been pushing for this thermonuclear reaction, this Super Bomb.

Los Alamos was a gathering of all the stars of nuclear physics, including many of the people who had been thrown out of, or fled from, Germany. They weren't that old either; they were maybe 10 or 15 years older than I. When I look back, I think it was astonishing. You know Robert Oppenheimer, the director of Los Alamos, was, I believe, 38. The head of the theoretical division was almost exactly the same age. They were all relatively young. It is amazing that such a big project was entrusted to such young people.

The Two Cultures about science and humanities.) It was decided that I would finish out three years and earn my degree.

As the end of the third year was approaching, a friend and I went to visit Sir Lawrence Bragg, who was head of the Cambridge Physics Department and a very famous X-ray crystallographer. He and his father had shared the Nobel Prize for physics in 1915 for their discovery of using X-rays to analyze crystal structures. We asked him if there was any possibility of getting a job at Cambridge after we got our degrees. He didn't make any commitments, but the man who was tutoring me in physics was a nuclear physicist who (unbeknownst to me at the time) was working on the British atomic bomb project. He took me on, so as soon as I graduated, I went into his lab to work on that research. I was working in a group measuring fission in the various isotopes that were candidates for use in bombs.

In 1944 you came to the United States and worked at Los Alamos. You were actually there at the time the atomic bomb was made. What was it like to be working there during this time?

The sheer excitement of the project itself was quite amazing. The people working on this project felt that they were definitely in a competition with Nazi Germany. Nobody knew what was happening over there, and the discovery of fission

After the war was over, you went back to Cambridge to teach. Was it during this time that your interest in *how* physics was taught started?

I think the idea of putting most of my time into physics education came later, after I came back to the States, in fact. I knew from a very early stage that I enjoyed the teaching side, including the tutoring on a one-to-one basis, and as a lecturer at Cambridge, of course, I was creating my own courses as I went along. You didn't usually just work from a prescribed textbook. But I was also devoting a lot of time to nuclear research. Then I went to the University of South Carolina. I had visited there and sensed that they were trying to build up a good quality physics department from scratch and I felt that I could make a difference by being there, whereas in an established place like Cambridge, it didn't matter so much whether you were there or not. So it was a kind of challenge. I joined the faculty there, and shortly afterward the department head died suddenly, and I became head of the department. I stopped working on nuclear physics at that time and was involved entirely with teaching and administration.

You were not the only one who felt an improvement was needed in science education. Did anyone have any ideas how to go about it?

There was a fellow called Jerrold Zacharias at MIT. In 1956—just before the Soviets launched Sputnik, he was always proud of pointing that out—he decided that something desperately needed to be done about science education in the United States.

He wrote to the president of MIT suggesting that there ought to be some major attempt to upgrade the teaching of physics in secondary schools. So, he involved a lot of college and university people and high school teachers in a joint project to

create a completely new physics course. This course (PSSC, for Physical Science Study Committee) was completed in about 1959, and has been continued as a published book (*PSSC Physics*) ever since, going through various editions. It was a major step forward in the quality of physics that was being offered for secondary school use. In order to prepare teachers for teaching this course, there was a number of special institutes, summer institutes and academic year institutes. I ran an academic year institute in South Carolina for teachers who lived in the state. Soon after that, I was invited to spend a semester at MIT because after the high school course had been finished, there was an interest in creating college level courses that would match.

How did the proposed courses differ? Were there drastic changes?

Compared to what MIT and many other places were doing then, they were rather drastic changes. The existing courses were very heavily geared to the future engineers or physical scientists, so they were rather narrow in content. We were very concerned with broadening the syllabus, to bring in a good deal of modern physics which was absent from the course in those days. Some of the people developed really good experiments that related to quantum physics and relativity, which had been totally absent before.

Based on your experience here at MIT, do you feel there will be more of an emphasis on experimental work in physics courses throughout the country?

I really hope so. I think the most desirable thing is to make physics a really first-hand experience to students, not just something they get out of a textbook. To establish something for yourself and understand what it means to make a measurement and critical analysis. If you have done some measurements on your own, you would have an idea of how seriously you should take it.

What current issues will have an impact on the way we teach physics?

There are various issues. One is that there is a lot more consciousness developing these days that perhaps the way we teach the subject is far from ideal. The standard sort of course in many places is particularly narrow in that the whole of the first year may not discuss anything beyond the 19th century. Let's compare this with what the biologists do. In the introductory biology courses these days, the biologists emphasize molecular biology and students are learning things that are on the frontiers of biological research. Physics will show up very poorly if we don't do something comparable.

It seems the methods of teaching physics have been scrutinized and revised for some time, at least since the late 1950's. How does the teaching of physics differ from other sciences, such as chemistry or biology?

I believe that, in an important sense, physics is the simplest of all the sciences. In physics, the fundamental ideas tend to be very easily stated and the applications we use in physics tend to be uncomplicated. Most chemists couldn't begin to explain, in detail, what goes on in an atomic reaction. They sort of use the results, very sophisticated results, of physics and say "here they are," and then have a more empirical set of rules as to how they apply to chemical processes. In biology, you just have to accept that "this is the way nature is" and you don't worry with the finer structure of biological molecules and try to explain them in terms of the basic physics that is involved. In that sense, physics is simpler. This gives us the opportunity to have students dig more deeply into the actual processes of understanding a scientific idea. Because we deal with problems which, from one point of view, are simpler, they do lend themselves to encouraging the thought process that one uses in attacking a problem. Now that's just a physicist's point of view and maybe a chemist or a biologist would see it differently.

One conviction of mine is that the teaching of basic physics does not have to involve a lot of mathematics. I am a great believer in arithmetic. I think there is an enormous amount one can do without fancy mathematics. So that's something I try to stress in my own teaching—using a minimum of theoretical baggage in order to get a result.

What areas of physics are likely to be most exciting in the future?

Primarily, I believe astronomy and astrophysics would be most exciting. Secondarily, at the other extreme, the innermost structure of matter would be of great interest. (Of course, in modern astrophysics and cosmology the two come together.) It seems to

me that these are two obvious frontiers which are still expanding and which, to me at least, are more fundamentally interesting than some of the other very remarkable studies— for example, in condensed matter physics—which lead to the creation of marvelous devices. These studies would excite some students, of course, those interested in learning what it is that underlies the physics of the compact disc player and so on.

Do you have any suggestions for students of physical science?

Perhaps the strongest single message I would want to give is that physics isn't something that just happens inside the walls of a classroom or lab. It's out there all of the time. Walk around with your eyes open. Think about the physical aspects of the ordinary experiences of life and read books. Play with ordinary objects and ask yourself why do they behave as they do.

Let's take, for example, something as simple and common as a flying insect. Why is it that small insects, like mosquitoes, have high pitched sounds? How does that relate to a direct measure of how fast they are flapping their wings? Think about the physics of the energy balance of your own body, the food you take in and the calories and how that converts into the work you can do or how far you can run. If you can relate principles of physics to what goes on in your body, I think that's a helpful ability to have. But don't forget the excitement of looking outside. Observe the basic physical principles through real world situations. There is almost an infinite variety of things that one can observe that way and many questions that you can ask.

Inside Tony French still lives the eager, 19-year-old student.

This interview was conducted by Kate Pachuta for Saunders College Publishing.

Our blue planet, as seen from space, is the only home most of us will ever know. As a result, it behooves us to have a basic understanding of scientific knowledge to help us preserve and protect our home in space. (Courtesy NASA)

C H A P T E R

1

Boundaries and Measurements in Physical Science

1.1
BOUNDARIES

Physical science is the study of matter and energy. This incredibly broad subject covers a range from the smallest particle of the atom to the largest collection of stars and galaxies, from the energy in the faintest ray of light to the primordial explosion of the entire Universe. The physical scientist looks back in time to the very formation of our Universe and forward to its ultimate fate. From the smallest to the largest, from the beginning to the end—these are the boundaries of this field of study. Within this realm, only life and living objects are omitted from the physical sciences. Living systems are covered by the biological sciences.

This book is an introduction to physical science. As an introduction to this introduction, let us quickly pass from boundary to boundary, from the smallest to the largest. The remainder of this section is broken into short paragraphs, each accompanied by an illustration. Each paragraph covers an object or entity that is 10, 100, or 1000 times larger than the object in the previous paragraphs.

The unit of distance used in this section is the centimeter (cm). One cm is about the width of your little fingernail. Although 1 cm is easy to visualize in human terms, other numbers used in physical science are not at all easy to visualize. Within these short paragraphs, a tremendous range in size is covered—from a scale of approximately 0.0000000000001 cm to 1,000,000,000,000,000,000,000 cm. Some of these quantities are so large or so small that it is often difficult to keep track of all the zeros. To make matters easier, the position of the decimal point in each number is noted in parentheses. In this system, the number 1 is taken as a reference. Now consider a large number such as 1000. The three zeros in 1000 indicate that the decimal point has been moved three places to the right of the number 1. Therefore, 1000 is followed by the reminder (+3 decimal places). In the number 0.1, the decimal point has been moved one place to the left of 1 and is followed by the reminder (−1 decimal place). Later in this chapter, an even more convenient shorthand for writing large and small numbers will be introduced.

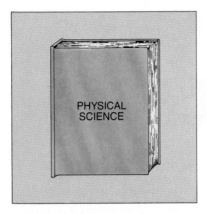

One cm in the picture represents about 10 cm (+1 decimal place) in real life. Many common objects that people normally deal with in their everyday lives are 10 to 50 cm in length, width, or diameter. This textbook, for example, is about 20 cm by 30 cm.

Advancing our outlook by a factor of 10 moves us up to the range of the largest objects built by humans. One cm in the picture now represents 10,000 cm (+4 decimal places). Oil tankers and skyscrapers, both about the same size, are shown superimposed on one another to represent this boundary of human endeavor.

(Courtesy NASA)

Our first jump brings us to objects approximately 100 times larger than your textbook. Therefore, the scale of our drawing has been changed, so that in this picture 1 cm represents 1000 cm (+3 decimal places). This first leap immediately takes us outside the range of things that can be picked up and carried about, but the scale is definitely within the realm of ordinary human experience. A tree, an apartment building, or a large truck could all be shown here, but instead a space shuttle was chosen because it stands as a symbol of our scientific era.

Another jump of 100 puts us at the one to a million scale. One cm in the picture represents 1,000,000 cm (+6 decimal places). The numbers are starting to get large, and so are the objects. On this scale, Mt. Everest would rise a little less than 1 cm above sea level, and a group of mountains with a large glacier flowing through them fits nicely on our 5 cm square.

(Courtesy NASA)

(Courtesy NOAO)

One cm in the picture represents 1,000,000,000 (one billion) cm (+9 decimal places). By the time our drawing reaches the one to a billion scale, the entire Earth comes into view. To a person on a ship in the middle of the ocean or on foot on one of the vast deserts, prairies, or ice caps of the planet, Earth may seem to be expansive or nearly boundless; but as our focus moves ever outward, this planet quickly recedes to a tiny speck in the cosmos.

One more jump of a factor of 10 brings the Sun into full view. One cm in the picture represents 100,000,000,000 cm (+11 decimal places). The diameter of the Sun is roughly 100 times the diameter of Earth and ten times the diameter of Saturn and its magnificent rings.

(Courtesy NASA)

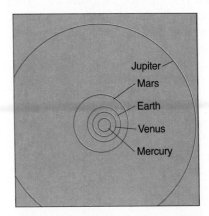

One cm in the picture represents 10,000,000,000 (ten billion) cm (+10 decimal places). A picture of Saturn and its most prominent rings fits nicely into our one to ten billion scale, of the order of magnitude of ten times larger than Earth. Photographs of Saturn's rings taken by spacecraft in the early 1980s have added greatly to our knowledge of this planetary system.

One cm in the picture represents 10,000,000,000,000 cm (+13 decimal places). In this scale the Sun appears as a small dot surrounded by the first four planets—Mercury, Venus, Earth, and Mars. The orbits of the planets appear as neat red lines in the picture, but it is important to realize that these are imaginary lines. If you were perched in a space capsule in a position to view the inner Solar System from this perspective, you would see the Sun shining in a black sky, and those planets that were visible would appear as small spots of reflected light.

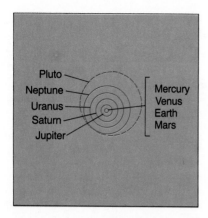

One cm in the picture represents about 1,000,000,000,000,000 cm (+15 decimal places). Here the entire Solar System comes into view. In this illustration, the Sun has become so small that it is impossible to draw it to scale and still see the dot. The dot is therefore larger than it should be. As you can see, in this drawing even the orbits of the inner planets are crowded together. The dominant feature now becomes the orbits of the five outer planets—Jupiter, Saturn, Uranus, Neptune, and Pluto.

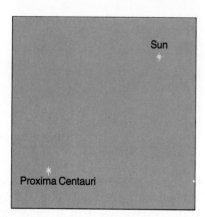

Expanding our field of vision another 1000-fold brings two small spheres into the frame of vision. One of the spheres represents the Sun; the other represents the closest star, Proxima Centauri. One cm in the picture now represents 1,000,000,000,000,000,000 (+18 decimal places). In this vastness of outer space, it is cumbersome to express distances in centimeters. Therefore, astronomers have adopted a unit called a light-year (ly). One ly is the distance that a beam of light travels in a year. Since light travels at 30,000,000,000 cm/s (in a vacuum), a light-year is also proportionally large, approximately 1,000,000,000,000,000,000 (+18 decimal places), and Proxima Centauri is about 4.2 ly away from our Solar System.

One cm in the picture represents about 100,000,000,000,000,000,000 cm (+20 decimal places) or 100 ly. Moving our perspective outward into space, a cluster of about 40 stars is shown. These form the local group that surrounds our Sun.

(Courtesy NOAO)

One cm in the picture represents about 10,000,000,000,000,000,000,000 cm (+22 decimal places) or 10,000 ly. The photograph shows a collection of some 100 billion individual stars. A collection of stars grouped together in this manner is called a **galaxy.** This picture is a galaxy like our own Milky Way galaxy. In order to imagine the scale of this drawing, think about the fact that a beam of light would have to travel some 100,000 years to traverse the Milky Way galaxy from edge to edge.

tured here, and therefore most of the field of vision would be empty blackness. Astronomers estimate that several hundred billion galaxies exist and that the farthest one from us is about 15 billion ly distant.

One cm in the picture represents about 1,000,000,000,000,000,000,000,000 cm (+24 decimal places) or about 1 million ly. In a scale in which 1 cm represents 1 million ly, about a dozen separate galaxies, each one containing some 100 billion stars, are brought into view. If you were to try to draw a picture of our Sun on this scale, the required dot would be smaller than a single atom.

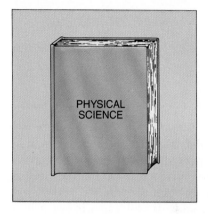

One cm in this picture represents 10 cm in real life. Before you lose yourself in the mysteries of intergalactic space, return for a moment to the ordinary, the book in your hand, in preparation for an imaginary journey into the realm of very small objects.

One cm in the picture represents 100,000,000,000,000,000,000,000,000 cm (+26 decimal places) or 100 million ly. In this last picture, each dot represents an entire galaxy, and many, many dots are shown. The picture is hardly drawn to scale, for in reality the spots of light would be much smaller than they are pic-

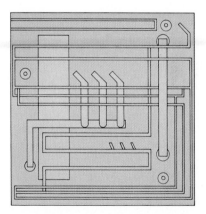

One cm in the picture represents 1/10 cm or 0.1 cm (−1 decimal place). A square 0.1 cm on a side would just about fit inside the letter o in the print used in this book. As small as this seems, this is the size of an electronic circuit inside a computer chip. These chips provide the working channels and the memories of various types of computers, which have initiated the electronic revolution that is causing rapid changes in our society.

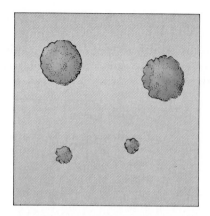

A 100-fold descent drops us into the range of objects that are barely visible. One cm in the picture represents 0.001 cm (−3 decimal places). Some day when you are indoors and sunlight is shining through the window, look at right angles to the sunbeam and you will see a haze of tiny pieces of dust about this size. This dust is made up of many different types of air pollutants: small particles of smoke and soot, some living organisms and pieces of dead ones, and small, oily droplets with various compounds dissolved in them or adhering to them.

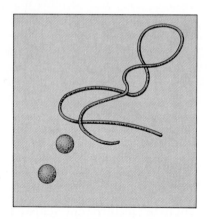

One more 100-fold descent brings the largest molecules into view. One cm in the picture equals 0.00001 cm (−5 decimal places). On the right side of the drawing you see a piece of a strand of DNA, the material that transmits genetic information from one generation to another. Thus, DNA is the molecular foundation for life as we know it. Next to the strand of DNA, two smaller molecules of starch are shown. Although starch

molecules are still very large compared with ordinary molecules such as those of water, oxygen, or sugar, they are much smaller than the strand of DNA, and less detail is seen.

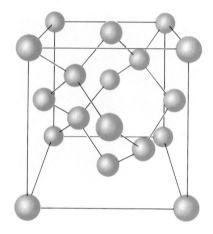

One cm in the picture represents 0.0000001 cm (−7 decimal places). Objects of this scale are below the field of vision of even the most powerful electron microscopes; but if you could look into a world of this size, you would see individual atoms. Pictured here is a tiny segment of a diamond, which is made up of a regular array of carbon atoms.

An additional 10-fold descent shows us an individual oxygen molecule floating in space. One cm in the picture represents 0.00000001 cm (−8 decimal places). If you wanted to draw a picture of a person on an equivalent scale, you would need a piece of paper approximately 200,000 kilometers (120,000 miles) long!

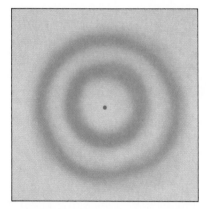

Another 100-fold drop brings us inside the atom itself. One cm in the picture represents 0.0000000001 cm (−10 decimal places). In this example, a carbon atom, pictured earlier as a part of the diamond, is shown. The dense central core, the nucleus of the carbon atom, is smaller than a period set in the print of this book. The rest of the space is essentially void, with an occasional electron whizzing about. Since electrons do not move in defined orbits as planets do, shading is used to picture the region in which the electrons are moving about.

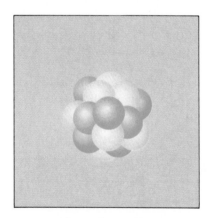

One cm in the picture represents 0.0000000000001 cm (−13 decimal places). In this picture, you see the nucleus of a carbon atom. Note that the scale has been magnified 10,000

times since two pictures ago when the oxygen molecule was in plain view. Thus, the nucleus, the dense central core of the atom, is much, much smaller than the atom itself. Yet the nucleus is made up of still smaller particles called protons and neutrons.

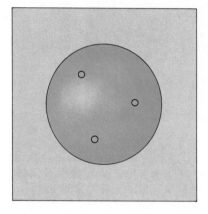

For a long time, protons and neutrons were considered to be elemental, indivisible particles, but today they are thought to be made up of even smaller particles. In this final picture, in which 1 cm represents 0.00000000000001 cm (−14 decimal places), an individual proton is shown to be composed of particles called quarks, which are surrounded by empty space. These particles are incredibly small, just as the Universe is incredibly large. No structure is shown here because no one really knows what a quark "looks like"; the very concept of "looking" at a quark is, in itself, meaningless.

The world we live in today is becoming increasingly affected by technological change. Automobiles, airplanes, computers, sophisticated weapons, nuclear power plants, factories, and oil refineries all profoundly affect the human condition. Although no one can hope to be an expert in all the different sciences that influence our lives, we can all understand the basic scientific concepts behind a great many different disciplines. Such a background will also help you to understand many issues of public policy that are related to scientific matters. For example, consider the question of nuclear power. Today many people

(b)

(a)

The instrument panel of (a) a Boeing 747 jet and (b) of the space shuttle are studded with gauges and measuring devices of all types.
(Shuttle photo courtesy NASA.)

are asking, "What are the advantages and disadvantages of a nuclear future?" Related to this question, one might ask, "How dangerous is a nuclear power station?" No one will become a nuclear engineer by reading this book, but the chapter on nuclear physics should provide a basic understanding of the concepts involved.

Beyond the realm of issues that directly affect our lives, there is another large body of science that is more theoretical in nature. Scientists ask questions like, "How old is the Earth?"; "What materials lie within its core?"; and "How was the entire Universe formed?" Although a search for the answers to these and other similar questions may not be of any immediate practical consequence, they remain an integral part of science just because human beings are curious.

1.2
SYSTEMS OF MEASUREMENT

Measurement is fundamental not only to experimental science but also to our whole technological way of life. Measuring devices are all around us. For example, clocks, thermometers, rulers, and bathroom scales are present in most homes, and automobiles are equipped with speedometers, fuel gauges, and a variety of other instruments.

A measurement must include both a quantity and a description of what is being measured. Thus, it is meaningless to say that an animal is 3. The quantity, by itself, does not give us enough information. The animal could be 3 years old, or

3 meters tall, or have a mass of 3 kilograms. Seconds (s), meters (m), and kilograms (kg) are all units. If you are asked to measure the distance between two points, your answer must include the magnitude of the length and the unit that you are using. For example, suppose you measure the length of a short pencil and find it to be 6.20 centimeters (cm). The same distance could also be expressed in other units, because 6.20 cm is equal to 2.44 inches (in), or 0.0620 m, or 0.203 feet (ft).

It is much easier for different observers to communicate with one another if everyone uses the same system of measurement. The SI (for Système International), which is the present-day descendant of the metric system, is internationally recognized and is now used in nearly all the nations in the world. The use of a universal measuring "language" facilitates international communications.

As shown in the following table, the SI has seven base units from which all other units can be derived.

Quantity	Unit	Symbol
Length	meter	m
Mass	kilogram	kg
Time	second	s
Temperature	kelvin	K
Electric current	ampere	A
Amount of substance	mole	mol
Luminous intensity	candela	cd

TABLE 1–1 Names and Symbols for SI Prefixes

Prefix	Symbol	Multiply by
tera	T	10^{12}, one trillion
giga	G	10^{9}, one billion
mega	M	10^{6}, one million
kilo	k	10^{3}, one thousand
deci	d	10^{-1}, one tenth
centi	c	10^{-2}, one hundredth
milli	m	10^{-3}, one thousandth
micro	μ	10^{-6}, one millionth
nano	n	10^{-9}, one billionth
pico	p	10^{-12}, one trillionth

As we move through this text, we will encounter all of these units except the candela; it is included just to complete the set.

The SI base and derived units are used throughout this text. In some instances, references to the British (or conventional) system of measurement are made for comparison.

The SI rules specify that the symbols or abbreviations for units are not followed by periods, nor are they changed in the plural. Therefore, scientists write, "The river is 10 m wide" (not "10 m. wide" or "10 ms wide").

Various combinations of these base units give us a large variety of **derived units.** A derived unit is made up of some combination of base units. For example, area is measured as a length times a length and can be expressed as (meter)(meter), or m^2. Similarly, volume is a length times a length times a length and can be expressed as m^3. As another example, speed is length divided by time and can be expressed as meters/second, or m/s.

The SI also considers units other than its official ones. Some of these units are accepted because they are so widely used, others are accepted only "temporarily," and still others are to be avoided. Hour (h), liter (L), and degrees Celsius (°C) are examples of accepted units. The calorie, sometimes used in the measurement of heat, is a unit that is now outside the SI.

The International System is particularly convenient because it provides a set of prefixes that express larger or smaller quantities than the standard units. The larger ones are multiples of 10, and the smaller ones are decimal fractions. The names and symbols of these prefixes are shown in Table 1–1. Thus, the prefix *kilo* means 1000, so 1

kilometer (km) is 1000 meters. Similarly, *milli* means one thousandth (0.001), so 1 milligram (mg) is one thousandth of a gram, or 0.001 g.

The first three base units, and some units derived from them, are discussed in the following paragraphs.

1.3 LENGTH

The SI unit of length is the meter (m). The meter originally was defined as one ten-millionth of the distance between the North Pole and the Equator. Since it is nearly impossible to calibrate ordinary measuring devices accurately against such a global standard, a more manageable meter was created by placing two marks on a platinum alloy bar (Fig. 1.1), which was called the International Prototype Meter. More recently, there has been a return to a "natural" standard for the meter, based on the speed of light. The current official definition of the meter is "the distance traveled by light in a vacuum during 1/299,792,459th of a second." Don't memorize this value; it is more important to be able to visualize the magnitude of the meter. For example, you might keep in mind that a kitchen cabinet or laboratory bench is almost a meter high, and that an average professional basketball player is about 2 m tall.

Area and volume are quantities derived from length. Area is the amount of space on a two-dimensional surface and is expressed in SI units in terms of square meters (m^2). Volume is the amount of space within a three-dimensional region and is expressed in terms of cubic meters (m^3) in SI units. For volume, the SI also recognizes the liter (L), which is $1/1000 \ m^3$, or $1000 \ cm^3$. A common unit derived from the liter is the milliliter (mL). One mL is 1/1000 of a liter, or $1 \ cm^3$.

1.4 MASS

The SI unit of mass is the kilogram (kg). The standard kilogram is the mass of a block of platinum alloy stored at the International Bureau of Weights and Measures in Sèvres, France, and it is almost exactly equal to the mass of 1000 cubic centimeters (cm^3) of water at a temperature of

FIGURE 1.1 The U.S. copy of the Standard Meter, Prototype Meter No. 27, was delivered from France in 1890.
(Courtesy of National Bureau of Standards.)

FIGURE 1.2 Prototype Kilogram No 20, the national standard of mass for the United States. It is a platinum-iridium cylinder 39 mm in diameter and 39 mm high.
(Courtesy of National Bureau of Standards.)

3.98°C (Fig. 1.2). Let us again point out that you do not need to memorize this exact definition of a kilogram. We are more concerned with your

ability to measure in these units and to visualize their magnitude.

1.5
TIME

The SI unit of time is the second, which was originally defined as 1/86,400th of a day. However, the spin of the Earth varies slightly over time, so the length of a day is constantly changing. Therefore, a second is now defined as the duration of 9,192,631,770 vibrations of a specific type of light emitted by a specific atom (cesium-133). Once again, this is not a fact to be memorized; it is introduced merely to give an example of the type of precision that is available to and required by the scientific community of the 1990s. Other base units and various derived units are discussed in appropriate places in the text and are summarized in Appendix A.

1.6
SIGNIFICANT FIGURES

One of the problems often encountered by a beginning student of physics is that of significant figures. Just because your calculator reads to 8 or more digits does not mean that you know the answer to a problem to that accuracy. For a discussion of significant figures and how to manipulate them, refer to Appendix B. Throughout the example problems and homework exercises in this text, unless otherwise specified, we shall assume that the given data are precise enough to yield an answer having three significant figures. Thus, if a problem states that a length is 5 m, it is understood that the length is actually known to be 5.00 m.

1.7
CONVERSION FACTORS

It is often necessary to change a quantity expressed in one unit to the same quantity expressed in another unit. For example, if you are traveling in France, your restaurant bill may come to 80 francs, and you might wish to know how much that is in U.S. dollars. Or if you are baking a pie, you might find it necessary to convert tablespoons to cups; in the laboratory you might measure something in centimeters and wish to express it in meters. The mathematical manipulations do not change the cost of your dinner or the size of the object; they change only the units in which they are expressed.

To start with very familiar units, consider the simple problem, "How many seconds are there in 1.5 minutes (min)?" You know the answer is 90 s, but it is instructive to study the reasoning process. First, write the equation

$$60 \text{ s} = 1 \text{ min}$$

Next, divide both sides of this equation by 1 min as

$$\frac{60 \text{ s}}{1 \text{ min}} = \frac{1 \text{ min}}{1 \text{ min}} = 1$$

or, more simply

$$\frac{60 \text{ s}}{1 \text{ min}} = 1$$

The fraction $^{60 \text{ s}}/_{1 \text{ min}}$ is called a "conversion factor" because it can be used to "convert" minutes to seconds. Since this factor has a value of 1, its reciprocal

TABLE 1–2 Handy Conversion Factors*

To Convert from	to	Multiply by the Conversion Factor
centimeters	inches	0.394 in/cm
feet	centimeters	30.5 cm/ft
	meters	0.305 m/ft
grams	pounds (avdp.)	0.0022 lb/g
inches	centimeters	2.54 cm/in
kilograms	pounds (avdp.)	2.20 lb/kg
kilometers	miles	0.621 mi/km
liters	quarts (U.S., liq.)	1.06 qt/L
meters	feet	3.28 ft/m
miles (statute)	kilometers	1.61 km/mi
pounds (avdp)	kilograms	0.454 kg/lb

*A more complete list is given in Appendix A.

$$\frac{1 \text{ min}}{60 \text{ s}}$$

also has a value of 1 and is an equally correct conversion factor that can be used to "convert" seconds to minutes. Multiplying or dividing anything by 1 does not change its value, so it is correct to carry out such operations. How do we decide which of these two factors gives the correct answer to the problem? Let us try both to show which is right and which is wrong:

$$1.5 \text{ min} \frac{60 \text{ s}}{1 \text{ min}} = 90 \text{ s}$$

Note that the units cancel as algebraic quantities, as shown by the slash marks. If you multiply by the wrong conversion factor, you end up with a meaningless answer, as shown below.

$$1.5 \text{ min} \frac{1 \text{ min}}{60 \text{ s}} = 0.025 \frac{\text{min}^2}{\text{s}}$$

This answer cannot be correct because the units are wrong. It is impossible for an incorrect use of a conversion factor to give a correct answer.

Table 1–2 gives some common conversion factors between SI and British units. A longer list is given in Appendix A. Use of the conversion Tables 1–1 and 1–2 is illustrated by the examples provided above.

EXAMPLE 1.1 Give them a thousand meters and they will take a kilometer

Long distance runners run a 10,000 m race. How far is that in **(a)** km, **(b)** miles (mi)?

Solution (a) The conversion from meters to kilometers is 1 km = 1000 m or

$$\frac{1 \text{ km}}{1000 \text{ m}} = 1$$

Therefore,

$$10{,}000 \text{ m} \frac{1 \text{ km}}{1000 \text{ m}} = 10 \text{ km}$$

(b) The conversion factor from Table 1–2 is

$$\frac{0.621 \text{ mi}}{\text{km}} = 1$$

Therefore,

$$10 \text{ km} \frac{0.621 \text{ mi}}{\text{km}} = 6.21 \text{ mi}$$

EXAMPLE 1.2 Looking for a gas station

A car can travel 500 km on a tank of fuel. How far is that in miles?

Solution The conversion factor, from Table 1–2, is

$$\frac{0.621 \text{ mi}}{\text{km}} = 1$$

Thus,

$$500 \text{ km} \frac{0.621 \text{ mi}}{\text{km}} = 311 \text{ mi}$$

EXAMPLE 1.3 A race car or a clunker?

Suppose that a race car travels at an average speed of 240 km/h. What is its speed in m/s?

Solution In this example, both units need to be changed, so two conversion factors are required. They are

$$\frac{1000 \text{ m}}{1 \text{ km}} \quad \text{and} \quad \frac{1 \text{ h}}{3600 \text{ s}}$$

We have

$$240 \frac{\text{km}}{1 \text{ h}} \frac{1000 \text{ m}}{1 \text{ km}} \frac{1 \text{ h}}{3600 \text{ s}} = 66.7 \frac{\text{m}}{\text{s}}$$

EXAMPLE 1.4 Getting a feel for SI

(a) A human who is 3 ft tall is probably a child. Is a person who is 3 m tall a child?

(b) A world-class runner can run a mile in a little under 4 minutes. Is it possible to run a kilometer in 4 minutes?

Solution These problems are designed to give you an intuitive feel for SI units We need not solve for exact numerical answers.

(a) Since there are approximately 3 ft in a meter, a person who is 3 m tall is approximately 9 ft tall and is not a child, but a giant.

(b) A kilometer is equal to about 0.62 mile and is therefore considerably less than a mile. Thus, an athlete could easily run a kilometer in 4 minutes.

1.8
SCIENTIFIC OR EXPONENTIAL NOTATION

During our introductory survey of the Universe from the largest objects to the smallest, very large and very small numbers were needed to express size. Recall that these numbers were cumbersome for the simple reason that it was tedious to count all the zeros. At that time, the number of decimal places in each number was noted in parentheses after the number. An even better system, called **scientific notation** or **exponential notation,** is used by scientists all over the world. This system is based on exponents of 10, which are shorthand notations for repeated multiplications or divisions. The end result is that all numbers are written as a number between 1 and 10 along with a "power of ten" which locates the decimal point.

A positive exponent is a symbol for a number that is to be multiplied by itself a given number of times. Thus, the number 10^2 (read "ten squared" or "ten to the second power") is exponential notation for $10 \times 10 = 100$. Similarly $3^4 = 3 \times 3 \times 3 \times 3 = 81$. The reciprocals of these numbers are

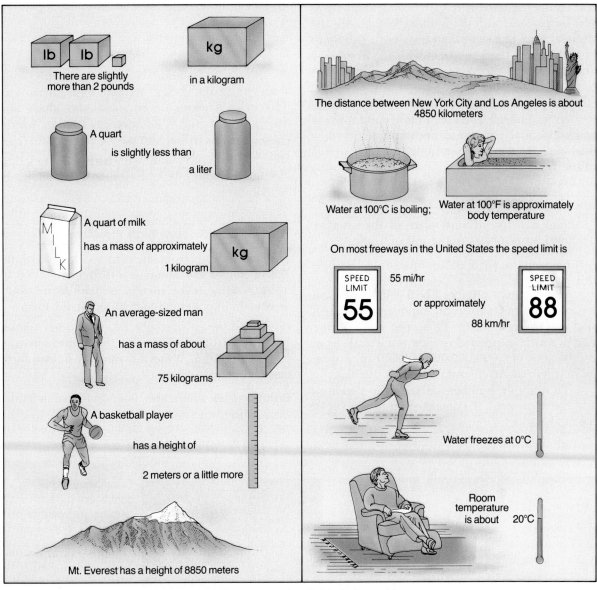

FIGURE 1.3 Since people in the United States must inevitably learn metric measurement to be compatible with the rest of the world, it is important to "think metric." Some useful rule-of-thumb relationships are shown.

expressed by negative exponents. Thus, $10^{-2} = 1/10^2 = 1/(10 \times 10) = 1/100 = 0.01$.

Positive and negative powers of 10 are shown below:

$10^4 = 10 \times 10 \times 10 \times 10 = \underline{10000}$ (4 places to the right)

$10^3 = 10 \times 10 \times 10 = \underline{1000}$ (3 places)

$10^2 = 10 \times 10 = \underline{100}$ (2 places)

$10^1 = 10 = \underline{10}$ (1 place)

$10^0 = 1$ (0 places)

$10^{-1} = 1/10 = 0.1$ (1 place to the left)

$10^{-2} = 1/(10 \times 10) = 0.01$ (2 places)

$10^{-3} = 1/(10 \times 10 \times 10) = 0.001$ (3 places)

$10^{-4} = 1/(10 \times 10 \times 10 \times 10) = 0.0001$ (4 places)

Notice that to write 10^4 in longhand form you simply start with the number 1 and move the decimal four places to the right, as 10000. Similarly, to write 10^{-4} you start with the number 1 and move the decimal point four places to the left to arrive at 0.0001.

It is just as easy to go the other way— that is, to convert a number written in longhand form to an exponential expression. For example, the decimal place of the number 1,000,000 is six places to the right of 1. Thus,

$$1,000,000 = 10^6$$

Similarly, the decimal place of the number 0.000001 is six places to the left of 1 and

$$0.000001 = 10^{-6}$$

What about a number like 3,000,000? If you write it $3 \times 1,000,000$, the exponential expression is simply 3×10^6. Thus, the mass of the Earth, which, expressed in long numerical form, is 5,980,000,000,000,000,000,000,000 kg, can be written more conveniently as 5.98×10^{24} kg.

To multiply numbers in scientific notation, add the exponents. To divide, subtract the exponents. Thus,

$$10^4 \times 10^3 = 10^7$$

and

$$\frac{10^5}{10^3} = 10^5 \times 10^{-3} = 10^2$$

Scientific notation is used frequently with numbers in the metric system. Thus, there are 100 cm in a meter or, in shorthand, there are 10^2 cm in a meter. A kilometer is 1000 m, which can be written 10^3 m. Similarly, a microgram (μg) is 0.000001 g, or 10^{-6} g. There are 1000, or 10^3, watts in a kilowatt and 1,000,000, or 10^6, watts in a megawatt.

EXAMPLE 1.5 How high the Moon?

(a) The distance from the center of the Earth to the center of the Moon is approximately 3.8×10^5 km. How many meters is that? Write your answer in exponential and in longhand form.

(b) The distance between two hydrogen nuclei in a hydrogen molecule is approximately 0.000000007 cm. Write this number in standard exponential notation.

Solution (a) There are 1000, or 10^3, m in a km. Converting,

$$(3.8 \times 10^5 \text{ km})\left(\frac{10^3 \text{ m}}{1 \text{ km}}\right) = 3.8 \times 10^8 \text{ m}$$

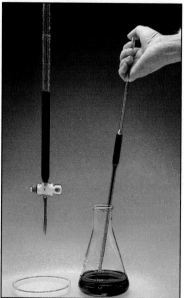

Scientists use the containers shown to measure volumes of liquids.

To convert from exponential to longhand form, move the decimal point eight places to the right:

380000000

This can be written as 380,000,000 m.

(b) As discussed in the text, the goal is to express the number using the digit 7 (as a number between 1 and 10), followed by an exponent showing the correct decimal place. Therefore, start with the 7 and count from right to left.

$$0.000000007 = 7 \times 10^{-9} \text{ cm}$$

SUMMARY

The introductory section reviews the scale of size within the Universe. The **International System of Units** (SI), an outgrowth of the metric system, is used in scientific circles throughout the world. A value in one unit can be converted to the same value expressed in another unit by multiplying by the appropriate conversion factor. Large or small numbers are conveniently expressed in terms of exponents of 10.

KEY WORDS

International System of Units (SI)
Base unit

Derived unit
Scientific or exponential notation

QUESTIONS

SYSTEMS OF MEASUREMENT

1. How many (a) milligrams are there in a gram, (b) centimeters in a meter, (c) grams in a kilogram, (d) micrometers in a meter, (e) meters in a kilometer?
2. Estimate the length or distance of the following: (a) The altitude in m of a commercial airplane in normal flight. (b) The length of a housefly. (Use appropriate SI units.)
3. Estimate the time interval of the following: (a) The time in s between heart beats. (b) The average age of a college student in s.
4. Estimate the mass of the following: (a) The mass in kg of a baseball. (b) The mass of a housefly. (Use appropriate SI units.)
5. (a) A certain power plant has an output of 20,000,000 watts. Express this output in megawatts. (b) A component in a TV set is rated as 0.00000012 farad. Express this rating in microfarads.
6. Give two reasons why the SI is easier to use than the British system.
7. Identify an object in your everyday experience that has: (a) a mass of 1 to 2 kg; (b) a mass of 1 to 2 g; (c) a length of 1 to 2 mm; (d) a length of 1 to 2 cm; (e) a length of 1 to 2 m; (f) a length of 1 to 2 km; (g) a volume of 1 to 2 mL; (h) a volume of 1 to 2 L; (i) a volume of 1 to 2 m³.

CONVERSIONS

8. A convenience store has several items on sale. Which are good buys and which are not? (a) Steak sells for $1.50 per kg. (b) Gasoline is $1.00 per liter. (c) Peanut butter is 10 cents per gram.
9. Fill in the blanks:
   2.5 in = _____ cm; 3.9 quarts = _____ L;
   100 g = _____ kg; 16 miles = _____ km;
   16 miles = _____ m; 80 km/h = _____ miles/h;
   3.8 m = _____ ft; 0.1 cm = _____ in.
10. An average NFL lineman has a weight of about 280 lb. If there are about 2.2 lb in a kg at the surface of the Earth, what is his mass in kg?
11. A European dress pattern calls for 3 m of fabric. How many yards would you have to buy to make the dress?
12. A football player is 6 ft tall. How tall is he in meters? In centimeters?
13. The distance between New York and San Francisco is approximately 3000 mi. How far is that in kilometers?
14. Suppose it takes you about 2 min to read a page in this textbook. How many milliseconds does it take?
15. (a) A world class cyclist can pedal 300 miles in a day. Would it be possible to pedal 300 km in a day? (b) Five people would not ride comfortably in a 10 ft boat. What about five people in a 10 m boat?

The speedometer on many new automobiles can be set to read either in conventional units or in SI. As an exercise for you, this car has been driven 62,980 km. How far is that in miles? Also, if this car is driven at 55 mi/h, how fast will the speedometer read this to be in km/h?

16. The speed of light is approximately 3.00×10^8 m/s. Express this number in terms of cm/s, mi/s, and mi/h.

17. The speed of sound is about 345 m/s. How fast is this in mi/h?

18. A tablecloth is 45 cm by 60 cm. What is its area in cm^2 and in m^2?

19. Until recently, the world's land speed record was held by Colonel John P. Stapp, USAF. He rode a rocket-propelled sled that moved at a speed of 282.5 m/s. What is this speed in mi/h?

20. A room in a house is 12 ft by 20 ft and has ceilings 8 ft high. What is the volume of this room in m^3?

EXPONENTIAL NOTATION

21. (a) Write the following numbers in scientific notation: 1573; 0.00589; 647; 6,300,000; 0.16. (b) Write the following in longhand form: 1.5×10^4; 6.34×10^{-3}; 9.02×10^{-8}.

22. (a) Write the following numbers in longhand numerical form: The diameter of a bacterium is about 10^{-4} cm. The radius of the Sun is 6.9×10^5 km. There are approximately 10^{11} stars in the Milky Way galaxy. (b) Write the following in exponential form: The temperature of the core of the Sun is approximately 15,000,000°C. The outer region of the Sun's atmosphere has a density of about 0.000000001 times that of the Earth's atmosphere.

23. (a) The mass of the hydrogen atom is 1.67×10^{-27} kg. Write this in longhand numerical form. (b) The mass of the Earth is about 6×10^{24} kg. Write this in longhand numerical form.

24. (a) The distance of the Earth to the most distant quasar (a strange object that will be discussed in the section on astronomy) is about 1.5×10^{26} m. Write this out in longhand numerical form. (b) The diameter of an atomic nucleus is about 8×10^{-15} m. Write this out in longhand numerical form.

25. There are approximately 32,000 aluminum cans in a ton of scrap. How many cans are there in 10 tons? Express your answer in exponential notation.

26. An average automobile is driven about 10,000 miles per year. How many centimeters is it driven in a year? Express your answer in scientific notation.

ANSWERS TO SELECTED NUMERICAL QUESTIONS

1. (a) 1000, (b) 100, (c) 1000, (d) 1,000,000, (e) 1000

5. (a) 20 MW, (b) 0.12 microfarad

11. 3.28 yd

12. 1.83 m, 183 cm

13. 4830 km

14. 120,000 ms

16. 3×10^{10} cm/s, 186,000 mi/s, 6.70×10^8 mi/h

17. 771 mi/h

18. 2700 cm^2, 0.27 m^2

19. 632 mi/h

20. 54.4 m^3

21. (a) 1.573×10^3, 5.89×10^{-3}, 6.47×10^2, 6.3×10^6, 1.6×10^{-1}; (b) 15,000, 0.00634, 0.0000000902

25. 3.2×10^5

26. 1.61×10^9 cm

Artist's concept of an aerospace plane. (Courtesy NASA)

CHAPTER

2

Motion

We will begin our discussion of physics by studying motion, a subject that is a common part of our daily existence. In fact, motion is so common in our everyday lives that most of the topics covered are already somewhat familiar to you. That is, words such as speed, velocity, and acceleration are currently a part of your vocabulary. Thus, one of the primary goals of this chapter is to be sure that your intuitive ideas about motion are correct and to place some of these ideas into a slightly more quantitative context than they may be at present.

In a larger sense, this chapter begins the study of a portion of physics referred to as *mechanics*. Mechanics is the study of motion and of the forces that cause motion. Thus, this chapter is divided into two parts. The first deals with the study of motion without regard to what causes the motion. This segment of mechanics is often referred to as *kinematics*. Following this, we shall investigate the relationship of forces and motion, commonly called *dynamics*.

2.1
EARLY IDEAS ABOUT MOTION

It is somewhat surprising to find that descriptions of motion were slow in developing. In fact, great scholars worked on this subject for thousands of years before our present-day ideas became firmly established. One of the first serious students of motion was Aristotle (384–322 B.C.), a famous philosopher in ancient Greece, who considered the movement of objects in terms of what he called "natural motion." He taught that an object moved according to the "nature" of the object. He stated that there are four elements—earth, air, fire, and water—that are the basic building blocks of the world around us. Also, according to his viewpoint, every object had a proper, or natural, place where it should be. A rock, for example, was composed of the earth element, and if it were tossed upward, it would attempt to return to the Earth. A lighter object, such as a feather, would do the same, but because it was not as heavy as the rock, it would not try as diligently to return to its proper place, and therefore it would fall more slowly to the Earth. On the other hand, smoke was composed of the fire element, and its natural place was above the air element. Therefore, smoke would drift upward from a smoldering fire.

These teachings of Aristotle, along with others that we will consider later, became a firmly entrenched part of people's outlook on nature, and, in fact, these ideas reflected everyday observations reasonably well.

The main difficulty with Aristotle's approach was that some of his theories were at odds with physical observations. Additionally, Aristotle presented his ideas without resort to experimentation. This means that his study of motion closely resembled that of a philosophical argument built on logic and debate without resorting to what Aristotle considered to be the lowly task of seeing if his laws really described events occurring in the natural world. Nevertheless, the authority of Aristotle was such that his ideas held sway for about 2000 years. Thus, his incorrect physics has been taught and believed far longer than has our present-day view of the subject. Aristotle's science was eventually challenged by Galileo Galilei (1564–1642), considered by many to be the dominant figure in leading the world of physics into the modern era. Among the changes wrought by Galileo was the introduction of the concept of time into the study of physics. For example, Aristotle held that the most important feature in determining the motion of an object was how far away it was from its proper place. Galileo, however, recognized that it was the time of fall, or the time for the motion to occur, that was an important missing link. With the introduction of time into physics, Galileo was able to develop the concept of accelerated motion. Let us now examine some of our present-day notions of motion.

2.2 SPEED

In some fashion, people have always been concerned with the subject of **speed.** For example, in today's world our busy schedules cause us to

Galileo Galilei

glance frequently at our watches to see how fast we must travel to make it to an appointment. Likewise, prehistoric people also must have pondered speed as they sought ways to capture a rapidly moving antelope.

Everyone can already give a basic definition of speed. In fact, when you tell someone how fast you are going in a car, you are defining speed. For example, suppose you are traveling at 55 mi/h. The units, mi/h, are essentially a definition of speed. That is, speed is the distance an object moves divided by the time required for the object to travel this distance. In equation form this is

$$\text{speed} = \frac{\text{distance}}{\text{time}}$$

or symbolically,

$$v = \frac{d}{t} \tag{2.1}$$

where d is the distance traveled and t is time required to cover this distance.

As an example, suppose a car moves 5000 m in 200 s. The speed is

$$v = \frac{d}{t} = \frac{5000 \text{ m}}{200 \text{ s}} = 25.0 \text{ m/s}$$

It is often useful to define two different types of speed, **average speed** and **instantaneous speed.** The definition presented in Eq. 2.1 is that of average speed. Average speed gives us information about the gross characteristics of a trip but omits many important details. For example, in the 5000 m trip above, it may well have been that the driver traveled very fast at the beginning, stopped for a traffic light, then made the last portion of the trip at a moderate speed because of a nearby police car. Nonetheless, the result was that the car traveled the 5000 m in 200 s for an average speed of 25.0 m/s.

The instantaneous speed of an object is defined as the speed of an object at any instant of time. The speedometer on your car tells you your instantaneous speed.

EXAMPLE 2.1 Strike the bum out!

Sound travels at a speed of approximately 345 m/s. If you are sitting in the stands at a baseball game, how long after a player 275 m from you hits a baseball do you hear the sound of ball striking bat?

Solution The equation for speed can be solved for time as

$$\text{time} = \frac{\text{distance}}{\text{speed}}, \quad \text{or} \quad t = \frac{d}{v}$$

If you need help with algebraic manipulations such as the above, consult Appendix E. We find

$$t = \frac{d}{v} = \frac{275 \text{ m}}{345 \text{ m/s}} = 0.797 \text{ s}$$

Let us give you the healthy warning that you should try calculations like this on your own. If you use a calculator, this device usually gives your answer out to eight or ten decimal places. However, our problems are always stated so that three-place accuracy is assumed. This means that you should consider the answer out to only three significant figures. (For a more in-depth explanation of significant figures, consult Appendix B.)

EXERCISE

Often we follow up one of our example problems with one for you to do to test your understanding. Our first one is nonmathematical.

A car travels 1000 m in 20 s. **(a)** How many

FOCUS ON . . . Equations

Equation 2.1 is the first equation to be introduced in this text. Equations are simple, concise statements of the relationships between physical properties. An equation not only summarizes a given relationship, but by omission it tells us what factors do *not* affect the others. For example, suppose someone asked you how the color of a vehicle affects its speed. Of course, you know that the speed of a vehicle is in no way affected by its color. This intuitive knowledge is supported by a look at the equation, for it contains no symbol for color. This example may sound trivial, but as more complex concepts are introduced, don't forget this generalization. Many students have become confused by making a problem more complex than it is and trying to include factors that are irrelevant.

---

average speeds does it have, and **(b)** how many instantaneous speeds does it have?

Answer **(a)** one, **(b)** an infinite number

2.3 VECTORS AND SCALARS

So far in this chapter you have encountered only three different physical quantities—distance, time, and speed—but as you continue your study, you will meet many more. However, all the physical quantities that you will run into can be placed in one of two categories—**vector** quantities and **scalar** quantities.

A scalar quantity is one that requires only a knowledge of its magnitude (and the units associated with the magnitude) to tell you all you need to know about it. Examples of scalar quantities are the number of pages in this textbook, the amount of fluid in a container, the temperature of a room, and the amount of change you have in your pocket or purse.

A vector quantity has both magnitude and direction. Force is a common physical quantity that is a vector. We have not discussed forces yet in this text, but your intuitive ideas about forces should be sufficient to enable you to understand why both a magnitude and direction must be specified for them. For example, suppose you walk up to a

This is a roadside "vector." As an exercise, decide what vector quantity this sign is asking the driver to change. Is it displacement, velocity, acceleration, or more than one of these?

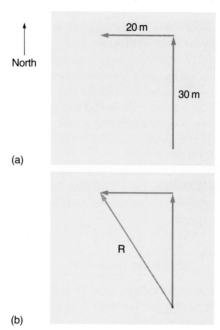

FIGURE 2.1 The triangle method for adding vectors.

door that is standing open and exert a 10 pound force on it. A simple statement of the strength of the force, 10 pounds, does not provide enough information to tell you what is going to happen to the door. The door moves differently if a 10 pound push is exerted on the door than it does if the force is a 10 pound pull. Thus, force is a vector quantity.

A second example of a vector quantity is **displacement.** *A displacement is the straight-line distance and direction from where motion begins to where it ends.* To see that displacements are vectors, consider the following situation. Suppose that someone tells you that a football player made a run of 30 m, stopped momentarily while he avoided a tackler, and then ran 20 m farther. How far is the player from his initial position? The answer that may pop into your mind is 50 m. However, you have assumed that both runs, or displacements, were in the same direction, which is not necessarily the case. The 20 m run could have been in the opposite direction to the 30 m run, in which case, the distance of the player from his starting point would be 10 m. Or the 30 m run could have been north and the 20 m run toward the west. How far away from the starting point is he in this case? The steps for answering this question are shown in Figure 2.1. First, a convenient scale, such as 1 cm equals 10 m, is used to draw the displacements, maintaining their relative directions with respect to one another. Figure 2.1a shows the resulting sketch. Then a vector is drawn which starts at the tail of the first and ends at the tip of the second. This is vector R in Figure 2.1b. The length of this vector can be measured from the figure and converted to an actual displacement by using the same scale that was used initially (1 cm = 10 m). If you try this for yourself, you will find this displacement to be approximately 36 m. This general process is referred to as *adding vectors,* and the specific technique used here is called the *triangle method for vector addition.*

EXAMPLE 2.2   I'm tired, and I'm weary, and I want to go home.

The triangle method for vector addition applies even if the displacements are not at right angles to one another as they were in our example above. For example, suppose that a jogger runs 20.0 m due north and then turns to run at an angle 45° north of east for 40.0 m. How far away from her starting point is the jogger at the end of this run?

Solution Figure 2.2 shows the steps for completion of the problem. A scale is chosen to draw the

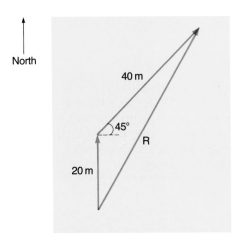

FIGURE 2.2 Example 2.2.

vectors, and the resultant is then sketched in and its length measured. Try it for yourself to find that the resultant displacement, or vector sum, is about 55.8 m in magnitude and north–northeast in direction.

2.4
VELOCITY

In everyday life, the terms "speed" and "velocity" are often used interchangeably, but in the world of physics the two have distinctly different meanings. The basic distinction between the two is that

Speed is a scalar quantity whereas velocity is a vector quantity.

For example, if we specify that a car is moving at 20 km/h northward, we have specified the velocity, which is a vector quantity. On the other hand, if all we tell you about the car is that it is traveling at 20 km/h, we have specified the speed, which is the magnitude of the vector without regard to its direction. Speed is important in many applications, but there are many practical situations in which we need to know more about an object than simply how fast it is traveling. For example, suppose your girlfriend tells you that she is going out for her morning jog. She tells you that she is going to pace herself so that she runs at a speed of exactly 3 km/h, that she will run in a straight-line path, and that she plans to run for exactly one hour. She then asks you to come pick her up

in your car. Can you satisfy this request if no more information than this is given? You know that she is exactly 3 km from her starting point, but if she does not tell you the direction of her run, you will have no idea of where to go to give her a ride. Velocity is a vector quantity that requires a knowledge of the magnitude and direction; thus, if she is to completely specify her velocity, she will do so by saying that she will run at, say, 3 km/h due north on Maple Street.

To find the magnitude of the velocity, you use exactly the same equation as specified earlier for speed:

$$\text{velocity} = \frac{\text{displacement}}{\text{time}} \qquad (2.2)$$

However, it is convenient to shorten this equation such that we let v = velocity, d = displacement, and t = time. Thus, we have

$$v = \frac{d}{t}$$

2.5
ACCELERATION

As you travel along a highway in your car, you seldom travel at a constant velocity for long periods of time. Red lights, the flow of traffic, or a nearby state policeman will cause you to alter your velocity. **Acceleration** is defined as *the change in velocity of an object divided by the time it takes the change in velocity to occur.* In equation form, this can be expressed as

$$\text{acceleration} = \frac{\text{final velocity} - \text{initial velocity}}{\text{time}}$$

or

$$a = \frac{v_f - v_i}{t} \qquad (2.3)$$

Recall that velocity is a description of the speed and the direction of an object. Therefore, if an object moving at a constant speed changes direction, the velocity also changes. Since acceleration is a change in velocity, a body moving at a constant speed around a curve is accelerating.

This concept may seem confusing at first until you think of it in human terms. You will learn in a later section that an object will accelerate only if forced to do so. Think of a situation in which you are riding in a car. When you speed up suddenly, the seat pushes against your back. Similarly, if you slow down suddenly while you are leaning against the dashboard, you feel the dashboard pushing against you. Now, what happens when you turn a sharp corner? You may feel the side of the vehicle press against your shoulder as various forces cause you to change your direction and therefore your velocity. Starting, stopping, and turning are all forms of acceleration.

There are some points that need to be clarified about the equation for acceleration, and these features can best be demonstrated with an example. For that reason, go through the following example carefully; do not skip past it.

EXAMPLE 2.3   Rev it up!

(a) A motorist traveling in his car causes his car to change from a velocity of 5.00 m/s due north to a velocity of 15.0 m/s, also north, in a time of 5.00 s. What is his acceleration?

Solution Let us call the direction of travel of the car, north, the positive direction. This means that both the velocities listed above have + signs associated with them. Thus, we have

$$\text{acceleration} = \frac{15.0 \text{ m/s} - 5.00 \text{ m/s}}{5.00 \text{ s}}$$

$$= 2.00 \, \frac{\text{m}}{\text{s}^2}$$

First note the units. An acceleration of 2.00 m/s² means that on the average the velocity of the car is increasing by 2.00 meters per second every second.

(b) The motorist, now traveling at 15.0 m/s, decides that he will lower his velocity to 5.00 m/s in a time of 5.00 s. (He is still traveling due north.) What is his acceleration in this case?

Solution From the definition of acceleration, we find

$$\text{acceleration} = \frac{5.00 \text{ m/s} - 15.0 \text{ m/s}}{5.00 \text{ s}}$$

$$= -2.00 \, \frac{\text{m}}{\text{s}^2}$$

The negative sign indicates that the direction of the acceleration vector is south and thus the car is slowing down at an average rate of 2.00 meters per second every second. Such an acceleration is usually called a *deceleration*.

2.6
ACCELERATION DUE TO GRAVITY

The acceleration of a falling object intrigued scientists for hundreds of years. Aristotle taught that heavier objects fall faster than do lighter ones. To illustrate Aristotle's point of view, consider a baseball that is 20 times heavier than a tennis ball. According to his approach, if the two are dropped simultaneously, when the baseball hits the ground, the tennis ball will only have fallen 1/20th as far. Before you read further, why don't you try it for yourself? If the two are allowed to fall for only a short distance, such as from eye level, you will find that they strike the ground at the same time. You should recall that Aristotle believed that natural laws could be understood by logical reasoning alone and that experimentation was largely unnecessary. In light of your simple experiment, it is surprising that Aristotle's teachings on falling bodies could have been accepted for about 2000 years.

Galileo questioned the conclusions of Aristotle and decided to test them by direct experiment. According to popular legend, Galileo took two objects, a heavy one and a light one, to the top of the Leaning Tower of Pisa and dropped them together. They both hit at the same time! It should be obvious that if they are dropped from rest and hit at the same time, their motions are alike in all ways as they fall. Because they are accelerated, they therefore must fall with the same acceleration. It should be noted here that Galileo recognized that a fluffy object like a feather or a piece of paper falls more slowly than a denser, more compact object like a rock. He concluded correctly that air resistance is the primary factor causing the difference in the time of fall in such cases. It is to his credit that he was able to make the leap of imagination to recognize that in the absence of such resistive forces, all objects, heavy or light, would fall with the same acceleration. On August 2, 1971, a vivid demonstration of this fact took place on the Moon. Astronaut David Scott simultaneously released a hammer and a feather, and they fell with the same acceleration to the lunar surface. This demonstration would surely have pleased Galileo. *We shall define an object that falls without any resistance to its motion to be a freely falling object.*

Galileo's experiments with freely falling objects were important in that they established that objects fall with the same acceleration, but more importantly he introduced the process of experimentation as a necessary part of scientific investigation. Granted, there had been experimentation before the time of Galileo. For example, some wealthy scientists had their slaves perform the menial task of doing experiments for them, and many astronomers had made and recorded observations. Galileo's work, however, led to the idea that experimentation is the only way to truly test the validity of a scientific proposition.

Galileo suggested that objects fall with the same acceleration everywhere, but actually the acceleration that a falling object experiences varies slightly at different locations. Measurements show, however, that the acceleration due to gravity is approximately 9.8 m/s² everywhere on Earth. This means that its speed changes at a rate of 9.8 m/s every second.

It is a simple matter to observe the speeds at various times to enable us to find a simple equation for the instantaneous speed, v, of a body that is falling from rest. This is

velocity = (9.8 m/s²) time

or

$$v = (9.8 \text{ m/s}^2) \, t \qquad (2.4)$$

You should note that this equation is just our old equation for the definition of acceleration, Eq. 2.3, in the form v = at, where we have set the

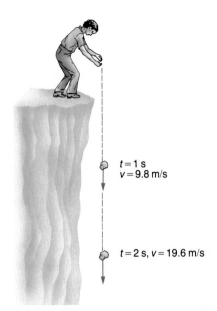

FIGURE 2.3 The speed of an object in free fall increases at the rate of 9.8 m/s for every second it falls.

initial velocity equal to zero and have used 9.8 m/s² for the acceleration. Often, in future equations, we will use the symbol g to stand for the acceleration due to gravity. Thus, Eq. 2.4 could be written as v = gt.

Thus, as shown in Figure 2.3, and indicated by Eq. 2.4, an object dropped from rest off a building will be traveling at 9.8 m/s after 1 second, at 19.6 m/s after 2 seconds have passed, and so forth.

If a ball is thrown upward, the acceleration due to gravity remains the same, but the effect of gravity is now to slow the object at a rate of 9.8 m/s every second. For example, if the ball has a speed upward of 29.4 m/s at the instant it leaves the hand, it will have decreased in speed by 9.8 m/s after one second. Refer to Figure 2.4. Thus, after 1 second, it will be traveling upward at a speed of 19.6 m/s. In another second, its speed will have declined by another 9.8 m/s, so its speed will be 19.6 m/s − 9.8 m/s = 9.8 m/s. Finally, after one more second, the speed will have been reduced to zero. It then comes back down as though it had been dropped from rest and accelerates, or increases its speed, at a rate of 9.8 m/s².

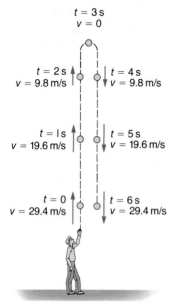

FIGURE 2.4 As a body in free-fall rises, its speed decreases at a rate of 9.8 m/s every second. While falling, its speed increases at a rate of 9.8 m/s every second.

EXAMPLE 2.4   Look out below!

A student drops his textbook off a high-rise dormitory. After 1.40 seconds of fall, how fast is the book traveling?

Solution   We use $v = (9.8 \text{ m/s}^2)\, t$ to find

$$v = (9.8 \text{ m/s}^2)(1.40 \text{ s}) = 13.7 \text{ m/s}$$

EXERCISE

How much time must pass before the textbook reaches a speed of 26.7 m/s?

Answer   2.72 s

2.7
MOTION WITH CONSTANT ACCELERATION

The most common case of an object moving with constant acceleration is that of a freely falling body just discussed. However, cars or other objects often travel with constant acceleration as well (at least for short distances). As a result, let us discuss this important kind of motion more completely.

At the end of the previous section, we found that the velocity of a falling object can be found at any instant of time via the equation

$$v = (9.8 \text{ m/s}^2)\, t$$

We can generalize this equation to apply to objects moving with a constant acceleration, other than falling objects, by substituting a, the acceleration, for 9.8 m/s^2 in the equation above. We have

$$v = at \tag{2.5}$$

You should note that this equation applies only if the object under consideration starts from rest.

Galileo was able to find a relationship between the distance covered by an object moving with constant acceleration and the elapsed time. His method was to roll balls down a slight incline and to measure the distance they moved during successive time intervals. The purpose of the incline was to slow down the movement of the ball sufficiently to enable him to take accurate measurements of time. The relationship he found was

$$d = \frac{1}{2}at^2 \tag{2.6}$$

where d is the displacement, a is the acceleration, and t is the elapsed time. You should note, again, that this equation is valid only when the object under consideration starts from rest.

EXAMPLE 2.5   Looking for the checkered flag

A jalopy starts from rest and moves in a straight line with a constant acceleration of 2.0 m/s^2 for a time of 5.0 s

(a) Find the velocity of the car at the end of this time interval.

Solution   From $v = at$, we have

$$v = (2.0 \text{ m/s}^2)(5.0 \text{ s}) = 10 \text{ m/s}$$

(b) Find the distance covered during the 5.0 s time interval.

Solution From d = $\frac{1}{2}$at^2 we have

$$d = \frac{1}{2}(2.0 \text{ m/s}^2)(5.0 \text{ s})^2 = 25 \text{ m}$$

EXERCISE

At the end of the 5.0 s time interval given above, the car stops accelerating and begins to move at constant velocity. What is its velocity after another 5.0 s have passed and what distance has it covered during this second 5.0 s interval?

Answer v = 10 m/s, d = 50 m.

2.8
FORCES AND MOTION

Thus far, we have investigated some terms common to the study of motion, including "speed," "velocity," and "acceleration." However, we have not answered questions such as: What causes a car to accelerate? What causes a car to move at a constant velocity? Under what conditions does a car move at constant acceleration? The answer to these questions can be found from one of three laws of motion discovered by Isaac Newton (1642–1727). Newton was born at Woolsthorpe, England, on Christmas Day in the same year that Galileo died. He is considered by many to be the most brilliant scientist who has ever lived. During an 18-month period in his early twenties, he formulated the law of universal gravitation, invented the mathematical concept of calculus, discovered the three laws of motion, and proposed theories on light and color. Even before this period, he had gained a measure of immortality by inventing the reflecting telescope. He was buried in Westminster Abbey with the following epitaph: "Mortals, rejoice that so great a man lived for the honor of the human race."

As noted above, Newton's contributions to physics were many and varied. However, as numerous as were his ideas and concepts, we shall devote the remainder of this chapter to those that apply to the study of mechanics. These are the famous first, second, and third laws of motion.

2.9
FORCE

Most people have an intuitive understanding of what is meant by "force"; it is usually thought of as a push or pull on an object. For our purposes at present, this inherent idea is sufficient, but we shall return to this topic again later in this chapter and place our ideas concerning forces on a more rigorous basis.

In this country, the unit used to measure a force is the pound. For example, when you say that your weight is 150 lb, this means that the Earth is pulling down on you with a force of 150 lb. In scientific usage, and in most foreign countries, the most commonly used unit of force is the newton. To be precise, the relation between a pound and a newton is

$$1 \text{ N} = 0.224 \text{ lb}$$

As an exercise, try calculating your weight in newtons. To see if you are on the right track, the weight of a 150 lb individual is about 670 N.

Forces are vectors

We pointed out earlier that physical quantities like velocity and displacement are vectors. We have seen how to add vectors, such as two displacement vectors, to find the resultant displacement, and we add forces to find the resultant force by the same technique. In fact, among the most important tasks that one must perform to analyze the motion of an object is to be able to recognize all the forces that act on the object and then to find the resultant, or net, force. The **resultant force,** or **net force,** is defined as the vector sum of all the forces acting on an object. Let us examine these steps for a couple of situations.

Consider an overhead view of a safe as shown in Figure 2.5. We see two not-so-bright burglars attempting to move the safe by exerting forces on it in the directions indicated. They both

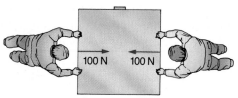

FIGURE 2.5   Each burglar pushes on the safe with a force of 100 N, but the net force on the safe is zero.

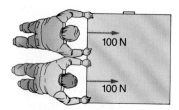

FIGURE 2.6   Forces in the same direction add. The net force on this safe is 200 N.

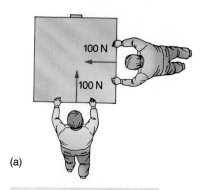

(a)

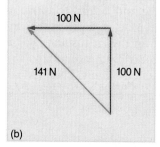

(b)

FIGURE 2.7   (a) Two forces not along the same direction must be added by the triangle method. (b) The triangle method yields a resultant of 141 N.

push with a force of 100 N, but since the forces are in opposition to one another, the resultant force on the safe is zero.

If the two burglars exert forces as shown in Figure 2.6, the resultant force is just the sum of the two individual pushes, and the safe moves as though a single force of 200 N were being exerted on it. (We ignore friction in all these cases.)

However, what about the situation shown in Figure 2.7a? In this case, the two burglars are exerting forces on the safe at right angles to one another. The net force can be found by the triangle method of vector addition, as shown in Figure 2.7b. You can draw the forces to scale for yourself and measure the resultant force to be about 141 N.

2.10
THE PYTHAGOREAN THEOREM

In our discussion of the addition of vectors, we have found the resultant of two vectors at right angles by the use of a graphical technique. There is an alternative to this approach that allows you to find the resultant by mathematical techniques alone. This approach may be traced back to the Greek thinker Pythagoras, who died about 500 B.C. According to the Pythagorean theorem, the length of the sides of a right triangle, shown in Figure 2.8, are related as

$$R = \sqrt{A^2 + B^2}$$

This says that the length of the longest side R, called the hypotenuse, is equal to the square root of the sum of the squares of the lengths of sides A and B. To see how to use this equation in a practical application, let us suppose that we have two displacements at right angles, as shown in Figure 2.9. The length of the vector forming side A is 4 m, and the length of the vector forming side B is 3 m. What is the resultant of these two vectors? To find the answer, we proceed as follows:

$$
\begin{aligned}
R &= \sqrt{(4 \text{ m})^2 + (3 \text{ m})^2} \\
&= \sqrt{16 \text{ m}^2 + 9 \text{ m}^2} \\
&= \sqrt{25 \text{ m}^2} \\
&= 5 \text{ m}
\end{aligned}
$$

To verify the Pythagorean theorem result, you might check this answer with the graphical techniques discussed earlier.

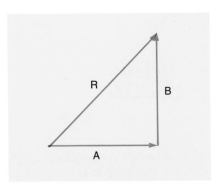

FIGURE 2.8 The Pythagorean theorem provides an alternative to the graphical process of vector addition. The resultant vector R is found from R = $\sqrt{(A^2) + (B^2)}$.

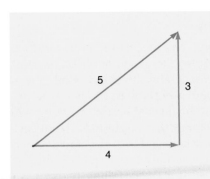

FIGURE 2.9 Use of the Pythagorean theorem shows us that the length of the resultant is 5 units.

Recognizing forces

It is often necessary in solving mechanics problems to be able to find the net force on an object by use of the method given above. However, even before one can apply the techniques of vector addition, one must be able to pick out all the forces that are being exerted on an object. Usually this is a trivial exercise that can be done easily, but occasionally it is a little difficult to do. Because of this, let us give you a few tips on how to recognize when a force exists.

As we observe the world around us, forces seem to fall into two convenient categories. These are **action-at-a-distance** forces and **contact** forces. *Action-at-a-distance forces are those which one object exerts on another even when there is no physical contact between the two.* In the world of physics

there are not many of these. The most common is the force of gravity. As you sit in your room reading this, the Earth is pulling on you with a force that is directed toward the center of the Earth and that has a magnitude equal to your weight. If you jump off a building, this force acts on you while you are falling. This means that the force of gravity is present even in those cases where there is no physical contact between you and the Earth. As we venture further into physics, we will find that there are other action-at-a-distance forces. Two charged objects exert electrical forces of attraction or repulsion which are of the action-at-a-distance type. Likewise, a magnet can attract a piece of iron or other magnetic material across space without the two touching. In our day-to-day experience, these are the only three action-at-a-distance forces we encounter. In fact, as far as the study of mechanics is concerned, the only one you need to be concerned with is the force of gravity.

Contact forces arise when there is actual physical contact between two or more objects. For example, when a batter hits a baseball, the bat exerts a force on the ball only during the time when the bat and ball are actually in contact. Likewise, a punter exerts a force on a football only while his toe is in contact with the football. As a general rule, contact forces can arise anytime there is physical contact between two objects. It should be noted that these forces, which we are calling contact forces, can be shown to be action-at-a-distance forces when we consider them on a submicroscopic level. For example, the actual origin of the contact force exerted by a bat on a ball arises from electrical interactions among atoms of the two objects. However, on a large-scale basis, these electrical forces are not obvious. Instead, we recognize their presence in terms of a gross force that we call a contact force.

Before we continue with our study, let us pause to get some practice in recognizing the presence of forces.

EXAMPLE 2.6

Figure 2.10a shows a simple situation in which a book rests on a table. Find all the forces that act on the book.

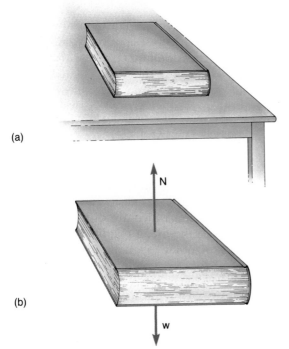

(a)

(b)

FIGURE 2.10 (a) A book at rest on a table. (b) There are two forces on the book; w its weight, and N the normal force.

Solution To approach this problem, draw a sketch of the book by itself. Figure 2.10b is our sketch. Now, let us begin to find the forces that act on it. First, are there any action-at-a-distance forces present? Of course there is always one, the force of gravity, which is manifested as the weight of the object. We have sketched this in Figure 2.10b as w, and we have indicated its direction toward the center of the Earth. There are no other action-at-a-distance forces to be considered for our situation, so we must now attempt to recognize what, if any, contact forces are acting. To determine what may be exerting a contact force on the book, we must first ask whether there is anything touching the book. The only object that is in contact with the book is the table, and it does, indeed, exert a force on the book. This force is directed perpendicular to the surface along which the book and table make contact. This force is often called the *normal force*, and is pictured as N in Figure 2.10b. Since nothing else is in physical contact with the book, there are no other contact forces present and our diagram is complete.

FIGURE 2.11 An object in free fall has only one force acting on it. This force is its weight.

EXAMPLE 2.7

Suppose the book of the preceding example falls off the table. Draw a diagram of the book that shows all forces acting on it while it is falling to the floor.

Solution The diagram is shown in Figure 2.11. The weight, the action-at-a-distance force, is always present, and this is indicated as w in the figure. There is nothing in contact with the book, so there can be no contact forces acting. (We are ignoring forces that might arise from such factors as air resistance.) Thus, the only force acting on the book as it falls is its weight. In fact, this is the definition of a freely falling body; it is one that has only one force, the weight, acting on it. As you recall, freely falling bodies have an acceleration of 9.8 m/s².

2.11 NEWTON'S FIRST LAW OF MOTION

Take a look at a book lying at rest on your desk. Why is the book not moving? Imagine the Rock of Gibraltar at rest at the mouth of the Mediterranean Sea. Why does the Rock of Gibraltar not move with respect to the Earth? **Newton's first law of motion** gives us an insight into why things that are at rest stay at rest, and what must happen to them if they are to begin to move. We can state Newton's first law of motion in partial form as:

An object at rest will remain at rest unless a net force acts on it.

We saw in Example 2.6 that only two forces act on a book resting on a table—the force of

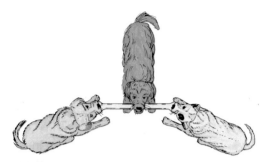

FIGURE 2.12 If the bone does not move, the net force on it is zero.

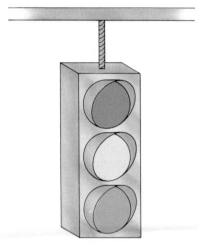

FIGURE 2.13 Find the tension in the cable if the traffic light weighs 150 N.

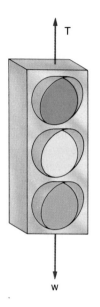

FIGURE 2.14 The forces on the traffic light are its weight w and the tension in the cable T.

the tension in the cable, as shown in Figure 2.14. (If you are not sure about this, re-read the section on recognizing forces.) Since Newton's first law tells us that the net force must be zero, the only way this can be possible is if the tension in the cable is also equal to 150 N.

Newton's first law applies as shown above to objects that are at rest, but as we shall now see, it also applies under certain conditions to objects that are moving. To investigate motion via the first law, consider the following experiment. Take a book and slide it across a floor. The book slides for only a very short distance and then comes to a stop. The ancient Greeks chose to concentrate on this aspect of matter. They said that the natural state of matter was for it to be stopped. When an object such as the book is set into motion, it naturally comes to a stop. But take this same book and slide it across a smooth, highly polished floor. What happens now? Again, it comes to a stop, but not nearly as soon.

It was Galileo who looked at motion and concluded that scientists were considering it from an unproductive viewpoint. He said that the natural state of matter is to remain in motion once it is set in motion. Objects do not "naturally" stop; they "naturally" continue to move. In our everyday world, this does not seem to make sense, because every object that we see moving eventually stops.

gravity and the normal force. Thus, according to Newton's first law, these two forces must cancel one another to give a net force of zero if the book is to remain stationary. Figure 2.12 shows a bone being tugged by three hungry dogs. If we watch the bone and observe that it is not moving, we know that the dogs are pulling such that the vector sum of all the forces being exerted by the dogs is zero.

EXAMPLE 2.8

A traffic light weighing 150 N hangs from a cable, as shown in Figure 2.13. Find the tension in the cable.

Solution To work this problem, we must first be able to recognize all the forces acting on the traffic light. There are two of these, the weight and

The reason that the book does stop in its slide across the floor is that there is a force of friction acting on it, and the reason that the book slides farther on a polished floor is because the force of friction is smaller than it is on a rough surface. On an air hockey table, it would slide even farther before friction brought it to rest. Now, let your imagination leap to a floor so highly polished that the force of friction could be reduced to zero. In this case, the book would slide forever, because there would be no force to stop it.

This new idea of motion was formalized by Isaac Newton as part of his first law. He said

An object in motion will remain in motion at the same velocity (same speed and same direction) unless a net force acts on it.

Thus, when you see a car moving down a road at a constant speed and in a straight line, according to Newton's first law the net force on the car must be zero.

EXAMPLE 2.9   Frequent flyer tickets accepted

A jet airplane is flying a straight course at a constant speed of 500 mi/h. If the engines of the jet exert a forward thrust on the plane of 40,000 N, what is the force of air resistance acting on the plane?

FIGURE 2.15 If the jet is to fly at a constant velocity, the forward thrust by the engines T must be equal to the force of air resistance f.

Solution If the plane is flying at a constant speed in a straight line, the net force on it must be zero. The only way this can be possible is if the force of air resistance, f in Figure 2.15, is equal to the forward thrust of the engines, T. Thus, f = 40,000 N.

We can now form a complete statement of Newton's first law as:

An object at rest remains at rest, and an object in motion with constant velocity continues in motion with the same velocity unless a net external force acts on it.

2.12 INERTIA

Newton's first law of motion is often referred to as the law of inertia. To see why it has acquired this name, consider some passengers standing in a bus. If the bus is not moving, the passengers can easily stand up, but what happens if the bus suddenly begins to accelerate forward? You know from your own experience that the passengers will fall backward in the bus. From the insights gained by our brief study of the first law, let us see if we can understand why they fall. From the viewpoint of the first law, the passengers were initially at rest, and they will remain at rest until a net force acts on them to change their motion. *The tendency for matter to remain in whatever state of motion that it is in is called* **inertia.** Thus, since the passengers were at rest, their tendency is to stay at rest, and they fall because the bus accelerates out from under them. Of course when they regain their footing and are holding onto a strap, the strap exerts a force on them to make them accelerate along with the bus. After the bus has reached a constant velocity of 80 km/h, the passengers again can stand comfortably without support. Now suppose the driver throws on his brakes. You know that the passengers will pitch forward. From the inertia viewpoint, the passengers when traveling at a constant speed in a straight-line path will continue to move in this fashion unless a net force causes them to change this motion. If they are not prepared to grab hold of something, their inertia will cause them to continue to move and to fall forward as the bus stops under them.

These observations were placed on a quantitative basis by Newton with the definition of a quantity called **mass.** *Mass is a term used to describe how much inertia an object has.* A very massive object requires a large force to start it into motion or to stop it once it is started. As we saw in Chapter 1, in SI units, mass is measured in kilograms.

Often, beginning students of physics confuse weight and mass, but the two are distinctly different. The weight of an object is the force of attraction that the Earth exerts on it. At the surface of the Earth, an object with a large mass also has a great weight because the Earth is pulling on it very strongly. However, if this same object is taken thousands of kilometers into outer space, the pull of the Earth on the object has weakened greatly. Thus, its weight has decreased, but what about its mass? Since mass is a measure of how difficult it is to start an object into motion or to stop it once it is in motion, consider yourself to be in outer space and our massive object comes hurtling toward you like a meteor. Even though it has no weight, you will still find it very difficult to stop. Thus, the mass of an object does not change regardless of its physical location in space. If an object has a mass of 10 kg on Earth, it will have this same mass on the Moon or in the deepest reaches of outer space.

2.13
NEWTON'S THIRD LAW OF MOTION

Since Newton named his laws the first law, the second law, and the third law, it seems reasonable that we should discuss them in that order. However, of the three, the second law is a little more difficult for beginning students than are the first and the third. As a result, let us withhold our discussion of the second law for a few pages and examine the third law of motion next.

There is an age-old admonition that a parent often gives to a child while administering a spanking. The parent often says, "This hurts me as much as it does you." As we shall see, at least in principle, this description is absolutely correct from the standpoint of Newton's **third law of motion.**

Basically, what Newton discovered and presented with his statement of the third law of motion is the fact that forces always occur in pairs. There is no such thing as a single, isolated force. When discussing these force pairs, Newton chose to call one of them the *action* and the other the *reaction.* There is no real significance in these names. He could equally well have chosen to call one of them force A and the other force B, or he could have selected designations from a multitude of other possibilities. Also, when discussing these force pairs, there is nothing that signifies that one of them is the action and the other the reaction. If you choose to call a particular one of

these forces the action and the other the reaction, that is perfectly acceptable, and it is just as acceptable if someone else talking about these same forces chooses to use the opposite designations.

Newton's statement of the third law is:

For every action there is an equal and opposite reaction.

This simple statement of the third law is perfectly acceptable if one already understands all of the subtleties of the statement, but one must understand all the fine print. To recognize action-reaction pairs, first note that these pairs have three features. (1) The forces are equal in magnitude. (2) They are opposite in direction. (3) They always act on different objects.

As an example, suppose you are watching two boxers go at each other in the ring. One of them exerts a force of 10 N on the chin of his opponent. This is a force, and let us call this the action. Where is the second force, the reaction? An easy way to find the reaction is to make a simple declarative sentence concerning the action force and then invert the sentence to find the reaction. For our case above, we might state:

The action is the force that the fist exerts on the chin.

Now invert the sentence to find the reaction. The resulting sentence is

The reaction is the force that the chin exerts on the fist.

Thus, we see that if boxer A exerts a 10 N force on the chin of boxer B, the chin of boxer B also exerts a force of 10 N on the fist of boxer A. Now you can see that the statement, it hurts me as much as it does you, is valid, at least in principle. (Of course, chins are more sensitive to forces than are fists.)

Figure 2.16 shows these forces, and you should note that all three characteristics of action-reaction pairs are satisfied. The forces are equal in magnitude because both are 10 N forces. They act in opposite directions. This can easily be demonstrated by noting that the force exerted by the fist on the chin snaps the head of boxer B back, while at the same time, the force of the chin on the fist of boxer A acts to slow down the forward motion of the hand. Finally, note that the forces

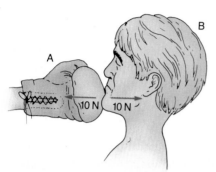

FIGURE 2.16 The force of 10 N on boxer B and the oppositely directed 10 N force on the hand of boxer A are an action-reaction pair.

FIGURE 2.17 A 500 N student sits in a chair. Find all the action-reaction pairs.

act on different objects: one was on the fist, the other on the chin.

EXAMPLE 2.10   I felt the Earth move!

Figure 2.17 shows a 500 N student sitting in a chair. We will consider the chair to be light enough that we do not have to consider its weight. Find all the action-reaction pairs.

Solution   In a sense, this problem is also a review of the first law. The first thing we must do is to find all the forces acting on the different objects—the student, the chair, and perhaps something else. We consider first the forces on the student, as shown in Figure 2.18. Force A in Figure 2.18a is the force that the chair exerts upward on the student. Force B is the weight of the student. From the first law, we recognize quickly that these two forces must be equal in magnitude and oppositely directed. Otherwise, the student

would either sink toward the floor or levitate upward. So the forces are equal in magnitude and oppositely directed, but are they action-reaction pairs? The answer is that they cannot be because they both act on the same object, the student, and we know that action-reaction pairs are always on different objects.

Before we go further with our analysis, let us find the forces acting on the chair. These are shown in Figure 2.18b. They are C, the downward force exerted by the student on the chair, and D, the force exerted upward on the chair by the Earth. (D is the normal force, which we have encountered in previous problems.) These also must be equal in magnitude and opposite in direction, but they are not action-reaction pairs since both act on the chair. However, we now have enough information at hand to begin to locate some action-reaction pairs. First, consider force A, the force of the chair on the student. Let us call this the action and make our declarative sentence concerning this force. We have

> The action is the force that the chair exerts on the student.

This is force A. Now turn the sentence inside out.

> The reaction is the force that the student exerts on the chair.

This is force C. Thus, A and C are action-reaction pairs.

Now the situation becomes a little more difficult, but only slightly so. If we call force B, the weight of the student, the action, what is its reaction? Again, form the sentence.

> The weight of the student, the action, is the force exerted on him by the Earth.

Inversion gives:

> The reaction is the force exerted on the Earth by the student.

This reaction force is shown in Figure 2.18c. The force, which acts upward in the figure, is one of 500 N, the same as the weight of the student. Thus, the student is exerting an attractive force on the Earth tending to pull it toward him, but don't expect a lot of movement on the part of the Earth because of its extremely large mass.

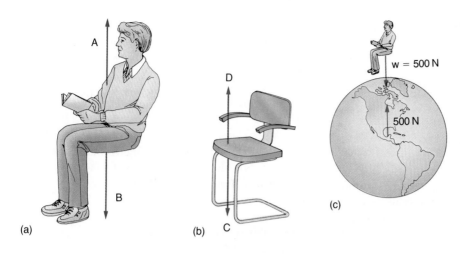

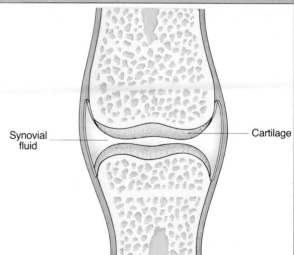

FIGURE 2.18 (a) The forces on the student are A, the force exerted on him by the chair, and B, his weight. (b) The forces on the chair are C, the downward force exerted on it by the student, and D, the normal force exerted on it by the surface supporting the chair. (c) The reaction force to the weight of the student is the gravitational force he exerts on the Earth.

FOCUS ON . . . Friction Forces in the Human Body

We take our bodies for granted and seldom pause to stand in awe of the marvelous engineering work that nature has done on all aspects of muscles, joints, and organs. For example, nothing is taken as much for granted as the ability to stand upright or to run when we so desire. Yet if it were not for some exceptional applications of the use of frictional forces in the joints, these tasks would be impossible. Consider the simple task of standing upright. If the frictional forces within the joints were too small, the bones would tend to slip across one another, and we would fall when they moved out of alignment. On the other hand, there are times when we do not want the frictional forces between the joints to be very large. An example of this would occur when you want to run or to do any exercise that requires rapid movement. If the frictional forces were too large within the joints, the bones would not be slippery enough to allow them to move freely across one another. As a result, there are times when we want the frictional forces within the joints to be large, and there are other times when we want these frictional forces to be relatively small. How does the body adjust to such varying demands?

The figure shows the construction of a typical joint in the human body. The ends of the bones are covered with cartilage, which is a sponge-like material, and the joint is encased in a capsule that contains a lubricating material called synovial fluid. When the bones are at rest with respect to an an-

A joint in the human body.

other, as they would be when we are standing stationary, the cartilage absorbs much of the synovial fluid; the ends of the bones become relatively dry, and the friction force between the two increases. On the other hand, when the bones are in relative motion, as they would be in a knee joint when a person runs, the bones press together more firmly, and this squeezes some of the synovial fluid out of the cartilage. This released fluid lubricates the ends of the bones and allows them to slide freely across one another. Truly, this is marvelous engineering.

We will leave it to you to find the reaction to the force D, the force exerted by the Earth on the chair. *Hint:* You may once again have to involve the Earth in this problem.

2.14
NEWTON'S SECOND
LAW OF MOTION

Newton's first law tells us what happens when the net force acting on an object is zero. The object either remains at rest if it is already at rest, or it moves with a constant velocity if it is in motion. But what happens if the net force on an object is not zero? Newton's answer to this is that the object is accelerated. The relationship between force, mass, and acceleration was summarized in his **second law of motion.**

The acceleration of an object is directly proportional to the net force acting on it and inversely proportional to the mass of the object.

If the proper units are selected, Newton's second law can be stated in equation form:

$$\text{acceleration} = \frac{\text{net force on object}}{\text{mass of object}}$$

or, rearranged and condensed, we have

Force = (mass)(acceleration)

or, in symbols,

$$F = ma \qquad (2.7)$$

In this equation F is the *net* force on the object, m is its mass, and a is the acceleration of the object. It should be noted here that the second law is often considered to be the most important law in mechanics because it stands as the first step in developing and understanding so many other concepts such as the conservation of momentum and the conservation of energy, which we will meet in the next chapter.

The newton

Newton's second law is used as the equation by which forces are defined. For example, let us apply the second law to a 1 kg object. We ask ourselves the question, "What force must we apply to this object, in order to give it an acceleration of 1 m/s^2?" To find the answer, we substitute into the second law as

$$F = ma = (1 \text{ kg})(1 \text{ m/s}^2) = 1 \frac{\text{kg m}}{\text{s}^2}$$

Thus, this force has a magnitude of 1. Its units, $^{\text{kg m}}/_{\text{s}^2}$, are lumped together and called a newton, N. Thus, we see that *a 1 N force is that force that will give a 1 kg object an acceleration of 1 m/s^2.*

The relationship between weight and mass

We demonstrated in an earlier section that weight and mass are two entirely different concepts, but there is a relationship between the two, and Newton's second law enables us to find this relationship. To do so, consider the situation shown in Figure 2.19, which shows a rock of mass m falling freely through space. The only force acting on the rock as it falls is its weight, w, as shown in the figure. Thus, let us substitute into Newton's second law that F = w, and a = g, where g is the acceleration due to gravity. We have

$$F = ma$$

$$w = mg \qquad (2.8)$$

This innocent-looking relationship between w and m is just a special case of the second law, but it will be used often and may be considered a conversion factor between w and m, as long as the object is on or near the Earth's surface. For example, a typical adult has a mass of about 75.0 kg.

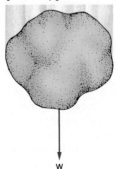

w

FIGURE 2.19   In the absence of air resistance, the only force on a freely falling rock is its weight w.

What is the weight of this individual in newtons? From w = mg, we have

$$w = mg = (75.0 \text{ kg})(9.8 \text{ m/s}^2) = 735 \text{ N}$$

2.15 EXAMPLES OF NEWTON'S SECOND LAW

Newton's second law requires a little more mathematical effort than do the other two of Newton's laws, and as a result, in this section we will give you a few worked-out examples to show you its application to real-world situations. It might be helpful to read the problem first and then attempt to work it out on your own before you look at our solution.

EXAMPLE 2.11   Race cars and stones

(a) A race car with a mass of 500 kg can accelerate from 10.0 m/s to 40.0 m/s in 4.00 s. How much force is required to cause this acceleration?

(b) Imagine that you had a stone with a mass of 500 g and threw it with a force of 100 N. What would be its acceleration while the force was acting on it?

Solution   (a) We will use F = ma. The acceleration is calculated according to the definition of acceleration as

$$a = \frac{\text{change in velocity}}{\text{time}}$$
$$= \frac{40.0 \text{ m/s} - 10.0 \text{ m/s}}{4.00 \text{ s}} = 7.50 \text{ m/s}^2$$

Substituting into Newton's second law:

$$F = (500 \text{ kg})(7.50 \text{ m/s}^2) = 3{,}750 \text{ N}$$

(b) In this case, we are asked to solve for acceleration. Newton's second law is used as

$$a = \frac{F}{m} = \frac{100 \text{ N}}{0.500 \text{ kg}} = 200 \text{ m/s}^2$$

Note that even though much less force is delivered to the stone than to the car, the acceleration of the stone is much greater. This occurs because the stone is much less massive than the car.

EXAMPLE 2.12   Watching your weight

A sled of weight 98.0 N is pulled across an icy, frictionless lawn by a boy who pulls with a force of 5.00 N on the sled. What is the acceleration of the sled?

Solution   Be careful here; you must use Newton's second law, F = ma, and in order to do so, you need to know the mass, m, of the sled. We are given the weight as 98.0 N, so we can easily find the mass by

$$m = \frac{w}{g} = \frac{98.0 \text{ N}}{9.8 \text{ m/s}^2} = 10.0 \text{ kg}$$

Now we can find the acceleration of the sled from the second law.

$$a = \frac{F}{m} = \frac{5.00 \text{ N}}{10.0 \text{ kg}} = 0.500 \text{ m/s}^2$$

EXERCISE

The problem above has a logical error in it, in the sense that it would be impossible for the boy to pull the sled. Why?

EXAMPLE 2.13   Get out of the way! I'm coming through.

How far does the sled travel in 3.00 s assuming that it starts from rest?

Solution   This is not really an example of the second law, but it does point out that you cannot forget the material studied earlier in this chapter. (Physics tends to be cumulative.) If you refer back to our discussion of motion with constant acceleration, you find that we can calculate how far an object travels in a certain period of time by use of the equation $d = \frac{1}{2}at^2$, as long as it starts from rest and has a constant acceleration. Thus,

$$d = \frac{1}{2}(0.500 \text{ m/s}^2)(3.00 \text{ s})^2 = 2.25 \text{ m}$$

EXAMPLE 2.14 Auld acquaintances and net forces are not to be forgotten.

The child continues to pull on the 10.0 kg sled of the preceding two examples with a force of 5.00 N. However, when he moves onto a portion of the lawn that is not quite as icy as before, he finds that there is a friction force of 3.00 N that begins to act on the sled. What is now the acceleration of the sled?

Solution You must recall here that F in F = ma is the net force that acts on the sled. Because the friction force opposes the motion, the forces that act on the sled are shown in Figure 2.20. Thus, the net force acting on the sled is

$$F = 5.00 \text{ N} - 3.00 \text{ N} = 2.00 \text{ N}$$

and the acceleration of the sled is

$$a = \frac{F}{m} = \frac{2.00 \text{ N}}{10.0 \text{ kg}} = 0.200 \text{ m/s}^2$$

2.16
THE NATURE OF FRICTION

We have mentioned the subject of friction briefly in the preceding sections, and there we relied on your understanding of forces of friction based on common experience. In this section, we will examine friction a little bit more carefully, because it plays such an important role in the world of physics and technology.

Friction forces arise when one object attempts to move across another. For example, as a baseball player slides along the ground while stealing a base, there is a friction force that the ground exerts on him. *Frictional forces always act in a direction such that they oppose motion.* Thus, if the ballplayer slides toward the left, the friction force exerted on him by the ground is to the right.

FIGURE 2.20 The forces on the sled are the 5 N pull and the 3 N friction force.

Determining the direction of a force of friction is not very difficult to do, but complications sometimes arise when one attempts to predict the strength of this frictional force. Factors such as the conditions of the surfaces involved and the speed of the object must be taken into consideration. Figure 2.21a shows a brick resting on a rough surface. In order to study the nature of frictional forces, let us begin to pull on it to the left, as shown, with only a very small force of 1 N. (A force of 1 N is approximately equal to the weight of a small apple.) From your own experience, such a small force exerted on the brick will not set it into motion. Figure 2.21b shows why the brick does not move. When motion is impending toward the left, a frictional force arises to oppose it, and this frictional force f is toward the right. Since the object is not moving, Newton's first law tells us that the strength of the frictional force must be equal to that of your pull, or 1 N. *Frictional forces like this, which arise even when there is no motion, are called forces of static friction.*

Now, let us pull slightly harder with a force of 2 N. Still the object does not move, so we conclude that the frictional force must have risen also and is now equal to 2 N.

Let us pull harder yet, with a force of 3 N. Still the object does not move, but now we find that if we pull only a fraction of a newton harder, the object *will* begin to move. From these observations, we see that the force of static friction between two surfaces has a maximum value that it can reach. If one pulls on an object with a force greater than the maximum value of this frictional force, the object begins to move. To see the origin of the force of static friction, consider Figure 2.22, which shows a greatly magnified view of the surface of the brick in contact with the floor. Note that little bumps on each surface have settled against one another. For the brick to slide, it must either rise over the bumps or break them off, and either action requires force. Even very smooth objects have tiny, molecular or atomic-sized irregularities. In fact, as one object is moved across

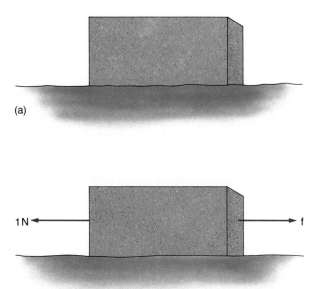

(a)

1 N ◄─────────────────────────► f

FIGURE 2.21 (a) A brick resting on a rough surface. (b) When a small force of 1 N is exerted toward the left, it is opposed by a 1 N friction force to the right.

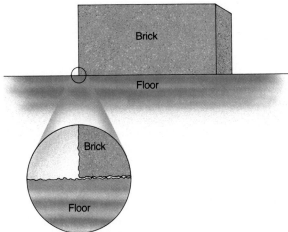

FIGURE 2.22 Magnified view of surfaces of a floor and a brick.

another, the atoms themselves can cling together and oppose the motion.

Once an object is set into motion, the bumps of one surface do not have time to fully settle into the holes in the other, and the friction force is less than when resting. Thus, if a force of 3 N is required to overcome the force of static friction, a smaller force, say 2 N, is large enough to keep the object moving at constant velocity once it begins to move. *The force of friction that acts on a moving object is called the force of kinetic friction.*

The preceding explanation of the origin of friction forces also explains why friction is reduced after rough objects have been rubbed back and forth against one another. This wears down the irregularities, resulting in smoother surfaces. Friction is also reduced by lubricating the surfaces with oil. The film of oil separates the surfaces so that there is no intimate contact between the irregularities of each.

2.17 DENSITY

There is an age-old trick question that we would like for you to try. It goes like this, "Which has

more mass, a kilogram of feathers or a kilogram of lead?" The answer is, of course, that they both have the same mass, one kilogram. The reason that this question often catches people off their guard is that there is a tendency to think of a given volume of each. For example, you might try to imagine how much mass of feathers a small paper bag would hold, and then think of that same bag filled with lead. As a result, the off-the-cuff response would be that the lead has more mass. The point of this example is to indicate that it is frequently important to discuss not just how much mass something has, but to also indicate in some way how this mass is compressed or spread out. To this end, a quantity called **density** is defined as follows. *Density is the mass of a substance divided by the volume of the substance.* In equation form this is

$$\text{density} = \frac{\text{mass}}{\text{volume}}$$

or, in symbolic form

$$\rho = \frac{m}{V} \qquad (2.9)$$

FOCUS ON . . . When Is Free Fall Not Free Fall?

We have said in our discussion of free fall that objects accelerate downward with an acceleration of 9.8 m/s². This, of course, is true when the falling objects are in a vacuum with no force of air resistance to impede their motion. It is also almost true when the distance of fall is relatively short, because the force of air resistance does not have time to affect the motion appreciably. For example, try dropping a small, heavy rock and a tennis ball from eye level. Chances are great that you will not be able to tell that the rock hit the ground first, but if you take these objects to the top of a tall building and drop them off, you will find that there is a difference in the time of fall. Apparently, the two do not fall with the same acceleration, and the reason for this difference is the force of air resistance acting on the two. The Figure shows the forces acting on an object falling through air, or any fluid. If the only force were w, the weight, the object would fall with an acceleration of 9.8 m/s², but the presence of the force of air resistance, f, means that the net force acting on the object is smaller than w. Thus, from Newton's second law, we see that when the net force, F, decreases, the acceleration also decreases.

There are two factors that are important to consider in determining how large the force of air resistance is that acts on an object. They are (1) the area presented by the object to the air, and (2) the speed at which the object falls. Let us look at these in turn. You can do a simple experiment to convince yourself that the area presented to the air affects the speed of a falling object. Take a sheet of paper and drop it. You will find that the sheet tends to flutter

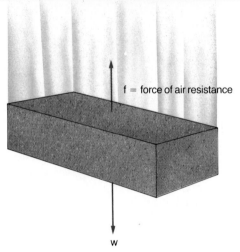

f = force of air resistance

w

The forces on a falling body when air resistance is present.

In a spread-eagle position, air resistance increases and terminal speed decreases.

where ρ is the symbol for density, m is the mass of the substance, and V is its volume. In the SI, density has units of kg/m³.

One primary advantage that density has in the discussion of substances is that under the same conditions of temperature and pressure every sample of a particular substance has its own characteristic density. That is, all iron has the same density, as does all gold, silver, and so forth. Table 2–2 gives the density for several representative substances. Note that the densities of solids are normally about 1000 times greater than densities for gases. This means that the spacing between molecules in the gaseous state is considerably greater than in the solid state.

TABLE 2–2 Densities of Selected Substances

Substance	Density (kg/m³)
Ice	0.917×10^3
Aluminum	2.70×10^3
Iron	7.86×10^3
Copper	8.92×10^3
Silver	10.5×10^3
Gold	19.3×10^3
Mercury	13.6×10^3
Water	1.00×10^3
Air	1.29
Oxygen	1.43
Hydrogen	8.99×10^{-2}

All values are for atmospheric pressure and 0°C.

to the ground. However, if you take the same sheet of paper and crumple it up into a ball, it falls much more rapidly. The reason that it falls more slowly when opened up as a sheet is that the area presented to the air is great, and air molecules that the sheet encounters must be pushed around it in order for the paper to move through the air. Each of the air molecules that strikes the sheet exerts a tiny force on it, but trillions of these air molecules are pushed aside as the sheet falls, so the cumulative effect of all these forces produces a large total force of air resistance on the paper. We can see that the force of air resistance also depends on the speed of fall by noting that as the speed of a falling object increases, the number of molecules of air that it runs into each second also increases. This means that more molecules have to be pushed out of the way, and since each exerts a force on the falling object, the net force of air resistance must increase.

With these facts concerning air resistance in mind, let us examine the motion of a sky diver as he falls toward the ground. What can he do to alter the speed at which he falls? If he spread-eagles himself, with both arms and legs extended, he presents a large area to the air, and he falls more slowly than if he moves into a diving position so that little area is presented to the air. It is of interest to note that without the effects of air resistance, a rain shower would be an event to be feared. In the absence of air resistance, a raindrop would reach speeds of about 200 mi/h before striking the ground, and a simple sprinkle could cause great devastation to foliage.

Let us consider a falling object in the presence of air resistance in terms of Newton's second law of motion, as shown in the figure. Initially, the force of air resistance is fairly small because its speed is small. However, as the object continues to fall, the speed increases, and so does the force of air resistance (by Newton's third law). The net force, F, on the object is w − f, where w is the weight and f is the force of air resistance. As f increases, the net force decreases. Thus, from F = ma, we see that as F, the net force, decreases, the acceleration a must also be decreasing. If the acceleration is decreasing, this says that the velocity is not increasing at as rapid a rate. Finally, there comes a time at which the force of air resistance has risen to the point where it is exactly equal to the weight of the falling object. At this point, F becomes equal to zero and, consequently, so does a. When the acceleration goes to zero, the speed no longer changes. From this point on, the force of air resistance does not change, and the object continues to fall with a constant speed. This speed of an object when the force of air resistance is equal to the weight of the object is called the **terminal speed.** Obviously, the purpose of a parachute is to decrease the terminal speed of a parachutist to the point where he can strike the Earth without endangering himself. Without a parachute, the terminal speed of a sky diver with a closed parachute is about 125 mi/h, while a sky diver with an open parachute reaches a terminal speed of only about 11 mi/h.

EXAMPLE 2.15   Rooking the King?

A king buys a block of material which a salesman says is gold. The block has a mass of 60.0 kg and measures 0.20 m by 0.20 m by 0.10 m. The king feels he may have been gypped, since the salesman operated out of the back of his car. Is the material truly gold?

Solution   The volume of the gold brick is given by V = (length)(width)(height). Thus, we find

$$V = (0.20 \text{ m})(0.20 \text{ m})(0.10 \text{ m}) = 0.0040 \text{ m}^3$$

The density is found to be

$$\rho = \frac{m}{V} = \frac{60.0 \text{ kg}}{0.0040 \text{ m}^3} = 15000 \text{ kg/m}^3$$

Thus, the density of the "gold" is considerably less than the density of 19,300 kg/m³ characteristic of pure gold. Off with his head!

EXAMPLE 2.16   The density of planets

The mass of Earth is 5.98×10^{24} kg, and its average radius is 6.37×10^6 m. Saturn is much more massive, 5.68×10^{26} kg, but its average radius is also considerably larger, 5.85×10^7 m. Consider each of them to be spheres ($V = \frac{4}{3}\pi r^3$) and calculate the density of each ($\pi = 3.142$).

Solution The volume of Saturn is

$$V_S = \frac{4}{3}\pi r^3 = \frac{4}{3}(3.142)(5.85 \times 10^7 \text{ m})^3$$
$$= 8.39 \times 10^{23} \text{ m}^3$$

and the volume of Earth is

$$V_E = \frac{4}{3}\pi r^3 = \frac{4}{3}(3.142)(6.37 \times 10^6 \text{ m})^3$$
$$= 1.08 \times 10^{21} \text{ m}^3$$

We are now able to calculate the density of each as

$$\rho_E = \frac{m}{V} = \frac{5.98 \times 10^{24} \text{ kg}}{1.08 \times 10^{21} \text{ m}^3}$$
$$= 5.54 \times 10^3 \text{ kg/m}^3.$$

and

$$\rho_S = \frac{m}{V} = \frac{5.68 \times 10^{26} \text{ kg}}{8.39 \times 10^{23} \text{ m}^3}$$
$$= 0.677 \times 10^3 \text{ kg/m}^3$$

It is interesting to note that the density of Saturn is less than the density of water. This means that Saturn would float in a bathtub filled with water—if such a large tub could be found for it.

2.18 PRESSURE

We saw in the last section that knowing the mass of an object is important, but that sometimes it is more important to know how that mass is distributed. To take care of this difficulty, we defined density as the mass of an object divided by the volume over which this mass is distributed. A similar problem arises with forces. Certainly they are important, but it is also important to know how they are applied to a surface. To indicate how this can become important, consider a 600 N man standing on a linoleum floor. The force he is applying downward on the linoleum is equal to his weight, but this force does no apparent damage to the surface. However, if he happens to be wearing a pair of golf shoes with numerous metal cleats protruding from the soles, one finds that he does considerable damage to the floor by standing on it. Why? In both cases, the force ap-

plied to the floor is 600 N, yet the effect of this weight has been different in the two cases. The answer is that in the first case, the applied force is spread out over a wide area of contact between the sole of the shoe and the floor. In the second case, even though the same force is applied to the floor, the area of contact is considerably smaller. The only area in contact with the floor is the cross-sectional area of the tips of the metal cleats. Because of the importance of considering the area of contact in many practical situations, we define **pressure** as

$$\text{pressure} = \frac{\text{force}}{\text{area}}$$

or

$$P = \frac{F}{A} \tag{2.10}$$

where P is the pressure, F is the applied force, and A is the area over which this force is distributed. In the SI, the units of pressure are N/m^2, but as we shall see in Chapter 15, there are some frequently used alternatives to these units.

EXAMPLE 2.17 The pressure exerted by a brick

A brick has a weight of 20 N and dimensions of 0.200 m by 0.080 m by 0.050 m. Find the pressure exerted by the brick on a surface **(a)** when it is resting with its largest face in contact with the surface, and **(b)** when it is resting so that its smallest face is in contact.

Solution **(a)** The area of contact when the largest face is in contact with the surface is A = (0.200 m)(0.080 m) = 0.016 m². Thus, the pressure is given by

$$P = \frac{F}{A} = \frac{20 \text{ N}}{0.016 \text{ m}^2} = 1250 \text{ N/m}^2$$

(b) When the smallest face is in contact, the area is A = (0.080 m)(0.050 m) = 0.0040 m², and the pressure exerted by the brick on the surface is

$$P = \frac{F}{A} = \frac{20 \text{ N}}{0.0040 \text{ m}^2} = 5000 \text{ N/m}^2$$

SUMMARY

Speed is a scalar quantity and is defined as distance divided by time. A **vector quantity** is described by magnitude and direction whereas a **scalar quantity** is specified only by its magnitude. **Velocity** is a vector quantity that describes speed and the direction of travel. **Acceleration** is change in velocity divided by time; an object is accelerating when it is speeding up, slowing down, or turning. The acceleration of a falling object near the surface of the Earth is a constant equal to 9.8 m/s². As a result, the velocity of a falling object is constantly increasing in magnitude as it falls and decreasing as it rises. Newton's three laws of motion are: (First) **A body at rest remains at rest, and a body in uniform motion in a straight line remains in such motion, unless acted upon by a net force.** (Second) **The net force on an object equals the mass times its acceleration (F = ma).** **Mass** is a quantitative measure of the inertia of an object and is related to the weight of the object by w = mg, where g is the acceleration due to gravity. (Third) **For every force there is always an equal and opposite force.** Frictional forces always arise when one object tries to slip over or pass another, and they always act in a direction to oppose such motion. The **density** of an object is its mass divided by its volume. **Pressure** is defined as the force per unit area acting on a surface.

EQUATIONS TO KNOW

(1) $\text{speed} = \dfrac{\text{distance}}{\text{time}} = \dfrac{d}{t}$

(2) acceleration
$= \dfrac{\text{final velocity} - \text{initial velocity}}{\text{time}}$

(3) $v = at$ (if the initial velocity is zero)

(4) $d = \dfrac{1}{2}at^2$ (if the initial velocity is zero)

(5) $F = ma$

(6) $w = mg$

(7) $\text{density} = \dfrac{\text{mass}}{\text{volume}} = \dfrac{m}{V}$

(8) $\text{pressure} = \dfrac{\text{force}}{\text{area}} = \dfrac{F}{A}$

KEY WORDS

Average speed	Velocity	Action-at-a-distance force	Friction
Instantaneous speed	Acceleration	Contact force	Weight
Scalar quantity	Acceleration due to	Inertia	Density
Vector quantity	gravity	Mass	Pressure
Displacement	Force	Net force	

QUESTIONS

EARLY IDEAS ABOUT MOTION; SPEED

1. Can the average speed of an object ever be greater than its instantaneous speed? Can the reverse be true?

2. A speed limit sign along an interstate highway says 65 mi/h. Is this referring to average or instantaneous speed?

3. (a) When a police officer gives someone a ticket for speeding, is the officer concerned with the driver's average speed or instantaneous speed? Explain. (b) On the entrance to many turnpikes, drivers are given a card that indicates the place of entry so the proper toll can be levied at the exit. The time of entry is also noted on the card. At the exit station,

the toll officer can determine the driver's speed during the journey, since both the elapsed time and the distance traveled are known. Can the officer determine the driver's instantaneous speed, the average speed, or both? Explain.

4. What is the speed of a horse that travels 20.0 km in 3.00 h?

5. The speed of sound in air is about 345 m/s. The speed of light is so fast that at first approximation it seems to travel distances of 100 km or so instantaneously. If you see a flash of lightning and hear thunder 8 s later, how far away is the lightning storm? Express your answer in meters and in kilometers.

6. An electron in the picture tube of a TV set may travel at a speed of about 2×10^6 m/s. If so, how long does it take the electron to travel the 0.20 m length of the tube?

VECTORS AND SCALARS

7. Which of the following are vectors and which are scalars: (a) The length of one class period of the physical science class in which you are enrolled. (b) The displacement of a car from one city to another. (c) The temperature of your classroom. (d) The results of a poll taken on the popularity of the President.

8. Look around you in the room where you are reading this textbook and identify what you believe to be several physical quantities. Classify them as vectors or scalars.

9. Use the triangle method of vector addition to find the resultant displacement for a pool ball that moves 60 cm down a table, hits the bumper, and rebounds along the same path for a distance of 15 cm before stopping.

10. A shopper in a supermarket starts at the end of an aisle and pushes his cart for a distance of 20 m. He then turns at an angle of 90° and moves for a distance of 10 m. Use the triangle method of vector addition to find the resultant displacement of the shopper from his starting point. Check your result with the Pythagorean theorem.

11. Suppose the shopper of the previous problem made a 20° turn with respect to his initial direction of travel instead of a 90° turn. What then is his resultant displacement?

VELOCITY

12. Explain what is incomplete about the statement, "The velocity of the automobile was 80 km/h."

13. A device on the dashboard of your car is called a speedometer. Why isn't it called a velocity-meter?

ACCELERATION

14. Which of the following can alter the acceleration of a car? (a) The gas pedal, (b) The brakes, (c) The steering wheel.

15. A bicyclist can accelerate from 0 km/h to 10 km/h in 2 s. Similarly, an automobile can accelerate from 80 km/h to 90 km/h in 2 s. Which vehicle is accelerating at a faster rate? Explain.

16. Find the acceleration of a car for the following circumstances: (a) The car increases its velocity from 10 km/s northward to 15 km/s northward in a time of 3 s. (b) The car changes its velocity from 10 km/s northward to 5 km/s northward in 3 s.

17. Can a car ever have an instantaneous velocity of zero and an acceleration that is not zero at the same instant?

18. Can a car ever have an instantaneous velocity of 10 km/s and a zero acceleration at the same instant?

ACCELERATION DUE TO GRAVITY

19. If a ball is dropped from the top of a tall tower, and at the same time a bullet is fired horizontally from the same tower, which object will strike the ground first?

20. You have two bricks, one heavy and one light. The heavy one is dropped, and its time of fall is measured; then the experiment is repeated for the light one. Finally, the two are glued together so that they fall as a unit, and the time of fall is measured for this combination. Discuss the expected results of this experiment from the point of view of Aristotle. Of Galileo.

21. The acceleration of gravity is stated as 9.8 m/s² in the text. Use conversion of units to show that this is equivalent in conventional units to 32 ft/s².

22. A box is dropped off a high building. At what time after release is it falling with a speed of 15 m/s?

23. A ball is thrown upward from the ground such that it reaches a height of 15 m. (a) What is its acceleration on the way up at a point 10 m above the ground; (b) on the way down at a height of 10 m above the ground; (c) at a height of 15 m above the ground?

MOTION WITH CONSTANT ACCELERATION

24. Imagine that you are sitting in an airplane cabin but do not have the seatbelt fastened. Describe your motion with respect to the rest of the cabin and the forces acting on you when the aircraft is (a) climbing upward, (b) flying level, (c) accelerating downward at a rate of less than 9.8 m/s², (d) ac-

celerating downward at a rate equal to 9.8 m/s^2, (e) accelerating downward at a rate greater than 9.8 m/s^2.

25. A rock dropped down a deep well falls for 5.00 s before striking the water. (a) How fast was it going when it hit? (b) How deep is the well?

26. A car is moving north with a velocity of 10.0 km/h when it begins to accelerate in the same direction at a rate of 5.0 km/h^2 and does so for a period of 2.00 s. Can the equation d = ½at^2 be used to find its displacement during the 2.00 s time interval? Explain.

27. A model airplane accelerates from rest with an acceleration of 5.00 m/s^2. What distance does the plane cover in a time of 3.00 s?

28. A snail starts from rest and covers a distance of 3.00 cm in 20.0 min with a constant acceleration. Find this acceleration in m/s^2.

FORCE

29. Find the weight in newtons of a 120 lb person.

30. Find the weight in lb of a 500 N person.

31. Can a force directed vertically on an object ever cancel a force exerted horizontally?

32. A person holds a textbook at arm's length. Identify all the forces acting.

33. A baseball is struck by a bat and flies toward an outfielder. While the ball is in flight, identify all the forces acting on it.

34. A traffic light is supported by two cables. Identify all the forces acting on the light.

35. A force of 40 N is exerted northward on an object, and a force of 30 N is directed eastward. Use the triangle method of vector addition to find the resultant force on the object. Check the magnitude by use of the Pythagorean theorem.

36. Repeat the preceding problem if the 30 N force is directed 20° east of north.

NEWTON'S FIRST LAW OF MOTION AND INERTIA

37. The word "inertia" is used in many nonscientific applications. Write a physical and a nonphysical definition for "inertia" and show that the spirit of the word is similar in both cases.

38. Is an object's inertia more closely related to its mass or to its weight? Explain.

39. An astronaut traveling at a steady speed along a straight path from Mars to Jupiter has to go outside the ship to do a repair. While she is working, she lets go of her wrench. Does the wrench fall far behind and get lost in space? Explain.

40. You can remove the dust from a coat by shaking it. Could you shake out the dust in free space in the absence of gravity? Explain.

41. A magic trick often used by children is to yank a tablecloth out from under a place setting of dishes. Use the concept of inertia to explain how this can be done without breaking the dishes.

42. A passenger in the rear of an airplane claims that he was injured when the plane suddenly decelerated, sending a suitcase flying toward him from the front of the plane. If you were the judge, would you let this case come to court?

43. In a rear-end collision, the passenger in the front car often suffers a whiplash injury because his neck is rapidly bent backward. Explain why the neck is snapped backward in this way.

44. It is possible for you to lie in a prone position with a concrete block resting on your chest and allow someone to break the block with a sledge hammer without causing any damage to you. Explain how this can happen.

45. A car with the emergency brake on is parked on a hill. Does gravity exert a force on the car? If so, why doesn't the car roll downhill?

NEWTON'S THIRD LAW

46. If you were to try to drive a large nail into hard wood with a light hammer, the hammer would bounce up at each blow. Explain this in terms of Newton's third law.

47. Identify the action-reaction pairs in the following situations: (a) a kangaroo jumps, (b) a pool ball glances off another ball, (c) a football player catches a punt, (d) a steady wind blows against the sail of a ship.

48. A person holds a 5 kg bag of groceries at arm's length. Identify all the forces acting and separate them into action-reaction pairs.

NEWTON'S SECOND LAW

49. An astronaut floating in free space is handed two objects and asked to determine which is more massive. Outline a simple procedure for solving the problem.

50. Consider two possible devices for measuring mass: a spring scale and a two-pan balance. In the spring scale, the object to be measured is placed on a spring, and the compression of the spring is recorded. In the two-pan balance, the object to be measured is placed on one pan, and calibrated masses are placed on the other until the two sides balance. Which device will show the same result on Earth as on the Moon? Would either device be useful in a space lab?

51. What is the mass in kg of an object having a weight of 20 N?

52. What is the weight in N of an object having a mass of 20 kg?

53. A bullet with a mass of 0.015 kg is propelled from the breech of a rifle with a force of 3000 N. What is the acceleration of the bullet while the force is acting on it?

54. A ball with a weight of 10 N is thrown forward with a force of 20 N. What is its acceleration while the force is acting?

55. In which of the following situations is the net force on the object in question equal to zero: (a) a car rolling along a road at constant velocity, (b) a car slowing down for a stoplight, (c) a bridge supporting a load of 300 automobiles, (d) a baseball leaving the pitcher's hand, (e) a person sitting in a chair, (f) a ball perched on a sea lion's nose, (g) a tree swaying in the wind? Explain.

56. It is often said that Newton's first law is just a special case of the second law. What is meant by this statement?

57. If gold were sold by weight, would you rather buy it in Denver or in Death Valley? If sold by mass, which site would be your choice?

58. A small 70 kg motorcycle carrying a 75 kg passenger has a force of 250 N acting on it. (a) What is its acceleration? (b) How far will it travel, starting from rest, in 4 s?

FRICTION

59. A moving company places a washing machine on a truck without tying it down, yet the machine makes the entire trip without moving on the bed of the truck. Explain carefully how this is possible. What would the driver of the truck have to do in order to make the machine move on the bed? (A move like this can be dangerous. Tie down your washing machine.)

60. A small 70 kg motorcycle carrying a 75 kg passenger has a forward force of 250 N acting on it exerted by the road, while a force of air resistance of 75 N acts on the two. What is the acceleration of the unit?

61. Why can rubbing your hands together in cold weather warm them up?

62. What is wrong with this statement? A baseball player sliding toward second base was tagged out because friction pulled him toward home plate.

DENSITY

63. (a) What volume of gold would have a mass of 10 kg? (b) What volume of air would have this same mass? (Refer to Table 2–2.)

64. Some of the gold bricks at Fort Knox have dimensions of 25 by 10 cm by 5 cm. What is the mass of one of the bricks?

65. A traveling salesman wants to sell you a solid silver block at a tremendously reduced price. The block has a mass of 0.27 kg and dimensions of 0.1 by 0.1 by 0.01 m. Is this silver? If not, identify this substance from the values listed in Table 2–2.

PRESSURE

66. (a) Find the maximum pressure that the gold brick of problem 64 can exert on the surface of a table by placing one of its faces flush against the surface. (b) Find the minimum pressure.

67. A king-size water bed is 250 cm long, 250 cm wide, and 30 cm deep. Calculate the mass of water in the bed and the pressure it exerts on the floor if it lies flat on the floor.

68. Indian Fakirs do stretch out on a bed of nails for a nap. Explain how this is possible.

69. Why are nails made with sharp points rather than blunt ends?

ANSWERS TO SELECTED NUMERICAL QUESTIONS

4. 6.67 km/h

5. 2760 m, 2.76 km

6. 10^{-7} s

9. 45 cm

10. 22.4 m at an angle of 26.6° with respect to the initial direction of travel

11. 29.6 m at an angle of 6.64°

15. The acceleration is 5 m/s^2 for both

16. (a) 1.67 m/s^2 north, (b) 1.67 m/s^2 south

22. 1.53 s

25. (a) 49 m/s, (b) 123 m

27. 22.5 m

28. 4.17×10^{-8} m/s^2

29. 536 N

30. 112 lb

35. 50 N at 53.1° north of east

36. 68.9 N at 8.51° east of north

51. 2.04 kg

52. 196 N
53. 2×10^5 m/s^2
54. 19.6 m/s^2
58. (a) 1.72 m/s^2, (b) 13.8 m

60. 1.21 m/s^2
63. (a) 5.18×10^{-4} m^3, 7.75 m^3
66. (a) 47,200 N/m^2, (b) 9450 N/m^2
67. 1880 kg, 2950 N/m^2

Top: The energy of natural waterfalls, such as this one at Cumberland Falls in Kentucky, has been utilized in many worthwhile ways ranging from the grinding of corn to the production of electricity. (Courtesy Bill Schulz.) *Bottom: Momentum is conserved in all collisions.* (Courtesy NASA.)

C H A P T E R

3

Conservation Laws

Newton's laws of motion, which we discussed in Chapter 2, revolutionized the world of physics and altered man's outlook on his environment. Phenomena that had defied explanation were now understandable by people with only modest scientific knowledge. With the use of only these laws, aspects of nature as diverse as the motion of a ball rolling along the ground and the motion of a spacecraft can be explained. Additionally, they were important to physics because they led to the explanation and derivation of other laws of physics. In this chapter we will look at two of these outgrowths of Newton's laws of motion, the conservation of momentum and the conservation of energy. Both of these concepts were known, at least to a limited extent, before the advent of Newton's work, but a true understanding and development of them was possible only following Newton.

The conservation of momentum is most frequently used in situations in which objects collide.

Thus, after studying this chapter we hope you will have a better understanding of why two pool balls move as they do when they bump together on a billiard table or why a gun recoils when a shot is fired.

The concept of energy is one of the most important in science. In everyday usage, we think of energy in terms of the cost of fuel for transportation and heating, of electricity for lights and appliances, and of the foods we consume. However, these ideas do not really define energy. They tell us only that fuels are needed to do a job and that those fuels provide us with something we call energy. We shall investigate in this chapter the many different forms in which we find energy, and we shall discover that one form of energy can be converted into another form. In fact, this transformation of energy from one form to another is an essential part of the study of physics, chemistry, biology, geology, and astronomy. Thus, a study of energy is essential to all the fields of science, but you may also be surprised at the extent to which an understanding of energy is important to your day-to-day existence, regardless of your field of study.

3.1
WHAT IS A CONSERVATION LAW?

If we are to study conservation laws, it behooves us to understand exactly what is meant when we say that something is conserved. In its simplest explanation, when something is conserved, it is never lost. If we have a certain amount of a conserved quantity at some instant of time, we find that we have exactly the same amount of that quantity at any later time. This does not mean that the quantity cannot change from one form to another during the elapsed time, but if we consider all the forms that can be taken, we will find that we always have the same amount. Let us consider the amount of money that you now have. As you are aware, your checking account balance is not a conserved quantity; it is very likely to decrease during the course of a month. However, for the moment, let us assume that your money *is* conserved. This would mean that if you have a dollar in your pocket, you will always have that dollar. However, you may find that it changes form. That is, one day it may be in the form of a nice crisp bill; the next day, you may have 100 pennies, and the next day an assortment of silver. But when you total it up, you always find exactly one dollar. It would be nice if money behaved in this way, but of course it doesn't. However, momentum and energy are conserved, and the purpose of this chapter is to examine the ramifications of this statement.

3.2
MOMENTUM

Frequently on the nightly news you hear a statement such as, "Candidate Blunderbuss predicts victory based on the fact that he is gaining momentum in the polls." In such everyday usage, the term momentum expresses the idea that candidate Blunderbuss is moving forward rapidly, but let us examine the concept of momentum to see how it is defined and used in physics. **Momen-**

tum *is defined as the product of the mass of an object and its velocity.* In equation form, with m the mass and v the velocity, we have

$$\text{momentum} = mv \qquad (3.1)$$

From Eq. 3.1, we see that a large object such as a locomotive can have a large momentum even if it is moving very slowly. On the other hand, a small object such as a fly can also have a large momentum if it is moving fast enough. You should also note that because velocity is a vector quantity, so is momentum. Thus, to specify momentum completely, we also must give its direction, which is the same as that of the velocity.

EXAMPLE 3.1 To swat or not to swat

A 10,000 kg locomotive is rolling leisurely along a track with a velocity of 0.25 m/s. With what speed would a 10^{-4} kg fly have to move in order to have the same momentum as the locomotive?

Solution The momentum of the locomotive is

$$\text{momentum} = (10,000 \text{ kg})(0.25 \text{ m/s})$$
$$= 2500 \text{ kg m/s}$$

This must also be equal to the momentum of the fly, so the velocity of the fly must be

$$v = \frac{\text{momentum}}{m} = \frac{2500 \text{ kg m/s}}{10^{-4} \text{ kg}}$$
$$= 2.5 \times 10^7 \text{ m/s}$$

This corresponds to a speed of 25 million meters per second. In a science fiction story our fly might have this velocity, but not in real life.

3.3
CONSERVATION OF MOMENTUM

The real beauty and usefulness of momentum lie with the fact that *momentum is conserved in collisions between objects.* For example, consider a hockey puck sliding across the frictionless surface of ice. The puck has mass and velocity; thus it has momentum. Now suppose that this puck collides head on with another hockey puck, at rest on the ice, and after the collision both pucks go zooming

FOCUS ON . . . Transfer of Momentum in a Rocket Engine

It is, of course, correct to say that the upward thrust of a rocket engine is the "equal and opposite reaction" to the downward thrust of the exhaust gases. But some students ask, what actually pushes the rocket skyward? The exhaust gases don't push a rocket anywhere. To understand the answer to this question, think about what happens inside the engine itself. A rocket engine is a chamber in which gases burn at a rapid but controlled rate. As the rocket fuel burns, hot (that is, fast-moving) molecules of gas shoot out in all directions. Those that hit other molecules exchange momentum with them. Those that hit the walls of the firing chamber exchange momentum with the rocket. The molecules that hit the left wall (see figure) are balanced by those that hit the right, so the rocket gains no net momentum in either horizontal direction. The molecules that hit the top of the firing chamber push the rocket upward. What about the exhaust gases, those that travel straight downward? They do not hit any part of the rocket at all, so they do not affect its motion directly. Since the rocket is forced upward, but not downward, the net momentum is skyward, and up it goes.

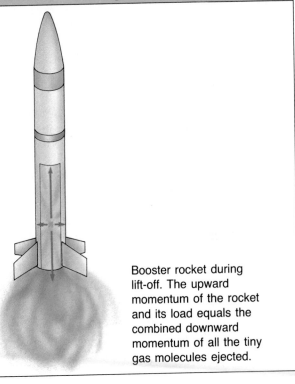

Booster rocket during lift-off. The upward momentum of the rocket and its load equals the combined downward momentum of all the tiny gas molecules ejected.

away. If we measure the momentum of the initial hockey puck and find the total momentum of the two pucks after the collision, the net momentum before is equal to the net momentum afterward. In fact, regardless of how complicated the collision may seem, momentum is always conserved in collisions between two or more objects. As an example of what seems to be a slightly more complicated collision, consider our hockey puck again, but this time assume that it collides with and sticks to a piece of mud as in Figure 3.1. After the collision, the puck and mud move as a unit; nevertheless, the momentum of the puck before the collision is exactly the same as the momentum of the combined puck-mud combination after the collision.

In equation form, we can express the conservation of momentum as

> momentum before = momentum after

(3.2)

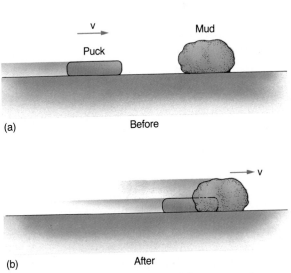

FIGURE 3.1 (a) The hockey puck has momentum before the collision. (b) When the puck picks up the mud after collision, the velocity of the combination is less, but the mass is greater. The momentum is the same before and after.

EXAMPLE 3.2   Batter up

A 50 kg baseball pitching machine is at rest on a smooth playing surface. A 0.30 kg baseball is fired toward the right from the machine at a speed of 30 m/s. Find the recoil velocity of the machine.

Solution   This is an example of conservation of momentum. In this problem, however, the initial momentum is zero because the pitching machine and the baseball are both initially at rest. After the ball is thrown, the ball has momentum toward the right, which we will call the positive direction, and the machine recoils to the left. We have

momentum before = momentum after
$$0 = (50 \text{ kg})v + (0.30 \text{ kg})(30 \text{ m/s})$$

Solving for v, we find

$$v = -0.18 \text{ m/s}$$

The negative sign indicates that the pitching machine moves to the left, in the negative direction.

EXAMPLE 3.3   Road hogs beware

A car of mass 2000 kg traveling at a speed of 20 m/s collides with another car of the same mass which is stopped for a traffic light. The two cars become entangled and move together as a unit after the collision. What is the velocity of this wreckage?

Solution   We use conservation of momentum as

momentum before = momentum after

The momentum before the collision is solely that of the oncoming automobile. Thus,

$$
\begin{aligned}
\text{momentum before} &= mv \\
&= (2000 \text{ kg})(20 \text{ m/s}) \\
&= 40{,}000 \text{ kg m/s}
\end{aligned}
$$

The momentum after is the product of the mass of the combined wreckage and the velocity v of this wreckage. We have

momentum after = $(4000 \text{ kg})v$

Conservation of momentum yields

$$40{,}000 \text{ kg m/s} = (4000 \text{ kg})v$$

From which,

$$v = 10 \text{ m/s}$$

EXAMPLE 3.4   Daredevil pool balls

Pool ball A of mass 0.30 kg rolls to the right across a table with a velocity of 3.00 m/s and collides head on with a 0.20 kg ball B moving toward the left with a speed of 2.0 m/s. After the collision, the lower mass ball has a velocity of 1.50 m/s toward the right. Find the velocity of ball A.

Solution   This is a straightforward application of conservation of momentum. We have

$$
\begin{aligned}
(0.30 \text{ kg})(3.00 \text{ m/s}) &+ (0.20 \text{ kg})(-2.00 \text{ m/s}) \\
&= (0.30 \text{ kg})v + (0.20 \text{ kg})(1.50 \text{ m/s})
\end{aligned}
$$

Note that we have selected the positive direction for velocity to be toward the right. As a result, the velocity and momentum of the 0.20 kg ball before the collision are negative.

We solve the above for v to find

$$v = 0.667 \text{ m/s}$$

The positive value indicates that the velocity of the 0.30 kg ball is still to the right, in the direction we have selected as positive.

EXAMPLE 3.5   I felt the earth move again.

A student hurls a piece of mud toward the right at a stationary brick wall. When the mud strikes the wall, it sticks to the wall and stops. Discuss conservation of momentum as applied to this situation.

Solution   At first glance, it looks like conservation of momentum has failed us in this application. Certainly, before the collision occurred, the mud had momentum toward the right, but after the collision nothing appears to be moving. This seems to be in violation of the conservation of momentum because if something has momentum to the right before, something must have the same momentum to the right afterward. What has happened?

The answer is that something does indeed have momentum to the right after the collision.

The wall is firmly affixed to the ground, and it is the entire Earth that picks up the net momentum to the right after the collision. Don't expect the Earth to change velocity very much, however, because of its huge mass.

EXERCISE

In light of the previous discussion, consider the following situation. You are standing at rest in a room, and then you begin walking toward the right. Discuss conservation of momentum in this situation. Before the walk, there was no momentum because nothing was moving, but when you took your first stride, you had momentum to the right. For the net momentum to be zero, something has to gain momentum to the left. What is it?

3.4 WORK

Imagine holding a heavy bucket of water straight out from your body at arm's length for a couple of hours. It isn't a pleasant way to spend an afternoon, but if you ever choose to do it, you will probably be quite tired at the end and you will think that you have done a lot of work. Even though your tired muscles indicate that a lot of work has been done, we shall shortly see that according to the physics definition of work, you have not done any at all. Occasionally we encounter a word that has a totally different meaning in physics than it does in everyday usage. Perhaps of all such cases, none differs quite as much as does the definition of work from that used in nonscientific situations.

Work is defined as the product of a force, F, and the distance, d, over which the force acts. In equation form this is

Work = (force)(distance)

and in symbolic form

$$\text{Work} = \text{Fd} \qquad (3.3)$$

where F is the force and d is the distance moved along the direction of the force.

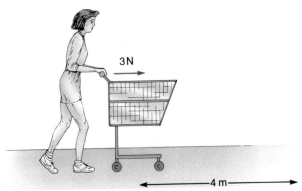

FIGURE 3.2 How much work is done by the shopper moving a grocery cart with a force of 3 N for a distance of 4 m?

As an example, in Figure 3.2 a grocery cart is being pushed by a shopper with a force of 3.0 N. If the cart moves a distance of 4.0 m, how much work is done? We find

Work = Fd = (3.0 N)(4.0 m) = 12 N m

The units of work are N m, and this combination of units is lumped together and given the name joule (symbol J). Thus, the work done is 12 J.

Now, return to the example with which we started this section. When you are holding the bucket of water at arm's length, you are certainly exerting a force on it, but as long as there is no movement of the bucket, d is zero and the work is also zero.

A question that has probably already occurred to you is why we have paused to discuss work. This chapter is supposed to be about momentum and energy—why this apparent digression? The reason is that energy is defined in terms of work. More specifically, *an object has **energy** if that object has the ability to do work.* We will examine this relationship between work and energy more closely in the following sections.

EXAMPLE 3.6 More about work

A person supports a 20 N bucket as shown in Figure 3.3 as he walks a distance of 3.0 m. How much work is done by the bucket-carrier?

Solution It is tempting but *incorrect* to substitute into the defining equation for work to find

FIGURE 3.3 How much work is done by the farmer carrying the 20 N bucket for a distance of 3 m?

$$\text{Work} = Fd = (20 \text{ N})(3.0 \text{ m}) = 60 \text{ J}$$

The reason that this is incorrect lies with the definition of the distance d. Note in the definition given earlier, the distance d is the distance moved *in the direction of the force*. Granted, the person exerts a force, 20 N, and he moves a distance of 3 m, but the 3-meter movement is horizontal while the force is vertical. There is no movement in the direction of the force, so the work done is zero.

3.5
KINETIC ENERGY

Imagine that you have a nail driven part way into a vertical board, as shown in Figure 3.4a. You would like to drive the nail farther into the board, but no hammer is handy. How can you accomplish this task? If a rock is nearby, one way to drive the nail is to throw the rock at it, as shown in the figure. Let us examine this nail-driving episode from the standpoint of work and energy. Recall that in the last section, we said that an object has energy if it has the ability to do work. Does our moving rock have energy? It does, because when it strikes the end of the nail the rock

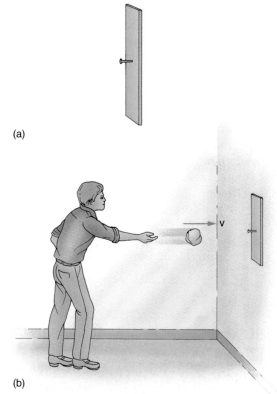

FIGURE 3.4 (a) A nail in a board. (b) One way to drive the nail into the board.

exerts a force, F, on the nail and the nail is moved some distance, d, into the plank. Thus, there is a force on the nail, a movement of the nail in the direction of the force, and work is done. As a result, we can say that the moving rock must have energy. *The energy that an object has because of its motion is called **kinetic energy**.*

The amount of kinetic energy that a moving object possesses can be shown to depend upon two factors: the mass of the moving object and its speed. It should seem reasonable that mass is a factor because a more massive rock exerts a greater force on the nail and drives it deeper into the plank than does a less massive rock. Likewise, if one throws the rock with a high speed, the nail is driven deeper than if the rock moves slowly. In fact, the kinetic energy of an object depends on the square of the speed. The kinetic energy of an object is given by

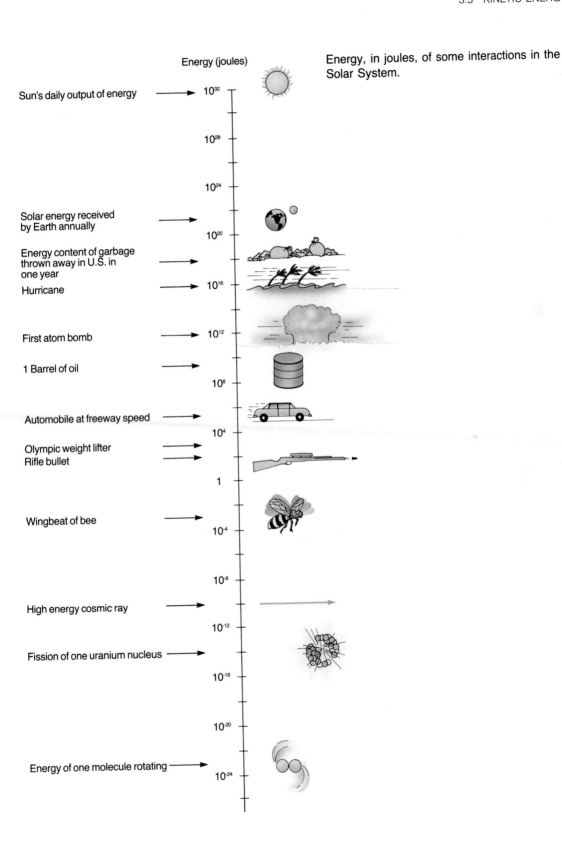

Energy (joules)

Energy, in joules, of some interactions in the Solar System.

Sun's daily output of energy → 10^{32}

10^{28}

10^{24}

Solar energy received by Earth annually → 10^{20}

Energy content of garbage thrown away in U.S. in one year →

Hurricane → 10^{16}

First atom bomb → 10^{12}

1 Barrel of oil →

10^{8}

Automobile at freeway speed →

10^{4}

Olympic weight lifter →
Rifle bullet →

1

Wingbeat of bee →

10^{-4}

10^{-8}

High energy cosmic ray →

10^{-12}

Fission of one uranium nucleus →

10^{-16}

10^{-20}

Energy of one molecule rotating → 10^{-24}

$$KE = \frac{1}{2}mv^2 \tag{3.4}$$

where m is the mass of the object and v is its speed. A derivation of the kinetic energy equation appears at the end of the chapter.

EXAMPLE 3.7   Duck!

A 0.20 kg baseball is thrown with a speed of 30 m/s. How much kinetic energy does the baseball have?

Solution   This is a straightforward substitution into the defining equation for kinetic energy. We have

$$KE = \frac{1}{2}mv^2 = \frac{1}{2}(0.20 \text{ kg})(30 \text{ m/s})^2 = 90 \text{ J}$$

As anyone who has ever played baseball is aware, this is enough energy to do considerable injury to someone standing in the flight path.

3.6 GRAVITATIONAL POTENTIAL ENERGY

As a review, recall that an object is said to have energy if it has the ability to do work. We saw in the last section that a rock can have energy because of its motion, since the moving object can do work. Let us now examine a different type of energy that an object can possess, called **gravitational potential energy**. *Gravitational potential energy is the energy that an object has because of its location in space.* To examine this concept more carefully, consider once again our task of driving a nail. As shown in Figure 3.5a, the plank is now horizontal, and we are going to drive the nail by dropping our rock onto it. It should be obvious to you that the dropped rock will exert a force, F, on the nail and cause the nail to move a distance, d, into the plank. Thus, the rock can do work on the nail, and from our definition the rock must have energy. This energy arises because of the location of the rock in space above the nail.

Consider lifting an object of mass m a distance h above the surface of the Earth. To lift the

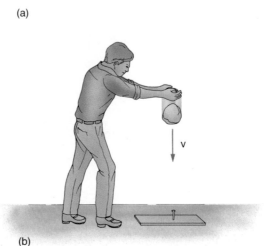

(b)

FIGURE 3.5   (a) A nail in a horizontal plank. (b) Another way to drive the nail.

object you have to do work on it, and the amount of work you do is given by

$$W = Fd = mgh$$

where mg is the weight of the rock. You must exert a force at least equal to the weight of the rock in order to lift it through the height h. This work that you have done on the system can be returned to you. Thus, we define the gravitational potential energy, PE, of the rock when at the height h above the ground to be

$$PE = mgh \tag{3.5}$$

It should seem reasonable that the potential energy depends on the mass of the object. For example, in our case of dropping the rock, it is obvious that a more massive rock does more work on the nail than a lighter rock. Also, a given object drives the nail farther into the plank, or does more work, if it is dropped from a great height than it does if it falls only a short distance.

EXAMPLE 3.8   Climbing higher and higher

(a) A secretary in a highrise office building holds a 2.0 kg file folder 0.50 m above a desk, at posi-

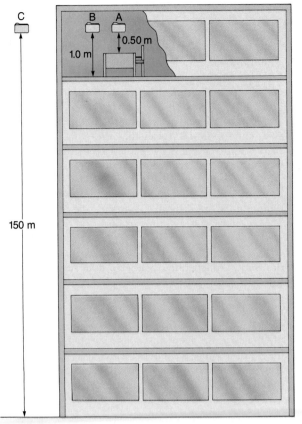

FIGURE 3.6 The reference level for measuring gravitational potential energy is arbitrary.

tion A in Figure 3.6. Find the gravitational potential energy with respect to the top of the desk.

Solution The potential energy can be easily found from PE = mgh as

PE = (2.0 kg)(9.8 m/s²)(0.50 m) = 9.8 J

(b) If he moves slightly to B so that the file folder is not above the desk, a more convenient reference level might be the surface of the floor, which is 1.0 m below the file folder. What is the potential energy with respect to the floor?

Solution Note that h in PE = mgh is relative to whatever level you choose. As a result, we find the PE as

PE = mgh = (2.0 kg)(9.8 m/s²)(1.0 m)
= 19.6 J

(c) Finally, our secretary moves to an open window and holds the file folder out the window at C

such that it is 150 m above street level. How much potential energy does the folder have relative to ground level?

Solution Again, we calculate the potential energy as

PE = mgh = (2.0 kg)(9.8 m/s²)(150 m)
= 2.9 × 10³ J

In all three cases above, the potential energy was measured with respect to some reference level. This reference level is completely arbitrary, and you can choose the one that you like. However, in a given problem usually one particular reference level stands out as being the most suitable one to use. We shall have more applications later, which will help you in making judicious selections of these reference levels.

3.7
CONSERVATION OF ENERGY

The true beauty and usefulness of the concept of energy arose when it was discovered that energy is conserved. We shall see that a problem that might be very difficult to solve by the use of Newton's laws of motion becomes quite simple when approached from the viewpoint of the conservation of energy. As noted earlier, when a quantity is conserved, we find that we have as much of it at a later time as we do initially. This does not mean that the quantity cannot change form between the two times. As an example, consider the situation in which we hold a rock above the ground at position A in Figure 3.7. At this location, all of the energy of the rock is in the form of gravitational potential energy. However, when we release the rock, its gravitational potential energy decreases as the rock falls, because its height above the ground becomes less. We find, however, that its speed, and hence its kinetic energy, is increasing. Finally, just before striking the ground, all of the rock's initial potential energy is completely converted to kinetic energy. Observations such as this lead to one of the most fundamental laws of physics, the **law of conservation of mechanical energy,** which states

In the absence of friction, the sum of the kinetic energy and the gravitational potential energy of a system is constant.

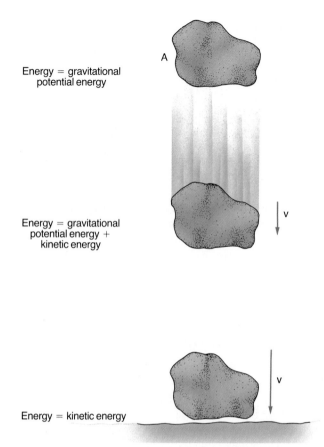

A

Energy = gravitational
potential energy

v

Energy = gravitational
potential energy +
kinetic energy

v

Energy = kinetic energy

FIGURE 3.7 The energy of a falling object changes
from gravitational potential energy to kinetic energy.

In equation form, this may be stated as

$$KE + PE = \text{constant} \qquad (3.6)$$

EXAMPLE 3.9 Good toss for a rookie

A baseball player throws a 0.15 kg baseball
straight up into the air with a speed of 15 m/s.
How high does the ball rise above the level at
which it is released.

Solution The situation is shown in Figure 3.8, in
which we have called point 1 the position at which
the ball is released from the thrower's hand and
point 2 the highest point reached by the ball. Initially, all of the energy of the ball is in the form of
kinetic energy, assuming that we set the zero level

FOCUS ON . . . Energy in the Human Body

The ultimate source of all energy used by the
human body is plant life, and since plants gain their
energy from the Sun, all the energy used by us is
ultimately traceable to the energy of sunlight. Before you begin to protest about the hamburger that
you had for lunch, recall that the cow dined on
plants, and the energy it gained from that source
was ultimately transferred to you. Much of the energy supplied to your body is used just to keep the
machine operational. That is, it maintains the body
temperature, it keeps all the organs operating, and
it allows us to walk, to move around, and to perform
work.

The process by which a plant stores energy from
the Sun is called photosynthesis. The steps are as
follows: The foliage of a green plant stores carbon
dioxide and water, and these two would forever
remain separated if it were not for the presence of a
substance called chlorophyll in the plants. Chlorophyll acts as an agent that allows sunlight striking
the plant to cause a chemical reaction in which carbon dioxide and water are changed to a form of
sugar called glucose, with some oxygen left over.
The reaction is symbolized as

(carbon dioxide) + (water)
  + (energy) $\longrightarrow$ (glucose) + (oxygen)

Long chains of these glucose molecules join together to form cellulose, but nevertheless this stored
energy is retained and, under the proper conditions, can be reclaimed. One way to reclaim this energy is to burn a piece of wood. At a high enough
temperature, the chemical reaction just discussed
can proceed in the opposite direction. The glucose
stored as cellulose can combine with oxygen from
the air and be changed back to carbon dioxide and
water. In this process, the energy originally absorbed from the sunlight is released, primarily in
the form of heat. When an animal eats the plant, the
energy stored in the glucose molecules is released to
the animal. This release of energy takes place in
mitochondria, small bodies found in cells. These
mitochondria contain enzymes that break up the
glucose into simpler molecules, which then react
with oxygen to form carbon dioxide, water, and
energy. The energy released is then utilized by the
body to perform its various functions.

FIGURE 3.8 Energy changes from kinetic to gravitational potential energy as a ball rises into the air.

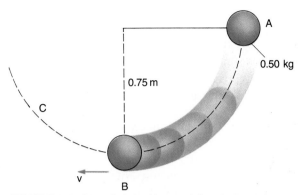

FIGURE 3.9 How fast is the pendulum ball moving at point B? How high does it rise as it moves along path C?

for potential energy at the release point, point 1. As the ball rises, it slows down to finally come to a stop at the top of its flight path, point 2. At this location, all of the energy of the ball is gravitational potential energy. We find its initial total mechanical energy as

$$KE_i + PE_i = \text{constant}$$

or

$$\frac{1}{2}(0.15 \text{ kg})(15 \text{ m/s})^2 + 0 = 16.9 \text{ J}$$

The final total mechanical energy is

$$KE_f + PE_f = 0 + (0.15 \text{ kg})(9.8 \text{ m/s}^2)h$$

where h is the maximum height reached. Because mechanical energy is conserved, we have

$$(0.15 \text{ kg})(9.8 \text{ m/s}^2)h = 16.9 \text{ J}$$

from which we find

$$h = 11.5 \text{ m}$$

(You should note that in order to cancel units in the equation above, J must be changed back to its equivalent of kg m/s².)

EXAMPLE 3.10 How to be a swinger

(a) An object of mass 0.50 kg is attached to a string of length 0.75 m, to make a pendulum as

shown in Figure 3.9. The string is held horizontal and released at position A. How fast is the object going when it reaches the bottom of its arc, position B?

Solution Initially, all of the energy of the pendulum is in the form of gravitational potential energy, if we set the zero level for the measurement of potential energy at the bottom of the arc. Thus, at position A, the total mechanical energy is

$$mgh = (0.50 \text{ kg})(9.8 \text{ m/s}^2)(0.75 \text{ m}) = 3.68 \text{ J}$$

At B all of this potential energy has been converted into kinetic energy, and conservation of mechanical energy gives us

$$3.68 \text{ J} = \frac{1}{2}(0.50 \text{ kg})v^2$$

Or, solving for v we find

$$v = 3.83 \text{ m/s}$$

(b) At position B, the speed of the object causes it to move through the bottom of the arc and continue along its path as shown in Figure 3.9. Find how high it rises above the bottom of the arc.

Solution In this case, we find that all of the initial energy of the object at B is in the form of kinetic energy. We found in part (a) that the kinetic energy that it has is 3.68 J. When the object climbs to the same height as point A, all of this kinetic energy is converted again into gravitational potential energy. Thus, it is a simple calculation for you to show that the object will climb to a height of 0.75 m before stopping.

From this, we see that in the absence of friction, the object on the end of the string will continue to swing back and forth forever, always rising to the same height at the end of each swing and having the same speed at the bottom of the arc. You know from your own experience that this does not occur; the pendulum soom comes to a stop. In the next section, we shall examine why the motion eventually ends.

3.8 FRICTION FORCES AND THE CONSERVATION OF ENERGY

We pointed out in our discussion of the motion of a pendulum (Example 3.10) that a difficulty seems to have arisen with our approach to the conservation of energy. If the pendulum continues to interchange energy between kinetic and potential, it will swing forever and never lose any of its total mechanical energy. A pendulum never does this, so it is obvious that we have omitted something from our discussion—namely friction.

As the pendulum moves through the air, it experiences two important frictional forces. One of these arises at the point where the pendulum string is attached to the ceiling. The second is air resistance against the swinging pendulum. These frictional forces cause a decrease in the energy of the pendulum such that as the pendulum swings from an extreme position to the bottom position, the amount of kinetic energy gained is not quite equal to the amount of gravitational potential energy lost. And as the pendulum moves up toward the end of its swing, it does not end up with the same amount of potential energy at the top as the kinetic energy it had at the bottom. As a result, the pendulum does not return to the same height that it reached on its previous swing. Over a period of time, friction robs all of the mechanical energy from the system, and the pendulum ceases to swing.

The question now is, "Where does this lost energy go?" The answer is that this energy is not really lost. Instead, it appears as a third form of energy, which we will call **thermal energy**. In order to see how this form of energy enters the picture, imagine the effect that the swinging object has on the air molecules in its path. A molecule just in front of the object is in a position similiar to a golf ball as the golf club approaches. When the pendulum hits the air molecules, it knocks them away, just as a golf club knocks away the ball. Other molecules are hit by these molecules as they fly away, and in general the billions of billions of molecules near the pendulum gain speed from the stirring effect of the pendulum's motion. Different molecules are affected differently, but the important thing to note is that, on the average, the air molecules gain speed as a result of the swinging object.

Individual molecules have mass; thus, a speed gain by the molecules means a gain in their kinetic energy. This, then, is where some of the energy of the pendulum goes. As the pendulum swings through the air, it stirs up air molecules and increases their kinetic energy. *This energy of motion of the molecules or atoms of any substance is referred to as thermal energy.* If one were able to measure the increase in energy of the molecules during one swing of the pendulum, one would find that it would account for *some but not all* of the decrease in the energy of the pendulum. The remainder of the energy loss can be accounted for by friction at the point of support. Constant rubbing at this location causes an increase in thermal energy of the support and of the string attached to it. We will examine thermal energy in more detail in a later chapter.

3.9 POWER

Imagine a designer of heavy equipment who builds a crane that is capable of lifting a 4000 kg mass to a height of 30 m. With only this much information, it would be difficult to sell this device to a practical-minded contractor, because he wants to know not only how much the crane can lift but also how long it takes to lift it, because he may have to lift several such objects during the course of a work day at his construction site. Thus, in many practical applications, we need to know more than just how much energy output a

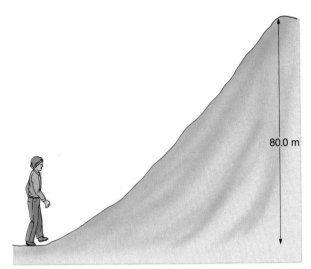

FIGURE 3.10 Discuss the energy transformations that take place as the student climbs a hill that rises 80 m vertically.

device has. We must also consider *time,* a factor that has not appeared in our analysis of energy thus far.

Power is defined as the amount of energy converted from one form to another divided by the time required to make this conversion.

This is

$$\text{Power} = \frac{\text{energy}}{\text{time}} \qquad (3.7)$$

The units of power are J/s, and this collection of units is referred to as a watt, W. An alternative form of measuring power is the horsepower, hp, chosen to equal 550 ft lb/s. Based on this choice the relation between the hp and the watt is

1 hp = 746 W

Consider the definition of power in terms of a 100 W light bulb. When we say that the power of a light bulb is 100 W, we mean that energy is being converted from electrical energy to other forms at the rate of 100 J every second. These other forms are light energy and thermal, or heat energy. Note that we are accustomed to discussing our pieces of electrical equipment in terms of power measured in watts, and we are accustomed to discussing mechanical equipment, such as the

output of an automobile engine, in terms of horsepower. However, there is no fundamental reason for this; it is by convention only that this occurs. We could equally well refer to a 0.2 hp light bulb or to a 60,000 W car motor.

EXAMPLE 3.11 Not eating would be easier

A 70.0 kg student decides to lose a little weight by climbing a hill that rises vertically by 80.0 m, as shown in Figure 3.10.

(a) Discuss the energy transformations that occur during the climb.

Solution The final gravitational potential energy of the student relative to his position before the climb is

$$mgh = (70.0 \text{ kg})(9.8 \text{ m/s}^2)(80.0 \text{ m})$$
$$= 54,900 \text{ J}$$

This energy is supplied by energy transformations inside the body in which stored energy from food is converted by chemical processes to gravitational potential energy.

(b) What is the power output of the student if he makes the climb in a time of 200 s?

Solution We have

$$\text{Power} = \frac{54900 \text{ J}}{200 \text{ s}} = 275 \text{ W}$$

Thus, the student is converting chemical energy to gravitational potential energy at the rate of 275 J each second.

3.10 SIMPLE MACHINES

In this section, we will take a look at two classifications of **simple machines,** the inclined plane and the lever. Regardless of how simple or complex a machine may be, we will find that the principle that underlies its operation is the conservation of energy. A machine is a type of tool that alters the magnitude or direction of an applied force. Thus, a machine can change a small force to a large force or, alternatively, a large force to a small force.

In order to understand machines, it is essential to remember the difference between force and energy. A force is any push or pull. As we will learn, it is relatively easy to amplify or reduce an applied force. But energy is different, because it can be neither created nor destroyed. Thus, no machine can create energy. The difference between force and work can be emphasized by the following example. Recall that work, which is related to energy, is defined as

Work = (force)(distance)

Imagine that 12 J of potential energy is available to you. You can design a machine to utilize those 12 J to exert a force of 1 N for 12 m. Another machine could be built to use the 12 J of energy to transmit a force of 12 N for 1 m. (We have assumed here that frictional forces are zero.) Of course, it would be impossible to build a machine that would use 12 J of energy to exert a force of 12 N for 12 m, or even for 1.0001 m, for either process would create energy. Thus, a given amount of energy can be used to exert a large force for a small distance or a smaller force for a larger distance.

Inclined planes

Suppose two people are given the task of lifting a heavy iron ball onto a platform 2 m off the ground. If the ball weighs 400 N, the work required to lift it will be

W = Fd = (400 N)(2 m) = 800 J

Thus, 800 J of energy are required to lift the ball—no more and no less. But there are no restrictions on how this work is to be performed. The stronger person may lift the ball straight up. In this case, the force exerted will be 400 N and the work performed will be W = (400 N)(2 m) = 800 J as indicated above.

However, a weaker person who is unable to lift the ball might build a ramp and roll the ball up, as shown in Figure 3.11. In this example, the ramp is 10 m long. Thus, the weaker person has to exert a force over a longer distance. But because the total work must be the same, the magnitude of the force is less. Since 800 J are per-

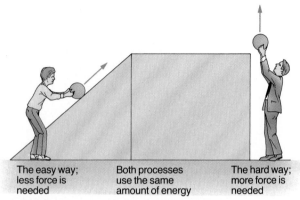

FIGURE 3.11 Less force is needed to roll a ball up an inclined plane than to lift it straight up. However, the total energy required is the same in both instances.

The easy way; less force is needed   Both processes use the same amount of energy   The hard way; more force is needed

formed over a distance of 10 m, the force needed can be calculated as

$$F = \frac{W}{d} = \frac{800 \text{ J}}{10 \text{ m}} = 80 \text{ N}$$

Thus, the weaker person is able to raise the ball to the required height by using only 80 N of force. The ramp, called an **inclined plane**, reduces the force required to do a given amount of work by increasing the distance. Don't forget, however, that the total energy needed to perform the task is the same in both instances. An inclined plane is a simple machine. In general, if there are two ways to perform a given task,

work done by one method = work done by any other method

and

(large force)(small distance) = (small force)(large distance)

Levers

There are a great many other clever ways of performing work. Think of a simple seesaw, supported in the middle as shown in Figure 3.12a. If a person on one end pushes downward, the other end will naturally move upward. Since the board is supported in the middle, both ends will travel the same distance—when one end goes down

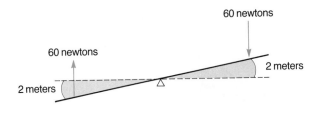

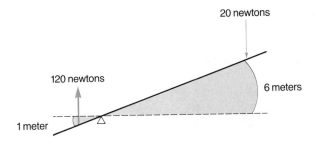

(a) 60 newtons x 2 meters = 60 newtons x 2 meters
120 joules = 120 joules

(b) 120 newtons x 1 meter = 20 newtons x 6 meters
120 joules = 120 joules

FIGURE 3.12 A seesaw.

2 m, the other end goes up 2 m. If one end is pushed down with a force of 60 N, 120 J of work are performed. Since energy is conserved, it is possible to exert an equal force of 60 N on the other side for an equal distance of 2 m. But now imagine that the seesaw is positioned so that it is supported not in the middle but off to one side, as shown in Figure 3.12b. Because one side is longer than the other, it will move upward farther. Let us say, for example, that whenever the long end travels 6 m, the short end travels only 1 m. Now what happens if the long side is pushed down with a force of 20 N, depressing it the full 6 m? The work performed is W = (20 N)(6 m) = 120 J. Since energy is conserved, 120 J of work must be performed on the other side. But, since the other side travels a smaller distance (only 1 m), it must exert more force if it is to do equal work. The force, as found from W = Fd, is

$$F = \frac{W}{d} = \frac{120 \text{ J}}{1 \text{ m}} = 120 \text{ N}$$

Thus, once again, we see that a given amount of work can be performed by exerting a small

force over a large distance or a large force over a small distance.

Any device analogous to a seesaw that pivots over a fixed point is called a **lever.** The pivot point is called the fulcrum. For convenience, the force applied to one end of the lever is called the input force, and the force developed at the other side of the lever is called the output force. For any lever, we can make use of the following relationship

(Input force)(distance from input force to
fulcrum)
= (output force)(distance from output
force to fulcrum)   (3.8)

It is immediately obvious how useful this relationship is. Suppose you are asked to lift a 5000 N stump, but you are not strong enough to pick up a stump that weighs more than 500 N. Figure 3.13 shows how you might accomplish the task. As shown, a long rigid bar is used as a lever, and a rock is used as a fulcrum. Let us suppose that you have arranged your rod so that the distance from the fulcrum to the point where you push down is 2 m and that you push down with a force of 500 N. What distance d can the weight of the stump be from the fulcrum so that you can lift it? From eq. 3.8, we have

(500 N)(2 m) = (5000 N)(d)

From which we find

d = 0.2 m

EXAMPLE 3.12   Row, row, row your boat

The oars of a rowboat are simple levers that may be used as shown in either Figure 3.14a or 3.14b. In which of these two configurations is the most force generated against the water, and in which arrangement does the oar travel the farthest distance through the water?

Solution   Figure 3.15a and b demonstrate the results of the rowing process for each situation. When the distance from the hand to the fulcrum is small, as in Figure 3.15a, a large input force is applied over a small distance, and the result is a

FIGURE 3.13 An attempt to move a stump with a lever. Can you locate the position of the fulcrum?

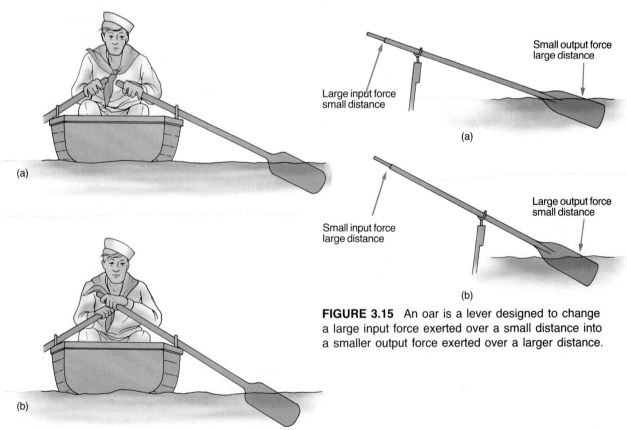

(a)

(b)

FIGURE 3.14 An oar as a lever. The oarlock is the fulcrum.

Large input force small distance

Small output force large distance

(a)

Small input force large distance

Large output force small distance

(b)

FIGURE 3.15 An oar is a lever designed to change a large input force exerted over a small distance into a smaller output force exerted over a larger distance.

Is this a lever in action? If so, identify the input force, the output force, and the fulcrum.

small output force over a large distance. For the case of Figure 3.15b, a small input force applied over a long distance produces a large output force over a small distance.

Typically, a rower is not concerned about applying a lot of force as much as he is about sweeping the water over a large distance. Thus, the configuration of Figure 3.15a is the more desirable rowing mode.

3.11
DERIVATION OF KINETIC ENERGY EQUATION

A moving body has energy of motion, or kinetic energy. Energy is the capacity to do work, and a moving body can do work by colliding with a sta-

tionary body and forcing it to move. Let us consider how much energy a stationary body can gain as it is accelerated at a uniform rate; this energy is its kinetic energy.

Let m = mass of body
0 = initial velocity
v = final velocity
d = distance the body travels as its velocity changes uniformly from 0 to v
t = time

Then,

change in velocity = final velocity − initial velocity = $v - 0 = v$

The acceleration of the body is given by

$$a = \frac{\text{change in velocity}}{\text{time}} = \frac{v}{t}$$

and the average velocity of the body is

$$v_{av} = \frac{\text{final velocity + initial velocity}}{2} = \frac{v}{2}$$

The distance traveled by the object, d, is found by multiplying the average velocity by the time. We find,

$$d = \frac{v}{2}t$$

Dividing both sides of the above equation by t, we have

$$\frac{d}{t} = \frac{v}{2} \tag{a}$$

The kinetic energy of the body is the energy it gains from the work done on it. Thus,

Kinetic Energy = (force)(distance)
$$= (ma)(\text{distance}) = m\frac{v}{t}d = (mv)\frac{d}{t}$$

But, $d/t = v/2$ (from equation (a) above). Therefore,

Kinetic Energy $= (mv)\left(\frac{v}{2}\right) = \frac{1}{2}mv^2$

SUMMARY

Momentum is defined as the product of the mass of an object and its velocity. It is found that momentum is conserved in collisions between objects; this is the **law of conservation of momentum**. **Work** is defined as the product of a force and the distance over which the force acts. An object has **energy** if it has the ability to do work. Energy exists in many forms, two of which are **kinetic energy**, which is the energy an object has because of its motion, and **gravitational potential energy**, which is the energy of an object because of its location

in space. In the absence of friction the sum of the kinetic energy and the gravitational potential energy of an object is constant; this is the **law of conservation of mechanical energy**. Frictional forces convert kinetic energy and gravitational potential energy into **thermal energy**. **Power** is defined as the energy converted from one form to another divided by the time to make the conversion. A **simple machine** is a device that alters the magnitude or direction of an applied force. Two examples are the inclined plane and the lever.

EQUATIONS TO KNOW

Momentum = mv

Work = force × distance = Fd

Kinetic Energy = $\frac{1}{2}mv^2$

Gravitational PE = mgh

Power = $\frac{energy}{time}$

For levers: (Input force)(distance from input force to fulcrum) = (output force)(distance from output force to fulcrum)

KEY WORDS

Conservation laws
Momentum
Law of conservation of
    momentum
Work

Energy
Joule
Kinetic energy
Potential energy
Gravitational potential energy

Power
Watt
Law of conservation of
    mechanical energy

QUESTIONS

MOMENTUM AND CONSERVATION OF MOMENTUM

1. Consider a situation in which two guns are to fire identical shells with the same muzzle velocity. Would you prefer to withstand the kick of the heavier or the lighter of the two guns?
2. What is the momentum of a 4000 kg truck traveling at 80.0 km/h?
3. A student is stranded in the middle of a frozen, frictionless pond with nothing except his physical science book along for comfort. How could the student use the law of conservation of momentum to get off the ice?
4. As a bullet leaves the muzzle of a gun, the gun is propelled backward, and momentum is conserved. What actually forces the gun backward? The bullet? The exploding gases? Describe in detail the

forces within the gun barrel as the bullet is being fired.
5. You are standing on top of a table and drop off to collide with the Earth. Discuss how momentum is conserved in your collision.
6. A bullet with a mass of 50.0 g is fired out of a gun with a muzzle velocity of 400 m/s. If the gun has a mass of 3.50 kg, what will be its velocity after the bullet is fired, assuming that the gun is free to recoil?
7. A charging rhino moving at 2.00 m/s collides with a stationary Volkswagen. If the mass of the rhino is three times that of the VW, what is the velocity of the entangled wreckage after the collision? (Neglect friction.)
8. A puck of mass 20.0 g moves with a velocity of 1.30 m/s toward the right on an air hockey table. It collides with a puck of mass 60.0 g moving to the

left with a speed of 2.00 m/s. After the collision the lower mass puck moves to the left with a speed of 0.90 m/s. What is the velocity of the heavier puck?

9. What happens to the momentum of an object if its velocity is tripled?

10. Gases are often stored in cylinders at very high pressures. Signs attached to these cylinders warn that they can become very dangerous projectiles if punctured. Use the law of conservation of momentum to explain this.

11. A sailor is becalmed on a lake, but he has with him a battery-powered fan. He finds that when he directs the wind from the fan against the sail he does not move. Why not? He also finds that if he turns the fan around so that it is blowing away from the sail he will move. Why?

WORK

12. A crane exerts a force of 40,000 N over a distance of 40 m. How much work has it performed?

13. You stand all day holding a book stationary at arm's length. At the end of the day, you may feel that you have done a lot of work, but from the standpoint of our definition of work, have you done any?

14. A truck with a mass of 5000 kg is rolling along a level road at constant velocity. If there were absolutely no friction or air resistance, how much work would be required to continue to move the truck for a distance of 1.00 km?

15. A horse is pulling a wagon down a hill. As the wagon tends to roll faster than the horse is walking, the animal must hold back the wagon. Does the horse perform work on the wagon as they move downhill? Explain.

16. (a) The engine of a model airplane exerts a forward force of 4.00 N on a plane as it moves through the air for a distance of 6.00 m. (a) How much work does the engine do? (b) If a force of air resistance equal to 2.00 N is exerted on the plane during this short flight, how much work does this force do? (*Hint:* When the force and the displacement are in opposite directions, the work is negative.) (c) Find the net work done on the plane during the flight.

KINETIC ENERGY

17. What happens to the kinetic energy of an object if its speed is tripled?

18. A carpenter is trying out hammers of different sizes. The heaviest hammer has a mass of 1 kg but is too heavy to swing rapidly. The carpenter finds that if he uses a 0.5 kg hammer, he can swing it twice as fast. With which hammer can he generate more kinetic energy, or will it be the same with both? Explain.

19. A football player has a mass of 100 kg and can run at a rate of 7.0 m/s. (a) What is his kinetic energy? (b) If another player is lighter, having a mass of only 80 kg, how fast must he run to maintain an equal kinetic energy?

20. Can the kinetic energy of an object ever be a negative number?

21. You are sitting at rest in a car traveling down the highway. Do you have any kinetic energy?

22. A ball is thrown straight up into the air. At what point(s) is its kinetic energy a maximum? Is the momentum also a maximum at this same time?

GRAVITATIONAL POTENTIAL ENERGY

23. A roller coaster with a mass of 500 kg is sitting on top of the highest incline. The incline is 30 m above the ground and 15 m above the dip in the track below. (a) What is the potential energy of the machine with respect to the ground? (b) With respect to the track below?

24. A rock climber scaling a nearly sheer cliff reaches a ledge 500 m above the valley floor. Then he climbs 10 m above the ledge. If he falls from this point, he will land on the ledge. If he has a mass of 75 kg, what is his potential energy with respect to a fall at this point?

25. A baseball is thrown straight up into the air. At what point(s) is the gravitational potential energy a maximum?

26. A high diver weighing 700 N leaps from a 10 m tower. (a) What is his potential energy initially? (b) At 4 m above the surface of the water? (c) At the surface of the water?

CONSERVATION OF ENERGY AND FRICTION FORCES: THE LAW OF CONSERVATION OF ENERGY

27. A bicycle rider coasts down a hill and up another one. Use the law of conservation of energy to explain why she cannot go as far up the second hill as her initial height on the first.

28. A physics teacher constructs a pendulum by tying one end of a bowling ball to a wire and attaching the free end to the ceiling. The teacher then pulls the bowling ball aside and up against his chin. When he releases it, should he step back to keep from being hit by the ball after it swings through its complete arc?

29. Find the kinetic energy of the high diver of problem 26 at each of the positions listed.

30. A high jumper wants to clear a 2 m bar. What must be his speed upon takeoff to reach this height?

31. A baseball is thrown straight up into the air and reaches a height of 18 m. What was the speed when it left the thrower's hand?

32. In the absence of air resistance, an object thrown into the air will return to ground level with the same speed as it had when thrown upward. Use the conservation of energy principle to verify this statement.

33. A diver drops off a 5 m high diving board. How fast is he going when he strikes the water?

34. (a) Water towers to supply water to a city are placed either on very high hills or on large elevated supports. Why? (b) A person living on the top floor of a building often finds that the water runs very slowly out of his faucet, but a person on the ground floor finds that it comes out forcefully. Why?

35. Would we find the water cooler or warmer at the base of a waterfall than at the top?

36. A car is braked to a stop from a very high speed. What happens to its kinetic energy?

POWER

37. What is energy? How are the concepts of heat and power related to energy?

38. Three farmers are faced with the problem of hauling a ton of hay up a hill. The first makes 20 trips, carrying the hay himself. The second loads a wagon and has his horse pull the hay up in four trips. The third farmer drives a truck up in one load. Which process—manpower, animal power, or machine power—has performed more external work? Which device is capable of exerting more power?

39. Find the hp rating of a 100 W light bulb.

40. Find the wattage rating of a 90 hp car engine.

41. When taking an exam, you have about the same energy output as a 200 W light bulb. Find the total energy in joules you would release during a one-hour exam.

42. Two 70 kg students are on their way to a class on the second floor of a building 2.8 m above ground level. One strolls up in a time of 60 s, while the second sprints up in 4 s. Compare the energy output and power output of each.

SIMPLE MACHINES

43. Suppose you had to lift a heavy iron ball 2 m in the air. You could either lift the ball outright or roll it up a ramp. Which route would require more work? More applied force? Explain.

44. A screw can be considered to be an inclined plane wrapped around a shaft. Explain why less force is needed to turn a screw into a piece of wood than is needed to hammer a nail into an equivalent piece of wood.

45. Suppose you wish to exert a force of 2000 N against a piece of machinery to slide it across the floor. In order to do the job, you decide to use a lever and position the fulcrum 0.25 m from the load. If you are capable of exerting a force of 400 N, how long will the lever have to be to enable you to move the machine?

46. The muscle and bone structure of your arm is shown in the figure. If your muscle exerts a force of 100 N, is more or less than 100 N delivered to a load in your hand? If your muscle moves 1 cm, does your hand move more or less than 1 cm? Explain.

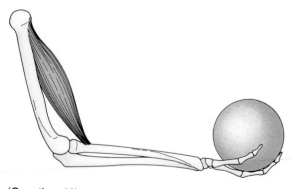

(Question 46)

ANSWERS TO SELECTED NUMERICAL QUESTIONS

2. 320,000 kg km/h
6. −5.71 m/s
7. 1.50 m/s
8. −1.27 m/s
12. 1.60×10^6 J
16. (a) 24.0 J, (b) −12.0 J, (c) 12.0 J

19. (a) 2450 J, (b) 7.83 m/s
23. (a) 147,000 J, (b) 73,500 J
26. (a) 7000 J, (b) 2800 J, (c) 0
29. (a) 0, (b) 4200 J, (c) 7000 J
30. 6.26 m/s
31. 18.8 m/s

33. 9.90 m/s
39. 0.134 hp
40. 67,100 W
41. 7.2×10^5 J
45. 1.25 m

The circular motion of the atmosphere is evident in this cyclonic storm north of Hawaii as photographed by the crew of Apollo 9. (Courtesy NASA)

CHAPTER

4

Newton's Law of Gravity and Some Special Kinds of Motion

Legend has it that Isaac Newton was sitting under an apple tree and was inspired to formulate the law of gravity when he saw an apple fall to Earth. It is not certain that this event actually happened, (although Newton himself told the story), but it is a fact that Newton was the first to realize that objects in the heavens respond to gravitational forces. The Moon, the Earth, every object feels the gravitational tug of every other object.

In this chapter, we shall examine this universal law of gravitation, and we shall also delve into some special kinds of motion. The two particular types of motion that we shall examine are that of a projectile and that of an object moving in a circular path.

4.1
EARLY IDEAS ABOUT GRAVITATION

We already examined a few of the early ideas concerning gravitation when we studied the motion of a freely falling object in Chapter 2. We saw there that the thread of thought that eventually led to our modern ideas of gravitation can be traced to the ancient Greek civilization and to the ideas of Aristotle (384–322 B.C.). For example, we found that one of his ideas concerning falling bodies was that heavier objects fall faster than lighter ones. Aristotle's philosophy saw the Universe as heavily goal oriented. From this point of view, the "natural" place of material objects was "down" and objects sought their natural place, their goal. Heavier objects sought the Earth more than did lighter ones, so they would fall faster than lighter ones. Also, as we have seen, Galileo refuted this theory of Aristotle and opened the door to a more satisfactory approach that was eventually fully developed by Newton. We have seen that Newton's three laws of motion provide the foundation for describing and interpreting all types of motion. For example, Newton stated that an object naturally moves in a straight line with a constant speed unless caused to change this motion by a net force (the first law). But when you throw a ball toward a friend, it does not go in a straight line; it curves down toward the Earth.

FOCUS ON . . . Artificial Gravity

There is a well-known science fiction story called "Ring World" in which an artificial world is found in space which is shaped like a huge ring that is rotating. People inside that world can move around just as though they were on the surface of a planet. Somehow, the rotation of the ring produces a form of artificial gravity. Let us see how this can be done.

Consider a bug riding on the surface of a record on a turntable. If the record is turned on at a low speed such as 33⅓ revolutions per minute, the bug may be able to turn along with the record. If it moves in a circular path, there must be a centripetal force acting on it, and this force is provided by friction between the bug's feet and the record. At a slow speed, this frictional force may be large enough to cause the bug to rotate right along with the record. However, if the record is moved to the next highest speed, 45 rpm, the bug may slide off the edge of the record. This occurs because the frictional force is not large enough to provide the necessary centripetal force for it to turn in the circular path at this higher speed. However, suppose that the record is constructed so that it has a wall built around its rim as shown in part (a) of the Figure. When the bug slams into this wall, it can now move in the circular path because the wall exerts a force on it toward the center of the circular path. This force provided by the wall is now the centripetal force. Let us now turn the speed up to 78 rpm. At this speed, the wall must exert an even larger force to provide the necessary centripetal force to keep the bug turning with the record. The wall is capable of doing so, and the bug rotates. In fact, if the bug desires, he may reorient his position so that he is standing against the wall as shown in part (b). Or, if the bug is so inclined, he could take a stroll around the rim provided, walking just as though a gravitational force were acting on him. In actual practice, the bug on our turntable would stand slightly sideways because the Earth is

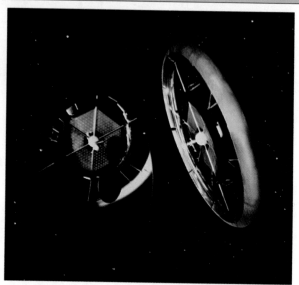

An artist's conception of an artificial space station. (Courtesy NASA)

also exerting a force on him, but in outer space this extra force would not be present.

This imaginary experience for a bug could be used in a practical way for travelers who some day may have to spend long periods of time in space capsules flying through the Solar System or even on flights to nearby stars. On such a journey, prolonged weightlessness could become inconvenient and perhaps even unhealthful. Therefore, scientists envision some sort of spinning space station as shown above. In this case, the centripetal force acting on the travelers in the space vessel would act as an artificial gravity. "Ring World" is correct; such artificial worlds are possible.

Aristotle would have explained that the object curves downward so that it can reach its goal, the Earth. Newton, however, concluded that there is an unbalanced force on the ball as it moves through the air which causes it to deviate from a straight-line path. The force is the object's weight, and the weight results from the phenomenon we call gravitation. In the next section, we shall present Newton's theory of gravitation,

which explains not only how a ball moves while in flight but also the movements of heavenly objects.

4.2
THE LAW OF GRAVITY

Newton stated his law of gravity as follows:

Every object in the Universe is attracted to, and

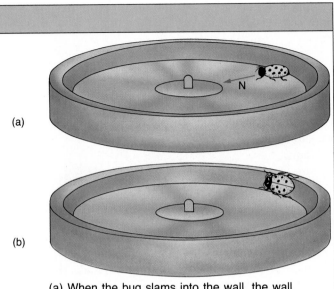

(a)

(b)

(a) When the bug slams into the wall, the wall exerts a force on it causing it to follow its circular path. (b) The bug under artificial gravity.

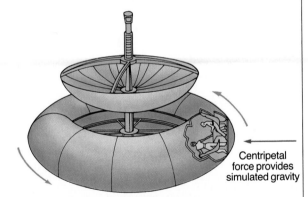

(c)

(c) Travelers in a spinning space station would experience a force pressing them to the outer rim. This force is a type of simulated gravity. A stationary observer outside the station would see this as a centripetal force. However, people in the station would feel that they are being pushed outward by a centrifugal force.

attracts, every other object in the Universe by a force which we call the force of gravity.

At first look, this law does not seem to relate to reality, because if you hold two pencils in your hands, you do not feel the gravitational tug that one exerts on the other. Yet, according to Newton's law of gravity, they are exerting this force one on another. However, if you actually calcu-late the magnitude of this force by use of the equation given below, you find that for ordinary pencils this force has a magnitude of about 2×10^{-11} N, hardly measurable. In fact, if you go to larger objects such as two automobiles, you still find that the force of gravitational attraction between the two is so small as to be unnoticeable. Thus, if you have a fender-bender with another car, don't try to use the law of gravity as your defense with the arresting officer.

In equation form, the law of gravity may be stated as

$$F = G \frac{m_1 m_1}{r^2} \qquad (4.1)$$

where G is the universal gravitational constant determined experimentally to have a value of 6.67×10^{-11} N m^2/kg^2 in SI units. The quantities m_1 and m_2 are the masses of the two attracting objects, and r^2 is the square of the distance between the centers of the two objects.

EXAMPLE 4.1 Dueling pencils

A pencil of mass 0.10 kg is held at a distance of 0.20 m from a second identical pencil. Find the force of gravitational attraction between the two.

Solution The force of gravitational attraction is found by direct substitution into Eq. 4.1. We have

$$F = G \frac{m_1 m_2}{r^2}$$
$$= (6.67 \times 10^{-11} \text{ N m}^2/\text{kg}^2)$$
$$\frac{(0.10 \text{ kg})(0.10 \text{ kg})}{(0.20 \text{ m})^2} = 1.67 \times 10^{-11} \text{ N}$$

EXAMPLE 4.2 I'm climbing higher and higher

(a) The typical mass for an adult is about 70 kg. Find the force of gravitational attraction exerted on a person of this mass by the Earth (mass = 5.98×10^{24} kg) at the surface of the Earth. The radius of the Earth is about 6.38×10^6 m. This force exerted on the individual by the Earth is the weight of the person.

(b) Find the weight of the person if he is in a spaceship at a height above the Earth equal to the radius of the Earth.

Solution **(a)** The weight of the person is found from Eq. 4.1 as

$$F = w = G \frac{m_1 m_2}{r^2} = (6.67 \times 10^{-11} \text{ N m}^2/\text{kg}^2)$$

$$\frac{(5.98 \times 10^{24} \text{ kg})(70 \text{ kg})}{(6.38 \times 10^6 \text{ m})^2} = 686 \text{ N} = 154 \text{ lb}$$

(b) At a height above the Earth equal to its radius, $r = 1.28 \times 10^7$ m, and from the law of gravity we find the weight as

$$F = w = G \frac{m_1 m_2}{r^2} = (6.67 \times 10^{-11} \text{ N m}^2/\text{kg}^2)$$

$$\frac{(5.98 \times 10^{24} \text{ kg})(70 \text{ kg})}{(1.28 \times 10^7 \text{ m})^2} = 170 \text{ N} = 38 \text{ lb}$$

Notice the pattern indicated in this example. When you are twice as far away from the center of the Earth in (b) as in (a), your weight is only one-fourth as much. In a similar fashion, if you move three times farther away, your weight will be only one-ninth as much. A relationship between variables that produces this result is called an "inverse square" relationship.

EXAMPLE 4.3   Gravity and distance

(a) Use Newton's law of gravity to find an expression for the acceleration due to gravity as a function of height above the surface of the Earth.

(b) Use your expression to find the acceleration of gravity at a height of one Earth radius above the Earth's surface.

Solution **(a)** At first thought, this seems to be a difficult problem, but let us assume that we have an object of mass m a distance r away from the center of the Earth of mass M. As we have seen above, in this case Eq. 4.1 gives us the weight of the object. Thus, we have

$$F = w = G \frac{mM}{r^2}$$

But, we also know from Chapter 2 that the weight of an object is given by $w = mg$, where g is the acceleration due to gravity. Thus, we have

$$mg = G \frac{mM}{r^2}$$

or, upon canceling the common term m from both sides, we have

$$g = G \frac{M}{r^2}$$

You should note that there is nothing special about this equation that makes it apply only to the Earth. It applies equally well to any planet. Thus, we can generalize as

$$g_{planet} = G \frac{M_{planet}}{r^2_{planet}}$$

Thus, we see that the more massive the planet, the greater the acceleration due to gravity at its surface. However, be careful because in general the more massive the planet, the larger is its radius, and an increased radius tends to make the force of gravity smaller.

(b) Using the values for the mass of the Earth and the distance found in Example 4.2, we find

$$g = (6.67 \times 10^{-11} \text{ N m}^2/\text{kg}^2) \frac{(5.98 \times 10^{24} \text{ kg})}{(1.28 \times 10^7 \text{ m})^2}$$

$$= 2.43 \text{ m/s}^2$$

EXERCISE

Find the acceleration due to gravity at the surface of Jupiter, which has a mass of 1.90×10^{27} kg and a radius of 6.99×10^7 m.

Answer 25.9 m/s^2.

4.3
WEIGHTLESSNESS

As we have seen, as an object moves higher above the surface of the Earth, it weighs a little less. Thus, you weigh less standing on the top of Mt. Everest than you do at sea level. Similarly, astronauts become lighter as they fly skyward, for they travel farther and farther from the center of the Earth. However, if their weight when they are in orbit is actually calculated, a curious fact becomes evident. The Skylab spaceship launched by the United States in 1974 orbited approximately 320 km above the surface of the Earth. Since the Earth is 6400 km in diameter, the astronauts

FOCUS ON . . . The Tides

As you sit in a comfortable chair, studying a subject such as the movement of the oceans, it becomes easy to lose touch with the natural power and majesty of this phenomenon. To appreciate how people first began to think about such systems, imagine that you are working on a fishing boat off the coast of Alaska. You would notice that the level of the ocean rises and falls in a cycle of approximately 12 hours. Thus, if the water were low at noon, it would reach maximum height at about 6:00 P.M. and be low again near midnight. These vertical displacements are called **tides.** You would soon find out that low tide does not recur at the same time each day. If today's low tide occurred at 4:43 A.M., tomorrow's might not occur until 5:29 A.M., and the next day's would be still later, until eventually low tide would not occur until noon. If you were observant, you would notice that each day the tides were delayed approximately 40 to 50 minutes, or about the same amount of time that the moonrise is delayed from day to day.

To understand tides, consider the Earth-Moon system, as shown in the Figure. For simplicity, imagine a situation in which there are no continents on the Earth and the surface is one giant ocean. At any one instant of time, one section of this giant ocean (marked A in the figure) lies just under the Moon, while all other regions are farther away. Since gravitational force is greater for objects that are closer together, the part of the ocean closest to the Moon is

attracted with a force greater than the force on the center of the Earth, point B in the figure, and the force exerted on the Earth at B is greater than the force exerted on that portion of water at C. The relative magnitudes of these forces are shown by the arrows in the Figure. The large force on the water at A causes the ocean to bulge outward toward the Moon, resulting in high tide. At this same time, the center of the Earth, B, is being more strongly attracted toward the Moon than is the water at C, and as a result, the Earth is "pulled" away from the water. This means that the ocean at point C is left behind a little, so a bulge is formed at C. This bulge is the high tide 180° away from the Moon. Thus, the tides rise and fall two times a day. At a given time of day, the tide is high at a point on the Earth directly facing the Moon, and it is simultaneously high at a point exactly on the opposite side of the Earth.

Following this reasoning, it is easy to explain why the high tide appears later every day. An observer on Earth located just under the Moon at noon of one day sees that the gravitational pull of the Moon causes the ocean to bulge and the tide to rise. Twenty-four hours later, the Earth has made one revolution; but because the Moon has traveled some distance in its orbit during that time, the Earth must spin a little farther for an observer to be again directly under the Moon. Thus, the tide reaches a maximum a little later each day.

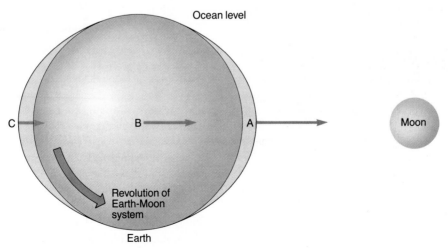

Schematic view of tide formation. (Magnitudes and sizes are exaggerated for emphasis.)

(a)

(b)

(a) Astronauts J. Hoffman and M. Seddon demonstrate weightlessness using a Slinky in Discovery's main deck, April 1985. (b) Astronaut Truly has a slight problem handling a computer print-out while in a weightless condition on the 1981 space shuttle. (Courtesy NASA)

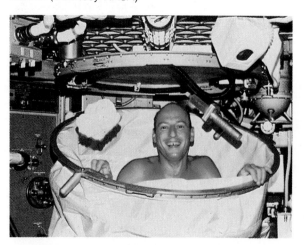

As an exercise, discuss the problems this Skylab astronaut encountered while trying to take a shower in a weightless condition. (Courtesy NASA.)

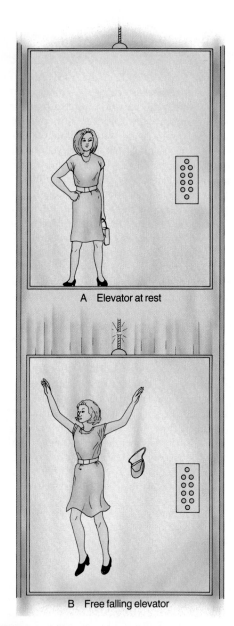

A   Elevator at rest

B   Free falling elevator

FIGURE 4.1 Weightlessness in a freely falling elevator.

were located about 6720 km from the center of the Earth. A person who weighed 650 N (146 lb) on Earth would have weighed 587 N (131 lb) in the orbiting Skylab. But how could that be? Television broadcasts from space showed the astronauts floating weightlessly about the capsule.

To answer this perplexing question, let us set aside the spaceship for a moment and think about a woman carrying a pocketbook as she steps into an elevator, as shown in Figure 4.1. Assume that just as the luckless woman steps in, the cable suddenly breaks and the elevator starts to fall freely down the shaft. Just at that moment, the scared rider lets go of her pocketbook. There are now three objects (the elevator, the rider, and the pocketbook) falling down the shaft. Remember that the acceleration of gravity is equal for all objects. Thus, assuming that air resistance is negligible, all three accelerate downward at exactly the same rate. For every meter that the elevator falls, the pocketbook and the woman also fall 1 m. Therefore, if the pocketbook were $\frac{1}{2}$ m above the floor of the elevator initially, it would remain at that height relative to the elevator during the entire descent. If someone photographed the scene, the pocketbook would appear to be suspended, weightless, in the air. Similarly, if the rider had jumped 10 cm in the air just before the cable snapped, she would remain at that height, as if she were weightless. This is only an apparent weightlessness because at any instant she would weigh just as much as she would if the elevator were motionless. Obviously, she has to have weight, for otherwise she would not fall along with the elevator.

The apparent weightlessness of an astronaut in an orbiting spacecraft is similar to the apparent weightlessness of the luckless woman in the falling elevator. Rocket action carries the craft aloft until it is, say, 320 km above the Earth, then redirects the capsule and accelerates it horizontally. Finally, the engines shut off. At this point, the satellite begins to fall, but as we shall see in a later section in this chapter, if the arc of the fall is the same as the arc of the Earth's surface, the satellite orbits rather than crashing into the Earth. The vehicle is falling freely all the time, however. The astronauts in the capsule are like the rider and the pocketbook in the falling elevator. They are falling, but since their enclosure is falling at the same rate, they appear to be weightless when in fact they are not.

4.4 PROJECTILE MOTION

The types of motion that we have looked at thus far have been those in which the object under consideration moved along a straight-line path. For example, when we looked at freely falling

FIGURE 4.2 The only force on a projectile in flight is its weight w.

objects, they were either moving vertically downward or upward along a straight-line path. We will now turn our attention in the rest of this chapter to the motion of objects moving in a plane. The first kind of motion that we shall look at is called **projectile motion.** This is the kind of motion followed by a batted baseball or a kicked football.

Before we look at this motion in detail, we ask you to repeat a mental exercise discussed in Chapter 2, namely that of finding all the forces that act on an object. Here is the situation. A punter kicks a football—your task is to find all the forces acting on it while it is in flight, ignoring air resistance. Before you read further, think about it a moment and remember that there are only two types of forces in nature—contact forces and action-at-a-distance forces.

If your answer is only one force, you are absolutely correct. There are no horizontal forces at all acting on the football once it is in flight. Obviously, while the football was being kicked, there was a force acting on it exerted by the punter as long as his toe was in contact with the ball. However, once the ball has lost contact with his foot, there is nothing else touching the ball to exert a contact force on it. Thus, the only remaining possibilities are the action-at-a-distance forces, of which there is only one that we need worry about, the force of gravity, or the weight. This force is shown in Figure 4.2.

It is important to examine the forces acting on the football, because such an analysis enables us to determine some features of the motion of any object moving as a projectile. To do so, we shall consider the motion of a projectile in two

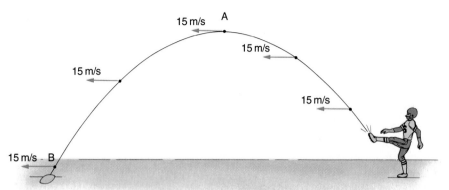

FIGURE 4.3 The horizontal velocity of a projectile is a constant.

parts: the horizontal motion and the vertical motion. Let us look at each of these in turn.

Horizontal Motion There are no forces acting on a projectile along the horizontal direction. Consider this statement from the point of view of Newton's second law, F = ma. If we consider F to be the net force along the horizontal, we find

$$0 = ma$$

but from this we must conclude that the acceleration along the horizontal is also zero. If the acceleration along the horizontal is zero, this means that the object can neither speed up along this direction nor slow down. *Thus, the motion along the horizontal is one of constant velocity.* This aspect of the motion of a projectile is pictured in Figure 4.3. For our specific example, a punter has kicked a football such that at the instant it left his toe, the ball was moving along the horizontal with a velocity of 15 m/s, and this velocity remains the same at all points. That is, it is traveling horizontally with a speed of 15 m/s at the top of its path (position A), at the end of its path just before it strikes the ground (position B), and at any of the infinite number of other intermediate points.

Vertical Motion The motion along the vertical direction is exactly like a type of motion that we examined in Chapter 2, free fall. As we have seen, the characteristic of a freely falling body is that the only force acting on it is its weight, and that is the case for a projectile. This means that as the object rises, its vertical velocity slows at a rate of 9.8 m/s every second, and after it passes the apex of its flight path and begins to descend, it

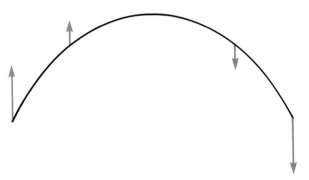

FIGURE 4.4 The vertical velocity of a projectile decreases as it rises and increases as it falls in the same way as that for a freely-falling object.

speeds up at a rate of 9.8 m/s every second. The vertical velocity of a projectile follows a pattern like that shown in Figure 4.4. Note that initially it has a large upward velocity that gradually decreases as it rises. This is indicated by the fact that the length of the velocity vectors decrease with height. Finally, at the top of the path, its vertical velocity reaches zero for an instant. However, it continues to accelerate with the acceleration due to gravity; as a result, it now begins to increase in speed in the downward direction as it falls, as shown in Figure 4.4.

To understand some of the characteristics of projectile motion, consider the following problem. You are a punter for your college team, and you have the fantastic ability to kick the ball so that it always leaves your toe traveling at the same speed. You would like to kick the ball so that it travels the maximum distance. The question is, at what angle with the horizontal should you punt

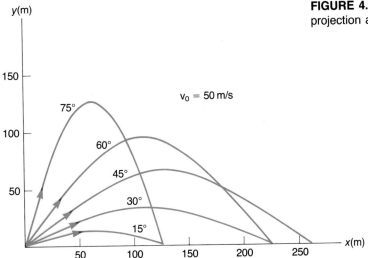

FIGURE 4.5 The flight of a projectile with different projection angles.

the ball? Figure 4.5 shows the result of such an analysis for balls kicked at several different angles. In the absence of air resistance, the ball travels the maximum distance when kicked so that its initial direction of travel is at an angle of 45° with respect to the horizontal. Also, note from Figure 4.6 that whatever the angle, in the absence of air resistance, the path is symmetrical about a line through the apex of the path and perpendicular to the ground. If air resistance is important, as it would be in real-world situations, the path deviates from this symmetrical pattern. The actual result is that the path is much like that shown in Figure 4.6. As you would expect, the ball will travel a shorter distance with air resistance present than it would without, as Figure 4.6 indicates.

EXAMPLE 4.4 Which will hit first?

A baseball is thrown horizontally at the same instant an identical baseball is dropped straight down, as shown in Figure 4.7. Which of the two will hit the ground first and why?

Solution They will both hit at exactly the same instant. This occurs because both balls have identical vertical motions. That is, they both fall vertically with the acceleration due to gravity. As another example of this, suppose that a baseball outfielder is sitting on a fence when a batter hits the ball directly at him. If the outfielder is to catch the ball, all he has to do is drop off the fence. He will accelerate downward at the same rate as the falling baseball, so they will both reach

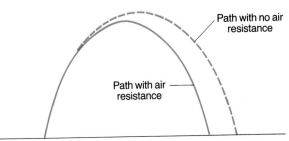

FIGURE 4.6 A projectile travels a shorter distance when air resistance is present than it does in its absence.

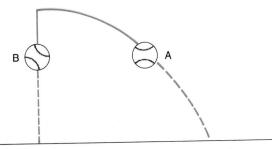

FIGURE 4.7 Ball A is thrown horizontally at the same instant that ball B is dropped. Which will hit the ground first?

FIGURE 4.8 Example 4.5.

ground level at the same time. This assumes that the baseball has sufficient horizontal speed to reach him.

EXAMPLE 4.5   A new kind of bowling alley

Imagine that a flat bowling alley is built on the edge of a cliff, as shown in Figure 4.8. A bowler rolls a ball down the alley and off the cliff as shown. Which of the diagrams best describes the motion of the ball after it has rolled off the edge? (Neglect the effects of friction and air resistance.)

Solution   The ball would follow path b. Initially the ball is propelled (forced to accelerate) by the bowler's arm. The ball moves in a straight line at constant speed until it reaches the edge of the cliff. What happens the instant the ball goes over the edge? There are no horizontal forces acting on the ball. Therefore, the horizontal velocity remains unchanged because there are no horizontal forces to speed it up or slow it down. There is, however, a net vertical force, its weight, which produces a net downward acceleration. The resultant motion of the ball, shown in Figure 4.8b, is the combination of these two independent components, a constant horizontal velocity and a simultaneous acceleration due to gravity. As an exercise, explain why a, c, and d are incorrect.

FIGURE 4.9 Example 4.6.

EXAMPLE 4.6

Imagine that you are running at a constant velocity across a field and you drop a ball while continuing to run, as shown in Figure 4.9. Again assuming that air resistance and friction are negligible, will the ball fall (a) in front, (b) behind, or (c) right under your hand?

Solution   This problem is very similar to the previous one. Just before the ball is dropped, it has a horizontal velocity equal to the velocity of the runner. When the ball is released, it starts to fall but still continues to move forward with its original horizontal velocity. At this point, it is analo-

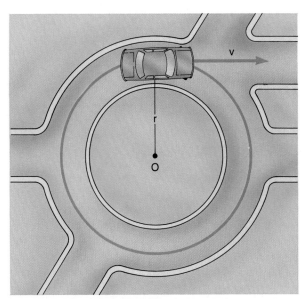

FIGURE 4.10 Does the car moving at a constant speed have an acceleration?

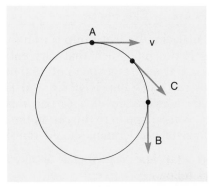

FIGURE 4.11 The direction of the velocity of an object moving in a circular path changes as shown.

gous to the bowling ball just after it has gone over the edge of the cliff. Since the horizontal velocities of both the runner and the ball are constant and equal, the ball remains directly under the runner's hand throughout its fall.

4.5
CIRCULAR MOTION

Here is a thought question for you to ponder before you read too deeply into this section. Figure 4.10 shows a car traveling in a circular path with a constant speed. Does the car have an acceleration? At first thought, it seems obvious that the answer should be no, but as is so often the case, what seems obvious may be incorrect. The car does indeed have an acceleration. Let us see why.

In Chapter 2, we examined the definition of acceleration, and found it to be

$$a = \frac{\text{change in velocity}}{\text{time for change to occur}}$$

A superficial look at this equation tells us that the change in velocity of the car is zero because the car is traveling at a constant speed. However, take a second look. This equation actually says

that acceleration is equal to the change in *velocity* divided by the time over which this change occurs. This means that since velocity is a vector quantity there are two ways in which it can change. (1) *The magnitude of the velocity, the speed, can change.* This is what has been happening in all of the various types of motion that we have examined in this text thus far. (2) *The direction of the velocity can change,* and that is what is happening in the present situation. Figure 4.11 shows how the direction of velocity is changing. At point A, the direction is toward the east; at point B, it is toward the south. For a point in between, such as C, it points in the direction shown.

It is a relatively straightforward derivation to find the change in velocity for the car, but we shall not do so here. Instead, it suffices to say that it can be shown that *the direction of the acceleration of an object moving in circular motion is always toward the center of the circular path followed* and has a magnitude given by

$$a_c = \frac{v^2}{r}$$

where v is the magnitude of the velocity and r is the radius of the circular path followed by the object. The acceleration, a_c, is called the **centripetal acceleration.** The word "centripetal" means center-seeking, and this designation arises because the direction of this acceleration is always toward the center of the circular path.

EXAMPLE 4.7   Exit ramps

A particular circular exit ramp off an interstate highway is designed such that cars entering it should be traveling at a speed of 15.0 m/s. The radius of this exit ramp is 40.0 m. **(a)** Find the centripetal acceleration of a car on this ramp. **(b)** What would happen to the car on the ramp if this centripetal acceleration should vanish?

Solution   (a) The centripetal acceleration is found as follows

$$a_c = \frac{v^2}{r} = \frac{(15.0 \text{ m/s})^2}{40.0 \text{ m}} = 5.63 \text{ m/s}^2.$$

(b) The centripetal acceleration of the car arises because the direction of its velocity is changing. This means that if a car has zero centripetal acceleration, the direction of travel of the car must *not* be changing. Thus, if a_c should go to zero, the car would have to travel in a straight-line path. As we will see later, this can happen to the car when the exit ramp is icy.

EXAMPLE 4.8   Centripetal acceleration on the Earth

The Earth has a radius of 6.38×10^6 m at the equator and turns once on its axis in a day, 86,400 s. Find the centripetal acceleration of a bug resting at the equator due to the rotation of the Earth.

Solution   First, we must find the speed at which the bug is turning. This is the speed at which the Earth turns, which can be found from v = distance/time.

The distance moved by the bug in a time of 86,400 s is the circumference of the Earth given by

$$\text{distance} = 2\pi r = 2(3.142)(6.38 \times 10^6 \text{ m})$$
$$= 4.01 \times 10^7 \text{ m}$$

and thus, the velocity of the bug is

$$v = \frac{\text{distance}}{\text{time}} = \frac{4.01 \times 10^7 \text{ m}}{86400 \text{ s}}$$
$$= 4.64 \times 10^2 \text{ m/s}$$

From this, we are able to find the centripetal acceleration of the Earth as

$$a_c = \frac{v^2}{r} = \frac{(4.64 \times 10^2 \text{ m/s})^2}{6.38 \times 10^6 \text{ m}} = 0.0337 \text{ m/s}^2.$$

This centripetal acceleration is virtually negligible in comparison to the acceleration due to gravity of 9.8 m/s^2.

4.6
CENTRIPETAL FORCE

In the preceding section, we found that an object that travels in a circular path always has an acceleration toward the center of the path given by $a_c = v^2/r$. Let us now examine this statement from the standpoint of Newton's second law of motion, F = ma. The second law tells us that if an object has an acceleration, a, it also must have a net force, F, acting on it. Thus, since an object moving in circular motion has an acceleration toward the center of the path that it follows, there must also be a resultant force toward the center of the circular path. This force is actually the force that makes the object follow the circular path and is called a **centripetal force.**

Often, when beginning students of physics encounter the term "centripetal force," they tend to think of it as some kind of mystical force that arises to keep an object in its circular path. Avoid this. Centripetal forces are just like all the other forces you have seen so far. That is, they are forces such as the tension in a string, a force of gravitational attraction, friction, or a host of other possibilities. Let us take a look at a few of these situations.

Figure 4.12a shows a rock attached to the end of a string being whirled in a vertical circle. Identify the centripetal force in this instance. The force which is acting toward the center of the circular path is a contact force, the tension in the string. To see that this force does indeed cause the rock to follow its circular path, consider what would happen if the string should break, as in the b part of Figure 4.12. At the instant the string breaks, the object is traveling toward the left, and it will fly off in a straight-line path toward the left. As it leaves, however, gravity causes it to move as a projectile. Thus, it follows the path shown in the figure.

Figure 4.13 shows the Moon following its almost circular path about the Earth. Identify the

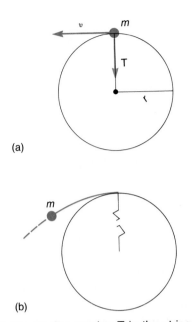

(a)

(b)

FIGURE 4.12 (a) The tension T in the string causes the rock to follow its circular path. (b) If the string breaks, the ball moves away tangent to the circular path and becomes a projectile. The path followed by the projectile assumes the rock is being swung in a vertical circle.

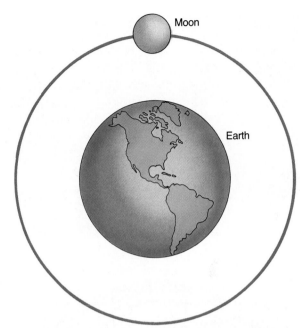

FIGURE 4.13 Identify the centripetal force acting on the Moon.

centripetal force acting on the Earth. Remember that centripetal forces are no different from the other forces we have run into so far. Their only particular characteristic is that they point toward the center of the circular path followed by the object. The force which acts on the Moon and causes it to move in its circular path is the force of gravitational attraction exerted on it by the Earth. Thus, the centripetal force is given by

$$F = G \frac{m_e m_m}{r^2}$$

where m_e is the mass of the Earth, m_m is the mass of the Moon, and r is the distance from the Earth to the Moon.

Note that when a problem involves only centripetal forces and centripetal accelerations, Newton's second law, F = ma, reduces to

$$F_c = m \frac{v^2}{r} \tag{4.2}$$

where F_c is the centripetal force acting on the object and v^2/r is its centripetal acceleration.

EXAMPLE 4.9 The Red Baron in action

A model airplane of mass 0.75 kg flies in a circular path of radius 10 m at a speed of 15 m/s. The plane is held in its circular path by a cable. Find the tension in the cable.

Solution The centripetal force acting on the plane is the tension T in the cable, and this can be found from

$$F_c = m \frac{v^2}{r}$$

$$T = (0.75 \text{ kg}) \frac{(15 \text{ m/s})^2}{10 \text{ m}} = 16.9 \text{ N}$$

EXERCISE

If the cable in the exercise has a breaking strength of 20 N, what is the maximum speed at which the airplane can fly?

Answer: 16.3 m/s

(a)

(b)

(a) As an exercise, identify the centripetal force acting on these roller coaster riders as they make one of the circular turns at the bottom of the ride. (b) Also, use your knowledge of centripetal forces to explain why these roller coaster cars, and their riders, can stay on the track at the top of the loop.
(Courtesy Kings Island)

EXAMPLE 4.10 Moon over Miami

Find the centripetal force acting on the Moon.

Solution The Moon is held in its circular orbit by the force of gravitation exerted on it by the Earth. This is found as

$$F = G \frac{m_e m_m}{r^2}$$

where m_e is the mass of the Earth, m_m is the mass of the Moon, and r is the distance from the Earth to the Moon. We have

$$F = (6.67 \times 10^{-11} \text{ N m}^2/\text{kg}^2)$$
$$\times \frac{(5.98 \times 10^{24} \text{ kg})(7.36 \times 10^{22} \text{ kg})}{14.6 \times 10^{16} \text{ m}^2}$$
$$= 2.01 \times 10^{20} \text{ N}$$

This problem re-emphasizes the idea that there is nothing special about centripetal forces. When searching for them, you should look for old friends such as tension in strings, gravitational forces, and so forth.

4.7
ORBITING THE EARTH

It is sometimes confusing to students to find out that an object moving in a circular path has an acceleration directed toward the center of the path. The question often asked is, if the object is accelerating toward the center of the path, why does it not fall in toward the center? To see why it doesn't, let us give an example used by Newton himself to explain how an object can be made to orbit the Earth. It may be argued that Newton anticipated today's satellites.

Newton considered what could be done with a very powerful cannon and a mountain that extended above the Earth's atmosphere. If a cannonball were fired from the summit, its path would be somewhat as shown in Figure 4.14, path A. The cannonball behaves exactly like you expect in that it accelerates toward the Earth and strikes the surface. If, now, one uses a greater charge of gunpowder to shoot the cannonball at a greater speed, the ball would follow path B. Even more powder yet, and you get path C. With careful adjustment (and a powerful cannon) you should be able to shoot the ball such that as it falls toward the surface of the Earth, the natural curvature of the Earth causes its surface to curve out

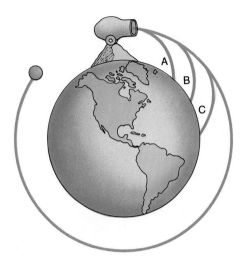

FIGURE 4.14 Newton's idea for placing a cannonball in orbit.

from under the ball as it falls. The result is a cannonball in orbit. Speeding cannonball and speeding Moon are equivalent.

4.8
CENTRIFUGAL FORCES

Of all the concepts of physics, perhaps the one that is most frequently misunderstood by students is that of **centrifugal** forces. The word "centrifugal" is derived from Latin roots meaning center-fleeing. Therefore, the word relates to forces that are directed away from the center of the circular path that an object follows. Because of the subtle nature of centrifugal forces, we shall have to spend some time talking about what they are and what they are not.

Consider a car turning a corner as shown in Figure 4.15. If the car is to make the circular turn on the exit ramp, there must be a centripetal force acting on it. This centripetal force is the frictional force exerted on the car by the roadway. On a very icy road where friction is minimal, this frictional force might be so small that the car would not be able to negotiate a sharp curve. According to Newton's first law, the natural tendency of the car is to maintain its motion along a straight-line path unless some external force, such as the frictional force, causes it to deviate from this straight line. Thus, on an icy road where the necessary centripetal force is not avail-

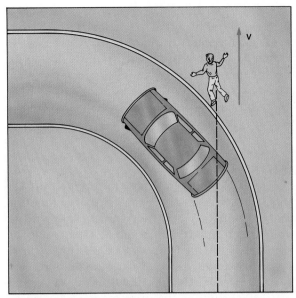

FIGURE 4.15 All moving objects have an inertial tendency to travel in a straight line. If a person were sitting on the roof of a car, and the car made a sharp turn, the person would be likely to continue to travel in a fairly straight line and fall off the car. The reason is that there is no force of sufficient strength to cause the person to follow a curved path.

able, the car does not move in the circular path; it continues on its straight-line course and goes off the road. Now, consider a passenger in the car. If he is to deviate from the straight-line path he is following before the turn, a centripetal force must act on him also. What is the origin of this force? This force arises from the frictional force exerted on the person by the car seat. If the turn is sharp enough, however, or made at a high speed, this frictional force may not be large enough to deviate him from his straight-line path. As a result, he slides across the seat until he strikes the door. The door then provides a push on him toward the center of the circular path. Thus, in this case the centripetal force is the force that the door of the car exerts on him. In Figure 4.15, we see what would happen to a person riding on the smooth roof of a car as it rounds a corner quickly. Frictional forces would not be large enough to produce the required centripetal force; there is not a door for the person to slide into to produce the necessary centripetal force, and as a result, he would continue to travel in a straight-line path and fly off the roof of the car.

Thus, the person slides off the roof of the car because of the *absence* of a centripetal force. However, if you were in the position of the person riding on the roof of the car, you would describe your experience in totally different terms than those presented above. Let us examine the most common explanation for what happens and see why it is incorrect.

If you were the person on the roof of the car when it begins to turn under you, you would suddenly find yourself flying off into space. Why? A person on a rooftop watching the event would say that you are obeying Newton's first law of motion and continuing on in your straight-line path as you should. But, your point of view would be completely different. As you went flying off into space, you might glance back at the car and see it turning in its circular path. Glancing quickly toward the center of the circular path followed by the car, you would notice yourself getting farther and farther away from this center point. Flipping through your memory bank of past experiences, you would ask yourself what has caused such occurrences in the past. You could explain your motion by saying that some force must be acting on your body to pull you away from the center of the circular path, and you could give this force a name, a centrifugal force. This is a fictitious force created only to explain your motion. It obviously cannot be a real force because there is nothing to create it. There are no action-at-a-distance forces that would act in this way, and as you fly through the air, there is nothing touching your body which could produce a contact force. Thus, you move as you do, not because of a centrifugal force, but because of the absence of a centripetal force to cause you to make the turn with the car.

Let us consider another example to show how people often invent centrifugal forces to describe common observations. When you wash clothes, at the end of the cycle the drum of the washing machine goes into a spin cycle during which it rotates rapidly in order to throw excess water off the clothing. Why is the water thrown off? The common answer is that a centrifugal force is exerted on the water droplets pulling them off the fabric. However, as you probably might guess, this fictitious force is not present. Let us consider the clothes as they rotate at a low rate of speed. A particular drop of water is held

FIGURE 4.16 The centripetal force exerted on the can by the string causes it to travel in a circular path.

to the fabric by molecular forces, and at a low turning rate, these forces are large enough to provide the centripetal force to allow the drop to turn with the clothes. However, as the rotational rate increases, there comes a time when the molecular forces are not great enough to provide the required centripetal force. When this occurs, the drop flies off in a straight-line path until it hits the outside of the drum and disappears down the drain. If a very intelligent amoeba were in the drop of water, it might explain what is happening to the drop by saying that it was pulled off the clothing by some centrifugal force. On the other hand, a person standing beside the washing machine would explain the phenomenon correctly by saying that the drop containing the amoeba was separated from the clothing simply because there was not a large enough centripetal force acting on it to cause it to turn along with the drum and clothing.

Now that we have seen what centrifugal forces are not, let us see what they are. Centrifugal and centripetal forces are action-reaction pairs. To explain this with an example, consider a situation in which a child attaches a string to a can and swings it around her head in a circular path, as indicated in Figure 4.16. The force that is causing the can to move in its circular orbit is the tension in the string. This is the centripetal force. To find out the reaction to this action, let us formulate a sentence as follows: The action is the force exerted on the can by the string. To find the reac-

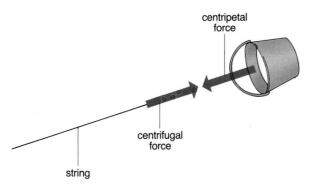

FIGURE 4.17 The centripetal force acts on the bucket, the centrifugal force is on the string.

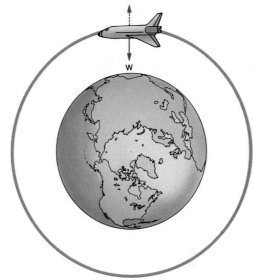

FIGURE 4.18 What is wrong with this figure?

tion, we invert the sentence as follows: The reaction is the force exerted on the string by the can. This is the centrifugal force. Thus, the action-reaction pair is as indicated in Figure 4.17. Note that the centrifugal force is not on the can; it is on the string.

EXAMPLE 4.11   Why doesn't it fall?

Back in the early days of the space program, government officials were often confronted with the question of how a satellite can stay out there. Why doesn't it fall to Earth? In searching for a simple answer, they devised an explanation that uses centripetal and centrifugal forces as shown in Figure 4.18. They said that there was a force on the satellite directed toward the center of the Earth, its weight, and this was the centripetal force. If this were the only force acting, they said incorrectly, the satellite would fall. They then said that there was another force, a centrifugal force, which acted on the ship in a direction away from the center of the Earth. This force is shown as a dashed force in Figure 4.18 to emphasize that it really isn't there. The explanation went on to say that the reason the satellite did not fall was because the two forces were action-reaction pairs, and since they are equal in magnitude and opposite in direction, they would cancel, and the object would sail happily along in its orbit. From what you know about Newton's laws of motion, you should be able to verify that this explanation violates all three laws. Try it. Explain why this violates **(a)** the third law, **(b)** the first law, and **(c)** the second law, and finally, explain **(d)** why the object doesn't fall.

Solution **(a)** Recall that action-reaction pairs have three fundamental properties. They are equal in magnitude and opposite in direction, and they act on different objects. These two proposed forces are equal in magnitude and opposite in direction according to space officials, but they both act on the satellite. Thus, they cannot be an action-reaction pair.

(b) If these two forces are equal in magnitude and opposite in direction, they would cancel one another, and produce a net force on the satellite equal to zero. According to Newton's first law, any object which has no net force acting on it will either not move at all or it will move at a constant speed in a straight line. Because the satellite is doing neither of these, the first law would be violated.

(c) We know that any object that has the direction of its velocity changing has an acceleration, which we have called the centripetal acceleration. Our satellite therefore has a centripetal acceleration, but, if so, it also must have a net force acting toward the center of its circular path. As noted in part (b), their explanation depends on there being no net force acting on the satellite. Thus, in $F = ma$, you would have an acceleration without a net force, and so the second law is also violated.

(d) The object does fall. As pointed out on several occasions in this chapter, the satellite *is falling* toward the Earth in an arc such that the curvature of the Earth causes its surface to bend away from the falling satellite to exactly the same extent that the satellite is bending down to try to meet the Earth.

SUMMARY

Newton's law of universal gravitation states that every object in the Universe exerts a force of gravitational attraction on every other object. The apparent weightlessness of objects in orbit occurs because they are all "falling" toward the Earth with the same acceleration.

Projectile motion is characterized by two separate types of motion. Motion along the surface of the Earth takes place at a constant velocity, while the vertical motion is one of constant acceleration.

Centripetal force is an inward force that causes an object to move in a curved path. Every object moving in a curved path has an acceleration toward the center of that path called a **centripetal acceleration.** If the centripetal force is the action, the reaction is an outwardly directed **centrifugal force.**

EQUATIONS TO KNOW

$$F = G \frac{m_1 m_2}{r^2} \text{ (Newton's law of universal gravitation)}$$

$$a = \frac{v^2}{r} \text{ (centripetal acceleration)}$$

$$F = m \frac{v^2}{r} \text{ (centripetal force)}$$

KEY WORDS

Gravity
Newton's law of universal gravitation

Apparent weightlessness
Projectile motion
Centripetal acceleration

Centripetal force
Centrifugal force

QUESTIONS

THE LAW OF GRAVITY

1. Is there a gravitational attraction between you and this book? If so, why don't you feel the pull?
2. In a desperate effort to lose weight, you decide to move either to Denver or to Death Valley. Which would be your best choice and why?
3. According to the law of universal gravitation, how far from the Earth would you have to get before its gravitational pull on you dropped to zero?
4. In this text, we frequently refer to the mass of a planet but never to its weight. Why not?
5. A man driving his car (total mass 1500 kg) crashed into a telephone pole (mass 500 kg). He used the defense in court that the gravitational pull on his car by the pole when he was 10 m from it was so great that he was unable to control the car. Calculate the force and rule on the case.
6. Find the acceleration due to gravity at the surface of (a) Mercury (mass = 3.18×10^{23} kg and radius = 2.43×10^6 m), (b) and Saturn (mass = 5.68×10^{26} kg and radius = 5.85×10^7 m).

7. A baseball is dropped from a height of 10.0 m on the surface of Mercury. (a) Use the results of problem 6 to find the speed of the ball just before it strikes the ground. (b) Repeat your calculation for Saturn.
8. Recently, some experimental evidence seems to indicate that the value of G in the law of universal gravitation is decreasing with time. What consequences could this have?

WEIGHTLESSNESS

9. Would it be possible for a person to appear to be weightless 2 km above the surface of the Earth? Explain.
10. Cyrano de Bergerac describes the following method by which he could reach the Moon:
    "Finally—seated on an iron plate,
    To hurl a magnet in the air—the iron
    Follows—I catch the magnet—throw
    again—And so proceed indefinitely."
Do you think this plan could work? Why or why not?

11. Science fiction writers imagined flights to the Moon long before human flight was a reality. Edgar Allan Poe wrote of a journey to outer space by balloon, Jules Verne had his characters fired to the Moon out of a mammoth cannon, and H.G. Wells used an antigravity machine. Which of these devices (if any) could, in theory, send a spacecraft to the Moon? If any are theoretically possible, why have they not been used in modern space programs?

12. If you jump off the top of a building, are you weightless on the way down?

13. If you punch a hole in the bottom of a container filled with water, the water drains out. However, if you drop the container while the water is coming out, the water ceases to flow out while the container is falling. Why?

PROJECTILE MOTION

14. Three children stand on the edge of a tall building. One drops a rock from rest over the side of the building, one throws an identical rock downward at 10 m/s, and the third throws an identical rock horizontally at 10 m/s. Which has the greatest acceleration downward?

15. Someone is angry with you, and you hide in a tree in order to escape a confrontation. Nevertheless, you are seen, and the other person throws a rock such that it is heading directly toward you. To avoid being hit, should you remain seated on the limb or drop out of the tree?

16. A person standing at the side of a road drops a coin, and at that same instant a person driving by in a car drops an identical coin. (a) Which of the coins has the greater acceleration? (b) Which of the coins will strike the ground first? (c) Which of the coins will have the greater speed when it strikes?

17. An airplane moving horizontally at a constant velocity drops a load of supplies to some stranded explorers. In the absence of air resistance will the supplies strike ahead of the plane, below it, or behind it?

18. A baseball is struck so that it leaves the bat with a vertical velocity of 10 m/s and a horizontal velocity of 8 m/s. (a) What are the values of the vertical and horizontal velocity at the maximum height reached by the ball? (b) What are the values of the vertical and horizontal velocities at the end of the flight just before it strikes the ground?

19. A jet pilot in horizontal flight fires a rocket directly forward. He then goes into an evasive dive. Describe what he would have to do, if anything, to shoot himself down.

20. You are riding in a car when you flip a coin straight up into the air as seen by you. Describe the motion of the coin as seen by an observer standing by the side of the road.

CIRCULAR MOTION; CENTRIPETAL FORCE; CENTRIFUGAL FORCE

21. The gravitational pull of the Sun holds the planets in orbit. Is that pull acting as a centripetal or as a centrifugal force? Explain.

22. If the gravitational field of the Sun disappeared magically and instantly, what would happen to the planets? If the gravitational field of the Earth disappeared magically and instantly, what would happen to you? What would happen to the Moon?

23. A person puts a stone inside an open tin can, ties a string to it, and spins the can around vertically. Why doesn't the stone fall out of the can?

24. Show that the expression for centripetal acceleration v^2/r has units of acceleration.

25. A car is moving around a circular racetrack when the centripetal acceleration suddenly goes to zero. What happens to the car?

26. A car is moving around a circular racetrack at a speed v. The speed is quickly increased to $2v$. What must happen to the centripetal force to keep the car on the track at the same radius? Where does this centripetal force come from?

27. A person swings a 2.00 kg object attached to the end of a 1.50 m long string around his head in a horizontal circular path at a speed of 3.00 m/s. What is the centripetal force acting on the object and what produces this force?

28. Why might a jet pilot tend to black out when pulling out of a steep dive?

29. A centrifuge is a device used in hospitals to separate heavy chemical substances from lighter ones. In these devices, a solution of the materials is placed in a test tube and whirled at very high speeds in a circular path. The heavy materials migrate to the bottom of the test tube, and the lighter ones appear at the top of the solution. Why?

30. If the breaking strength of the string in question 27 is 33 N, how fast can the object move?

ANSWERS TO SELECTED NUMERICAL QUESTIONS

5. 5.00×10^{-7} N
6. (a) 3.59 m/s^2, (b) 11.1 m/s^2
7. (a) 8.47 m/s, (b) 14.9 m/s
27. 12.0 N
30. 4.97 m/s

Solar flares from the Sun. (Courtesy NASA)

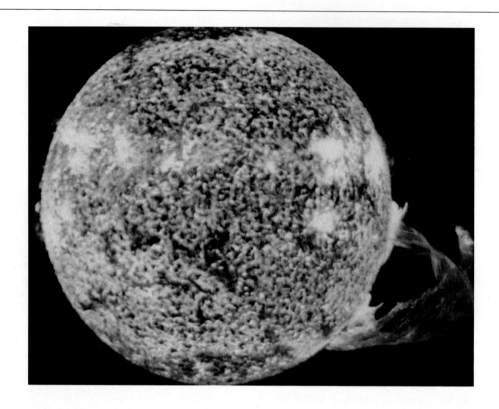

C H A P T E R

5

Thermal Physics

Our study thus far has dealt with a brief introduction to that portion of physics known as mechanics. We have examined such concepts as motion, Newton's laws, momentum, and kinetic energy. We now move to a second subdivision, that of thermal physics. Here we shall be concerned with such topics as temperature and heat.

5.1 TEMPERATURE

One of the first sensations of childhood is that of the relative hotness or coldness of objects. But why does one object feel hot while another may be cold to the touch? We will withhold the answer to this question for a few pages, because scientists had learned how to measure temperature long before they understood the internal difference between two bodies at different temperatures. The key experimental observation that enabled people to measure temperature is the fact that most materials expand when heated and contract when cooled. For example, if you blow up a rubber balloon and place it in a refrigerator, you will find that the size of the balloon diminishes as the temperature of the air inside the balloon decreases. In general, a change of size with a change of temperature is a property of all materials, regardless of whether they are in the form of solid, liquid, or gas. This property forms the basis for the most common of all temperature-measuring devices: the mercury (or alcohol)-in-glass thermometer.

A common mercury thermometer is illustrated in Figure 5.1. It consists of a glass tube with a very small inside diameter from which all the air has been removed. This tube is connected to a small bulb filled with mercury. When placed in contact with a hot object, both the mercury and the glass expand, but the increase in volume of the glass is not as great as the increase in volume of the mercury. The result is that the mercury gradually creeps up inside the stem. The temperature is measured by how high the mercury rises in the stem, as indicated by a scale on the glass.

5.2 TEMPERATURE SCALES

A standard and convenient way to calibrate a thermometer is by first inserting it into a mixture of ice and water, noting the height of the mercury, and then inserting it in boiling water while again noting the height of the mercury. A scale can then be engraved on the glass by use of these reference points. For example, in this country the most commonly used temperature scale is the **Fahrenheit** scale, in which the freezing point of water is chosen to be 32°F and the boiling point to be 212°F. This particular scale was originally set

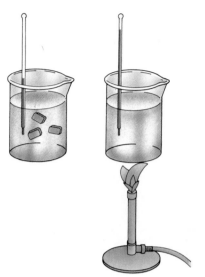

FIGURE 5.1 As the temperature rises, the mercury expands more than does the glass, and the level of the mercury is used as a measure of the temperature.

up to relate roughly to the lowest temperatures reached on our planet during the winter and to the hottest temperatures reached during the summer months. That is, the originator of this scale wanted the coldest days to have a temperature of about 0°F and the hottest about 100°F.

Although the Fahrenheit scale remains the one most often used in everyday life in this country, in the scientific community and throughout most of the world an alternative scale called the **Celsius** scale is the choice. This scale sets the temperature of the ice point to be 0°C and the temperature of boiling water to be 100°C.

Many weather forecasts give the temperature in both Celsius and Fahrenheit units. If you would like to check their accuracy, the equation that relates these two temperature scales is

$$T_C = \frac{5}{9}(T_F - 32) \tag{5.1}$$

where T_C is the Celsius temperature and T_F is the Fahrenheit temperature. Use this equation to verify that a Fahrenheit room temperature of 72°F corresponds to a Celsius temperature of approximately 22°C.

FOCUS ON . . . Comparing Fahrenheit and Celsius Temperatures

The British system of measurement uses the Fahrenheit scale to measure temperature. Gabriel Daniel Fahrenheit developed the scale that bears his name. He chose 0°F to be the temperature of a mixture of ice and common table salt. This was the coldest temperature that was easily reproducible in the laboratory at that time. He then arbitrarily defined the body temperature of a healthy human to be 96°F. Using this scale, he observed the boiling point of water to be 212°F. However, the bore of his glass thermometer must have been uneven, for body temperature is now recognized to be approximately 98.6°F. The table below gives the comparison between the two scales for several recognizable events.

	°C	°F
Arctic winter	−40	−40
Ordinary winter temperature	−10	14
Melting point of ice	0	32
Brisk autumn temperature	10	50
Normal room temperature	20	68
Summer day	30	86
Body temperature	37	98.6
Boiling point of water	100	212

How cold can a substance be cooled? As any material is cooled, the molecules that are its building blocks gradually lose kinetic energy, and when this energy has decreased as much as possible, the temperature can drop no lower. This theoretical lower limit of temperature is reached at approximately −273°C (−459°F). In our day-to-day activities we never approach this lowest limit of temperature. Even if we should travel to the planet Pluto, the planet farthest from the Sun in our solar system, we would find that the temperature there is −233°C. In fact, theory predicts that it is impossible to reach this lowest limit of temperature, but experiments on Earth have come very close. In general, these experiments are based on the principle that when a substance is magnetized it heats up, and when it is demagnet-

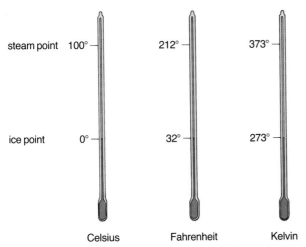

steam point 100°   212°   373°

ice point 0°   32°   273°

Celsius   Fahrenheit   Kelvin

FIGURE 5.2 Comparison of the Celsius, Fahrenheit, and Kelvin temperature scales.

ized it cools. In one experiment, copper was magnetized and then placed in liquid helium, thus dropping its temperature to about −272°C. When the copper was demagnetized while in the helium, its temperature dropped to a point only 0.000001 degree above the lowest theoretical temperature.

In view of the fact that there exists an absolute minimum of temperature, another scale, called the **Kelvin** scale, has been developed which sets its zero point at this minimum temperature. (This lowest theoretical temperature is thus often referred to as **absolute zero.**) On the Kelvin scale, the freezing point of water occurs at 273 K and the steam point of water at 373 K. The difference between the two temperatures is the same on this scale as on the Celsius scale. The Celsius, Fahrenheit, and Kelvin temperature scales are compared in Figure 5.2.

The relationship between the Kelvin scale and the Celsius scale is given by

$$T_K = T_C + 273 \qquad (5.2)$$

EXAMPLE 5.1   A unique temperature

(a) Convert the temperature −40°C to the Fahrenheit scale.

FOCUS ON . . . Comparing Celsius and Kelvin Temperatures

The difference between Celsius and Kelvin temperatures is 273°. Let us take a look at a few temperatures on these two scales.

$$0\ K = -273°C$$
$$100\ K = -173°C$$
$$1000\ K = 727°C$$
$$1{,}000{,}000\ K = 999{,}727°C$$

Note that at low temperatures, the difference between the two scales is significant. However, at very high temperatures the magnitude of the temperature is about the same whether it is expressed in kelvins or degrees Celsius. We will see in our study of astronomy that astronomers predominantly use the Kelvin scale. They frequently talk about extremely large temperatures such as 20,000,000. Does it really matter whether or not they explicitly express the particular scale, Celsius or Kelvin, that is used?

(b) Repeat for the Kelvin scale.

Solution **(a)** Equation 5.1 can be solved for the Fahrenheit temperature to give

$$T_F = \frac{9}{5}\,T_C + 32$$

Thus, if we substitute −40 for T_C, we find

$$T_F = -40°F$$

The temperature −40 is unique in that it is the same on both the Celsius and the Fahrenheit scales.

(b) The temperature of −40 on the Celsius scale can be converted to the Kelvin scale by use of Eq. 5.2.

$$T_K = T_C + 273$$

or,

$$T_K = -40 + 273 = 233\ K$$

This answer is read as 233 kelvins.

While boring cannons, Count Rumford wondered, "Where does all the work go after the hole is drilled?"

5.3 EARLY IDEAS ABOUT HEAT

When a hot object is placed in contact with a cold object, the two eventually reach a common temperature intermediate between the two initial temperatures. When such processes occur, we say that heat is transferred from the object at the higher temperature to the one at the lower temperature. But what is it that is being transferred? Early investigators believed that heat was an invisible, colorless, weightless material to which they gave the name **caloric** and that when two objects at different temperatures were placed in contact, caloric was transferred from one to the other.

The first experimental observation suggesting that caloric does not actually exist was made by Benjamin Thompson at the end of the Eighteenth Century. Thompson, an American-born scientist, emigrated to Europe during the Revolutionary War because of his Tory sympathies. Following his appointment as director of the Bavarian arsenal, he was given the title Count Rumford. While supervising the boring of artillery cannon in Munich, Thompson noticed the great amount of heat generated by the boring tool, indicated by the fact that the water used to cool the tool had to be replaced continually as it boiled away. On the basis of the caloric theory, he reasoned that the ability of the metal filings to retain heat should decrease as their size decreased. These heated filings, in turn, would transfer caloric to the water, causing it to boil. To his surprise, Thompson discovered that the amount of water boiled away by a blunt boring tool was comparable to the quantity boiled away by a sharper tool. This meant that even when no, or few, metal shavings were produced, the water was still being heated. As a result, he concluded that the size and number of the shavings had nothing to do with the boiling of the water. He then reasoned that if the tool were turned long enough, he could produce an almost infinite amount of heat. Where was all this caloric coming from? Based on these observations, Thompson rejected the caloric theory of heat and suggested that heat is not a substance but rather some form of motion that is transferred from the boring tool to the cooling water. In another experiment, he showed that the heat generated was equal to the mechanical work done by the boring tool.

Although Thompson's observations provided evidence that brought the caloric theory of heat into question, it was not until the middle of the Nineteenth Century that the modern model of heat was developed. In this view, heat is treated as just another form of energy, one that can be transformed into mechanical energy.

5.4 MEASURING HEAT

If heat is just another form of energy, it seems obvious that we should measure it in joules. Until recently, however, this has not been the case. In most present-day applications, heat *is* indeed measured in joules, but there is an alternative method that harkens back to the early misunderstandings of the nature of heat, as exemplified by the caloric theory. Before a correct understanding of heat came about, units in which to measure it had already been developed. These units were so widely used and had become so ingrained in the world of science that they are still often used today.

One of these units is the **calorie (cal),** defined as *the amount of heat necessary to raise the temperature of 1 g of water from 14.5°C to 15.5°C.*

As an example of the use of this unit in practical situations, you will find that dieters often think in terms of counting calories. But be careful here, because the energy unit used in describing the energy equivalent of food is a Calorie, spelled with a capital C. The Calorie is actually 1000 of the calories defined above.

An alternative old way of measuring heat units is the **British thermal unit, Btu.** *This is defined as the heat required to raise the temperature of one pound of water from 63°F to 64°F.* Heating and air conditioning systems often have their capabilities specified in Btu. For example, a 20,000 Btu air conditioning unit is capable of removing 20,000 Btu of heat energy from a room every hour of operation.

Even though these units are still around and are frequently used, most of our work in future chapters will specify heat as measured in joules. The relationship between the calorie and the joule was first established by James Prescott Joule and is given by

$$1 \text{ cal} = 4.186 \text{ J}$$

5.5 THERMAL ENERGY AND TEMPERATURE

Atoms and molecules are always moving, even in a block of ice or in a drop of liquid nitrogen. The combined energy of motion of all the particles in a sample is called the thermal energy. If a beaker of cold water is placed on a block of hot iron, the temperature of the water rises. Energy from the iron has been transferred to the water. Energy transfers of this type occur all around us all the

FOCUS ON . . . Evaporation and Boiling

We found in our discussion of thermal energy and temperature that temperature is a measure of the average kinetic energy of the atoms or molecules of a substance. Obviously, by random processes some of the atoms of a substance end up with kinetic energies that are higher than the average, and some have energies lower than average. With this thought in mind, let us consider a common experience and see why it occurs. If you set a beaker of water out on a table, after a few hours some of the water will have evaporated. To see why this occurs, consider the molecules near the surface of the liquid. From time to time, one of them receives a higher than average kinetic energy via collisions and interchanges of energy between it and other molecules in the liquid. In fact, it often receives a large enough kinetic energy to escape from the surface of the liquid. Thus, the molecule has gone from a condition in which it is considered a molecule of a liquid to one in which it is considered to be in vapor form.

Let us take a look at this same process from a slightly different point of view. After swimming, you surely have noticed that your wet body feels cooler than it did when it was dry. The reason for this cannot be related to the temperature of the liquid because the same sensation is felt even after a hot bath. A deeper look at the process of evaporation provides an explanation. Note that we have said above that the molecules that escape a liquid by evaporation have a greater-than-average kinetic energy. But if the molecules leaving have a lot of energy, the ones left behind have, on the average, less energy than before. Thus, the remaining liquid is cooler. Therefore, evaporation is a cooling process, and the evaporation cools the swimmer.

If one adds heat to a liquid, the average kinetic energy of the molecules increases. Finally, at some temperature, the molecules attain enough energy to break free of one another even below the surface of the liquid. When this happens, bubbles of vapor form in the liquid and rise to the surface. We call this phenomenon **boiling.**

time. *Heat is defined as the energy that is transferred from one system to another when the two systems are in contact and at different temperatures.*

The relationship between heat and energy can be readily understood if we think about atoms and molecules, rather than beakers of water or masses of iron. For example, a drop of

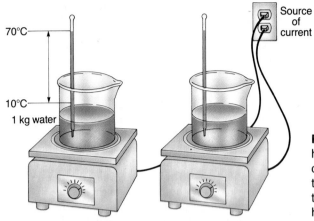

FIGURE 5.3 If two beakers are placed on identical hot plates at the same time, and both beakers contain the same mass of water, the temperature of the two will increase identically. Thus, the temperature change is proportional to the amount of heat added.

mercury consists of a large number of mercury atoms, and a drop of water consists of a great number of water molecules. The atoms of mercury or molecules of water move about and collide with one another. When thermal energy is transferred to a liquid, the atoms or molecules speed up. Now think about one molecule. To speed it up, a force must be applied to it, just as a force must be applied to a bowling ball or a dump truck to speed it up. When a force is applied over a distance, work is done. On a microscopic level, *heat is an expression of the work required to change the speeds of a collection of atoms or molecules.*

To see that heat energy and temperature are different, consider the following experiment. Suppose you add a measured amount of heat to a 200 liter container of water. You will find that the temperature of the water will change by a certain amount. But add this same amount of heat to a small cup of water, and its temperature may rise by several more degrees. The same amount of heat is transferred in both instances, but in the first case an individual molecule receives, on the average, only a small portion of this energy, whereas in the second case, an individual molecule receives a comparatively large amount of this energy. The end result is that the molecules in the smaller sample of water have large speeds, and consequently large kinetic energies, whereas the molecules of the larger sample have much smaller speeds and kinetic energies. *Temperature is a measure of the average kinetic energy of the atoms or molecules of a substance.*

5.6 SPECIFIC HEAT

Let us do a series of thought experiments that will enable us to discover a few facts relative to what happens to the temperature of an object as we add heat to it. These are referred to here as thought experiments simply because most of these observations should be familiar to you. If they are not, why don't you try them as actual experiments.

1. Imagine two equal sources of heat energy, such as identical hot plates plugged into the same source of current (Fig. 5.3). For our first exercise, we will place a beaker with 1 kg of water on both of the plates, measure the temperature of the water, and turn the hot plates on for the same period of time so that the *same* amount of heat energy is added to *both* beakers. At the end of this interval, we will again measure the temperature of both beakers. It should not come as a surprise to you that the temperatures of both containers of water have risen by the same amount. Now let us turn on the hot plates once again and watch the temperature of each for another period of time. Again, we will find that at the end of this interval both have risen in temperature by the same amount. The result of this simple observation is that the amount of temperature increase ΔT

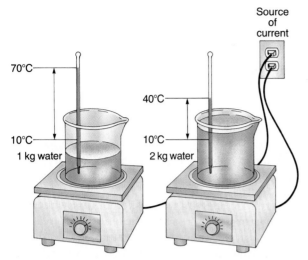

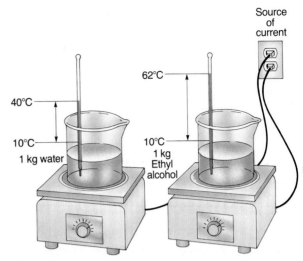

FIGURE 5.4 If two beakers are placed on identical hot plates, and one beaker has twice as much water as the other, the temperature of the larger volume of water will rise half as much as the temperature of the smaller volume of water.

FIGURE 5.5 If two beakers are placed on identical hot plates, and both contain equal quantities of different substances, generally the rise in temperature of the two liquids will be different.

is proportional to the amount of heat Q added. That is,

$$\Delta T \propto Q$$

Our experiment will work the same way if we reverse the process by extracting heat from the system. That is, for a given amount of a substance, the temperature decrease is directly proportional to the amount of heat removed.

2. Let us alter our experiment somewhat by placing twice as much water in one beaker as in the other (Fig. 5.4). With 2 kg in one beaker and 1 kg in the other, we again place our beakers on the identical hot plates and add energy for a given period of time. In this case we will find that the temperature increase of the 2 kg beaker will be only half that of the 1 kg beaker. (As above, this process is reversible. The temperature decrease of the 2 kg beaker would be half that of the 1 kg beaker if we remove equal amounts of heat energy.) The result of this experiment indicates that the temperature increase is inversely proportional to the quantity of matter in the sample. To be a little more specific, the temperature increase is inversely proportional to the *mass* of the substance.

Thus, we can write

$$\Delta T \propto \frac{1}{m}$$

3. Finally, let us add one last observation to our series of investigations. Let us place a beaker with 1 kg of water on one hot plate and a beaker with 1 kg of ethyl alcohol on the other, as shown in Figure 5.5. If the same amount of heat energy is added to each beaker, the temperature of the ethyl alcohol will rise about twice as fast as the temperature of the water. Thus, the same amount of heat energy added to equal masses of different substances will produce a greater temperature change for one than it does for another. We conclude that the temperature change depends on the substance to which we add the heat. We include this in our proportionality observations by use of a constant called c, defined as the **specific heat** for a particular material. As a proportionality, we express this observation as

$$\Delta T \propto \frac{1}{c}$$

Thus, the larger is c the smaller is the temperature increase.

TABLE 5–1 Specific Heats of Common Materials

Material	Specific Heat (J/kg°C)
Aluminum	901
Copper	387
Glass	838
Ice	2100
Iron	448
Silicon	704
Lead	128
Mercury	138
Steam	2010
Water	4190

Let us collect all of our observations into a single relationship. We also shall choose the value of c such that we have an equation rather than a proportionality. We have

$$\Delta T = \frac{Q}{mc} \qquad (5.3)$$

or

$$Q = mc\Delta T$$

Table 5–1 lists the specific heats for a variety of common materials in units of J/kg°C.

EXAMPLE 5.2   Pour on the heat

(a) Calculate the heat energy required to raise the temperature of 0.50 kg of water by 15°C.

(b) Repeat for 0.50 kg of sand, which has a specific heat of about 1500 J/kg°C.

Solution   (a) This is a straightforward application of $Q = mc\Delta T$. We find for water

$$Q = (0.50 \text{ kg})(4190 \text{ J/kg°C})(15°C) = 31,400 \text{ J}$$

(b) For sand, we have

$$Q = (0.50 \text{ kg})(1500 \text{ J/kg°C})(15°C) = 11,300 \text{ J}$$

EXERCISE

Let us slightly reverse the problem above. We found that adding 31,400 J of heat energy to 0.50 kg of water would increase its temperature by 15°C. Now suppose that you add 31,400 J of energy to 0.50 kg of sand. What would be its temperature increase?

Answer 42°C

5.7 SPECIFIC HEAT AT WORK IN NATURE

In Exercise 5.2, you were asked to find what happens when you add a specific amount of heat energy to equal amounts of water and sand. We found that the temperature of the water would increase by a small amount, 15°C, whereas the temperature of the sand would rise considerably more, by about 42°C. Undoubtedly, you have observed this effect on a trip to the beach. You have found that you can walk comfortably in the water, but if you try to walk across the sand-covered beach, you may find that it is almost hot enough to burn your feet. Why? In light of our exercise, we see that the Sun has been pouring down equal amounts of heat energy on both water and sand, but these equal amounts of heat added to each warms the water only slightly, while the sandy beach soars in temperature. (You should note that there actually are other factors involved in this practical application besides the differences in the specific heats of the substances. Water mixes while sand does not, and as a result, the Sun is actually heating a larger mass of water than of sand.)

While we are at the beach, let us take a look at how the high specific heat of water compared to that of land is responsible for the pattern of airflow at the seaside. During the daylight hours, the Sun adds roughly equal amounts of heat to both the beach and the water, but the lower specific heat of sand causes the beach to reach a higher temperature than the water, as noted above. Because of this, the air above the land reaches a higher temperature than that over the water; as a result, the air near the surface of the sand decreases in density and rises as cooler, denser air over the water rushes in to displace it, as shown in Figure 5.6a. This movement of air

HEAT

(a)   (b)

FIGURE 5.6 (a) On a hot day, the air above the warmer land becomes hot and rises as cooler air above the water moves in to replace it. The breeze is toward the shore. (b) At night, the warmer air is over the water, and the breeze blows away from the land.

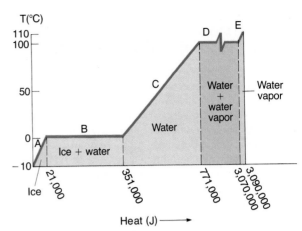

FIGURE 5.7 A graph of the temperature change of 1 kg of water initially at $-10°C$ (ice) as heat is added, changing it to steam at 110°C.

produces a breeze from sea to land during the day. The heated air gradually cools as it rises and thus sinks to set up a circulating pattern, as shown in the figure. During the night, the land cools more quickly than the water; thus, the circulating pattern of air flow reverses because the hotter air is now over the water, as shown in Figure 5.6b. You should not expect to observe this pattern every time you go to the beach, because prevailing winds caused by other factors often obscure this effect.

The high specific heat of water is also responsible for moderating the temperatures of regions near large bodies of water. As the temperature of a body of water decreases during the winter season, it gives off heat to the air, which carries it landward when prevailing winds are favorable. For example, the prevailing winds off the west coast of the United States are toward the land, and the heat liberated by the Pacific Ocean as it cools keeps coastal areas much warmer than they would be otherwise.

5.8
HEAT AND CHANGES OF PHASE

Let us suppose that we have a 1 kg ice cube at $-10°C$. If what we would like to have is 1 kg of steam at a temperature of 110°C, we can make the required transition with the addition of some heat. Let's examine the changes involved, one at a

time; as we do, some surprising features will appear. Figure 5.7 shows a graph of the temperature of the material versus the amount of heat added under conditions of standard atmospheric pressure. Let us examine each portion of the curve separately.

Part A During this portion of the curve, we are changing the temperature of the ice from $-10°C$ to 0°C. Since the specific heat of ice is 2100 J/kg°C, we can calculate the amount of heat added as follows:

$$Q = mc\Delta T = (1.0 \text{ kg})(2100 \text{ J/kg°C})(10°C)$$
$$= 21,000 \text{ J} = 2.1 \times 10^4 \text{ J}$$

Part B When the ice reaches 0°C, it remains at this temperature—even though heat is being added—until all the ice melts. The change in a substance from a solid to a liquid form or from a liquid to a gaseous form is called a **change of phase,** and the amount of heat to accomplish a phase change from solid to liquid is given by

$$Q = mL_f$$

where L_f is the so-called **latent heat of fusion** of the substance. The value of L_f depends on the

The temperature at which a liquid boils is affected by the pressure exerted on it. In general, as the pressure on a liquid is increased, the boiling point also increases. This is reasonable to expect from the molecular interpretation of boiling given in the box on evaporation and boiling. An increase in pressure on the surface of a liquid makes it more difficult for molecules to escape the liquid; hence, the boiling point increases.

The opposite is also true; a reduction of pressure reduces the temperature at which a liquid boils. Let us consider an interesting experiment that produces some unique results. Imagine that water is placed inside a container from which the air can be gradually pumped out. We start with the water at room temperature—hence far below its boiling point. As the air is evacuated, the pressure on the water is gradually reduced and so is the boiling point of the liquid. Finally, when the pressure has been reduced by about a factor of 95 percent, the boiling point is reduced to room temperature and the water boils.

We pointed out in the box on evaporation and boiling that the boiling process is actually a case of rapid evaporation, which is a cooling process. Hence, as the water in the container boils, it is also being cooled. If you reduce the pressure just a little bit more, the water will boil a little more vigorously and its temperature will decrease a little bit more. Finally, the temperature of the liquid will reach the freezing point (even though it is boiling). *Thus, you can simultaneously boil water and freeze it.*

particular material. For example, if ice is changing from ice to water at 0°C, $L_f = 3.34 \times 10^5$ J/kg. This means that it takes 3.34×10^5 J of heat to transform 1 kg of ice at 0°C to water at 0°C, or

$$Q = mL_f = (1.0 \text{ kg})(3.34 \times 10^5 \text{ J})$$
$$= 3.3 \times 10^5 \text{ J}$$

Part C Between 0°C and 100°C, there are no phase changes, and all the heat added to the water is being used to increase its temperature. We can find the heat required as

$$Q = mc\Delta T = (1.0 \text{ kg})(4190 \text{ J/kg°C})(100°C)$$
$$= 4.2 \times 10^5 \text{ J}$$

Part D At 100°C we have another phase change occurring as the water at 100°C changes to steam at 100°C. The heat required for this transition is found from

$$Q = mL_v$$

where L_v is called the **heat of vaporization.** For water, $L_v = 2.26 \times 10^6$ J/kg. We have

$$Q = mL_v = (1.0 \text{ kg})(2.26 \times 10^6 \text{ J})$$
$$= 2.3 \times 10^6 \text{ J}$$

Part E Finally, we must change the steam at 100°C to steam at 110°C. The amount of heat required for this is

$$Q = mc\Delta T = (1.0 \text{ kg})(2010 \text{ J/kg°C})(10°C)$$
$$= 2.0 \times 10^4 \text{ J}$$

Thus, the total amount of heat to change our ice at −10°C to steam at 110°C is

$$Q = (2.1 \times 10^4 \text{ J}) + (3.3 \times 10^5 \text{ J})$$
$$+ (4.2 \times 10^5 \text{ J}) + (2.3 \times 10^6 \text{ J})$$
$$+ (2.0 \times 10^4 \text{ J}) = 3.1 \times 10^6 \text{ J}$$

It should be noted that this process is reversible. That is, if we remove 3.1×10^6 J of heat from 1.0 kg of steam at 110°C, we will end up with 1.0 kg of ice at −10°C.

One of the features of the above that you surely noted is that it takes a tremendous amount of heat to accomplish a phase transition. For example, we saw that in order to transform 1 kg of ice at 0°C to water at 0°C, 3.3×10^5 J of heat energy had to be added. To understand why so much heat energy is necessary, we must realize that a phase transition produces a rearrangement of the molecules of the substance. Consider the liquid-to-gas transition. The molecules of a liquid are close together, and the forces between them are stronger than they are in a gas, in which the molecules are far apart. As a result, in order to move the molecules farther apart by some distance d, we must exert a force. This means that work must be done on the liquid to overcome

TABLE 5–2 Heat of Fusion and Vaporization

Substance	Heat of Fusion	Heat of Vaporization
Ethyl alcohol	1.04×10^5 J/kg	8.54×10^5 J/kg
Mercury	0.12×10^5 J/kg	2.72×10^5 J/kg
Water	3.34×10^5 J/kg	22.6×10^5 J/kg
Lead	0.25×10^5 J/kg	8.71×10^5 J/kg
Silver	0.88×10^5 J/kg	23.4×10^5 J/kg
Gold	0.65×10^5 J/kg	15.78×10^5 J/kg

these attractive molecular forces. The amount of heat required to vaporize a substance is equal to the amount of work that must be added to the fluid to accomplish this separation. A similar explanation applies to the large amount of heat required to change a solid to liquid. However, because the average distance between molecules in the gas phase is much larger than in either the liquid or the solid phase, we might expect that more work is required to vaporize a given amount of a substance than to melt it. Thus, the heat to vaporize, the heat of vaporization, is much larger than the heat required to melt a given substance, the heat of fusion. The heat of vaporization and the heat of fusion of several common substances are given in Table 5–2.

EXAMPLE 5.3 Physics gothic

On nights when the temperature is expected to fall below freezing, farmers often protect fruits and vegetables stored in a cellar by placing large vats of water in the cellar with the produce. Explain how this works.

Solution As the water freezes at 0°C, each kilogram of ice that forms releases about 3.3×10^5 J of heat to the surroundings. This helps to keep the cellar temperature high enough to prevent damage to the stored food.

5.9
HEAT TRANSFER

Heat energy can be transferred from one place to another by three different processes: conduction, convection, and radiation. Regardless of the process, there can be no heat transfer from one substance to another unless the two objects are at different temperatures. In this section, we shall briefly examine these three methods of transferring heat energy.

Conduction

Each of the three processes of heat transfer can be examined by considering the various ways you can warm your hands over an open fire. For example, if you stick a poker into a roaring campfire, you will soon find that heat is transferred from the flame along the poker to your hand. In this instance, the heat reaches your hand through the process of **conduction.** There are two factors that are responsible for this heat transfer. Let us look at these in turn.

1. Before the poker is inserted into the flame, the atoms of the metal are vibrating back and forth about their equilibrium positions. As the flame heats the rod, the vibration of those atoms near the flame becomes larger and larger. Because they are swinging farther and farther away from their equilibrium positions, they collide with neighboring atoms and transfer some of their energy to them. This causes these neighbors to vibrate with larger amplitudes also, and they, in turn, collide with atoms farther up along the poker. This increase in the amplitude of vibration of the atoms gradually works its way along the rod until the end being held is reached. We perceive this increase in the amplitude of vibration (increase of kinetic energy) as a temperature increase of the rod, and thus our hand is warmed as our molecules vibrate with a larger amplitude.

2. The explanation of the transfer of heat by conduction, as given above, is only part of the story. Some heat is indeed transferred by the vibration of the atoms of the substance. But all substances contain atoms, so why are all substances not equally good conductors of heat. It is well-known that some materials like copper conduct heat extremely well, whereas others like asbestos do not. Those materials that allow heat to travel through them easily are called good **conductors,** and those that inhibit the

FOCUS ON . . . The Hot Water Versus Cold Water Race

One of the most commonly asked questions about heat transfer problems is, "Does hot water freeze faster than cold water?" The answer to the question is simple: no, it does not. However, it is not as simple as one might believe to prove the assertion, and in fact, many common experiences tend to show just the opposite. Let us look at one and see if we can find a simple explanation of why it occurs.

Suppose you have two pipes against the wall of your home, and the temperature drops below freezing. Invariably, it seems as though it is the hot water pipe that you have to thaw out. If cold water freezes faster than hot, why did the hot water pipe freeze before the cold? The answer has nothing really to do with physics; instead it relies simply on the fact that you use more cold water than you do hot. The water in both pipes decreases in temperature when against the cold wall, but before the cold water gets to the freezing point, someone flushes a commode, washes his hands, and so forth. As a result, the cold water moves on to be replaced by another batch of water to be cooled. Since you have not used any hot water, however, the water in that pipe stays in place against the cold wall and continues to drop in temperature. Thus, there is an alternative explanation for this commonly observed phenomenon.

To see why the cold water freezes faster than the hot, think of a situation in which you are removing heat from a pan of water at 50°C and from another pan at 20°C. The hot water may cool faster at first because it is at a greater temperature difference with its surroundings and thus will radiate more heat to the atmosphere. However, when it cools to 20°C, it is in the same state as the pan of cool water and will cool down in the same way as it did until it reaches 0°C.

It would seem to be a simple matter to test the statement that the cold water freezes faster than the hot by placing two pans in the freezing compartment of a refrigerator. However, there are other factors that influence the results of your experiment. First, if you place the hot pan and the cold pan in the compartment at the same time, there is an interchange of heat between the two such that the cool warms up a little and the hot cools down. Thus, they affect one another and the results of your experiment. Another factor that you have to watch for is the contact of the pans with the walls of the freezing compartment. If there is a frost build-up on the walls, the hot pan will melt through this frost, allowing it to make intimate contact with the colder walls of the compartment so that heat can be conducted away from it faster than from the cold pan, which remains somewhat insulated from the cold walls by the layer of frost.

---

flow of heat are called **insulators.** The reason that metals are good conductors is because they have a large number of free electrons that can aid in transferring heat energy throughout the material. As the end of the poker is heated, the electrons in the hot end begin to move more rapidly. This increase in kinetic energy is transferred by collision to electrons farther up the rod, and we perceive this increase in energy of the electrons as an increase in temperature of the rod. Thus, heat energy is transferred throughout a substance by the increase in the amplitude of vibration of the atoms and by the increase in kinetic energy of individual electrons of the material.

It is quite easy to recognize the difference between a good conductor of heat and a poor conductor. Suppose you remove an ice tray and a package of frozen vegetables from a freezer. You will easily notice that the ice tray feels considerably colder than does the cardboard package, even though you know that they are both at the same temperature. This can be explained by noting that the metal ice tray is a good conductor, whereas the cardboard is a fairly good insulator. The metal conducts heat from your hand better than does the cardboard package; as a result, the tray feels colder than the carton. By use of a similar argument, you should be able to explain why a tile floor feels colder to your bare feet than does a carpeted floor.

The ability of a given material to conduct heat depends on its cross-sectional area and its length. Figure 5.8 shows a bar placed between two objects at different temperatures. *The rate*

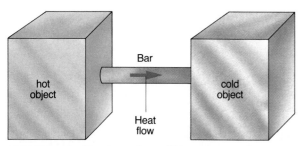

FIGURE 5.8 The heat flow will increase as the cross-sectional area of the bar increases and will decrease as the length of the bar increases.

FIGURE 5.10 Warming your hand by convection. Rising convection currents of air pass over your hands and warm them.

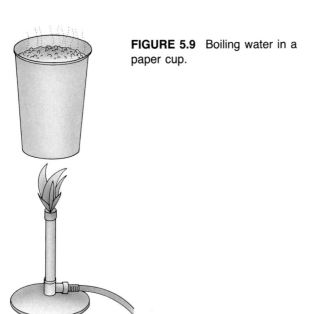

FIGURE 5.9 Boiling water in a paper cup.

at which heat energy flows through the bar increases as the cross-sectional area increases and decreases as the length of the bar increases.

EXAMPLE 5.4   Boiling water in a paper cup

If you hold a paper cup filled with water over an open flame as in Figure 5.9, you can bring the water to a boil before the temperature of the cup becomes high enough to burn. Explain this phenomenon by use of the fact that as the length of a material decreases, the rate at which it is able to transfer heat increases. (*Hint:* If you try this,

make sure that your cup does not have a rim around the bottom, or else it will burn. An alternative suggestion is to fill a paper bag with water rather than a cup.)

Solution The length to be considered is the thickness of the bottom of the paper cup. Because this thickness is so small, the heat from the flame is rapidly transferred through it to the water.

Convection

In the process of heat transfer by conduction, thermal energy is moved from one location to another via the vibration of molecules inside a substance (and, in the case of metals, by the transfer of kinetic energy between electrons). When energy is transferred by way of **convection,** *there is an actual movement of the material from one location to another.* To illustrate this, imagine warming your hands over an open flame as shown in Figure 5.10. In this situation, the air directly above the flame becomes heated, expands, and becomes less dense than nearby, cooler air. As a result, the heated air rises, and the cooler air moves in to replace it. Your hand is warmed as the rising hot air moves past it.

You should note that we have encountered convection in nature in an earlier section when we were describing the air flow patterns at a beach. This same air flow pattern takes place in a room heated by a radiator, as shown in Figure 5.11.

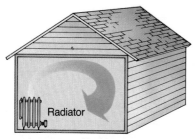

FIGURE 5.11 Warm air above the radiator rises and is displaced by cool air drawn in from the room.

FIGURE 5.12 Warming your hands by radiation.

Radiation

A third method of transferring thermal energy from one place to another is via the process of **radiation.** To illustrate this effect, consider Figure 5.12. In this case, someone is warming his hand by holding it in front of an open fire. Since there is nothing in contact with the flame and the hand, there is no way that heat can reach the hand via conduction. (There is, of course, air between the hand and flame, but because air is a very poor conductor of heat it will not allow much heat transfer by conduction.) Also, because the hand is not in the path of the rising hot air above the flame, heat is not transferred by convection. The process that remains is that of heat transfer by radiation.

We shall find in a later chapter that energy can be transmitted by way of a type of wave called an **electromagnetic wave.** Light is one example of electromagnetic radiation, radio waves are another, and X-rays are yet another. For the moment, it suffices to say that there is another type of radiation called **infrared radiation,** which is associated with the transfer of heat. Through electromagnetic radiation, approximately 1340 J of heat energy strikes each square meter of the top of the Earth's atmosphere every second. Some of this energy is reflected back into space and some is absorbed by the atmosphere, but enough reaches the surface of the Earth each day to supply all of our energy needs on this planet hundreds of times over—if it could be captured and used efficiently. The growth of the number of solar homes in this country is one example of an attempt to utilize this free energy.

All warm objects emit infrared radiation. You should note that "warm" is a relative term because every object that is at a temperature other than absolute zero emits infrared radiation, and the hotter the object, the more radiation it emits.

If your home is heated by a fireplace, the principal way that heat enters the room is by radiation. The chimney provides an outlet for convection currents to escape; thus, the room does not receive an appreciable amount of heat by this mechanism. In fact, cooler air drawn into the room to replenish the air supply exhausted through the chimney competes with the heat supplied to the room via radiation. As a result, the temperature of a drafty room may actually decline when a fireplace is used as the primary source of heat.

As an example of a common experience with infrared radiation, consider what happens to the atmospheric temperature on a winter night. If there is a cloud cover above the ground, the water vapor in the clouds reflects back a portion of the infrared radiation emitted by the ground; as a result, the temperature remains at moderate levels. On the other hand, in the absence of a cloud cover, this radiation escapes into space, and the temperature drops lower than it would if it were cloudy.

5.10 HINDERING HEAT TRANSFER

A common device to keep cool liquids cool or hot liquids hot is the thermos bottle. Let us examine its design, as shown in Figure 5.13, in order to see how it is able to perform this task. The walls of the bottle are made of glass and are silvered on the inner surface. The highly reflecting silvered surfaces reflect radiation, hindering heat transfer by radiation either into or out of the bottle. In order to prevent heat transfer by conduction or

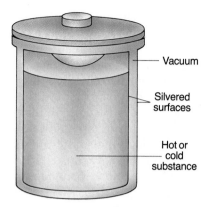

FIGURE 5.13 Design of a thermos bottle.

convection, the bottle has a double-walled construction with the air removed in the space between the two walls. A true vacuum conducts no heat at all; also, if there is no material medium there, convection currents cannot form in this enclosure. Thus, if all the air could be exhausted in this space, no heat would be transferred in or out by these two mechanisms. In practice, it is impossible to completely evacuate this interior space, so some heat is transferred by these two means.

Another situation in which one would like to minimize heat transfer occurs in the insulation of buildings. To prevent heat transfer between the interior of a home and its surroundings, insulating materials are placed in the walls and above the ceilings. These materials, such as fiberglass, are poor conductors of heat, so little heat transfer occurs by this process. Additionally, the insulating material is loosely packed so that pockets of air are trapped in cavities within the insulation. Air does not readily conduct heat, and, when air is trapped, convection currents cannot be set up. This same basic principle is used to keep the body warm with wool sweaters or down jackets. Both of these trap air near the body and hence reduce heat loss from the body by both convection and conduction.

5.11 THE GREENHOUSE EFFECT

The **greenhouse effect,** as its name implies, would certainly be of interest to a florist or a gardener, but the drought of 1988 brought this term into the popular press as an example of a catastrophe waiting to happen to the world. The greenhouse effect provides a good example of the principles of heat transfer and of ways to prevent heat transfer, but it is even more important in that it has global significance for life on this planet.

Let us examine the greenhouse effect first from the point of view of the glass enclosure so familiar to you at florists or plant nurseries. The underlying principle behind these buildings is that *glass allows visible light to pass through, but it does not allow infrared radiation to be transmitted as readily.* During the day, sunlight passes through the windows of a greenhouse and is absorbed by the earth, walls, and vegetation inside the structure. These objects then re-radiate this energy in the form of infrared radiation, but this radiation is now trapped because it cannot escape through the glass. As a result, the temperature of the interior rises.

An additional factor that causes the temperature to rise is the fact that convection currents are also inhibited in a greenhouse. This means that the heated air cannot circulate rapidly past those surfaces of the building that are exposed to the colder outside air. This prevents heat loss to the surroundings by this mechanism. In fact, many experts consider this to be an even more important effect in a greenhouse than the effect of the trapped infrared radiation.

Now let us turn to a phenomenon that occurs in the Earth's atmosphere which plays a role in determining the temperature of our planet. The primary constituent of the atmosphere that we need to consider is carbon dioxide. Carbon dioxide acts somewhat like the glass in a greenhouse in that it readily allows incoming visible light from the Sun to pass through to the surface of the Earth, but it does not allow infrared radiation to pass as readily. Thus, incoming visible light is absorbed at the surface of the Earth and is re-radiated in the form of infrared radiation, but this infrared radiation is absorbed and trapped by the carbon dioxide. This trapped heat energy causes the temperature of the surface of the Earth to be warmer than it would be if this infrared radiation could escape. This overall effect by which the temperature of the surface rises because of the trapping of infrared radiation by carbon dioxide is called the greenhouse effect. An example of a planet on which the greenhouse effect has run wild is Venus. Venus has an atmosphere rich in carbon dioxide, and because of the amount of trapped heat energy in this atmosphere, Venus is our warmest planet, approximately 850°F, even though it is not the closest planet to the Sun.

As fossil fuels (oil, coal, and natural gas) are burned on Earth, large amounts of carbon dioxide are released into our atmosphere. This, of course, causes the atmosphere to retain more heat by virtue of the greenhouse effect. Many scientists are convinced that the 10 percent increase in the amount of atmospheric carbon dioxide in the last 30 years could lead to drastic changes in world climate. The drought of summer 1988 led many scientists to speculate that these worldwide temperature increases are already on the way. It has been estimated that if the average global temperature should rise by only 2°C, this would be sufficient to melt the polar ice caps, thus causing flooding and the destruction of many coastal areas, an increase in droughts, and a reduction of already low crop yields in tropical and subtropical countries. Present-day agricultural areas such as the wheat belt in the midwest would move northward into Canada. The jury is still out as to whether or not a runaway greenhouse effect is indeed in control of this planet. However, it is an important problem that all nations must address.

SUMMARY

Three common temperature scales are the **Celsius,** the **Fahrenheit,** and the **Kelvin** or absolute scale.

A **calorie** is the amount of heat required to raise the temperature of 1 g of water from 14.5°C to 15.5°C.

Heat is a measure of the work required to change the speeds in a collection of atoms or molecules. **Temperature** is a measure of the average kinetic energy of the atoms or molecules of a substance.

The **specific heat** of a substance is the ratio of the amount of heat added to a substance divided by the mass of the substance and by the temperature change of the substance.

The change in form of a substance from solid to liquid, from liquid to gas, or vice versa, is called a change of phase. The amount of heat to change a unit mass of a substance from solid to liquid at the melting point is called the **heat of fusion.** The amount of heat to change a unit mass of a substance from liquid to vapor at the boiling point is called the **heat of vaporization.**

The three mechanisms of heat transfer are **conduction, convection, and radiation.** In the conduction process, heat is transferred by the motion of the atoms and electrons within the substance. Heat transfers by convection involve the movement of a heated body of liquid or gas. Heat transfer by radiation involves the absorption or emission of infrared radiation.

EQUATIONS TO KNOW

$T_C = \dfrac{5}{9}(T_F - 32)$ (converting between Fahrenheit and Celsius temperatures)

$T_K = T_C + 273$ (converting between Celsius and Kelvin)

$Q = mc\Delta T$ (heat transferred in the absence of a phase change)

$Q = mL_f$ (heat transferred during solid-liquid phase change)

$Q = mL_v$ (heat transferred during liquid-vapor change)

KEY WORDS

Temperature	Calorie	Change of phase	Convection
Celsius	Btu	Heat of fusion	Radiation
Fahrenheit	Heat	Heat of vaporization	Conductor
Kelvin	Specific heat	Conduction	Insulator
			Greenhouse effect

QUESTIONS

TEMPERATURE AND TEMPERATURE SCALES

1. If you place a thermometer in a container of hot water, you will often note that the level of the mercury falls slightly before it begins to rise. Explain why this happens.

2. Could a thermometer be made to work if the glass expanded more than the liquid trapped inside?

3. Galileo reputedly constructed one of the first thermometers by allowing water to rise and fall in a column of glass. Why do you suppose this type of thermometer did not catch on?

4. If the weatherman reports that the temperature is going to rise by 10° but does not refer to the type of scale he is using, in which of the following cases will the day end up the hottest? If he meant (a) 10 C°, (b) 10 F°, or (c) 10 K.

5. The temperature may reach −10°F on a cold day and 100°F on a hot day. Convert these temperatures to Celsius and to kelvins.

6. You are using a Celsius thermometer to read your body temperature. It says 38°C. Do you have a fever?

7. Aluminum expands more upon heating than does copper. What would happen if you heated a bar that consisted of a sheet of aluminum firmly bonded to a sheet of copper? How could such a bar be used as a thermometer? Can you think of any other use for such a device?

MEASURING HEAT AND SPECIFIC HEAT

8. (a) How many calories are needed to change the temperature of 1000 g of water by 40°C? How many calories are needed to warm 10,000 g of water by 4°C?

9. Repeat problem 8, except determine your answers in joules.

10. (a) How many joules are needed to heat 1.00 kg of copper 40.0°C? (b) How many joules are needed to heat 1.00 kg of iron 40.0°C?

11. How much energy in J will be released if 50.0 g of mercury is cooled by 25.0°C?

12. 500 cal of heat energy are added to a 5.00 kg mass of iron. What is the temperature increase of the iron?

SPECIFIC HEAT AT WORK IN NATURE AND HEAT AND CHANGES OF STATE

13. In general, cities have a higher average temperature than the surrounding countryside. There are many factors involved in this, but one of them is the difference in specific heat of soil and concrete. Which do you think has the higher specific heat?

14. Steam at 100°C will cause a more severe burn than water at 100°C. Why?

15. It takes longer to heat a cold room full of furniture than it does to heat an identical empty room. Explain. If the furnace is turned off when both rooms are warm, which one will cool off more quickly?

16. If a house is built with many large windows facing south, the sunlight passing through the glass will heat the interior, even on cold winter days. Imagine that two houses are identical except that one is constructed of wood and the other of much more massive stone and concrete. Both have many south-facing windows. If the furnace is turned off, (a) which one will be warmer during the sunlight hours and (b) which will be warmer during the evening? Explain.

17. Lead has a melting point of 327°C and a boiling point of 1750°C. Find the amount of heat in cal and J required to convert 1.00 kg of solid lead at 327°C to a vapor at 1750°C.

18. In winter, pioneers often stored barrels of water in their food storage cellars. Why?

19. In general, a city is hotter than the countryside. Would you expect breezes to blow toward or away from the city?

20. State whether your body would gain or lose heat from each of the following changes of phase. (a) An ice cube melts in your mouth. (b) Molten candle wax falls on your finger and solidifies.

21. Which of the following requires more energy: (a) melting 1 g of ice (at its freezing point) or boiling 1 g of water (at its boiling point); (b) vaporizing 1 kg of water or vaporizing 1 kg of alcohol; (c) vaporizing 1 g of water or vaporizing 5 g of mercury (both are at their boiling point).

22. How much heat in joules is needed to (a) melt 20.0 g of solid mercury; (b) vaporize 10.0 g of liquid mercury?

HEAT TRANSFER AND HINDERING HEAT TRANSFER

23. A piece of paper is wrapped around a rod made of wood, while another piece of paper is wrapped around a rod made of copper. When held over a flame, the paper wrapped around the wood burns first. Why?

24. A piece of metal feels colder than a piece of wood when they are both at the same temperature. Why?

25. Often you see highway signs which warn that bridges freeze faster than roadways. Why does this occur?

26. You need to boil some water quickly and you have a choice of two vessels in which to perform the task. One is a pan with a thin bottom, and the second is of the same material but has a thick bottom. Which would you choose and why?

27. Windowpanes in some houses actually consist of two pieces of glass with an air space separating them. Why are these much more efficient at holding heat in the house than single pane glass?

28. Heat radiation will not pass through glass. Why then should you not have more wall area of your house covered with glass to keep the inside warm?

29. A potato can be baked more quickly by inserting a nail through it. Why?

30. Builders insulate ceilings well but the floors not so well. Why?

31. You will find that outside temperatures often drop to their lowest on nights when there is no cloud cover. Which of the three mechanisms of heat transfer do the clouds protect against?

32. Of the three possible mechanisms for heat transfer, which is the most important for maintaining the temperature balance of the Earth?

33. When your body gets cold, the blood vessels in your skin contract. Why?

34. Why are dark-colored clothes more suitable for winter wear than white clothes?

ANSWERS TO SELECTED NUMERICAL QUESTIONS

5. $-23.3°C$, 249.7 K; $37.8°C$, 310.8 K
6. $= 100.4°F$, yes
8. (a) 40,000 cal, (b) 40,000 cal
9. 1.67×10^5 J
10. (a) 1.55×10^4 J, (b) 1.79×10^4 J

11. 173 J
12. $0.935°C$
17. 257,000 cal, 1.078×10^6 J
22. (a) 2.40×10^2 J, (b) 2.72×10^3 J

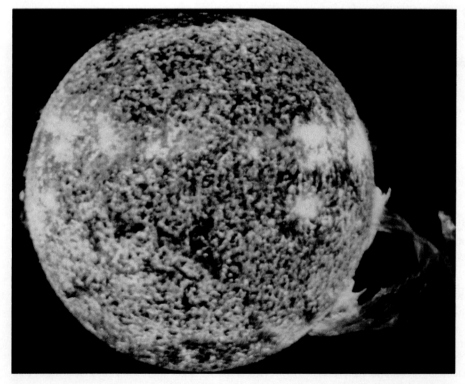

Solar flares from the Sun.
(Courtesy NASA)

This device, called Hero's engine, was invented around 150 B.C. by Hero in Alexandria. When water is boiled in the flask, which is suspended by a cord, steam exits through two tubes at the sides of the flank (in opposite directions), creating a torque that rotates the flask.
(Courtesy of CENCO)

From steam engines, to internal combustion engines, to the engine in this jet . . . they all obey the laws of thermodynamics. (Courtesy NASA)

C H A P T E R

6

Thermodynamics

Thermodynamics is a part of physics that falls loosely within the category of heat-related concepts. Basically, *thermodynamics deals with those processes related to the use of heat in practical applications.* Let us briefly look at some of the details that will concern us in this chapter.

Consider any mechanical device that is made of metal or other materials. After years of use or neglect, the device may wear out, become obsolete, break down, or rust. However, even though the device may have no inherent value in its present form, the atoms of the original metal are still there. If one has need of these raw materials, the bent and broken machinery can be remelted, and the materials that are reclaimed can be recycled. Likewise, the rust can be chemically reconverted to iron, which can then be recast to make some new device. Therefore, at least in principle, the world's supply of metals will never be depleted. They can be used again and again forever. But does this also apply to energy? Does the Earth have a stockpile of energy that will always be available, or is it possible that the world is slowly running down in regard to usable energy?

Coal is a storehouse of energy that can be released to do work for us. But what becomes of the energy when used? If we can find used metals and reuse them, can we also find used energy and recycle it in some way so that it can also be used repeatedly? A second question that we must consider is: Is it possible that we could somehow add some energy to our stockpile "for nothing"? After all, energy is not a material substance like a metal. We have defined energy as the ability to do work, and as such, energy is not something that you can touch and feel or store in a warehouse in the same sense that you can store several bars of metal. Is it possible that we could somehow do a conversion process on energy that would actually give us more in a usable form than we had at the beginning? These two questions plagued early scientists for a long time, and the search for answers led to the study of heat-motion, or thermodynamics.

6.1
WORK AND INTERNAL ENERGY

In order to consider the first law of thermodynamics, we must re-examine a topic that we have already covered—namely, work. Additionally, we must investigate a new concept, that of internal energy.

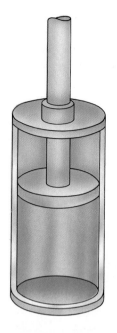

FIGURE 6.1 A gas in a cylinder fitted with a moveable piston.

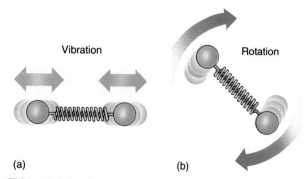

FIGURE 6.2 The rotational and vibrational motions of the molecules add to the internal energy of some gases.

Work

Often the materials that we must consider when discussing applications of thermodynamics are fluids, such as a gas or a liquid. As a result, we must briefly consider exactly what is meant when we say that we do work on a fluid, or what is meant when we say that a fluid does work for us. To aid us in our understanding, consider a cylinder filled with gas and fitted with a movable piston, as shown in Figure 6.1. When the system is in equilibrium, the gas holds the piston at some stationary height above the bottom of the container. However, we can change this height in a variety of ways. For example, suppose we play a flame over the container such that the enclosed gas is heated. As the gas warms, it expands and this causes the piston to rise in the container. The expanding gas thus exerts a force on the piston, and this force causes the piston to move through some distance. Thus, the gas does work on the piston. In turn, a shaft leading from the piston could be used to lift a heavy weight for us. As a result, work can be done *for us*.

When a system is doing work for us, we say that the work done is positive.

On the other hand, we could do work on the gas enclosed in the container. One way to do this is to exert a force on the piston and compress the gas.

When we do work on a system, we say that the work is negative.

Note that the volume of the gas must change when it does work or when work is done on it. When the system does work, the gas expands; when work is done on the system, it is compressed. When the volume remains constant, no work is done on or by the system.

Internal energy

If we could see the atoms of a gas, we would find them moving rapidly in random directions, colliding with the walls of the container and with one another. Thus, because of their motion, the atoms of the gas have kinetic energy. *The **internal energy** of a substance is the total energy (both kinetic and potential) of all the atoms or molecules that make up the substance.*

The simplest kind of gas is one composed of the so-called noble gases, such as argon, helium, and neon. In these gases, the individual components are single atoms that are not bound to other atoms. As a result, the only kind of energy that they can have is the kinetic energy of their random motion. However, in a more complicated gas, the rotational and vibrational motions of the molecules also contribute to the internal energy of the gas (Fig. 6.2). The atoms of the gas are connected to other atoms by forces that act as though the atoms were connected by small "springs" attached between the particles. The energy associated with the stretching or compressing of these small springs is a potential energy, and this potential energy contributes to the total internal energy of the gas (Fig. 6.2a).

Also, for these more complicated molecules, the group can rotate as shown in Figure 6.2b. The kinetic energy associated with this rotation is another part of the total internal energy of the gas.

6.2 THE FIRST LAW OF THERMODYNAMICS

*The **first law of thermodynamics** is an extension of the law of conservation of energy to include possible changes in the internal energy of a system.* The law is valid in all circumstances, and it can be applied in a wide variety of situations. In equation form, the first law is stated as

$$\Delta U = Q - W \qquad (6.1)$$

In this equation, Q is the heat added to or subtracted from a system. We will use the convention that Q is *positive* if heat is *added* to a system and *negative* if the system *loses* heat. W is the work done on or by the system. Work is *positive* if the system *does work* and *negative* if work is *done on* the system. The quantity ΔU is the change in internal energy of the system.

Thus, the first law of thermodynamics relates heat, work, and internal energy. The relationship expressed by Eq. 6.1 has been verified experimentally in an enormous array of situations.

EXAMPLE 6.1   A constant pressure process

A gas is enclosed in a container fitted with a piston. The piston is allowed to move such that the pressure of the gas always stays at the same value. (A constant pressure process is often called an isobaric process.) Heat is slowly added to the gas; as a result, the piston is pushed upward. Discuss this process from the point of view of the first law of thermodynamics.

Solution We are adding heat to the gas, so Q is a positive quantity. Likewise, the gas is doing work on its surroundings because it is pushing the piston upward. As a result, the sign of the work is positive. The sign of ΔU is found as

$$\Delta U = Q - W = (+) - (+)$$

Thus, the sign of ΔU could be either positive or negative depending upon which is larger, Q or W. However, in most processes of this nature, Q is greater than W. Thus, the sign of ΔU would be positive, indicating that the internal energy of the gas has increased.

EXAMPLE 6.2   A constant volume process

Heat is added to the container of gas in Example 6.1, but the piston is clamped in position so that it cannot move. Under such conditions, the pressure of the gas will not remain constant. Discuss the change in internal energy of the system for this process.

Solution Because the piston is clamped, it cannot move, and as a result, no work can be done either on or by the gas. (A force is exerted on the piston, but there is no movement.) Since heat is added to the container, the sign of Q is positive, and the first law becomes.

$$\Delta U = Q - W = (+) - (0)$$

or

$$\Delta U = Q$$

Thus, the sign of ΔU is positive, the same as that of Q, indicating that all the heat added to the system goes into increasing the internal energy of the gas.

6.3 HEAT ENGINES

One of the oldest applications of thermodynamics, and one that is still important today, is the study of heat engines. A heat engine is any device that converts heat energy to other useful forms of energy, usually mechanical or electrical energy. A heat engine is found to follow a three-step pro-

FOCUS ON . . . A Flywheel-Operated Subway Train

The first law says that energy can never be found for free. However, it is possible to build machines that operate very efficiently. Many modern devices are highly inefficient and could do the same job using less fuel if clever engineering practices were employed. For example, a subway train must accelerate away from a station and brake to a halt within a few minutes, and then repeat the process many times during the operating day. Acceleration requires large amounts of energy. The brakes of a subway train operate by friction; they slow the train by rubbing a brake pad against a wheel. When the brakes are applied, the energy of the motion of the train is converted to heat generated in the braking system. Thus, most of the energy used to accelerate the train is dissipated as heat. In the summertime, the braking mechanism raises the temperature of the cars and creates a need for increased air conditioning. Since the air conditioners also use energy, the whole system is highly inefficient.

An alternative subway system has been proposed, as shown in the Figure. This design employs a large, heavy flywheel mounted under each car. This flywheel is designed so that it can be connected to the wheels of the train or to an electric generator, also mounted under the chassis of each car. Instead of applying conventional friction brakes, the train engineer would pull a lever connecting the wheels of each car to a flywheel. A great deal of energy would be required to set all the flywheels spinning. This energy would come from the kinetic energy of the train. In other words, instead of dissipating the kinetic energy of the train as heat, the engineer could convert the kinetic energy of the train to rotational energy of the flywheels and thereby stop the train. While the subway train rests at the station to pick up and discharge passengers, the flywheels would be spinning rapidly. When power is needed to accelerate the train, the shafts of the flywheels could be connected to their electric generators, electricity could be produced, this electricity would then be used to power the train, and the train would speed up while the flywheels would slow down again. The train could produce some of its own electric power and in doing so would diminish the need for air conditioning as well.

Naturally, there would be some heat losses in such a system, and outside power would be needed, but much less than is needed at present.

Schematic of a flywheel-operated subway train.

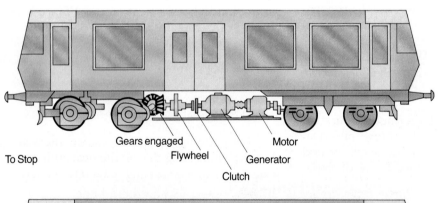

To Stop

Gears engaged Motor
Flywheel Generator
Clutch

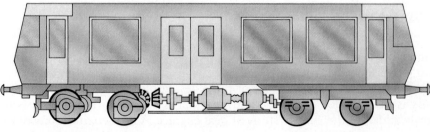

To Start

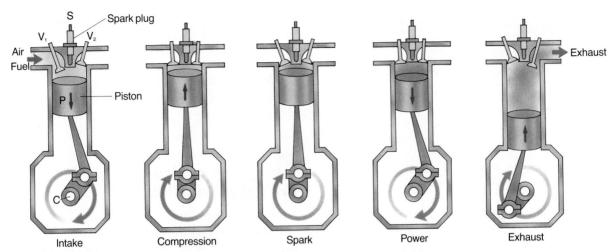

FIGURE 6.3 Schematic diagram of a gasoline internal combustion four-stroke cycle.

cess during one of its cycles. (1) Heat is absorbed from some source at a very high temperature, (2) work is done by the engine, and (3) heat is expelled from the engine at a lower temperature. For example, the internal combustion engine in an automobile extracts heat from a burning fuel and converts a fraction of this energy to mechanical energy. Some heat is then expelled from the system through the exhaust to the atmosphere. In the next section, we shall examine the gasoline engine more completely.

6.4
THE GASOLINE ENGINE

Our civilization has reached a point where engines are in constant use around us. Hardly anyone walks anymore. Most people usually move about with the aid of a motor, or engine, in an automobile, bus, or car. Similarly, most homes rumble quietly with the vibrations of furnace motors and blowers, refrigerators, freezers, and perhaps washing machines, dryers, and numerous other household devices. The design and development of the engines that have become so much a part of our lives provided much of the impetus for the study of thermodynamics. For this reason, let us examine a conventional type of engine.

The most common type of engine used in our automobiles is the four-stroke internal com-

bustion engine. The principal parts of the four-stroke engine are shown in Figure 6.3. The piston, P, is a piece of metal that can slide freely up and down inside a cylindrical chamber. V_1 and V_2 are valves that allow fuel to enter the cylinder and burned gases to leave. S is the spark plug, and C is the crankshaft. An up-and-down motion of the piston causes the crankshaft to rotate, and this rotation is transmitted by gears to the wheels of the car. The steps in a single cycle of the engine are best understood by examining what happens during each of the four strokes of the engine.

1. **The intake stroke:** The first stroke of the cycle is called the intake stroke. As Figure 6.3 shows, the piston moves down and, as it does so, valve V_1 opens. The gradually increasing space inside the cylinder causes the pressure there to drop. This drop in pressure draws a mixture of gasoline vapor and air through the open valve and into the cylinder.

2. **The compression stroke:** At the end of the intake stroke, the piston begins to move up in the cylinder. This starts the compression stroke, as shown in Figure 6.3. In this stroke, both valves are closed, and the mixture of air and gasoline vapor is compressed as the piston rises.

3. **The power stroke:** The third stroke begins when the piston reaches the end of the compression stroke and the gas mixture is fully compressed. At this instant the spark plug

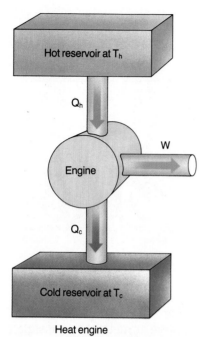

Heat engine

FIGURE 6.4 A representation of a heat engine. The engine (the circle) absorbs heat Q_h from a source of energy and rejects heat Q_c to the environment. In the process an amount of work W is done.

fires. As its name indicates, this device causes an electric spark inside the chamber, which ignites the fuel. The small explosion produced inside the chamber has sufficient force to push the piston downward, as shown in Figure 6.3. This stroke, the power stroke, generates the driving force that causes the crankshaft to rotate. This is the only one of the four strokes in which energy is transferred from fuel to crankshaft. The linear motion of the pistons is eventually converted to rotational motion of the wheels.

4. **The exhaust stroke:** At the end of the power stroke, when the piston is at the bottom of its travel, valve V_2 opens. The upward motion of the piston forces the burned fuel mixture out through the exhaust system, as shown in Figure 6.3. At the end of this exhaust stroke, the cycle is repeated.

Since the power stroke took only one-fourth of the total time of the cycle, the crankshaft would have to coast through the other three strokes if only one piston and one cylinder were

in the automobile. As a result, the engine would not run smoothly. In practice, most cars have four to eight cylinders and pistons to smooth out the drive of the engine and to produce significantly more power. Lawn mowers do have only one cylinder, whereas airplane engines may have sixteen or more.

6.5
THE SECOND LAW OF THERMODYNAMICS

In order to formulate our statement of the **second law of thermodynamics,** we must examine a heat engine in terms of the first law. To do so, it will be convenient to represent the engine schematically, as shown in Figure 6.4. The engine, represented by the circle, absorbs a quantity of heat Q_h from the heat reservoir at a temperature T_h. (For the gasoline engine, Q_h is supplied at a high temperature when the fuel is burned.) The engine then does work W. (This work is done during the power stroke of the gasoline engine.) Finally, the engine gives up heat Q_c at a temperature T_c. (For the gasoline engine, Q_c is exhausted to the atmosphere at the temperature of the atmosphere, T_c.)

You should take particular note of the fact that the internal energy change of the gas is zero for the complete cycle. This occurs because we started with a compressed gas ready to absorb heat during the burning process, and at the end of the four strokes, we have been through a cycle in which we have a compressed gas again ready to absorb heat. Thus, following a complete cycle, the gas is in the same state as it was when it entered the cycle. As a result, the change in internal energy, ΔU, is zero. Now let us apply the first law to the complete cycle. Work W is done by the gas; hence, W is a positive quantity. The net heat transferred is $Q_h - Q_c$. We have used a positive sign for Q_h because this is heat absorbed by the gas and a negative sign is used for Q_c because this is heat lost from the gas. Thus, we have

$$\Delta U = Q - W$$
$$O = Q_h - Q_c - (+W)$$

or

$$W = Q_h - Q_c \qquad (6.2)$$

There are a variety of ways of stating the second law of thermodynamics, but one of the more common is based on Eq. 6.2. This form of the second law is useful in understanding the operation of heat engines.

It is impossible to construct a heat engine that absorbs an amount of heat from a reservoir and performs an equivalent amount of work.

In light of Eq. 6.2, the second law says that it is impossible for W to equal Q_h. Another way of looking at this is to note that the second law says that some heat must always be exhausted to the environment. An instructive assessment of our circumstances is that the first law says we cannot get more work out of a process than the amount of energy we put in. The second law goes further and says that we cannot break even.

Let us examine the statement of the second law a little more carefully. Suppose a mass of coal is burned in a steam locomotive. The heat is converted to useful work, and the engine travels from Paris to Amsterdam. When the engine arrives in Amsterdam, the coal is gone. What happened to the energy? Could you somehow find it, save it, and use it to drive the train back to Paris? The answer is no. The heat from the coal was spread out into the environment. The air between Paris and Amsterdam was warmed slightly, but the locomotive cannot extract enough energy from the warm air to drive back to Paris. Thus, the energy from the coal cannot be recycled. This observation is a general one and explains why energy, once used, cannot be reused to perform work efficiently. In brief, materials can be recycled but energy cannot. The energy that was dissipated into the environment has not been destroyed—it still exists somewhere, but it is lost in the sense that it is no longer available to do work. Ingenious scientists have tried to invent heat engines that can convert all of the energy of a fuel into work, but they have always failed. It was found, instead, that a heat engine could be made to work only by the sets of processes indicated by our statement of the second law as given above. (1) Heat must be absorbed by the working parts from some hot source. The hot source is generally provided when some substance such as water or air is heated by the energy obtained from a fuel, such as wood, coal, oil, or uranium.

(a)　　　　　　(b)

(c)　　　　　　(d)

FIGURE 6.5 The second law of thermodynamics states: Any undisturbed system will naturally tend toward maximum disorder. If a drop of ink is placed in a glass of water, the ink will always disperse until it is evenly distributed.

(2) Waste heat must be ejected to an external reservoir at a lower temperature.

6.6
ALTERNATIVE FORMS OF THE SECOND LAW

The formulation of the second law, like that of the first, arose out of a long series of observations. If a hot iron bar is placed on a cold one, the hot bar always cools, while the cold one becomes warmer, until both pieces of metal are at the same temperature. No one has ever observed any other behavior. Similarly, if a small quantity of black ink is dropped into a glass of water, the ink disperses until the solution becomes uniformly black, as shown in Figure 6.5. A reverse process, such as one in which a jar of uniformly black ink spontaneously becomes black on one side of the jar and colorless on the other side, is never observed.

Carnot and the Efficiency of Engines

In 1824, a French engineer named Sadi Carnot (1796–1832) described a theoretical engine, now called a Carnot engine, that is of great importance from both a practical and a theoretical standpoint. He was able to prove that his engine is the most efficient engine possible. Thus, the Carnot engine sets an upper limit on the efficiency of all engines.

His imaginary engine consisted of a cylinder fitted with a piston and containing an ideal gas. (We shall define an ideal gas explicitly in our study of chemistry.) The walls of the cylinder and the piston are assumed to be made of a material that does not conduct heat. You should now begin to see why no one has constructed a Carnot engine. First, there are no gases that behave as an ideal gas at the temperatures and pressures that would be encountered in a real engine. Secondly, there are no materials that are perfect insulators from which you could build your cylinder and piston.

Carnot imagined that his engine would work in four stages, somewhat like the gasoline engine described in the text. (1) In stage one, the base of the cylinder is replaced by a wall that can conduct heat. In this way, the gas enclosed inside the cylinder is placed in contact with a large reservoir of heat at a temperature T_h. The gas expands at this constant temperature T_h. During this time, the piston rises and work is done by the gas. (2) The base of the cylinder is replaced with the nonconducting substance and the gas continues to expand. No heat can enter or leave the system during this stage, but because the gas is expanding, its temperature decreases from T_h to T_c. The piston continues to move up during this time, and work is done by the gas. (3) Once again the base of the cylinder is replaced by a conducting material, and the gas is allowed to come into contact with a reservoir at a temperature of T_c. The piston now moves down to compress the gas, and thus work is done on the gas. (4) In the final stage, the base of the cylinder is again replaced by the nonconducting material, and the gas is compressed until the temperature T_h is reached. There is work done on the gas in this stage.

If this sounds to you like a very complicated pro-cess, you are correct—it is. Nevertheless, Carnot was able to show that the efficiency of this engine was given by the simple equation

$$\text{eff} = \frac{T_h - T_c}{T_h} \, 100\%$$

where the temperatures T_h and T_c are Kelvin temperatures.

Let's take a look at the efficiency of a Carnot engine. Consider a steam engine that has a boiler that operates at 400 K. Water in the boiler is converted to steam, and this steam is used to drive the piston of the engine. The exhaust temperature is that of the outside air, approximately 300 K. What is the maximum efficiency of such an engine? The answer is provided by Carnot's expression for the efficiency, with $T_h = 400$ K and $T_c = 300$ K. We have

$$\text{eff} = \frac{T_h - T_c}{T_h} \, 100\%$$
$$= \frac{400 \text{ K} - 300 \text{ K}}{400 \text{ K}} \, 100\% = 25\%$$

This says that the highest efficiency that one could obtain would be achieved by running a Carnot engine between these high and low temperatures. Such an engine would have an efficiency of 25 percent. A real engine would have an efficiency considerably less than this value.

The expression for the efficiency also gives designers insight on how to increase the efficiency of an engine. For example, the efficiency can be increased either by lowering T_c or by raising T_h. In most cases, it is the latter that engineers attempt, because T_c is usually room temperature. Try it for yourself to show that if T_h is raised to 500 K while maintaining T_c at 300 K, the efficiency is increased to 40 percent. As an exercise, use the expression for the efficiency of a Carnot engine to show what one would have to do to get an efficiency of 100 percent. The answer is that T_c would have to be 0 K, absolute zero. Obviously, even on theoretical grounds, an exhaust temperature this low is impossible.

Thus, there appears to be a natural drive toward sameness, or random disorder. If there are two blocks of iron at different temperatures in a system, or a drop of ink in a glass of water, there is a differentiation of physical properties. Such differentiation results from some kind of orderly arrangement among the individual parts of the system. This is a subtle but important

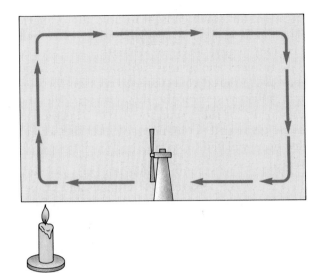

FIGURE 6.6 A simple heat engine. The flame sustains a convection current, and the moving air turns the blades of the toy windmill.

point. In your experience, how is order different from disorder? The answer is that order is characterized by repeated separations. Your room is orderly if all the books are separated from your socks—books on the shelves, socks in the drawer. It is disorderly if books and socks are all mixed up in both places. Similarly, if a small spot of black ink is separated from colorless water, the system is orderly. Which system is more natural (that is, more probable)? That's easy—if you neglect your room, does it naturally become more orderly or more disorderly? Disorderly, of course! The reason is that there are always more ways to be disorderly than to be orderly (or there are more ways to break rules than to follow them). Therefore, any system, if left alone, tends toward disorder.

Entropy is a thermodynamic measure of disorder. It is possible to state the second law of thermodynamics in terms of entropy as follows: It has been observed that *the entropy of an isolated system always increases during any spontaneous process;* that is, the degree of disorder always increases. Thus, if you drop a spot of ink in water, the ink spreads out evenly throughout the liquid. It becomes disorderly. If you don't clean your room regularly, it becomes messy. Similarly, a boat engine designed to extract heat from lake water as a source of energy and to release cool water as exhaust would cause a separation of hot and cold.

This is a more orderly arrangement. Any separation leads to a decrease of entropy of the system, and this is impossible to do without adding energy from some outside source.

Let us return now from the impossible to heat engines that use fuel, where the situation continues to be discouraging. In order to understand this concept, let us examine a very simple heat engine. Imagine that you have a box, and one side of the box is filled with hot air and the other side is filled with cold air. If left undisturbed, this system spontaneously becomes disorderly; that is, heat is exchanged until the temperature is uniform throughout the box. But now imagine that energy is continuously added to the system. Suppose a flame is placed under one side of the box, as shown in Figure 6.6. The flame heats the air above it. In turn, the hot air rises and cold air blows along the bottom of the box, creating a convection current, as shown in the diagram. This moving air has kinetic energy and has the capacity to perform work. If a well-balanced toy windmill is placed in the box, it rotates. Thus, the heat of the flame is converted to work.

The amount of work performed on the windmill is related to the velocity of the moving air. In turn, the air moves most rapidly if the temperature difference between the hot air and the cold air is greatest. Think of it this way: If you have very hot air and very cold air, the difference in density between the two types of air is great, and the hot air rises rapidly. In turn, this rapidly rising air initiates a rapid convection current, and the windmill is forced to rotate rapidly. Conversely, if the hot air is only slightly warmer than the cold air, it rises slowly, and the convection current moves slowly. The weak convection current transmits little force, and the work performed is small. This reasoning leads to another alternative statement of the second law as: *The quantity of work performed by a heat engine depends on the temperature difference between the hot reservoir and the cold reservoir.* Heat energy can be converted efficiently into work only if large differences in temperature are available.

EXAMPLE 6.3 Hot outside, cool inside

A refrigerator causes a separation of hot and cold; cold air is maintained inside the unit, while

FOCUS ON . . . Planetary Winds

Planetary wind systems are another type of heat engine. For example, the Sun warms the Earth at the equator more than at higher latitudes. Thus, the air rises vertically upward over the equator, as is shown schematically in the Figure. This rising air is replaced by cooler air moving along the surface. The air that moves across the surface is simply the wind and can indeed push ships across the ocean or cause windmills to rotate. Obviously, a convection current is greatest if the difference between the hot part of the system and the cold portion is large. Thus, the energy created by the wind is related to the temperature difference of the two parcels of air.

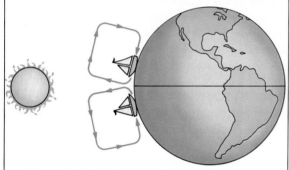

The global wind systems are heat engines powered by the Sun.

the kitchen remains warm. Is this separation a violation of the second law?

Solution   No, the second law is not violated. The second law states that the entropy of an isolated or undisturbed system always increases during any spontaneous process. A refrigerator operating in a room is neither undisturbed nor spontaneous. An outside source of electrical energy is added to this system to cause the separation of hot from cold.

EXAMPLE 6.4   Telling grapes from oranges

Imagine a closed system consisting of a mixture of wet oranges and grapes suspended by a shelf made of paper. Below the shelf there is a second shelf made of metal screen. The mesh size of the screen is large enough to permit grapes to pass

through, but small enough to hold the oranges. The box is sealed. After a while the paper shelf becomes soggy, loses its strength, and breaks. The oranges and grapes fall and are separated into two ordered collections, grapes on the bottom and oranges on the metal screen. Has the total order of the system increased? Has the entropy decreased spontaneously? Has the second law been violated? Explain.

Solution   The second law has not been violated. In a sense, the separation of oranges and grapes in this system is similar to the separation of hot from cold in a refrigerator. In a refrigerator, the potential energy in some fuel is used to generate electricity, and the electricity creates a separation of hot from cold. When the total system—fuel plus refrigerator—is examined, the system becomes more disordered than ordered, and the entropy increases. The same is true with this system. At the start, the fruit has potential energy of position. When the shelf breaks, the gravitational potential energy is lost, and in the process the oranges and grapes are separated. Thus, energy is exchanged for order, just as in a refrigerator. The system is not an isolated or undisturbed one, because gravity is doing work on it; therefore, the second law is not violated.

6.7 CONSEQUENCES OF THE SECOND LAW

There are three important practical consequences of the second law. The first is that since the quantity of work performed by any heat engine depends on the temperature difference between the hot working parts and the cooler surroundings, engineers must design an engine with not only a hot working substance but also an adequate cooling system. The second consequence is that whatever fuel is used to operate a heat engine, some waste heat is discharged into the environment, and it is therefore impossible to convert all of the potential energy of the fuel to useful work. Finally, as stated above, a fuel can be used only once; after it has been burned and work performed, the energy cannot be completely recycled.

FIGURE 6.7 Schematic view of an electric generator.

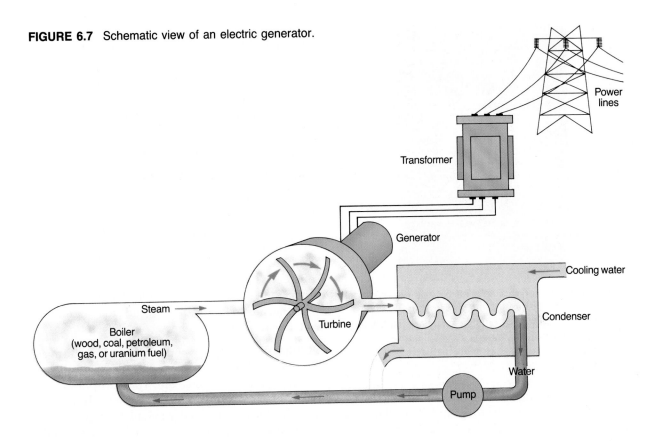

6.8
THE SECOND LAW AND THE GENERATION OF ELECTRIC POWER

Most of the electricity generated in the world today is produced in steam-driven power plants. Here water is boiled using the heat available from coal, oil, gas, or nuclear fuel. The hot steam is allowed to expand against the blades of a turbine. A **turbine** is a device that spins when air or water is forced against it. You can think of it as a kind of enclosed windmill. The hot, expanding steam forces the turbine to spin, and the spinning turbine then operates a generator, which produces electricity. After the steam has passed through the turbine, it is cooled, liquefied, and returned to the boiler to be re-used, as shown in Figure 6.7. Normally, the steam is cooled with river, lake, or ocean water. The cooling action of the condenser is essential to the whole generating process.

To understand this important concept, let us return to the discussion of the second law of thermodynamics. Recall that a windmill in a box of air

moves when the air on one side of the box has been heated. Also remember that the greatest amount of work is performed if the temperature difference between the hot air and the cold air is maximized. A steam turbine operates on the same general principle. Water is boiled, the steam is heated by some fuel, and the hot steam expands against the turbine blades. If the exhaust gases are cooled, then the temperature difference between the hot gas and the exhaust is increased, and more work can be performed. Therefore, some provision must be made to cool the steam at one end of the engine.

In practice, cooling is accomplished by circulating water around the condenser. It is obvious that maximum efficiency is reached with very hot steam and a very cool condenser. In practice, the nature of the metals in the turbine limits the temperature to a maximum of about 540°C. The low temperature is limited by the cheapest coolant, which is generally water. Within the constraints of these two limits, an efficiency of 60 percent is

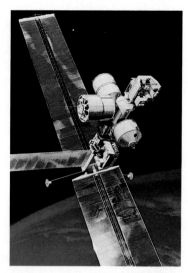

The solar panels on this satellite provide the electrical energy to drive its motors and transmitters. Solar power may in the future provide a substantial fraction of this country's energy needs.
(Courtesy NASA)

theoretically possible, but uncontrollable variations in steam temperature and miscellaneous heat losses reduce efficiency to about 40 percent, even in the best installations. Considering the power plant as a whole, this level means that for every 100 units of potential energy derived from the fuel, 40 units of electrical energy are available as useful work and 60 units of energy are dissipated to the surroundings as heat.

Although a modern electric power plant is only 40 percent efficient, it is more efficient than other common heat engines. A large diesel is 38 percent efficient, the gasoline engine in an automobile is 25 percent efficient, and a steam locomotive is only 10 percent efficient. Therefore, use of electricity is an efficient way to perform work. Less fuel is needed to operate an electric lawn mower or car than a gas-powered machine. But the generation of electricity to be used for heating is fundamentally inefficient despite the fact that advertisements encourage homeowners to "live better with clean, efficient electric heat." True, an electric heater is 100 percent efficient *in the home*, but that is only part of the system. At the power plant, heat must first be converted to work, and 60 percent of the available energy is lost during the work-to-heat conversion. (An ad-

ditional 7 percent of the original energy is lost during transmission, so that a total of 67 percent of the fuel energy is lost before the electricity is delivered to your home.) On the other hand, a direct fuel heater, such as a gas stove, delivers the heat directly where it is needed, and there is very little waste.

EXAMPLE 6.5 Evaluate this engine

In the tropics, the surface of the ocean is some 20°C warmer than the deep layers. Imagine an electric generator that operates on this temperature difference. The engine of such a device would use the warmer surface water to heat a gas, causing it to expand against the blades of a turbine. The exhaust gases would then be cooled with the subsurface water. Would such an engine violate the first law? The second law?

Solution It violates neither. Energy is derived from warm water and not "created" magically, so the first law is obeyed. The system operates on the temperature difference between a hot working substance and a cooler exhaust, so the second law is also obeyed. However, such a device would not be very efficient because the temperature difference is small. In fact, an experimental power plant of this kind has been built and tested near Hawaii. Although electric power has been produced, the cost of building and operating such a plant is so high that it is not economical to run, even though the energy is free.

6.9 THERMAL POLLUTION

Large amounts of heat are released from modern electric generating stations. For example, a 1000-megawatt facility, running at 40 percent efficiency, would heat 60 million liters of water by 8.5°C every hour. In many systems, lake, river, or ocean water is used as a coolant. It is not surprising that such large amounts of heat, added to aquatic systems, cause ecological disruptions. The term **thermal pollution** has been used to describe these heat effects.

What happens when the outflow from a large generating station raises the water temperature of a river or lake? Fish are cold-blooded ani-

FOCUS ON . . . Entropy and Living Systems

Superficially, living organisms appear to run counter to the second law by creating order out of disorder. Consider a frog developing from a tiny egg. The embryo builds large, highly ordered structures, such as muscles and other organs, from smaller building blocks such as protein molecules. However, living things are not really exempt from the second law of thermodynamics. They are not isolated systems, and they must use an outside supply of energy—food—to synthesize the molecules they need, to move substances around, and to combat the universal tendency toward increasing disorganization. The second law does not say that it is impossible to create ordered systems—there is order all around us in buildings, machinery, mountain ranges, trees, and so on. It simply states that, if left undisturbed, a system will spontaneously become disordered. If energy is added, the system can indeed be altered in any manner.

mals, which means that their body temperature increases or decreases with the temperature of the water. In natural systems, marine life is adapted to the prevailing water temperatures and their seasonal changes. When the temperature is raised above its normal level, all the body processes of a fish (its metabolism) speed up. As a result, the animal needs more oxygen, just as you need to breathe harder when you speed up your metabolism by running. But hot water holds less dissolved oxygen than cold water. Therefore, cold-water fish may suffocate in warm water. In addition, warm water can kill fish by disrupting their nervous systems. In general, not only fish but also entire aquatic ecosystems are rather sensitively affected by temperature changes. For example, many animals lay their eggs in the spring when the water naturally becomes warm. If a power plant heats the water so that some organism starts reproducing at the wrong time and the eggs are hatched in midwinter, the young may not find the food needed to survive.

The second law of thermodynamics assures us that it is impossible to invent a process to avoid the production of waste heat in steam-fired turbines. It is possible, however, to dispose of the waste heat with minimal disruption to the environment or, better yet, to put it to good use. Some techniques will now be described.

How to use the atmosphere as a heat sink

One approach to the problem of thermal insult to our waterways is to dispose of heat into the air. Air has much less capacity per unit volume for absorbing heat than does water, so the direct action of air as the cooling medium in the condenser is not economically feasible. For this reason, power plants must still be located near a source of water, the only other available coolant. However, the water can be made to lose some of its heat to the atmosphere and then can be recycled into the condenser. Various devices are available that can effect such a transfer.

The two cheapest techniques are based on the fact that evaporation of water is a cooling process. Many power plants simply maintain their own shallow lakes, called **cooling ponds.** Hot water is pumped into the pond, where evaporation as well as direct contact with the air cools it, and the cooled water is drawn into the condenser from some point distant from the discharge pipe. Water from outside sources must be added periodically to replenish evaporative losses. Cooling ponds are practical where land is cheap, but a 1000-megawatt plant needs 1000 to 2000 acres of surface, and the land costs can be prohibitive.

A cooling tower, which can serve as a substitute for a cooling pond, is a large structure, often about 180 m in diameter at the base and 150 m high (Fig. 6.8). Hot water is pumped into the tower near the top and sprayed downward onto a mesh. Air is pulled into the tower by either large fans or convection currents and flows through the water mist. Evaporative cooling occurs, and the cool water is collected at the bottom, as shown in Figure 6.9. No hot water is introduced into aquatic ecosystems, but a large cooling tower loses over 4 million liters of water per day to evaporation. Thus, fogs and mists are more common in the vicinity of these units than they are in the surrounding countryside. These fogs and mists are less ecologically disruptive than the thermal pollution of waterways, but they do affect the quality of the environment. Many new electrical generating stations are using cooling towers to dispose of waste heat.

FIGURE 6.8 A cooling tower at a nuclear reactor site in Oregon.
(Courtesy of U.S. Department of Energy.)

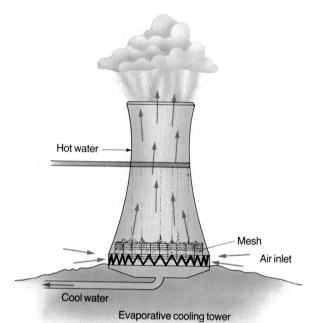

FIGURE 6.9 Schematic view of a wet cooling tower.

FOCUS ON . . . The Drinking Bird

When you think of an engine, you probably think in terms of the kind used in your car, as discussed in the text. However, any device that is capable of transforming heat into work meets the criterion of being an engine. One of the most common of these is the drinking bird toy that you have undoubtedly seen. The Figure shows how one of these toys works. The inside of the bird is constructed with two cavities, A and B, separated by a fluid, usually ether. To start the bird into motion, a piece of fabric that covers the beak of the bird is wet with water, as shown in (a). This water begins to evaporate from the beak, and since evaporation is a cooling process, the ether vapor inside the head of the bird also begins to cool down. As this gas cools, the pressure inside chamber A becomes less than that in chamber B, and the liquid begins to creep up inside the tube, as shown in (b). At some height, the weight of the bird's head becomes greater than the weight of its body, and the bird tips forward, as shown in (c). In this way, its beak becomes wet once again. Also, in this position, chambers A and B are no longer separated, so the pressures in the two come to the same value, and the fluid can also drain back into the body of the bird. Thus, the bird now tips to the upright position, as shown in (d) and the cycle begins anew. This continuous motion of the bird could be used to drive a (small) piston and to do some work. Thus, the drinking bird qualifies as an engine, but you are grasping at straws if you ever expect one to run your automobile for you.

A complete cycle of a "drinking bird" engine.

Use of waste steam (cogeneration)

No matter how well they work, the cooling systems described above are still elaborate means of throwing energy away. However, some of this discarded energy can still be useful. Waste steam is too cool to produce work efficiently, but it is hot enough for many other industrial processes. For example, the steam can be used to cook food in a cannery, to heat wood pulp in a paper mill, or to process petroleum in an oil refinery. In the United States, most of the waste steam for electric generation is discarded, leading to thermal pollution, but in Europe much of the waste steam is sold to other industries. Energy is utilized efficiently, and pollution problems are reduced (Fig. 6.10). Some companies in the United States do utilize their excess heat efficiently. One example is the relationship between the Baywood, New Jersey, refinery of Humble Oil and Refining Company and the Linden, New Jersey, generating station. The Linden power plant is capable of producing electricity at 39 percent efficiency. This efficiency is lowered by less than optimum cooling of the condenser, and some of the waste heat is sold as steam to Humble. If the two-plant operation is considered as a single energy unit, the overall efficiency of power production is raised to a level of 54 percent. The process is beneficial to many; the companies save money, fuel reserves are conserved, and thermal pollution of waterways is reduced.

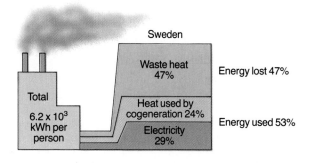

Sweden

Total 6.2 x 10³ kWh per person

Waste heat 47% — Energy lost 47%

Heat used by cogeneration 24% — Energy used 53%

Electricity 29%

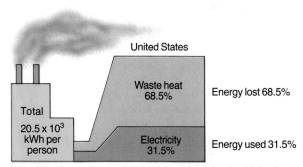

United States

Total 20.5 x 10³ kWh per person

Waste heat 68.5% — Energy lost 68.5%

Electricity 31.5% — Energy used 31.5%

FIGURE 6.10 Use of a fuel to produce electricity in Sweden and the United States. (kWh is an abbreviation for kilowatt hour, a unit of energy.)

SUMMARY

When a system does work on its environment, the work is defined as a positive quantity; when the environment does work on a system, the work is defined as negative. Heat is considered a positive quantity when it is added to a substance and a negative quantity when it is removed. The **internal energy** of a substance is the total energy of all the atoms or molecules that compose the substance. The **first law of thermodynamics** is an extension of the law of conservation of mechanical energy to include internal energy.

An **internal combustion engine** is a four-stroke engine—the intake stroke, the compression stroke, the power stroke, and the exhaust stroke.

The **second law of thermodynamics** can be stated in a variety of ways: (1) It is impossible to construct a heat engine that absorbs an amount of heat from a reservoir and performs an equivalent amount of work. (2) The **entropy** of an isolated system always increases during any spontaneous process. (3) The quantity of work performed by a heat engine depends on the temperature difference between the hot reservoir and the cold reservoir.

EQUATIONS TO KNOW

$\Delta U = Q - W$ (the first law of thermodynamics)

KEY WORDS

Work

Internal energy

The first law of
  thermodynamics

Engine

The second law of
  thermodynamics

Entropy

Thermal pollution

Cogeneration

QUESTIONS

WORK AND INTERNAL ENERGY

1. A gas is contained in a cylinder fitted with a piston. The gas is then taken through the following processes. In each case, state whether work is done on or by the gas and whether heat is added to or removed from the gas. What is the sign of the work and the heat transferred in each case? (a) The piston is clamped firmly in position and a flame is played across the base of the container. (b) The cylinder is covered with a material that will not conduct heat, and the gas is allowed to expand. (c) The gas is compressed while the nonconducting wall is still in place.

2. Discuss how the internal energy changes of the following gases might manifest themselves. That is, does an increase of the internal energy produce a change solely in the random kinetic energy of the atoms, do internal molecular rotations or vibrations play a role, and so forth? (a) A monatomic gas (composed of individual atoms), (b) a diatomic gas (composed of two atoms bound together).

FIRST LAW OF THERMODYNAMICS

3. An engineer designed and built the roller coaster shown in the Figure. He was fired. Why?

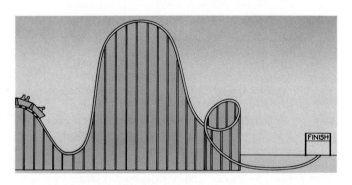

FINISH

4. Two bicycles are coasting down one big hill and up another. Both bicycles are identical except that one cyclist has a small electric generator attached to the rear wheel to power a portable cassette tape player. Which cyclist will coast farther uphill? Explain.

5. Imagine an isolated system that consists of a battery and an electric motor connected through pulleys to a set of weights. When the battery is connected to the motor with properly insulated wires, the motor can do 50,000 J of work lifting the weights before the battery runs down. If shoddy wires are used and some partial short circuits develop, the system will spark and sputter, and a fully charged battery will do only 25,000 J work. Does this observation disprove the first law of thermodynamics? Explain.

6. A car designer has the idea to position large flywheels inside a conventional automobile. He reasons that if the flywheels can be connected to the wheels while the car is coasting downhill, energy can be conserved for the next uphill pull. Do you think that this is a good idea? Explain.

7. A gas is compressed so that 10,000 J of work are done on it; at the same time 5000 J of heat are released to the environment. What is the internal energy change of the gas?

8. During an expansion process of a gas, 3000 J of heat are added to the gas, and it does 2000 J of work. What is the internal energy change of the gas?

9. In what way is the statement, "You can't get something for nothing," related to the first law of thermodynamics?

10. The first law of thermodynamics is a modification of a conservation law encountered earlier in your study in this text. What is that conservation law and how are they related?

HEAT ENGINES AND THE GASOLINE ENGINE

11. Why is an automobile engine called an "internal combustion" engine?

12. What is a cylinder? A piston? A spark plug? What is the function of each?

13. Explain what happens inside an automobile engine during the power, exhaust, intake, and compression strokes. During which parts of the cycle is the piston moving upward? Downward? At what point in the cycle do the valves open? When do they close? When does the spark plug fire? How many valves are there in a four-cylinder engine?

14. What would happen if a small hole were burned in the face of one intake valve so that gases could leak through even when the valve was closed? Explain.

15. What would happen if a small hole were burned in the face of one exhaust valve so that gases could leak through even when the valve was closed? Explain.

16. The piece of metal connecting the piston to the crankshaft (which ultimately drives the wheels) is called the connecting rod. What would happen to a car if a connecting rod broke?

17. The piston of an internal combustion engine is surrounded by a ring that presses tightly against the piston and the wall of the cylinder. Explain why a car with worn piston rings loses power.

SECOND LAW OF THERMODYNAMICS AND ALTERNATIVE FORMS OF THE SECOND LAW

18. In what way is the statement, "You can't even break even," related to the second law of thermodynamics?

19. Write three statements of the second law of thermodynamics.

20. If a room is the same temperature as the surrounding air, and an air conditioner is turned on in the house, the room will get cooler while the outside is heated. Is this a violation of the second law of thermodynamics? Explain.

21. Suppose a refrigerator is placed in the middle of a closed room, turned on, and the refrigerator door left open. Will the temperature of the room rise, fall, or remain the same? Explain.

22. Mountain ranges slowly erode and crumble, ultimately weathering down to flat land. Is entropy increasing or decreasing during this process? Explain.

23. (a) Automotive engineers are experimenting with building engine blocks out of ceramic materials instead of metals. Since ceramics can withstand greater thermal stress without failing, ceramic engines operate at higher temperatures than ordinary metal ones do. All other factors being equal, will a ceramic engine be more, or less, efficient than a metal engine? Explain. (b) All other factors being equal, will a car give better gas mileage in the summer or in the winter? Explain.

24. One of the early pioneers in the science of thermodynamics, Julius Mayer, started thinking about work and heat while he was a physician on board a trading ship. Dr. Mayer noticed that the sailors ate less when they were in the tropics than they did when they were in colder regions, yet they performed the same amount of work regardless of location. Was this an observation of the first law or of the second law? Explain.

25. Does water have more entropy when frozen or when liquid?

26. When rolling dice, it is far more probable that one will roll a seven than a two. Defend the statement that the entropy of a system of dice is higher for seven than it is for two.

THE SECOND LAW AND THE GENERATION OF ELECTRIC POWER

27. Consider two power plants, one that uses 500°C steam as its working substance and one that uses 150°C steam. In all other respects, both are identical. Both use cooling water at 30°C and produce the same amount of electricity. Which power plant discharges more thermal energy to the environment? Which uses less fuel? Explain.

28. Why is electric heat less efficient than a small propane or oil furnace that is installed in a house? Would it be possible to generate electric heat as efficiently as it is to produce heat by burning fuel directly in the home?

29. In a hydroelectric generating system, a turbine is driven by a stream of falling water. The water itself may be cool, and no auxiliary cooling system is required, yet these facilities are nearly 99 percent efficient. No fossil fuels are used. Do hydroelectric power plants violate the laws of thermodynamics? Review the entire cycle, answering the questions: What is the ultimate source of energy of the system? How is the second law obeyed?

30. Would it be practical to increase the efficiency of a power plant by cooling the condenser with a giant refrigerating unit? Explain in terms of the second law.

THERMAL POLLUTION

31. Define thermal pollution. How does it differ in principle from air or water pollution?

32. Explain why nuclear-fueled power plants require more cooling water than fossil-fueled plants.

33. Since marine life is abundant in warm tropical waters, why should the warming of waters in temperate zones pose any threat to the environment?

34. Describe some difficulties with the use of waste steam for home heating. Describe the potential benefits.

ANSWERS TO SELECTED NUMERICAL QUESTIONS

7. 5000 J   8. 1000 J

From the soft, entrancing music of an orchestra, to the beat of a marching band, to a rock star, to a child playing with a tin whistle, all sounds have a vibrating object as their source. (Courtesy: Greg Perry)

C H A P T E R

7

Wave Motion: Sound

7.1
WAVE MOTION

When most people think of a wave, they visualize water waves traveling across the surface of a lake or crashing against the seashore. A water wave is made up of a collection of moving particles. However, wave motion is different from the other types of motion that we have studied so far, in that the wave itself takes on an identity separate from that of the moving particles of which it is composed. In fact, the wave pattern has its own special properties and behavior.

There are many different types of waves; three common ones are water, sound, and light waves. In one sense, these three are so different that they hardly seem to resemble each other at all. However, there are certain features of waves that are common to all the various types that one may encounter in nature. In this chapter, we shall attempt to determine some characteristics of waves that are common to all the various types, and in doing so, we shall focus much of our attention on one particular type of wave, the sound wave.

7.2
DESCRIBING A WAVE

Longitudinal and transverse waves

All of the different types of waves that we will discuss in this text can be placed into one of two categories. They are either **longitudinal** waves or **transverse** waves. Let's look at the difference between these two categories.

A wave can be set up on a rope by tying one end to a wall and moving the other up and down to form a pattern moving along the rope. In Figure 7.1, a bump has been created on the rope by one upward motion of the hand. The bump moves away from the hand and down the rope toward the right. Let us now examine the motion of the rope a little more carefully as the bump passes by a particular location. We shall do so by dropping a piece of paint onto a small section of the rope and focusing our attention on the movement of the painted segment. We find that the motion of the segment is up and down while the bump moves to the right. *Waves of this kind, in which the moving particles vibrate at right angles to the direction in which the wave travels, are referred to as*

135

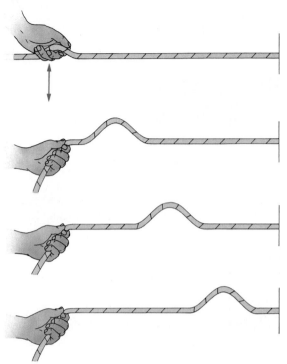

FIGURE 7.1 A transverse wave. Any given point on the string moves up and down, but the wave moves horizontally.

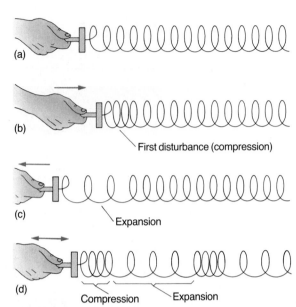

FIGURE 7.2 (a) The spring is held at rest. (b) Moving the hand toward the right compresses the coils near the hand. (c) Moving the hand to the left creates an expansion in the coils. (d) To-and-fro motion of the hand sets up a series of compressions and expansions.

transverse waves. Waves traveling along the surface of a pool of water are basically transverse waves. This can be demonstrated by considering the motion of a cork placed in the water. As the wave passes by, the cork bobs up and down. (For a water wave, there is a small back-and-forth motion of the cork, but the primary motion is up and down.) Individual water drops follow a path in the water like that followed by the cork. We shall see in a later chapter that light waves are transverse waves.

A different kind of wave pattern can be set up by attaching one end of a spring to a wall and moving the other end in a back-and-forth motion along the direction of the spring, as shown in Figure 7.2. As the hand moves to the right in Figure 7.2b, the coils of the spring near the hand are compressed, and this compression begins to move along the spring. When the hand moves to the left, as shown in Figure 7.2c, the coils near the hand move farther apart, and this expansion then begins to move to the right. If this to-and-fro motion of the hand is continued, as in Figure

7.2d, a series of compressions followed by expansions moves along the spring toward the right, forming a wave. Note that the individual coils of the spring do not travel along the spring. It is the wave—the disturbance—that travels. Also, note that if you focus your attention on one particular coil, its motion is a to-and-fro motion that basically follows the motion of the hand. *When the individual particles that constitute the wave move to-and-fro along the same direction as the wave travels, the wave is called a longitudinal wave.* We shall see in the next section that a sound wave is a longitudinal wave.

Graphical representation of a wave

A general pattern that is used to represent a wave is shown in Figure 7.3. (This type of pattern is called a sine curve.) It should be noted that this representation is similar in appearance to an actual transverse wave on a rope. However, this

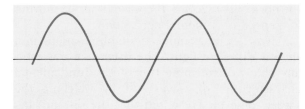

FIGURE 7.3 A wave can be represented graphically by a sine curve.

same figure can be used to represent a longitudinal wave. To see this, refer to Figure 7.4. We note that at those points where the coils of the spring are compressed in part a of the figure, the sine curve of part b is at a crest. Likewise, at those points where the coils are far apart in part a, the curve in part b is at a trough. Thus, although a transverse and a longitudinal wave are distinctly different, the same kind of curve can be used to visualize both.

The speed of a wave

Imagine yourself resting on a beach and watching the waves come rolling in. If you focus your attention on one particular crest of a wave, you will see it come moving in toward the shore with a speed of perhaps a few feet per second. Thus, a wave has a velocity in that it travels with some speed and in some given direction.

As we saw in Chapter 2, the speed of a car is relatively easy to measure. All that you have to do is to measure the distance the car moves and divide by the time it takes to move this distance. The speed of a wave can be measured by exactly the same technique. In this case, you would focus your attention on, say, a crest of a wave, watch how far it moves in a given time interval, and divide this distance by the time interval.

The frequency of a wave

While watching a wave at the beach, you might decide to focus your attention on a fisherman's cork bobbing up and down in the water. An examination of the motion of this cork would enable you to find yet another important feature of a wave, its **frequency.** *The frequency of a wave is the number of complete vibrations that the wave makes each second.* Since the cork is following the vibrations

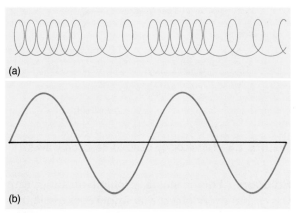

FIGURE 7.4 Representing a longitudinal wave by a sine curve.

of the water, its frequency of vibration is the same as that of the water. As another example, suppose you attach one end of a rope to a wall and vibrate the other end up and down with your hand. A wave travels down the rope, with each segment moving such that it emulates the movement of the hand. That is, if your hand makes three up-and-down cycles each second, each point of the rope will also make three up-and-down vibrations each second. In this case the frequency of the wave is

$$f = 3 \; \frac{\text{vibrations}}{\text{s}} = 3 \; \frac{\text{cycles}}{\text{s}}$$

Note the different ways in which frequency can be expressed. Units often used are vibrations per second or cycles per second. In calculations, only the "per second" is used, as we will see in our example problems yet to come. The official name given to the unit is the **hertz** (abbreviated Hz). Thus, we can express the frequency of the wave on the rope as

$$f = 3 \; \text{Hz}$$

EXAMPLE 7.1 Tock tick goes the erratic clock

A clock repairman is attempting to repair an erratic grandfather clock. He notes that the pendulum of the clock makes 12 complete vibrations every 4.0 s. What is the frequency of motion of the pendulum?

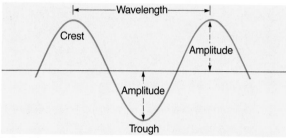

FIGURE 7.5 Terminology used to describe waves.

Solution This is not a problem dealing with waves, but every object that undergoes oscillatory motion, such as the clock pendulum, has a frequency of vibration. We can calculate this frequency by noting that the definition of frequency is the number of complete vibrations made every second. Thus,

$$f = \frac{\text{vibrations}}{\text{s}} = \frac{12 \text{ vibrations}}{4.0 \text{ s}} = 3.0 \text{ Hz}$$

The wavelength of a wave

Figure 7.5 shows a snapshot of a wave on a string. Another important characteristic of a wave, called the **wavelength,** is shown in the figure by the horizontal arrows. (The symbol used to represent wavelength is the Greek letter lambda, λ.) *The wavelength of a wave is defined as the distance between two consecutive points on a wave that are behaving identically.* Thus, a wavelength could be the distance between two consecutive crests on a wave, between two consecutive troughs, or between any two identical, consecutive points.

The relationship between speed, wavelength, and frequency

We will develop a relationship between the speed of a wave, its wavelength, and its frequency by use of a thought experiment. Let us consider waves moving under a fishing dock. Suppose fishing is slow and a fisherman starts wondering about the waves he sees rolling in toward him. He estimates that the distance between crests is about 3 meters, and he notes that a crest hits the dock once every second. What, he wonders, is the speed of these waves? If each wave is 3 meters long and one hits

the dock every second, the waves must be moving at a speed of 3 meters per second.

Now let us examine these observations in a slightly different light. The fisherman was actually able to determine two of the fundamental characteristics of waves just by watching the waves come in. He noted that the distance between consecutive crests was 3 meters; thus, he found the wavelength ($\lambda = 3$ m). He noted that the waves were hitting the dock at a rate of one each second; thus, he found the frequency ($f = 1$ Hz). From his observations, he was able to find the speed by multiplying the wavelength times the frequency as

speed = (wavelength)(frequency)

or symbolically

$$v = \lambda f \tag{7.1}$$

where we use v for the speed of the wave.

EXAMPLE 7.2   Sound waves

A tuning fork produces a 256 Hz note, corresponding to middle C on a piano. It is found that this fork produces a sound wave having a wavelength of 1.30 m. From this information determine the speed of sound.

Solution Equation 7.1 is valid regardless of the type of wave under consideration. Thus, we can find the speed of sound as

$$v = \lambda f = (1.30 \text{ m})(256 \text{ Hz}) = 333 \text{ m/s}$$

As we shall see, the speed of sound depends on temperature, but this is a reasonably accurate value for the speed of sound on a winter day when the outside temperature is 0°C.

EXAMPLE 7.3   Light waves

Light travels at a speed of 3×10^8 m/s in a vacuum and, for all practical purposes, through air. The wavelength of a particular shade of yellow light is 580 nm (1 nm = 10^{-9} m). Find the frequency of this light.

Solution Equation 7.1 can be used to find the frequency as

$$f = \frac{v}{\lambda} = \frac{3 \times 10^8 \text{ m/s}}{580 \times 10^{-9} \text{ m}} = 5.17 \times 10^{14} \text{ s}^{-1}$$
$$= 5.17 \times 10^{14} \text{ Hz}$$

The amplitude of a wave

Shown on the sketch of Figure 7.5 is one last term that is characteristic of a wave, the **amplitude.** Note the horizontal line in the figure, which we have called the reference line. If the wave under consideration were a wave on a string, this center line would be the straight-line shape taken by the string in the absence of a wave. If the wave were, instead, a water wave, the reference line would be the undisturbed surface of the water in the absence of the wave. Thus, from the figure we see that *the amplitude of a wave is the distance that the wave has been raised (or lowered) from the undisturbed position.*

You might try a simple experiment on your own to discover the relationship between the speed of a wave and its amplitude. Run a bathtub full of water and wait until the surface becomes reasonably smooth. Then, wiggle your finger in the center of the pool to set up a small amplitude wave on the surface. Watch the wave and try to estimate its speed. Now repeat the process, except this time hit the water more violently so that a large amplitude wave is set up. Again, try to estimate the speed of this wave. Admittedly, it is a little difficult to do this experiment with any accuracy, but we hope that you will see that both waves seem to have the same speed. If you were to do this more accurately, you would find that *the speed of a wave and the amplitude of the wave are not related.*

The amount of energy carried by a wave determines the amplitude of the wave. This last statement should be obvious to you from your experiences at a beach. On a day when the water is relatively calm, you are able to play in the surf and allow the small amplitude waves to break against you with no difficulty. However, if the wave amplitudes become large because of a storm, you may find that you are unable to stay on your feet when they surge into you. The difference occurs because the larger amplitude waves are carrying more energy.

7.3
SOUND WAVES

So far in our discussion of waves, we have focused our attention on terms and concepts that are common to all types of waves. From now on in this chapter, we will narrow our point of view and concentrate exclusively on one important type of wave, the sound wave. We shall begin our discussion by considering how a sound wave is produced. Regardless of whether you are listening to a heavy-metal rock star's guitar or the melodious sound of a violin concerto, the sounds have the same basic source. Whether you are listening to the sweet trill of a robin or to the harsh sound of an angry shout, the sound waves originate in the same basic way. *The source of any sound is a vibrating object.* When the rock star strums his guitar, the source of sound is the vibrating string; likewise, the sound from a violin comes from a vibrating string. When the robin trills, its vibrating vocal cords are the source of the sound, just as they are for the angry shouter. You pick your own sound; whether it is pleasant or irritating, the source of the sound can always be traced to a vibrating object.

When an object vibrates, it disturbs the air near it, and this disturbance moving through the air is what is referred to as a sound wave. Let us examine exactly how a sound wave is produced by considering how a tuning fork produces a sound. A tuning fork like that shown in Figure 7.6 consists of two metal prongs that, when struck, vibrate back and forth. Consider Figure 7.7 to see how the sound wave is produced. When one prong of the tuning fork moves to the right, as in Figure 7.7a, air molecules near the fork are forced more closely together than normal. We call these regions **compressions.** When the prong swings to the left, as in Figure 7.7b, the air molecules to the right of the prong have room into which they can move; hence, the air molecules are not squeezed together as closely as normal. Such a region is called a **rarefaction.**

As the tuning fork continues to vibrate, additional compressions and rarefactions are pro-

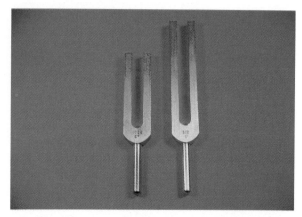

FIGURE 7.6   Tuning forks.

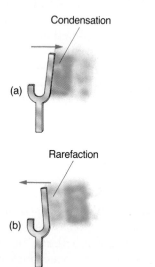

Condensation

(a)

Rarefaction

(b)

FIGURE 7.7   (a) When the tuning fork prong moves to the right, air molecules in front of it are pushed close together. (b) A rarefaction is produced when the prong swings back to the left.

duced, and these spread out through the air like ripples in a pool of water. After a short period of time, the air might look as shown in Figure 7.8, if one were able to actually see the molecules. We are able to represent the sound wave by our familiar curve, as shown by a comparison between parts a and b of Figure 7.8. We see that where the sound wave has a compression in part a, the curve in part b is shown to be at a crest. Likewise, at points where the sound wave has a rarefaction, the curve is seen to be at a trough.

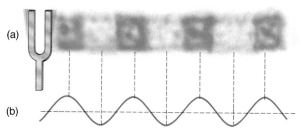

(a)

(b)

FIGURE 7.8   (a) After several vibrations, the air in front of the tuning fork is filled with compressions (condensations) and rarefactions moving away from the fork. (b) A graphical representation of the sound wave. Note that a compression is represented by a crest in the sine curve and a rarefaction by a trough.

7.4
SOUND: A LONGITUDINAL WAVE

We can determine whether sound is a longitudinal wave or a transverse wave by once again considering what a tuning fork does to the air near it. In Figure 7.9, we will focus our attention on one particular air molecule as the prong of a tuning fork moves to the right and then back to the left. As we see in Figure 7.9a, as the prong swings to the right, the air molecule is also forced to the right, and in Figure 7.9b as the prong swings back to the left, the air molecule follows its motion and also moves to the left. We know that the sound wave is spreading out from the tuning fork, toward the right in our figure. Thus, we see that as the sound wave moves outward toward the right, an individual air molecule vibrates to and fro from right to left. If you go back and check the definition of a longitudinal wave, you will see that such a wave is characterized by the fact that the vibrations of the material are along the same direction as the wave travels. Thus, sound is an example of a longitudinal wave.

7.5
THE SPEED OF SOUND

The speed of sound in air at 0°C is about 331 m/s (1090 ft/s, or 741 mi/h). This is a very fast speed, but it is very small when one considers that the speed with which light travels through air is 3.00×10^8 m/s. Everyone has at one time or another observed vivid demonstrations of the enor-

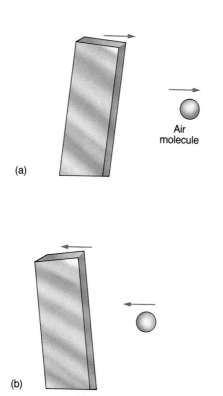

(a)

Air
molecule

(b)

FIGURE 7.9 (a) As the prong of the fork swings to the right, the air molecule shown moves to the right. (b) When the prong moves to the left, the air molecule also moves to the left.

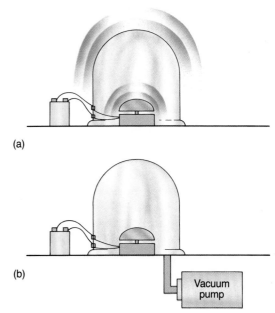

(a)

(b)

Vacuum
pump

FIGURE 7.10 (a) With air in the container, we can hear the bell, but (b) when the air is removed, sound cannot travel through a vacuum, and we cease to hear the bell.

mous differences between these two speeds. For one example, imagine yourself in the centerfield bleachers at a baseball game, some 300 m from home plate. When a batter hits a baseball, the sound of bat striking ball will take about 1 second to reach you, but the speed of light is so fast that you observe the batter make contact almost instantaneously. Thus, you see the batter hit the ball and a brief moment later you hear the contact.

The speed of sound also depends on the type of material through which the sound wave is traveling. In general, sound travels fastest through a solid, next fastest through a liquid, and slowest through a gas. Sound has to have a material to travel through; it does not move through a vacuum at all. This can be demonstrated by setting an electric bell ringing inside a vacuum chamber, as shown in Figure 7.10. When air is in the container as in Figure 7.10a, you can hear the bell ringing quite well, but note what happens when a

vacuum pump is attached to the container and the air is pumped out. As the air is evacuated, the sound of the bell gradually diminishes, until when the air is almost all gone, the sound of the bell disappears completely (Fig. 7.10b). Because light travels through a vacuum, you can still see the bell ringing inside the container—you just can't hear it.

The reason that sound travels differently through different types of materials depends on a property of a material called its **elasticity.** A very elastic material quickly springs back to its original shape when something distorts it. In such a material, the individual atoms act as though they were attached to their neighbors by springs. This makes it easy for a disturbance of one molecule to be transferred to its neighbors. Thus, when one atom is caused to swing to and fro by a sound wave, this motion is quickly transmitted to nearby atoms, and the sound wave spreads rapidly through the material. Solids are more elastic than liquids, which in turn are more elastic than gases. Thus, sound travels with decreasing speed through solid, liquid, and then gas.

A final factor affecting the speed of sound in air is the temperature of the material. The reason for this is that as the temperature of a substance increases, the speed of the individual atoms in the material increases. As a result, the atoms can collide more often; thus, a disturbance of one atom is transmitted more quickly to its neighbors. For a sound wave in air, it is found that the speed of sound increases with temperature according to the following equation

$$v = v_0 + (0.61)T \qquad (7.2)$$

where v_0 is the speed of sound at 0°C, 331 m/s, and T is the temperature in degrees Celsius.

EXAMPLE 7.4   The happy wanderer

A hiker glances at his Acme wrist thermometer and notes that the temperature is 21°C. What is the speed of sound through air at his location?

Solution   The speed can be found from Eq. 7.2. We find

$$v = v_0 + (0.61)T$$
$$= 331 \text{ m/s} + (0.61)(21°C) = 344 \text{ m/s}$$

EXAMPLE 7.5   The happy yodeler

The happy hiker of Example 7.4 emits a sound and finds that an echo returns to him a time of 2.00 s. How far away is the mountain that reflected the sound?

Solution   From the definition of speed, we find that the distance traveled by the sound wave is

$$d = vt = (344 \text{ m/s})(2.00 \text{ s}) = 688 \text{ m}$$

However, the sound wave had to travel to the mountain and back. Thus, the actual distance to the mountain is 688/2 = 344 m.

7.6 FREQUENCY AND WAVELENGTH OF SOUND

A longitudinal disturbance traveling through the air is called a sound wave only if its frequency is between 20 and 20,000 Hz, because it is only within this range that the normal human ear is sensitive. Waves having a frequency greater than 20,000 Hz are called **ultrasonic** waves and can be heard by some animals, including dogs.

Ultrasonic waves have important medical applications. For example, these waves are often used to clean objects. The object to be cleaned is placed in a liquid through which an ultrasonic beam is then passed. These waves set the contaminants on the surface of the object into a rapid vibration that shakes them free. A second important medical application of ultrasonic waves is as an imaging tool to examine internal organs or to observe a fetus. In the latter application, an instrument that is both a source and a detector of ultrasonic waves is passed across the mother's abdomen. An ultrasonic wave emitted by the instrument penetrates the body of the mother and is reflected to the receiver by the fetus. The reflected waves are converted to an electrical signal that is then used to produce an image on a phosphorescent screen. Certain birth defects such as spinal bifida are easily detected by this technique.

Another interesting application of ultrasonic techniques is the ultrasonic ranging unit designed by the Polaroid Corporation. This device is used to determine the distance from the camera to the object to be photographed. A burst of ultrasonic waves is emitted from the camera and travels toward the subject, which reflects part of it back to the camera, where it is picked up by a detector. The time interval between emission and the detection of the echo can be used to find the distance to the object to be photographed, using the same technique demonstrated in Example 7.5.

In everyday life, the terms "frequency" and "pitch" are often used interchangeably. For example, the flute is an instrument that characteristically emits sounds having a high frequency, whereas some horns emit sounds of low frequency. When speaking of these, it is common to refer to the flute has having a high pitch and to the horn as having a low pitch.

EXAMPLE 7.6   Speed and frequency

Imagine yourself at a symphony and use your experience with what you hear when you listen to the performance to convince yourself that the speed of sound does not depend on the frequency of the sound.

Solution If you are to enjoy a symphony, the musical notes played must reach you at the precise instant they should, relative to other notes. Suppose that speed did depend on frequency, such that high frequencies travel faster than low frequencies. If that were the case (and it isn't), the notes from a high-frequency instrument would reach you in the audience more quickly than would the notes from a bass instrument. Instead of being in perfect timing and coordination, the sound would end up being a jumble and not soothing at all. Thus, the speed of sound is completely independent of the frequency of the sound. This seems somewhat surprising in view of the equation $v = \lambda f$, because this equation seems to indicate that as the frequency, f, goes up, so should the speed v. However, this is not the case; as one moves to higher frequencies, the wavelength decreases such that the product of λ and f always remains equal to the speed of sound.

7.7 LOUDNESS OF SOUND AND THE DECIBEL SCALE

We noted earlier that the energy carried by a wave depends on the amplitude of the wave. Thus, a loud sound causes large amplitude vibrations of the molecules in the air, whereas a soft sound produces small vibrations. Be careful here, however, because your definition of a large vibration may not be in agreement with what actually happens in a sound wave. The faintest sound that the human ear can detect, at a particular frequency, is called the **threshold of hearing,** and the loudest sound that the ear can tolerate is called the **threshold of pain.** If one could examine an individual molecule of air when a sound equal to the threshold of hearing is passing by, one would find that the molecule moves with an amplitude of vibration of only about 10^{-11} m. This is an incredibly small number, so we see that the ear is an extremely sensitive detector of sound. On the other hand, suppose that a sound at the threshold of pain passes by. In this case, the amplitude of vibration of the molecule increases

TABLE 7–1 Decibel Level of Common Sounds

Noise Level	Source of Sound
0	threshold of hearing
10	rustle of leaves
30	soft whisper
40	mosquito buzzing
50	average home
60	ordinary conversation
70	busy street traffic
100	power mower
120	threshold of pain
130	rock concert
150	jet engine (at 30 meters)
180	rocket engine (at 30 meters)

to about 10^{-5} m. This is still a very small amplitude of vibration, but it is extremely large in comparison to the vibration amplitude at the threshold of hearing.

In order to discuss the loudness of sounds, the **decibel** scale was devised. This scale is constructed so that the lowest perceptible sound for the normal ear is assigned a 0 decibel level. A sound carrying 10 times as much energy as the 0 decibel sound is said to have a decibel level of 10. A reasonable guess is that a 20 decibel sound carries 20 times as much energy as the zero decibel scale, but that is incorrect. The scale is devised so that the 20 decibel level sound carries 100 times the energy of the 0 decibel sound and 10 times the energy of the 10 decibel sound. Thus, an increase of 10 decibels means that the energy content of the sound increases by 10 times. For example, a 50 decibel sound carries 10 times more energy than a 40 decibel sound. On this scale, a sound at the threshold of pain has a decibel level of 120. Table 7–1 indicates the decibel level of several common sounds.

7.8 INTERFERENCE

It is a characteristic of all types of waves that they can interact with one another. To get an idea of how this phenomenon works, let us consider two rivers, as in Figure 7.11, coming together and melding into a single river. We are going to assume that these are perfect rivers in that we don't

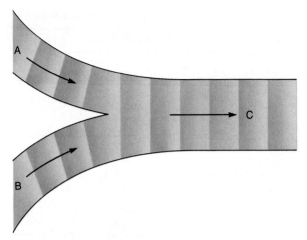

FIGURE 7.11 When the waves in rivers A and B meet in river C, what does the resulting wave look like?

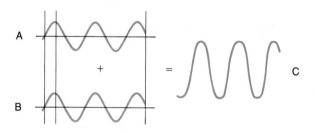

Constructive interference

FIGURE 7.12 When two waves interfere such that crest meets crest and trough meets trough, constructive interference occurs.

have to worry about sand bars, turbulence, or any other effect that would disturb a wave moving along the surface of these rivers. With all these restrictions placed on our example, it should be obvious to you that you are not going to be able to observe these effects on a real river. The phenomenon *can* be demonstrated with water waves, but it takes a more ideal setting, such as a laboratory, to produce it.

Let us consider identical waves moving along branches A and B of the river in Figure 7.11. These waves have the same frequency, wavelength, and amplitude. The question is: What happens to these waves when they come together in river C? The answer is that the resulting disturbance in C depends on the phase relationship

between the waves in A and B. For example, suppose at some instant of time, the wave in river A and the wave in river B match up as shown in Figure 7.12. Waves like this, such that crest matches up with crest and trough matches up with trough are said to be *in phase*. When these waves come together in river C, the resultant disturbance of the water looks like that shown in the figure. The two waves, A and B, are said to have undergone **constructive interference**. The resul-

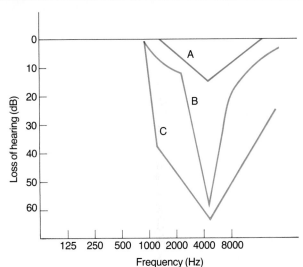

Patterns of hearing loss from exposure to industrial noise. (A) Temporary loss of hearing. (B) After 20 years. (C) After 35 years.

quency. At a 2000 Hz frequency, for example, it is estimated that occupational exposure to 95 decibel noise (about as loud as a power lawn mower) will depress one's hearing ability by about 15 dB in 10 years. Occupational noise, such as that produced by bulldozers, jackhammers, diesel trucks, and aircraft, is deafening many millions of workers.

What about extremely loud noises? Recent concern over exposure of people to rock music stems from the fact that such music is often played very, very loudly. Sound levels of 124 decibels have been recorded in some nightclubs and concerts. Such noise is at the edge of pain and is unquestionably deafening. Noise levels as high as 135 decibels should never be experienced, even for a brief period, because the effects can be instantaneously damaging. Such an acoustic trauma might occur, for example, as the result of an explosion. If the noise level exceeds about 150 or 160 decibels, the eardrum may be ruptured beyond repair.

A living organism, such as a human being, is a very complicated system, and the effects of a stress or disturbance follow intricate pathways that may be very difficult to understand. Many investigators believe that loss of hearing is not the most serious consequence of excess noise. The first effects are anxiety and stress reactions or, in extreme cases, fright. These reactions produce body changes such as increased rate of heart beat, constriction of blood vessels, digestive spasms, and dilation of the pupils of the eyes. The long-term effects of such overstimulation are difficult to assess, but scientists do know that in animals it damages the heart, brain, and liver and produces emotional disturbances. The emotional effects on people are, of course, also difficult to measure. One known effect is that work efficiency goes down when noise goes up.

moved from his noisy environment? That will depend on how noisy it has been and how long he has been exposed. In many cases, his chances for almost complete recovery will be fairly good for about a year or so. However, if the exposure continues, hearing loss becomes irreversible, and eventually he will become partially deaf. Look at curves B and C in the figure to see a typical downward progression caused by prolonged exposure to industrial noise.

In general, noise levels of about 80 decibels or higher can produce permanent hearing loss, although, of course, the effect is faster for louder noises, and it is somewhat dependent on the fre-

tant wave, C, has the same frequency and wavelength as A and B, but its amplitude is the sum of the amplitudes of A and B.

On the other hand, if the phase relationship is like that shown in Figure 7.13, the crest of one matches up with the trough of the other, and we say that the two waves are *180° out of phase.* When these two waves come together in river C, they cancel one another and the surface of the water remains calm. This occurs because when one of

the waves is trying to produce a crest by pulling the water upward, the other wave is attempting to produce a trough by pulling the water downward. The result is no motion of the water at all. In this case, we say that the waves are undergoing **destructive interference.**

Interference effects are important in understanding many day-to-day occurrences, and particularly in our understanding of most musical instruments. It should be noted that if the two

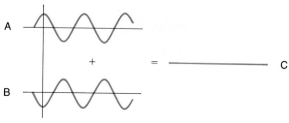

Destructive interference

FIGURE 7.13 Destructive interference occurs when crest meets trough.

FIGURE 7.14 Interference of waves with different frequencies.

waves interfering have different frequencies, the situation becomes more complicated, for they interfere constructively part of the time and cancel part of the time, as shown in Figure 7.14.

EXAMPLE 7.7   Unequal amplitudes

Two waves having a frequency of 10 Hz interfere constructively. One of the waves has an amplitude of 2 ft and the other has an amplitude of 1 ft. What are the frequency and amplitude of the resultant wave?

Solution The resultant wave has the same frequency as that of the two individual waves that produce it. Thus, the frequency of the resultant wave is 10 Hz.

To find the resultant amplitude when constructive interference is occurring, we must add the two amplitudes together. As a result, the resultant wave has an amplitude of 3 ft.

EXERCISE

Repeat this example assuming the waves undergo destructive interference.

Answer frequency = 10 Hz, amplitude = 1 ft

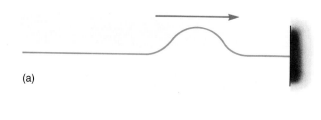

(a)

(b)

FIGURE 7.15 (a) A bump sent down a rope which is tied to a wall is (b) reflected so that it turns over on the rope.

7.9
STANDING WAVES

In the preceding section, we examined what happens when two waves both traveling in the *same* direction come together. However, many of the interference effects important to musical instruments occur when two waves traveling in *opposite* directions interfere. One way in which such a set of conditions can occur arises when a wave is sent down a string that is attached to something at the end toward which the wave is traveling. For example, the string could be tied to a wall, or it could be attached to one of the standards on a guitar. When the wave hits the fixed end, it is reflected back, and the reflected wave interferes with the oncoming wave.

Before we can truly understand the kind of interference pattern that results on the string, we must consider what happens to a wave when it is reflected. You can find out on your own by way of a simple experiment. Tie one end of a rope to a wall and send a pulse (a small portion of a wave) down the rope by shaking the other end briefly. You might end up with a situation like that shown in Figure 7.15a, with the bump traveling on top of the rope. The bump travels along the rope, hits the wall, and is reflected. You would find that upon reflection, the bump turns over on the rope such that it returns on the bottom of the rope, as in Figure 7.15b.

FOCUS ON . . . Resonance

Every object that is free to vibrate has a particular set of frequencies at which it "prefers" to vibrate. For example, a guitar string prefers to vibrate at its fundamental or one of its overtones. When we pluck or push an object in such a way that our pushes match one of the object's preferred frequencies, we say that resonance exists. Under resonance conditions, the amplitude of vibration of the object can become extremely large. Opera singers have demonstrated this vividly by breaking crystal goblets with their powerful voices, as shown in the photograph. In this case, resonance occurs when the wavelength of the sound wave emitted by the singer is the same length as the distance around the rim of the glass. If the singer is able to sustain the note, the amplitude of vibration can increase to the point where the glass shatters.

Our vocal cords disturb very small masses of air, but the vibrating cords cause resonant vibrations within various cavities in a person's neck and head. If you ever have a bad cold and some of these cavities are filled with fluids, the quality of your voice changes because the fluids alter your natural resonant frequencies.

Wine glass shattered by the amplified sound of a human voice. (Courtesy Memorex Corporation)

Resonance can occur in mechanical situations as well. Any structure, such as a building or a bridge, naturally sways in the wind. If such a structure is made to vibrate at its natural frequency, the vibrations increase in amplitude. In 1940, the Tacoma Narrows Bridge was set into a resonant vibration by a wind blowing down the canyon that it spanned. The normal oscillation increased to the point that the bridge collapsed, as shown.

The destruction of the Tacoma Narrows Bridge, caused by resonant vibrations activated by the wind.

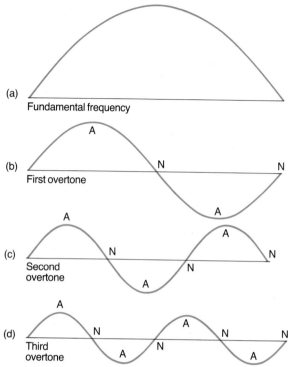

FIGURE 7.16 Fundamental frequency and overtones. The points labeled A have large amplitude vibrations, while at the points labeled N, there is no motion of the string at all.

While you have the rope tied to the wall, try another experiment to see what happens when you shake the rope at different frequencies. In this case, you are causing a wave to move toward the wall; likewise, a wave reflects off the wall and returns toward you. If you are careful and shake the rope at just the right frequency, you will find that you can cause ongoing crests to meet reflected troughs at the same point on the rope, resulting in a cancellation of motion at these points. The various patterns that you could set up on the rope are shown in Figure 7.16. The frequency that produces the pattern shown in Figure 7.16a is called the **fundamental frequency** (or first harmonic). A higher shaking frequency would produce a pattern like that in Figure 7.16b. This is called the **first overtone** (or second harmonic). The pattern in Figure 7.16c corresponds to the second overtone, and so forth. All of these patterns are characterized by the fact that the wave does not appear to move along the string at all. There are large amplitude vibrations at points like A in Figure 7.16b, whereas other points like those labeled N have no vibration amplitude. However, the overall wave pattern does not move along the rope at all. As a result, this kind of wave pattern is referred to as a **standing wave.**

The relationship between these patterns is that if you have to vibrate the string at a frequency of 100 Hz to produce the fundamental frequency, you will have to cause it to vibrate at 200 Hz to produce the first overtone, at 300 Hz to produce the second overtone, and so forth. If f_f is the frequency of the fundamental, the overtone frequencies are given by

$$f_0 = nf_f \qquad (n = 2, 3, 4, 5 \ldots)$$

This says that the frequency of any overtone is some integral multiple of the fundamental frequency.

A standing wave pattern on a string.

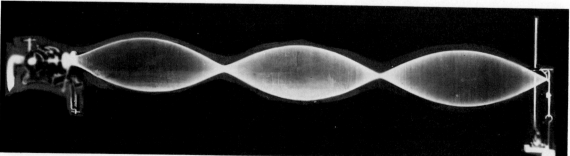

FOCUS ON . . . Beats

Many interference effects occur when two waves of the same frequency attempt to travel through the same region of space in opposite directions. This is how standing waves on a guitar string are formed. But what happens if the two waves moving through the same region of space travel in the same direction and have slightly different frequencies? The answer is that the waves move into constructive interference at some instant of time, and a brief instant later they undergo destructive interference. Consider part (a) of the figure. There we see two waves that we can consider as having been emitted by two tuning forks having slightly different frequencies. Part (b) shows the interference pattern formed by these two waves. At some particular time indicated by t_a, the two waves interfere destructively. This occurs because one of the forks is emitting a compression at the same instant that the other is emitting a rarefaction. At some later time, t_b, however, the vibrations move into step with one another, and constructive interference occurs. This means that at this instant, the two forks are simultaneously emitting compressions and rarefactions. As time continues, the two forks continually move into and out of step because of their differing frequencies. Consequently, a listener hears an alternation in loudness, known as **beats.**

The number of beats heard per second is known as the beat frequency. The beat frequency is found to be equal to the difference in frequency between the two sound sources. One can use beats to tune a stringed instrument, such as a piano, by listening to the beats produced between a string and a tuning fork. The string can be loosened or tightened to change its frequency and bring it into tune with the accurate fork. For example, suppose a particular string on a piano is supposed to emit a frequency of 440 Hz but is not doing so. To bring the string into its desired niche, a tuning fork of frequency 440 Hz is sounded. Let us suppose that two beats per second are heard. What is the frequency of the piano string? The beat frequency of two beats per second means that the string and the tuning fork differ by 2 Hz in frequency. Thus, the string could have a frequency of either 438 Hz or 442 Hz. The string has probably worked loose, which would cause it to have a lower than normal frequency. Thus, its frequency is probably 438 Hz.

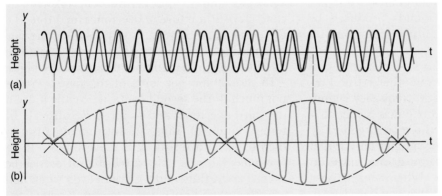

(a) Two sound waves of slightly different frequencies are shown traveling in the same direction. (b) The interference pattern produced by these two waves. Note that the amplitude of the resultant wave oscillates in time, producing alternating loud and soft sounds.

7.10 MUSICAL SOUNDS

When a guitar string is plucked, the waves produced on the string reflect back and forth between the end supports, and thus standing wave patterns are set up. Similarly, standing waves are set up on other stringed instruments such as the violin or cello when they are bowed. Surprisingly, perhaps, standing waves also produce the musical sounds when a wind instrument, such as an

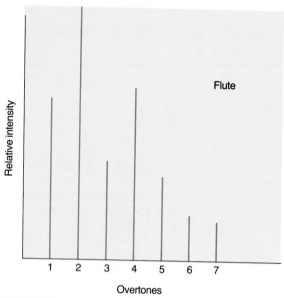

FIGURE 7.17 The characteristic mixture of overtones for a flute.

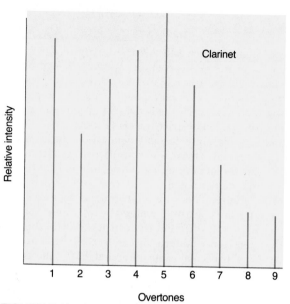

FIGURE 7.18 The characteristic mixture of overtones for a clarinet.

organ, is played. In this case, sound waves bounce back and forth between the ends of a pipe, and these reflecting waves interfere to produce standing waves. However, the sounds produced by musical instruments are quite complex, as we shall see in what follows.

If you are very careful, you can cause a guitar string, or the string of any other musical instrument, to produce a single frequency sound. For example, if you distort a guitar string to a shape that looks like any of those pictured in Figure 7.16, you will find that it vibrates only at that particular frequency. Such a pure tone, however, seldom happens in practice. When a string on a guitar is plucked, the sound that you hear is basically that of the fundamental, but the string does not vibrate with a simple pure frequency. Instead, the string vibrates such that it emits not only the fundamental but several overtones as well. This means that if we try to cause a string to vibrate at a fundamental of 220 Hz, the resultant sound will really be a 220 Hz sound with a little 440 Hz mixed in, a little 660 Hz, and so forth. All of these frequencies add together to produce a complicated sound.

As another example, if you try to play a particular note on a flute, the resultant sound is com-posed of a certain amount of the fundamental, a lot of the first overtone, not much of the second overtone, and so on. Figure 7.17 displays the characteristic mixture of overtones for a flute. On the other hand, if you try to play the same note on a clarinet, the amount of each overtone mixed in with the fundamental is different. As Figure 7.18 shows, you get some of the first overtone, not much of the second overtone, a little more of the third overtone, and so forth. Because all different kinds of musical instruments produce their own characteristic mix of overtones with the fundamental, it is easy to tell the difference between, say, a flute and a clarinet even when both are attempting to play the same fundamental frequency. A note from any instrument contains a mixture of the fundamental and its overtones. In a quality instrument, this mixture is more complex and therefore fuller than in an inferior one. Thus, it is easy to distinguish a cheap piano from a quality grand piano.

The musical scale now in use is an example of survival of the fittest among many competing patterns. The present arrangement seems to be the one that produces the most satisfying tones to the ear. The scale of a typical piano keyboard is shown for reference purposes in Figure 7.19.

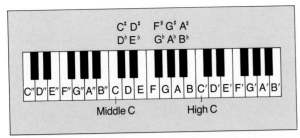

FIGURE 7.19 A typical piano keyboard.

FIGURE 7.20 The observers at A and B hear the same frequency when the sound source is at rest.

The notes on the scale are labeled from A to G with 12 keys, both black and white, separating these two extremes. Let us start with middle C to follow the pattern of frequencies. Middle C has a basic frequency of 261.6 Hz. Each key upward on the scale has a frequency of 1.0595 times the frequency of the one that precedes it. Thus, the next key is the black key, labeled either C sharp (C♯) or D flat (D♭), and the frequency is given by (261.6 Hz)(1.0595) = 277.1 Hz. The next key is D, and its frequency is 1.0595 times the frequency of the black key, (277.1 Hz)(1.0595) = 293.6 Hz. If you continue this process on up the scale, you find that at the end of the basic 12-key segment, you reach high C, which has a frequency of exactly twice that of middle C, 2 × 261.6 Hz = 523.2 Hz. The basic 12 keys are called an octave, and as the numbers indicate, the same notes in two adjacent octaves have frequencies such that the higher frequency is twice that of the lower.

EXAMPLE 7.8   Give me a C

(a) Use the fact that each key on a piano scale is 1.0595 times the frequency of the one that precedes it and the fact that the frequency of middle C is 261.6 Hz to find the frequency of A.

Solution We will not go through all the arithmetic for you, but start with middle C with its given frequency of 261.6 Hz and work your way up the scale by multiplying by 1.0595 for each key until you reach A on the scale. You will find that A has a frequency of 440.1 Hz.

(b) What is the frequency of C below middle C?

Solution This is one octave below middle C, and

since the corresponding frequencies in consecutive octaves have a ratio of 2 to 1, the frequency of low C is 261.6/2 = 130.8 Hz.

7.11
THE DOPPLER EFFECT

Have you ever stood beside a train track and noticed the change in frequency of a train whistle as the train approached and then passed you? The siren of an ambulance approaching and then receding from you is another situation in which you may have encountered this phenomenon. This change in frequency was first explained by Christian Doppler (1803–1853) and bears his name: the **Doppler effect.** He tested his mathematical explanation by placing trumpeters on a flatcar and having it pulled repeatedly past listeners chosen because of their ability to accurately estimate frequencies.

To understand the cause of this frequency change, consider a sound source, such as a tuning fork, emitting a constant-frequency sound. When the tuning fork is at rest, as in Figure 7.20, waves spread out from it uniformly in all directions. We picture the crests of the sound waves by the dark blue lines in the figure. Notice that the distance between consecutive crests is the same on each side of the fork, and as a result, observers A and B hear the same frequency. However, if the fork is set into motion toward the right, as in Figure 7.21, the situation changes dramatically because, as the figure shows, the crests are spread further apart behind the fork and are closer together in front of it. To see that this is reasonable, consider crests 1 and 2 behind the fork. Between the times when 1 and 2 are emitted by the fork, the fork

FOCUS ON . . . Sonic Booms

To visualize a sonic boom, think of something moving rapidly through water. If the object is traveling faster than the speed of the waves it creates, it therefore leaves its waves behind (see photo of the ducks). Moreover, the wave energy is being continuously reinforced by the forward movement of the object. The result is a high-energy wave, called a wake, that trails the object in the shape of a V and that slaps hard against other vessels or against the shoreline. The sonic boom is a high-energy air wave of the same type. The tip of the wake moves forward with the airplane, while the sound itself moves out from the wake at the speed of sound in air. The faster the airplane, the more slender is the wake.

To understand the geometry of the wake, consider the next Figure. A stationary object (A) remains in the center of the circular waves it generates. The waves from a moving object crowd each other in the direction of the object's motion (B). The object is, in effect, chasing its own waves. Recall that sound travels in air at sea level at a speed of about 345 m/s. When the speed of the object equals the speed of the wave, the object will not see any waves before it; it will just be keeping up with them (C). Such high speeds are usually expressed in Mach numbers after Ernst Mach (1838–1916), a physicist who made important discoveries in sound. The Mach number is defined as

$$\text{Mach number} = \frac{\text{speed of object}}{\text{speed of sound}}$$

If an object is traveling at the speed of sound, then the numerator and the denominator of the equation are the same, and the Mach number equals 1. Mach 2 is twice the speed of sound, Mach 3 is three times the speed of sound, and so forth.

Figures (D) and (E) show the wave pattern at supersonic speeds. Note that the object is always ahead of its waves. A passenger in a supersonic transport would therefore not hear the sound of its motion. Instead, the waves crowd each other, and the effect is a significant elevation of pressure at the advancing boundary of the overlapping wave fronts.

Of course, an airplane travels within its medium and not, like a boat, on the surface of its medium. Therefore, the outlines shown in (D) and (E) are two-dimensional projections of what are really coni-

Notice the vee-shaped bow wave behind the ducks. This is similar to the shock wave behind a supersonic airplane, except the shock wave is three-dimensional, in the shape of a cone.

cal shapes. Furthermore, an airplane is more than a point in space; therefore, a whole series of such cones is generated. It is sufficient to consider only the nose and the tail of the plane and to represent the entire space between the forward and rear cones as the volume of the disturbance, as shown in the next figure. As the conical shock wave strikes the ground, people within the volume of the shock wave hear the sonic boom.

To be struck unexpectedly by a sonic boom can be quite unnerving. It sounds like a loud, close thunderclap, which can seem quite eerie when it comes from a cloudless sky. Depending on the power it generates, the sonic boom can rattle windows or even shatter them.

It is important to avoid the misconception that the sonic boom occurs only when the aircraft "breaks the sound barrier," that is, passes from subsonic to supersonic speeds. On the contrary, the sonic boom is continuous and, like the wake of a speedboat, trails the aircraft during the time that its speed is supersonic. Furthermore, the power of the sonic boom increases as the supersonic speed of the aircraft increases.

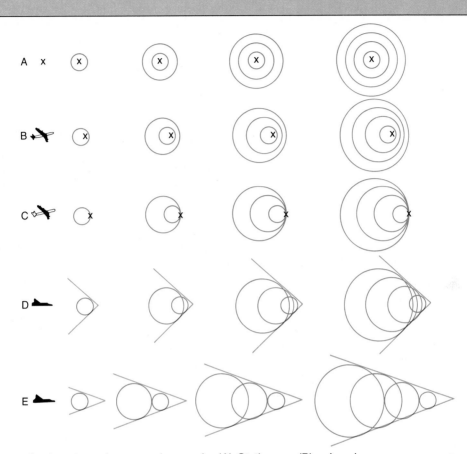

Wave patterns of subsonic and supersonic speeds: (A) Stationary; (B) subsonic, (C) sonic, Mach 1; (D) supersonic, Mach 1.5; (E) supersonic, Mach 3. From left to right, the waves are shown at equal time intervals as they expand.

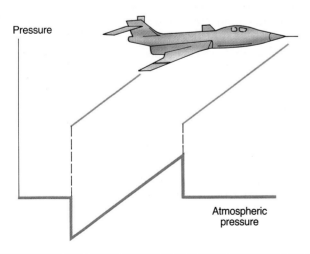

Sonic boom. Shock waves originate from both the front and rear of a supersonic plane.

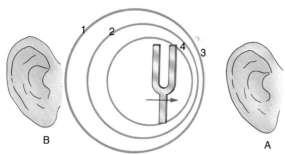

FIGURE 7.21 When the sound source is in motion, the observer in front of the source at A hears a higher frequency than he did when the source was at rest, and the observer behind the source at B hears a lower frequency.

itself has moved toward the right. As a result of this movement, 2 is farther behind 1 when emitted than it would be if the source were still. On the other hand, the distance between consecutive crests, such as 3 and 4, in front of the moving fork is decreased because in the interval between emission of the two, the fork has moved to the right. This means that the motion of the fork helps 4 to keep up with 3. A listener to the right finds more crests reaching him per second than when the fork was at rest, and as a result, his ear correctly interprets the sound as having a higher frequency. Observer B behind the fork (to the left) detects fewer crests per second reaching him and interprets this as a decrease in sound frequency.

Doppler correctly predicted that a similar effect should be observed for light waves, but he was unable to demonstrate it because of the difficulty in his time of working experimentally with the high frequencies of light waves. The present-day science of astronomy uses Doppler effects to determine many important facts about distant astronomical objects. For example, if we look at the light coming to us from a distant galaxy, we might note that it contains a particular frequency component of, say, 5.16×10^{14} Hz. If we then compare this to light sources here on Earth, we might find that a particular substance emits a light having a frequency component of 5.17×10^{14} Hz. Thus, the light from the galaxy also came from the same kind of source, but the frequency is slightly lower. From the Doppler effect, we know that the lowering of frequency means that the source is moving away from the Earth. In the visible portion of the spectrum, red light has the lowest frequency; thus, because the light from the galaxy has been shifted toward the lower, or red, end of the visible spectrum, we refer to the light as having been red-shifted. The fact that the light from distant galaxies is red-shifted is the rule rather than the exception. On this observation rests the now-accepted theory that the Universe is expanding.

EXAMPLE 7.9   The Doppler effect and the Sun

When one observes the Sun, it is found that the light from one side of the Sun is red-shifted and the light from the other side of the sun is blue-shifted. What does this piece of information tell us about the Sun?

Solution   The side of the Sun that emits the red-shifted light is moving away from us, while the side that emits the blue-shifted light is moving toward us. The only way that this could happen is if the Sun were in rotation, with one side rotating away and the other side rotating toward the Earth.

SUMMARY

A wave in which the vibrations are at right angles to the direction of travel of the wave is a **transverse wave.** If the vibrations are in the same direction as the direction of travel of the wave, the wave is a **longitudinal wave.**

The **wavelength** of a wave is the distance between successive **crests** or **troughs.** The speed of a wave is the rate at which the disturbance moves, and the **frequency** is the number of repetitions, or cycles, per second. The **amplitude** is the intensity of the disturbance.

A sound wave is a **longitudinal** wave and is manifested as a succession of compressions and rarefactions of a medium, such as air. The speed of a sound wave is about 331 m/s at 0°C, and audible frequencies for a normal human ear range from 20 to 20,000 Hz.

The **decibel** scale used to measure the energy content, or loudness, of a sound wave (a) starts at zero, which represents the softest audible sound, and (b) represents each tenfold increase in sound intensity as an additional 10 dB.

Waves exhibit **interference.** When they come together such that crest meets crest, **constructive interference** is produced. When crest meets trough, **destructive interference** occurs.

Standing waves are produced when two waves of the same frequency attempt to travel in opposite directions through the same region of space. Musical tones contain a fundamental frequency and some combination of overtones.

When an observer moves toward a source of sound, or if the source moves toward the observer, the frequency of the sound increases. If the relative motion is such as to separate the source and listener, the frequency of the sound decreases. This is the **Doppler effect.**

EQUATIONS TO KNOW

$v = f\lambda$ (speed = frequency times wavelength for all types of waves)

$v = v_0 + (0.61)T$ (the speed of sound as a function of temperature)

KEY WORDS

Longitudinal wave
Transverse wave
Sine curve
Crest
Trough
Wavelength
Speed

Frequency
Amplitude
Compression
Rarefaction
Ultrasonic waves
Decibel
Interference

Fundamental frequency
Standing waves
Overtones
Resonance
Doppler effect

QUESTIONS

DESCRIBING A WAVE

Unless otherwise specified, use 340 m/s as the speed of sound in air.

1. The speed of light is 3.00×10^8 m/s. Calculate (a) the wavelength of a gamma ray with a frequency of 10^{22} Hz; (b) the frequency of a radio wave with a wavelength of 30 m (gamma rays and radio waves travel at the speed of light); (c) the wavelength of a sound wave with a frequency of 1000 Hz.

2. Can you change the wavelength of a wave without simultaneously changing its frequency?

3. While observing waves at a beach, you note that the distance from the crest of a wave to the next consecutive trough is 3 m. What is the wavelength of this wave?

4. You hold one end of a spring in your hand with the other end attached to a wall. How would you move your hand to set up (a) a longitudinal wave in the spring? (b) A transverse wave?

5. For the sketch shown in the Figure, what is (a) the amplitude of the wave? (b) The wavelength of the wave? (c) If its speed is 30 m/s, what is its frequency?

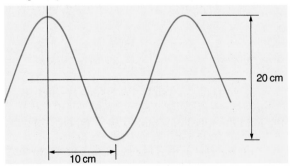

10 cm

20 cm

6. A metronome is set so that it makes 20 complete vibrations in 25 s. What is the frequency of the metronome?

SOUND WAVES

7. When someone says that a sound wave can be represented by a sine curve, which of the following statements is implied? (a) The molecules move along in a wavy motion like water waves. (b) A graph of air density versus distance will be a sine curve. (c) A graph of the degree of increase or decrease of air pressure above or below atmospheric pressure versus distance will be a sine curve. (d) A graph of pressure versus time at any one point will be a sine curve. (e) The molecules move back and forth like the inking on a piece of paper used to draw the sine curve. (f) Since the air does not ripple, there is no real sine function, but rather we are using a figure of speech to describe a sound wave.

8. The distance between the compression of a sound wave and the next consecutive rarefaction is found to be 1 meter. Can this wave be heard as an audible sound wave?

9. If sound travels at 340 m/s, what is the range of wavelengths to which a human ear can respond if the person can hear all sounds with frequencies between 20 Hz and 20,000 Hz?

10. What happens to the frequency of a sound wave if the wavelength is tripled?

11. What happens to the speed of a sound wave if the wavelength is tripled?

12. What happens to the speed of a sound wave if the amplitude is tripled?

13. Devise an experiment to prove that the speed of sound does not depend on the frequency of the sound.

14. Devise an experiment that would enable you to measure the speed of sound in (a) water, (b) aluminum.

15. A tugboat lost in the fog in a harbor sounds its whistle, and the captain hears an echo returned from the surface of an oil tanker 6 s later. How far away is the tanker? Don't forget the sound has to travel to the tanker and return.

16. Sketch a sine curve to represent two waves described as follows: They have equal amplitudes but one has twice the frequency of the other.

17. The sound waves used in normal conversation are usually between 500 Hz and 5000 Hz. What are the wavelengths of these sounds?

18. Find the speed of sound in air at (a) 0°C, (b) 22°C, and (c) 50°C.

19. Find the speed of sound in air at (a) 0°F, (b) 22°F, and (c) 100°F.

20. Some animals can hear sounds as high as 100,000 Hz. What is the wavelength of this wave?

LOUDNESS OF SOUND AND THE DECIBEL SCALE

21. A person hears a cry in the woods that is 1000 times the intensity of the faintest audible sound. What is the sound level in decibels?

22. (a) How much more sound energy is carried by a 50 decibel sound than a 0 decibel sound? (b) How much more sound energy does the 50 decibel sound carry than a 30 decibel sound?

23. Can there be a negative decibel level?

24. A dog can hear sounds inaudible to a human. Suppose a dog could just hear a sound whose intensity is 100 times less than the faintest audible sound for humans. What is the minimum decibel level to which the dog is sensitive?

25. Estimate the decibel level of the following sounds: (a) a train passing by while you are near the tracks, (b) a pin dropping on a wooden table, (c) the background noise in a church.

26. The sound intensity of a motorcycle at a distance of 8 m is 90 dB. How many times greater is the energy level of this sound than the faintest audible sound?

27. (a) The sound intensity of a garbage disposal unit is 80 dB. How many times greater is the energy content of this sound than the faintest audible sound? (b) How many times greater is the energy content in a sound of a 120 dB thunderclap than the sound of a garbage disposal unit?

INTERFERENCE

28. Imagine that two equal pure tones are generated by speakers mounted on a wall and situated 5 m apart. What will you observe if you stand in the center of the room and move a sensitive decibel meter slowly back and forth in front of them? Explain.

29. Two speakers emitting identical sound signals are placed side by side. You start to move one of them back and a listener in front of the speakers listens for an interference effect to be produced. (a) At what minimum distance would you move the speaker to hear destructive interference? (b) At what minimum distance would you have moved the speaker to hear constructive interference?

30. Can a transverse wave and a longitudinal wave interfere?

31. Two water waves are interfering to produce constructive interference with a wave of amplitude 2.2 m. One of the individual waves has an amplitude of 1.6 m. What is the amplitude of the other?

32. If the two waves of problem 31 were moved into destructive interference, what would be the amplitude of the wave?

STANDING WAVES AND MUSICAL SOUNDS

33. Sketch the wave pattern for a string vibrating in its fourth overtone.
34. The fundamental frequency of a string is 125 Hz. What is the frequency of the fifth overtone?
35. A string driven by a tuning fork is found to vibrate in its third overtone when the frequency of the fork is 400 Hz. What is the frequency of the fundamental?
36. If the frequency of middle C is 262.6 Hz, find the frequency associated with the key labeled G.
37. What is the wavelength emitted when middle C is sounded?
38. What is the frequency of the first overtone of middle C?
39. If a standing wave pattern is set up in a small tank of water, will a cork placed in the water move at all? Explain.

THE DOPPLER EFFECT

40. Certain types of stars pulsate in size. (a) If you observe the light from such a star while it is growing, would you see the light shifted toward higher or lower frequencies? (b) Repeat part (a) for the case in which the light is observed while the star is diminishing in size.
41. You are standing at the end of a city block observing a fire engine at the other end of the block. The engine has its siren going and it is driving in a circular path. Describe what you would hear when the engine comes toward you, moves away from you, and drives parallel to your line of sight.
42. When a fire truck is at rest with respect to you, you hear a sound from its siren having a frequency of 500 Hz. Late at night you hear the same engine emitting a frequency of 502 Hz. Is the engine coming toward you or away from you?
43. A wind is blowing toward you in the same direction as a fire engine moves with its siren sounding. In what way would the wind affect the sound you hear?
44. In some movies a spaceship accelerates to near the speed of light and suddenly the stars disappear. How could the Doppler effect account for this?
45. How could the Doppler effect be used to prove that our Sun is rotating?
46. A sound source and an observer are both traveling in the same direction at the same speed. Will a Doppler shift in the sound be heard?
47. A binary star system consists of two stars in revolution about one another. Describe how the Doppler effect will change the light reaching us from these stars.

ANSWERS TO SELECTED NUMERICAL QUESTIONS

1. (a) 3×10^{-14} m, (b) 10^7 Hz, (c) 0.34 m
3. 6 m
6. 0.8 Hz
8. yes, f = 170 Hz
9. 0.017 m to 17 m
15. 1020 m
17. 0.68 m and 0.068 m
18. (a) 331 m/s, (b) 344 m/s, (c) 362 m/s
19. (a) 320 m/s, (b) 328 m/s, (c) 369 m/s
20. 3.4×10^{-3} m
21. 30 dB
22. (a) 10^5, (b) 100
24. -20 dB
26. 10^9
27. (a) 10^8, (b) 10^4
31. 0.6 m
32. 1 m

CHAPTER

8

Electricity

We are sure that every person reading this book has had elderly people, such as grandparents, tell them about the way things were in their childhood. They can remember when there were no radios, TVs, stereos, and a whole host of other electronic devices. Things have changed a lot in the lives of these people, but there have also been many changes in electrical equipment that have occurred in your own memory. Many of you can recall when an electronic appliance was a large, bulky, tube-type item. Likewise, such accepted modern-day devices as the hand-held calculator, the computer, the VCR, the compact disc player, and so forth were either not available or available only at a prohibitive cost. Thus, you too will have stories to tell your grandchildren.

Much of the impetus for these devices arose with the advent of microminiaturization of electronic components. This led to the production of circuits filled with transistors, diodes, resistors, and capacitors on a chip so small that its individual parts can be seen only with a microscope. Although these devices are quite fantastic, they all obey the basic laws of physics. The trail that has led to these modern conveniences started with the Greeks about 500 B.C. The interest of these early Greek thinkers was in static electricity.

Their results were modest, and it is safe to say that they could never in their wildest dreams have envisioned where their investigations would lead.

8.1
STATIC ELECTRICITY

The first observations of static electricity occurred when someone noticed that a waxlike substance called amber would attract small objects after it had been rubbed with wool. We now know that this phenomenon is not restricted to amber and wool but that a similar effect can be observed (to some extent) when almost any two nonmetallic substances are rubbed together. To describe this change in the physical properties of these objects, scientists said that they had been given an **electric charge.** For example, when hard rubber is rubbed with wool, both the rubber and wool become charged. Likewise, rubbing a glass rod with silk charges both objects. It is very simple to show that there are two different kinds of charge, which were given the names **positive** and **negative** by Benjamin Franklin (1706–1790). Figure 8.1 shows one experimental technique that can be used to demonstrate the two different kinds of charge and to show some additional properties of

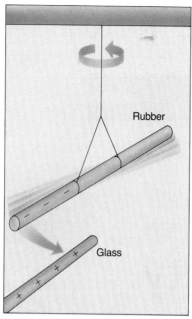

FIGURE 8.1 A negatively charged rubber rod is attracted by a positively charged glass rod.

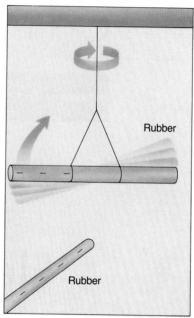

FIGURE 8.2 Two rods having the same type of charge repel one another.

charged objects. In the figure, a rubber rod that has been charged is suspended by a string so that it can swing freely, and a charged glass rod is brought nearby. It is found that the rubber rod swings toward the glass rod, indicating that the charges on the two rods are attracting one another. It has become customary to say that the glass rod has a positive charge and the rubber rod has a negative charge. Based on our experiment, we can state that *unlike charges attract one another.* A similar experiment, illustrated in Figure 8.2, shows that a negatively charged rubber rod is repelled by another negatively charged rubber rod. This is also true for two positively charged glass rods. Thus, we can state that *like charges repel one another.*

In order to understand what is happening when an object becomes charged, let us digress briefly to examine a model of the atom, often called the planetary model because of its similarity to our Solar System. (It should be noted here that this planetary model of the atom has now been replaced by a more accurate quantum mechanical model, which will be discussed in Chapters 12 and 13.) In the Solar System, the planets

orbit the Sun, bound into their paths by the gravitational attraction exerted on them by the Sun. Likewise, in the planetary model of the atom, small, light, negatively charged particles called **electrons** circle the relatively massive central core of the atom, called its **nucleus.** The nucleus has two fundamentally different types of particles in it. One of these is called a **neutron** and is so named because it carries no electric charge. The second type is called a **proton,** and the *proton is nature's basic carrier of positive charge. The planet-like electrons are nature's basic carriers of negative charge.* Under normal circumstances, there is an electron circling a nucleus for every proton in the nucleus. The magnitude of the electrical charge on an electron is identical to the magnitude of the charge on the proton. Thus, an atom is electrically neutral. Likewise, ordinary material made of neutral atoms also has no net charge. Thus, even though a block of copper may contain literally trillions of charges, it exhibits no electrostatic effects because the net charge is zero.

When atoms are assembled such that they form a solid object such as a rubber rod, it is found that the atoms are basically fixed in a given

FOCUS ON . . . How Does a Photocopying Machine Work?

In past generations, if you wanted copies of a document, you could either write it over and over with, perhaps, a quill pen, or, after vast technological developments occurred, you could use carbon paper to do the task. However, the job is now done quickly and effortlessly by a Xerox machine. The basic process for producing Xerox copies was discovered by Chester Carlson in 1940, and in 1947 the Xerox corporation utilized his methods to produce the now familiar office machines.

The Figure illustrates the various steps taking place in a machine when you press the "copy" button. In part (a) a drum deposits a coating of selenium or a selenium compound onto the surface of the paper, and via friction the coating is simultaneously given a positive electric charge. A beam of light is then directed toward the paper to be copied and a lens focuses the image of the original on the selenium-coated paper as seen in part (b). Selenium is a substance that is photoconductive. This means that when light strikes it, the lit portion becomes a conductor of electric charge. Thus, those positive charges on the selenium surface are neutralized when struck by light, and only the dark portions, the writing, that light could not penetrate retain a charge. This means that a copy of the original is formed on the paper by an arrangement of positive charges that duplicate the print on the original. Next, a powder called a toner, which has a negative charge, is spread across the paper as seen in part (c). The toner sticks to the paper at the places that still have a positive charge. In this way, the writing becomes visible. Finally, heat is applied, which causes the toner to melt and stick to the paper to form the final, permanent copy.

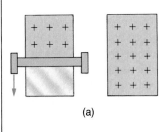
(a)

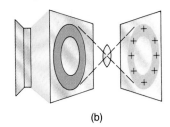
(b)

(c)

(a) A positively charged coating of selenium is deposited on a piece of paper. (b) Light neutralizes some of the positive charge. (c) A toner adheres to the positively charged image.

location. Individual atoms may vibrate back and forth about some equilibrium position, but they do not migrate freely throughout the solid. On the other hand, some of the electrons of the material can get loose from the nucleus and become relatively free to wander throughout the body of the substance. Thus, when two objects are charged by friction, the basic mechanism that occurs is that electrons have been transferred from one of the objects to the other. *It is always the electrons that move around in a solid and never the protons.* For example, when a rubber rod is rubbed vigorously with wool, electrons move from the wool to the rubber. Since the rubber now has an excess of electrons, it has a net negative charge, and since the wool has lost electrons, it now has a net positive charge. All materials have their own characteristic ability to hold onto their electrons; for the case above, it is the rubber rod that has the greater ability to hold onto its electrons than does the wool. As an exercise, explain to yourself what happens when a glass rod is rubbed with silk.

Note that in the process of charging by friction, no charge is ever created or destroyed. Thus, we can say that *electric charge is conserved.* An object becomes charged negatively by gaining electrons from some other object; electrons are not created out of thin air. Also, an object becomes charged positively by losing electrons; positive charges are not being created nor are negative charges being destroyed.

8.2 MEASURING CHARGE

When we discussed the measurement of length and mass, we decided that some fundamental unit of comparison was needed for the measurement process. Thus, we adopted the basic unit of length to be the meter and the basic unit of mass

FOCUS ON . . . Lightning and Lightning Rods

The common phenomenon of lightning is nature's large-scale display of electrical energy. Lightning occurs when charges jump from one cloud to another or between a cloud and the Earth. The method by which clouds become charged has never been completely explained, but the process is somewhat as follows.

As rain falls, updrafts of air pass the falling raindrops, and these air currents seem to cause large drops to split into two drops, one large and one smaller. The larger one carries a net positive charge after the split, while the smaller has a net negative charge. The larger drop continues to fall under its own weight, but the smaller one is swept upward by the rising air currents. The result is a separation of charge between layers of the clouds. The overall effect is very similar to that encountered in charging a rod by friction.

When a positively charged cloud drifts near a negatively charged one, the charges are attracted, and they may leap together in a jagged bolt of lightning. If a cloud charged, say, positively moves over a house, electrons from the Earth are attracted by the cloud and spread out all over the surface of the house, giving it a net negative charge. A lightning discharge occurs when bursts of charge move downward from the cloud toward the Earth. These short strokes, called step-leaders, overcome the resistance of the air to the flow of charges. A powerful return from the ground along the path of the step-leaders is the visible lightning stroke. Protection against destructive discharge can be effected by lightning rods, which are pieces of metal with sharp points on the ends. These are attached to the house in such a way that they extend higher than any other portion of the building. Wires leading from the rods are connected to metal posts driven several feet into the earth, so that if a discharge does occur, it is safely carried to ground away from the structure. Charges tend to accumulate at sharp points and to be sprayed off of these points. Thus, as charges build up on the house, the sharp pointed ends of the rods spew the charges off into the air before a level of charge sufficient to produce a lightning bolt accumulates.

FIGURE 8.3

Particle	Charge	Mass
electron	-1.6×10^{-19} C	9.11×10^{-31} kg
proton	1.6×10^{-19} C	1.67×10^{-27} kg
neutron	0	1.68×10^{-27} kg

unit is a little more difficult for beginning students to become comfortable with than are units such as meters and kilograms, because a coulomb of charge is not something that you talk about in everyday life. To give you a slight feel for the unit, it is found that an electron has a charge of -1.6×10^{-19} C (C = coulombs) and a proton has a charge of 1.6×10^{-19} C. Some of the important properties of the components of an atom are shown in Figure 8.3. (Note that the proton and neutron have about the same mass, and that this mass is about 1840 times greater than the mass of the electron.)

It should be obvious to you after a moment's reflection that you cannot give an object any charge that you like. This is considerably different from measuring the length of an object. For example, you may measure the length of one object to be 1.39869 m long, while the length of another can be 1.39868 m long. A length can be subdivided in an infinite number of ways. This is not the case for electric charge. The smallest negative charge that an object can have occurs when that object gains a single electron. In that case, its charge would be -1.6×10^{-19} C. The next smallest negative charge would arise if the object gains two electrons, leaving it with a charge of -3.2×10^{-19} C. Never will you find an object that has a charge of, say, -2.0×10^{-19} C.

As a point of interest, in modern-day theoretical study of matter, it has been proposed that there should exist particles that are even more fundamental than the neutron and proton. These particles are called **quarks,** and they are predicted to have charges that are either 1/3 or 2/3 the magnitude of the charge of the electron or proton. Theoretically, neutrons and protons are actually bound units of three quarks of different types. Many experimental investigations are now underway in an attempt to discover these elusive particles. The results of these investigations have thus far proven fruitless, but the

to be the kilogram. It is also necessary to be able to measure charge, and the basic unit by which charge is measured is called the **coulomb.** This

search continues, for there are good theoretical grounds for their existence.

8.3 COULOMB'S LAW

We have seen that charged objects exert forces of either attraction or repulsion on other charged objects. In 1789 Charles Coulomb investigated these forces in an attempt to find an equation that would predict their strength. He found that the force is given by an equation that is quite similar to an equation you have already encountered, the force of gravitational attraction. This force law is referred to as **Coulomb's law,** and is given in equation form by

$$F = k \frac{q_1 q_2}{r^2} \qquad (8.1)$$

where q_1 is the charge on one of the objects and q_2 is the charge on the other. The quantity r is the distance of separation of the charges, assuming that they are small enough to be considered to be located at a point, or if the charges are spread out over the surface of a sphere, it is the distance between the centers of the spheres. The constant k is found experimentally to have the value 8.99×10^9 N m^2/kg^2, but we shall round this off to 9.0×10^9 N m^2/kg^2.

As we have noted, there are similarities between Newton's law of universal gravitation and the Coulomb force law, Eq. 8.1. However, there are also differences. For example, Newton's law predicts that the force between objects having mass is always one of attraction, whereas the force between charged objects, given by Coulomb's law, can be either attractive or repulsive. Also, as we shall see in the example below, electrical forces are considerably stronger than gravitational forces.

EXAMPLE 8.1 Comparing gravitational and electrical forces

In a simple hydrogen atom, a single electron circles a single proton such that their distance of separation is 0.53×10^{-10} m. **(a)** Find the magni-

tude of the gravitational force of attraction between the two, and **(b)** find the magnitude of the electrical force of attraction.

Solution **(a)** Figure 8.3 gives us the masses of the electron and proton. The gravitational force of attraction is given by Newton's law of universal gravitation as

$$
\begin{aligned}
F_g &= G \frac{m_1 m_2}{r^2} \\
&= (6.67 \times 10^{-11} \text{ N m}^2/\text{kg}^2) \\
&\quad \times \frac{(1.67 \times 10^{-27} \text{ kg})(9.11 \times 10^{-31} \text{ kg})}{(0.53 \times 10^{-10} \text{ m})^2} \\
&= 3.61 \times 10^{-47} \text{ N}
\end{aligned}
$$

(b) We use the charge of the electron and proton to find the force of attraction between the two from Coulomb's law as

$$
\begin{aligned}
F_e &= k \frac{q_1 q_2}{r^2} \\
&= (9.0 \times 10^9 \text{ N m}^2/\text{kg}^2) \frac{(1.6 \times 10^{-19} \text{ kg})^2}{(0.53 \times 10^{-10} \text{ m})^2} \\
&= 8.20 \times 10^{-8} \text{ N}
\end{aligned}
$$

Thus, we can easily see that electrical forces between electrons and protons in an atom are considerably stronger than gravitational forces between the two.

8.4 THE CONCEPT OF FIELDS

In our study of mechanics, we ran across a type of force called an action-at-a-distance force. This was the force of gravitational attraction between two objects that have mass. These were in distinction to the more common contact forces. To refresh your memory on these forces, contact forces arise when there is actual physical contact between the objects. For example, when a fighter punches his opponent, there is actual physical contact between his fist and the chin of the challenger. On the other hand, the Earth exerts a force of gravitational attraction on the Moon even though there are some 240,000 miles of empty space between the two. There is no physical contact between the two at all, and forces of this kind are called action-at-a-distance forces. We have now encountered our second action-at-

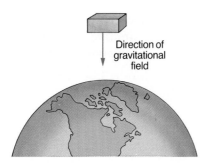

FIGURE 8.4   A brick in the gravitational field of the Earth.

FIGURE 8.5   A representation of the gravitational field of the Earth.

a-distance force, the electrical force between two charged objects. These forces are somewhat unusual and rare; in fact, we will meet only one more example of this class of forces in our study—magnetic forces. There are many ways to discuss this classification of forces, but a method developed by Michael Faraday (1791–1867) is of such practical importance that we shall pause to examine his procedure.

Gravitational fields

Consider a brick held above the surface of the Earth as in Figure 8.4. What happens to the brick when it is released? The answer, of course, is that it falls. But why does it fall? A complete answer to the question is very difficult and is still a topic of theoretical investigation. However, a partial answer might be that it falls because the Earth exerts a gravitational force on it, which pulls it downward. An alternative to this answer is provided by the field approach. In this method, it is said that the Earth somehow alters the space around it such that a field is set up in this space; we call this field a gravitational field. When any object that has mass, such as our brick, moves into this gravitational field, it finds a force exerted on it. In this case, the force is one of attraction, which pulls the brick toward the Earth. This alternative approach does not produce much new insight into gravitational forces; therefore, we did not introduce the concept of gravitational fields into our earlier discussions of gravity. However, because of the importance of the field concept in electricity and magnetism, let us continue with the idea of gravitational fields. A field is defined to be a vector quantity. Thus, it must have magnitude and direction.

The direction of a gravitational field is defined to be in the direction of the gravitational force on an object placed in the field.

Thus, in Figure 8.4, we see that our brick has a force on it directed toward the center of the Earth when the brick is in the gravitational field of the Earth. Thus, at the point where the brick is located, the gravitational field is in the direction indicated in the figure. However, regardless of where we go above the surface of the Earth, the force of gravitational attraction is always toward the center of the Earth, so we could use lines to represent this gravitational field as shown in Figure 8.5.

The strength of the field is defined as the magnitude of the gravitational force of attraction on the brick divided by the mass of the brick. In equation form,

$$\text{Gravitational Field} = \frac{F_g}{m_b} \qquad (8.2)$$

where F_g is the force of attraction exerted on the brick by the Earth and m_b is the mass of brick.

The electric field

Let us now parallel our discussion of gravitational fields as presented above in an attempt to find an alternative way of looking at electrical forces. When a small positively charged object is near an object having a considerably larger positive charge, what happens? Figure 8.6 shows us that the small object is repelled from the larger object. But why? One answer is that the larger object exerts a force on it given by Coulomb's law; thus, the object is pushed away by this force. However,

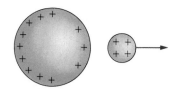

FIGURE 8.6 The small charge is repelled by the larger charge.

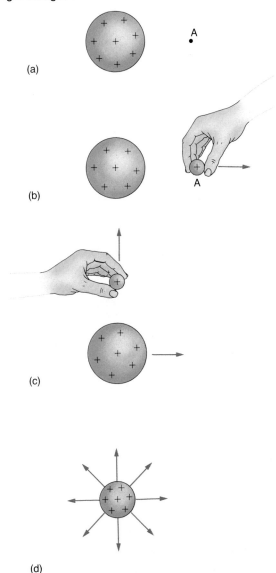

(a)

(b)

(c)

(d)

FIGURE 8.7 (a) To find the electric field at point A, we (b) place a tiny positive charge at that point. The direction in which an electrical force acts on that charge is the direction of the field. (c) The electric field can be found at other points to show that the field (d) radiates away from the center of the large charged object.

our objective here is to examine the field concept. In this alternative viewpoint, the large charged object somehow alters the space around it such that an **electric field** is set up in this space. According to this outlook, when another charged object is in this electric field, it finds an electrical force acting on it and moves accordingly.

As for gravitational fields, we must define a direction for electric fields:

The direction of an electric field at any point is in the direction of the force that would be exerted on a small **positively** charged object if it were placed at that point.

Let us consider this definition for a moment. In Figure 8.7a, a large positively charged object is shown. The question is: What is the direction of the electric field at the point A? To find this direction, in our imagination we place a tiny positive charge at this location and ask in what direction would there be an electric force acting on it? In Figure 8.7b, we have placed this charge at A, and since positively charged objects repel one another, the electrical force on the test charge is toward the right. Thus, we say that the direction of the field at A is to the right, or it is away from the center of the larger object that sets up the field. If we move all around the large object as in Figure 8.7c, determining the direction of the field at each point, we find that we can represent the field as shown in Figure 8.7d. The electric field of a positively charged object consists of lines radiating away from the center of the object.

Can you use the definition of the direction of electric field lines to show that the electric field around a large negatively charged object looks like that shown in Figure 8.8?

By analogy with our discussion of gravitational fields, the magnitude of the strength of an electric field is given by the strength of the electrical force exerted on a small object with a charge q divided by the strength of the charge. Thus,

$$E = \frac{F_e}{q} \tag{8.3}$$

The concept of electric fields is very important in the study of electricity, and one important

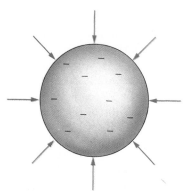

FIGURE 8.8 The electric field around a negatively charged object.

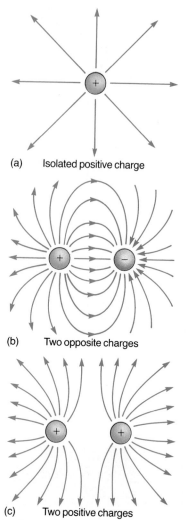

(a)   Isolated positive charge

(b)   Two opposite charges

(c)   Two positive charges

FIGURE 8.9 Lines of force surrounding electrical charges.

value is indicated by the way that sketches of electric fields are drawn. When a small positive charge is placed very close to a positively charged sphere, the small charge will be pushed away with a very great force. This force of repulsion gradually diminishes as the small charge moves farther and farther away. However, if the force F_e is diminishing in strength, Equation 8.3 shows us that the strength of the electric field is also diminishing. Thus, the electric field is strong close to the large sphere and becomes weaker as we move away from it. One advantage of the electric field concept is that we can represent the strength of the field by the way in which we sketch the electric field lines. For example, in Figure 8.9a, we see that the lines representing the field are closely spaced near the large positively charged object and become farther apart as they move away. Thus, the spacing of the lines is used to convey the strength of the field. The more closely spaced the lines, the stronger the field.

Figure 8.9b illustrates the field between a positive charge and a negative charge. What can you say about the strength of the field between the two charges? Is it relatively large or small? (Answer: Since the lines are relatively closely spaced in this region, the field is strong.) Figure 8.9c shows the field lines between two positively charged objects. Describe the field in the region between the two charges. (Answer: Since the lines are not very close together in this region, the field must be relatively weak.) Some other properties

of the field concept are indicated by the example problems that follow.

EXAMPLE 8.2   Which way does the charge move?

(a) You walk into a room carrying a small positive charge. When you release it, you discover that it moves from east to west because of an electrical force acting on it. What is the direction of the electric field in the room?

Solution Recall that the direction of an electric field is the direction in which an electrical force is exerted on a small positive charge. Thus, the direction of the field is from east to west. A conclusion that will be important to us later is that *a positive charge always moves in the direction of an electric field.*

(b) You walk into the room of part a and find that all you have with you is a small negative charge. When you release it, which way will it move?

Solution When you released a positive charge in a, you noticed that it moved from east to west. This must have occurred because of a force of repulsion exerted on the positive charge by the charged object setting up the field. However, a little thought should convince you that a negative charge should behave exactly opposite a positive charge. Thus, the negative charge will move from west to east. The conclusion that we can draw is *a negative charge always moves in a direction opposite to that of an electric field.*

We shall discover later that a battery is capable of setting up electric fields inside wires attached to it. Thus, if there are free positive charges in the wire, they move in the direction of this field, and if there are free negative charges in the wire, they move in a direction opposite to that of the electric field.

EXAMPLE 8.3 Speeding up a proton

A proton is released at rest in a region where an electric field of magnitude 500 N/C exists.

(a) Find the magnitude of the force on the proton.

Solution The force on the proton (charge = 1.6×10^{-19} C) can be found from Eq. 8.3 as

$$F = qE = (1.6 \times 10^{-19} \text{ C})(500 \text{ N/C})$$
$$= 8.0 \times 10^{-17} \text{ N}$$

This is an extremely small force, but remember that it is acting on an extremely light particle. Thus, it can produce large accelerations and large speeds in a very short time. Let us investigate this.

(b) What will be the speed of the proton after it has traveled a distance of 1.00 cm? Assume that the proton starts from rest.

Solution The force is constant. Thus, from Newton's second law, we see that the acceleration is also a constant. The acceleration is

$$a = \frac{F}{m} = \frac{8.0 \times 10^{-17} \text{ N}}{1.67 \times 10^{-27} \text{ m/s}}$$
$$= 4.79 \times 10^{10} \text{ m/s}^2$$

To find the speed, we can use the equations of motion with constant acceleration. First, let us find the time for the proton to travel 1 cm from $d = \frac{1}{2} at^2$. We have

$$0.01 \text{ m} = \frac{1}{2} (4.79 \times 10^{10} \text{ m/s}^2)t^2$$

from which

$$t = 6.46 \times 10^{-7} \text{ s}$$

We can now find the speed from $v = at$. We have

$$v = at = (4.79 \times 10^{10} \text{ m/s}^2)(6.46 \times 10^{-7} \text{ s})$$
$$= 3.09 \times 10^4 \text{ m/s}$$

8.5
THE VAN DE GRAAFF GENERATOR

Many of the concepts dealing with electric charges, electric forces, and electric fields can be demonstrated with a device designed and built in 1931 by Robert J. Van de Graaff. His device, called a **Van de Graaff generator,** serves as an excellent teaching tool when the device is constructed on a small scale, but larger models are still in use in nuclear physics research laboratories where they are used to investigate the secrets of the atomic nucleus. A classroom-size demonstration unit like that shown in Figure 8.10 may be about 1 meter tall, but those in use in nuclear research may be as tall as a two-story building. Regardless of the size, the basic principles of operation remain the same.

Figure 8.11 shows the schematic details of such a generator. A motor in the base drives a pulley, which causes a rubber belt to turn as indi-

FIGURE 8.10 A small-scale Van de Graaff generator that can be used for classroom demonstrations.

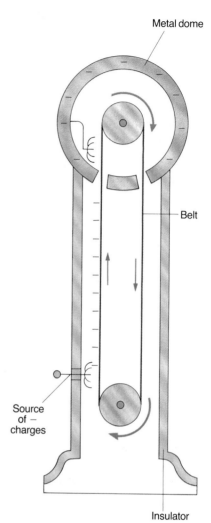

FIGURE 8.11 Cut-away view of a Van de Graaff generator.

cated by the arrows in the figure. In a small-scale device, the rubber belt rubs against a material that produces a negative charge on the belt by a process that is much like that of charging by friction, discussed earlier in this chapter. This negative charge is carried upward by the belt, and at the top of the device, a comb-like row of points is suspended near the belt. Since the negative charges are repelled by one another, they move as far away from one another as possible, so they run off the belt and onto the metal contacts. From there they spread out all over the surface of the metal dome. In this fashion, it is possible to give the dome quite a large negative charge. (In larger machines, the charge is sprayed onto the rubber belt by devices called ion sources. These are capable of giving the belt either a negative or a positive charge; hence, the dome can end up with a charge of either sign.)

If a person stands on an insulating stool and places his or her hand on the top of the dome while it is being charged, the person becomes a part of the Van de Graaff dome and charge spreads out all over the body, as in Figure 8.12. (The purpose of the insulating stand is to prevent charge from leaking off the person through the feet to the floor.) The fact that the person's body is covered with a large charge can easily be seen if the person has straight, fine hair. The charges spread out all over the hair, and the repulsion between these charges causes the hair to stand out away from the head.

In a more important application, these devices are used in physics laboratories to produce

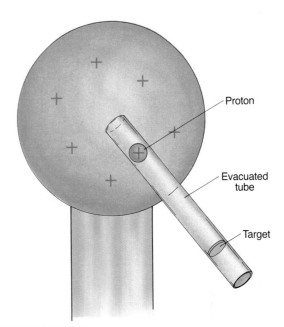

FIGURE 8.12 A person touching the charged dome also becomes charged.

FIGURE 8.13 Protons are repelled by the positively charged dome. When they strike the target, nuclear reactions may be produced.

As an exercise, explain why touching the generator causes this person's hair to stand on end.

nuclear reactions. For our example, let us suppose that the dome has been given a positive charge, as shown in Figure 8.13. A tube from which the air has been removed is attached to the dome, and protons are released inside the tube at a location near the dome. The repulsion between the positively charged dome and the positively charged protons causes the latter to move away from the dome through the evacuated tube. At the end of the tube, the protons strike a target made of the atoms that the investigator wishes to study. The nuclear reactions that occur will be discussed in detail in Chapter 13.

8.6 A BLACK-AND-WHITE TELEVISION RECEIVER

The essential pieces of a black-and-white picture tube are shown in Figure 8.14. (It should be noted here that a common piece of electronic equipment to view electrical signals is the oscilloscope, and the method of operation of the tube in one of these devices is very similar to that of a picture tube in a television receiver.) In the neck of the picture tube, a metal cylinder A is heated

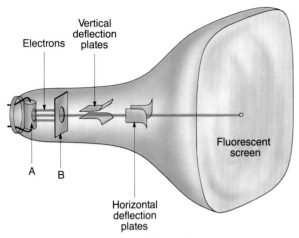

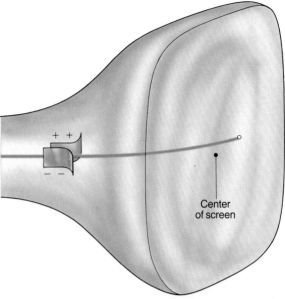

FIGURE 8.14 A TV picture tube. Electrons from A are accelerated toward plate B and then move toward the fluorescent screen. The charged plates deflect the beam from point to point on the screen.

FIGURE 8.15 Gradually increasing the charge on the plates as shown would cause the electron beam to move to the right side of the screen.

The "picture tube" in a test instrument called an oscilloscope is very much like that in a television receiver.

by an electric current until the temperature of the cylinder becomes hot enough to boil off electrons. Located a short distance down the tube is a plate B that has been given a strong positive charge by electrical circuits not shown in the figure. The negatively charged electrons are attracted toward this plate and gain a very high speed just before they reach it. Most of the electrons strike this plate, but a few pass through a hole in its center and continue on in a straight-line path until they strike the front surface of the picture tube. The front of the tube is painted with small phosphorescent dots that emit white light

when struck by electrons. Thus, an observer watching the screen would see a bright spot of light when electrons strike it.

Watching a single spot of light like this does not make for stimulating programming, so to enhance the excitement, two sets of plates are installed in the neck of the tube to make the electron beam, and hence the spot of light, move around. These plates are the horizontal and vertical deflection plates. To understand how these deflection plates work, let's consider them separately.

Initially, let's assume that the electron beam travels directly down the center of the tube and forms a spot of light at the center of the screen. External circuits can change the amount and the sign of the charge present on the horizontal plates. Assume that positive charge is gradually placed on one of the plates and negative charge on the other. This charge produces an electric field in the space between the plates, which causes the electron beam to be gradually deflected from its straight-line flight path. You can follow the movement of the electron beam by watching the movement of the spot of light on the face of the screen. For the situation shown in Figure 8.15, you would see the spot gradually move to the right side of the screen. Here is a question for you to answer on your own. What would happen if the following steps were taken in sequence? (1) The horizontal plates are charged so that the beam starts at the extreme left side of the screen. (2) Now the charge on the plates is gradually changed so that the beam slowly moves to the right side of the screen. (3) The electron beam is turned off when the spot of light reaches the right side of the screen, and (4) the charge on the plates is changed so that the beam is again focused at the extreme left position when it is turned back on. (5) The steps are continually repeated in the order given.

The answer is that an observer would see a spot of light continually sweeping from left to right on the screen face. Still this is nothing to get excited about, so let's include the vertical plates in our discussion. Combined action of the horizontal and vertical plates causes the electron beam to sweep across the tube face to form a line at the top of the tube. As the beam sweeps across the top, the number of electrons leaving the electron gun is controlled by varying the heating current. Thus, the line is neither all dark nor all light, but shadings of intensity of the light are observed. The plates are then charged such that the beam moves downward slightly and sweeps out another line across the screen. In a television set, 30 complete sweeps across the *total* face of the tube are made each second. By varying the strength of the heating current, the varying shades of dark and light can produce a picture on the screen. The rapid scanning of the beam across the screen produces a series of pictures so rapidly that the eye cannot follow the production of an individual one. Thus, a series of pictures is flashed before the eye, each slightly different from the preceding one, and the perception of motion is produced. The strength of the electron beam current is controlled by a signal to the antenna sent out by a TV transmitting station and received by the antenna.

8.7 VOLTAGE

The study of practical electric circuits can be made easier by referring to certain analogous

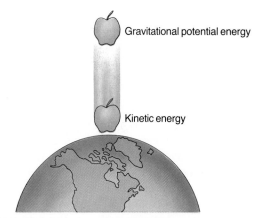

FIGURE 8.16 The gravitational potential energy of the apple is converted to kinetic energy as it falls toward the Earth. The falling apple is capable of doing work.

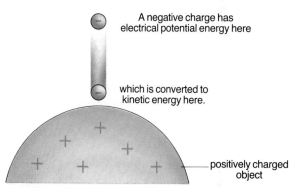

FIGURE 8.17 Charged objects can have electrical potential energy.

topics covered when we examined mechanics. Think of an apple near the surface of the Earth. If we grasp the apple and lift it into the air, we do work on it, and as we saw in our study of energy, the potential energy of the apple is increased. If the apple is then dropped, the potential energy is converted into kinetic energy, as shown in Figure 8.16. In turn, the moving apple can perform work for us. The same situation arises for electric charges. Suppose you have a large positively charged object (analogous to the Earth) and a small negatively charged object (analogous to the apple), as shown in Figure 8.17. In order to pull the negatively charged object away from the positively charged one, you must do work. When the two are separated, the small charge has an energy due to its position, and we call this kind of energy

electrical potential energy, by analogy to our use of the term "gravitational potential energy." If the negative charge is released, it accelerates toward the positive charge, and the potential energy is converted into kinetic energy. Once again work can be done for us.

The concept of gravitational potential energy is an extremely important one in the study of mechanics. Thus, one might think that electrical potential energy would be just as valuable in our study of electricity. Granted, electrical potential energy is important to us now, but it turns out that another related concept is of even greater importance. We shall find that in the study of electric circuits, a quantity called **potential difference,** or in colloquial terms, **voltage,** is more frequently used. *The difference in potential between two points is defined as the work performed by a charge q as it moves between these points divided by the magnitude q of the charge.* An alternative form of this same definition is that *the potential difference between two points is equal the energy that a charge q transforms from electrical energy to other forms of energy divided by the magnitude of the charge q.* In equation form this is expressed as

$$\text{Potential Difference} = \frac{\text{electrical energy transformed}}{\text{charge}} \quad (8.4)$$

or symbolically

$$V = \frac{\Delta(\text{energy})}{q} \quad (8.5)$$

where V is the change in potential difference, Δ(energy) is the energy transformed, and q is the charge in coulombs. As Eq. 8.5 indicates, the units of potential difference are joule/coulomb,

$$V = \frac{1\,J}{1\,C} = 1\,V$$

where the units of joule per coulomb, J/C, have been defined as the volt, *V*.

If you are a typical student beginning the study of physics, the concepts of potential difference and voltage seem somewhat obscure. In order to get a better feel for how voltage can be of importance, we will begin our consideration of electric circuits in the next section.

8.8 ELECTRIC CIRCUITS

Figure 8.18 shows a very simple electric circuit consisting of a battery connected to a light bulb and a switch. Under normal operating conditions, charges move through the wires and the bulb, and the bulb lights, as shown in Figure 8.18a. If the switch is opened, leaving a gap in the circuit, the charges cannot cross the air gap, and the bulb does not glow, as shown in Figure 8.18b. In order for the charges to move there must be a closed conducting path for them to move through. For convenience in sketching our circuits, we shall use certain circuit symbols to represent the various pieces that can go into a circuit. For example, a battery is represented by the symbol |ı. *A device that hinders the motion of charge through a circuit is called a resistor* and is represented by the symbol -\/\/\/-. A switch is represented as —/—. These symbols enable us to draw the circuit of Figure 8.18b as shown in Figure 8.18c.

Let us consider the battery in our circuit shown in Figure 8.19, and let us assume that it is a 12 *V* battery. When we say that a battery has a voltage of 12 *V*, we mean that chemical reactions occurring inside the battery maintain one of its terminals at a potential 12 *V* higher than that of the other. The long line drawn in the symbol for a battery is used to indicate which of the two terminals is being held at the higher potential. This high-voltage terminal is often called the positive terminal, while the other is often called the negative terminal. For convenience in our discussion, let us assume that the negative terminal is held at a voltage of 0 *V*. Any point in a circuit that is held at a voltage of 0 *V* is said to be grounded, and a ground location in a circuit is indicated by the symbol shown in Figure 8.19.

Now let us consider the motion of charges in our circuit, and also for convenience in our discussion, let us assume that positive charges move through the circuit. (We shall see in a later section that it is actually negative charges that move in a circuit.) We have already noted that our battery is

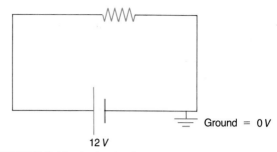

FIGURE 8.19 A simple circuit.

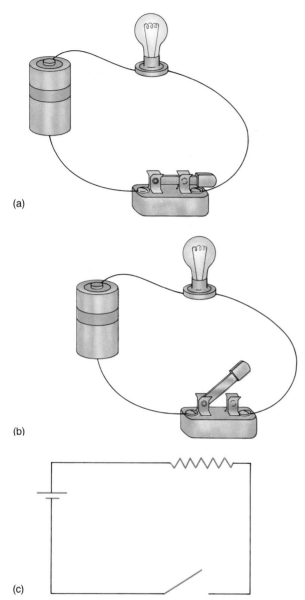

FIGURE 8.18 (a) The bulb glows if there is a complete circuit. (b) The bulb does not glow if the switch is open. (c) Schematic diagram of the circuit in part (b).

a 12 *V*, or a 12 J/C, battery. This means that every 1 coulomb of charge that leaves the positive terminal carries with it 12 J of electrical energy. This energy is furnished to the charges by the chemical reactions occurring inside the battery. As our 1 coulomb of charge moves around the circuit, it gives up this electrical energy to the pieces of equipment that make up the circuit. For our example, the only piece of equipment that we have included is the light bulb, indicated as a resistor. In the resistor, the 12 J of electrical energy supplied by the 1 coulomb of charge is converted to other forms of energy, namely light and heat. When the 1 coulomb of charge returns to the negative terminal, it returns with zero electrical energy. It has given up all of its energy to the devices it encounters. The charge then moves through the battery, where chemical reactions replenish its electrical energy to a 12 J level once again, and the charge is ready to begin another transit of the circuit. In this way, charges moving through a circuit continually supply power to light bulbs, stereos, radios, or whatever else the battery is connected to, until the chemicals inside the battery are used up and the battery "runs down."

It is often helpful to compare a circuit like that of Figure 8.19 to a mechanical system like that of Figure 8.20. In Figure 8.20, water is caused to circulate through a pipe by means of a water pump. The pump is analogous to a battery, and the motion of the water is analogous to the motion of charges in the electric circuit. (Loosely speaking, a battery is a charge pump.) The water then drops into a funnel filled with loosely packed sand. The sand, which inhibits the flow of water, is analogous to the resistance of an electric circuit.

8.9 ELECTRIC CURRENTS

In the previous section, we discussed the motion of charges in an electric circuit as though positive

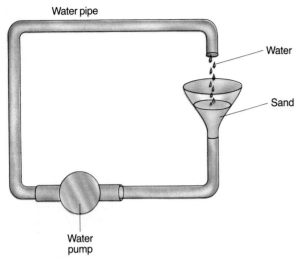

FIGURE 8.20 A water pump analogy for the electric circuit of Figure 8.19.

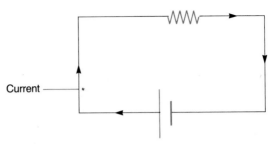

FIGURE 8.21 The sense of an electric current is from the positive terminal of a battery toward the negative terminal.

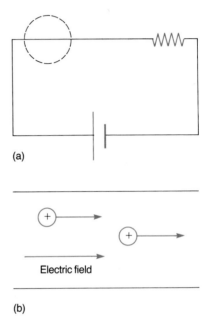

FIGURE 8.22 (a) The dashed circle is the part of the circuit we will investigate. (b) In this portion of the circuit, there is an electric field directed as shown. This field would cause positive charges to move from left to right.

charges do the moving. This is not the case; it is actually the electrons that move. However, when electric circuits were first studied, it was not known whether it was the positive charges or the negative charges that were in motion. As a result, early investigators discussed circuits in terms of the motion of positive charges and defined the sense of a current in terms of positive charge movement. In Figure 8.21, we show the direction selected for the sense of the electric current. We assume that the charge exits the positive terminal, moves around the circuit, and re-enters the battery at the negative terminal. Thus, *we define the sense of a current to be in the direction of motion of positive charges.* We know that this cannot be the true state of affairs because, as we noted in our discussion of charging by friction, the positive charges are rigidly fixed in place in a wire, and it is the electrons that are free to move around. What actually happens is that electrons leave the negative terminal, move around the circuit, and re-enter the positive terminal to have their energy replenished. Don't be confused by our choice for the sense of a current. Just recall that this "opposite" direction is chosen because of a historical misunderstanding of the true nature of current. Our assumption that it is positive charges in motion in a circuit will cause us no difficulties in our analysis of electric circuits.

Now imagine that you shrink to a size small enough to fit into a wire and that you are carrying with you a device to "count charge" and a watch. The segment of the wire that you have dropped in on is outlined in Figure 8.22a. The first thing that you notice when you enter the wire is that there is an electric field present that has the direction shown in the figure. A battery sets up an electric field in a wire that originates on the positive terminal of the battery and stops on the negative terminal. Thus, in the magnified segment of the wire shown in Figure 8.22b, you see an elec-

tric field directed from left to right. You now re-call that positive charges tend to move in the di-rection of an electric field, so you see positive charges moving past you from left to right. Thus, the sense of the current in the wire is from left to right in Figure 8.22b.

The quantity of charge moving through a cross-sectional area of the wire divided by the time re-quired for the charge to move through this area is defined as the magnitude of the current in the wire.

This is

$$I = \frac{Q}{t} \qquad (8.6)$$

where I is the symbol used to represent current, Q is the charge moving through a cross-sectional area, and t is the time required for the charge to move through the area. The units of I are cou-lombs per second, C/s, and these units are grouped together and referred to as an amp, symbol A.

$$I = \frac{1\,C}{1\,s} = 1\,A$$

Let us consider one more detail concerning the motion of charges through a wire. If one could determine the actual speed with which charges move through a wire, one would find it to be surprisingly small. Usually, the speed is such that an individual charge moves only a small frac-tion of a centimeter in one second. If this is the case, why is it that when we turn on a wall switch, a light bulb comes on almost instantaneously? If the electrons are moving very slowly, it seems that it might take several hours for one to leave the wall socket and finally reach the light. To under-stand what happens, consider the analogy shown in Figure 8.23. Here we see a group of people lined up to view the Grand Canyon. A new arrival pushes against the back of the line in an effort to move forward, and this push is transmitted down the line very rapidly. The unfortunate result is that the person nearest the rim is shoved off into the canyon. This same effect happens to the

FIGURE 8.23 Analogy of charge motion in a circuit. The force of a push by a person at the end of the line is transmitted to those at the front of the line.

charges in a wire. As a charge enters the wire from the wall outlet, it pushes on those in front of it in the wire, and these pushes cause a charge to leave the wire at the other end. Thus, the move-ment of charges in the wire is established almost instantaneously. As we noted, it is the presence of an electric field in the wire that is causing the charges to move, and when you turn on a switch in a circuit, the electric field inside the wire is set up at the speed of light, 3.00×10^8 m/s. The elec-tric field moves through the wire far more rap-idly than do the individual charges.

EXAMPLE 8.4 Getting a feel for currents

In a circuit, 3.0×10^{-3} C of charge moves past a cross-sectional area every 2.0 s. Find the magni-tude of the current in the wire.

Solution The current can be found from Eq. 8.6 as

$$I = \frac{Q}{t} = \frac{3.0 \times 10^{-3}\,C}{2.0\,s} = 1.5 \times 10^{-3}\,A$$
$$= 1.5\,mA$$

where the symbol m = milli = 10^{-3}

EXAMPLE 8.5   Counting electrons

The current in a circuit is 3.0 A. How many charges, each having a charge of 1.6×10^{-19} C, pass through a cross-sectional area of the wire every second?

Solution   Let us first find the amount of charge which flows past the cross-sectional area each second. We do this by use of the definition of current, Eq. 8.6.

$$Q = It = (3.0 \text{ A})(1 \text{ s}) = 3.0 \text{ C}$$

But if each charge has a magnitude of 1.6×10^{-19} C, the total number of positive charges is

$$\text{number} = \frac{3.0 \text{ C}}{1.6 \times 10^{-19} \text{ C/charge}}$$
$$= 1.9 \times 10^{19} \text{ charges}$$

8.10
RESISTANCE

We have said that certain devices placed in an electric circuit impede the motion of charges through the circuit. This hindrance to the motion of charges is called **resistance.** To understand what produces the resistance in a circuit, let us consider the motion of a single charge as it moves through a wire that has resistance. When the charge leaves the positive terminal of the battery, it has a high electrical potential energy, and it is moving toward the negative terminal of the battery where it will have zero electrical potential energy. Thus, from the conservation of energy, we see that the kinetic energy of the charge will increase as it moves. Periodically, however, the charge collides with the atoms of the wire, and when one of these collisions occurs, the moving charge is slowed. The kinetic energy of the moving charge is given to the struck atom, which causes the atom to vibrate about its fixed location with a larger amplitude of vibration than it had before the collision occurred. After the collision, the moving charge begins to speed up once again, until it collides with another atom. Thus, it is the collisions with the fixed particles of the material that hinder the motion of a charge through the wire, and this hindrance is called the resistance of the wire.

The collisions inside a wire carrying a current cause the atoms of the wire to vibrate with larger than normal amplitudes, and we recognize this fact by a temperature increase of the wire. Thus, inside a material that has resistance, kinetic energy of the moving charges is being transformed into heat energy. If the material gets hot enough it will glow, first with a dull red light and then, as it gets hotter, with a white light. Inside a light bulb, the charges move through a piece of tungsten wire, called the filament of the bulb, which has a very high resistance to the motion of charges. A current of about 1 amp through the filament causes its temperature to reach about 2000°C. At this temperature, the filament becomes white-hot and emits light.

Materials that have a very high resistance to the motion of charges are called **insulators.** Examples include asbestos, rubber, and glass. Other materials, such as the metals gold, silver, and aluminum, allow charges to move through them freely. These materials are called **conductors.** It is convenient for us that there exists a wide range of resistances in the materials in our environment. Thus, it is easier for an electric current to pass through thousands of kilometers of wire in an overhead transmission line than through the few centimeters of insulating matter that stands between the wire and its supporting tower.

8.11
OHM'S LAW

It is debatable which of the laws that we studied during our investigation of mechanics stands as the most important, but many scientists would pick Newton's second law. Likewise, if one had to choose the most important law in the study of electricity, the one we will look at in this section, Ohm's law, would finish high on the list. The law, discovered by George Simon Ohm (1787–1854), relates the voltage supplied to a circuit to the current in the circuit and the resistance of the circuit. In equation form, **Ohm's law** is

$$V = IR \tag{8.7}$$

where V is the applied voltage, I is the current,

FOCUS ON . . . Superconductivity

In the last few years, a lot of attention has been directed toward new advances in the study of a class of materials called **superconductors.** Superconductors are materials that behave basically as ordinary substances above a temperature called the critical temperature. However, as the temperature is decreased to and below the critical temperature, they behave as though they have zero resistance. This phenomenon was discovered by Kamerlingh-Ohnes in 1911 when he was working with mercury. He found that when the material was cooled below 4.2 K, its resistance dropped essentially to zero. Since that time, thousands of other materials have been found to exhibit superconducting properties. Aluminum, tin, lead, zinc, and indium are some common examples. The feature of superconductors that has kept them from having many practical applications and more widespread use is that the critical temperature is extremely low for all of them. But in 1986 an oxide of barium, lanthanum, and copper was discovered to be superconducting at 30 K by George Bednorz and K. Alex Muller, working at the IBM Zurich Research Laboratory in Switzerland. The two were awarded the Nobel Prize in 1987 for their research.

In 1987, scientists at the University of Alabama in Huntsville and the University of Houston reported a critical temperature of 92 K for an oxide of yttrium, barium, and copper. Later in 1987, the critical temperature went to 105 K for an oxide of bismuth, strontium, calcium, and copper, and in 1988 an oxide containing thallium was discovered to be superconducting at 125 K. Thus, the rise in critical temperature toward room temperature is increasing, and it is not out of the question that such a goal could be reached in a few years.

If room temperature superconductors could be developed, the electronics industry would be revolutionized. Superfast computers could be developed, and all electronic items would feel the impact of the new technology. One of the main problems with the distribution of electrical energy across country is that there are unavoidable power losses in the transmission lines caused by the resistance of the wires. If these wires could be constructed of zero-resistance materials, this problem would be solved, and the cost of electrical energy could drop drastically.

FOCUS ON . . . Microphones and Telephone Receivers

There are many different types of microphones in use today, but we shall concentrate here on the type that is found in the mouthpiece of a telephone, the carbon microphone. The Figure shows the design of such a microphone and the details of how it is connected to a circuit. The primary resistance in the circuit is the carbon granules packed into a box inside the mouthpiece. When a sound wave strikes the steel diaphragm on one side of the box, it flexes in and out following the compressions and rarefactions of the sound wave. When the carbon granules are compressed tightly together, they make intimate contact with each other, and the resistance of the circuit is low. When the diaphragm flexes outward because of a rarefaction, the granules become more loosely packed, and the resistance in the circuit is high. The current in the circuit, produced by the battery, follows these changes in resistance, fluctuating between a large value and a small value. These variations in current are sent through a transformer and into the transmission lines of the telephone company. (Transformers will be discussed in Chap-

ter 9.) These changing currents are converted by the listener's earpiece back into sound waves.

Carbon microphones are fine for telephone transmission because they reproduce sound frequencies quite well up to 4000 Hz, and most of the frequencies of normal conversation are below this value. At higher frequencies, they are not very efficient. Thus, if someone tries to play a flute solo over your telephone, your enjoyment may not be very satisfactory.

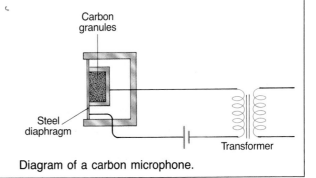

Diagram of a carbon microphone.

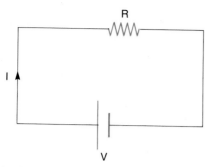

FIGURE 8.24   What could be done to this circuit to change the current, I?

and R is the resistance. From this equation, we see that the units of resistance are V/A. These units are grouped together and called an ohm.

$$R = \frac{V}{I} = \frac{1\ V}{1\ A} = 1\ \Omega$$

Resistances are thus stated in terms of ohms, where the symbol Ω (Greek upper case omega) is used in place of the word ohm. Thus, a good insulator might have a resistance of $10^8\ \Omega$, and a good conductor might have a resistance of $10^{-3}\ \Omega$.

To see that Ohm's law agrees with common sense, let us consider a simple circuit consisting of only a battery and a resistor as shown in Figure 8.24. Solving Ohm's law for the current, I, we find

$$I = \frac{V}{R}$$

Now, ask yourself the question: What can we do to alter the current in this circuit? To increase the current, it seems reasonable that we could do one or more of the following: We could increase the applied voltage, that is, use a larger battery; we could decrease the resistance of the circuit; or we could do both. Isn't this what Ohm's law says we should do? Likewise, we could decrease the current by doing the opposite of the suggestions above.

EXAMPLE 8.6   The philosophical implications of making toast

The heating element inside a toaster has a resistance of 12 Ω. In use, a toaster is connected to a wall outlet which supplies 120 V. (This voltage is an alternating voltage, but for our example, the direction in which the current moves is of no significance.) Find the current in the heating element of the toaster.

Solution   From Ohm's law we find the current to be

$$I = \frac{V}{R} = \frac{120\ V}{12\ \Omega} = 10\ A$$

EXERCISE

Find the resistance of a light bulb that draws 0.50 A when connected to a 120 V source.

Answer   240 Ω

8.12 DIFFERENT TYPES OF ELECTRICAL CIRCUITS

If several appliances are to be wired into a single circuit, they can be connected either in **series,** as shown in Figure 8.25a, or in **parallel,** as shown in Figure 8.25b. The characteristics of parallel and series circuits are very different from one another. One obvious difference can be noted by a casual glance at Figure 8.25a and b. In a series circuit, connections are made such that there is only one path for charges to take as they move around the circuit. On the other hand, in a parallel circuit, there are branches in the circuit that allow some charges to follow one branch while other charges follow another pathway. Let us study each of these classifications of circuits in turn.

Series circuits

Let us build a series circuit by starting with one that contains only a single 100 Ω resistor, say a light bulb, as in Figure 8.26a. Now let us add additional resistors such as a stereo and a toaster as in Figure 8.26b. As these appliances are added, the total resistance connected to the battery increases. (This is like adding more and more sand inside a water pipe.) Now let us examine what we have done by use of Ohm's law, written as I = V/R. The voltage V connected to the circuit has not changed, because we have not changed the

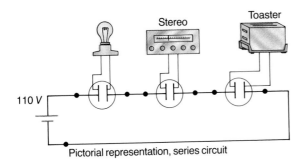

Pictorial representation, series circuit

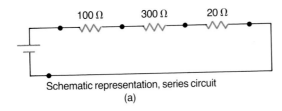

Schematic representation, series circuit
(a)

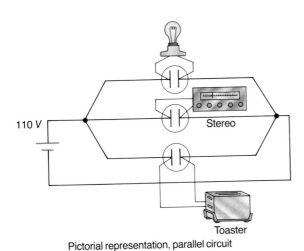

Pictorial representation, parallel circuit

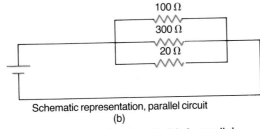

Schematic representation, parallel circuit
(b)

FIGURE 8.25 (a) A series circuit. (b) A parallel circuit.

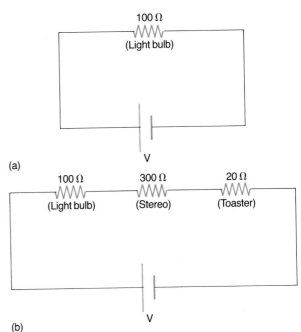

(a)

(b)

FIGURE 8.26 (a) A series circuit with one resistor. (b) As more resistors are added, the current must decrease, but the current will still have the same value at all points in the circuit.

battery; however, the resistance has increased as more and more appliances are added. Thus, we see that the current in the circuit must decrease. Series circuits are integral parts of every piece of

electronic equipment that you own, but a little thought shows that for household wiring, this arrangement is unsatisfactory for several reasons.

1. Each new resistor that you add decreases the current in the circuit. As a result, if you add a new lamp to a room, the new light bulb decreases the current supplied to that room. This would cause the light bulbs already present to glow a little less brightly, your toast would be light rather than dark, and so forth.

2. In a series circuit, there is only one pathway for the charges to move along. This means that the current in all parts of the circuit has the same value. For example, in Figure 8.26b, if the current in the light bulb is 0.5 A, the current in the stereo is also 0.5 A, and the toaster likewise has the same current. This has to be true because the charge motion through the circuit is like that of water flowing through a pipe that does not leak. If 12 gallons per minute are flowing through a pipe at one location in a series plumbing connection, there must be 12 gallons per minute flowing at all other loca-

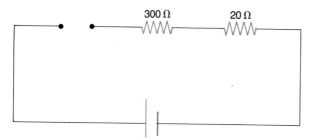

FIGURE 8.27 Series circuit with light bulb burned out.

tions in the piping. If not, either water would be escaping from the pipes or it would be building up at some location. Such a set of circumstances does not occur for water, and it certainly does not occur for charges flowing through an electric circuit. We have already discussed the fact that charge is conserved, and if the current were different at different parts of a series circuit, this would imply that charges are being either created or destroyed at some location in the circuit. Household circuits would not be effective at all if they were designed such that all devices connected in the line had to have the same current in them. Some devices such as an electric heater require large currents, whereas others such as a stereo may require only a small fraction of an amp. Also, certain devices such as a computer are designed for current at a constant, specified value. Think what would happen if you were working on your computer and you added a lamp to your series wiring arrangement. The addition of the light would decrease the current in the entire circuit, and your computer could be very unhappy with the change.

3. A series circuit has yet another disadvantage that is perhaps even more important than the others mentioned, although they are bad enough as is. Imagine that the light bulb in Figure 8.26b burns out. When a light bulb burns out, the filament in the bulb actually breaks. Therefore, the circuit is interrupted as shown in Figure 8.27, and no current at all can exist because there is not a complete, uninterrupted pathway through which the charges can move. This means that if your house were wired in series and one light bulb burned out,

all the lights would go out. In addition, your toaster, stereo, and any other appliances would also stop. Of course, you wouldn't know which light bulb blew originally, so you would have to grope around to find the culprit that broke the circuit. Some Christmas tree lights are wired in series, so perhaps you already know from practical experience the agony of trying to find the bad bulb in a chain of 25 or so.

Parallel circuits

We can construct a parallel circuit beginning as shown in Figure 8.28a, where we have connected a light bulb to our battery. As we progress from Figure 8.28b to c and beyond, we are connecting other devices to the battery. As each new device is added, entirely new pathways are formed for charges to move through. For example, in Figure 8.28c, charges leaving the battery have a choice of pathways when they reach point A. Some charges move through the light bulb to point B and then back to the negative terminal of the battery. Simultaneously, other charges follow the pathway from A through the stereo to B and back to the battery, and still others move from A through the toaster to B. Thus, each new device that we add forms an independent route from the battery through the circuit. An arrangement of this nature is ideal for a household electric circuit, because all the difficulties associated with a series arrangement discussed in the previous section are avoided here.

1. When additional appliances are added to a parallel circuit, the current through each one is independent of the current through the others. The circuit of Figure 8.28c is not affected when, say, the light bulb is turned off because there is still a pathway for charges to move through all the other devices.
2. Since each appliance is independent of the others, the current drawn by each is dictated solely by the resistance of the appliance.
3. If one appliance in a parallel circuit burns out or is unplugged, as shown in Figure 8.29, the other pathways are unaffected, and the appliances in the other branches operate normally.

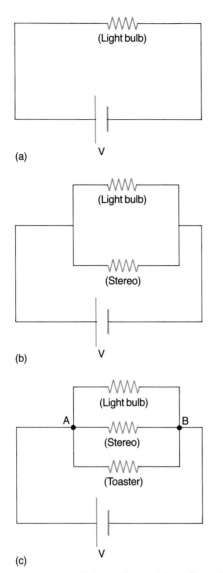

(a)

(b)

(c)

FIGURE 8.28 A parallel circuit provides alternative pathways for charges to move.

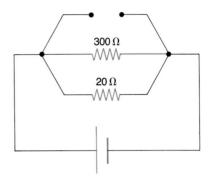

Parallel circuit with light bulb burnt out

FIGURE 8.29 If one appliance in a parallel circuit burns out, the other pathways still operate normally.

Thus, if one light bulb burns out, the others work just as well, and it is easy to find the dark one.

EXAMPLE 8.7   Resistors in series

A certain light bulb has a resistance of 144 Ω.

(a) Find the current through this light bulb when it is connected to 120 V.

Solution Ohm's law provides the answer as

$$I = \frac{V}{R} = \frac{120\ V}{144\ \Omega} = 0.833\ A$$

(b) If two of these light bulbs are connected in a series circuit, find the current through each.

Solution The total resistance connected to the voltage source will now be 288 Ω. Thus, the current will be

$$I = \frac{V}{R} = \frac{120\ V}{288\ \Omega} = 0.417\ A$$

Note that the supplied voltage does not change. Thus, the current supplied to each will be one-half the value that is supplied to one alone. As a result, the bulbs will glow less brightly as more and more are added.

EXAMPLE 8.8   Parallel circuits and the need for fuses

The two light bulbs of the previous example are connected in parallel as shown in Figure 8.30a.

(a) Find the current through each and the total current drawn from the voltage supply.

Solution Each of the light bulbs is connected directly across the 120 V voltage supply. Thus, the current through each is

$$I = \frac{V}{R} = \frac{120\ V}{144\ \Omega} = 0.833\ A$$

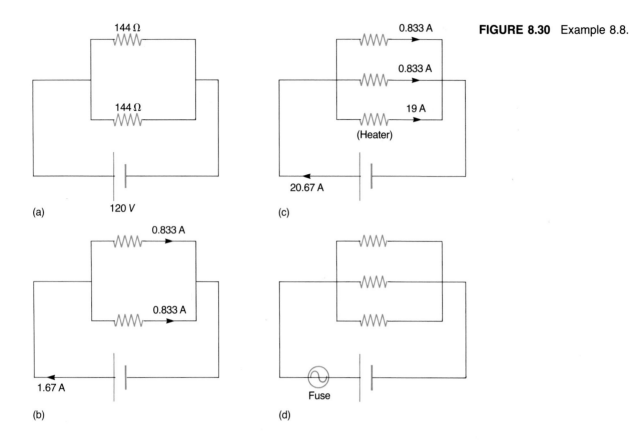

FIGURE 8.30 Example 8.8.

Thus, we see from Figure 8.30b that the total current drawn from the voltage supply is 1.67 A.

(b) An electric heater is now added to the circuit as in Figure 8.30c. The resistance of this heater is such that it will draw a current of 19.00 A when turned on. Find the current now drawn from the supply voltage and discuss the implications of what is happening.

Solution The total current drawn from the voltage supply is

$$I_T = 1.67 \text{ A} + 19.00 \text{ A} = 20.67 \text{ A}$$

The supply lines running through the house have a low resistance, but they still do have *some* resistance. As a result, these wires get hotter as the current through them increases. 20.67 A is a dangerously high level for them to carry, because there is a danger that they can overheat and cause a fire. In order to ensure that the current does not exceed safe levels, fuses are inserted in the supply line as shown in Figure 8.30d. Fuses are protective devices that allow some maximum cur-

rent, say 20 A, in the circuit before they "blow," causing all current to stop. A typical fuse is shown in Figure 8.31. It consists of a metal strip through which the current passes. This strip is made of a metal that melts when the current reaches a certain level. In modern homes, alternative devices called circuit breakers are used instead of fuses. Most of these devices are designed to use electromagnets, but the basic purpose remains the same— to cut off the current before it reaches a dangerous level.

8.13
POWER

In our study of mechanics we defined power as the rate at which work is done, or alternatively, as the rate at which energy is transformed from one form to another. The equation used is

$$\text{Power} = \frac{\text{energy transformed}}{\text{time for the transformation}} \quad (8.8)$$

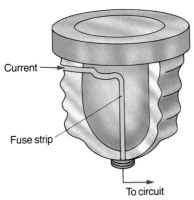

FIGURE 8.31 A simple fuse.

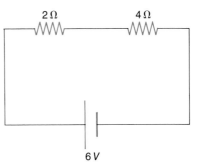

FIGURE 8.32 Example 8.7.

and the units are Power = J/s = W, where W is the symbol for a watt.

For electrical circuits, power is easily calculated if one knows the voltage and the current. Let us reason through the process.

Current tells us how much charge passes through a device each second. Voltage tells us how much energy each of these charges gives up to the device. Let us examine what the units have to say for the product of current times voltage, IV.

$$IV = \left(\frac{\text{charge passing through device}}{\text{time}} \right) \times$$
$$\left(\frac{\text{energy given to device}}{\text{charge}} \right)$$
$$= \frac{\text{energy given to device}}{\text{time}} = \text{Power} \quad (8.9)$$

Thus, we can find the power supplied to a device by finding the product of the current drawn by the device and the voltage supplied to it.

The equation Power = IV can be used regardless of the kind of circuit element you are considering. That is, you could find the power supplied to a circuit by a battery or the power converted to heat by a resistor. On the other hand, there is an alternative equation that applies *only* to resistive elements in a circuit, such as a light bulb. This alternative form is found by substituting V = IR for V in Eq. 8.9. We find

$$\text{Power} = IV = I(IR) = I^2R \quad (8.10)$$

EXAMPLE 8.9 Finding the power

(a) In the circuit of Figure 8.32, find the current drawn by the two resistors.

Solution The total resistance of the circuit is $2\,\Omega + 4\,\Omega = 6\,\Omega$, and the total voltage supplied to the circuit is that of the 6 V battery. Thus, the current is

$$I = \frac{V}{R} = \frac{6\ V}{6\ \Omega} = 1\ A$$

(b) Find the power supplied to the circuit by the battery.

Solution The power supplied is found from Power = IV. We have

$$\text{Power} = IV = (1\ A)(6\ V) = 6\ W$$

(c) Find how much power is consumed by each resistor.

Solution The power consumed by a resistor is found from Power = I^2R. The current through both resistors is the same, since this is a series circuit. Thus, the power consumed by the 2 Ω resistor is

$$\text{Power} = I^2R = (1\ A)^2(2\ \Omega) = 2\ W$$

and the power consumed by the 4 Ω resistor is

$$\text{Power} = I^2R = (1\ A)^2(4\ \Omega) = 4\ W$$

Thus, we see that the device that has the higher resistance converts more electrical energy into heat than does the lower resistance device, when the circuit is a series circuit.

8.14
THE COST OF ELECTRICAL ENERGY

Each month, an electric utility company presents you with a bill for the amount of electrical energy that you have converted from electrical form to other forms, such as heat, light, or sound. Let us examine the procedure that an electric company uses to determine your bill. The amount of energy that you have converted to your needs can be determined from the total power supplied to your home and the time that it has been supplied. The definition of power shows us how to do this calculation.

Energy converted
= (power)(time for conversion)

However, the electric company computes the power in units of kilowatts rather than watts, and they measure the time in hours rather than seconds. Thus, the units for the energy converted in your home are

Energy converted = (kW)(h)
= kWh (called a kilowatt-hour)   (8.11)

To see how your bill is determined, let us consider an example problem.

EXAMPLE 8.10   Turn off that light bulb

Suppose you accidentally leave a 100 W light bulb burning in your basement for a 24 h period. How much money will the utility company charge you for the energy consumed by the bulb if electricity costs 8 cents per kWh?

Solution   A 100 W light bulb is a 0.100 kilowatt bulb. Thus, the energy consumed in kWh is

Energy converted = (0.100 kW)(24 h)
= 2.4 kWh

and if electricity costs 8 cents per kilowatt-hour, your oversight in leaving the bulb on will cost you

(2.4 kWh)(8.0 cents/kWh) = 19 cents

EXAMPLE 8.11   Read the fine print

All electrical devices are required by federal law to have certain pieces of information supplied with them, usually on a metal plate on the back of the device. This plate gives you enough information to calculate how much the device will cost to operate. For example, a plate on a black-and-white TV set says that the TV will draw 0.75 A of current when connected to a voltage source of 120 V. Find the cost of watching 6.0 h of soap operas on this TV in a location where electricity costs 9.0 cents per kWh.

Solution   The power consumed by the device is found as

Power = IV = (0.75 A)(120 V) = 90 W

Thus, the device is a 0.090 kilowatt TV set, and the energy consumed in kWh is

Energy consumed = (0.090 kW)(6.0 h)
= 0.54 kWh

At 9.0 cents per kWh, the total cost is

(0.54 kWh)(9.0 cents/kWh) = 4.9 cents

8.15
ELECTRICAL SAFETY

Everyone who has ever been around electricity has at one time or another shocked themselves. Most of the time a shock produces no effect other than an unpleasant tingling sensation, but if the electrical contact is made in the wrong way, death can occur. Thus, let us digress briefly to discuss some of the dangers of electricity.

Electricity causes damage to the body in two ways: It can cause the muscles of a vital organ to contract and stop its life-sustaining function, or it can burn the tissues of the body. The primary consideration in the first instance is the amount of current that passes through the muscles of the vital organ. It is difficult to generalize concerning the amount of current that will cause various effects on the body, because factors such as the physical condition of the person involved and the exact current path are important. Generally, a current of about 0.001 A will be felt as an electric shock, and a current of 0.01 A *can* be fatal if it passes through a vital organ such as the heart. Currents of 0.02 A can paralyze the respiratory muscles and result in suffocation. A current of the order of 0.1 A passing near the heart is almost certainly fatal because it can cause the heart muscles to contract irregularly.

At first thought, it seems that these are extremely small currents to produce such dramatic effects. For example, consider a 3 V flashlight battery connected in a circuit to a 6 Ω bulb. A simple Ohm's law calculation shows that the bulb draws a current of I = 3 V/6 Ω = 0.5 A. Is it, then, dangerous to touch the circuit or to deal with it in any way? The answer is obviously no because you have touched such circuits many times without even feeling a hint of a shock. Why not? The reason is that the human body has a very high resistance. For example, dry skin has a normal resistance of perhaps 200,000 Ω or greater, depending on where the contact is made. Thus, if you connect yourself directly across the battery, the current through your body is only about I = 3 V/200,000 Ω = 0.000015 A, well within the safe range. Frequently, when people are killed by electricity, the reason is because their skin is wet, a factor that may reduce the resistance of the body to as low as a few hundred ohms. Thus, we see why it is especially dangerous to handle electrical equipment while you are in a bathtub or while you are standing barefoot on wet ground.

In order to help protect consumers, electrical equipment manufacturers now use electrical cords that have a third wire, called a case ground. To see how this works, let us consider the drill being used by the "fixit" person in Figure 8.33. A two-wire device has one wire, called the hot wire, that is connected to the high-voltage (120 V) side of the input power line to the home, and the second wire is connected to ground, to 0 V. Under normal operating conditions, the path of the current through the drill is as shown in Figure 8.33a. However, the cord inside the device can become frayed and a "short circuit" can occur if the high-voltage wire comes into contact with the case of the drill, as shown in Figure 8.33b. In this circumstance, the pathway for the current is from the high voltage wire through the consumer and to the earth—a pathway that can lead to death for the user. Protection is provided by the third wire, which is connected to the case of the drill, as shown in Figure 8.33c. In this situation, if a short occurs, the easiest path for the current is from the high-voltage wire through the case and back to

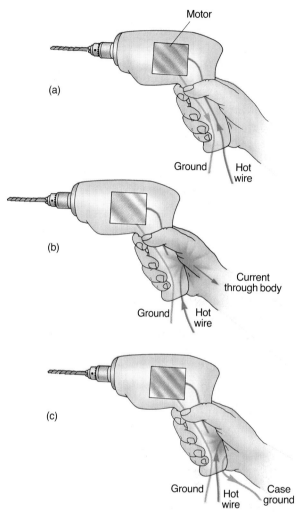

FIGURE 8.33 (a) Normal path of current is from the hot wire through the motor, to ground and back to the power supply. (b) If the hot wire shorts to the case of the drill, the current path is from the hot wire, to case, through the body, and to ground. (c) If a case-ground is provided, the current goes from the hot wire, to case, through the case-ground, and back to the power supply.

ground through this third wire. The high current produced can blow a fuse or trip a circuit breaker before the consumer is injured.

EXAMPLE 8.12   Bird on a wire

Explain why a bird can perch on a 10,000 V power line without being electrocuted.

Solution   If you cannot answer this question, try a simpler one: Why doesn't a person get hurt by stepping off a curb in Denver, Colorado, at elevation 1.6 km? The answer is simple—even though Denver lies far above sea level, there is no way to fall that distance, and a person stepping off the curb falls only a few centimeters, which is hardly lethal. Just as height per se is not dangerous, but only differences in height are, so high voltage is not dangerous, only differences in voltage are. What is of concern to the bird is not the potential of the line, but the potential difference between its two feet, that is, the voltage directly across its body. If the bird stands with its feet a few centimeters apart, the resistance of the copper wire separating his feet is only about 10^{-6} Ω, and if the wire carries a current of 1 A, the potential difference between the feet of the bird is

$$V = IR = (1\ A)(10^{-6}\ \Omega) = 10^{-6}\ V$$

The bird is safe because there is so little voltage applied directly across his body.

If, on the other hand, the bird straddled two wires, one at 10,000 V and the other at 0 V, the potential difference between its feet would be the full 10,000 V, and the bird would die instantly. Therefore, if the bird, or a human for that matter, touched one foot to the high-voltage line and the other foot to the ground, the resulting shock would be lethal (Fig. 8.34).

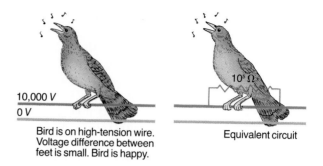

Bird is on high-tension wire. Voltage difference between feet is small. Bird is happy.

Equivalent circuit

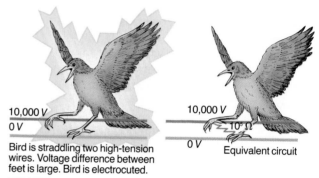

Bird is straddling two high-tension wires. Voltage difference between feet is large. Bird is electrocuted.

FIGURE 8.34   A bird on a high-voltage wire.

SUMMARY

If two electric charges are the same $(++)$ or $(--)$, they repel each other; opposite charges attract. Ordinary matter is made up of **atoms.** Every atom has one or more **electrons,** which are negatively charged, and **protons,** which are positively charged, along with **neutrons,** which have no charge. The magnitude of the charge on a proton and an electron is 1.6×10^{-19} C **(coulombs).** The magnitude of the force between charged particles can be found from **Coulomb's law.**

An **electric field** is said to exist in a region of space if a force of electrical origin is exerted on a charged particle brought into that region. The direction of an electric field is defined to be the direction that a positive charge would move under the action of the field.

Voltage is the difference in potential energy between two points, divided by the magnitude of the charge. An **electric current** is any concerted movement of electric charge. A current of 1 **ampere** is said to exist in a circuit if 1 coulomb per second of charge passes a cross-sectional area of the circuit.

Ohm's law states that the voltage drop across a resistor is equal to the current times the resistance. In a **series circuit,** the resistors are aligned so that there is a single pathway for charges to move along; in a **parallel circuit,** the pathway is branched and charges move through all the branches simultaneously. Electric **power** is equal to current times the voltage and is measured in **watts.**

EQUATIONS TO KNOW

$F = k \dfrac{q_1 q_2}{r^2}$ (Coulomb's law)

$E = \dfrac{F_e}{q}$ (definition of electric field)

$V = \dfrac{\Delta(\text{energy})}{q}$ (definition of potential difference)

$I = \dfrac{Q}{t}$ (definition of electric current)

$V = IR$ (Ohm's law)

$P = IV = I^2 R$ (power)

KEY WORDS

Electric charge
Electron
Proton
Neutron
Coulomb's law
Coulomb
Electric field
Van de Graaff generator

Voltage
Potential difference
Direct current
Electric current
Ampere
Resistance
Electric circuit
Ohm

Ohm's law
Parallel circuit
Series circuit
Electric power
Watt
Kilowatt-hour

QUESTIONS

STATIC ELECTRICITY AND MEASURING CHARGE

1. An object is given a net positive charge via friction. Does the mass of the object increase, decrease, or stay the same?
2. A thin metal ball cannot be given a static charge by rubbing it against another material. Why not?
3. Often when you try to charge something by friction on a humid day, your experiment proves unsuccessful. Why?
4. How many electrons would be required to produce a charge of -1 C?
5. If the people who first designated the electron as the basic carrier of negative electricity and the proton as the basic carrier of positive electricity had reversed their decision, how would our world have been affected?
6. A thin piece of aluminum foil is draped over a piece of wood. When a rod with a positive charge is touched to the foil, the leaves stand apart. Why do they repel one another?
7. One object is given a positive charge and another a negative charge. What happens when these two objects are placed in contact with one another?

COULOMB'S LAW

8. Compare and contrast Coulomb's law and the law of universal gravitation.
9. In the text, it is said that gravitational forces are weaker than electrical forces. Yet, when you try to pick up something heavy, you recognize gravitational force instantly, but you seldom are affected by electrical forces. Why not?
10. Two protons are placed 1 meter apart. Could another proton be placed at some point in space so that it would feel no net electrical force from these two? Where would this point be, and why?
11. Isolated charges as large as 1 coulomb are seldom encountered in nature. (a) Find the electrical force of repulsion between two 1 C charges separated by 1 m. (b) Charges actually encountered are usually of the order of 10^{-9} C. Find the force of repulsion between two of these charges separated by 1 m.
12. An electrical force of 10^{-11} N is exerted on an electron. (a) Find its acceleration. (b) Repeat for a proton.

THE CONCEPT OF FIELDS

13. A proton is released in a room where an electric field is present, and you discover that electrical forces move it upward. What is the direction of the field in the room?

14. If the charge released had been an electron in problem 13, would your answer for the direction of the field still be the same?

15. An electron is placed 1 meter away from an object having a charge of 10^{-9} C. (a) What is the acceleration of the electron? (b) What is the strength of the electric field set up by the object having the 10^{-9} C charge?

16. A proton is released from rest in an electric field of strength 100 N/C. (a) What is the acceleration of the proton in magnitude and direction? (b) How fast will the proton be moving after 10^{-3} s?

17. Electric field lines never cross one another. Why not? (*Hint:* What would happen to an electron if it was moving along a field line and reached the point of intersection?)

THE VAN DE GRAAFF GENERATOR

18. (a) When a metal object is given a charge, all the charge resides on the outside surface of the metal. Why? (b) If the object is hollow, the electric field inside the metal is zero. Why?

19. In light of problem 18, why would it be safe to be on the inside of a car if it is struck by lightning?

20. Would it be safe to be inside the dome of a Van de Graaff generator when it is given an extremely large charge?

21. Suppose you were wearing clothes affected by static cling. How would the clothing be affected when you come near a Van de Graaff generator?

22. When a demonstration is done in which a person's hair is made to stand on end by having him touch a Van de Graaff generator, the person is asked to stand on an insulating stool. Why?

VOLTAGE AND ELECTRIC CIRCUITS

23. Distinguish carefully between potential and potential energy.

24. A high-voltage transmission wire is at a potential of 120,000 V, local power lines are at a potential of 2200 volts, and household wiring is at 110 V. Assume these voltages can be treated as though they were direct current voltages and find the potential difference between (a) the high-voltage line and the local power line; (b) the local power line and the ground; (c) the high-voltage line and the household wiring.

25. Some devices operate directly off 110 V, whereas others operate off a small transformer that converts the voltage to, say, 5 V before it is applied to the device. Based on the definition of potential difference, what do you suppose would happen to the 5 V device if it were connected directly to 110 V?

26. Two identical electrical devices are connected one after the other and to a 12 V battery. Discuss the energy transformations that occur in this circuit.

ELECTRIC CURRENT

27. We have defined the sense of an electric current to be in the direction of motion of positive charges rather than in the direction of motion of electrons, which are the charges that really move in a circuit. This is often defended by saying that "an electron moving right to left has the same effect as an equal positive charge moving left to right." Defend this position.

28. A current of 10^{-3} A exists in a circuit. How long a time would have to pass before 1 coulomb of charge moved past a cross-sectional area in the circuit?

29. A current of 10^{-3} A exists in a circuit. How many electrons move past a given cross-sectional area of the circuit in 10 minutes?

RESISTANCE AND OHM'S LAW

30. (a) Calculate the resistance of a simple circuit in which a 40 V power supply produces a current of 4 A. (b) Calculate the voltage of a battery if it can push 1.5 A through an 8 Ω resistor.

31. It is found that the resistance of a material increases with temperature. On the basis of what is happening inside a wire as charges move through it, explain why you would expect this increase in resistance.

32. Suppose you insert batteries into a flashlight such that they oppose one another. On the basis of Ohm's law, explain why you would not expect to get a current in the circuit.

33. An air conditioning unit has a compressor that is connected to 120 V and draws 90 A when first turned on. What is the effective resistance of the compressor at this instant?

34. The resistance of 10 cm of a certain wire is about 0.006 Ω. How many meters of this wire would you have to connect to a 12 V battery in order to limit the current to 3 A?

DIFFERENT TYPES OF ELECTRICAL CIRCUITS AND POWER

35. (a) Explain why there must always be two wires

leading to an electric light bulb if it is to draw current and light up. (b) In an automobile, the wire from one terminal of the battery (the "ground") generally is connected directly to the metal chassis. The other terminal is said to be "hot." Therefore, any electrical device, such as a light bulb, can be energized by first connecting one of its wires to the chassis and then running the other wire to one of the terminals of the battery. Explain. From which terminal, the ground terminal or the "hot" terminal, should the wire be run? Which terminal, the positive or the negative, should be grounded, or does it matter?

36. Calculate the resistances of (a) a 25 W light bulb, (b) a 50 W light bulb, (c) a 50 W stereo amplifier, (d) a 100 W light bulb, (e) a 250 W toaster, and (f) an 800 W space heater, assuming all are designed to operate at 110 V.

37. During a "brown-out" in a big city, fuel shortages force a power company to supply 90 V potential instead of the usual 110 V potential. If a 250 W

pure copper wire has a lower resistance than heater elements, why doesn't the wiring in your house or the wire in an extension cord get hot when it is drawing current?

39. A customer brought her new portable radio into the repair shop, complaining that although the radio operated well, a set of batteries lasted only a short time before running down. Speculate in a general way on the nature of the problem. Would you think the radio was drawing excess voltage or excess current? Discuss the reasons for your answer.

40. Explain why Christmas tree lights are generally wired in parallel. If they were strung together in series and one bulb blew out, would it be hard or easy to find that bulb? Explain.

41. Are the lights in your automobile wired in series or parallel? Defend your answer.

42. Calculate the current in the wire and the power delivered to each resistor in each of the diagrams in the Figure.

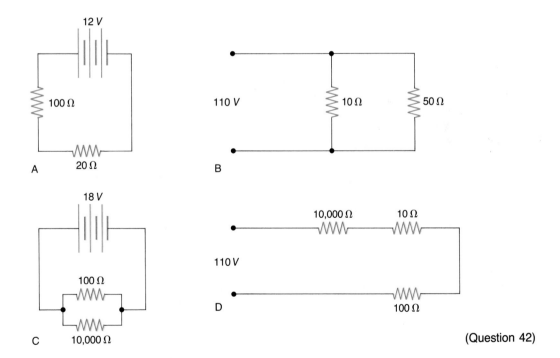

(Question 42)

toaster is plugged in during such a brown-out, how much power will it actually deliver?

38. If two different heaters are compared, the one with the lower resistance uses more power. Since

43. Assume that you have a 12 volt battery and a small appliance with a resistance of 6 ohms. The appliance manufacturer states that the appliance will burn out if more than 15 watts are drawn. Can the appliance be safely connected to the battery? If not, design a circuit using the 12 volt battery that will operate in such a manner as to supply 15 watts to the appliance.

44. If a household fuse burns out, the circuit can be re-established by inserting a penny into the socket. Explain why this works but why this is a very dangerous practice that should never done.

45. All automobile circuits are provided with fuses. What would happen if an automobile were wired without fuses and a short circuit developed?

46. A certain light bulb has a resistance of 200 Ω. (a) What is the current in this bulb when connected to a 12 V battery? (b) If three of these light bulbs are connected in series to this same battery, what is the current?

47. The three light bulbs of the preceding problem are connected in parallel to the 12 V battery. What is the current in each bulb, and what is the potential difference across each?

48. Two 100 Ω resistors are connected in series to a 30 V battery. (a) What is the power supplied to each? (b) What is the power supplied to each if the two are connected in parallel?

THE COST OF ELECTRICAL ENERGY

49. A black-and-white TV has a power rating of about 90 W. If electricity costs 8.0 cents per kWh, how much will it cost you to watch TV 4 h a day for a year?

50. Repeat problem 49 for a 300 W color set.

51. Suppose you have a job that requires you to load 600 N bags of fertilizer on a platform that is 1 m above ground level. You lift one bag every 30 s. (a) How much work do you do in an 8 h day? (b) What is your power output in kilowatts? (c) If you are paid at the rate of 10 cents per kWh, how much money do you make in a day?

52. A certain electric heater draws 10 A of current when operated at a voltage of 110 V. If electricity costs 8.0 cents per kWh, how much does it cost to run this heater for an 8-hour day?

ELECTRICAL SAFETY

53. If it is the current that passes through your body that does the damage, why do you see warning signs that say, Danger: High Voltage?

54. Why is cutting off the third prong on a plug so that it can be used in a two-prong outlet a dangerous thing to do?

55. You leap off the top of a tall building and grab a high-voltage wire on the way down. (a) Are you in danger while you are swinging from the wire? (b) Suppose the wire breaks. Are you in danger while you are falling while holding onto the wire? (c) While still holding onto the wire, you touch the ground. Are you injured?

56. A power line falls on your car while you are in it. Are you in danger while in the car? Are you in danger if you open the door and step out of the car?

57. Why is it suicidal to use a hair dryer while in the bathtub?

ANSWERS TO SELECTED NUMERICAL QUESTIONS

4. 6.25×10^{18} electrons
11. (a) 9.0×10^9 N, (b) 9.0×10^{-9} N
12. (a) 1.10×10^{19} m/s², (b) 5.99×10^{15} m/s²
15. (a) 1.58×10^{12} m/s², (b) 9×10^{-9} N/C
16. (a) 9.58×10^9 m/s², (b) 9.58×10^6 m/s
28. 1000 s
29. 3.75×10^{18} electrons
30. (a) 10 Ω, (b)12 V
33. 1.33 Ω
34. 66.7 m

36. (a) 484 Ω, (b)242 Ω, (c) 242 Ω, (d) 121 Ω, (e) 48.4 Ω, (f) 15.1 Ω
37. 167 W
46. (a) 0.06 A, (b) 0.02 A
47. 0.06 A, 12 V
48. (a) 2.25 W, (b) 9 W
49. $10.51
50. $35.04
51. (a) 5.76×10^5 J, (b) 2×10^{-2} kW, (c) 1.6 cents
52. 70.4 cents

The production of electric power and the distribution of this power is possible only because of a connection between magnetic effects and electrical effects.

C H A P T E R

9

Electromagnetism

The first magnetic materials discovered were in the form of mineral deposits found in Magnesia in Asia Minor. The mineral was given the name lodestone or magnetite. The term "lodestone" is derived from the Saxon word *laedan,* meaning "to lead." The association between leading and the mineral arises because ancient navigators learned that if a bar of this mineral was mounted so that it could rotate freely, one end would always point north. Scientists now know that these materials are permanent magnets and that they align themselves with the magnetic field of the Earth.

The study of magnetism began with the investigation of the properties of these naturally occurring permanent magnets. The use of magnets as compasses to find direction opened up new trade and exploration routes and led to alterations in the life of the people of that time. However, the impact of magnetism on society became even more profound when it was discovered that an electric current can produce magnetic effects in the space around the wire. Because of the connection between electricity and magnetism, we shall embrace all of the phenomena discussed in this chapter under the single heading of electromagnetism.

9.1 MAGNETS

Experiments show that every magnet, regardless of its shape, has two poles, called **north** and **south.** The names north and south pole arose from the use of magnets as direction finders. The end of a magnet that points toward the north of the Earth is called the "north seeking" end, or more simply the north pole of the magnet. The interaction between the poles of two magnets is similar to that observed between charged objects. For example, like electric charges repel one another, and so do like magnetic poles. Also, as unlike charges attract, so do unlike poles. Because of this correspondence between electric forces and magnetic forces, it is tempting to carry the analogy a step farther and say that magnetic effects are produced in a material by tiny isolated north and south poles, somewhat like isolated positive and negative charges. If true, then perhaps what is happening when a material becomes magnetized is that there is a separation of these tiny poles, with the north species accumulating at one end of the magnet and the south species at the other.

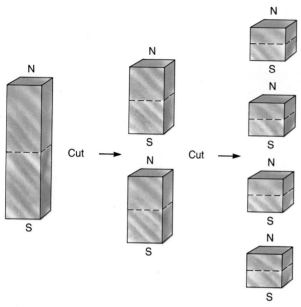

FIGURE 9.1 If a permanent magnet is cut in half, two smaller magnets are produced. Is it impossible to isolate separate north and south magnetic poles by repeated cutting?

If such tiny magnetic entities do exist, it seems that it would be easy to separate them. One way would be to start with a bar magnet and break it in half, leaving one end with an excess of tiny north poles and the other with an excess of tiny south poles. However, if you attempt this experiment, you find that cutting a magnet in half produces two smaller magnets, *each* with its own north and south pole. Slice again, and you get four magnets, and so on. If your instruments were fine enough, you could keep cutting until you had an enormous collection of atomic-sized magnets, but each would still have a north pole and a south pole as in Figure 9.1. It should be noted here that attempts to detect an isolated magnetic monopole (an isolated south or north pole) have been unsuccessful, but the search continues. However, if such isolated monopoles are indeed found someday, our basic understanding of ordinary magnetic effects will not be drastically altered. As we shall see, an understanding of what makes a magnet a magnet is explained by the fact that the ultimate source of magnetic effects is the motion of charged particles at the atomic or the subatomic level. We shall return to

this topic after we have investigated the relationship of electricity and magnetism a little more carefully.

There are other similarities between magnetic and electric effects. For example, we saw earlier that two objects can become electrically charged by rubbing them vigorously together. This charging by friction process leaves one object with a positive charge and the other with a negative charge. In a similar fashion, an unmagnetized piece of iron can become magnetized by stroking it with a bar magnet. After the stroking, the iron is found to be magnetic, with a north pole and a south pole. There are other ways in which an unmagnetized piece of iron can be magnetized. One way is to place the iron near a strong magnet and either hammer the iron or heat it. The hammering and the heating accelerate the process, but neither is necessary. The iron will also become magnetized if it is left alone near a strong magnet; however, the time period for the magnetism to occur will be longer. For example, iron fences left standing for many years often become magnetized because of the presence of the magnetic field of the Earth. Also, the naturally occurring magnetic material, lodestone or magnetite, became magnetized because of the field of the Earth.

9.2 MAGNETIC FIELDS

The concept of electric fields was introduced earlier to provide a way to discuss the forces acting on charged particles. It is also convenient to discuss forces that are magnetic in origin in terms of a field concept. One of the tasks that must be considered when discussing a field is to define a process by which we determine the direction of the field. For a magnetic field this is done by defining the *direction of a magnetic field to be in the direction that the north pole of a compass needle points when placed in the field.* Figure 9.2a shows how a magnetic field line of a bar magnet can be traced with a small compass. The compass is first placed near the north pole of the magnet and the direction in which the compass needle points is noted. The compass is then displaced slightly in the direction indicated by the needle, and the direction in which the needle points is once again noted. Fig-

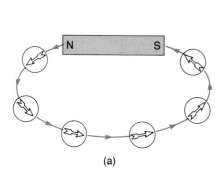

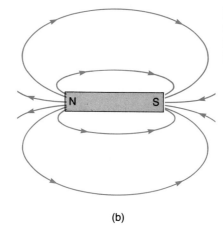

FIGURE 9.2 (a) Tracing the magnetic field of a bar magnet. (b) Several field lines of a bar magnet.

(a) (b)

ure 9.2b shows several magnetic field lines around a bar magnet.

One of the characteristics that we observed for electric fields is that an electric field originates on a positive charge and terminates on a negative charge. Superficially, it appears that we might be able to make a similar statement for magnetic fields, because in Figure 9.2b it appears that magnetic field lines originate on north poles and terminate on south poles. However, this is not the case. *Magnetic field lines have no starting or stopping points. Magnetic field lines always close on themselves.* For example, if we could trace the magnetic field lines of the bar magnet through the magnet itself, we would find the lines continue as shown in Figure 9.3.

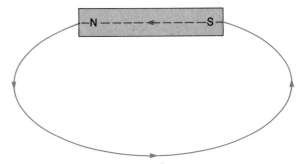

FIGURE 9.3 Magnetic field lines never start or stop.

EXAMPLE 9.1 The spacing of magnetic field lines

When we studied electric fields, we found that sketches of the fields convey information about the strength of the field by the way in which the lines are spaced. At those locations where the lines are drawn closely packed, the field is strong, and where the lines are far apart, the field is weak. This same technique is used to represent magnetic fields. With this in mind, **(a)** discuss the strength of the magnetic field in the region of space between two bar magnets aligned as shown in Figure 9.4a. **(b)** Repeat (a) for two bar magnets aligned as shown in Figure 9.4b.

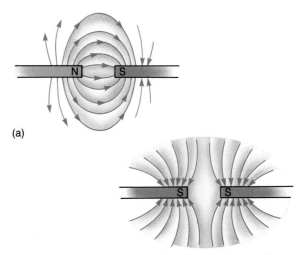

FIGURE 9.4 (a) The field lines between two unlike poles, and (b) the field lines between two like poles.

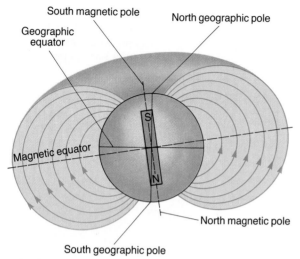

FIGURE 9.5 A schematic view of the magnetism of the Earth. The north-seeking pole of a compass is attracted toward the north geographic end of the Earth, which is a magnetic south pole. Imagine a bar magnet positioned as shown, but, as explained in the text, there is no bar magnet in the center of the Earth. Note that the magnetic south pole does not coincide exactly with the geographic north pole.

Solution **(a)** The magnetic field lines are drawn closely spaced in the region of space between the two magnets, indicating that the magnetic field is strong in this region.

(b) The magnetic field lines are not closely spaced in the region between the two south poles, indicating that the field is weak there.

9.3
THE MAGNETIC FIELD OF THE EARTH

As mentioned previously, a compass is merely a magnetic needle that is suspended so that it can rotate freely. Since navigation relies on the fact that a compass aligns itself at least roughly in the north-south direction of the Earth, it is obvious that the Earth itself must possess some sort of magnetic field. In fact, *the magnetic field of the Earth is similar to the field that would be produced if a giant bar magnet were situated inside the planet with its south magnetic pole near the north geographic pole* (Fig. 9.5). However, several independent obser-

vations convince us that the magnetic field of the Earth cannot be due to large masses of permanently magnetized material buried beneath the Earth's surface. For one thing, geologists believe that sections of the interior of our planet are hot and liquid, and a solid magnet would either melt or lose its magnetism under such conditions. Secondly, the position of the Earth's north magnetic pole moves measurably from year to year. In fact, geologists have determined that the field has switched *direction* from time to time in the past. Evidence for this is obtained by examination of basalt, a type of igneous rock that contains iron, which is spewed forth by volcanic activity on the ocean floor. As the lava cools, it becomes magnetized and leaves a permanent record of the Earth's magnetic field at the time it solidified. These rocks can be dated by their radioactivity to provide the evidence for these periodic reversals of the Earth's field. No theory can explain how a solid magnet inside the Earth could produce this effect.

It is believed that the source of the Earth's field is charge-carrying convection currents in the molten core of the Earth. As noted earlier, charges in motion can produce a magnetic field, and this field would be similar to that set up by a bar magnet. No one knows for sure what processes are at work inside the Earth to produce its field, because no one knows exactly what is happening at the Earth's core. The core is too far beneath the surface to be observed directly, and any laboratory or computer models of this region represent only approximations of reality.

Whatever the exact mechanisms, planetary magnetic fields are quite common. The magnetic field of the Earth is hardly unique. Several other planets and moons in our Solar System, our Sun, and most of the other stars all produce magnetic fields. There is some evidence to indicate that the rate of rotation of a planet has an effect on the magnetic field set up by that planet. For example, Jupiter rotates faster than Earth, and recent space probes indicate that Jupiter's magnetic field is stronger than ours. Venus, on the other hand, rotates more slowly than the Earth, and its magnetic field is found to be weaker.

If a compass needle is allowed to swing freely in both the horizontal and vertical directions, it is found that the needle is horizontal with respect to

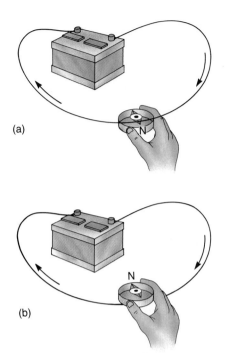

(a)

(b)

FIGURE 9.6 (a) A compass needle points as shown when held below the wire, and (b) it reverses direction when placed above the wire.

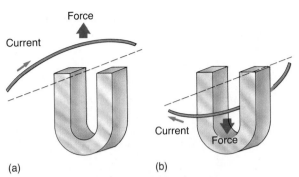

(a)   (b)

FIGURE 9.7 (a) A wire carrying a current is deflected upward when the current direction is as shown. (b) It is deflected downward when the current is reversed.

the Earth's surface only when it is near the equator (Fig. 9.5.). As the compass moves northward, it points more and more toward the surface of the Earth, until at a point just north of Hudson Bay in Canada, the needle points directly downward. This location was first found in 1832, and it is considered to be the site of the south magnetic pole of the Earth. This site is approximately 1300 mi from the Earth's geographic north pole. Similarly, the Earth's north magnetic pole is about 1200 miles away from the Earth's geographic south pole. This means that a compass needle points only roughly toward the geographic poles of the Earth. For example, along a line through Florida and the Great Lakes, a compass indicates true north, but in Washington State, the needle aligns 25° east of true north.

9.4
THE CONNECTION BETWEEN ELECTRICITY AND MAGNETISM

Early scientists did not recognize the connection between electricity and magnetism. Magnetic compasses were used by navigators probably more than 2000 years ago, and scientists have experimented with electric currents for several hundred years, but the first investigation of the interrelationship between the two phenomena was conducted by Hans Christian Oersted in 1820. Oersted's simple experiment was to hold a magnetic compass needle under a wire carrying a direct current, as shown in Figure 9.6a. He noticed that the needle of the compass was always affected in the same way. When the compass was placed below the wire as shown in Figure 9.6a, the needle lined up at right angles to the wire with its north end pointing as shown. When the compass was placed above the wire, the needle again aligned so that it was perpendicular to the wire, but the needle reversed direction, as shown in Figure 9.6b.

In a second experiment, Oersted had a flexible wire running through the poles of a horseshoe magnet, as shown in Figure 9.7a. When a current passed through the wire, the wire bent upward or downward, depending on the direction of the current, as shown in Figure 9.7b. When the wire was held in line with the direction of the magnetic field lines of the magnet, the wire was not affected. The sections that follow will tell us why these effects were observed.

The magnetic field set up by a current-carrying wire

Whenever an electric current exists in a wire, a magnetic field is set up in the space around the wire. Thus, if a compass is brought into the field, it is affected in accordance with Oersted's obser-

A large horseshoe magnet.

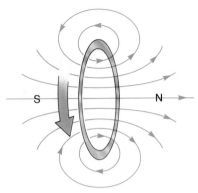

FIGURE 9.9 The magnetic field of a loop of wire carrying a current.

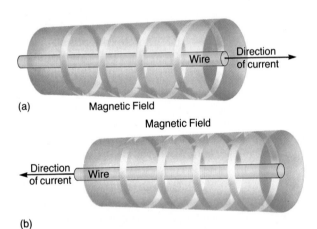

(a)   Magnetic Field

Magnetic Field

(b)

FIGURE 9.8 A magnetic field around a current-carrying wire.

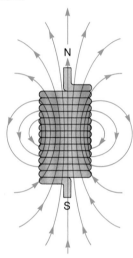

FIGURE 9.10 Magnetic field set up by a coil of wire.

vations. It is now known that a wire carrying a current produces a magnetic field such that the lines of the field form concentric circles around the wire, as shown in Figure 9.8. The direction of the field depends on the direction of the current, as shown in Figure 9.8a and b.

If the wire carrying a current is bent into the shape of a loop, the magnetic field pattern set up by it is like that shown in Figure 9.9. Finally, if several loops of wire are wound around a frame, the magnetic field set up is as shown in Figure 9.10. Several turns of wire can produce a magnetic field that is quite strong. Coils of wire used to produce magnetic fields are called **electromagnets,** and they have many practical applications.

Magnetic forces and moving charges

If a magnet is held near a wire that is carrying a current, a force is exerted on the wire, which can cause the wire to move as Oersted demonstrated. Oersted's discovery showed that electrical energy of the moving charges can be converted into mechanical energy, which causes the wire to move. This result ultimately led to the development of the electric motor. Subsequent investigation of these forces by Oersted and others showed that the moving charges do not have to be confined to a wire.

Any charge that is moving in a magnetic field has a magnetic force acting on it.

A large coil of wire used to demonstrate magnetic effects of electric currents.

Figure 9.11 shows a positive charge moving upward on the page in a region in which a magnetic field exists. The crosses are used to indicate that the direction of the magnetic field is into the paper. (It is easy to remember the meaning of the crosses if you think of the magnetic field as being represented by arrows with feathers at the back of the arrow. Thus, the crosses indicate the arrow moving away from you, and a dot indicates the tip of the arrow coming toward you.) When a charge moves through a magnetic field in this manner, a force is exerted on the charge by the magnetic field. The force, however, is not in the direction

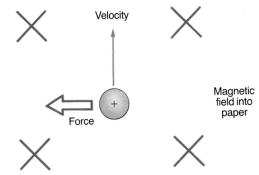

FIGURE 9.11 A charged particle moving perpendicularly to a magnetic field has a force on it as shown.

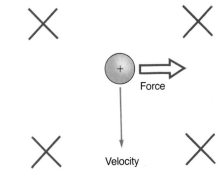

FIGURE 9.12 Reversing the direction of the velocity of the charge reverses the direction of the force.

of the field, nor is it in the direction of the motion of the charge. In fact, the direction of the force is perpendicular to both the field and the velocity of the charge. For the situation shown in Figure 9.11, the direction of the force is toward the left. If the motion of the charge is reversed as shown in Figure 9.12, the direction of the force is also reversed. On the other hand, if the motion of the charge is along the direction of the magnetic field lines as in Figure 9.13a, or in the direction opposite to that of the field lines as in Figure 9.13b, there is no force acting on the charges.

Moving charges in a magnetic field have a force acting on them whether they are in air or in a material. Since an electric current is a motion of charges through a wire, it is expected that there would be a force on the wire. For example, in Figure 9.14a, we see a portion of a wire carrying a current upward on the page and in a magnetic field directed into the paper. As Figure 9.14b shows, each charge has a force acting on it toward the left, and this force is communicated to the atoms of the wire and to the "sides" of the wire, producing a net force toward the left on the wire.

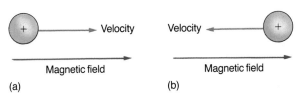

FIGURE 9.13 If the motion of the charge is (a) in the same direction as the magnetic field or (b) in the opposite direction, there is no magnetic force on the charge.

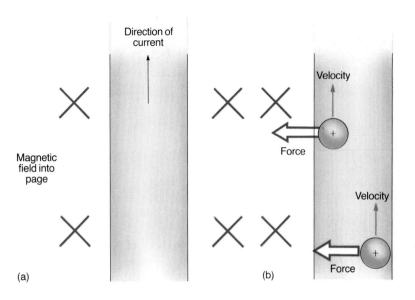

Direction of current

Velocity

Force

Velocity

Force

Magnetic field into page

(a)

(b)

FIGURE 9.14 (a) A wire carrying a current upward in a magnetic field directed into the page. (b) Each moving charge has a force on it to the left. This produces a net force on the total wire which is to the left.

Experimentally, it is found that *the strength of the force acting on a current-carrying wire is dependent on the magnitude of the current and on the strength of the field.* The larger the current and the magnetic field, the stronger the force.

EXAMPLE 9.2   Which way does the compass point?

(a) Figure 9.15 shows a wire carrying a current coming out of the page and the magnetic field pattern set up by this current. If a compass needle is placed at A and then at B in the figure, which way will the needle point at each location?

Solution A compass needle will align in the direction of the magnetic field. Thus, it will point as shown in Figure 9.16.

(b) Figure 9.17 shows the magnetic field set up by an electromagnet. Which end of the electromagnet is the north end? Also, discuss the strength of the field inside the coil, assuming that the field lines inside the electromagnet are drawn correctly.

Solution Magnetic field lines always appear to exit from the north end of a magnet. Thus, the left end of the electromagnet is the north pole. Since the field lines are drawn evenly spaced

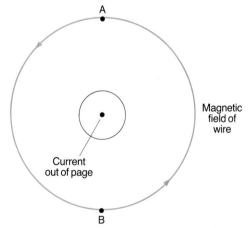

FIGURE 9.15 Which way will a compass point at locations A and B?

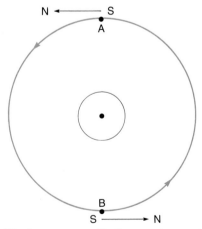

FIGURE 9.16 A compass will align as shown at locations A and B.

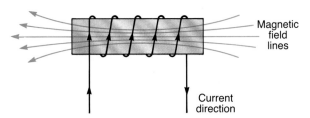

FIGURE 9.17 Which end of the electromagnet is the north end?

throughout the interior of the coil, the spacing indicates that the magnetic field has the same value at all points in the coil. The fact that a coil can set up a uniform field in its interior is an important feature of the device in many applications.

EXAMPLE 9.3   Steering charged particles

In Chapter 8, we discussed how a black-and-white television set works. We saw that the charges in the tube could be moved from point to point on the screen by use of electrical forces produced by horizontal and vertical deflection plates. Discuss how this steering could also be done with electromagnets. In fact, in practice the steering of the moving charges *is* done with electromagnets.

Solution   Figure 9.18 shows how the electromagnets are arranged at the neck of the tube. When a current is sent through the coils, a magnetic field is produced that exerts a force on the electrons when they pass through the field. This force deflects the electrons from their path, and by varying the current in the electromagnets and hence

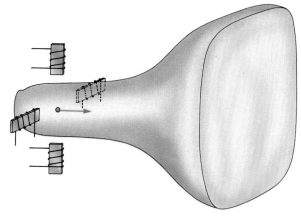

FIGURE 9.18   Electromagnets used to steer electrons in a TV picture tube.

the strength of the field, the beam can be moved to any location on the face of the screen.

9.5
WHAT MAKES A PERMANENT MAGNET MAGNETIC?

Although it is possible to isolate electrically positive particles and to separate them from negative ones, no one has ever isolated particles with only a north or a south magnetic pole. As we have seen, any object that is magnetic has both a north and a south pole. A clue to the ultimate cause of magnetism is given by the kind of magnetic field that is produced by a loop of wire as shown in Figure 9.19a. We see that the field produced is quite similar to that produced by a bar magnet as shown in Figure 9.19b.

FIGURE 9.19   (a) The magnetic field set up by a loop of wire and by (b) a bar magnet are very similar.

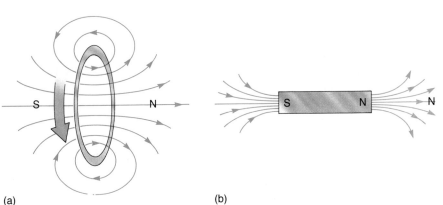

(a)   (b)

The explanation of why magnetic materials are magnetic lies in the structure of iron. This magnetism arises in two ways: (1) An individual atom can be magnetic. (2) The individual electrons of the atom possess an intrinsic magnetism. To see why an atom is magnetic, again consider the fact that a coil of wire produces a magnetic field. If a coil can produce a magnetic field, then any charge moving in a circular path should also produce a magnetic field. That is precisely what is happening in an atom as an electron circles the nucleus of the atom. Each negatively charged electron circles about the nucleus, thus constituting a motion of charge, or an electric current. From this, we see that a net magnetism for the total iron bar could be produced if we could somehow cause several of these atomic magnets to align inside the material. This does occur, but only to a limited extent. The reason that this is not a more important factor in magnetic materials is that the magnetic field produced by one electron in an atom is often cancelled by the magnetic field produced by another nearby revolving electron in the same atom. The net result is that the magnetic field produced by the electrons orbiting the nucleus is either zero or very small for most materials.

The most important reason for the magnetic effects exemplified by materials such as iron lies in an intrinsic property of the individual electrons. To understand this property, let us think about the Earth orbiting the Sun. As it revolves around the Sun once each year, the Earth is also rotating on its axis once each day. In like manner, as an electron circles the nucleus, it is also rotating on its axis. These rotating electrons also produce a magnetic field as they turn. In four materials (iron, cobalt, nickel, and gadolinium), the magnetic field of one (or more) electrons produces a net magnetic field for an individual atom. In turn, several of these atoms act in unison to produce a small cluster of atoms with a net magnetic field for the cluster. These groups of cooperating atoms are called **magnetic domains.** In an unmagnetized piece of material, we can picture these domains as pointing in random directions, as shown in Figure 9.20a. When the material becomes magnetized, these domains have been caused to align as shown in Figure 9.20b. The alignment of domains explains why the strength

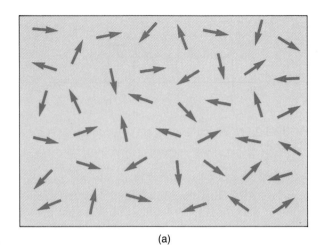

(a)

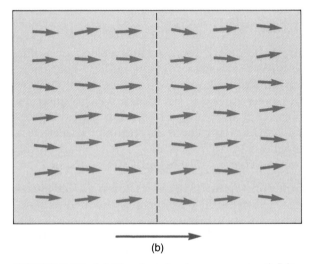

(b)

FIGURE 9.20 (a) The domains in an unmagnetized substance point in random directions. (b) When the substance is magnetized, the domains tend to align.

of an electromagnet is increased dramatically by the insertion of an iron core into the magnet's center. The magnetic field produced by the current in the loops of wire causes the domains to align and thus to produce a strong field. Figure 9.20b shows why one ends up with two identical magnets when a magnet is broken in half. Breaking the magnet along the dashed line in the figure results in two pieces that still have north and south poles at each end. There is no way to ever break a magnet into small enough pieces to isolate a free north pole and a free south pole.

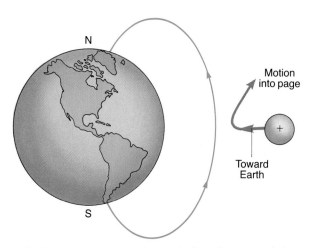

FIGURE 9.21 A charged particle heading toward the surface of the Earth is deflected by the Earth's magnetic field as shown.

9.6
COSMIC RAYS AND THE EARTH'S MAGNETIC FIELD

The Earth's magnetic field provides a protective shield around our planet. The Earth is constantly subjected to bombardment by many different types of particles and radiation from outer space. The most energetic particles, known as **cosmic rays,** originate from interstellar space and are believed to be ejected when dying stars explode. Other high-speed particles originate from huge storms on the surface of the Sun. Energetic particles from either source can produce disruptions in our lives and environment that range from the extremes of producing damage in living cells and altering genetic information carried in the reproductive organs of all living things to the more minor disruptions of television and radio communication.

When charged particles approach the Earth, they come under the influence of its magnetic field. Since these field lines are roughly parallel to the surface of the Earth over much of its area, an incoming particle is likely to approach the Earth in a direction that is perpendicular to this field. Now, recall that when the velocity of a particle and a magnetic field are at right angles, a magnetic force is exerted on the particle, which in this case deflects the particle sideways as shown in Figure 9.21. Thus, most of the particles are deflected away from the surface of the Earth. Of

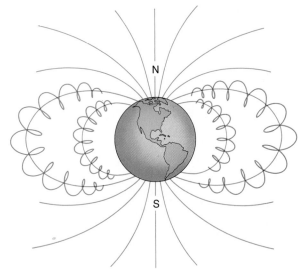

FIGURE 9.22 A charged particle is trapped by the Earth's field and spirals along one of the field lines.

course, some do make it through, and these constitute a part of the background radiation to which all life has become adapted. If the Earth's field were to disappear, all life would be subjected to a much higher number of incoming particles. It is likely that more mutations would occur, with results that cannot be predicted. It is interesting to note that evolution has not been steady during geologic time. Instead, there have been periods of rapid change and other periods when life forms have changed little. Some scientists believe that the periods of rapid evolutionary change may coincide with times when the Earth's magnetic field was weakened or nonexistent.

With the advent of orbiting satellites, a most astounding discovery was made in 1957 by James Van Allen of the University of Iowa. Instruments carried aloft in spacecraft found two belts high in the atmosphere in which many charged particles were trapped. (The discovery of these belts is considered to be the first major scientific discovery of the Space Age.) These belts are filled by particles coming in from outer space such that their direction of travel would bring them in almost parallel with the Earth's magnetic field. In such a case, it can be shown that a particle would be trapped by the field and would spiral around a magnetic field line, as shown in Figure 9.22. As the particle approaches one of the Earth's poles,

The Aurora Borealis is a common sight in Alaska, but a rare event as far south as Arizona. On April 12, 1981, the red glow of the aurora in the northern sky was seen near Tucson, Arizona. (Courtesy NOAO)

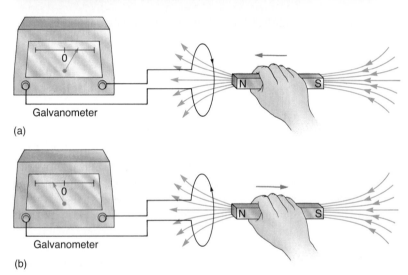

FIGURE 9.23 Relative motion between a magnet and a coil of wire produces an electric current in the coil.

the field gets stronger, and the spiral winds in on itself. Finally, near the pole the particle reverses its direction of travel and spirals back toward the other pole. A proton trapped in this way may travel from pole to pole once every few seconds and may remain trapped for several hundred years.

There are two belts that seem to favor this trapping process. One is about 2000 miles above the Earth's surface, and the second is about 10,000 miles above the Earth. For the most part, these trapped particles go unnoticed by us in our day-to-day lives. However, the particles occasionally build up to such an extent that large numbers of them spill out of their to-and-fro orbits above

one of the poles. These spilled particles lose energy in collisions with air molecules and produce the beautiful **auroras** that sometimes light the northern and southern skies.

9.7 ELECTROMAGNETIC INDUCTION

Shortly after Oersted's experiments, people began to speculate that since a current produces a magnetic field, perhaps a moving magnet would produce an electric current. Eleven years after Oersted's experiments, Joseph Henry of the United States and Michael Faraday of England

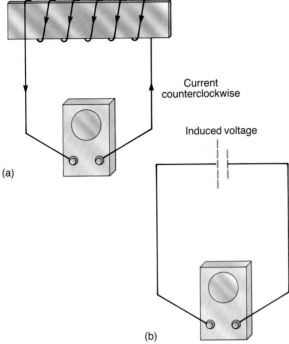

Current counterclockwise

Induced voltage

(a)

(b)

FIGURE 9.24 (a) The current produced by the coil is in the same direction as it would be if (b) a battery were in the circuit. The voltage produced is called an induced voltage.

electricity in which there has been a current set up in a wire. However, in all those cases, there has *always* been a power supply, such as a battery, present to produce the current. In the present case, there is a current with no apparent source. In these circumstances, we say that *the effect of the magnetic field has been to induce a current in the coil.* It is as though the changing field lines passing through the coil somehow produce a phantom battery in the coil. For example, if the current is counterclockwise, as shown in Figure 9.24a, this battery would have to be connected as shown in Figure 9.24b to produce the current. This voltage that is produced is referred to as an **induced voltage.**

In performing this experiment yourself, you could easily demonstrate several facts:

1. There is an induced current in the coil only when there is relative motion between the coil and the magnet. That is, a current is induced when the magnet is moved closer to the stationary coil or pulled farther away, or a current is induced when the coil is moved closer to or farther away from a stationary magnet.
2. The faster the relative movement, the greater the current.
3. The number of turns of coil also affects the current. If the coil has several turns, the magnet affects each individually; thus, twice as many turns result in twice as much current.

Think of the simplicity of these experiments of Oersted, Henry, and Faraday and of their significance in world history. Oersted discovered that electrical energy can be converted to mechanical energy. Then Henry and Faraday discovered that mechanical energy can induce an electric current. The way was paved for the development of electric motors, generators, and the entire electric age.

independently tried to produce an electric current by moving a magnet up and down through a coil of wire. The experiments performed by these men were relatively simple, yet on these modest grounds lie the basic principles used to generate the enormous amounts of electrical energy produced throughout the world today. Let us examine what they did.

In Figure 9.23a, a bar magnet is placed close to a coil of wire such that some of the magnetic field lines from the magnet thread through the coil. Also, a sensitive current detector is placed in the coil. Now, if you move the magnet toward the coil (Fig. 9.23a) so that more magnetic field lines thread through the coil, you will note that while the motion of the magnet takes place, the current detector deflects toward the right, indicating the presence of a current in the coil. On the other hand, if you move the magnet away from the coil (Fig. 9.23b) so that fewer field lines thread through, you will note that the needle deflects in the opposite direction. This result is astonishing. We have seen many examples in our study of

---

EXAMPLE 9.4 Is there motion or not?

Figure 9.25 shows an electromagnet connected to a battery and an open switch. Nearby is a coil of wire with a current detector A. When someone closes the switch, is there an induced current in the coil? Why or why not?

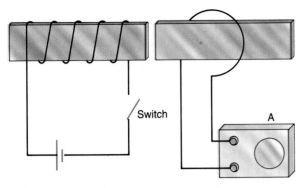

FIGURE 9.25 When someone closes the switch, is there a current in detector A?

9.8
TAPE RECORDERS

The common reel-to-reel tape recorder that is a part of every music enthusiast's collection provides some interesting examples of practical uses of the principles we have discussed thus far. The design of a typical recorder is shown in Figure 9.26a. Tape moves from one side of a cassette reel past a recording head and a playback head to a take-up reel in the same cassette. The tape itself is a plastic ribbon coated with a metallic oxide that can become magnetized in the presence of a magnetic field.

The steps in recording are illustrated in Figure 9.26b. A sound wave is translated by a microphone into an electric current, which is then amplified and passed through a wire wound around a doughnut-shaped piece of iron. The iron ring and the wire form an electromagnet that constitutes the recording head of the instrument. The varying current in the wire produces a varying magnetic field in the iron, and this field is completely inside the iron except at the point where a slot is cut in the ring. At this location, the magnetic field fringes out of the iron and magnetizes the small pieces of metallic oxide on the tape as the tape is pulled past the head by a motor. The pattern of magnetization preserved on the tape simulates the frequency and loudness of the sound signal originally used to produce the magnetization. Thus, *the recording process uses the fact that a current passing through an electromagnet produces a magnetic field.*

To reconstruct the sound signal, the tape is allowed to pass through the recorder again, this time with the playback head in operation. Figure

Solution At first thought, the answer might be no because there does not seem to be any relative motion between the two coils. However, there *is* a deflection of the current detector when the switch is closed, indicating the presence of a current. Let us see why this occurs.

With the switch open, there is no current in the electromagnet, and hence it produces no magnetic field. As a result, the single coil finds no magnetic field lines threading through it. When the switch is closed, there is a current; hence, the electromagnet produces a magnet field. Now the coil does have a magnetic field threading through it. Thus, even though there has been no relative motion, the net effect is as though the coil were moved from far away where the field is zero to very close to the electromagnet where the field is strong. It is as though one had "tricked" the coil into believing that either it or the electromagnet had moved. Thus, there *is* an induced current.

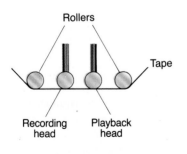

(a)

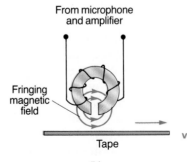

(b)

FIGURE 9.26 (a) The essential elements of a tape recorder. (b) As the tape is drawn by the recording head, the fringing magnetic field of the head magnetizes the tape.

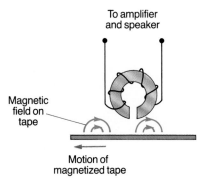

FIGURE 9.27 Playback of a magnetic tape in a tape recorder.

9.27 shows that this head is very similar to the recording head in that it consists of a doughnut-shaped piece of iron with a wire coil wound around it. When the magnetized tape is pulled past this head, the varying magnetic field pattern present on the tape produces changing field lines through the wire coil on the playback head. These changing lines induce a current in the coil that corresponds to the current in the recording head that produced the tape originally. This changing current can be amplified and used to drive a speaker. *The playback process is thus an example of induction of a current by a moving magnet.*

9.9
ELECTRIC MOTORS

An **electric motor** is a device that converts electrical energy into mechanical energy. The first conversion of this kind was performed by Oersted when he held his compass near an electric current. The magnetic field produced by the wire caused the compass to turn. Thus, electrical energy was converted to mechanical energy of motion. However, this was not a very useful or efficient motor, for once the needle turned so that it was along the direction of the magnetic field of the current-carrying wire, the needle stopped moving. If an electric drill were constructed in this manner, it would turn only half a revolution and then stop. In order to keep a motor running, some way must be found to alternate the current flow so that the shaft of the motor is forced to turn continuously. To understand this concept, refer back to Oersted's original experiment, redrawn in Figure 9.28. When the current is in one direction as shown in Figure 9.28a, the compass

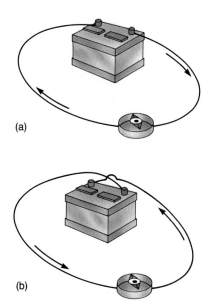

FIGURE 9.28 Demonstration of the principle behind an electric motor. (a) The compass needle aligns as shown when the current is right to left, and (b) it reverses direction when the current direction reverses.

needle moves such that it points in a direction perpendicular to the wire, and there it stops. But if the wires connected to the battery are reversed at this precise instant, the current changes direction as in Figure 9.28b, the magnetic field produced by the wire also changes direction, and the needle rotates 180° as shown. If the wires are once again reversed, the needle rotates again to the position of Figure 9.28a, *ad infinitum.* Thus, the needle rotates continuously if some way can be found to switch the wires rapidly and in synchronization with the compass needle.

The essential parts of a slightly more realistic motor are shown in Figure 9.29. Here a coil of wire is mounted so that it is free to rotate, and a shaft connected to it turns as the coil turns. Connected to the coil are two semicircular pieces of metal called brushes. In order to see how the motor works, refer to Figure 9.30, where we show a battery as the source of current in the coil. In Figure 9.30a, current leaves the positive terminal of the battery, moves through brush A around the coil in the direction indicated, through brush B, and back to the battery. The side of the coil labeled JK is in a magnetic field, and it is carrying a current. Thus, there is a magnetic force acting on it. The direction of this force is upward, as

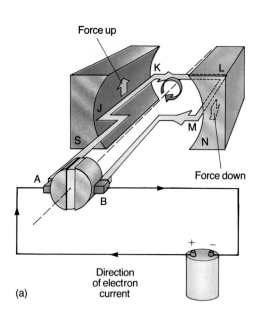

Force up

Force down

Direction
of electron
current

(a)

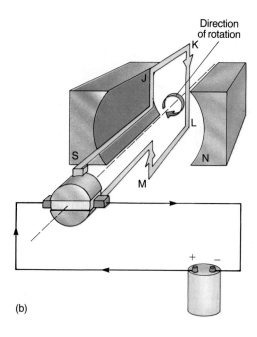

Direction
of rotation

(b)

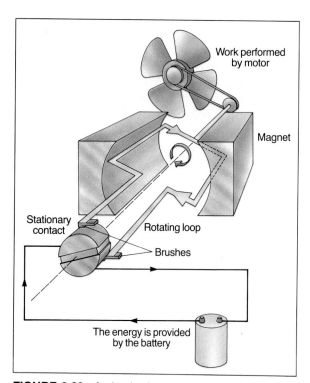

Work performed
by motor

Magnet

Stationary
contact

Rotating loop

Brushes

The energy is provided
by the battery

FIGURE 9.29 A simple d.c. motor.

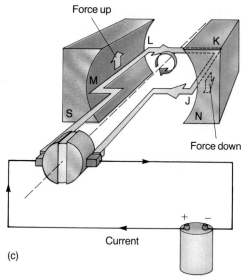

Force up

Force down

Current

(c)

FIGURE 9.30 Several stages in the motion of the
coil of a d.c. motor.

FOCUS ON . . . Magnetohydrodynamics (MHD)

A mechanical generator is not the only way to produce electricity. Recall from the text that an electric current can be induced if a coil of wire is moved in a magnetic field. In a broader sense, the forces exerted on charged particles moving through a magnetic field can also be used to generate a voltage and a current. In a magnetohydrodynamic (MHD) generator, there are no mechanical moving parts. Some source of heat is used to strip electrons off of ordinary atoms to create a hot mixture of free electrons and positive ions, called a **plasma.** This plasma expands through a nozzle and travels at supersonic speeds through a magnetic field. When the charged particles are in the magnetic field, magnetic forces push electrons upward and positive charges down-

ward (see Fig). These charges strike the plates, which are in turn connected to an external circuit. These charges create a voltage and a current that is supplied to the light bulb shown. MHD generators suffer no frictional losses, because they have no mechanical parts. In addition, the plasma can be maintained at a very high temperature, resulting in high efficiency. Finally, after the plasma has passed through the magnetic field, it is still hot and the thermal energy can be used to boil water, create steam, and run a secondary mechanical generator, as indicated in the figure. For all these reasons, MHD generators are more efficient than mechanical installations and may become the system of choice in the future.

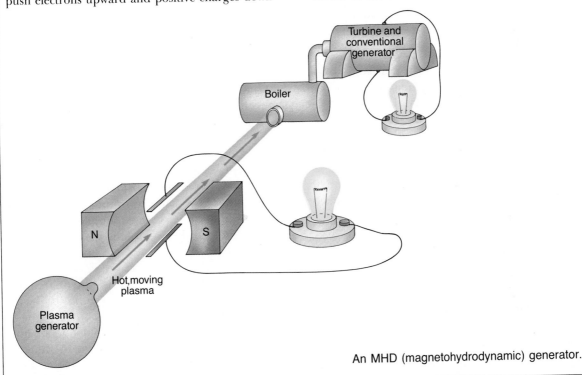

An MHD (magnetohydrodynamic) generator.

shown. Likewise, the wire labeled LM has a force directed downward acting on it. These forces acting on the coil cause the coil to rotate until it reaches the position shown in Figure 9.30b. At this position, the brushes momentarily become disconnected from the rings and, as a result, no current passes through the coil. Thus, there are no forces acting on the sides JK and LM. How-

ever, the motion of the coil causes it to continue to rotate under its own momentum until it reaches the position shown in Figure 9.30c. Sides JK and LM have now changed position relative to that shown in Figure 9.30a, and the forces on these wires are now in the direction indicated. Thus, the coil continues to rotate, always in the same direction.

FOCUS ON . . . Motors, Generators, and Electric Cars

Motors and generators are so similar that one device can be used for both functions. If electric cars are to become popular in the future, they will probably contain a motor-generator unit. When power is needed, electrical energy is drawn from the batteries to drive the motor. However, when the driver wishes to slow down or is rolling downhill, a control unit converts the motor into a generator, and the kinetic energy of the car is used to generate electricity and recharge the batteries. Since no system is 100 percent efficient, the batteries need periodic recharging from an external source, but the motor-generator system is still much more efficient than a simple motor.

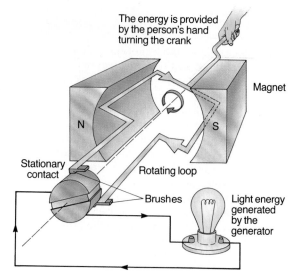

FIGURE 9.31 A simple generator.

The motor discussed here is not likely to be very powerful. Recall that the magnitude of the force on the segments JK and LM depends on the strength of the current in the wires and on the strength of the magnetic field that these wires are in. A practical motor differs from our simple picture in two respects. First, one loop of wire as shown in our diagram carries only a small amount of current, so the core of any useful electric motor must have many loops of wire rather than one. Secondly, we have produced a magnetic field by use of a permanent magnet. These magnets are not suitable for a rugged, long-lasting motor because they are extremely heavy and may easily become demagnetized. As a result, the magnetic field is provided by an electromagnet.

9.10 ELECTRIC GENERATORS

An electric motor operates on the principle that whenever a current-carrying wire is in a magnetic field, a sideways force acts on the wire. A **generator** utilizes the principle that a current is induced in a coil of wire when the number of magnetic field lines passing through the cross-sectional area of the coil changes with time. A generator looks very much like a simple motor. The difference between the two is often expressed by saying that a motor uses a current to produce rotary motion, whereas a generator uses rotary motion to produce a current, as indicated in Figure 9.31. A cycle of the motion of a simple generator is depicted in Figure 9.32. The basic parts of a generator consist of a coil of wire, a means to make the coil rotate, and two rings that rub (or brush) against stationary contacts. The brushes allow for connections to be made to some external circuit, which could be an entire city but in our figure is a single light bulb. Let us follow what happens as the coil is caused to rotate.

When the coil is horizontal, as in Figure 9.32a, no magnetic field lines pass through it, but a small rotation in the direction indicated by the arrows causes several lines to pass through its cross-sectional area. The rapid change in lines through the coil induces a current in the coil. This current flows around the coil in the direction indicated, through the brushes and rings to the light bulb. As the coil moves from the horizontal position to a vertical one, as shown in Figure 9.32b, the current decreases. This occurs because when the coil is almost vertical, a slight rotation of it does not change appreciably the number of field lines threading through it. In fact, the current drops to zero when the coil is exactly vertical. In Figure 9.32c, the coil has again rotated through the vertical and back to a hori-

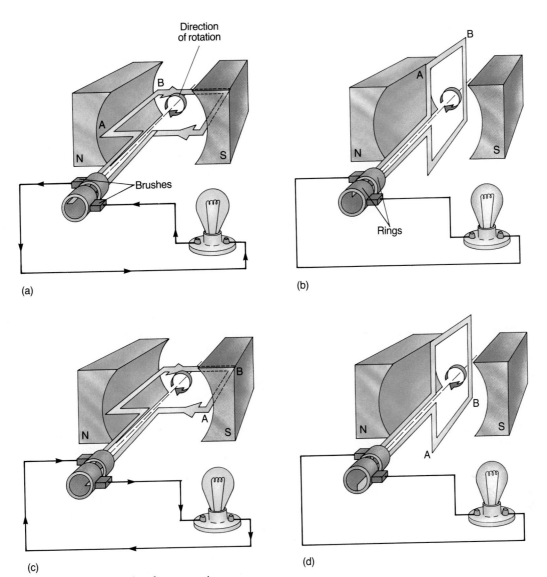

FIGURE 9.32 A cycle of a generator.

zontal position. During this rotation, the current becomes large again, but a careful analysis shows that the current through the light bulb has reversed direction. A continued rotation of the coil again takes it to the vertical position as in Figure 9.32d, and again the current drops to zero.

Note an important result brought forth in our analysis. The current through the light bulb is considerably different from that which would be produced by an ordinary battery. A battery produces a constant current that always moves in the same direction through the bulb. The generator produces a current that has a large value in

one direction, decreases to zero, increases to a large value in the opposite direction, and again decreases to zero, as shown in Figure 9.33. A current of this nature, one that changes direction and magnitude regularly, is called an **alternating current** and is the type provided to your home by a utility company. The alternating current produced by generators in this country varies at 60 Hz. That is, the current goes through a complete cycle 60 times per second.

The problem of causing the coil to rotate is solved in a variety of ways; a common method is to direct falling water against the blades of a de-

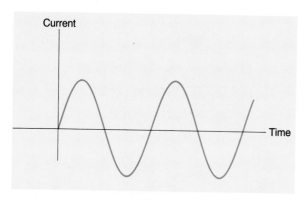

FIGURE 9.33 Alternating current.

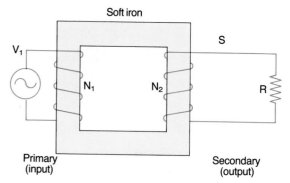

FIGURE 9.34 A simple transformer.

vice called a turbine. The water striking the blades produces rotational motion of a shaft on which the coil of the generator is mounted. Coal, oil, and nuclear power plants use turbines in a similar manner. However, in these cases the heat generated by the burning fuel is used to convert water to steam, which is then directed at the blades of the turbine.

9.11 TRANSFORMERS AND TRANSMISSION OF ELECTRICITY

Most of the electricity utilized today is generated in large, centralized power plants, and in North America there is an increasing trend toward situating these plants near sources of fuel, especially coal mines, rather than near urban centers where most of the electricity is needed. The reason for this choice is that it is cheaper to transmit the electricity than it is to transport the coal. As a result, long-distance transmission lines have become common. An important problem that utility companies face is that of transmitting the electrical energy over large distances to the consumer without losing much of the energy in the process. Let us briefly examine this problem.

Electrical energy is lost in the transmission lines primarily because some of the electrical energy is converted to heat in the lines. These heat losses occur because the wires have resistance, and the equation Power = I^2R tells us the rate at which energy is lost through resistive heating. Utility companies have the choice of transmitting

their electrical energy either at high voltage and low current, or at high current and low voltage. As Power = I^2R shows, less energy is lost if the electricity is transmitted at low current and high voltage. As a result, modern long-distance transmission lines operate at 120,000 V or more. However, it would be disastrous to feed electricity into household circuits at this potential; appliances would spark uncontrollably, wires would burn out, and a slight mishap would be fatal. Thus, the high voltages used to transmit the electricity from the source to the site of consumption must be "stepped down" to the normal voltages encountered in a home, approximately 120 V. A device that accomplishes this conversion is called a **transformer.** A simple transformer is diagrammed in Figure 9.34. It consists of a number of turns of wire wound around a piece of iron. One set of windings, called the **primary** coil, is connected to a source of alternating voltage that is to be changed. A second coil of wire, called the **secondary,** is also wrapped on the same iron core. As the direction and magnitude of the current in the primary change, the magnetic field set up by the primary also changes simultaneously. The iron core provides a pathway for the magnetic field lines produced by the primary to be led through the secondary coil. Thus, the secondary coil finds a changing magnetic field threading through its cross-sectional area. As we have seen, a changing magnetic field through the secondary produces an induced current and voltage in this coil. To see how such a device can be helpful to us, let us assume that there is one loop of wire in the primary

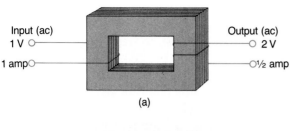

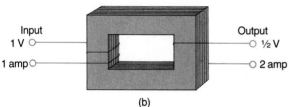

FIGURE 9.35 (a) A 1 volt potential in one loop of the primary coil produces 1 volt in each loop of the secondary. Therefore, the potential in the secondary coil is amplified to 2 volts, but the current is reduced by half. (b) In this situation, the voltage is cut in half, and the current is doubled.

and two loops in the secondary. A changing magnetic field in the primary induces a voltage in each loop of the secondary. Thus, if the input signal to the primary has an amplitude of 1 *V*, the output from the secondary will be 2 *V* (Fig. 9.35). When a transformer is used in this way to step up the voltage, it is appropriately called a *step-up transformer*. If the coil is constructed such that the secondary has one loop while the primary has two loops, the voltage at the secondary will be decreased by a factor of two, and we have a *step-down transformer*. The relationship between the number of turns and the voltages is given by

$$N_p V_s = N_s V_p \qquad (9.1)$$

where N_p and N_s refer to the number of turns on the primary and secondary, respectively, and V_p and V_s are the voltages at the primary and secondary.

If the voltage at the secondary is greater than that at the primary for a step-up transformer, does this mean that we are getting something for nothing? Since we know that energy is conserved,

(a)

(b)

(c)

(a) A small demonstration transformer which shows the primary and secondary windings and the magnetic core which links them. (b) A stepdown transformer used near a home to reduce the voltage supplied to the house to about 120 V. (c) A stepdown transformer used near a city to reduce the voltage from the utility company down to about 3500 V.

obviously we are not. Let us see why. Since the primary and secondary are operating together for the same amount of time, the power supplied to the primary also appears at the secondary. Since P = IV, this equation tells us that if the voltage at the secondary is greater than that at the primary by a factor of, say, two, the current at the secondary is stepped down by a factor of two. Thus, an increase of voltage at the secondary is accompanied by a simultaneous decrease in the current by the same amount. The relationship between the power at the primary and at the secondary is given by

$$(IV)_{secondary} = (IV)_{primary} \qquad (9.2)$$

Large transformer centers near every big city step down the 120,000 V arriving from the generating plant to about 2200 V for transmission along conventional "telephone pole" routes to residential and commercial buildings. The 2200 V must be stepped down again before electricity is brought into a house or a business. Therefore, another transformer, one that steps down the voltage to about 110 V, is placed near these establishments.

EXAMPLE 9.5   Stepping down

Imagine that you wish to step down a high-voltage line at 120,000 V to 2200 V for local transmission.

(a) If there are 100,000 coils of wire in the primary circuit of the transformer, how many coils will be needed in the secondary?

(b) If the high-voltage line carries 100 A, how many amps will be present in the secondary circuit?

Solution   (a) In this situation, a step-down transformer is needed. The number of turns on the secondary can be found from Equation 9.1 as

$$N_s = N_p \frac{V_s}{V_p} = (100,000 \text{ turns}) \frac{2200 \ V}{120,000 \ V}$$
$$= 1830 \text{ turns}$$

(b) To find the current in the secondary, let us use the fact that power is conserved between the primary and the secondary as

$$(IV)_{secondary} = (IV)_{primary}$$

or

$$I_s(2200 \ V) = (100 \ A)(120,000 \ V)$$

from which we find

$$I_s = 5450 \ A$$

SUMMARY

Like magnetic poles repel and unlike poles attract. The direction of a magnetic field is determined by the direction in which the north pole of a compass needle points. Magnetic field lines have no stopping or starting points.

The magnetic field of the Earth is created by some sort of electric generator in the Earth's fluid interior.

When an electric current travels perpendicular to a magnet field, a sideways force is generated. When a magnet is forced to move perpendicularly to an electric wire, or when a wire or any charged particles are forced to move through a magnetic field, a current is generated.

Magnetism is produced by individual domains, and a permanent magnet is created when these domains are aligned. Whenever the number of magnetic field lines passing through a coil is changed, an electric current is induced in the coil.

In an **electric motor,** a current is forced to travel perpendicularly to a magnet, and mechanical work is produced. A **generator** operates on the reverse principle; namely, a coil of wire is mechanically turned within a magnetic field and electricity is generated. **Transformers** can be used to alter the voltage and current (but not the energy) in an alternating current circuit.

EQUATIONS TO KNOW

$N_pV_s = N_sV_p$ (voltage-turn relationship for a transformer)

$(IV)_{secondary} = (IV)_{primary}$ (power relationship for a transformer)

KEY WORDS

North pole (of a magnet)
South pole (of a magnet)
Magnetic field
Magnetic domain

Electromagnetic induction
Electromagnet
Electric motor

Electric generator
Transformer
Plasma

QUESTIONS

MAGNETS AND MAGNETIC FIELDS

1. Are magnetic fields more similar to electric fields or to gravitational fields? Defend your answer.
2. You have two pieces of metal; one is magnetized and one is not. By using *nothing* else except the two pieces of metal, how could you determine which is the one that is magnetized?
3. You have two equivalent iron magnets and keep one in the refrigerator and the other in a hot oven for the same length of time. Which one do you think will preserve its magnetism more completely? Defend your answer.
4. Based on what you know about the magnetic field of the Earth, draw a sketch that represents this field. Your pattern of lines should also continue through the interior of the Earth.

THE MAGNETIC FIELD OF THE EARTH

5. What is the source of energy for the magnetic field of the Earth?
6. The Moon has virtually no magnetic field. Does that information tell us anything about the structure of the interior of the Moon? Defend your answer.
7. Sketch a diagram of the Earth showing the direction a compass would point at (a) its south pole, (b) its north pole, and (c) the equator.
8. Airline pilots have to reset their compasses to correct for irregular variations in the magnetic field of the Earth. What could cause these variations?
9. During the history of our planet, its magnetic field has undergone periodic reversals. Suppose one of these reversals should take place now. How would our lives be affected?

THE CONNECTION BETWEEN ELECTRICITY AND MAGNETISM

10. Modern ships and airplanes use both conventional compasses and sophisticated electronic equipment for navigation. Would it be possible for the electronic equipment to damage the effectiveness of the compass? Do you think that precautions would be needed to shield the compass from the ship's electronics? Explain.
11. If you rub a balloon on your hair to charge it and then place the charged balloon near a compass, will the compass needle deviate from pointing northward? What will happen if you wave the charged balloon back and forth across the top of the compass? Explain.
12. Imagine that you are a space explorer who discovers a rain of high-energy, fast-moving, glowing particles in the atmosphere of some faraway planet. Design a simple experiment to determine whether or not these particles are electrically charged.
13. Will a magnet held near the screen of a TV tube affect the picture? Why? (It is not advisable to try this as a home experiment unless you enjoy paying for TV service calls.)
14. A beam of protons is moving across the room when the north pole of a magnet is pointed at the beam

such that the magnetic field and the motion of the particles are perpendicular. Will the beam be deflected toward the magnet, away from it, sideways, or not at all?

15. How could a magnet be used to determine whether a beam of charged particles is positive, negative, or neutral?

16. Explain why an electromagnet made with ten loops of wire will be stronger than one with five loops of wire. How much stronger will it be?

17. If you were building an electromagnet to lift scrap iron, would you use alternating current or direct current, or would it make any difference? Explain.

WHAT MAKES A PERMANENT MAGNET MAGNETIC

18. An iron nail will be attracted to either end of a bar magnet. Why?

19. The magnetic pattern around a magnet such as a horseshoe magnet can be traced out by placing a piece of plastic over the magnet and sprinkling iron filings on the plastic. The filings align along the field lines. Why?

20. A magnet is capable of supporting long chains of paper clips, each clip dangling from one above it. Explain on the basis of magnetic domains what is happening in each clip.

COSMIC RAYS AND THE EARTH'S MAGNETIC FIELD

21. Would the magnetic field of the Earth deflect high-speed neutrons from outer space? Explain.

22. As mentioned in problem 9, the Earth has undergone periodic reversals in its magnetic field. If the field should drop to zero, what effect would it have on (a) the Van Allen belts? (b) Auroras? (c) The number of charged particles striking the Earth from outer space?

ELECTROMAGNETIC INDUCTION

23. If a powerful electromagnet is suspended by a long cable and held stationary over a coil of wire, will a current be induced in the wire? If a wind comes up and the magnet sways in the breeze, will this induce a current in the wire? Explain.

24. A popular classroom demonstration consists of dropping a light but strong magnet down the length of a piece of copper pipe. Instead of falling rapidly through the pipe, the magnet floats down the pipe. Explain how this could occur?

25. A coil of wire is placed close to the north pole of a magnet, and the flexible coil is rapidly crushed to zero cross-sectional area. Is there a current in the coil while it is being crushed? Explain.

26. A bar magnet is placed on the turntable of a record player and set into rotation at 45 rpm. A coil of wire near the turntable is found to have a current induced in it. Describe how this current would vary with time.

ELECTRIC MOTORS AND GENERATORS

27. Electric motors can be constructed to operate from a direct or an alternating power source, but inside the motor there must always be alternating current. Explain.

28. As the price of copper wire continues to increase, do you think that small portable electric motors might someday be built with permanent magnets rather than electromagnets? Defend your answer.

29. The motor described in the text and illustrated in Figure 9.29 has a loop of wire spinning in a magnetic field. Would it be possible to build a motor in which a permanent magnet rotates in an alternating electric field? Explain.

30. When you start an electric motor, there is an initial rapid surge of current, followed by a rapid approach to constant current flow. Explain why this current surge occurs. Can the surge be harmful to the motor itself? If so, how?

31. Does a generator create energy? Explain.

32. Would a generator work if the coil were held stationary and the magnet rotated? Explain.

TRANSFORMERS AND TRANSMISSION OF ELECTRICITY

33. If the primary coil of a transformer has 100 loops of wire carrying 10 amps at 2000 volts, how many loops are needed in the secondary coil to produce 4000 volts? What about 200 volts? 2000 volts? What would the current be in each case?

34. A transformer can be built to step up the potential of a line from 50 volts to 500 volts. Is this a violation of the conservation of energy? Explain.

35. Slot cars and electric trains operate at about 10 to 15 volts. The 110-volt household current is stepped down to lower voltage in a transformer. The speed of the toys can be changed by changing the voltage applied to them. Draw a labeled diagram of a transformer that can supply a variable output voltage.

36. Explain why direct current is not generally supplied in household circuits.

37. Why is a transformer constructed such that an iron core passes through both coils?

38. The primary of a transformer is connected to a battery, and a direct current flows in the coil. If a switch in the primary circuit is opened, will there be a voltage induced in the secondary?

39. A transformer is constructed with 100 turns in the primary and 400 turns in the secondary. If a voltage of 20 V is applied at the primary and 2 A of current exists in the primary, find (a) the current and (b) voltage in the secondary.

40. According to the text, power losses from generating stations are reduced when the voltage is stepped up to large values. Why doesn't the company then step the voltage up to even higher values?

ANSWERS TO SELECTED NUMERICAL QUESTIONS

33. (a) 200 turns, (b) 10 turns, (c) 100 turns, (d) 5 A, 100 A, 10 A

39. (a) 0.5 A, (b) 80 V

CHAPTER

10

Properties
of Light

Throughout this book, there has been a consistent attempt to relate the concepts of physics to everyday experiences. This was reasonably easy to do in our study of mechanics, because the motion of objects and the forces that cause this motion are readily apparent to our senses. The task became less easy when we delved into electricity and magnetism. We can understand how the principles of electricity and magnetism are applied on a large scale to provide us with electrical energy to heat our homes and do a myriad of other wonderful things for us. However, electricity *per se* is elusive. To understand all of these large-scale phenomena, we must search for answers in the invisible world of electrons and atoms.

As we begin our study of light, we will find that it becomes even more difficult to get our hands on the fabric of light. Although we will be able to explain many everyday occurrences related to light, the question of what it is will be quite difficult to answer. The nature and behavior of light are quite different from anything that can be explained in terms of experience with natural phenomena. It is no wonder that many great physicists of the past have struggled with light.

Newton, for example, considered light to be a stream of particles. According to his approach, a source of light emitted tiny particles, something like a stream of BBs, and we are able to see objects when these little bullets bounce off them and enter the eye. Other scientists found that many of the properties of light could be explained only by considering it to be a form of wave motion. Thus, when we see an object, we do so because light from it breaks like a wave on the retina of our eye. Most of the concepts discussed in our study of light will be explained by the wave model of light. However, keep the fact stored away that sometimes the behavior of light can be explained only from the Newtonian viewpoint of it being a stream of particles. This so-called dual nature of light, in that it sometimes behaves as a wave and sometimes as a particle, is unique among the topics and concepts that we have studied, but there are other fascinating characteristics of light.

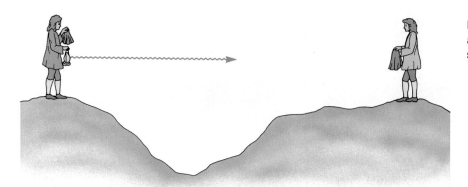

FIGURE 10.1 Galileo's attempt to measure the speed of light.

10.1
THE SPEED OF LIGHT

We pointed out in the opening section that light has many unusual characteristics. One of them is its enormous speed, which we now know to be 3.00×10^8 m/s, or 186,000 mi/s. Think of the enormity of this value. If you turn on a flashlight and its light could spread out through space without being absorbed or diminished, an observer 186,000 miles distant would detect the light beam one second later. It is no wonder that if you turn on an electric light bulb in your room, or throw a switch that turns on a light across the road or even on a distant hilltop, the rays seem to appear almost instantly. How is something as fast as the speed of light measured?

The first recorded experiment in which an attempt was made to determine the speed of light was conducted about A.D. 1600, and it was done by a person first encountered in our study of mechanics, Galileo Galilei. He attempted to measure the speed of light by positioning two people on hilltops separated by a distance of about 1.5 km, as shown in Figure 10.1. Both experimenters carried shuttered lanterns. One person quickly removed the cover from his lantern and began to record the time. As soon as his partner on the neighboring hill saw the light beam reach him, he uncovered his lantern. When the first observer saw that the second lantern had been uncovered, he stopped his measurement of time. From these observations, Galileo hoped to be able to calculate the speed of light from v = distance/time. However, as you might guess, Galileo found that human reaction time was so slow compared with the speed of light that all he was really able to measure was the time required to uncover the lanterns. As far as he could tell, the light made the round trip instantaneously, or at least with an unmeasurably fast speed.

Several techniques have been used to measure the speed of light. Since it is impossible to turn switches on and off fast enough to time the speed of a light ray as one would time the speed of, say, a human runner, all earthbound measurements have been performed with the aid of some continuously moving device. The most famous experiment of this sort was performed by Albert Michelson in 1880.

His apparatus, shown in Figure 10.2, consisted of a rotating eight-sided mirror, an intense source of light, a flat mirror approximately 35 km from the rotating one, and an observer located as shown. With the mirror originally at rest, the light source and observer were aligned so that the light would shine on face A of the eight-sided mirror, reflect to the distant mirror, bounce back to side B, and enter the eye of the observer, as shown in Figure 10.2a. The mirror could be rotated at different speeds, and it was found that light was reflected into the eye only at certain rates of rotation. If the mirror is rotating just right, the light will reflect off side A and return at the instant side C has moved into place to reflect the beam into the observer's eye. However, if the rotation rate is either too slow or too fast, C will not be in the proper position to reflect light into the eye, and the light will bounce off in some random direction, as indicated by Figure 10.2c and 10.2d. Thus, by accurately adjusting the speed of rotation of the mirror until the light was reflected to the observer, Michelson was able

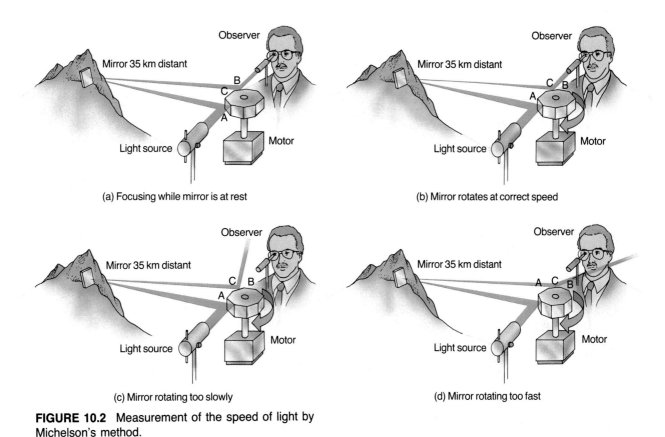

FIGURE 10.2 Measurement of the speed of light by Michelson's method.

to calculate the speed of light. His experimental value was 2.9992×10^8 m/s.

Various other methods have been used to measure the speed of light since this attempt by Michelson. The currently accepted value for the speed of light is 2.997924574×10^8 m/s. For convenience in calculation, we shall round off this value to 3.00×10^8 m/s.

Scientists now know that the speed of light in a vacuum is a universal constant.

At any time and any place anyone chooses to measure the speed of light in a vacuum, it will travel at 3.00×10^8 m/s, never faster and never slower.

If a beam of light travels in an uninterrupted path for an entire year, it will cover a distance of 9.5 trillion km (9.5×10^{12} km). This distance is called a **light-year.** But numbers can be misleading: On an astronomical scale, the speed of light no longer seems so great. For example, the closest star to our Solar System is about 40×10^{12} km (40 trillion km) away. A light ray leaving that star for Earth must travel for about 4 years before it reaches us. If our focus of interest is expanded even farther out into space, even a distance of 4 light-years can seem small. The most distant galaxies now known are about 10 billion light-years from Earth. This means that if one of those galaxies had exploded 9 billion years ago, we still wouldn't know about it for another billion years to come.

EXAMPLE 10.1 Could you work a little faster?

In view of our currently known value for the speed of light, how far apart in miles would Galileo's hills have needed to be in order for the round-trip flight of the light to take 1 second?

Solution In 1 second the light would travel a distance of 186,000 mi. Thus, the hills would have to have been 93,000 miles apart for 1 second to elapse for the round trip.

EXAMPLE 10.2 What happened yesterday?

(a) Explain why astronomers often say that when they are looking through a telescope, they are looking backward in time.

(b) If the Sun is 93,000,000 miles away, how far back in time are you looking when you glance at the Sun?

Solution (a) When we look at a distant object, we see it not as it is now, but as it was when the light left it. For example, when we look at a star 10 light-years away, we see that star as it was 10 years ago, because it has taken 10 years for that light to reach us. Thus, if an observer 70,000,000 light-years away should at this instant focus an absurdly powerful telescope on the Earth, he would not see us; instead, he would see dinosaurs peacefully grazing on a now ancient meadow.

(b) The time for light to reach us from the Sun is found from the definition of speed as

$$t = \frac{d}{v} = \frac{93000000 \text{ mi}}{186000 \text{ mi/s}} = 500 \text{ s} = 8.33 \text{ min}$$

Thus, we see the Sun not as it is now, but as it was approximately 8 minutes ago.

10.2
RAYS OF LIGHT

When an object emits light, the waves spread out uniformly in all directions. We will discuss the motion of these waves in terms of rays. *A ray is a line drawn in the direction in which waves are traveling.* For example, a sunbeam passing through a darkened room traces out the path of a ray. Figure 10.3 shows a small object emitting light waves. We have drawn the figure such that the circles indicate crests of the wave. The same type of picture could be used to represent ripples spreading out from a vibrating object placed in a pool of water. Several rays of light are indicated in the

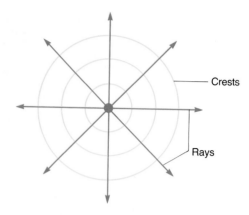

● = Source of light

FIGURE 10.3 A small source emitting rays of light.

figure. Observations indicate that *these rays of light always travel in straight lines, as long as the light does not move from one material into another.* As we shall see later, when light moves from one material into another, it changes direction.

10.3
THE LAW OF REFLECTION

Most of the objects around us are not emitters of light, but instead act as reflectors of the light that falls on them. For example, if light moving through air strikes a mirror, the wave bounces back from the mirror. When waves bounce back from an object, they are said to be reflected. Light is reflected from the surface of wood, paper, metal, soil, and anything else that is described as opaque (nontransparent). Transparent substances such as glass or water allow most of the light to pass through them, although they do reflect some of the light.

To discover the law of reflection, refer to Figure 10.4, where we see an incoming beam of light reflecting off a surface. At the point where the incoming beam of light strikes the surface, point A, we sketch in a reference line drawn perpendicularly to the surface. The angle between the incident beam of light and the perpendicular is called the **angle of incidence.** When the light ray is reflected, it bounces off on the other side of the perpendicular line as shown. The angle between the perpendicular and the reflected ray is

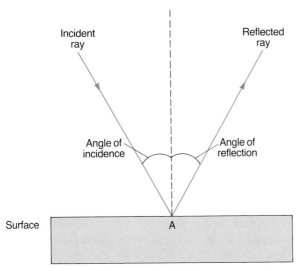

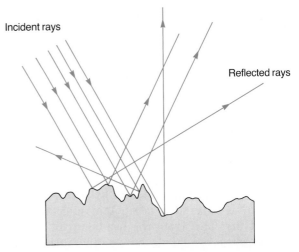

FIGURE 10.4 When light is reflected from a surface, the angle of incidence is equal to the angle of reflection.

FIGURE 10.5 Scattered reflections from a rough surface produce diffused light.

called the **angle of reflection.** A careful measurement in an actual experiment would show that *the angle of reflection equals the angle of incidence*. Additionally, it is found that *the incident ray, the reflected ray, and the perpendicular line all lie in the same plane.* These pieces of information are referred to as the **law of reflection.** In Figure 10.4, we have sketched the reflecting surface as perfectly flat

and smooth, but what happens if the surface is rough, as shown in Figure 10.5? As a group of rays approach the surface, different ones strike at different angles of incidence, yet all obey the law of reflection. The result is that the reflected rays are scattered in all directions as shown. Light rays reflecting in this manner are said to be diffusely reflected, and such light is called diffuse light.

(a) The mountainous terrain is clearly reflected in a still, calm lake. The smooth water is an effective mirror. (b) When the lake surface is disturbed by waves, reflected light is scattered, and the image is blurred. Rough surfaces may reflect light but do not make good mirrors.

(a)

(b)

EXAMPLE 10.3 Stay home when it rains

It is common experience that it is more difficult to drive at night after a heavy rain than it is when the roadway is dry. Use the law of reflection to explain why this is so.

Solution When the roadway is dry, the light from an oncoming car is scattered in random directions off the rough pavement. Thus, light reaches your eye from all directions. However, after a rain, microscopic irregularities in the pavement surface are filled in by water, and the roadway is covered by a smooth reflecting surface. Thus, light from the headlights of an oncoming car reflected off the smooth water surface comes to your eye from only one direction, making it more difficult to see a wide expanse of the road, which reduces the ability to drive safely.

10.4 THE FORMATION OF IMAGES

The fact that light forms images is useful to us, and it also provides us with recreational opportunities. One example of a useful image is that formed in a flat mirror. A glance at your image tells you whether you look as good as you would like before meeting the public. On the other hand, the image formed on a movie screen provides us with entertainment. The image in a mirror and the image on a theater screen are both examples of images, but the two types are distinctly different. Before we look at some optical devices that form images, it is useful to pause briefly to discuss how images are formed.

Images are formed in two ways: *(1) An image is formed when light rays actually intersect, and (2) an image is formed when light rays appear to intersect.* We shall look shortly at the actual processes of forming images by these two methods. For now, remember two more facts about these images. An image formed when light rays actually intersect is called a **real image.** A characteristic of real images is that they can be caught on a screen. Thus, the image formed by a movie projector is a real image. An image formed when light rays only appear to intersect is called a **virtual image.** A virtual image cannot be caught on a screen. This

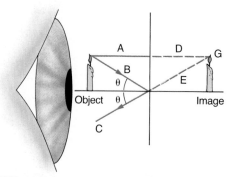

FIGURE 10.6 Tracing rays of light to find the location of an image formed by a flat mirror.

means that if you hold a sheet of paper at the location where the image appears to be, no image is formed on the paper. When you look at your image in a flat mirror, the image appears to be formed some distance behind the mirror. However, if you move a sheet of paper around behind the mirror, no image is ever formed on the paper.

10.5 FLAT MIRRORS

The kind of mirror that you probably have in your bathroom is a flat mirror, which means that the glass surface of the mirror is not curved. In order to see how such a mirror forms an image of an object placed before it, we will have to follow at least two different rays of light that leave the same point of the object.

In Figure 10.6, we follow two rays leaving the tip of a candle placed before the mirror. One of these rays, labeled A, leaves the candle tip and strikes the mirror head-on. In this case, the angle of incidence is zero, and the ray of light is reflected such that the angle of reflection is also zero. This means that it reflects back directly on itself. The next ray, labeled B, leaves the tip of the candle at any arbitrary angle and strikes the mirror. This ray, too, obeys the law of reflection and follows the path labeled C after reflecting. These two reflected rays enter the eye of an observer placed as shown in the figure. (The eye has been enlarged for clarity.) When these rays enter the eye, it traces both rays back along a straight-line path to the point from which they appear to originate. Thus, as far as the eye is concerned, the reflected ray A appears to have come from some point along the line labeled D in the figure, and

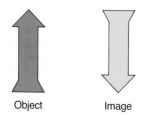

FIGURE 10.7 If the image is turned over with respect to the object, it is said to be an inverted image.

the ray C appears to have come from some point along the path labeled E. The dashed lines drawn for D and E show that these lines intersect at point G; thus, this is the point from which the eye believes the rays originated. At point G, where the rays appear to intersect, is the location of the image. If you refer back to the last section, you will find that the image formed is like that discussed as case two, and the image is a virtual image. You could follow rays of light from all points on the candle to see where the image of these points is found, and you would end up with a complete image formed at the position G indicated in Figure 10.6.

You could draw Figure 10.6 to scale and measure all the angles and distances carefully to discover some facts about the images formed by flat mirrors. Some of the features that you would find are:

1. The image is as far behind the mirror as the object is in front of it. This means that if a book is held 20 cm in front of a flat mirror, the image will be formed 20 cm behind the mirror.
2. The image is unmagnified. This means that if the object is 2 meters tall, the image will also be 2 meters tall.
3. The image is erect. This means that if an object in the shape of an arrow is placed in front of a flat mirror such that the tip of the arrow points toward the ceiling, the image formed of the arrow will also point toward the ceiling. An image is said to be inverted if it points in a direction 180° away from that pointed by the object. Such a situation is shown in Figure 10.7.
4. The image has right-left reversal. This means that if you raise your right hand while standing

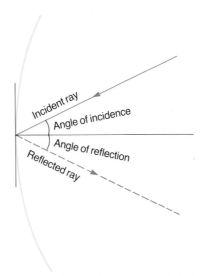

FIGURE 10.8 Reflection from a curved surface. The angle of incidence equals the angle of reflection with respect to a line drawn perpendicular to the tangent to the surface, as shown.

in front of a flat mirror, your image will seem to be raising its left hand.

10.6 CURVED MIRRORS

Curved mirrors are used in many practical applications. If light strikes a curved surface, it is reflected so that the angle of incidence equals the angle of reflection, just as with a flat surface, as shown in Figure 10.8. There are many different types of curved mirrors, but for our present discussion, let us choose a curved mirror that is a segment of a sphere. For example, suppose you cut out a segment of a round Christmas tree ornament and use it as a mirror. You would find that you could use it in two ways. If you allowed the light to reflect off the inward curving side, as in Figure 10.9a, the type of mirror that you have is called a **concave mirror.** On the other hand, if you are interested in the light reflected from the other side of the segment, as in Figure 10.9b, you have a **convex mirror.** For convenience, let us examine only one type of mirror, the concave mirror, and see what kind of image(s) it can form.

Figure 10.10 follows two rays of light that leave an object in the shape of an arrow placed at

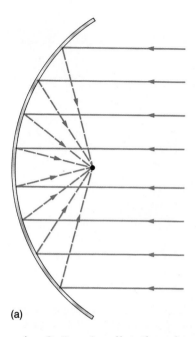

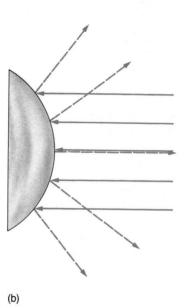

(a) (b)

FIGURE 10.9 (a) A spherical concave mirror, and (b) A spherical convex mirror.

point O. Ray A strikes the mirror at the center of the curved segment and reflects, obeying the law of reflection, to follow path B. The second ray of light leaving the object follows path C, hits the mirror head-on, and reflects back on itself. These two rays *actually* intersect at point I. Thus, the image formed at I is a real image. The figure has been drawn to scale, so other pertinent facts about the image can be determined by inspection: (1) The image is smaller than the object, and (2) it is inverted.

Figure 10.11 follows two rays from an object at O that reflect off a convex mirror. We will leave it for you to follow the path of the rays leaving the object and show that an image is formed at point I. Note the characteristics of the image formed:

(1) The image is erect, and (2) it is smaller than the object.

When describing two apparently similar types of mirrors, a distinguishing characteristic called the **focal length** is often used. *The focal length of a mirror is the distance of the image from the mirror when the object is placed an infinite distance away.* For example, when you use a mirror to form an image of the Sun, for all practical purposes the image is formed at the focal point of the mirror. This is true because the Sun is so far away that it can be considered to be at infinity. Figure 10.12 diagrams this situation. Because the object is so far away, by the time the light rays reach the mirror they are traveling parallel to one another

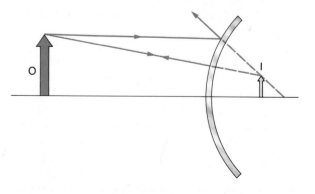

FIGURE 10.10 Formation of an image by a concave mirror.

FIGURE 10.11 Image formation by a convex mirror.

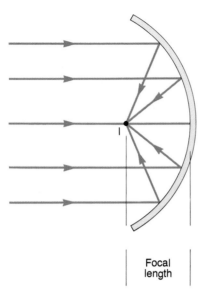

FIGURE 10.12 The focal point of a concave mirror.

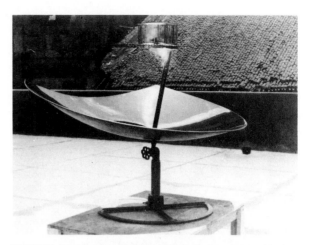

FIGURE 10.13 A reflecting parabola can concentrate enough light energy to cook food.
(Courtesy of Dan Halacy, University of Wisconsin)

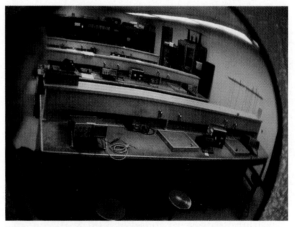

Note that the image formed in this convex mirror is diminished and erect.
(Courtesy Bill Schulz)

as shown. After reflection, the light rays come to a focus at I, and the distance from the image to the mirror is called the focal length.

Curved mirrors are often used in telescopes, except that the shape of the mirror is usually not a part of a segment of a sphere. In this application, the shape is usually chosen to be one called a **parabola.** Parabolic shapes are better than spherical shapes because the light reflected by a parabola is much more sharply focused at the image than it would be by a spherical segment, which smears the light out somewhat at the image position. Figure 10.13 shows a small parabolic reflector that can concentrate enough sunlight to boil water, cook a hamburger, or bake a chicken. In France, there is a multistory parabolic mirror that can concentrate proportionally larger amounts of solar energy. This solar tower can produce enough steam by heating water to power a small electric generating station. Similarly, parabolic antennas that reflect radio waves are used to collect signals from objects in outer space that emit radio waves. In recent years, many important astronomical discoveries have been made with these radio telescopes.

EXAMPLE 10.4 Objects may be closer than they seem

The rearview mirrors in most modern automobiles are designed to use convex spherical mirrors. These mirrors have a cautionary warning printed on them which reads, "Objects may be closer than they seem." Explain why this message is necessary.

Solution Note the characteristics of the image formed by a convex mirror, as shown in Figure 10.11. The key feature here is that *the image is always demagnified.* Thus, even if a car is right on your rear bumper, the small image formed could mislead you into believing that the car is farther away than it really is. Thus, a dangerous driving maneuver might be undertaken.

(a) Sled analogy

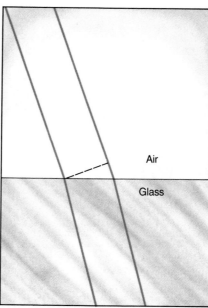

(b) Refraction of light

FIGURE 10.14 (a) A sled turns upon striking a paved roadway at an angle, because one runner hits the roadway and slows down before the other does. (b) A ray of light behaves in the same manner.

10.7 REFRACTION

A light ray travels in a straight-line path as long as it does not move from one material into another. However, if the ray moves from one transparent material into another, the ray undergoes a change in direction. This bending is called **refraction.** The reason that light bends is because it has a different speed in the two materials. Let's see why this causes bending.

Light travels at a speed of 3.00×10^8 m/s in a vacuum or, for all practical purposes, in air. However, when light moves into another material, it slows down. For example, the speed of light in glass is about 2.00×10^8 m/s and the speed in water is about 2.25×10^8 m/s. To see why this change in speed causes bending, consider an analogy from common experience. Suppose you are sledding down a snowy hill and you hit a roadway head-on, perpendicular to the roadway. The sled will not move as fast on the roadway, but because you hit it head-on, you will not change direction. On the other hand, if you hit the road at an angle, one runner will reach the pavement before the other, as shown in Figure 10.14a. The runner that hits first slows down, while the other runner, which is still on the snow, continues to

travel at a rapid speed. The result is that the sled turns abruptly and follows the path shown in the figure. Light acts in a comparable manner. If light traveling in air enters a transparent liquid or solid head-on, it slows down, but its direction of travel does not change. However, if it moves at an angle from one material into another in which it travels more slowly, it turns (undergoes refraction) just as the sled did when it hit the road. See Figure 10.14b.

To describe the refraction process, consider the situation shown in Figure 10.15, where a ray of light in air is incident on the surface of a pool of water. At the point where the incident ray strikes the water, we construct a reference line perpendicular, or normal, to the surface. The angle between the incident ray and the normal is called the angle of incidence, and the angle between the ongoing ray and the normal is called the angle of refraction. Figure 10.15a has been drawn in a manner that demonstrates a characteristic of the refraction process for this situation; namely, *as light moves into a material in which it travels more slowly, its direction of travel bends toward the perpendicular.* The reverse of the situation also holds. For example, suppose the source of light was beneath the surface of the pool of water. A

FOCUS ON . . . The Rainbow

A rainbow can be seen anytime you are between the Sun and a rain shower, and the concepts of how a rainbow are formed can be understood by means of the laws of refraction. To see how this works you must first understand that white light from the Sun is really a composite of all the colors from red through violet. (The subject of color spectra will be discussed more thoroughly in the next chapter.) Consider a ray of light from the Sun passing over your head and striking a raindrop, as shown in the Figure. The white light passing into the drop is bent out of its original direction of travel, as we know it will be from our study of refraction. However, we also find that some colors are bent more than others. Specifically, violet light is bent the most and red the least. These red and violet rays, and all the ones in between, continue until they hit the back side of the drop, where much of the light is reflected back toward the front surface. When these colors strike the front of the drop, they are bent a little more, and they end up separated as shown.

To see how this separation of the colors of light leads to the formation of a rainbow, consider the man shown in the Figure. If he observes a raindrop high in the sky, the red light from it reaches his eye, whereas the violet light passes over his head. The observer would interpret the color of this drop to be

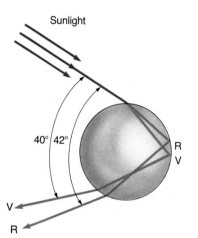

Light refracted by a drop of water is separated into its component colors, thus forming a rainbow.

red. A drop lower in the sky would direct violet light toward his eye and would be seen as having that color. The red light from this drop would strike the ground and not be seen. Colors between the red and violet extremes would be sent to the eye by drops located between these two, and thus the rainbow would be red on the outside and violet on the inside.

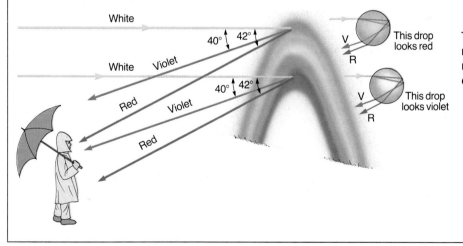

The formation of a rainbow by light refracted in water drops.

ray directed toward the surface at an angle would follow the path indicated in Figure 10.15b. Note the identification of the angles of incidence and refraction for this situation. The general rule is that *when light moves from a material into one in which* *it travels faster, the ray is bent away from the perpendicular.*

The refraction of light explains many common optical phenomena. For example, when you look at a fish in a pool of water, the fish appears

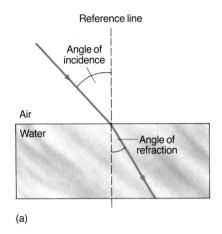

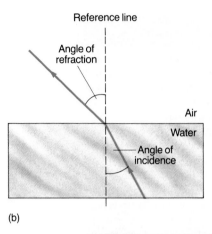

FIGURE 10.15 (a) Light moving into a material in which it travels more slowly is bent toward the perpendicular reference line. (b) When light moves into a material in which it travels faster, it is bent away from the perpendicular reference line.

to be closer to the surface than it really is, because of refraction. Figure 10.16 shows why. We follow two rays leaving the fish, passing through the surface, and bending. When these rays enter the eye of an observer, the eye traces the rays back to the point where they appear to have originated. This point is point I, where an image of the fish is formed. Thus, the fish appears to be lifted toward the surface of the water. The apparent bending of a pencil when inserted in water is also explained in the same way. Each point of the pencil that is beneath the surface appears to be lifted toward the surface, just as was the fish, and the pencil appears to be bent (Fig. 10.17).

A mirage in the desert is a well-known optical illusion caused by refraction. Travelers in the desert have reported "seeing" water only to find

(a)

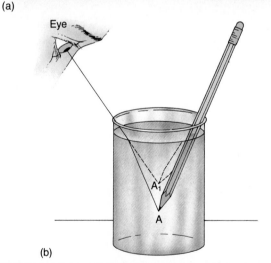

(b)

FIGURE 10.17 (a) The pencil appears to be bent because of refraction. (b) Solid lines show actual position of pencil and actual path of light ray. Dashed lines show observed position of pencil and the mentally projected light ray.

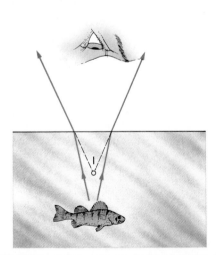

FIGURE 10.16 A fish appears to be closer to the surface than it actually is because of refraction.

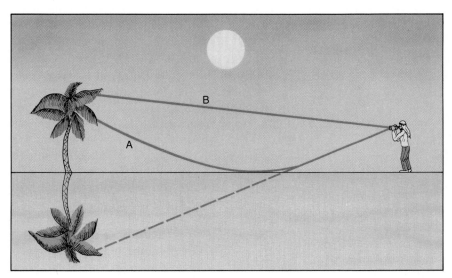

FIGURE 10.18 A desert mirage occurs when light rays coming from a distant object are bent as they pass through air of different densities.

out that no such pool ever existed. On a hot day in the desert, the earth is very warm, and thus the air next to the earth is also warm. Warm air is less dense than cool air, so a ray of light headed toward the desert surface, such as ray A in Figure 10.18, is moving through a material that continuously becomes less dense as the ray approaches the ground. As the density of the air becomes less, the speed of the ray also changes, and this leads to a continual refraction that causes the ray to follow the curved path indicated in the figure. An observer thus sees rays of light from the tree reaching him from two different directions. One is the direct path B and the other is the curved

path A. The eye of the observer traces these rays back to the points where they appear to originate. The brain is fooled and assumes the rays coming from the ground originate below the surface, while the direct rays produce the expected view of the tree. Thus, the viewer sees the tree and an inverted image of the tree. Based on prior experience, the observer would interpret the inverted image of the tree to have been produced by reflection in a pool of water in front of the real tree.

EXAMPLE 10.5 Follow the bending light

A ray of light in air strikes the surface of a glass block as shown in Figure 10.19. The ray is then observed to exit the block at some point along the bottom surface. Trace a possible path for the ray through the block, based on refraction.

Solution At the top surface, we construct a perpendicular line and identify the angles of incidence, i, and refraction, r, as shown in Figure 10.20. Because the speed of light is slower in glass than in air, the light is shown to have been bent toward the perpendicular. While in the glass, the light follows a straight-line path, but it will refract again when it exits into air. This occurs at the lower surface, where we have again constructed a perpendicular line and identified the angles of incidence and refraction for this surface. In this case, the light will bend away from the perpendicular as it moves into the air.

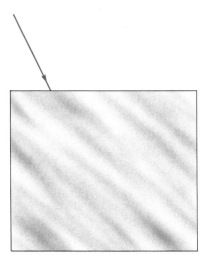

FIGURE 10.19 Trace a possible path for the ray.

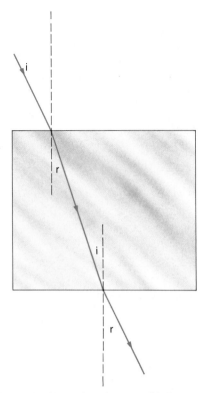

FIGURE 10.20 Following the ray of light.

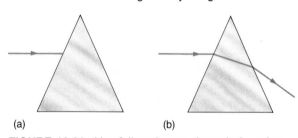

(a)   (b)

FIGURE 10.21 You follow the ray through the prism.

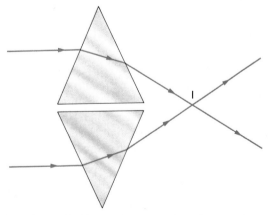

FIGURE 10.22 Two prisms would produce an image at I.

10.8
LENSES

In order to see how lenses work, let us consider another problem similar to that of Example 10.5. In this case, we will trace a ray of light through the prism shown in Figure 10.21a. The ray is incident on the left surface and is observed to exit the glass at the right surface. By now, you should be able to identify the angles of incidence and refraction at each surface and, by use of the law of refraction, convince yourself that a possible path for the ray of light is like that shown in Figure 10.21b.

Being able to trace a ray of light through a prism is important because one useful type of lens, called a **converging lens,** can be made by stacking two prisms base to base, as shown in Figure 10.22. In this case, incoming rays of light follow the path shown and intersect at I. Since the rays are actually intersecting at I, a real image is formed at this location. Also note what effect our two stacked prisms have had on the incoming parallel beams of light. The rays have been deviated from their path such that they come together, or converge—thus, the name converging lens. In actual practice, a converging lens is produced by grinding the shape out of a single piece of glass and rounding off all the sharp corners to produce a shape like one of those shown in Figure 10.23. *Any lens that is thicker at its center than it is at the rim will act as a converging lens.*

The transformation shown in Figure 10.24a will produce another important type of lens, called a **diverging lens.** Note that incoming rays of light are diverged, or spread apart, after having passed through the lens. Figure 10.24b shows a variety of different types of diverging lenses, all of which have the characteristic that they are *thinner at the center than at their rim.*

FIGURE 10.23 Several types of converging lenses.

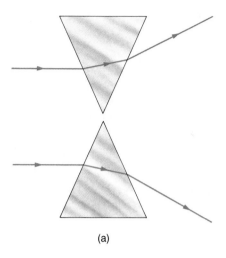

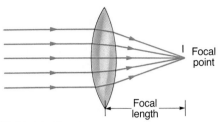

FIGURE 10.25   The focal point and focal length for a converging lens.

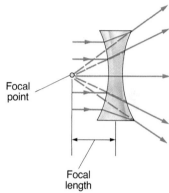

FIGURE 10.26   The focal point and focal length for a diverging lens.

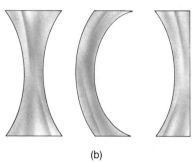

FIGURE 10.24   (a) Another way to stack prisms. (b) Several types of diverging lenses.

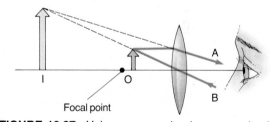

FIGURE 10.27   Using a converging lens as a simple magnifier.

An important descriptive feature of a mirror is its focal length, and this same terminology is also used to describe lenses. Figure 10.25 shows incoming parallel rays of light from a distant object such as the Sun. The lens converges these rays such that an image is formed at I. *The point at which parallel rays are converged to an image by a converging lens is called the focal point of the lens, and the distance of this point from the lens is called the **focal length.*** These concepts are illustrated in Figure 10.25.

If incoming parallel rays of light strike a diverging lens, they are bent as shown in Figure 10.26. *The point from which the rays appear to originate after bending by a diverging lens is called the focal*

*point, and the distance of this point to the lens is called the **focal length.***

10.9 A SIMPLE MAGNIFIER

As you are no doubt aware, an important use of a converging lens is as a simple magnifier. To see how this magnification occurs, consider Figure 10.27. Here we see an object placed at O, and we follow two rays of light that leave the object, move

FOCUS ON . . . Vision correction with lenses

One of the most common applications of lenses is in the correction of imperfect eyesight. In order to follow how this is done, consider the Figure, which shows the important parts of a typical eye. The front of the eye is covered by a transparent membrane called the cornea. This is followed by a clear liquid region (the aqueous humor), an aperture (the iris and pupil), and the lens. When light enters the eye, it is bent slightly by the cornea, but most of the bending takes place as it passes through the lens. The object of the game for the cornea and the lens is to bring light rays together such that an image is formed on the back surface of the eye at the retina. The surface of the retina contains millions of light-sensitive receptors, called rods and cones, which generate electrical impulses when struck by light. These impulses are transmitted via the optic nerve to the brain, where the image is interpreted. It should be obvious from this brief discussion that distinct vision is possible only when the image is formed directly on the retina.

In the defect commonly called nearsightedness (myopia), the eye is too deep or the lens too strong; as a result, the image is brought to a focus in front of the retina (see Figure, part (a)). The distinguishing feature of this imperfection is that distant objects are not seen clearly. To correct this problem, an ophthalmologist or an optometrist fits the eye with a diverging lens, which prevents the light from coming to a focus until it reaches the retina, as shown in part (b) of the Figure.

A second common defect of the eye is called farsightedness (hyperopia). Here distant objects are seen clearly, but near objects are indistinct. The lens of the eye is unable to bring diverging rays of light to a focus on the retina and instead tries to form an image behind it. Figure (a) shows the problem, and Figure (b) shows how a converging lens can correct the difficulty.

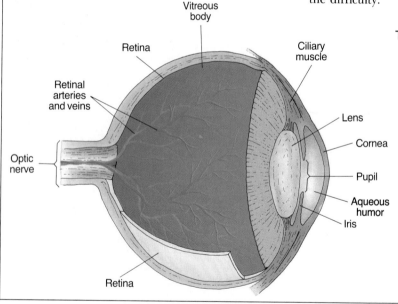

The primary parts of a human eye.

through the converging lens, and pass into the eye of an observer. In this case, unlike any case that we have discussed so far, the object to be viewed is placed very close to the lens. When the object is close to the lens, the lens converges the rays of light as usual, but it is not able to converge them enough that they intersect. Thus, a real image is not formed. As shown in Figure 10.27, the rays A and B *have* been converged somewhat by the lens, but not enough to cause them to cross. When these rays enter the eye of the observer, the eye follows these two rays back to the point where they appear to have originated; this is I in the figure. Note that the image at I is much larger than the object. Thus, it is magnified. Is the image real or virtual?

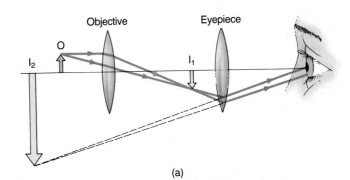

(a)

FIGURE 10.28 (a) Image formation by a compound microscope. (b) A compound microscope.
(Courtesy Bill Schulz)

(b)

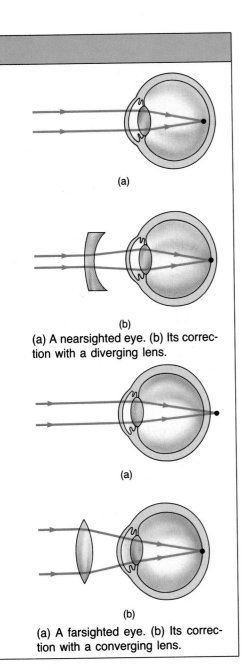

(a)

(b)

(a) A nearsighted eye. (b) Its correction with a diverging lens.

(a)

(b)

(a) A farsighted eye. (b) Its correction with a converging lens.

EXAMPLE 10.6   A compound microscope

A simple magnifier provides only limited assistance in inspecting the minute details of an object. Greater magnification can be achieved by combining two lenses in a device called a compound microscope. Figure 10.28 shows the construction details of such a device. Also, Figure 10.28 traces rays of light that leave the object O and move through the lens system. Use your knowledge of refraction and image formation to explain qualitatively how this device works.

Solution We will follow two rays of light from the specimen at O through both lenses to see how magnification occurs. The rays first pass through a lens called the objective lens, which converges the rays and causes them to intersect at point I_1. Examination of I_1 shows that this image is real, inverted, and slightly enlarged. The second lens in the system is called the eyepiece, and it acts as a simple magnifier. This lens is used to examine the image I_1 and to magnify it. Thus, a final image is formed by the eyepiece at I_2. Because both lenses have produced magnification, the final image is much larger than it would be if only a single lens were used.

10.10 TELESCOPES

There are two fundamentally different kinds of telescopes, both of which have the same basic purpose: to aid us in viewing distant objects such as the planets in our Solar System. The two classifications are (1) the **refracting telescope,** which uses a combination of lenses to form an image, and (2) the **reflecting telescope,** which uses a curved mirror and a lens to form an image. Let us examine these in turn.

The refracting telescope

The first type of telescope to be constructed was the refracting telescope, and it was first used in a systematic way for observation of the heavens by Galileo. With it, Galileo observed mountains on the Moon, the phases of Venus, the stars in the Milky Way, and the moons of Jupiter. These observations were important to the development of astronomy because, as we shall see later, when these observations were made, there were two competing views of our Universe. One of these, called the **geocentric model,** held that the Universe was Earth-centered. This meant that the Earth was considered to be at the center of the Universe, and all planets, moons, stars, and so forth were thought to revolve about the Earth. In the 1500s, a competing view, called the **heliocentric model,** was developed by Copernicus, a Polish astronomer. Copernicus was not the first to hold the belief that the Sun was at the center, but he was the first to put this model on a firm scientific footing. In 1543, he published a book detailing his observations and beliefs while on his death bed; however, the book did not receive wide circulation and probably would have faded into obscurity for many years had it not been for the work of Galileo. Many of Galileo's observations with the telescope could be explained only by the heliocentric model, and as a result, he became an outspoken advocate of this view of the heavens. His ability to communicate caused the heliocentric viewpoint to become the accepted belief by many educated people during his lifetime. Let us now examine the refracting telescope to see how it is used to bring the heavens closer to Earth.

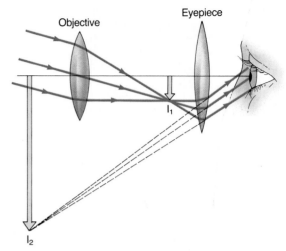

FIGURE 10.29 Diagram of a refracting telescope.

Two lenses are used in a fashion similar to that used in a compound microscope. Figure 10.29 shows the construction details and the image formation processes. Light from a distant object enters a large diameter lens, called the **objective lens** (or just the objective), as parallel rays, and the rays are converged to form an image at point I_1. The second lens, called the **eyepiece,** is then used as a magnifier to form an image at I_2. We shall not derive this result here, but the magnification, M, of either a refracting or a reflecting telescope is given by

$$M = \frac{f_o}{f_e} \qquad (10.1)$$

where f_o is the focal length of the objective and f_e is the focal length of the eyepiece. The largest refracting telescope in the world, located at Yerkes Observatory in Williams Bay, Wisconsin, is designed such that the diameter of the objective lens is 1 m.

Surprisingly enough, when a telescope is used as a tool in astronomy, the magnification is seldom of importance, because stars are so far away that regardless of how much magnification you attempt to use, they always appear as simple points of light. No telescope can magnify a star enough to enable one to see any detail on its surface. The primary uses for an astronomical tele-

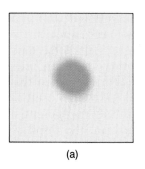 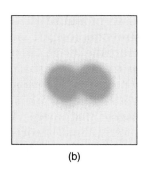 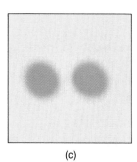

(a)   (b)   (c)

FIGURE 10.30 An image seen through a telescope with (a) a poor resolving power, (b) better resolving power, and (c) the best resolving power.

scope are stated in terms of three types of powers: (1) the **light-gathering power,** (2) the **resolving power,** and (3) the **magnifying power.**

The light-gathering power of a telescope is important because the objects that are of astronomical interest are very distant; hence, the light reaching Earth is very faint. Thus, it is important to collect as much light as possible from them in order to form as distinct an image as possible. Imagine a rain storm occurring, and someone sets you the task of going outside in the rain and collecting as much falling water as possible. Won't you achieve the most success by going outside with a container having as large a diameter as you can find? Likewise, the light from a distant astronomical object is falling on the Earth like rain, and the job for an astronomer is to collect as much of this light as possible. The obvious solution is to use a telescope with a large diameter objective. Thus, *the light-gathering power of a telescope depends on the diameter of the objective.*

The resolving power of a telescope refers to the ability of a telescope to distinguish between objects that are close to one another. For example, suppose you use a telescope that does not have a good resolving power to view some distant object in space. You see something that looks like Figure 10.30a. You then use a telescope with a slightly greater resolving power, and the image looks as shown in Figure 10.30b. You then use a telescope with a good resolving power, and you see an image like that shown in Figure 10.30c. Thus, what you saw as a single object with the poorly resolving telescope is found really to be two close-together objects when observed with a telescope with a good resolving power. *The characteristic of a telescope that determines its resolving power*

is the diameter of the objective. Thus, the larger the diameter, the greater is the ability of the telescope to distinguish the fine details in an image.

The final power, the magnifying power of a telescope, is given by Eq. 10.1. From this equation, we see that it is the focal lengths of the objective and the eyepiece that determine the magnifying power of the instrument. As noted earlier, magnification is not important when viewing objects beyond our Solar System, but it can be of importance when viewing nearby objects such as planets.

EXAMPLE 10.7   Be a wise telescope shopper

A shopper for a telescope is unable to decide between two models. The two telescopes are described below.

Tele-scope	Focal Length of Eyepiece	Focal Length of Objective	Diameter of Objective
A	2.4 cm	150 cm	12 cm
B	0.06 cm	100 cm	8 cm

(a) If the main interest of the shopper was in light-gathering power, which would you advise him to buy?

Solution Light-gathering power is determined by the diameter of the objective lens. Thus, he should buy telescope A.

(b) If the main interest of the shopper was in resolving power, which would you advise him to buy?

Solution Resolving power also depends on the diameter of the objective. Thus, telescope A is once again the choice.

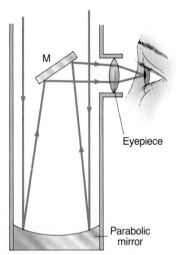

FIGURE 10.31 Image formation by a reflecting telescope. The parabolic mirror collects light over a large area and focuses it toward the center of the tube. Before the image is formed, a small flat mirror redirects the image to the eyepiece.

(c) If the main interest of the shopper was in magnification, which would you advise buying?

Solution We will have to calculate the magnification of each in order to answer this question. This is done for A as

$$M_A = \frac{f_{oA}}{f_{eA}} = \frac{150 \text{ cm}}{2.4 \text{ cm}} = 62.5$$

and the magnification of B is

$$M_B = \frac{f_{oB}}{f_{eB}} = \frac{100 \text{ cm}}{0.06 \text{ cm}} = 1670$$

Thus, telescope B is the scope of choice.

Reflecting telescopes

Even before Isaac Newton gained immortality via his discovery of the three laws of motion and of the law of universal gravitation, he had achieved some fame in scientific circles because he invented the reflecting telescope. The heart of the reflecting telescope is a parabolic mirror mounted at the base of a tube. The mirror collects light from a large area and focuses it as shown in Figure 10.31. Incoming light from an astronomical object converges toward a focus, but before the image can be formed, a small flat mirror reflects the light toward the side of the telescope barrel. A real image is formed when this light converges, and an eyepiece is then used to observe this image. To illustrate the potential power of such a collection system, the pupil of your eye collects light available in a circular area a few millimeters across, but the light-gathering mirror in the Hale telescope at Mt. Palomar, California, is about 5 m in diameter.

All new astronomical telescopes at observatories around the world are reflectors. This occurs because there are certain features in the design and construction of large refractors that are difficult or impossible to overcome. Let us look at some of these difficulties and see how they are either not present in a reflector or can be easily overcome with a reflector.

1. If one is to collect as much light as possible, it is necessary to make the objective lens in a refractor very large. The grinding and polishing required for the front and back surfaces of this lens are quite difficult and very expensive. A reflector, on the other hand, has only one surface to be ground, since light bounces off the objective mirror rather than passing through it.

2. The already faint light from a distant object is dimmed even more as it passes through the objective lens of a refractor, because the glass absorbs a percentage of the light passing through. Additionally, bubbles or imperfections in the interior of the lens will scatter some of the light, thus removing it from the portion eventually collected. Both of these factors are unimportant in a reflector, because the light does not pass through any glass on its way to the eyepiece.

3. It is very difficult to rigidly support the objective lens in a refracting telescope. Since the light must pass through the lens, the only way that it can be held in place is by supports around the rim of the lens. This leads to sagging in the very heavy glass, and can cause distortions in the final image. On the other hand, the mirror in a reflector can rest on a support at the base of the telescope.

FOCUS ON . . . Fiber Optics

An important application of the principle of total internal reflection lies in the burgeoning field of **fiber optics.** To understand what is happening here, consider light entering one end of a small, plastic fiber, as shown in the Figure. If the fiber is not bent at very sharp angles, the light will always strike the surface of the plastic such that it undergoes total internal reflection. Thus, the fiber can transmit the light from one location to another as if it were a "wire" for light.

The analogy between a wire and a light-transmitting fiber can be taken one step farther. An unvarying direct current in a wire carries no information and cannot, say, drive a speaker or transmit data from one computer to another. But if the current is turned on and off, information can be conveyed. The simplest form of electronic communication is Morse code, in which letters and numbers are carried by a series of dots, which are merely short bursts of current, and dashes, which are longer bursts of current. Obviously, a light-carrying fiber can carry information in this manner.

Of course, modern communications have gone far beyond the use of Morse code. Voice and other forms of information are carried by rapid oscilla-tions in an electrical signal. (See AM and FM transmission in Chapter 11.) The same effect can be achieved by using fiber optics. There are many advantages in transmitting information via light, because the density of information that can be impressed on light waves is considerably higher than that on conventional electrical distribution systems. A typical fiber optics system, for example, can transmit 300,000 telephone conversations simultaneously through a single glass fiber. By comparison, a single copper wire can carry only 24 different voice channels. In addition, the light rays are not affected by static from lightning or other electrical signals. Glass fibers are not only better than copper wires for carrying telephone conversations but also are cheaper. As a result, it seems inevitable that fiber optics will be used in the future both for conventional telephones and for more advanced communication systems. The first major fiber optics transmission system became operational in February of 1983 for a telephone link between Washington, D.C., and New York City. At present, many large cities have changed completely to fiber optics systems, and all new installations are of this type.

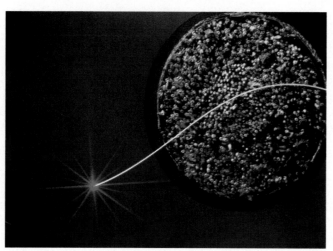

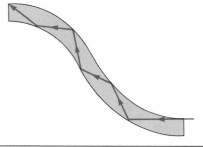

Light is seen emerging from the end of a single optical fiber. The cable behind the fiber contains hundreds of these fibers which are used for transmission of telephone signals in place of copper wires.
(Courtesy Corning Incorporated)

Reflection of light inside a transparent medium such as a piece of glass or plastic. Note how light which travels in straight line paths can turn a corner by repeated total internal reflections.

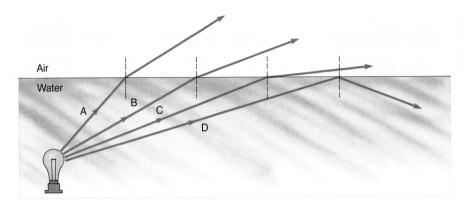

FIGURE 10.32 A ray such as D that strikes the surface at a large angle of incidence is totally internally reflected.

10.11
TOTAL INTERNAL REFLECTION

The phenomenon of total internal reflection is one that embodies facts related to both reflection and refraction. Let us examine how the principle works and then investigate some of its practical applications. To understand what happens, consider Figure 10.32. Shown are four rays of light leaving a source that is beneath the surface of a pool of water. Ray A strikes the surface and bends away from the perpendicular, as it must. Ray B strikes the surface, and it also is bent away from the perpendicular, but since its angle of incidence is greater than that for A, it is bent away from the perpendicular much more. Finally, consider ray C, which strikes the surface at an angle of incidence such that when it passes out of the water, it skims right along the surface. Any ray of light, such as D, striking the surface farther out simply cannot get out of the water at all. It is reflected from the surface just as though the surface were a perfectly reflecting mirror. Such rays are said to have undergone **total internal reflection.** *Total internal reflection can occur only when light rays are attempting to pass from a material in which they travel slower to one in which they travel faster.*

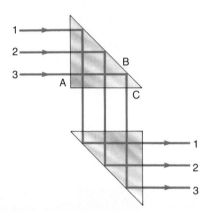

FIGURE 10.33 A periscope.

Figure 10.33 shows a way in which this phenomenon can be used in a practical situation. Shown are rays of light incident on the face of a prism having angles of 90°, 45°, and 45°. The rays strike side A and pass into the glass without being bent. They then strike side B, where they undergo total internal reflection and are reflected toward side C, where they exit the prism. A combination of two such prisms arranged as indicated in the figure will produce a change in direction of the light as shown. This combination could be placed in a tube and used as a periscope.

SUMMARY

The speed of light is 3.00×10^8 m/s and was originally measured with an apparatus using rotating mirrors.

When a ray of light is **reflected,** the angle of incidence equals the angle of reflection.

Images are formed when light rays actually intersect or appear to intersect. When the rays actually intersect, the image is called a **real** image; when they only appear to intersect, the image is **virtual.** An image that is turned over relative to the original object is said to be **inverted;** otherwise, it is **erect.** The image formed by a flat mirror is as far behind the mirror as the object is in front of it, unmagnified, erect, and left-right reversed. Curved mirrors may be either **concave** or **convex.** In either case, the position of an image for a distance object is called the **focal length** of the mirror. Curved mirrors may be used to form **reflecting** telescopes.

Refraction occurs when a light ray bends as it passes at an angle from one medium to another.

A **lens** is a piece of glass shaped such that it uses refraction of light to form images. Lenses are of two types: **converging** and **diverging.** In either case, the distance from the lens to the point where the image of a distant object is formed is called the **focal length** of the lens. Lenses may be used as simple magnifiers or, in combination, to form a **refracting** telescope.

When light attempts to pass from a material in which it travels at a certain speed into one in which it travels faster, **total internal reflection** can occur. In this situation, no light escapes, and the surface acts like a perfectly reflecting mirror.

EQUATIONS TO KNOW

$$M = \frac{f_o}{f_e} \qquad \text{(magnification of a telescope)}$$

KEY WORDS

Light-year	Real image	Focal length	Light-gathering power
Rays of light	Erect image	Lenses	Resolving power
Reflection	Inverted image	Converging lens	Magnifying power
Refraction	Flat mirror	Diverging lens	Total internal reflection
Image	Convex mirror	Refracting telescope	
Virtual image	Concave mirror	Reflecting telescope	

QUESTIONS

THE SPEED OF LIGHT

1. In another version of Galileo's experiment, he is said to have quickly uncovered a lantern and tried to measure the time required for light to travel to a mirror 1.5 km away and back. How long does it take for light to travel that distance? Explain why this attempt also failed.

2. If Michelson's mirror had had six sides instead of eight, would he have had to spin it faster or slower to measure the speed of light? Explain.

3. Find the length of a light-year in miles.

4. Our Sun is approximately 93 million miles away.
   (a) Find the distance to this star in light-seconds.
   (b) Find the distance to the Moon, about 240,000 miles distant, in light-seconds.

THE LAW OF REFLECTION

5. Draw a diagram showing light reflecting from a mirror when the angle of incidence is 0°.

6. Aristotle said that the Moon is smooth and polished as a looking glass. If the Moon truly were this way, what would it look like to the naked eye?

7. A mirror A is placed flat on the surface of a table, and a second mirror, B, is placed against it, but vertical. A ray of light strikes A at an angle of incidence of 60° and bounces off it and finally off of B. Draw a sketch of this situation, and find the angle of incidence and reflection at mirror B.

8. Repeat problem 7 for the case in which the mirrors are placed at an angle of 120° with respect to one another, rather than 90°.

THE FORMATION OF IMAGES AND FLAT MIRRORS

9. The word AMBULANCE is written in a very

strange way on the front of such a vehicle. It is written so that when you look in your rearview mirror, you are able to read the word. Write the word ambulance as it should be printed on the front of the vehicle. Use a mirror to look at what you have written, in order to see if you did it correctly.

10. A person 2 m tall stands in front of a flat mirror. What is the minimum height of the mirror so that the person can see all of himself, but no more. Be careful, the answer is not 2 m.

11. If you are standing 3 m in front of a mirror, you see yourself essentially as another person would if he were standing how far from you?

12. Often, when you look out a window at night, you see two images formed by the glass. How are these images produced?

13. You have a flat mirror, but you don't know whether it is silvered on the front or the back. How could you use a beam of light bounced off the mirror to determine how it is silvered?

CURVED MIRRORS

14. Curved mirrors in amusement parks can make a person appear to be fat, skinny, or even wiggly. Using a diagram, explain briefly how a curved mirror distorts an image.

15. Convex mirrors are often placed in the corners of stores so that observers can watch for shoplifters. Why do you think convex mirrors are selected?

16. A small light bulb is placed 10 cm from a concave mirror, and a real image is formed 20 cm from the mirror. Is the focal length of the mirror 20 cm? Why or why not?

17. A dish antenna to receive television signals from satellites is made in the shape of a concave mirror. Why do you think this shape is used rather than the shape of a convex mirror?

18. All the properties of concave mirrors were not discussed in the text. Let us take an example of a use of one of these mirrors to let you figure out some of their characteristics on your own. A dentist uses a small concave mirror to examine teeth. The mirror must be held closer to the tooth than the focal length of the mirror. Under these circumstances do you believe (a) the image is real or virtual, (b) erect or inverted, (c) magnified or unmagnified?

REFRACTION

19. If you place a pencil straight down into a glass of water and look at it from above, it does not appear to be bent. But if you place it in the water at an angle and eye it obliquely, it does appear bent. Explain.

20. A ray of light bends as it enters the body of a lens and bends again when it emerges. Explain.

21. The Earth's atmosphere is denser near the surface than at high altitudes. If a scientist is trying to locate the exact position of a weather balloon flying in the upper atmosphere, is the density gradient of the atmosphere a factor to consider? Explain.

22. Mirages are common in the arctic as well as in the desert. Explain why an arctic mirage might appear and how it is different from a desert mirage.

23. Light travels faster in water than it does in glass. (a) Sketch the path followed by a ray of light moving obliquely from water into glass. (b) Repeat for the ray moving from glass into water.

24. We are able to see the Sun for a short period after it has actually sunk below the horizon. Explain why refraction in the Earth's atmosphere could produce this effect.

LENSES AND A SIMPLE MAGNIFIER

25. Explain why either a converging lens or a concave mirror could be used to build a solar cooker. Which would be more effective? Defend your answer.

26. Lenses used in eyeglasses may be either converging or diverging, depending on the defect of vision. One of the problems of lenses used in eyeglasses is that they must be ground such that the lash of the eye does not rub against them as the wearer blinks. (a) Show the general shape that a converging lens might have to avoid this problem. (b) Repeat for a diverging lens.

27. A small light bulb is placed 10 cm from a converging lens and an image is formed 20 cm from the lens. Is the focal length of the lens 20 cm? Why or why not?

28. (a) If you were trying to use the light from the Sun and a lens to set a fire, what kind of image would you be forming, a real image or a virtual image? (b) When you use a lens as a simple magnifier, is the image real or virtual?

TELESCOPES

29. Sometimes distant galaxies are so faint that they cannot be seen even with a powerful reflecting telescope. However, if a galaxy is followed for several hours while a piece of photographic film is placed at the eyepiece of the telescope, the galaxy can be detected. Explain how the film can detect an object in space whereas the eye cannot, even though film, *per se*, is no more sensitive than a person's eye.

30. Two telescopes have the properties listed below.

Telescope	Focal Length of Eyepiece	Focal Length of Objective	Diameter of Objective
A	25 mm	1250 mm	5 cm
B	6 mm	500 mm	8 cm

(a) Which has the greater resolving power?
(b) Which has the greater light-gathering power?
(c) Which has the greater magnification?

31. List the advantages of reflecting telescopes over refracting telescopes.

TOTAL INTERNAL REFLECTION

32. Light travels faster in water than in glass. (a) Can total internal reflection occur when light traveling in water is incident on glass? (b) Can it occur when light traveling in glass is incident on water?

33. Rearrange the prisms used in the periscope example given in the text so that you could see where you have been rather than where you are going.

34. Recall that the path followed by a light ray is reversible, and use the principles discussed in the section on total internal reflection to describe how a fish would view the world outside the water.

ANSWERS TO SELECTED NUMERICAL QUESTIONS

1. 10^{-5}s
3. 5.87×10^{12} mi
4. (a) 500 light-sec, (b) 1.29 light-sec
7. 30°
8. 60°
10. 1 m
11. 6 m

C H A P T E R

11

The Nature of Light

We noted at the beginning of Chapter 10 that light sometimes acts like a wave and sometimes like a particle. Thus far, we have been able to explain optical phenomena by adhering strictly to a picture derived from its wave nature. In this chapter, however, we look a little more closely at the dual nature of light. We shall examine some of the experiments that have been done that illustrate its wave nature, and we shall also look at some that require a particle explanation. Along the way, we shall attempt to answer the question: Where does light come from? We shall find that light is only one kind of wave from a spectrum of other similar types of radiation collectively referred to as electromagnetic waves.

11.1
IS IT A WAVE OR A PARTICLE?

What exactly is the nature of light? Two conflicting answers to this question were proposed in the late 1600s and early 1700s. Robert Hooke and Christian Huygens argued that light travels in waves, whereas Isaac Newton postulated that light rays consist of streams of particles—so-called packets of light. As we have already noted, this disagreement was ultimately resolved in a happy manner. Hooke and Huygens were correct—light exhibits wave behavior—and Newton was also right—light acts as if it is composed of a stream of particles. But how can light be two things at the same time, both a wave and a particle?

In a sense, this is an unfair question to ask. In the world around us, it is easy to distinguish between waves and particles. A thrown baseball unquestionably behaves like a particle, and a breaker crashing in on a beach unquestionably acts like a wave. However, in the submicroscopic world of electrons, protons, and light waves the distinction between waves and particles is not as sharply drawn. We must be content with saying that light is light, and it happens to exhibit properties of both waves and particles. There is no

245

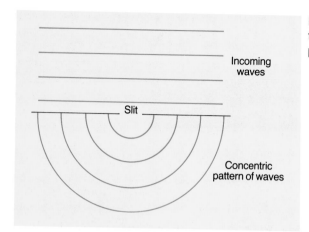

FIGURE 11.1 Incoming waves, from above in the figure, spread out in a concentric circular pattern after passing through a slit.

FIGURE 11.2 (a) Two point-source waves will interfere with each other, producing alternating patterns of wavy and quiet water. (b) An artist's graphical representation of the alternating pattern produced when two point-source waves interfere.

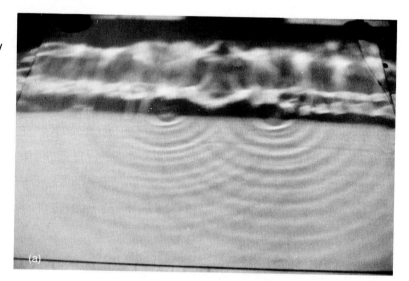

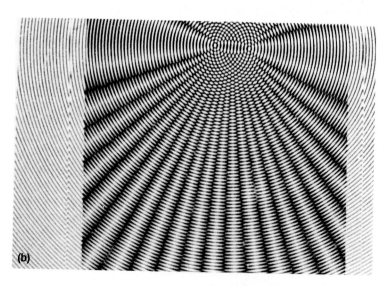

fundamental reason why it should be like ocean waves or speeding bullets. This blurring between wave-like and particle-like behavior is not unique to light. In fact, we shall soon see that electrons, protons, and other elementary particles also exhibit a dual nature in their behavior.

11.2
THE WAVE NATURE OF LIGHT— YOUNG'S DOUBLE-SLIT EXPERIMENT

For many years, scientists had attempted to establish whether or not light travels in waves. Then, in 1801, Thomas Young, a British physicist, decided that the best way to resolve this question

would be to determine whether or not light could exhibit interference. If so, its behavior would resemble that of known waves, such as water waves. Young knew that if a wave such as a water wave passed through a narrow opening, the opening would act as a source of new waves. Figure 11.1 is a sketch of what would happen. Incoming waves from above would spread out after passing through the slit and form a concentric pattern of circular waves below. If two narrow openings are placed side by side, two independent sets of waves would be produced. As these waves met, at some locations they would be in phase and constructive interference would occur, while at other locations they would be 180° out of phase and destructive interference would be produced. As a result, alternating patterns of quiet water and moving water would be produced, as shown in Figure 11.2. This works for water, but would it work for light? Young's experiment provided the answer.

The details of **Young's double-slit experiment** are indicated in Figure 11.3. He scratched two fine lines on a painted piece of glass and allowed light from a small source to pass through the openings (Fig. 11.3a). The light spread out from each slit, and interference between the two sources occurred. Along certain lines, labeled C in Figure 11.3b, the waves of light emerging from the two slits interfered constructively, while along other lines, midway between these, destructive interference occurred. This means that if the eye is placed in the position shown, alternating bright and dark lines are seen, as shown in Figure 11.4.

This demonstration of interference gave the wave model of light a strong boost. It was inconceivable that particles of light coming through these slits could cancel each other in a way that would explain the regions of darkness. Today, we still use the phenomenon of interference to distinguish wave-like behavior in any observation.

Although we will not discuss the details of the calculations, it is of historical interest to note that Young's experiment was the first to provide a way of measuring the wavelength of light. White light is composed of all the colors and hues ranging from red to violet. Measurements on these individual components of white light showed that the wavelength of red light is approximately 750 nm and violet light has a wavelength of about 400 nm (1 nm = 10^{-9} m).

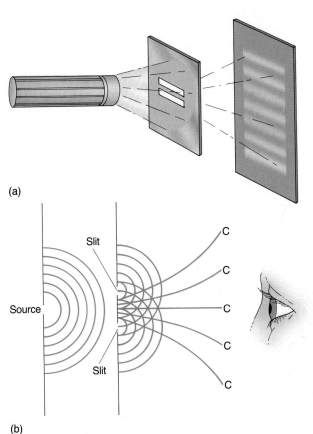

(a)

(b)

FIGURE 11.3 (a) Young's experiment was simple, yet it became a milestone of physics. He passed a beam of light through two narrow slits. When alternate patterns of light and dark appeared on a viewing screen, he deduced that light must exhibit wave behavior. (b) Constructive interference occurs along lines labeled C, and destructive interference occurs along lines midway between these.

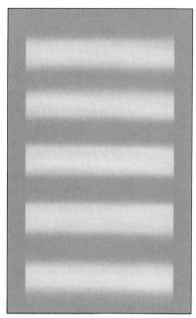

FIGURE 11.4 The alternating light and dark bands produced in Young's double-slit experiment.

EXAMPLE 11.1 Visible light frequencies

If the wavelength of red light is 750 nm and that of violet light is 400 nm, find the frequencies of these two extremes of visible light.

Solution In our study of wave behavior in Chapter 7, we found that all waves obey the equation $c = \lambda f$, where the symbol c represents the speed of light, 3.00×10^8 m/s, λ is the wavelength of the light, and f is its frequency. Thus, the frequency of red light is

$$f = \frac{c}{\lambda} = \frac{3.00 \times 10^8 \text{ m/s}}{750 \times 10^{-9} \text{ m}} = 4.00 \times 10^{14} \text{ Hz}$$

and that of violet light is

$$f = \frac{c}{\lambda} = \frac{3.00 \times 10^8 \text{ m/s}}{400 \times 10^{-9} \text{ m}} = 7.50 \times 10^{14} \text{ Hz}$$

11.3
THE PARTICLE NATURE OF LIGHT— THE PHOTOELECTRIC EFFECT

Approximately 100 years after Young's experiment demonstrated the wave nature of light, a

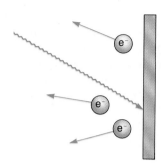

FIGURE 11.5 Incoming radiation dislodges electrons from certain metallic surfaces. This is the photoelectric effect.

series of experiments was performed that showed that light sometimes acts as though it is composed of a stream of particles. There are many processes in nature in which light exhibits its particle-like behavior, but the one that we shall consider is the **photoelectric effect.** The photoelectric effect is a process in which electrons are observed to be emitted by certain metals when light is shined on the metal. The process is indicated in Figure 11.5.

Wave theory can explain how light can knock electrons off a metal surface. After all, water waves can dislodge pebbles from a sandy beach, and a loud noise can knock down a delicate house of cards, so why shouldn't light waves dislodge electrons from a metal surface? However, further analysis of the photoelectric effect led to some results that could not be explained satisfactorily by assuming that light acts as a wave. Let us look at the key observations and show how the wave theory failed.

1. *Electrons are released by the metal when high-frequency light strikes the metal but not when low-frequency light strikes.* According to the wave model of light, both high and low-frequency light waves carry energy. Therefore, it should make no difference what frequency we use; electrons should be ejected for all types.

2. *Dim high-frequency light can produce the effect, but even very bright low-frequency light cannot.* Presumably, dim light would carry less energy than bright light. As a result, bright low-frequency light should eject electrons much more easily than dim high-frequency light.

FOCUS ON . . . Applications of the Photoelectric Effect

Practical applications of the photoelectric effect usually make use of an electronic component called a phototube, shown in the Figure. A phototube acts much like a switch in an electric circuit in that no current can flow through it when it is in the dark, but a substantial current can flow if the tube is exposed to light. The curved plate inside the tube of the Figure is made of a photoelectric material that will emit electrons when exposed to light. Thus, when light shines on this surface, the electrons emitted move through the vacuum in the tube to the collector (shown in black); these electrons constitute a current in the external circuit. One application of such a phototube as a burglar alarm is shown. When light shines on the phototube in part (a) of the figure, there is a current present in the external circuit, and this current energizes the electromagnet. The magnet then attracts the pole of a switch, and no current is allowed to flow in the portion of the circuit that contains the alarm. However, if a burglar interrupts the light beam, the current to the electro-

magnet is cut off, and the switch pivots to the right as shown in part (b). The switch in the circuit containing the alarm is closed and the alarm sounds. Obviously, these devices work better if ultraviolet light is used to activate the phototube so that its pathway will not be obvious to the burglar.

The second Figure shows how the photoelectric effect is used to produce the sound information on a movie film. The sound track is impressed on the film as an alternating pattern of dark and light lines along the side of the film. When light falls on this track from a light source, it can penetrate at some locations and activate the phototube shown. Likewise, at points where the sound track is dark, no light penetrates and the phototube is not illuminated. These changes in the amount of light reaching the phototube produce a fluctuating current in its circuit. This fluctuating current can be used to re-create the original sound signal by driving a speaker connected as shown.

A burglar alarm.

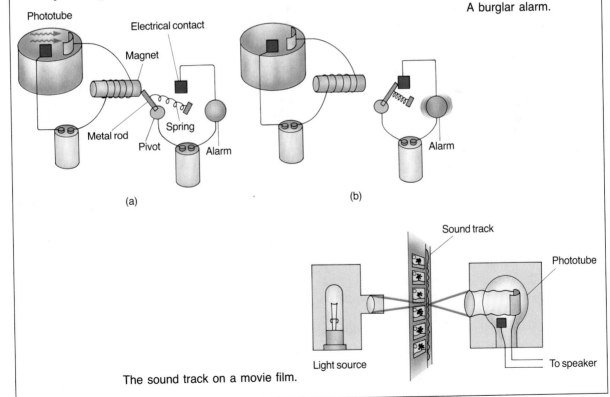

The sound track on a movie film.

3. *The electrons begin to leave the surface as soon as the light strikes the surface—there is no time lag at all, even for very dim light.* Calculations can easily be done to find out how much energy the incoming light carries and how much energy is required to eject electrons from the metal. These calculations show that light would have to shine on the metal for several minutes before individual electrons could gain sufficient energy to leave the metal.

4. *The higher the frequency of the incoming light, the greater are the speed and kinetic energy of the ejected electrons.* Again, it would be expected that the energy of the ejected electrons would depend on the brightness of the light but not on its frequency.

To indicate the extremely unusual behavior of the electrons emitted by the photoelectric effect, let us consider an analogy with water waves striking a beach and dislodging rocks from hard-packed sand. The amplitude of the water waves is analogous to the brightness of the incident light, and, of course, the frequency of the water waves is analogous to the frequency of the light. Consider only one of the observations above, number two, to see just how different the photoelectric effect is from our experiences with water waves. Observation two says that dim high-frequency light produces the photoelectric effect but very bright low-frequency light does not. This is the same as saying that large-amplitude water waves crashing into a beach will not release any pebbles if the frequency of the waves is low. However, a gentle, small-amplitude wave lapping against the shore will scatter rocks away from the shore easily if the waves have a high frequency. We know that water waves do not behave this way; large-amplitude waves dislodge rocks easily and push them away from the shore with large kinetic energies. As an exercise for you, use this water wave analogy for the other experimental observations for the photoelectric effect to see that they too are not in tune with common sense.

The results of the experiments on the photoelectric effect were most puzzling, but in 1905 Albert Einstein published a theory that explained the observations. Five years earlier, another physicist, Max Planck, had postulated that light is emitted discontinuously from a source of light.

By this he meant that the light coming from any source is not in the form of a continuous wave. Instead, the light is emitted in tiny bundles or packets. The original name for one of these discrete bundles of energy was the **quantum** (plural quanta), but the present-day accepted name is the **photon.** The amount of energy carried by an individual photon can have only a certain value and no others. The energy carried by a photon was determined by Planck to have a value given by

energy = h(frequency of the light)

or symbolically

$$E = hf \tag{11.1}$$

where h is Planck's constant = 6.626×10^{-34} J s. Note that Eq. 11.1 says that the higher the frequency of the light, the higher is the energy carried by a photon. Thus, photons of violet light carry more energy than do photons of red light.

With this photon picture of light in mind, let us see how Einstein was able to explain the photoelectric effect. Figure 11.6 indicates the model used by Einstein. The incoming light acts like a stream of bullets, with each bullet being an individual photon (Fig. 11.6a). An electron in the metal interacts with only one of these photons by absorbing the photon and taking on all of its energy; in this process the photon disappears (Fig. 11.6b). The electron now has an abundance of energy, which it uses in two ways. Part of it is used to simply get out of the metal. In other words, the surface of the metal acts like a fence that the electron must jump over in order to be freed from its "corral," as shown in Figure 11.6c. Any excess energy that the electron has after it uses some to escape from the metal now appears as kinetic energy, causing the electron to move away at a high speed if it has a lot of energy remaining or at a low speed if it has used most of it to escape (Fig. 11.6d).

With this picture in mind, let us examine all the experimental observations once again.

1. High-frequency photons have more energy than low-frequency photons. Thus, apparently, low-energy photons may not have enough energy to allow electrons to "jump the fence" and escape from the metal.

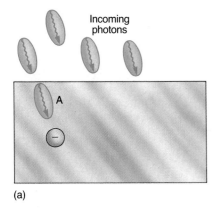

Incoming
photons

A

(a)

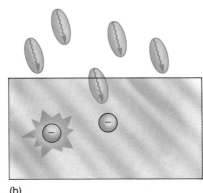

(b)

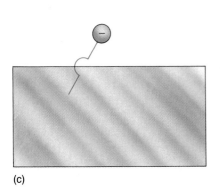

(c)

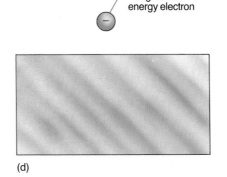

High kinetic
energy electron

(d)

FIGURE 11.6 (a) A stream of incoming photons strikes a metal. (b) The photon at A in (a) is absorbed by an electron and disappears. (c) The electron uses part of the photon's energy to escape the metal, and (d) the remainder of the photon's energy appears as kinetic energy of the electron.

2. A very bright beam of light would have a large number of photons in it, while a dim light would carry far fewer photons. Bright, low-frequency light would have many photons, but none of them would carry enough energy to cause the electrons to be ejected. Dim, high-frequency light has few photons, but each of them carries enough energy to free an electron from the metal.

3. Time does not enter the picture at all. As soon as a photon is absorbed by an electron, the electron instantly either has enough energy to escape the metal or it does not. There is no need to wait for the electron to gradually accumulate energy. Either a single photon does it or it isn't going to be done.

4. When a photon is absorbed by an electron, only a certain amount of the energy is needed by the electron to escape the metal. Any excess energy left over after the escape appears as kinetic energy of the freed electron. Thus, when an electron absorbs a high-frequency (or high-energy) photon, it has more energy left over than it does if it absorbs a low-frequency (low-energy) photon.

So what is light anyway? Is it a wave traveling through space, or is it a series of particle-like photons? This is a question that cannot be answered. Instead, all we can do is describe how light acts. It does exhibit wave behavior such as interference. Furthermore, as with other waves, there is a frequency associated with light. But it also acts like a series of particles. Since these particles can strike objects, be absorbed, and carry energy, when they strike an object they can cause the object to behave as though it had been struck by a particle. Thus, light acts as though it had a split personality—sometimes it acts like a wave, sometimes like a particle. Luckily for us, it never acts like both in the same experiment.

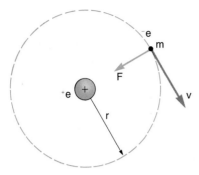

FIGURE 11.7 The hydrogen atom consists of a single negatively charged electron circling a single positively charged proton.

EXAMPLE 11.2 The energy of a red and a violet photon

(a) Find the energy carried by a "red" photon having a frequency of 4.00×10^{14} Hz.

Solution The energy is given by Eq. 11.1 as

$$E = hf = (6.626 \times 10^{-34} \, \text{J s})(4.00 \times 10^{14} \, \text{Hz})$$
$$= 2.65 \times 10^{-19} \, \text{J}$$

(b) Repeat part (a) for a violet photon of frequency 7.50×10^{14} Hz.

Solution

$$E = hf = (6.626 \times 10^{-34} \, \text{J s})(7.50 \times 10^{14} \, \text{Hz})$$
$$= 4.97 \times 10^{-19} \, \text{J}$$

EXAMPLE 11.3 Analyzing the photoelectric effect for a particular metal

It is found that in order for an electron to escape from a certain metal, the electron must gain 3.00×10^{-19} J just to get out.

(a) Will a beam of either the red photons or the violet photons in Example 11.2 have enough energy to cause the photoelectric effect to occur?

Solution The red photons have an energy of 2.65×10^{-19} J, but 3.00×10^{-19} J is required to eject electrons from the metal. So if the beam is red light of this frequency, the photoelectric effect will not occur. The violet photons, with an energy of 4.97×10^{-19} J, have enough energy to eject the electrons, so the photoelectric effect will occur.

(b) What is the kinetic energy of the ejected electrons when violet light shines on the metal?

Solution The energy of a photon is 4.97×10^{-19} J, and when this energy is absorbed by an electron, 3.00×10^{-19} J is used to escape from the metal. The remaining energy, 1.97×10^{-19} J, appears as kinetic energy of the electron.

11.4 THE SOURCE OF LIGHT

We have seen that a beam of light can be considered to be composed of literally trillions of small particle-like packets of energy called photons. But where do these photons come from? In this section, we shall find that photons are produced when electrons inside an atom are jostled around. To explain exactly what is happening, we shall consider the simplest atom, the hydrogen atom, which has one proton in its nucleus and one electron orbiting about the nucleus. This state of affairs is pictured in Figure 11.7. Under normal circumstances, the electron always orbits the hydrogen atom at a specific average distance from the nucleus and with a specific amount of energy. The electron is said to be in its **ground state** under these conditions. Now suppose that you want to move an electron farther away from the proton. Since the two are being held together by electrical forces, it should seem reasonable to you that you will have to do work on the electron to move it—that is, you will have to add energy to lift the electron farther out into space. However, there are some complications that we must address.

In 1913, Niels Bohr developed a model of the atom that was successful in explaining, among other things, how light is emitted by an atom. One of the assumptions made by Bohr was that there are only certain orbits in an atom in which an electron could orbit, and an electron will never be in any other orbit except one of these. This state of affairs is pictured in Figure 11.8. The electron has its lowest energy in the ground state, and it has higher energy as it moves to orbits farther from the nucleus. When an electron is in one

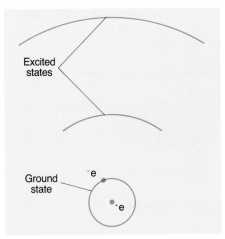

FIGURE 11.8 The ground state and a few excited states for the hydrogen atom.

of these higher orbits, the atom is said to be in an **excited state.**

This idea of only certain allowed orbits is a strange concept. If you get into your car and decide to drive in a circular path, there is nothing to prevent you from driving in a circle having any radius that you choose; there are no forbidden circular paths in between. But for the electrons in an atom, there are. Thus, if you want to move an electron from its ground state to an excited state, you must give the electron just enough energy to cause it to reach the excited level. (There are a variety of ways by which you could add energy to the atoms. You could heat a material made up of the atoms, pass an electric current through the material, or shine light on the material.)

If an electron has been caused to reach an excited level, it will not stay in that level for a long period of time. Instead, it will return very quickly to the ground state. However, in order to fall back to the ground state, the electron must release its excess energy. It does so by emitting a photon, as shown in Figure 11.9a. The energy carried away by this photon is exactly equal to the energy difference between the excited state and the ground state. You should also note that the electron does not have to move to the ground state in one single jump. Instead, if it is in a high excited state, it can stair-step down to the ground state, emitting photons in each jump, as shown in Figure 11.9b. The energy carried by a photon is given by $E = hf$; thus, if the energy is such that the frequency of the photon is in the visible range, we can see these emitted photons.

Therefore, according to the Bohr model of the atom, light is produced as electrons tumble down to the ground state from excited states. Specifically, in the hydrogen atom, it is found that visible light is produced when electrons fall from higher excited states down to the first excited state, as shown in Figure 11.10. Photons are released in other transitions also, such as from the first excited state to the ground state, or from the fourth excited state to the third excited state, but these photons do not have a frequency that places them in the visible portion of the spectrum. As we shall see later, this radiation can be ultraviolet (a radiation having a frequency greater than visible light) or infrared radiation (which has a frequency lower than visible light).

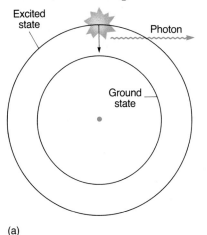

(a)

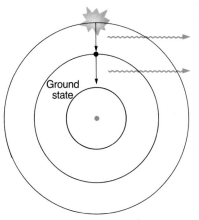

(b)

FIGURE 11.9 (a) When an electron jumps from an excited state to the ground state, it emits a photon. (b) In a high excited state, the electron can "stair-step" down to the ground state emitting photons with each step.

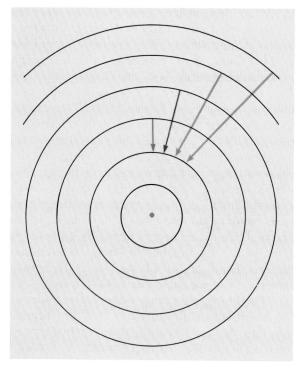

FIGURE 11.10 Visible light is produced from hydrogen by jumps which start on highly excited energy levels and end on the first excited state.

The origin of light has been discussed here in terms of the Bohr model of the atom, which pictures the atom as being much like our Solar System with electrons replacing planets and the nucleus of the atom replacing the Sun. This model is effective in helping one to gain a mental picture of the atom, but this simple model has been replaced by a more accurate quantum mechanical viewpoint, to be discussed in Chapters 12 and 13. We will find that the model of an atom with perfectly defined orbits and energy levels is not really the way that nature behaves.

EXAMPLE 11.4 Light and hydrogen atom

When the electron in a hydrogen atom is in the first excited state, it has a certain amount of energy. When the electron is in the second excited state, it has 3.02×10^{-19} J more energy than when in the first, and when it is in the third excited state, it has 4.08×10^{-19} J more than it does when in the first. Find the energy and the fre-

quency of all photons that could be emitted as the electron moves from the third excited state to the first excited state.

Solution If the electron jumps directly to the first excited state, it must get rid of its excess energy by emitting a photon with an energy equal to the difference in energy between these states. This energy is 4.08×10^{-19} J, and the frequency corresponding to this energy is

$$f = \frac{E}{h} = \frac{4.08 \times 10^{-19} \text{ J}}{6.626 \times 10^{-34} \text{ J s}}$$
$$= 6.16 \times 10^{14} \text{ Hz}$$

Note that this frequency is between 4.0×10^{14} Hz and 7.5×10^{14} Hz, the limits of the visible spectrum, and thus this photon will be seen as violet visible light.

However, the electron does not have to jump directly down to the first excited state from the third; instead, it can stair-step down by first jumping to the second excited state and then from there to the first. In the jump from the third excited state to the second, the energy given off is

$$E = 4.08 \times 10^{-19} \text{ J} - 3.02 \times 10^{-19} \text{ J}$$
$$= 1.06 \times 10^{-19} \text{ J}$$

corresponding to a frequency of

$$f = \frac{E}{h} = \frac{1.06 \times 10^{-19} \text{ J}}{6.626 \times 10^{-34} \text{ J s}}$$
$$= 1.60 \times 10^{14} \text{ Hz}$$

This frequency is not within the frequency limits of the visible spectrum. In fact, this frequency is lower than that of red light, and this radiation is referred to as infrared radiation.

Finally, the electron will jump from the second excited state to the first excited state. The energy emitted in this transition will be

$$E = 3.02 \times 10^{-19} \text{ J}$$

corresponding to a frequency of

$$f = \frac{E}{h} = \frac{3.02 \times 10^{-19} \text{ J}}{6.626 \times 10^{-34} \text{ J s}}$$
$$= 4.56 \times 10^{14} \text{ Hz}$$

This frequency corresponds to that of red light.

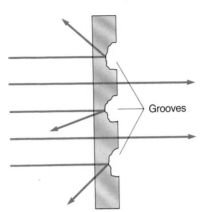

FIGURE 11.11 A side view of a diffraction grating.

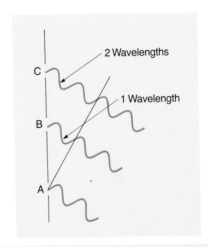

(a)

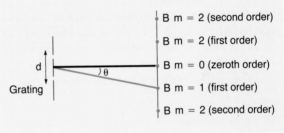

(b)

FIGURE 11.12 (a) Three rays from a diffraction grating which will undergo constructive interference. (b) The angle θ is measured above or below the central axis to a bright line B on the screen. The distance d is the distance between successive openings on the grating.

11.5
THE DIFFRACTION GRATING

We have stated at a number of different points throughout this book that white light is actually made up of all colors ranging from red through violet with all hues and shades appearing. As a child you probably verified this statement for yourself by using a prism to separate sunlight into its component colors. A device that is even better for breaking up light into its components is the **diffraction grating.** To understand how this device works, consider our discussion of Young's interference experiment. There we considered what happened when light waves interfered after having passed through *two* slits. The diffraction grating, on the other hand, allows light to pass through not just two but a large number of slits and to undergo interference.

Gratings are made by engraving closely spaced, parallel grooves on a piece of flat glass. A typical grating has about 6000 of these grooves per centimeter. As a result, the machine work is very detailed and precise. Figure 11.11 shows what happens when light is incident on a grating. Light passes through the glass unobstructed at locations between the grooves, but the light that strikes the grooves is either reflected or refracted to the side and is no longer a part of the beam.

Let us consider what happens to the unobstructed light when it falls on a screen. Figure 11.12a shows three rays of light with the same wavelength passing through adjacent openings and moving toward the same location on a screen. For convenience, let us refer to the openings as

slits. The light from slit A and the light from slit B move toward the screen, but we have selected a location on the screen such that the light from B will arrive exactly one wavelength behind the light from A. Thus, because of the extra distance B has to travel, it lags behind, but still when the two combine at the screen, crest will overlap crest and trough will overlap trough. This means that the light from these two slits will undergo constructive interference. The light from slit C also heads toward this location on the screen, but this light is one wavelength behind that from B and two wavelengths behind that from A.

This analysis can be continued for all the openings of the grating, but the result will be the same. Constructive interference will be occurring for all the waves headed toward this particular location on the screen. If we look at the screen at this position, we will see a bright line of light.

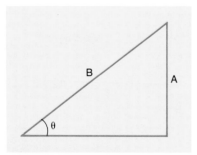

FIGURE 11.13 The sine of θ is the ratio of the length of A to the length of B.

Another bright line will be seen on the screen at a location such that the light from B has fallen two wavelengths behind that from A, and the light from C is two wavelengths behind B, or four wavelengths behind that from A. Thus, we will see alternating patterns of bright and dark on the screen.

The diffraction grating equation is

$$d \sin \theta = m\lambda \qquad (m = 0, 1, 2, \ldots) \quad (11.2)$$

Figure 11.12b defines the terms used in Eq. 11.2 and shows how to find the positions where bright lines occur. As the figure shows, bright lines occur at certain specified angles θ (Greek *theta*) from the central axis of the grating. λ (Greek *lambda*) is the wavelength of the light, and d is the distance between successive slits. Figure 11.13 defines the sine function for you. In trigonometry, the sine of an angle is defined in terms of two sides of a right triangle. The sine of an angle θ is the ratio of the side opposite θ, side A, to the hypotenuse, side B. It is not really necessary to understand the sine function in order to use Equation 11.2. Your calculator will give you the value for the sine. The quantity m is called the order number. A bright line occurs at a point along the central axis, as shown in Figure 11.12b. This line is called the **zeroth order** line, and m = 0 for it. The first bright line for a particular wavelength on each side of this central bright line is called the **first order** line (m = 1), the second bright line for a particular wavelength is the second order (m = 2), and so on.

You should take particular note here of the fact that for constructive interference to occur at

In this simple spectrometer, light passes through a set of focusing lenses so that the rays are parallel before striking the diffraction grating on the platform at the center. The resultant pattern is observed through the telescope. (Courtesy Bill Schulz)

some location, the light coming through an opening must fall an integral number of wavelengths behind the wave passing through a slit directly above it. Each of the colors that constitute white light has a different wavelength; therefore, a red bright line will not be formed at the same location on a screen as will, say, a green bright line. Thus, white light fans out into a spectrum of all the colors after passing through a grating.

As we shall see in a later section, the diffraction grating is an extremely important tool in the study of astronomy. No one has ever visited a star and sampled the material of which it is made. Yet astronomers have a good idea of the elements present in a star, the abundance of each element, the star's temperature, and so forth. This information is derived by studying the light from that star with a diffraction grating. In addition, an analysis of the diffracted light from a star or a galaxy can tell us whether that star is orbiting about another object or how fast the object is moving away from us or toward us. In short, the diffraction grating spectacularly enhances the ability of the telescope to give us information about astronomical objects.

EXAMPLE 11.5 Separating the colors

Two colors of light are incident on a diffraction grating having 6000 lines/cm. These colors are at the extreme ends of the visible spectrum; one is red light with a wavelength of 700 nm and the other is violet light of wavelength 400 nm. Find the angle at which one would have to move away from the central axis to observe the bright lines produced by each of these colors.

Solution First, we must find the distance between successive openings on the grating, d. This is done by noting that if there are 6000 lines/cm on the grating, then there are (1/6000) cm in a single line. Thus,

$$d = \frac{1}{6000} \text{ cm} = 1.67 \times 10^{-4} \text{ cm}$$
$$= 1.67 \times 10^{-6} \text{ m}$$

At the first occurrence of a red line, m = 1. The angle at which the red light is deviated is found as

$$\sin \theta_r = \frac{m\lambda}{d} = \frac{(1)(700 \times 10^{-9} \text{ m})}{1.67 \times 10^{-6} \text{ m}} = 0.419$$

and

$$\theta_r = 24.8°$$

The angle for the violet is given by

$$\sin \theta_v = \frac{m\lambda}{d} = \frac{(1)(400 \times 10^{-9} \text{ m})}{1.67 \times 10^{-6} \text{ m}} = 0.240$$

from which

$$\theta_v = 13.9°$$

Thus, if we start at the central maximum and gradually move our head to the side, we will first encounter a violet line at 13.9° and then a red line farther on at 24.8°. If the incoming light had contained all the colors of the rainbow, these colors would be found between these angles.

11.6 SPECTRAL ANALYSIS

The colors falling on a screen after white light passes through a diffraction grating is called a spectrum. This complete rainbow of colors is pretty and nice to look at, and it has some usefulness in the world of physics, but there are other kinds of spectra, and these have affected the world of physics and astronomy in a more far-

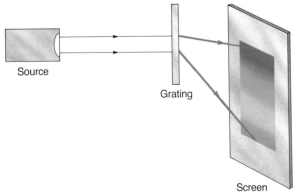

FIGURE 11.14 Observing spectra.

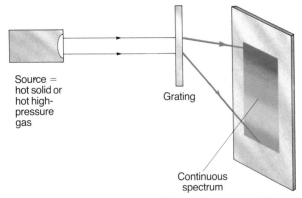

FIGURE 11.15 Production of a continuous spectrum.

reaching way than has that produced by sunlight. In this section, we will look at the three types of spectra that can be produced. They are the **continuous spectrum, the emission (or line) spectrum,** and the **absorption spectrum.**

The continuous spectrum

The general approach to observing spectra is shown in Figure 11.14. The light from a heated source is allowed to pass through a diffraction grating, which fans out its various colors and allows them to fall on a screen. For careful analysis, the screen is usually replaced with a sheet of photographic film so that the preserved images can be studied. That portion of Figure 11.14 consisting of the diffraction grating and screen or film is called a **spectroscope.** The type of pattern that appears on the film depends on the type of source that is used. If the object is a hot solid or a hot, high-pressure gas, as shown in Figure 11.15, the type of spectrum that appears on the film is one in which all the colors of the rainbow are present. A spectrum in which all the hues and shades appear is called a **continuous spectrum.**

In view of our discussion of how light is produced by the hydrogen atom, it should seem surprising that a heated solid produces all the colors in the rainbow. In an isolated gas atom, like the hydrogen atom, photons are produced when an electron jumps from an excited state toward the ground state. Since there are only a few pathways for jumps, it seems that photons of only a few different energies or frequencies would be emitted. Since frequency is the determining factor in the colors that we see, it would seem reasonable that only a few jumps would produce only a few distinct colors—not all the colors of the rainbow. The key to understanding why this vast array of colors does appear is to note that our energy level theory of the atom basically applies only for widely separated atoms, such as those in a gas. In solids or high-pressure gases, the atoms are packed so tightly together that neighboring atoms can influence one another. This alters the energy level pattern that they would have if they were isolated. Some of these closely packed atoms may have their first excited state quite far above the ground state, while another atom will have its first excited state quite close to the ground state, and other atoms will have all the possibilities in between these two extremes. As a result, there are possible jumps in the atoms that will produce a photon of virtually any energy within the visible spectrum, and consequently virtually any color. (Also included in the spectrum (but invisible) will be frequencies below red—the infrared—and frequencies above the violet—the ultraviolet.)

Emission spectra

A second type of spectrum, called an **emission spectrum,** can be produced by using a hot, low-pressure vapor as the source of light, as shown in Figure 11.16. In this case the spectrum that falls on the photographic film contains only a few distinct frequencies or colors instead of the rainbow found for a continuous spectrum. For example, if the heated vapor is sodium, only one color appears on the screen, a bright yellow. (Actually, the

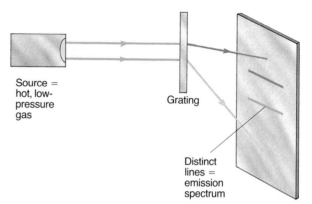

FIGURE 11.16   Production of an emission, or line, spectrum.

spectrum consists of two very closely spaced wavelengths, but, since the two are separated by only 0.6 nm, we shall refer to the two collectively as a single line.) The yellow light often used to illuminate streets is produced by sodium vapor lamps. If you return to the theory of the diffraction grating, you will find that if we measure the angle θ at which this color is deviated from the straight-through path, and if we know d, the slit spacing, we can accurately measure the wavelength, or the frequency, of this color.

We find that sodium is characterized by emitting only one frequency of light, and its emission spectrum contains only one line. On the other hand, if the heated vapor is mercury, it is found that four distinct colors, or frequencies, are observed. These are a particular frequency of yellow light, one of green, one blue, and one purple. (Actually, mercury contains several more lines that are very faint, but these would be seen only if long exposure times were used for the film.) In fact, it has been found that every chemical element produces its own characteristic pattern of spectral lines. In a sense, the lines produced by a heated vapor serve as fingerprints to identify the type of vapor that produced the lines.

The procedure of identifying materials according to the spectral lines they emit has been an important part of chemical analysis since it was first discovered by the German chemists Robert Bunsen and Gustav Kirchhoff in 1859, and the technique is still used in many laboratories, such as crime labs, to identify unknown substances.

When Bunsen and Kirchhoff developed this technique, they immediately began to catalog the spectra of all the known elements, and soon some surprising results turned up. For instance, when they were investigating the emission spectrum of mineral water, they found that an impurity in the liquid was producing an emission spectrum that had not been previously observed. These new lines were primarily in the blue region of the spectrum, and they named the element responsible for producing these lines cesium (from the Latin *caesium* for gray-blue). Later, while observing the vapor from a vaporized mineral sample, they found another pattern of previously unobserved lines. These lines were at the red end of the spectrum, and they named the element responsible for the production of these lines rubidium (from the Latin *rubidium* for red).

The explanation for why the elements emit the particular colors that are found in their emission spectra follows the basic interpretation used to explain why the hydrogen atom emits its particular frequencies, or colors. In a vapor, the atoms are so far apart that they do not influence one another as they do in a solid. Thus, only a small number of jumps are possible as electrons move from excited states toward the ground state. A few possible jumps mean that only a few possible energies exist for the emitted photons; so only a few possible frequencies are observed.

Absorption spectra

An absorption spectrum is produced by the technique shown in Figure 11.17. The light from a hot solid or high-pressure gas is first passed through a *cool gas*, and then the light is passed through a diffraction grating. The type of spectrum that is observed on the film appears at first glance to be a continuous spectrum. This is what you might expect because, as we saw above, the light produced by a heated solid or a high-pressure gas is a continuous spectrum. However, a more careful look shows that the spectrum is really not continuous at all. Instead, it is found that there are black lines scattered through the spectrum. Apparently, some particular frequencies, or colors, have been removed or absorbed out of the continuous background spectrum.

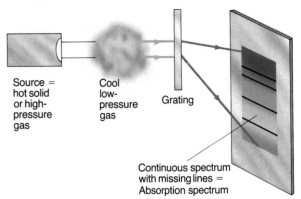

FIGURE 11.17 Production of an absorption spectrum.

A clue to finding out what is causing these absorption lines is found by noting the effect that the cool gas has on the absorption spectrum. For example, let us assume that the cool gas is sodium. We saw in the section on emission spectra that heated sodium vapor produces a single yellow line. If sodium vapor is used as the cool gas in Figure 11.17, it is found that all the colors of the rainbow are present in the resulting spectrum except one, and this missing color is identical to the one that would be produced by heated sodium vapor. That is, the missing color is the characteristic yellow line of sodium. Likewise, if we use mercury as our cool gas, we find that there are four lines missing in the absorption spectrum, and they are the characteristic yellow, green, blue, and purple lines of mercury.

To understand why the absorption spectrum looks as it does, we must examine the atomic processes that are responsible for it. An emission spectrum is produced as the electrons in excited atoms return to lower energy levels, emitting photons of light. As we have noted, there are many ways to excite atoms. Two ways are by heating the gas of atoms or by passing an electric current through the gas. However, there is yet another way, which is important to us now, and this process is somewhat like the photoelectric effect. Consider Figure 11.18. Shown in Figure 11.18a is a photon heading toward an atom. If this photon has just the right energy, it can be absorbed by an electron in the atom, and this excess energy of the electron raises it to an excited state, as shown in Figure 11.18b and c.

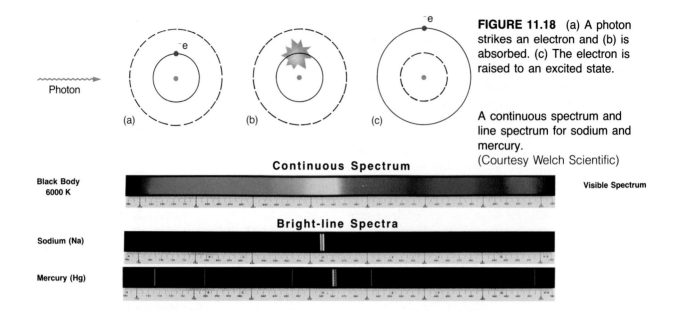

FIGURE 11.18 (a) A photon strikes an electron and (b) is absorbed. (c) The electron is raised to an excited state.

A continuous spectrum and line spectrum for sodium and mercury.
(Courtesy Welch Scientific)

Let us now return to our discussion of absorption spectra. In the white light coming from the heated solid, there are many yellow light photons. When these pass through the cool sodium vapor, these yellow photons are absorbed by the sodium atoms, because these photons have the precise amount of energy to lift an electron to an excited state. Thus, the cool vapor is removing from the beam precisely those photons that it would emit if it were already excited. These excited electrons will soon return to the ground state, emitting yellow photons as they go. However, these "new" photons can be sent out in any direction. As a result, very few photons of this frequency or color pass through the diffraction grating. This means that this color will be missing from the continuous spectrum when it is observed.

11.7
SPECTRA AND ASTRONOMY

No earthly being or human-made object has ever visited even one single star in the sky and brought back a sample for us to analyze here on Earth. Yet the store of known information about the stars is overwhelming. Astronomers can tell us the chemical composition of the stars, how fast they are moving toward us or away from us, whether they are orbiting other stars, and so forth. Certainly,

this fund of knowledge would be far less were it not for the telescope. But even with the telescope additional tools are necessary, and one of the most important, if not *the* most important, is the **spectroscope.** Most of our knowledge about the stars and other objects outside our Solar System comes from the study of the spectra produced by these objects.

As a brief quiz to see if you were paying attention in our discussion of spectra, what type of spectrum would be observed by looking at the light from a star such as our Sun? At first thought, it might seem that we should see a continuous spectrum, because in this case the source is a hot, high-pressure gas. However, that is not the case. The light produced initially by the star *is* a continuous spectrum, but before this light reaches an observer on Earth it must pass through two cool layers of gas. One of these is a relatively cool layer of gas surrounding the star like a halo, and the second is the atmosphere of the Earth. In both instances, certain frequencies are removed, so the spectrum observed is an absorption spectrum. The fact that the spectrum from our own star, the Sun, is an absorption spectrum was first noted in 1814 by Joseph von Fraunhofer. He found that the solar spectrum contained literally hundreds of dark lines missing from the background continuous spectrum. These missing lines, now called the Fraunhofer lines, provide the key that enable astronomers to determine the elements present on the Sun, as well as on any star.

For example, if the solar spectrum is compared to an absorption spectrum of sodium made in a lab here on Earth, it is noted that there are matching absorption lines in the Sun's spectrum. This tells us that sodium did the absorbing, but it actually does not tell us whether the sodium was present in the solar atmosphere or in our own, because the light must pass through both. We can make use of the Doppler effect to determine which atmosphere produced the absorption lines.

When we discussed the Doppler effect as applied to sound waves, we found that a sound has a slightly different frequency when it is moving toward us or away from us than it does if it is stationary with respect to us. This applies to all types of waves, including light waves. If we look at the edge of the Sun that is rotating away from us, we find that the light reaching us has been lowered in frequency. Let us now prepare a film of the absorption spectrum of sodium made in a laboratory here on Earth. The result would look like Figure 11.19a. Now let us look at the line corresponding to the absorption spectrum of sodium from light reaching us from the Sun. The pattern is identical, as shown in Figure 11.19b, except if the light is from the side of the Sun that is moving away from us, the lines are shifted slightly toward lower frequencies, toward the red end of the spectrum. These shifted lines must originate in the atmosphere of the Sun, because the atmosphere of the Earth is not moving with respect to us. By the same means we are able to examine the absorption spectra from stars and other astronomical objects and to determine what elements they contain. Our Sun is found to be composed of approximately 84 percent hydrogen, with most of the rest being helium. In fact, about two-thirds of the elements known on Earth have been found in trace amounts in the Sun. The remainder are probably also there, but in amounts so small that they cannot be detected by this procedure.

It is of interest to note that in 1868, astronomers found a pattern of absorption lines in the Sun that had never been seen in an element here on Earth. The conjecture at the time was that a

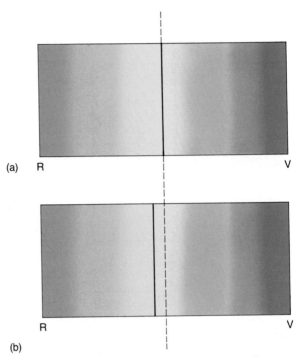

FIGURE 11.19 (a) Absorption spectrum of sodium made in a laboratory on Earth. (b) The spectrum line is shifted toward the red if the source is a star moving away from Earth.

new element unique to the Sun had been discovered. This element was named helium from the Greek word *helios*, meaning Sun. Of course, you know that this element has now been found on the Earth. However, it was only discovered in 1895 on the Earth when it was found as a gas emitted from uranium-containing minerals.

11.8
THE ELECTROMAGNETIC SPECTRUM

In our study of light, we have been concerned only with those waves that have frequencies detectable by the eye. Actually, this region of visible light constitutes only a very small portion of the **electromagnetic spectrum.** Electromagnetic radiation spans a wide range of frequencies (Fig. 11.20). Since each frequency has a different amount of energy per photon, waves of different frequencies affect us and our environment in different ways. Some warm our bodies, others can

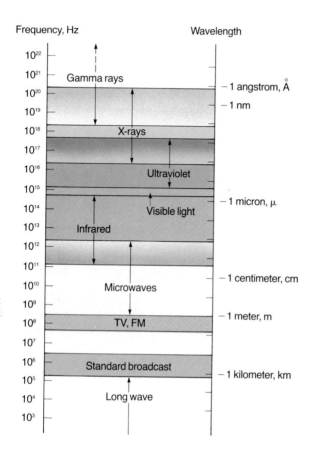

FIGURE 11.20 The electromagnetic spectrum.

kill germs, others destroy living tissue, and so forth. There are many sources of radiation in our environment. Sunlight, the ultimate source of most of the Earth's energy, contains a broad range of different frequencies. Some of this energy passes through the Earth's atmosphere and strikes the Earth, some is absorbed in the upper atmosphere and re-emitted at different frequencies, and some is reflected off the top of the atmosphere back into space. In addition to these natural radiations, there are technological sources, such as communication systems, radar, light bulbs, and X-ray machines, to name but a few. In order to appreciate the significance of electromagnetic energy in our environment and to understand how photons of different frequencies affect matter in different ways, we will take a brief look at all the different categories of waves that make up the electromagnetic spectrum.

FOCUS ON . . . Color

We have discussed in the text that white light consists of all the colors of the visible spectrum ranging from red to violet. When all of these colors are mixed together, the eye does not distinguish these individual components, but blends them together into what we refer to as white light. However, the eye also has the characteristic that all the colors of the spectrum can be produced by blending together only the three colors red, blue, and green. These colors are often called the **primary additive colors.** The color wheel, shown in the Figure, demonstrates how this mixing works. In fact, rather than take our word for it, you can do this on your own. Take three flashlights and cover the front of one of them with red cellophane, the face of another with blue cellophane, and the face of the last with green cellophane. If you now shine them on the wall in the pattern indicated by the Figure, you will find that various colors are produced when they overlap. For example, in the region where the red and blue overlap, the color magenta appears, and where the red and green overlap, yellow is produced. Note that in the center of the color wheel, white is produced in the region where all colors come together.

This technique of producing colors by mixing light beams is used to produce the image on the face of a color television set. If you examine the screen of a color TV closely, you will see that it is covered by a multitude of small dots painted on the glass. These dots are arranged in groups of three, where one of the three emits red light when struck by electrons, the second emits blue, and the third emits green light. Thus, if you are watching a western movie, at the location on the screen where the white hat of the hero is to appear, all the dots will be turned on by electron beams such that they glow with the same intensity. The three colors will mix together, and at a distance of a few feet from the screen, you will be unable to tell that the light is coming from closely spaced dots. Instead, you will see only the combined effect of white light. At the location where the black-hatted villain appears, no dots are turned on at the location of his hat. The yellow scarf of the school

Radio waves

Our discussion of the production of light by an atom revealed that visible light is produced when electrons in an atom jump from a state of high

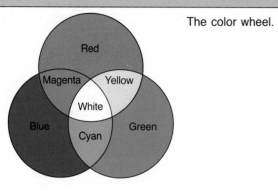

The color wheel.

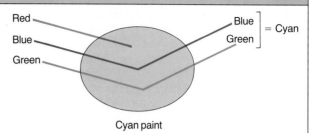

Cyan paint

Cyan pigment reflects blue and green light and absorbs red.

marm is produced by turning on only the red and green dots with equal intensity at the location of the scarf. How would you produce the pink hat of a hero of questionable character?

To understand why you see the colors of nature as you do, let us, for convenience, assume that white light consists of only the primary additive colors, red, blue, and green. The colors of objects around us are produced primarily by the colors that they reflect. For example, a red shirt appears red because it reflects the red light and absorbs all the other colors, assumed here to be blue and green. On the other hand, a yellow banana appears yellow because it will reflect the two primary colors red and green while absorbing blue light. These two colors mix together as yellow according to the color wheel. What color will an American flag take on in a room where we have only a green light bulb? To answer this question, let us assume the flag consists of a blue square in one corner and of alternating red and white stripes over the rest of the flag. Since there is only green light in the room, this color will be absorbed by the blue square and by the red stripes. Thus, since they are reflecting no color, they will appear black. Black is not really a color; it is the absence of color. The white stripes are capable of reflecting any color that falls on them, so since the only color present for them to reflect is green, these stripes will take on this color. Here is one more for you to try. A red rose with a green stem is brought into a room where a red light is present. What color does the rose appear to have. Try it to convince yourself that the rose petals will appear red, but the stem will be black.

Artists and others who study and work with color often state that the primary colors are not red, blue, and green as we have stated. Instead, they say that they are yellow, magenta (bluish red), and cyan (turquoise green). These colors are often called the **subtractive primaries,** because all the colors can be produced by selectively mixing pigments of these colors. The absence of these pigments will produce white paint, and the presence of all these pigments in equal proportions will produce black. To see how these pigments work, consider placing a drop of cyan paint on a piece of paper, as shown in the second Figure. When white light falls on the cyan paint, it reflects the colors blue and green while absorbing red (see the color wheel in the first Figure). The two colors blue and green are blended together by the eye, and the drop is interpreted as having its characteristic cyan color. Now here is an exercise for you to work on. What color will be produced when equal proportions of magenta paint and yellow paint are mixed? To answer this, recall that magenta will reflect only red and blue, while yellow will reflect only red and green. Thus, the combination will absorb every color except red, and the mixture appears red. Predict what color will be produced when magenta and cyan paint are mixed in equal amounts. The answer for you to work out for yourself is blue.

energy to a state of lower energy. There is an alternative way by which certain types of electromagnetic radiation can be produced. It is found that any time a charge is accelerated, the charge will release some of its energy in the form of electromagnetic waves. This is what happens in a radio antenna here on Earth. Charges in the antenna are caused to surge back and forth along a

FOCUS ON . . . Blue Skies and Red Sunsets

The sky on Earth takes on its characteristic blue color because of a resonance effect between the frequencies of the colors in the visible spectrum and that of molecular oscillators in the atmosphere of the sky. In our discussion of sound resonance, we found that any object that can vibrate prefers to vibrate at certain specific frequencies. If pushed at one of these frequencies, its amplitude becomes larger than if pushed at any other frequency. In the sky, the molecular oscillators have a preferred frequency of vibration that is the same as the frequency of blue light. Thus, when white light passes through the atmosphere, it is absorbed by the molecular oscillators, which in turn re-radiate it at the same frequency in all directions, as shown in the Figure. This phenomenon is called scattering. Thus, when we look in any direction in space, we see light coming to us from these molecules, and this light has the characteristic blue color of the sky. On the Moon, where there is no atmosphere, there are no oscillators to re-radiate skylight to an observer. Thus, a sky watcher on the moon would see a black sky above him.

At sunrise or sunset, the light reaching an observer on Earth has to travel a long distance through the atmosphere. Thus, before the light reaches the observer, the blue light is scattered out, but so also are most of the other colors from blue downward. The last color to be affected is that of the red light,

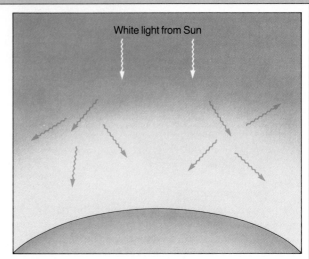

White light from the Sun has the blue portion of it scattered in all directions in the atmosphere of the Earth.

so it is still present in the light when it arrives. Thus, sunsets and sunrises are red. The beauty of these effects is enhanced by the fact that the red light is reflected off clouds. The water droplets in clouds are capable of reflecting all colors that strike them, and as a result, the clouds take on a pinkish hue as they reflect the red light from the Sun.

---

metal rod, and as they do so, electromagnetic waves of low frequency and long wavelength are produced. An AM (for *amplitude modulated*) radio station designs its antenna and associated electronic equipment such that the wave emitted by the station has a particular frequency between 530 and 1605 kHz, while the antenna of an FM (for *frequency modulated*) station emits a wave with a particular frequency between 88 and 108 MHz.

The frequency of the wave that a station is allowed to transmit is called the carrier wave. However, if the station did nothing more than emit that wave, it would not be a very exciting programming achievement. Some means must be provided to allow this wave to carry information. Let us examine this process for an AM station. Figure 11.21a is a representation of the carrier wave

emitted by the station. The sound signal to be sent out by the station at some instant of time is represented by the wave pattern shown in Figure 11.21b. Electronic equipment converts the sound signal to an electrical signal, and this signal is superimposed on the carrier wave, as shown in Figure 11.21c. Thus, the effect of the superimposed sound signal is to modify, or to modulate, the amplitude of the carrier wave. This modulated carrier wave travels through space until it is intercepted by an antenna in your home receiver. The electronic circuits in your radio work somewhat in reverse of those at the radio station. They separate the incoming wave (Fig. 11.21c) into two parts, the carrier wave and an electrical signal that emulates the audio wave. The carrier wave portion is tossed aside, and the electrical signal

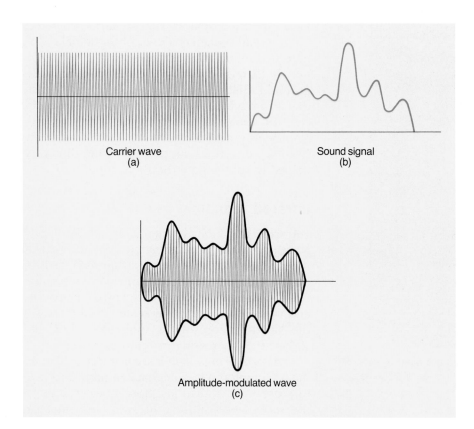

Carrier wave
(a)

Sound signal
(b)

Amplitude-modulated wave
(c)

FIGURE 11.21 Amplitude modulation.

carrying the information is sent to a speaker. This signal causes the cone of the speaker to vibrate to and fro, emulating the pattern shown in Figure 11.21b, thus reproducing the original sound wave.

EXAMPLE 11.6   Tune in your boom-box

(a) Find the wavelength of the carrier wave broadcast by an AM radio station assigned a frequency of 1430 kHz.

(b) Repeat for an FM station assigned a carrier wave frequency of 92.9 MHz.

Solution   **(a)** Regardless of the type of wave used, the equation that relates the speed of the wave c, the wavelength λ, and the frequency f, is given by $c = \lambda f$. We also must note that the speed for all the various types of electromagnetic waves in air is 3.00×10^8 m/s. Thus,

$$\lambda = \frac{c}{f} = \frac{3.00 \times 10^8 \text{ m/s}}{1430 \times 10^3 \text{ Hz}} = 209 \text{ m}$$

(b) For the FM wave, we find

$$\lambda = \frac{c}{f} = \frac{3.00 \times 10^8 \text{ m/s}}{92.9 \times 10^6 \text{ Hz}} = 3.23 \text{ m}$$

Microwaves

Microwaves are short-wavelength radio waves that have wavelengths between about 1 mm and 30 cm. One important use of these waves is in communication. For example, some cable television signals are transmitted as microwaves and received by antennas, as shown in Figure 11.22. In an application such as this, the frequencies used are selected so that the wave will not be absorbed appreciably by water molecules or other molecules in the atmosphere.

Another common use of microwave radiation is in the cooking of food. Water and oil molecules have a natural frequency of vibration that is in the microwave range. Thus, when you direct a beam of microwave radiation at the food, much

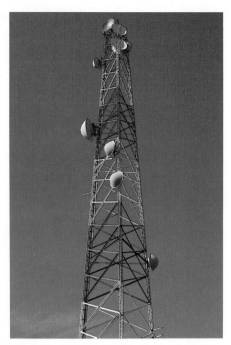

FIGURE 11.22 These antennas are used to receive cable television signals transmitted as microwaves.

of the radiant energy is absorbed by the water and oil molecules. This absorption is noted by the fact that these molecules begin to vibrate with larger amplitude, which is a signal of an increase in temperature of the substance. In fact, the increase in temperature is sufficient to cook the food. If there are no water or oil molecules in the substance, the microwaves either reflect off or pass through without absorption. This explains why glass or ceramic dishes do not become even lukewarm when they are placed alone in a microwave oven.

Recently, microwave radiation has given astronomers some insight into the question of how the Universe came into being about 20 billion years ago. The most widely accepted theory is that at the time of formation, the Universe was compressed into a ball of pure energy. Then an explosion occurred, which gives this theory its name as the **Big Bang** theory. Enormous amounts of radiation were produced at the time of the explosion, and from the remnants of the explosion came our stars, galaxies, planets, and all other parts of our Universe. Astronomers

have predicted that if this scenario is correct there should remain evidence of this initial radiation still around after all these years. The prediction is that the radiation, called **cosmic background radiation,** has changed its character such that now it is in the microwave region of frequencies. In 1965, radiation of the predicted frequency was detected striking Earth uniformly from all directions in the sky. This discovery helped to substantiate the Big Bang theory.

Infrared radiation

Infrared radiation is characterized by wavelengths between approximately 1 mm to that of the red end of the visible spectrum. Infrared radiation is colloquially called *heat radiation* because this radiation is the type emitted by warm objects, and when it is absorbed by materials, it increases the kinetic energy of the molecules of the absorber, thus increasing its temperature.

Human beings cannot see infrared radiation, but it is of interest to note that a rattlesnake is able to sense these radiations. Special sensors above the snake's eyes absorb infrared, and thus the snake "sees" warm animals even in the dark. Instruments utilizing infrared radiation can help us to "see" through the darkness, also. Since trees are generally at a different temperature from that of the earth, and nonliving objects such as houses or automobiles are at other temperatures, all these objects emit different frequencies of infrared radiation. Photographic film with an emulsion sensitive to infrared waves can detect these waves, as shown in Figure 11.23. Such photos are important in many applications. For example, the leaves of diseased trees are slightly cooler than the leaves of healthy ones, so infrared photographs can be used to pinpoint centers of disease in a forest before the unhealthy plants start to die. Similarly, if you take an infrared picture of a house on a cold winter night, you can map the insulating qualities of the structure; more heat escapes through poorly insulated regions than through well-insulated ones, and these differences in temperature show up as variations in brightness on the photo. Slightly revised versions of this process enable doctors to diagnose cancerous growths in the human body. An infra-

(a)

(b)

FIGURE 11.23 (a) A mountain scene photographed with film sensitive to visible light. (b) The same scene photographed with film sensitive to infrared radiation. Notice that while the black-and-white photograph of the tree and the tipi are significantly different, the infrared radiation emitted by the two is more similar.

red scan may reveal hot spots in the body, which are characteristic of malignancies.

Ultraviolet light

Just below the visible portion of the electromagnetic spectrum lies a region referred to as ultraviolet waves. These waves have wavelengths ranging from about 60 nm to 380 nm. As we moved through our discussion of the electromagnetic spectrum, we have moved from low frequency, low energy, radiation toward higher frequencies, and consequently higher energy photons. Ultraviolet light, which is more energetic than visible light, can excite electrons in most molecules and cause a multitude of chemical reactions. For example, the ultraviolet rays filtering through the Earth's atmosphere are largely responsible for several reactions on the surface of the skin. For example, they can cause a sunburn. An intense red or infrared light does not affect a person's skin, but even a moderate dose of ultraviolet light can produce a burn. Inhabitants of the Earth should be thankful that most ultraviolet radiation is absorbed in the upper atmosphere, or stratosphere, because larger doses reaching Earth could be harmful to humans. As we shall see later in our discussion of astronomy, there is fear that this protective covering is being depleted. A so-called ozone hole over the south polar regions is allowing ultraviolet radiation to penetrate to a much greater extent than in the past.

Ultraviolet light initiates many reactions in the atmosphere. For example, sunlight reacts with air pollutants to produce photochemical smog, and it is primarily the ultraviolet rays that produce these reactions.

Ordinary glass provides an effective shield against ultraviolet rays. As a result, you cannot get a suntan through a glass window. Some light sources such as mercury vapor lamps inherently emit a lot of ultraviolet. As a result, these lamps are usually enclosed in a glass housing to protect the user.

X-rays

X-rays have wavelengths in the range from about 10 nm to 10^{-4} nm. Thus, they have a high frequency and high-energy photons. In an X-ray generator, such as the one used in a hospital, electrons are boiled off a heated metal wire by a current passing through the wire, accelerated through a voltage of about 50,000 V, and directed toward a metal target, as shown in Figure 11.24. When the electrons strike the target, they slow down, and their kinetic energy is converted to electromagnetic energy in the form of X-rays.

Because of their high energy, X-rays have great penetrating power. As we all know, they can easily penetrate skin and flesh, but they cannot penetrate bone. As a result, if the X-rays are directed toward a photographic plate with an in-

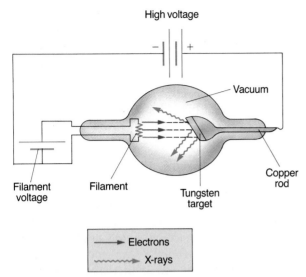

FIGURE 11.24 Diagram of an X-ray tube.

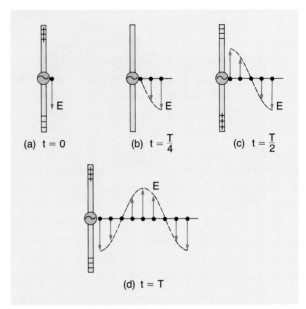

FIGURE 11.25 The electric field set up by oscillating charges in an antenna.

jured limb between the source and the film, the X-rays penetrate the flesh and expose the film, but the portion of the film covered by bone is not exposed. Thus, the presence of broken or injured bones is revealed when the film is developed. Because of their use as a diagnostic tool, X-rays have provided many medical benefits. However, X-rays can also seriously damage living cells. They can ionize molecules or even break bonds between atoms, and this damage can seriously disrupt the normal function of the cells. Studies show that high doses of X-rays lead to cancer and birth defects in animals. We shall discuss X-rays in more detail in a later chapter.

Gamma rays

Gamma rays have wavelengths ranging from about 0.1 nm to less than 0.00001 nm. Thus, gamma ray photons have very high frequencies and energies. Gamma rays are emitted by nuclei during some nuclear transformations. Under controlled usage, gamma rays are effective in combating cancer cells in the body, but in uncontrolled releases, they can severely damage the body and produce cancerous growths. Gamma rays will be discussed more completely later in our study of nuclear physics.

11.9
WHAT IS AN ELECTROMAGNETIC WAVE?

Earlier in this chapter, we discussed how light waves are produced when an electron in an atom jumps from an excited state to a lower energy level. We also pointed out in our discussion of radio waves that these kinds of waves can be produced by accelerated charges within an antenna. Regardless of the type of electromagnetic wave or how it is produced, its fundamental make-up is the same. For that reason, let us take a look at the production of an electromagnetic wave by an antenna to see exactly what constitutes such a wave.

In order to understand the fundamental principles of the production of a radio wave, consider Figure 11.25, which shows an alternating current generator connected to two metal rods. (The metal rods are the antenna.) The generator causes charges to move back and forth between the two rods. For example, at some particular instant of time, as shown in Figure 11.25a, the generator has forced an excess of positive charges into the top rod and an excess of negative charges into the bottom rod. This separation of charge sets up an electric field in the space surrounding the rod. As you recall from our discussion of electric fields in Chapter 8, this electric field is pointing downward, as shown in Figure 11.25a. Like a ripple in a smooth pond, this downward pointing electric field begins to move away from the antenna. However, as this field spreads out into space, the generator is changing the amount of charge on each of the rods, such that the downward-pointing field is decreasing in strength. This effect is pictured in Figure 11.25b. At some later time, the distribution of charge on the rods is such that the lower rod is positive and the upper one is negative. An upward-directed field is produced near the rod, as shown in Figure 11.25c. As the oscillations in the rods continue, the electric fields continue to spread away from the antenna, as shown in Figure 11.25d. The speed of this varying field through space is the speed of light. The name electromagnetic used to describe these waves now begins to make sense. As the "electro" portion of the name indicates, an electromagnetic wave is partially composed of a vibrating electric field.

The charges moving back and forth between the two rods constitute an electric current, and as we saw in our discussion of magnetism in Chapter 9, a current in a wire produces a magnetic field around it. Figure 11.26 shows the magnetic field pattern around the antenna during the portion of the cycle when the current is upward in the rods. Thus, at the same time that an electric field is being produced around the wire, a magnetic field is also being produced, and just as the electric field spread out away from the antenna like ripples on a pond, so does the magnetic field. Note that the direction of the magnetic field is always perpendicular to the electric field. In our diagrams above, the vibration of the electric field is up and down in the plane of the page, while the vibration of the electric field is into and out of the page. Thus, we see that *an electromagnetic wave consists of an electric field and a magnetic field vibrating at right angles to one another.* We also see that *the direction of travel of the wave is at right angles to the direction of vibration of both the electric field and the magnetic field.*

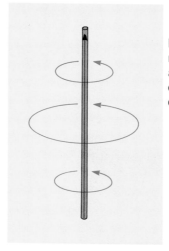

FIGURE 11.26 The magnetic field lines around an antenna carrying a changing current.

This discussion of how an antenna produces an electromagnetic wave is useful in that it provides a convenient way to show the make-up of these waves. However, the method discussed is only partially correct, because the fields produced in this manner would become extremely weak at short distances away from the antenna. The waves that actually reach us from an antenna utilize a principle that can be understood based on Lenz's law. Lenz's law says that a changing magnetic field can produce a changing electric field, which in turn can produce a current in a wire. It is also found that a changing electric field can induce a changing magnetic field. At large distances from the antenna, these effects predominate. Thus, the changing electric field and magnetic fields regenerate one another, and an electromagnetic wave is produced. Regardless of the method of production of a wave, however, the end result is the same: An electromagnetic wave is made up of a vibrating electric field and a vibrating magnetic field, and the two vibrations are perpendicular to the wave's direction of travel.

11.10 POLARIZED LIGHT

In our discussion of the production of an electromagnetic wave in the preceding section, we found that an electromagnetic wave meets the characteristics of a transverse wave—namely, that which vibrates, the electric field and the magnetic field, moves at right angles to the direction of travel. Figure 11.27 shows a representation of an elec-

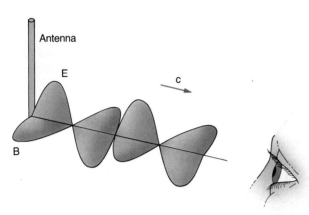

FIGURE 11.27 A representation of an electromagnetic wave leaving an antenna. The observer would see that the wave is polarized.

FIGURE 11.28 A polarized electromagnetic wave.

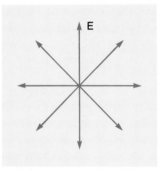

FIGURE 11.29 Representation of an unpolarized wave.

tromagnetic wave moving away from an antenna. Imagine yourself standing at the position shown in Figure 11.27, such that the wave is coming directly toward you. If you could actually see the electric field portion of the wave, it would consist of an up-and-down vibration as shown in Figure 11.28. Such a wave in which the vibrations of the electric field vector is back and forth along a particular direction is called a **polarized wave.** The vibrations emitted by a light source, such as an ordinary light bulb, are not of this nature, and the light is said to be **unpolarized.** To see why this is true, consider each atom in the filament of a light bulb to be a tiny antenna that is sending a wave toward you. The end result is a collection of trillions of waves, oscillating in all conceivable transverse directions. Such a mixture of waves is typical of all light sources, from the Sun down to the smallest light bulb. We picture an unpolarized light wave as shown in Figure 11.29, which indicates that the direction of vibration of the electric field could be in any direction as it comes toward you.

There are a variety of ways in which an unpolarized beam can be polarized, but we shall investigate only one of these. The process that we shall examine is that of selective absorption, which today is the most common method used. In 1938, E. H. Land discovered a material, which he called **polaroid,** that allows vibrations only along a particular direction to pass through. This mate-

rial is produced in sheets consisting of long-chain organic molecules, such as polyvinyl alcohol. During the manufacturing process, these sheets are stretched, which causes the long-chain molecules to align like pickets in a fence. The sheets are then dipped into a solution containing iodine, which causes the long-chain molecules to become conductors of electricity along the lengths of the chains. When light falls on these sheets, that portion of the light that has its electric field vibrations along the lengths of these chains is absorbed, while those vibrations at right angles to the chains are transmitted with little loss in intensity. Thus, the picket fence analogy is almost correct, except the light can pass through when the vibrations are perpendicular to the "pickets" but not when the vibrations are aligned with them.

The direction in the material that is perpendicular to the alignment of the long-chain molecules is called the transmission axis. Figure 11.30

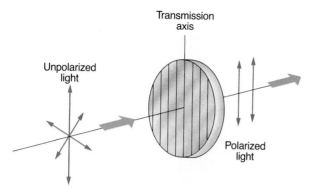

FIGURE 11.30 Unpolarized light becomes polarized after passing through a sheet of polaroid.

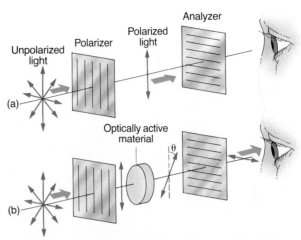

FIGURE 11.32 (a) The polarizer polarizes the light, which the analyzer then absorbs. (b) An optically active material rotates the polarized light so that some can pass through the analyzer.

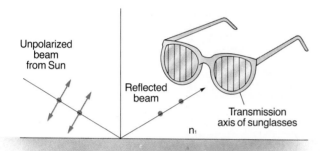

FIGURE 11.31 Unpolarized light can become polarized after reflection from a surface, such as that of water. Sunglasses, with transmission axis properly aligned, can cut down on the reflected light.

shows a beam of unpolarized light incident on a piece of polaroid material with its transmission axis aligned as shown. The light that passes through is polarized.

Polaroid lenses are popular in sunglasses to cut down on the glare from reflected light. To see how these glasses work, consider Figure 11.31, which shows unpolarized light from the Sun reflecting off a pool of water. When light is reflected off a surface, it becomes partly polarized, and if the reflecting surface is horizontal, the resulting polarization is also horizontal. (Before the advent of polaroid, the primary way of producing polarized light was by this reflection process.) As Figure 11.31 shows, the lenses of the polaroid sunglasses are oriented such that they will not let this horizontal vibration pass through.

An important use of polarized light involves the use of certain materials that are **optically active.** An optically active material rotates the direction of polarization of a beam of light passing through the material. Figure 11.32a shows a beam of unpolarized light incident on a piece of polaroid, which is called the polarizer. After passing through the polarizer, the light is polarized in the direction shown. The light now falls on a second piece of polaroid, one which has its transmission axis perpendicular to the vibration of the light. All of the light is absorbed by the analyzer, so no light reaches the observer shown. However, if a piece of optically active material is placed between the polarizer and analyzer, as shown in Figure 11.32b, the direction of vibration of the light is rotated slightly, and some light is able to move through the analyzer. This light can be blocked out once again by rotating the analyzer through the same angle as that through which the light beam has been rotated. It is found that the angle of rotation depends on the length of the sample and on the concentration if the substance is in solution. One optically active material is a solution of the common sugar, dextrose. A standard method for determining the concentration of the sugar solution is to measure the rotation produced by a fixed length of the solution.

(a)

(a) When the polaroids are uncrossed, light is easily transmitted, but (b) when one is rotated by 90°, they become crossed, and no light passes through. (Courtesy Bill Schulz)

(b)

SUMMARY

Light has a dual nature in that in some experiments it behaves like a wave and in others it behaves like a particle. It manifests itself as a wave in **Young's double-slit** experiment in that it undergoes interference, producing a series of alternating bright and dark lines.

Light manifests itself as a particle in the **photoelectric effect.** This effect is explained by assuming that light energy is carried in small packets called **photons** that are absorbed by electrons in certain metals, and the energy picked up by these electrons is sufficient to kick them out of the material.

Light is produced when electrons in **excited states** of atoms release their energy as they jump back to lower level states.

Light can be separated into its component frequencies by a **diffraction grating.** Such gratings are useful for observing the three different types of spectra, the **continuous,** the **emission,** and the **absorption.** The observation and analysis of spectra are important techniques in astronomy.

The **electromagnetic spectrum** consists of a vast array of waves, all of which are produced either by electronic transitions within atoms or by the acceleration of electric charges. The spectrum ranges from low frequency to high frequency as **radio waves, microwaves, infrared, visible, ultraviolet, X-rays,** and **gamma rays.**

EQUATIONS TO KNOW

E = hf (energy carried by a photon)

c = λf (basic wave equation)

KEY WORDS

Wave-particle duality
Young's double-slit
  experiment
Photoelectric effect
Photon

Ground state
Excited state
Diffraction grating
Continuous spectrum
Emission spectrum

Absorption spectrum
Electromagnetic spectrum
Radio waves
Microwaves
Infrared

Ultraviolet
X-rays
Gamma rays

QUESTIONS

YOUNG'S DOUBLE-SLIT EXPERIMENT

1. Would the bright lines of Young's double-slit experiment be farther apart for blue light than for red light? Defend your position.
2. If the incident light in Young's double-slit experiment were white light, what would the interference pattern look like?
3. When light waves interfere destructively and the light is destroyed, is energy also destroyed? Likewise, when light waves interfere constructively, the resultant pattern seems more intense than the simple sum of the two. Is energy being created?
4. Often when watching TV, you get a diminution of intensity on the screen caused by two beams reaching your antenna simultaneously. One is the direct beam; the other is from, say, an airplane flying overhead. Explain why the signal is reduced.

THE PHOTOELECTRIC EFFECT

5. Which of the following statements are true and which are false? Defend your answers. (a) A dim, high-frequency light will eject electrons with more kinetic energy than a bright, low-frequency light. (b) You can dislodge many high-energy electrons

from a metal surface with a bright, low-frequency light if you shine the light on the metal for a long time. (c) Light waves dislodging electrons from a metal behave like water waves dislodging rocks from a sandstone beach.

6. When the photosensitive cell of a light meter is exposed to light, a small electric current is produced that moves a needle to register the amount of light present. Could you obtain a true light meter reading near a powerful radio transmitter, or would the radio waves be likely to affect the instrument? Explain.
7. Sound can knock down a delicate card tower. If you are trying to destroy card towers with sound, would you be better advised to turn up the frequency or the amplitude? Explain. Light can knock electrons off the surface of a metal. If you are trying to dislodge electrons with light, would you be better advised to turn up the frequency or the brightness? Explain.
8. Would you expect the temperature of a metal to have any effect on the ease of removing electrons from it via the photoelectric effect?
9. If both visible light and ultraviolet light will eject

electrons from a metal, for which of the two would you expect the electrons to have the greater kinetic energy?

10. In a certain metal, photons of red light having an energy of 2.65×10^{-19} J will eject electrons with a maximum kinetic energy of 1.43×10^{-19} J. What is the amount of energy required to allow an electron to escape?

11. An electron in a certain metal has to have an energy of 2.35×10^{-19} J just to escape from the metal. What is the minimum frequency of the photon that will eject an electron? (*Hint:* What will be the kinetic energy of the ejected electron?)

WHERE DOES LIGHT COME FROM?

12. What is an excited state? How does an electron acquire the energy to be promoted into an excited state? What happens to the potential energy of an electron when it falls from an excited state to the ground state?

13. If a person does not have the strength to carry a rock up a hill, he may be able to roll it up slowly in stages and eventually accomplish the task. If one photon does not have the energy to raise an electron from one quantum level to the next, can many photons of the same energy move it up little by little? Explain.

14. The light emitted by hot sodium vapor consists of two colors very close together in frequency. What might the excited states responsible for this result look like?

15. According to the Bohr theory of the hydrogen atom, which has more energy, a photon emitted when an electron drops (a) from the second excited state to ground, or (b) from the third excited state to the second?

16. Is it possible for there to exist a single photon of white light?

17. The wavelength of the light emitted in a certain transition from a higher excited state to the first excited state of hydrogen is 656.3 nm. What is the energy difference between these two excited states?

18. An electron is in the fourth excited state of hydrogen. How many different energy photons could be produced as the electron returns to the ground state?

THE DIFFRACTION GRATING

19. Discuss the similarities and differences between the interference produced in Young's double-slit experiment and that produced by a diffraction grating.

20. Sodium emits two different wavelengths in the visible spectrum that are separated by 0.6 nm. That is, they are extremely close together in wavelength. Some diffraction gratings will show these as two separate lines, whereas others will not. That is, some will resolve them and some will not. What do you believe is the difference between a grating that will and one that will not?

SPECTRAL ANALYSIS

21. If a woman said that she could create electrons using only flint, steel, and a little dry tinder, would you believe her? If she said she could create photons using the same materials, would you believe her? Explain.

22. Describe the spectrum seen for the following situations. (a) The light from a hot tungsten filament is observed. (b) Light from a heated solid is passed through cool sodium vapor. (c) Hydrogen gas is heated until it emits light, which is then observed.

23. What kind of spectrum would be observed if hydrogen vapor is heated until it emits light, and then this light is passed through cool hydrogen vapor before being observed?

24. How might the spectrum from a star reveal the following properties about that star? (a) The composition of the star. (b) The speed of the star toward us or away from us. (c) The fact that the star is a binary star.

25. How could the spectrum from our Sun prove that it is rotating?

THE ELECTROMAGNETIC SPECTRUM

26. Which of the following objects would emit electromagnetic radiation? (a) A balloon with a slight negative charge on it falling to the earth. (b) An electron moving in a battery-operated circuit. (c) An electron oscillating to and fro in a wire as an alternating current. (d) A baseball leaving a pitcher's hand.

27. What is the electromagnetic spectrum? Is there a sharp distinction between radio and microwave radiation? Microwave and infrared radiation? Infrared and visible radiation? Explain.

28. List in order of increasing energy: ultraviolet, visible, X-ray, microwave, radio, infrared.

29. When the infrared picture shown in Figure 11.23 was taken, the photographer wished to filter out some of the more energetic visible light entering the camera. Do you think he used a filter that removed mostly blue light or mostly red light? Explain.

30. Photographers who work in dark rooms with pho-

tosensitive paper generally use a low-intensity red light in order to see. Would a dim blue light work just as well? Explain.

31. The flame of a welding torch is blue at the inner tip, white in the middle, and red on the outside. Which part of the flame is the hottest? Explain.

32. A fire inside a pot-bellied stove emits visible light. However, the stove itself does not change color when it gets hot. What has happened to the energy of the visible-frequency photons?

33. Imagine that you have two beams of light with the same number of photons in each beam, but one consists of ultraviolet frequencies and the other of visible light. Which beam carries more energy, or are they the same? Which requires more energy to generate, or are they the same? Explain.

34. Explain why X-rays are more apt to damage living tissue than are ultraviolet rays.

ANSWERS TO SELECTED NUMERICAL QUESTIONS

10. 1.22×10^{-19} J   **11.** 3.55×10^{14} Hz   **17.** 3.03×10^{-19} J

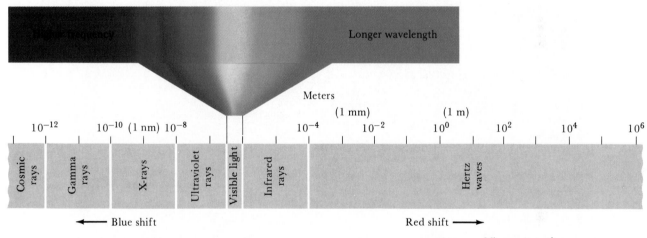

The range and regions of the electromagnetic spectrum.

The tracks of these subatomic particles made visible in a device called a bubble chamber give evidence of the unseen world of the atom and of the atomic nucleus. The curvature of the tracks is caused by an external magnetic field exerting a force on the charged particles. (Courtesy Lawrence Berkeley Laboratory)

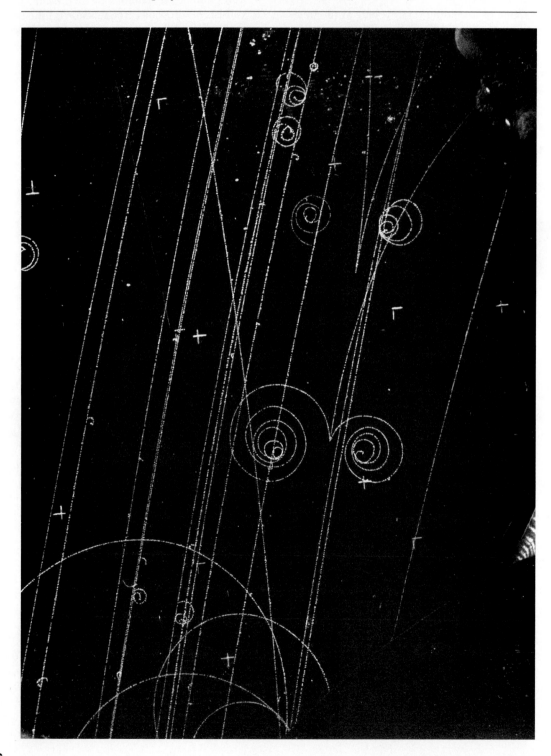

CHAPTER

12

Inside the Atom

When science is presented in a textbook, the various fields are neatly separated and packaged into units, chapters, and sections. In this book, for example, motion, energy, electricity, and magnetism have all been discussed as discrete topics. However, the history of the development of scientific thought is not nearly so orderly and compartmentalized. In many cases, seemingly different branches of science progress independently, and at first it seems as though the separate discoveries bear little relevance to each other. Then a key insight is brought forth that shows that the different components are really interconnected. Sometimes experimental technique is not refined enough to answer a crucial question until some apparently unrelated event occurs that enables a forward leap to be taken.

The study of the atom is one such agglomeration of many different types of research in different fields that were brought together to form one complete picture. During the Eighteenth and early part of the Nineteenth Centuries, various aspects of physics such as electricity, magnetism, and light were considered to be different from

each other, and the field of chemistry was thought to be pretty much unrelated to physics. Then in a relatively short period of time around the beginning of the Twentieth Century, scientists discovered that these seemingly different fields of inquiry are all intimately interrelated. These discoveries marked an exciting time in the history of science, for they changed our understanding of the physical world. In turn, this increased understanding created the foundation for the vast changes in human society that have occurred during the past few generations.

12.1 ATOMS

Think of a gold bar that has a mass of 1 kg. The volume of such a bar would be about 52 cm^3. Does this mean, however, that there is nothing in this volume but wall-to-wall gold with no empty space? If the bar is cut in half, the two resulting pieces still retain their chemical identity as solid gold. Imagine that the bar is cut again and again, indefinitely. Will the smaller and smaller pieces

Dalton and the Atom

Dalton's insight into the nature of the atom certainly is one of the great milestones of science, and he is remembered as a great scientist. However, he did make errors, and some of his key assumptions were wrong. In studying chemical reactions, he assumed "the rule of greatest simplicity," meaning that atoms would combine in the simplest possible manner. Thus, he assumed that water would be formed from the combination of one atom of hydrogen and one atom of oxygen and that the formula for water would be HO. Today we know that the correct formula for water is H_2O.

always be the same substance, gold? The ancient Greek philosophers thought about such matters, and two of them (Leucippus and Democritus) could not accept the idea that such cutting could go on forever. Ultimately, they speculated, the process must end when it produces a particle that no longer can be cut. In Greek, *a tomos* means "not cuttable," and from this we get the word **atom.** Thus, the word atom is used to describe the smallest, ultimate particle of matter.

Many centuries later, in 1803, John Dalton moved the atomic theory from a speculative basis to a scientific foundation. Dalton was a chemist interested in **chemical changes.** Chemistry is the subject of Chapters 15 through 17, but for now let us define a chemical change as the transformation of one material (or materials) to another. For example, consider hydrogen and oxygen, which are gases at room temperature. If the two gases are mixed and ignited, they combine and are transformed into another material—water. Dalton knew that hydrogen and oxygen always combine in fixed, constant proportions. For example, he found that 8 g of oxygen would always combine with 1 g of hydrogen, no more and no less. If an excess of oxygen is added to the container such that the 8:1 ratio of oxygen to hydrogen is exceeded, the excess oxygen will be left over after the chemical reaction. Similarly, if excess hydrogen is added, that excess will not enter into the reaction. Think how different this is from, say, baking a cake. You can bake a cake with one egg

or two, a lot of sugar or no sugar, and the result will still be a cake. Different cakes may taste different, but they will still be recognizable as cake, as long as you basically follow the recipe. However, water found anywhere on Earth is always found, within experimental error, to contain the same ratio of elements.

Dalton recognized that atoms are the fundamental units of matter. A collection of atoms of a single type forms a special class of substances known as **elements.** Gold, silver, hydrogen, and oxygen are all elements, because each of these substances is made up of only one kind of atom. Water, on the other hand, is not an element because it is made up of two different kinds of atoms, hydrogen and oxygen. In water, the different types of atoms are bound together as though they were connected by springs, and the combination of atoms is called a **molecule.** Any substance, such as water, made up of a fixed proportion of different atoms is called a **compound.**

The major contribution that Dalton made was that he recognized that all atoms of the same element have the same mass. To explain the transformation of hydrogen and oxygen into water, Dalton would have reasoned as follows. A specific mass of oxygen always combines with a specific mass of hydrogen because the atoms of oxygen (with their fixed mass) are combining with the atoms of hydrogen (with their fixed mass). The resultant compound, water, made up of billions of these atoms, always has a definite percentage, by mass, of these two atoms. If matter were continuous, Dalton reasoned, this definite ratio of mass would not always be found.

Furthermore, Dalton realized that chemical change involves transfers of whole atoms from one substance to another. During a chemical reaction, atoms are not created or destroyed, nor are they divided into parts; instead, the reactions merely change the way the atoms are bonded to one another. During chemical change, bonds between atoms are made or broken so that new compounds are formed and/or previously existing compounds are broken up. This subject will be investigated more thoroughly in our study of chemistry.

Dalton placed the atomic theory on a sound footing with his experimentation, but he never

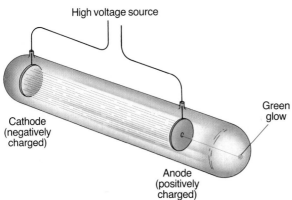

FIGURE 12.1 A simple cathode ray tube. If a high voltage is imposed between two electrodes in an evacuated tube, a current will pass between them. If a small slit is placed in the anode, some of the rays will pass through the slit and strike the glass. When they hit the glass, a greenish glow appears. Thus, the position of the rays can be located precisely.

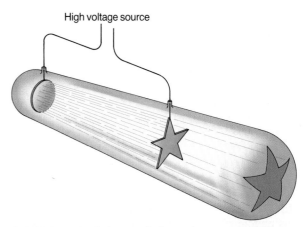

FIGURE 12.2 If the anode is cut in some specific shape, a shadow is observed behind the anode. This result shows that the cathode rays travel from the cathode to the anode. It also implies that the rays consist of a stream of particles that move in straight lines.

attempted to study the composition of an individual atom. Like the ancient Greeks, he still considered atoms to be ultimate, indivisible particles. Their make-up, or internal structure (if any), was not even considered.

12.2
THE DISCOVERY OF THE ELECTRON

The concept of a structureless atom began to change as a result of experiments with electricity. In 1752, Benjamin Franklin learned that lightning was a burst of electrical energy and therefore that electricity could travel through a gas. Other scientists found that the phenomenon of electrical discharge could be duplicated in the laboratory with a device called a **cathode ray tube.** This device is constructed by sealing two pieces of metal into a glass tube and then pumping most of the air out of the tube. The pieces of metal are then connected to a high-voltage source as shown in Figure 12.1. The positive plate is called the **anode,** and the negative plate is the **cathode.** When the high-voltage source is connected to these plates, an unusual phenomenon occurs. With a hole drilled in the anode, as shown in Figure 12.1, a green glow is seen on the end of the glass tube. The obvious question was: What is

causing this light to appear at the end of the tube? A series of experiments was devised in an attempt to answer this and other questions, and the results are discussed below.

1. Scientists observed that if the anode is placed in the center of the tube as shown in Figure 12.2, a shadow will be cast on the glass behind it. The existence of this shadow shows that something is being emitted by the cathode and is traveling toward the anode. As a result of this observation, the radiation became known as **cathode rays.**
2. Two charged plates were mounted inside the tube. When the ray passed between the plates, it was bent, or attracted, toward the positive plate and repelled by the negative one (Fig. 12.3). This experiment proved that the beam was negatively charged.
3. Similarly, the cathode rays could be bent by a magnetic field. Thus, the cathode rays behaved more like a stream of charged particles than a beam of light. By measuring how much the cathode rays were deflected by a given electric and magnetic field, physicist J. J. Thomson calculated the ratio of the charge to mass for the particles. He found this ratio to be 1.76×10^{11} C/kg.
4. The tubes were constructed using a variety of

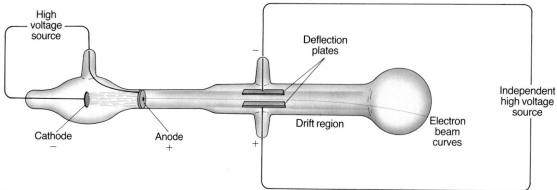

FIGURE 12.3 Thomson's cathode ray apparatus. J. J. Thomson's experiment was performed in the following manner: The cathode and the anode were placed in one end of a long evacuated tube, and a slit was cut in the anode. Some of the cathode rays would pass through the anode and drift through the tube. If no other fields were present, they would strike the center of the end of the tube and could be located by the green glow that appeared on the glass. Thomson then created an electric field perpendicular to the rays by placing two charged plates inside the tube and applying a high voltage between the plates. When this was done, the cathode rays were deflected toward the positive plate. Cathode rays were also deflected by a magnetic field. From these experiments, Thomson was able to measure the mass/charge ratio of the cathode ray particles.

materials for the cathode and a variety of residual gases inside the tube. (Although most of the gas inside the tube is pumped out, a small amount does remain.) Experiments always gave the same value for the charge-to-mass ratio.

Thomson's interpretation of the data was bold indeed. Since the cathode rays contained a fixed and constant ratio of charge to mass, he reasoned that the ray must consist of discrete particles. Today, these particles are called **electrons.**

At the time, Thomson had no way to prove that an electron was more or less massive than an atom. He assumed, without any definite proof at all, that electrons must be much less massive than atoms. Therefore, it followed that atoms must be made up of smaller, more fundamental particles. Thus, he concluded that atoms are not the smallest indivisible units of matter. Following this bold proposition, there arose an immediate interest in measuring the absolute value for either the mass or the charge of a single electron. This was finally accomplished some 15 years later in 1911. The most accurate modern measurements give the following values:

$$\text{charge of electron} = 1.60219 \times 10^{-19}\,\text{C}$$
$$\text{mass of electron} = 9.10953 \times 10^{-31}\,\text{kg}$$

By contrast, the mass of the lightest atom, hydrogen, is 1.67×10^{-27} kg, so the mass of an electron is only about 1/2000 that of the lightest atom. Electrons are indeed a small subdivision of the atoms, as Thomson had predicted.

Ordinarily matter is electrically neutral. Electrons are negatively charged. Therefore, it is obvious that matter must also contain positive electrical charge. What is the nature of this positive charge?

Electrons can be pulled out of atoms relatively easily. Thus, electrons can be removed from a piece of rubber by rubbing it with fur, and they can be pulled out of the cathode above by a high voltage. This ease of removal makes it possible to perform experiments such as Thomson's cathode ray studies that identified the electron and measured some of its properties. However, the positively charged portion of the atom is

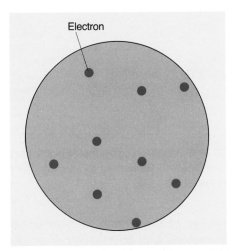

FIGURE 12.4 Thomson's plum pudding model of the atom.

much harder to study because it is difficult to separate and to isolate the positive charge from ordinary matter.

12.3
EARLY MODELS OF THE ATOM

The earliest model of the atom was that of a tiny, hard, indestructible sphere. However, when J. J. Thomson began to unravel the electrical nature of the atom, new models arose. One of these was the so-called plum pudding model suggested by Thomson. As shown in Figure 12.4, this model pictured the atom as a volume of positive charge with electrons embedded in it like plums in a plum pudding.

In 1911, an important experiment was conducted by Geiger and Marsden under the guidance of Lord Ernest Rutherford (1871–1937) that proved that the plum pudding model could not be correct. Before discussing the details of the experiment, let us ask, in general, how scientists can study something as small and seemingly unfathomable as the interior of an atom. To understand the approach, consider the following analogy. Imagine that you are a warrior in one of those fantasy novels where dragons, spells, and wizards are commonplace. As you wander across the mysterious countryside, you come upon an open valley. In the middle of the valley, you see a dense cloud hanging low over the ground. You cannot see into the cloud, and some magic spell prevents you from walking into it. As a guardian of your people, however, you feel that it is important to find out what lies inside the cloud. How would you go about your task?

Perhaps the first thing you would do is sit at the edge of the cloud and observe it. If an occasional burst of fire, steam, and smoke came forth, you might deduce that a dragon is hidden inside. If volleys of arrows flew out at you, you might suspect that hostile people lurk within.

A second and more active procedure would be to shoot something into the cloud. If you fired 100 arrows randomly into the cloud and all passed through and emerged on the other side, you might guess that there wasn't much solid matter hidden inside it. If 50 arrows passed through and 50 did not, then you would guess that there was a mixture of solid objects and open space. Other possibilities exist. The arrows might be absorbed, but new objects, say dragon scales, might fly back at you. Each observation would tell you something about the nature of the matter inside the cloud.

The procedure of shooting something into the cloud and observing what happens is analogous to what Rutherford and his colleagues did. As we shall see later, some atoms are radioactive, which means that they spontaneously emit certain kinds of radiation. One type of this radiation is a positively charged particle called an **alpha particle,** which is now known to be the nucleus of a helium atom. The experiment that Geiger, Marsden, and Rutherford conducted used alpha particles from radioactive atoms as arrows to bombard atoms in a thin metal foil, as shown in Figure 12.5. The results of the investigation were quite astounding to Rutherford and his colleagues. They found that some of the alpha particles were deflected through large angles, as shown in Figure 12.5; in fact, a few of them were deflected back along their initial direction of travel. To use Rutherford's words, "It was quite the most incredible event that had ever happened to me in my life. It was almost as incredible as if you fired a 15-inch shell at a piece of tissue paper and it came back and hit you."

If the plum pudding model were correct, large-angle scatterings would have been impossi-

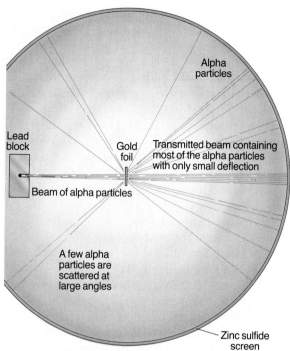

FIGURE 12.5 Rutherford's experiment with alpha particles. When some of the massive alpha particles were scattered at large angles, Rutherford deduced that the nucleus must be a small, dense, central core.

FIGURE 12.6 Head-on collisions of alpha particles with nuclei enabled Rutherford to estimate the size of a target nucleus.

ble. In such an atom, an alpha particle passing through the atom would be equally attracted and repelled in all directions, so the different forces would cancel and the alpha particle would pass right through the foil with little or no deflection.

To explain his observations, Rutherford theorized that the atom is mostly empty space, which accounts for the fact that most of the alpha particle arrows pass on through the foil without striking anything. However, to account for the occasional deflections, Rutherford also concluded that the positive charge and most of the mass of the atom must be concentrated in a very small volume, which he called the **atomic nucleus.** The electrons, he assumed, circled about the nucleus in orbits like planets about the Sun.

Rutherford's scattering experiments also led him to an estimate of the size of the nucleus. Figure 12.6 shows how he approached this problem. Rutherford considered only those alpha particles that were reflected back on themselves when they approached the nucleus. He knew the initial ki-

netic energy of the alphas, and conservation of energy enabled him to calculate how close the alpha could approach the nucleus before the Coulomb repulsion force exerted on the alpha by the nucleus would stop it and turn it around. He found that the alpha particles approached the nuclei to within 3.2×10^{-14} m when the foil was made of gold. Thus, the gold nucleus must be less than this value. For silver atoms, the distance of closest approach was found to be 2×10^{-14} m. From these results, Rutherford concluded that the nucleus has a radius no greater than 10^{-14} m. Other experiments had determined an approximate size for the entire atom, and the relative size of the nucleus to the atom was such that the atom as a whole was approximately 10,000 times larger in diameter than the nucleus. To understand the proportions in an atom, imagine that an atom is enlarged in size so that its diameter is equal to the length of a football field and one end zone (about 100 m). In such an atom, the nucleus would be a sphere about the size of a very small marble at the center of the field.

12.4
INSIDE THE NUCLEUS

Following Rutherford's discovery that the atom consists of a central nucleus with electrons swirling around it, the question arose as to whether or not the nucleus has structure. That is, is the nucleus a solid unit of charge or does it consist of a collection of individual particles? The exact composition of the nucleus has not been defined completely even today, but by the early 1930s, a model of the nucleus had evolved that is useful in

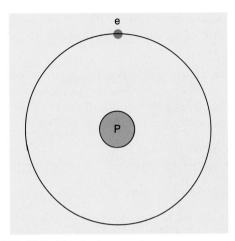

FIGURE 12.7 The hydrogen atom.

TABLE 12–1

Particle	Relative Charge	Relative Mass
electron	−1	0
proton	+1	1
neutron	0	1

about equal to that of the proton. This particle is the **neutron,** and its existence was verified conclusively by nuclear reaction experiments in 1932 by James Chadwick (1891–1974).

In our future discussions of the nucleus, we shall make use of the following quantities:

1. The **atomic number** of an atom is the number of protons, or units of positive charge, in the nucleus. It is also the number of electrons in the neutral atom. Thus, a hydrogen atom with its single proton has an atomic number of 1. An atom of carbon has an atomic number of 6, indicating that it contains 6 protons and, when neutral, 6 electrons. (We shall often refer to the charges on the electron and proton in terms of relative charges. For example, a proton has a relative charge of +1 and an electron has a relative charge of −1. This is just a shorthand way of saying that the proton has a charge of 1.6×10^{-19} C and an electron has a charge of -1.6×10^{-19} C.)

2. The **mass number** of an atom is the total number of protons and neutrons in the nucleus of the atom. For example, the most common form of carbon atoms has a mass number of 12 and an atomic number of 6. This means that since there are 6 protons in the nucleus, there must be 12 minus 6, or 6, neutrons. We shall also often use a relative scale for discussing masses of protons, neutrons, and electrons. Since the proton and the neutron have almost equal masses, we shall set their relative mass equal to 1, and because the electron is so much lighter than either of these two, we shall set its relative mass equal to zero. Table 12–1 indicates the relative charges and masses of the components of the atom.

It is convenient to have a symbolic way of representing nuclei that will show how many protons and neutrons are present in the nucleus. For

helping us to understand much about how it behaves. By that time, physicists had determined that the charges of the nuclei of the different elements were all whole number multiples of the charge on the electron. For example, a neutral iron atom has 26 electrons, and a neutral copper atom has 29 electrons. Consequently, the nucleus of an iron atom must have a charge equal but opposite to the charge of 26 electrons, the nucleus of a copper atom must have a charge equal but opposite 29 electrons, and so on. This observation led to the obvious conclusion that there must be particles inside the nucleus and that each particle bears a positive charge equal but opposite to that of the electron. These particles are, of course, protons, the existence of which was verified by Rutherford in 1920.

According to this model, the simplest atom, hydrogen, would consist of a single proton with a single electron swirling around it, as shown in Figure 12.7. From this model, we see that the mass of a proton must be equal to the mass of the hydrogen atom less the mass of the electron. This gives us a proton mass of 1.67×10^{-27} kg. However, complications arise as one moves to the next simplest atom, helium. Neutral helium must have in its nucleus two protons, yet the mass of a helium atom is about twice the mass of two protons. This led to the conclusion that there is another constituent part of the nucleus, a particle that must have no charge and that must have a mass

the element iron, this notation is $^{56}_{26}$Fe. Fe is the chemical symbol for iron; 56 is the mass number, and 26 is the atomic number. Thus, there are 56 total particles (protons and neutrons) in the iron nucleus, and 26 of these are protons. So, there are 30 neutrons.

12.5 ISOTOPES

All atoms of a particular element have the same number of protons in their nuclei. Thus, the nucleus of an element identified chemically as carbon always contains six protons. However, the nuclei of a given element may contain different numbers of neutrons. An alternative way of expressing this is to say that all atoms of a particular element have the same atomic number, but they may have different mass numbers.

Atoms that have the same atomic number but different mass numbers are said to be **isotopes.**

For example, in a sample of carbon, 98.6 percent of the carbon nuclei are the isotope $^{12}_{6}$C, and about 1.1 percent are of the form $^{13}_{6}$C. Found in even smaller percentages are the isotopes $^{11}_{6}$C and $^{14}_{6}$C. Even the simple hydrogen atom has three isotopes. They are $^{1}_{1}$H, ordinary hydrogen, $^{2}_{1}$H, deuterium, and $^{3}_{1}$H, tritium.

EXAMPLE 12.1	**Counting protons and neutrons**

(a) How many neutrons are in the nucleus of ^{37}Cl?

Solution Note that we have given you the mass number of chlorine, but we have not given the atomic number. All chlorine nuclei have the same atomic number, and you can find this in the table of elements (see the inside back cover). Find chlorine for yourself in this table and verify that it has an atomic number of 17. The number of neutrons is now easily found as

number of neutrons
= mass number − atomic number
= 37 − 17
= 20

(b) What are the symbol and the mass number of an isotope of an element containing 19 protons and 22 neutrons?

Solution If the nucleus contains 19 protons, its atomic number is 19. Search for atomic number 19 in the table of elements and prove for yourself that this nucleus is potassium, symbol K.

The mass number
= number of neutrons + number of protons
= 22 + 19
= 41

Thus, this isotope is represented as $^{41}_{19}$K, or more simply as ^{41}K.

12.6 THE BOHR ATOM

Taken together, the experiments performed by Thomson and Rutherford showed that an atom consists of a small, comparatively massive, positively charged nucleus surrounded by light, negatively charged electrons swirling in orbits around the nucleus. This picture, by itself, is a start, but there were some perplexing problems that this simple model could not explain.

1. It was known that if an electric charge is accelerated, it emits energy in the form of visible light or other forms of electromagnetic radiation. This presents a problem because an electron circling the nucleus has a centripetal acceleration. As a result, the electron continuously radiates energy, but if it loses energy, it should quickly spiral into the nucleus! In effect, the atom should collapse, and the predicted collapse time is quite short, of the order of 10^{-8} s after its formation. Thus, the question arises: How have atoms remained stable for the life of the Universe?

2. As we saw in Chapter 11, atoms of a particular element emit only certain specific frequencies of light that are characteristic of the particular element. Why? Also, we have seen that an atom will absorb only those frequencies of light that it will emit. Why? As noted in (1) above, the accelerated electron in an atom should emit light, but calculations indicate that the

Niels Bohr (1885–1962)

light emitted should consist of *all* visible frequencies, not distinct frequencies.

The first significant attempt to answer these questions was offered by Niels Bohr in 1913. We have already taken a brief look at the Bohr model of the atom in our discussion of spectra in Chapter 11. However, because of its importance in the world of physics, let us re-examine it here and explore its assumptions in greater detail. Bohr based his model of the atom on two postulates. They are:

1. Electrons orbit the nucleus in distinct paths, or orbits, similar to the orbits of the planets around the Sun. However, out of all the infinite number of possible orbits, only those with certain radii actually occur. Bohr didn't understand why this was so; in fact, this postulate contradicted the known laws of physics. In the gravitational analogy, a spacecraft can orbit the Earth at any radius, not just a few "allowed" radii. As part of this assumption, Bohr also assumed that the allowed orbits are characterized by the fact that they are *nonradiating*

orbits. This means that the electron can revolve about the nucleus indefinitely in these orbits without radiating away any of its energy. This gets us out of the predicament of having the electron continuously radiating energy and thereby spiraling into the nucleus.

2. If an electron is in a high (excited) orbit, it will fall back to its ground state, getting rid of its excess energy by emitting a photon. The photon carries an energy that is related to its frequency by $E = hf$. As we have seen, this explains why an atom emits only certain frequencies of light and no others. Conversely, if a photon of the precise frequency or energy is incident on an atom, the photon can be absorbed by an electron, causing it to move to an excited state. Thus, the atom will absorb only those photon energies that it will emit.

The validity of Bohr's assumptions about how the atom works could be determined only by how successful they were in agreeing with experimental observations. From his assumptions, Bohr was able to derive equations from which the frequencies of light that would be emitted by hydrogen could be calculated. The radius of the hydrogen atom could be calculated from his work, and the energy of the electron in each of its possible orbits could be found. These calculated values agreed with experimental findings. Thus, his work was enormously successful to scientists of the time. However, perhaps of even more importance was the fact that the Bohr theory gave us a model of what the atom looks like and how it behaves. Once a basic model is constructed refinements and modifications can be made to enlarge upon the concept and to explain finer details. His model did have limitations, and the picture of an atom with perfectly defined orbits and energies has been largely superseded today. We shall look at some of these innovations in the next section.

12.7
PARTICLES AS WAVES

We have examined the peculiar dual nature of light on several occasions throughout our study of physics. Beginning students of physics often find it troubling that light sometimes behaves like a wave and at other times like a particle. If you are one of these students, don't be concerned; many eminent physicists have shared your concern. However, in 1924, Louis de Broglie proposed the idea that *particles of matter also have wave properties.* Thus, the circle was complete, photons sometimes act like waves and sometimes like particles; now electrons, protons, and even baseballs share this dual nature in that they may at times act like waves.

De Broglie proposed that the wavelength of a particle could be found by the equation

$$\lambda = \frac{h}{mv} \tag{12.1}$$

where h is Planck's constant, $h = 6.63 \times 10^{-34}$ J s, and the product mv is the momentum of the particle, the mass times its velocity.

If de Broglie's theory is correct, then it should be possible to detect the wavelike behavior of electrons by passing a beam of them through something like a diffraction grating. After light passes through a grating, alternating patterns of light and dark are seen on a screen. Could a similar pattern be observed for electrons? The experiment was difficult to perform, because in order for a diffraction grating to work for light, the distance between the slits must be of the order of the wavelength of the radiation used. As we shall show in an example problem that follows this section, the wavelength of electrons is too short to enable an ordinary diffraction grating to work. However, crystals are made up of orderly arrays of atoms, and as it turns out, the atoms themselves are spaced at just the right intervals to provide a diffraction grating for electrons. Therefore, if moving electrons pass through certain crystals, they will be fanned out into alternating bands, providing they have a wave nature. In 1927, C. J. Davisson and L. Germer succeeded in measuring the wavelength of an electron by this process and thereby verified the wave nature of matter.

EXAMPLE 12.2 The wavelength of an electron and a baseball

(a) Calculate the wavelength of an electron (m = 9.11×10^{-31} kg) moving with a speed of 10^7 m/s.

Solution Eq. 1 gives

$$\lambda = \frac{h}{mv} = \frac{6.63 \times 10^{-34} \text{ J s}}{(9.11 \times 10^{-31} \text{ kg})(10^7 \text{ m/s})}$$
$$= 7.28 \times 10^{-11} \text{ m}$$

This wavelength is in the range of X-rays in the electromagnetic spectrum.

(b) A baseball pitcher throws a 0.150 kg baseball with a speed of 30.0 m/s. What is the wavelength of the baseball?

Solution Again, we use Eq. 12.1 to find

$$\lambda = \frac{h}{mv} = \frac{6.63 \times 10^{-34} \text{ J s}}{(0.150 \text{ kg})(30.0 \text{ m/s})}$$
$$= 1.47 \times 10^{-34} \text{ m}$$

This wavelength is so small that one could never find a diffraction grating with slits this small through which the baseball could pass. Thus, we could never observe interference effects with a large-scale object like a baseball.

12.8 WAVE MECHANICS AND THE HYDROGEN ATOM

One of the first successes of de Broglie's wave theory of particles was in explaining the arrangement of electrons in an atom. To understand de Broglie's model, we will have to recall some facts that we learned in our study of sound. There we found that we could set up standing waves on a string by tying one end of it to a wall and shaking the other end at just the right frequency. When the end of the rope is vibrated, a wave is sent down the rope and reflected off the wall. The two waves, traveling in opposite directions, interfere, and standing waves are produced. A standing wave on a string will have an integral multiple of half-wavelengths present on the string, as shown in Figure 12.8.

In de Broglie's application of wave theory to the hydrogen atom, he pictured an electron in orbit as having the properties of a circular standing wave. To understand his model, consider Figure 12.9. In Figure 12.9a, we have drawn three standing wave patterns: One contains a full wavelength, the second contains two full wavelengths,

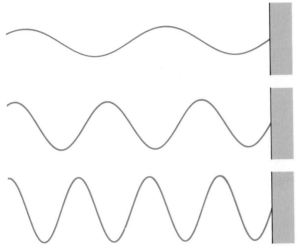

FIGURE 12.8 Several standing wave patterns on a string.

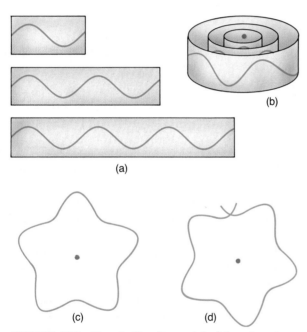

FIGURE 12.9 The de Broglie model of the atom can be visualized (a) by drawing 1, 2, 3, or more electron wavelengths on a flat piece of paper and (b) then wrapping the drawing around a nucleus. (c) An integral number of wavelengths is allowed. (d) If an orbital has a fractional number of wavelengths, the wave interferes with itself, and will be destroyed. Therefore, fractional orbitals do not exist.

and the third has three full wavelengths. Now imagine those wave patterns to be bent into a circular pattern as shown in Figure 12.9b.

FOCUS ON . . . Heisenberg's Uncertainty Principle

The Schrödinger theory talks about measuring the probability of finding an electron in a given region. Is it possible, even theoretically, to measure the exact position and momentum of a single electron? The answer is no. To understand this concept, think of the problem involved with locating an electron. Of course, it's absurd to try to measure an electron with a rule; it is far too small. What is needed is the tiniest probe to "touch" the electron with. The most delicate probe would be a single photon. Therefore, imagine trying to locate an electron by "bouncing" a photon off of it in much the same way that a ship can be located by using radar to bounce microwaves off of it. If all our measuring and detecting devices were perfect, could such a procedure measure the exact position and momentum of an electron? The answer is still no.

First of all, the photon moves as a wave, so at the very best, we could only locate the electron within the range of one wavelength of the photon. Second, a photon carries energy and momentum. Thus, when it bounces off the electron, conservation of momentum says that the electron will have its momentum changed. Thus, the act of measuring changes that which we are trying to measure, the momentum of the electron. As a result, there is an inherent inaccuracy in any attempt to measure the exact location and momentum of an electron. This principle is called the **Heisenberg uncertainty principle.**

De Broglie's viewpoint was that *one of Bohr's nonradiating orbits was formed when the circumference of the orbit was equal to an integral number of wavelengths.* The ground state orbit had a circumference of exactly one wavelength, the first excited state had a circumference of two wavelengths, and so forth. Figure 12.9c shows an excited state composed of five complete electron wavelengths. An orbit like that in Figure 12.9d would never be formed, because it does not have an integral number of wavelengths. A fractional wavelength orbit would annihilate itself because of destructive interference.

This initial effort by de Broglie was important to our understanding of the atom in that it provided a mechanism by which wave theory

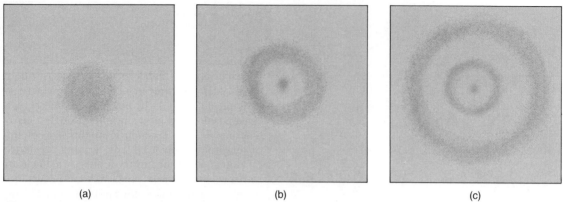

| (a) | (b) | (c) |

FIGURE 12.10 Three diagrams representing the shape of the electron cloud (a) in the ground state and (b) and (c) in two excited states of hydrogen. The densely shaded regions indicate places where an electron is likely to be found; the less densely shaded zones indicate regions where the probability of finding an electron is low. Note that these are not photographs of electrons, but rather models based on a mathematical theory of the atom.

could be applied in determining the details of atomic structure. Real success with wave mechanics, however, occurred two years later when an Austrian-German physicist Erwin Schrödinger developed a wave equation that described how particle waves change with position and time. This equation demonstrated convincingly that the wave nature of particles was a necessary feature in understanding the subatomic world. Schrödinger's equation has been successfully applied to the hydrogen atom and to many other submicroscopic systems.

Schrödinger's equation was mathematically very elegant. We shall describe it here by saying that what one is attempting to determine by solution of the equation is a quantity represented by the Greek letter Ψ (psi), called the **wave function.** The closest analogy between Ψ and ordinary attributes of a wave is that Ψ is most closely related to the amplitude of the wave. Its importance in wave mechanics lies in the fact that all that one can know about the behavior of a particle of matter can be determined from a knowledge of its wave function. For example, if Ψ is a wave function for a single particle, the value of Ψ^2 at some location at a given time is proportional to the probability of finding the particle at that location at that time. As a result of wave mechanics, our

viewpoint of the electron arrangements in atoms has now been modified in the following ways.

1. The electron is no longer required to remain at certain specific distances from the nucleus as in the Bohr theory. For example, the Bohr model predicts that the electron will always be found at a distance of 0.53×10^{-10} m from the nucleus when in the ground state. Wave mechanics says that it is most probable that the electron will be at this distance, but it also says that there is a probability of finding the electron at other distances from the nucleus.

2. The Bohr theory says that the electrons orbit the nucleus in a plane like the planets revolving around the Sun. Wave mechanics says that there is a probability of finding electrons anywhere in a spherical region around the nucleus.

As a result of these modifications, the currently accepted sketch of an atom in its ground state would be pictured as an electron cloud surrounding the nucleus, as shown in Figure 12.10a. Regions where the cloud is pictured as dense represent those locations where the electron is most likely to be found. Figure 12.10b and c represent electron cloud pictures of the orbits for excited states of hydrogen.

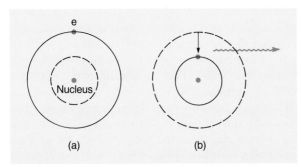

FIGURE 12.11 (a) An electron in an excited state. (b) When the electron drops to a lower level, a photon is emitted.

12.9
LASERS

The information and understanding that we have gained so far enable us to understand one of the most fantastic devices of the Twentieth Century, the **laser.** The word laser is an acronym for **l**ight **a**mplification by the **s**timulated **e**mission of **r**adiation. In principle, the acronym does a pretty good job of explaining how a laser works, provided that one understands the meaning of all the words.

Stimulated emission is a process that we have not considered previously, so let's examine how it works. We have seen in our earlier discussion that the electrons in atoms can be promoted to excited states by various processes. However, when the electron returns to its ground state, it gets rid of its excess energy by releasing a photon. This process is pictured in Figure 12.11. An electron typically does not spend very much time in an excited state before it de-excites; 10^{-8} s is a typical length of time. However, some atoms have excited states in which an electron seems to get

stuck for periods of the order 10^{-3} s to 10^{-2} s. Such states are called almost stable, or *metastable*, states. The electron will eventually return to the ground state from a metastable state and emit a photon. However, the process of stimulated emission can induce the electron to de-excite earlier than usual. Figure 12.12 pictures the stimulated emission process. In Figure 12.12a, we see an atom in an excited metastable state. In Figure 12.12b another photon comes along *that has exactly the same frequency as the photon that would be emitted by the excited atom if it were to de-excite.* This incoming photon "stimulates" the excited atom to de-excite, and the result is shown in Figure 12.12c. The initial incoming photon continues on along in its initial direction, but now it is accompanied by a photon from the de-excited atom. Note that both these photons have the *same frequency.*

Figure 12.12c shows another feature of the two photons. The two leave the atom with the *same direction of travel, and they are in phase.* Thus, when one is at a crest, so is the other. If these two photons come upon another atom that is in a metastable state, they can stimulate it to emit its radiation, and another photon joins the march, all photons in step and traveling in the same direction. Light composed of photons in step and moving in the same direction is said to be **coherent.** Coherent light is different from that emitted by an ordinary light bulb. The atoms in a light source operate independently of one another, in that once excited, they return to the ground state at random intervals and emit light of random frequencies, in random directions, and with no relationship between crests of one photon and that of another. The light of an ordinary source is often

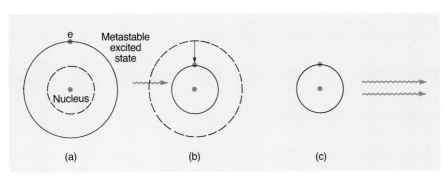

FIGURE 12.12 (a) An electron in a metastable state. (b) An incoming photon stimulates the atom to de-excite. (c) The photon emitted joins with the photon in part b to move away in step.

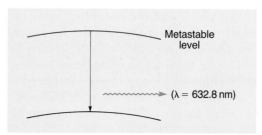

FIGURE 12.13 The metastable state of neon.

described as like waves created in a fountain. Little waves and ripples are created everywhere with no particular order or pattern between them.

Let us now apply these pieces of information to see how the light of a helium-neon laser is produced. The orbital arrangement of neon gas is shown in Figure 12.13. One of its levels is a metastable level from which an electron can descend to a lower orbit and in the process emit a photon with a wavelength of 632.8 nm. A mixture of helium and neon is confined to a glass tube sealed at the ends with mirrors, as shown in Figure 12.14a. One of the mirrors is completely silvered so that it reflects all light that strikes it, while the other is only partially silvered so that it will reflect some light and transmit some. A high-frequency power supply, not shown in the figure, is then connected to the tube such that it causes free electrons in the gas to sweep back and forth. These moving electrons collide with neon and helium atoms and excite them. Some of the neon atoms are excited to the metastable state. Helium is present in the

tube because it has the property of being able to efficiently transfer its energy to a neon atom in a collision between the two. As a result, we end up with a huge number of neon atoms in an excited metastable state. The details of what happens next are pictured in Figure 12.14a. Some of the excited neon atoms soon de-excite and emit photons. As shown, some of these (A and B in the figure) escape from the side of the tube and do not enter into the lasing action. However, other photons are emitted along the axis of the tube (C in the figure). One of these photons moving along the axis of the tube encounters an excited neon atom and causes it to emit a photon by the stimulated emission process. These two photons then stimulate other excited atoms to emit photons, and thus the avalanche grows (Fig. 12.14b). Light amplification is produced, starting with a single photon. In Figure 12.14c, the photon beam strikes the silvered end of the tube, is reflected and traverses the tube once again, with the photons stimulating as they go. When the photons strike the partially silvered end, some of them emerge from the tube, and this constitutes the laser beam.

The first experimental laser was operated in 1960, and today laser technology has become so widespread that it is entering into the realm of the commonplace. One obvious feature of the laser is that its beam is *highly directional*. Space scientists have been able to send laser light beams to the Moon from the Earth and reflect them off

FIGURE 12.14 (a) Neon atoms, shown as dots, are excited to a metastable level. These atoms eventually return to the ground state and emit photons. Most of these photons are lost through the sides of the tube (atoms A and B), but some move along the axis of the tube (atom C). (b) A photon moving along the axis of the tube encounters excited neon atoms and causes them to emit a photon by stimulated emission. This results in an avalanche of photons all traveling along the axis of the tube. (c) The avalanche of photons is reflected off the completely silvered mirror and continues to cause stimulated emission as it moves toward the partially silvered end. Those photons which emerge from the right end of the tube constitute the laser beam.

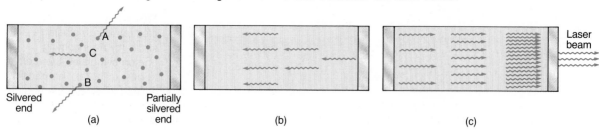

A laser beam arcs toward the Moon. The experiment bounced the light beam off a small 1-meter square mirror left on the lunar surface by Apollo astronauts, then received the beam approximately two seconds later at another telescope on Earth. This technique yields measurements of the Moon's distance.

mirrors left there by astronauts. The reflected beam was detected on Earth, and by this means the distance to the Moon has been determined to within a few centimeters. If conventional light sources had been used, the beam would have spread out to such an extent that its reflection would have been impossible to detect. In a similar experiment, scientists from two different continents have simultaneously bounced laser beams off the Moon. By accurately measuring the angle of the emitted beam with respect to the Earth and the time required for the round trip, and by using some simple trigonometry, the distance between continents has been measured with extreme accuracy. In more mundane applications, the highly directional feature of a laser beam has enabled engineers to dig tunnels precisely along a straight line, align bridges perfectly, and so forth.

The output of a laser is a very narrow beam that carries a lot of energy, and this energy can be concentrated into an even smaller area by lenses. As a result, when the beam is focused on a material, high temperatures can be achieved quite rap-

idly. A temperature of 6000°C can be reached in about half a millisecond with a powerful laser. Thus, a selected portion of metal can be melted and welded without destroying nearby areas. Eye surgeons are also using lasers to "weld" detached retinas back into position without disturbing the rest of the eye. Machinists have used the beams to melt tiny holes in small jewels used as bearings in watches or other precision machinery. Holes can even be drilled through a diamond, the hardest substance known. Unfortunately, many devices that can be used for the good of the human race can be used for destruction as well. A beam of energy that can create intense heat, that can be focused precisely, and that travels in a straight-line path at the speed of light has obvious uses in warfare.

The output of a helium-neon laser is at a single wavelength, 632.8 nm. However, more advanced lasers are tunable in that they can have an output beam at several different frequencies. One application of this is in determining the kinds of pollutants present in the atmosphere. Suppose you want to determine if a specific pollutant is present. If light energy of a specific wavelength is absorbed only by that pollutant molecule and not by normal air molecules, then the absorption of light from a laser beam operating at this wavelength becomes a method of detecting the presence of the pollutant. By measuring the amount of light absorbed, the concentration of that particular air pollutant can be determined.

The fact that the light is coherent means that very precise interference patterns can be produced with this device. Three-dimensional displays called **holograms** can be produced in this way. The applications of holography promise to be many and varied. For example, it is hoped that someday your television set will be replaced by one using the hologram photographic process. Instead of being flat, the picture you view will be in three dimensions.

One of the more recent applications of the laser that has found its way into the life of college students is the laser disk player that can be used in place of phonographs or tape players. Figure 12.15 shows the important pieces of one of these devices. The light from a laser is sent through

FOCUS ON . . . Making a Hologram

The Figure shows an arrangement that can be used to produce a hologram. Light coming from a laser is split into two parts by a half-silvered mirror H. One beam of light goes through the mirror and then through lens D, which diverges the beam, This spread-out beam then bounces off the subject and strikes a piece of photographic film. The second half of the beam is sent through lens A, which spreads it out and then bounces it off two mirrors, M1 and M2, and to the photographic film. The two beams of light then undergo interference at the location of the film, and it is this interference pattern that is really captured by the film. The pattern is so intricate that no vibration of the parts can be allowed while the hologram is being made. A move-ment or vibration as small as a fraction of a wavelength of the light used can ruin the interference pattern. The interference pattern produced relies on the fact that there is a constant phase relationship between the two beams during the period of time the film is being exposed. This constant phase relationship is possible only if the light is the coherent light from a laser.

When the photographic negative is produced, swirling patterns of lines are captured that cannot be interpreted by the eye. However, if laser light now is passed through the negative while one looks in the direction from which the light is coming, a three-dimensional view of the subject is seen hanging in space.

Making a hologram.

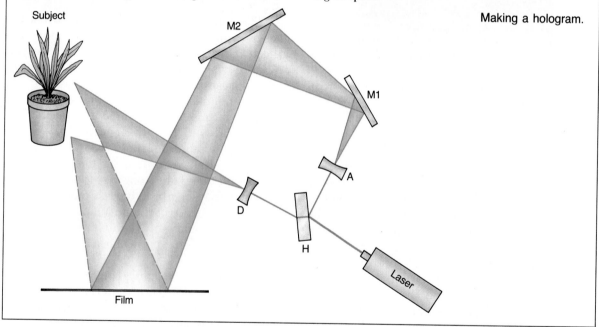

lens A, then toward a partially reflecting mirror. Part of the beam passes through this mirror, is redirected by another mirror through lens B and toward a laser disk. The light reflected off this disk retraces the original path until it reaches the partially reflecting mirror, where it is bounced toward a receiver. The receiver then uses the output it receives to drive a loudspeaker and to re-create the information that has been picked up off the laser disk. But how does the light retrieve information from the disk? In order to understand that, we will have to understand, to a limited extent, the **binary number system.** In ordinary arithmetic, we express the size of a quantity by representing its size on our ordinary decimal number system. Thus, a person 6 ft tall is taller than a 5 ft tall person. An alternative numbering system is the binary system. This system is the only one that can be understood by a computer, because it is a series of *on* and *off* signals. These on and off signals are segmented into groups of 8 individual signals, called a **byte.** Thus, in the binary system, the number one is represented as 00000001, this could be interpreted by a com-

FOCUS ON . . . Lasers in the Check-Out Line

Lasers are becoming more and more a part of everyday life, as a visit to your local supermarket will verify. In most larger stores of this genre, the cashier moves your grocery items across the beam from a helium-neon laser that senses information on the Universal Product Code, or bar code, label on a package. These bar codes are printed such that the spacing and darkness of successive lines can be interpreted by a computer as representing specific numbers. One of these bar codes is pictured in the Figure. The two thin lines at the far left are used to indicate to the computer that a new bar code is on the way to be interpreted. The next two lines, a heavy one and a thin one, are the basic code for the number 0. The number 0 is interpreted by the computer to indicate that the bar code for a grocery item is to follow. The next several lines are interpreted by the computer as the number 21140, which is the number that identifies the manufacturer. Two more thin lines then appear that indicate that the bar code for the item is on the way shortly. This pattern of lines is representative of the number 20786, which is the number assigned by the company to a can of green beans. As the beam of the laser moves across this bar code, the light is reflected back to a detector when a white space appears, but it is absorbed when

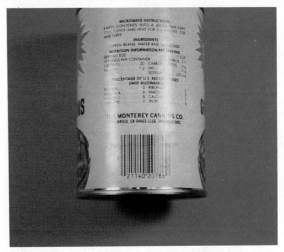

The bar code on a can of green beans.
(Courtesy Bill Schulz)

a dark space appears. The detector senses these varying shades of light and darkness as the numbers described above and sends this information to a computer that looks up the price for the product and rings it up on the cash register.

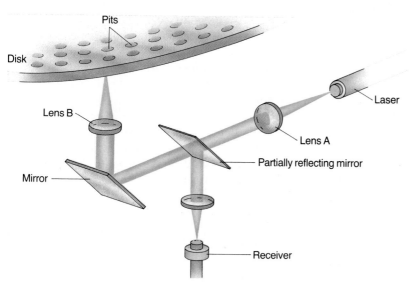

FIGURE 12.15 A laser disk player.

puter by turning a switch "off" for seven equal intervals of time then "on" for one of those time intervals. The number two in binary is 00000010. Turn your switch off for six equal intervals of time, on for one of these intervals, then off for one interval, and you have told the computer the number is 2. The decimal number three is 00000011, four is 00000100, five is 00000101,

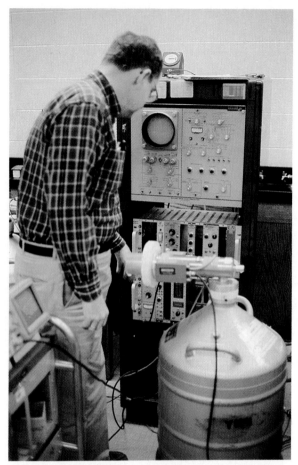

Physicist at work. Probing the secrets of the atomic nucleus requires sophisticated scientific equipment.

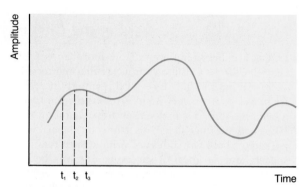

FIGURE 12.16 Converting the amplitude of a sound to binary.

Figure 12.16. The amplitude is sampled at regular intervals, and the loudness of the sound at any interval is then converted to a binary number. For example, suppose the amplitude of the sound signal in Figure 12.16 at t_1, t_2, and t_3 is decimal 4, 5, and 4, respectively. The computer will translate decimal 4, 5, and 4 to binary 00000100, 00000101, and 00000100, respectively. This pattern of binary numbers is then inscribed on a laser disk: A series of pits and smooth places are burned onto the disk such that each pit is smaller than the dot over an "i" in this textbook. When light from the laser in Figure 12.15 falls on one of these smooth places on the disk, it is reflected back to the detector, which interprets the presence of the reflection as an "on" signal or as a 1 in binary. When the light strikes a pit on the rotating disk, it is bounced off in some random direction and does not return to the detector. This is interpreted by the detector as an "off" or a 0 in binary. Electrical equipment connected to the detector then drives a loudspeaker according to the size of the binary number it receives. Thus, the original sound signal (Fig. 12.16) that was used to etch the disk is reproduced.

six is 00000110, seven is 00000111, eight is 00001000, ad infinitum. (We will leave it to you to determine how to relay these binary signals to the computer.)

The amplitude of a sound signal can be converted to a series of binary numbers, as shown in

SUMMARY

Dalton recognized that all atoms of the same element have the same mass and that chemical changes involve the transfer of whole atoms, not parts of atoms, from one substance to another. Atoms are not created or destroyed in ordinary chemical changes.

The negative charge in an atom is carried by elec-

trons, which are much less massive than the atom itself. Alpha-scattering experiments showed that an atomic nucleus occupies only a very small portion of the entire volume of an atom, even though it contains nearly all of its mass. **Atomic nuclei** consist of **protons** and **neutrons** (with the exception of hydrogen, which is a lone proton). A proton bears a positive charge equal but opposite to that of an electron. A neutron has approximately the same mass as a proton but is electrically neutral. The **mass number** of an atom is the total number of protons and neutrons. Atoms of the same element that have different mass numbers are called **isotopes.**

One of the first successful explanations of how an atom is constructed was the Bohr theory. This has largely been supplanted by a wave-mechanical model, in which the energy levels of electrons in atoms are described by a series of wave equations. An electron cloud diagram may be viewed as a description of the probability of finding an electron in any region of space around an atom.

We have found that light has a dual nature, but material objects also have a dual nature. The wavelength of a material object is given by the de Broglie equation.

Laser is an acronym for light amplification by the stimulated emission of radiation. The light from a laser is characterized by the fact that it is **coherent;** this means that all the photons maintain a fixed phase relationship with each other.

EQUATIONS TO KNOW

Mass number = number of protons + number of neutrons

$$\lambda = \frac{h}{mv} \quad \text{(de Broglie wavelength)}$$

KEY WORDS

Atom	Cathode ray tube	Atomic number	Wave mechanics
Chemical change	Cathode	Mass number	Wave function
Element	Anode	Isotope	Laser
Molecule	Proton	Nucleus	Stimulated emission
Compound	Neutron	Bohr atom	Coherence
Electron	Alpha particle	De Broglie wavelength	Hologram

QUESTIONS

ATOMS

1. The chemical formula for common table salt is NaCl, which means that sodium atoms are combined with chlorine atoms. Is this an element or a compound?

2. Microscopic particles of matter suspended in a fluid are seen to undergo a random, haphazard motion, darting one way and then another. This type of motion is referred to as Brownian motion and can be explained by the atomic theory of matter. What is your explanation?

3. Based on your explanation of Brownian motion in problem 2, why would a cork floating on the surface of the water not be affected?

4. An individual is lost in a deep forest. A bloodhound is brought in to track him. Based on the atomic theory of matter, how does the bloodhound do its job?

5. Some ordinary salt is dissolved in water. Is this a chemical change?

THE DISCOVERY OF THE ELECTRON

6. Explain the significance of the shadow cast in the cathode ray tube. In a cathode ray tube, the shadow is cast from the cathode to the anode. What conclusions would you draw if the shadow projected from the anode to the cathode? What conclusions would you draw if no shadow were cast at all?

7. What conclusions would you draw if the following observations were made during electrical experi-

ments in evacuated tubes? (a) The rays come from the anode and are attracted to the negative plate of a perpendicular electric field. (b) The rays come from both electrodes and are split into two discrete beams by a perpendicular electric field. (c) The mass/charge ratio of the cathode rays depends on the particular metal used as the cathode; that is, there is one value when the cathode is copper, another when it is nickel, and so on. Defend your answers.

8. It is also observed in discharge tubes that some "rays" do move toward the cathode, not away from it. What is the sign of the electrical charge on these "rays"? The properties of the particles that make up these "rays" depend on the identity of the residual gas that is left in the partially evacuated tube. What conclusion can you draw from this fact?

EARLY MODELS OF THE ATOM, INSIDE THE NUCLEUS, AND ISOTOPES

9. Describe a hypothetical atom consistent with the following experiments. In an evacuated tube, a high voltage pulls particles away from the anode. These particles are attracted toward a negative potential set at a right angle to their motion. Negative particles are scattered at large angles by the nucleus.

10. Why did Rutherford use alpha particles, rather than electrons, for his scattering experiment?

11. Explain the difference between the mass and the mass number of an atom.

12. Complete the accompanying table by substituting the correct numerical value where a question mark appears. (Use the table of elements on the inside back cover. Note that the table gives the symbols and atomic numbers of the elements, but not the mass numbers.)

Isotope in Nucleus	Atomic Number	Mass Number	Number of Neutrons
Oxygen-18	?	?	?
Strontium-90	?	?	?
Uranium-?	?	?	141
Iodine-?	?	131	?
?	17	35	?
?	?	226	138

13. What is the name of the element having atomic number 73?

14. Which one of the following has the most protons in its nucleus: iron (Fe), copper (Cu), or cadmium (Cd)?

15. What is the symbolic representation of (a) helium containing two protons and two neutrons, and (b) helium containing two protons and one neutron?

16. What is wrong with the following headline story, "Scientists have discovered a new isotope of hydrogen that contains two protons in its nucleus."

17. About how many times more massive is an atom of neodymium (Nd) than one of neon (Ne)?

18. Experiments showed that nuclear charges are whole number multiples of the charge on an electron. What conclusions would you draw if the smallest nuclear charge were half the charge of an electron? Explain.

THE BOHR ATOM

19. Explain why Bohr's model of the atom was an important step forward even though it was incomplete.

20. Discuss some differences and similarities between photons and electrons.

21. The transitions in hydrogen that produce visible light are those in which an electron in an excited state drops from its level to the first excited state. In these transitions, photons of wavelength 656.3, 486.1, 434.1, and 410.2 nm are emitted. How much more energy do the excited states from which these photons originate have than the first excited state?

22. An electron in a hydrogen atom drops from the first excited state to the ground state and emits a photon with a wavelength of 12.15 nm. (a) How much higher in energy is the first excited state than the ground state? (b) Is this an infrared, visible, or ultraviolet light photon? (c) What frequency photon could strike the electron in its ground state with just enough energy to kick the electron to its first excited state?

23. A hypothetical atom has its first excited state 3 units of energy higher than in the ground state, and in the second excited state, its energy is 3.5 units greater than in the ground state. What would happen to an electron in the ground state if struck by a photon having energy of (a) 2 units, (b) 3 units, (c) 3.2 units?

PARTICLES AS WAVES; WAVE MECHANICS; THE HYDROGEN ATOM

24. Describe an experiment that would verify that light acts as a wave and one that would verify that an electron acts like a wave.

25. Describe an experiment that would verify that light acts as a particle and one that would verify that an electron acts like a particle.

26. An electron and proton are both traveling with the same speed. Which has the longer wavelength?

27. Why are electron clouds used to describe the location of an electron in the wave mechanical view of the atom while specified orbits are used in the Bohr theory?

28. (a) What is the wavelength of a proton moving with a speed of 10^5 m/s? (b) What is the wavelength of a 70 kg person jogging at 2 m/s?

LASERS

29. What are the differences between laser light and ordinary light from a tungsten filament light bulb with respect to (a) the wavelength of the light, (b) the means of production, (c) the phase relationships within the beam?

30. If cost were not a factor, do you think that laser light bulbs could be used efficiently to illuminate your kitchen or shop? Explain why or why not.

31. Why does a laser emit only one color, as opposed to a continuous band of colors?

32. Laser light is often compared to a battalion of soldiers marching, while ordinary light compares to these soldiers milling around at a break in the march. Discuss these analogies.

33. In order for a laser to work, both ends of the tube must be sealed with a mirror that is at least partially reflecting. Why would an ordinary piece of glass not be sufficient?

ANSWERS TO SELECTED NUMERICAL QUESTIONS

21. 3.03×10^{-19} J, 4.09×10^{-19} J, 4.58×10^{-19} J, 4.85×10^{-19} J

22. (a) 1.64×10^{-18} J, (c) 2.47×10^{16} Hz

28. (a) 3.97×10^{-12} m, (b) 4.74×10^{-36} m

Nuclear reactions occurring deep in the interior of the Sun are responsible for its enormous production of energy. Shown here is a giant flare arching out from the surface of the Sun. (Courtesy NOAO)

CHAPTER

13

Nuclear Physics

In the preceding chapter, we discussed the experimental efforts of Rutherford and Bohr in establishing the planetary model of the atom. As we saw, this model has some imperfections, but it is still useful in that it is successful in many applications and its solar-system–like structure provides us with a simple visual picture. In this chapter, we shall examine the nucleus of the atom. We shall investigate many of its features such as radioactivity, and we shall also discuss nuclear reactions. Among the types of nuclear reactions that we will investigate are fusion reactions of the type that power the Sun and therefore are responsible for our life here on Earth. If fusion reactions are ever successfully harnessed, our energy needs on this planet will be satisfied for thousands of years.

13.1
NUCLEAR STABILITY

Why does a nucleus exist? At first thought, it seems quite improbable that a nucleus could possibly survive in nature. The reason for this assertion is that all nuclei, except for hydrogen, consist of positively charged protons packed closely together. As we saw in our study of electricity, charges of like sign repel one another via the Coulomb force, so why shouldn't the repulsion between protons be sufficiently great to cause the nucleus to fly apart? The reason that a nucleus can remain stable is because of a force that we have not yet discussed, called the **strong nuclear force.** Some of the characteristics of this force are illustrated in Figure 13.1. The force is an attractive force that acts between all nuclear particles. That is, as shown in Figure 13.1, the force acts between protons and protons, between protons and neutrons, and between neutrons and neutrons. This force of attraction is quite strong, much stronger than the Coulomb force of repulsion between charged particles. Thus, when protons and neutrons are assembled into a nuclear package, the force of attraction produced by the strong nuclear force binds the group together, overcoming the repulsive force between the protons. Other types of forces, such as the force of gravity and the Coulomb force, are noticeable in our everyday world because they are long-range

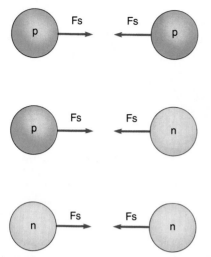

FIGURE 13.1 The strong nuclear force F_s is short range and attractive. It attracts protons to protons, protons to neutrons, and neutrons to neutrons.

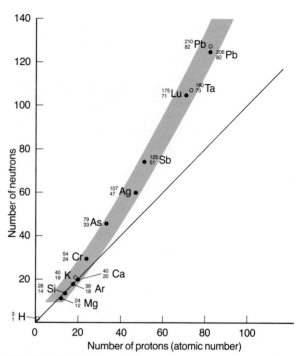

FIGURE 13.2 The stability region, approximately represented by the shaded area. Several naturally stable nuclei are given to serve as reference points, shown by the solid circles. A few naturally occurring radioactive nuclei are indicated by open circles.

forces that can act over considerable distances. The strong nuclear force, on the other hand, is a *short-range force* that becomes negligible in magnitude for distances in excess of nuclear dimensions, that is, greater than 10^{-14} m.

If you took a field trip around our planet collecting all the various forms of stable nuclei that you could find, you would return with approximately 400 different species. In your journey you would have found many nuclei that you would have had to reject—these are the radioactive nuclei. As we shall discuss in a later section, radioactive nuclei do not remain intact forever; instead, they spontaneously break up by emitting various types of radiation and becoming a nucleus of a different type. Let us plot a graph for our 400 or so particles as shown in Figure 13.2.

The graph is a plot of the number of neutrons in the nucleus, along the vertical axis, versus the number of protons in the nucleus, along the horizontal axis. Our plot shows that these nuclei cluster within the shaded region indicated in the figure. Our graph also reveals some surprising features. First, note that most of the light, stable nuclei contain equal numbers of protons and neutrons. One of the most common of this type of nuclei is the $^4_2$He nucleus, which contains two protons and two neutrons. However, as we

see from the graph, the heavy stable nuclei have more neutrons than protons. We can understand why this should be so by examining the characteristics of the strong nuclear force once again. As the nuclei become heavier, there are more and more protons that must be contained. As a result, the repulsive force that all the other protons exert on one individual proton becomes quite great, and it gets harder and harder to keep the nucleus from flying apart. To compensate for this repulsion, more neutrons must be added to the nucleus so that the attractive strong nuclear force between neutrons and protons can maintain equilibrium. There is a limit, however, to what the strong nuclear force can do. Eventually, the number of protons becomes so great that stability cannot be maintained by the addition of extra neutrons. This occurs when the number of protons reaches 83. All elements with atomic number greater than 83 do not exist as stable nuclei.

FOCUS ON . . . The Structure of the Proton

Recall that scientists once believed that the atom was a structureless, indivisible entity—the smallest indivisible particle imaginable. As we have noted, this picture has now been considerably modified. This development of the understanding of the atom suggests another line of questions. Is the proton itself really a structureless, indivisible entity, or is it composed of particles that are smaller yet? The answer is not yet in, but scientists are now theorizing that protons are made up of still smaller particles. This theory states that the particles are arranged so that there are dense regions and regions of empty space within the proton. The fundamental particles that make up protons and neutrons are called **quarks,** as shown in the Figure. Quarks were first postulated in 1963 by Murray Gell-Mann and George Zweig. In the original theory there were three types of quarks called the u, d, and s quarks, for "up," "down," and "strange." One of the unique features predicted for these particles was that they should have fractional electronic charges. That is, the u, d, and s quarks

have charges of $+\frac{2}{3}e$, $-\frac{1}{3}e$, and $-\frac{1}{3}e$, respectively. According to this scheme, a proton is composed of two u quarks and one d quark. Note that the charge adds up properly $(2 \times +\frac{2}{3}e -\frac{1}{3}e) = e$, the charge on the proton. The neutron is composed of two downs and one up. (You check this out to see if the charge adds up properly.) The strange quark is needed to produce other, more exotic nuclear particles.

Quarks provide a novel and important addition to our understanding of the world of physics, but several questions remain unanswered. Foremost among these is: Do they really exist? An attempt to answer this question is one of the paramount areas of research in high-energy nuclear physics today. If they do exist, why do they have fractional charges? Can isolated quarks exist in nature? By the time you read this, some of these questions may have been answered, but one can be sure that these will be replaced with other puzzling questions that may be even more fundamental.

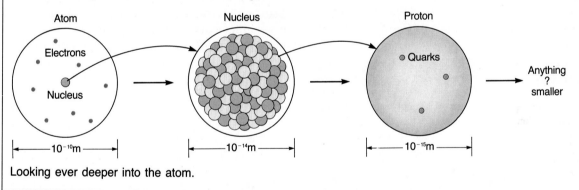

Looking ever deeper into the atom.

There are elements with atomic numbers greater than 83, of course, but all of these substances are radioactive.

A final feature of nuclear stability needs to be pointed out that may not be obvious from the graph of Figure 13.2. It is found that isotopes seem to be most stable when there are an even number of both protons and neutrons. There are very few stable nuclei with an odd number of both protons and neutrons. For example, there are 165 stable even-even nuclei and only four odd-odd nuclei. The reason for this occurrence is

that protons and neutrons tend to group together in pairs inside the nucleus, with protons pairing up with protons and neutrons pairing up with neutrons. (Protons do not pair up with neutrons in this way.) Thus, a nucleus with even numbers of both tends to be more stable. The four odd-odd nuclei are low mass number isotopes, $^{2}_{1}H$, $^{6}_{3}Li$, $^{10}_{5}B$, and $^{14}_{7}N$. On the other hand, there are 105 stable nuclei that are even-odd nuclei. This means that there are an even number of either protons and neutrons and an odd number of the other.

13.2
RADIOACTIVITY

In a sense, the study of nuclear physics actually began in 1896, when Henri Becquerel accidentally discovered radioactivity. He was working with a uranium-bearing ore when he discovered that a photographic plate in the lab had become exposed. In fact, he found that even if he wrapped the film in black paper so that no light could reach it, the film would still be exposed by the presence of the ore. After some careful experimentation, Becquerel found that it was the uranium that was producing the effect, and he proposed the idea that the uranium was spontaneously emitting some kind of radiation. *This emission of radiation from an element is now called ra-dioactivity.*

Several significant facts emerged from Becquerel's studies. The pure uranium compounds that were extracted from the mineral were less radioactive than the crude mineral itself. This difference implied that there were other more highly radioactive substances mixed with the ura-

nium. A series of careful, tedious separations carried out by Marie and Pierre Curie (wife and husband) resulted in the discovery of new radioactive elements, the most important of which was radium.

The Curies also learned that the radioactivity of substances is associated with the elements, not with their compounds. Thus, a gram of radium has the same radioactivity in the form of a pure metal, Ra, as in the form of any of its compounds, such as radium carbonate, $RaCO_3$. In any chemical bonding process such as this, it is the electrons of the atoms that are primarily responsible for holding the various atoms together. Thus, since the electrons did not seem to be playing a role in the radioactive decay of radium, scientists were led to the conclusion that

radioactivity is associated with the atomic nucleus.

When a naturally radioactive source such as uranium is placed at the bottom of a long, narrow hole in a block of lead, most of the radiation is absorbed by the lead, but a thin beam comes out of the hole. To investigate the nature of the radiation emerging from the hole, charged plates were

---

FOCUS ON . . . Radon Pollution

The Curies were the first to notice that the air in contact with radium compounds becomes radioactive. It was shown that this radioactivity came from the radium itself, and the product was therefore called "radium emanation." Rutherford and Soddy succeeded in condensing this "emanation" to a liquid and thus confirmed the fact that it is a real substance—the inert, gaseous element now called radon, Rn.

It is now known that the air in uranium mines is radioactive because of the presence of radon gas. The mines must therefore be well ventilated to help protect the miners. However, the fear of radon pollution now extends to our own homes. Many types of rocks, soils, as well as some brick and concrete, contain very small quantities of radium. Some of the resulting emissions of radon find their way into homes and other buildings. The most serious problems arise from leakage of radon from the ground into the house. A practical remedy is to exhaust the air through a pipe just above the underlying soil or gravel directly to the outdoors by means of a small fan or blower.

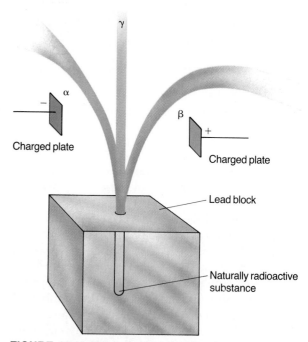

FIGURE 13.3 The behavior of alpha, beta, and gamma emissions in an electric field.

FOCUS ON . . . The Geiger Counter

Detection of radiation by allowing it to expose photographic film was one of the first methods for establishing the presence of radioactive decay. Most of the newer methods depend on the fact that as radiation passes through a gas (or any material) it ionizes some of the atoms of the gas. This means that it knocks electrons from their orbits about the nucleus and frees them from the atom. For example, suppose a beta particle, in passing through a gas, comes close to one of the electrons in an atom of the gas. Since both electrons have a negative charge, they repel one another, and the fast moving beta may knock the electron out of its orbit. Thus, in its trail, a beta particle leaves a wake of free electrons and positively charged atoms, called ions. The detection of this trail is the objective of radiation detectors such as the **Geiger counter.**

A Geiger tube, shown in the diagram, consists of a metal tube with a wire down the center. One end of the tube is covered with a very thin shield so that radioactive particles can pass into the tube easily. A high voltage is maintained such that the wire is held about 500 V higher than the grounded case of the tube. When a beta particle passes through argon gas in the tube, it leaves electrons and positive ions along its path. These charged particles respond to the pull of the high voltage, with the electrons moving toward the wire and positive ions toward the cylindrical housing. As each electron heads for the center, it gains more and more speed. During the trip, it encounters electrons in orbit about other gas atoms and knocks them loose. These electrons, in turn, free other electrons, and an avalanche is produced heading toward the center wire. When these electrons are collected by the wire, a burst of current is created in external circuits, and this current drives a counter to "count" the number of incoming beta particles. It is by this way that we are able to obtain half-life data such as that discussed in this chapter.

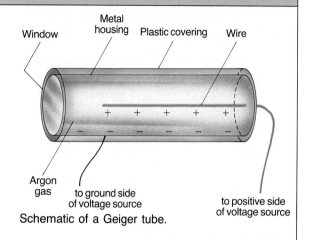

Schematic of a Geiger tube.

A hand-held Geiger counter.
(Courtesy Bill Schulz)

placed near the beam, as shown in Figure 13.3. It was found that some of the radiation is bent toward the positive plate, some is bent toward the negative plate, and some travels in a straight-line path, unaffected by the plates. As shown in Figure 13.3, these three different types of emission were initially given the names alpha (α) rays, beta (β) rays, and gamma (γ) rays.

The alpha beam is drawn toward the negative plate. This observation shows that the alpha rays consist of positively charged particles. Eventually, it was found that these particles would interact with electrons to form neutral helium gas. Thus, the alpha ray, or **alpha particle,** is the nucleus of a helium atom. Symbolically, we represent this particle as $^{4}_{2}\text{He}$.

FOCUS ON . . . Smoke Detectors

Most simple smoke detectors used in the home utilize radioactive sources to perform their task. To see how they work, consider the Figure. A battery is connected between two plates inside the chamber of the detector, and also in the circuit is a sensitive current detector and an alarm. A radioactive source near the two plates ionizes the air around it, and the charged particles created are drawn to the plates by the voltage of the battery. This sets up a small, but detectable, current in the external circuit. As long as this current flows, the alarm is deactivated. However, if there is a fire in the house, the smoke created drifts into the chamber of the smoke detector, and the ions become attached to the particles present in the smoke. These heavier particles do not drift as readily between the plates as do the lighter ions, and thus the current in the external circuit drops. The external circuit detects this decrease in the current and sets off the alarm.

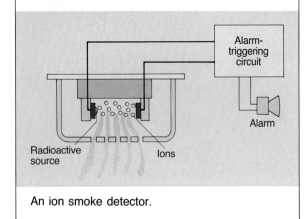

An ion smoke detector.

The beta rays are attracted toward the positive plate. Moreover, the beam of betas is more sharply bent than is the beam of alpha particles. These observations led to the conclusion that beta rays consist of negative particles that are very light. Further evidence shows that these particles are electrons. Symbolically we represent electrons as $_{-1}^{0}e$. This notation is indicative of the fact that the electron, e, has a negative charge equal to that of the proton (as shown by the subscript -1) and that its mass is so small relative to that of nuclear particles that it is negligible (as shown by the 0 superscript).

The gamma rays are not deflected at all by the charged plates; therefore, they have no charge. This beam could be composed of neutral particles, such as neutrons. However, some experimentation with the penetrating power of these various particles led to other conclusions. It was found that the alpha particles would be stopped by a sheet of paper or by a few inches of air. Beta particles would penetrate a thin sheet of steel. But the gammas could pass through several feet of concrete before being absorbed. The conclusion was drawn that gamma rays are a form of electromagnetic radiation, even higher in frequency and energy than X-rays.

13.3 THE DECAY PROCESS

As we noted in the last section, radioactive emissions are of three types, alpha particles, beta particles, and gamma rays. Let us examine in more detail the three decay processes that lead to these emissions.

Alpha decay

Let us pause for a moment and consider what happens to the atomic nucleus of radioactive elements. If a particle is emitted by the nucleus, the nucleus must break apart and lose its identity in the process. Thus, the nucleus of one atom is converted into the nucleus of another. This fact was discovered by Ernest Rutherford and Frederick Soddy in 1902. In particular, they were working with radium in an attempt to find what had happened to it chemically after an alpha particle emission occurred. Their chemical analysis showed that a lighter element, radon, was appearing in their initially pure sample of radium. We can write the nuclear process occurring as follows:

radium $\longrightarrow$ radon + alpha particle

It is of some historical interest to note that the results of their investigation produced a considerable amount of consternation in the two experimenters. This concern arose because **alchemy** had fallen into ill-repute by the 1900s. Alchemy is often called the chemistry of the Middle Ages. One of its fundamental assumptions was that ordinary metals could be *transmuted,* or changed, into gold. This was of obvious interest to kings and the well-to-do; therefore, much time was spent on scientific research devoted to determining how to make the desired transformation. In fact, Isaac Newton devoted much of his time and attention to alchemy before he gained fame via the law of universal gravitation and his laws of motion. The alchemists were never successful, because all the processes used were those in which chemicals were mixed together in various combinations. In such chemical reactions, the nucleus remains intact; as a result, a new element is not formed. However, Rutherford and Soddy had now found that one element, radium, was being transmuted into another, radon. According to a popular account, Soddy turned to his colleague and blurted, "Rutherford, this is transmutation!" Rutherford rejoined, "For Mike's sake, Soddy, don't call it transmutation. They'll have our heads off as alchemists." Rutherford and Soddy were careful to use the term "transformation" rather than "transmutation" in describing their results.

In any radioactive disintegration, the element that decomposes is called the **parent,** and the new element is called the **daughter.** Thus, in the decay of radium to radon, the parent is radium and the daughter is radon.

We can represent the radioactive decay of radium into radon in symbolic form by the notation

$$^{226}_{88}\text{Ra} \longrightarrow {}^{222}_{86}\text{Rn} + {}^4_2\text{He} \qquad (13.1)$$

This is a shorthand way of noting the changes that occur. The arrow indicates that the $^{226}_{88}\text{Ra}$ nucleus decays into a $^{222}_{86}\text{Rn}$ and an alpha particle, ^4_2He. Note the following about the decay process: (1) The total amount of charge (the atomic numbers) on each side of the arrow (before and after) is the same $(88 = 86 + 2)$. (2) The total mass number on each side of the arrow is the same $(226 = 222 + 4)$. Regardless of the decay process, we shall find that atomic number and mass number are conserved.

EXAMPLE 13.1 Time for a change

Uranium, $^{238}_{92}\text{U}$, decays by alpha emission. What is the daughter element formed?

Solution The decay can be written symbolically as

$$^{238}_{92}\text{U} \longrightarrow \text{X} + {}^4_2\text{He}$$

We have used the symbol X to represent the unknown daughter element. To identify the daughter, we first note that mass numbers must add up on each side of the arrow. Thus, the mass number of X must be equal to 234 $(238 = 234 + 4)$. The atomic number of X is found from the fact that the atomic numbers also must balance in the equation. The atomic number is thus found to be 90 $(92 = 90 + 2)$. Therefore, our reaction is

$$^{238}_{92}\text{U} \longrightarrow {}^{234}_{90}\text{X} + {}^4_2\text{He}$$

The periodic table on the inside front cover shows that the nucleus with atomic number 90 is thorium, Th. Thus, the process may finally be represented as

$$^{238}_{92}\text{U} \longrightarrow {}^{234}_{90}\text{Th} + {}^4_2\text{He}$$

Beta decay

The basic process in beta decay is one in which a parent nucleus emits an electron and a daughter nucleus is formed. The basic details discussed in alpha decay also apply here. That is, mass numbers and atomic numbers are conserved. A typical beta decay is that of an isotope of platinum, Pt, decaying into gold, Au, according to the following equation

$$^{197}_{78}\text{Pt} \longrightarrow {}^{197}_{79}\text{Au} + {}^{\ 0}_{-1}\text{e} \qquad (13.2)$$

You might note that this reaction represents the answer to the alchemists' dream of changing other metals into gold. A small difficulty is that ^{197}Pt does not occur naturally. We will discuss later how this isotope, as well as other artificially produced isotopes, are made.

At first thought, the beta decay process seems to contradict some of the facts that we have discovered about the nucleus. We have stated many times in the text that the nucleus is composed of protons and neutrons, yet in beta decay, an electron is emitted from the nucleus. From where does it come? The answer is that the electron is created inside the nucleus in a process by which a neutron is changed to a proton. Symbolically, this is

$$\ _0^1 n \longrightarrow \ _1^1 p + \ _{-1}^0 e \qquad (13.3)$$

On the basis of this transformation of a neutron into a proton, let us re-examine the beta decay process represented by Eq. 13.2. We note that the number of nuclear particles is the same before and after the reaction, 197. However, one of these nuclear particles has been changed from a neutron into a proton. Thus, the atomic number should be one greater after the process than before, and it is (78 before, 79 after).

EXAMPLE 13.2   The case of the disappearing sulfur

Sulfur-37 decays by beta emission. Find the daughter nucleus.

Solution   The decay process may be written as

$$\ _{16}^{37} S \longrightarrow X + \ _{-1}^0 e$$

where X is the unknown decay product. Balancing mass numbers, we find that X must have a mass number of 37 ($37 = 37 + 0$). It also must have an atomic number of 17 ($16 = 17 - 1$). Thus, the daughter has the representation $\ _{17}^{37} X$. In the periodic table, we find that the element with an atomic number of 17 is chlorine, Cl. Thus, the complete decay process is

$$\ _{16}^{37} S \longrightarrow \ _{17}^{37} Cl + \ _{-1}^0 e$$

Gamma decay

The emission of gamma rays by a nucleus is very similar to the emission of light by an atom. In an atom, photons are emitted when an electron in an excited state returns to the ground state. In its downward fall, the electron gets rid of its excess energy by emitting photons, which are usually in the infrared, visible, or ultraviolet portion of the electromagnetic spectrum. A nucleus also may have an excess of energy following a nuclear event, and it de-excites, or releases this pent-up energy, by emitting a photon. However, the amount of energy carried away by the photon is generally considerably greater than that released in an atomic process. Thus, the frequencies associated with the emitted photons are much higher, and they fall into the region of the electromagnetic spectrum called gamma rays.

A typical nuclear event that may lead to the emission of a gamma ray is that of alpha or beta decay. Often, following these types of decay, the nucleus is left in an excited state, and the alpha or beta decay is then shortly followed by the emission of a gamma ray. As an example, consider the beta decay of boron to carbon represented as

$$\ _5^{12} B \longrightarrow \ _6^{12} C^* + \ _{-1}^0 e$$

The asterisk following the symbol for carbon is used to indicate that the carbon nucleus is in an excited state after the decay. The carbon nucleus rids itself of the excess energy by a gamma ray emission as follows

$$\ _6^{12} C^* \longrightarrow \ _6^{12} C + \gamma$$

Note that gamma ray emission does not change the parent nucleus to a different daughter nucleus. The mass number and atomic number are the same before and after.

13.4 HALF-LIFE

Radioactive nuclei can decay by alpha, beta, or gamma emission, each of which produces drastically different effects on the original nucleus. For example, alpha decay reduces the mass number of the parent by four units and its atomic number by two units. Beta decay does not affect the mass number, but it increases the atomic number by one unit. Thus, the processes produce radically different results. But there is one feature of all types of decay that is common to the various forms of decay processes. *After a certain interval of time, half of the original number of radioactive nuclei will have decayed. This time is called the **half-life**.* Half-lives of radioactive substances vary from very long times, such as 4.5 billion years for uranium-238, to 10^{-21} s for lithium-5.

EXAMPLE 13.3   Counting carbon nuclei

A continual rain of particles, called cosmic rays, falls on the Earth each day. These particles come to us from nearby sources such as the Sun and from more distant heavenly objects. Cosmic ray activity in the upper atmosphere produces carbon-14 nuclei, which decay by beta emission with a half-life of 5730 y. If you start with a sample of 1,000,000 carbon-14 nuclei, how many will still be around in 22,920 y?

Solution In 5730 y, half of the original one million carbon-14 nuclei will have disappeared through beta emission. Thus, you will have 500,000 remaining. In another 5730 y (total elapsed time of 11,460 y) half of this number will have decayed, leaving you with 250,000 carbon-14 nuclei. In another half-life (total time = 17,190 y), the sample contains 125,000 carbons. Finally, in another half-life (total time = 22,920 y), there are 62,500 remaining.

Please be aware that the circumstances outlined above are ideal and unrealistic. Radioactive decay is governed by the laws of probability and is an averaging process. One million initial atoms seems like a lot, but in a real situation, there would be many more than this in a given radioactive sample. Thus, for the case above, if we were to actually be able to count the number remaining after one half-life, we would not get exactly 500,000. However, as the number in our original sample increases, the probability of getting extremely close to exactly one-half the original number remaining after one half-life increases dramatically.

13.5 NEUTRINOS

Let us briefly return to our discussion of beta decay, because there are some unusual features of it that initially brought into question one of the most fundamental laws of nature, the conservation of energy. The resolution of these difficulties led to the discovery of one of nature's most unusual creations, the **neutrino.**

In our study of the Bohr atom, electrons were shown to occupy various energy levels. These energy levels are easy to visualize because an electron in a higher energy level spends most of its time relatively far away from the nucleus, and an electron in a lower level is likely to be closer to the nucleus. Thus, the terms "falling from a higher energy level to a lower one" and "rising from a low one to a high one" bring obvious terrestrial analogies to mind. There are discrete energy levels in a nucleus as well. These are

a little harder to visualize, but they are real, nevertheless. To demonstrate that this is true, consider the process of alpha emission. Physicists observed that all the alpha particles emitted by a specific type of radioactive isotope travel outward from the nuclei with the same kinetic energy. This is illustrated in Figure 13.4a, where we see that all the alphas emitted carry the same amount of energy away from the nucleus as the nucleus de-excites. This also occurs in gamma emissions from nuclei. That is, when an excited nucleus de-excites via gamma emission, the gamma emitted has a specific amount of energy, as shown in Figure 13.4b.

However, radioactive decay via beta emission is different, as shown in Figure 13.4c. A specific type of nucleus decaying via the emission of beta particles was found to release the betas with a wide range of kinetic energies. For example, in one particular beta decay event, the fastest betas might carry away five units of energy, but most of the betas would carry less than this maximum amount. There would be some with four, some with three, down the line, and all fractional energies in between would also be seen. Thus, there would be some with 3.13, some with 3.14, and so forth. Theoretically, the fastest ones would carry the energy equal to the total energy released when the nucleus de-excited, but what about the slower ones? Was energy somehow being lost in the decay process? Some prominent physicists speculated that the law of conservation of energy might not be valid for beta decay. To compound the difficulties, it was also found that momentum was apparently not being conserved in beta decay processes.

In 1930, W. Pauli proposed that a third particle must be present to carry away the "missing energy" and to conserve momentum. This particle was later given the name **neutrino** (little neutral one) by Enrico Fermi. Thus, in all beta decay processes, the nucleus emits a neutrino (symbol v) in addition to the electron. As a result, the decay of platinum to gold via beta emission should be written to include the presence of the neutrino as

$$^{197}_{78}\text{Pt} \longrightarrow ^{197}_{79}\text{Au} + ^{0}_{-1}\text{e} + v$$

The neutrino is a curious and elusive particle. It was predicted to have the following properties.

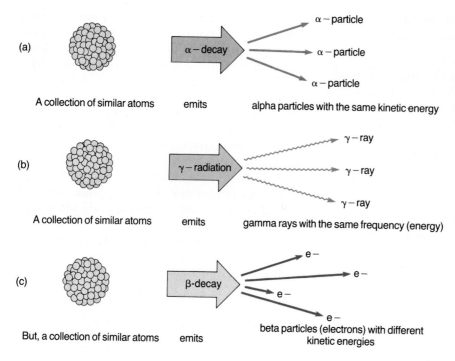

(a) A collection of similar atoms   emits   alpha particles with the same kinetic energy

(b) A collection of similar atoms   emits   gamma rays with the same frequency (energy)

(c) But, a collection of similar atoms   emits   beta particles (electrons) with different kinetic energies

FIGURE 13.4 All the alpha particles from the nuclei of a given radioactive isotope fly off with the same kinetic energy. Similarly, gamma rays from a particular type of decay all carry the same energy (frequency). However, beta particles from a specific decay have varying amounts of energy ranging from almost zero up to some maximum amount. This observation led to the discovery of neutrinos.

1. It is electrically neutral.

2. It has an extremely low mass. In fact, its rest mass may be zero, although recent theories suggest that it does have some mass.

3. It interacts very, very weakly with other particles of matter.

When one considers the ramifications of these properties, it is not surprising that the neutrino was very difficult to detect. Because it is neutral, it could not be deflected by electric or magnetic fields. Its small mass and the fact that it interacts very weakly with matter means that it could travel long distances without encountering another particle and being captured.

To appreciate the order of scale of the neutrino and of the structure of the atom, let us return to our analogy of shooting an arrow into a cloud in our fantasy world described in the preceding chapter. Imagine that there were a forest inside the cloud. If you shot an arrow blindly into this forest, the probability that the arrow would travel a given distance through the forest would depend upon (a) how closely the trees were packed together, (b) the diameter of the trees, and (c) the diameter of the arrow. To take this analogy to the atomic scale, consider atoms to be the trees and a neutrino to be the arrow. The question is: If you shoot a neutrino into a block of lead, how thick a piece of lead would you need to have a 50 percent chance of stopping, or trapping, a neutrino? The answer isn't measured in millimeters, centimeters, meters, or even kilometers. It is measured in light-years (Fig. 13.5). If it were possible to construct a tube of solid lead extending all the way from the Earth to the nearest star, four light-years away, and a neutrino were shot into the tube, chances are that it would emerge at the other end without having hit *anything at all* along the way. As elusive as is the neutrino, it was finally detected experimentally in 1950. Theorists believe that neutrinos are all around us all the time. For example, nuclear processes occurring inside the Sun produce neutrinos in abundance that escape from the Sun and go pouring out into space. There may be many more neutrinos in the Universe than there are protons and neutrons. These neutrinos are speeding through our Solar System, our planet,

FIGURE 13.5 If a solid lead tube extended from the Earth to the closest star, four light-years away, and a neutrino flew off into the tube, there is a high probability that it would pass through to the other end without hitting anything along the way.

and even through you as you read this, but they seldom touch anything along the way.

13.6 NUCLEAR REACTIONS

In 1919, Ernest Rutherford allowed alpha particles from a radioactive source to pass through nitrogen gas. He found that some particles were emitted from the gas which were more penetrating than the alpha particles. Finally, by passing these new particles between charged plates, he was able to show by the way that they were deflected that they were protons. Based on his observations, Rutherford concluded that a nuclear reaction had occurred that could be represented as follows:

$$\,^4_2\text{He} + \,^{14}_7\text{N} \longrightarrow \,^{17}_8\text{O} + \,^1_1\text{H}$$

This reaction equation is used to indicate that an alpha particle bullet, ^4_2He, strikes a nitrogen nucleus, $^{14}_7\text{N}$, and an interaction occurs such that the debris consists of a proton, ^1_1H, and an oxygen nucleus, $^{17}_8\text{O}$. This type of nuclear reaction is called a **bombardment reaction.** Note that mass numbers and atomic numbers add up on each side of the arrow, just as they did for radioactive decay processes. Since the time of Rutherford, thousands of nuclear reactions have been observed, particularly with the advent of particle accelerators in the 1930s. The study of nuclear reactions has been one of the most profitable techniques used by experimentalists in physics to probe the secrets of the nucleus. We shall examine many of these reactions in the remainder of this chapter.

EXAMPLE 13.4 The elusive neutron

(a) In 1932, a nuclear reaction was produced by James Chadwick which is of significant historical importance. In the experiment, Chadwick bombarded ^9_4Be with alpha particles. He found that there were two reaction products, one of which was $^{12}_6\text{C}$; what was the other?

Solution The reaction process may be represented as follows:

$$\,^4_2\text{He} + \,^9_4\text{Be} \longrightarrow \,^{12}_6\text{C} + \text{X}$$

Balancing mass numbers and atomic numbers, we see that X has the representation ^1_0X. The mass number of 1 indicates that the unknown particle has a mass approximately equal to that of a proton, but the atomic number 0 indicates that the particle has no charge. The unknown particle is a neutron, ^1_0n.

This was the first experiment to provide positive proof of the existence of neutrons. The experiment had the additional advantage that it added another type of "bullet" to those that could be used to bombard nuclei. Alpha particles used as bullets have the disadvantage of being positively charged. This means that when they are used as projectiles to cause reactions with heavy nuclei (with a lot of protons), the alphas must have a large energy to overcome the repulsive Coulomb force exerted on them by the nucleus. The neutral neutron, produced in reactions such as the one described above, does not suffer this disadvantage.

(b) A neutron-induced reaction that would have been of interest to the alchemists of the Middle Ages is shown below. What is the reaction product X?

$$\,^{1}_{0}\text{n} + \,^{198}_{80}\text{Hg} \longrightarrow \text{X} + \,^{2}_{1}\text{H}$$

Solution Balancing mass numbers and atomic numbers, we see that X is represented as $\,^{197}_{79}\text{X}$. A search through the periodic table of elements on the inside front cover reveals that this particle is gold (symbol Au). Thus, the complete reaction is

$$\,^{1}_{0}\text{n} + \,^{198}_{80}\text{Hg} \longrightarrow \,^{197}_{79}\text{Au} + \,^{2}_{1}\text{H}$$

The alchemists' dream has come true with this reaction, but be aware that the expense involved in producing gold in this way is prohibitive.

13.7
ARTIFICIALLY PRODUCED NUCLEI

In 1934, 15 years after Rutherford produced the first nuclear reaction, Irene and Frederic Joliot-Curie, Mme. Curie's daughter and son-in-law, converted boron to nitrogen-13, which is radioactive. This was the first *artificially produced* radioisotope. The reaction is

$$\,^{4}_{2}\text{He} + \,^{10}_{5}\text{B} \longrightarrow \,^{13}_{7}\text{N} + \,^{1}_{0}\text{n}$$

Uranium is the element with the largest atomic number that exists in nature in any appreciable amount. In 1940, researchers became able to extend the list of known elements beyond uranium. The first man-made element was produced by bombarding uranium with neutrons. The process that ensued is described by the following:

$$\,^{1}_{0}\text{n} + \,^{238}_{92}\text{U} \longrightarrow \,^{239}_{92}\text{U} + \gamma$$

The uranium-239 produced is radioactive and decays via beta emission as

$$\,^{239}_{92}\text{U} \longrightarrow \,^{239}_{93}\text{Np} + \,^{0}_{-1}\text{e}$$

Thus was produced neptunium, Np, the first man-made element. However, the process did not end there, because neptunium is also radioactive and decays by beta emission as

$$\,^{239}_{93}\text{Np} \longrightarrow \,^{239}_{94}\text{Pu} + \,^{0}_{-1}\text{e}$$

Following this reaction, the number of man-made elements beyond uranium had grown to two. Plutonium, Pu, proved to be of great importance in the history of mankind because, as we shall see, it played a role for the good through its use in nuclear power plants and for the bad in the production of nuclear weapons.

The reactions indicated above were the first to be performed in a laboratory setting to produce elements with an atomic number above that of uranium. Neutrons arrive from outer space as a part of the cosmic ray barrage. Infrequently, one of these neutrons strikes a uranium-238 nucleus and initiates the chain of events leading to the production of neptunium and plutonium. However, only trace amounts of these elements are found in nature, and the development of devices that use these elements as integral parts would have been impossible without the production by man-made avenues.

13.8
CARBON DATING

Much important information concerning the time line of life on this planet has been discovered because of a nuclear reaction that occurs high in the Earth's atmosphere. Let us first look at the sequence of events that occurs; then we shall examine how these events have been used to unravel many of the secrets about our past. The process begins with high-energy neutrons, which constitute a part of the cosmic ray bombardment falling on the Earth from outer space. The reaction that occurs when one of these neutrons strikes a nitrogen nucleus is

$$\,^{1}_{0}\text{n} + \,^{14}_{7}\text{N} \longrightarrow \,^{14}_{6}\text{C} + \,^{1}_{1}\text{H}$$

The end-products of the reaction are a proton and a radioactive isotope of carbon, carbon-14. Carbon-14 is radioactive and decays by beta emission as

$$\,^{14}_{6}\text{C} \longrightarrow \,^{14}_{7}\text{N} + \,^{0}_{-1}\text{e}$$

As far as scientists have been able to determine, carbon-14 has been produced at a fairly uniform rate in the upper atmosphere for at least the past

50,000 years, and just as regularly, this carbon has been decaying with a half-life of 5730 years to the stable isotope nitrogen-14. The rate of production of carbon-14 is such that there is about one nucleus of carbon-14 for every 10^{12} nuclei of the more common carbon-12 in our environment. Now, let us follow the processes that occur in nature which enable us to use carbon-14 as a clock to date the age of organic relics.

Carbon is a natural constituent of all organic materials; as a result, plants absorb carbon from the air and the earth. These plants then are eaten by animals, and the carbon becomes a part of their bones and flesh. The end result is that all organic materials, when alive, have the same ratio of carbon-14 to carbon-12 as that found in nature. When an animal dies, the stable carbon-12 stays around, but the radioactive carbon-14 begins to disappear by beta emission. Thus, the *ratio* of carbon-14 to carbon-12 in the remains begins to change. To understand how this fact enables us to date organic relics, consider the following scenario. At some time in the past, a hunter kills a deer and buries its bones near his campfire. An archaeologist then discovers the deer bones in the present day and analyzes them to determine the amount of carbon-14 still present. If he finds only half as much carbon-14 as that which would be found in a living deer, he knows that the deer died 5730 years ago.

There are limitations on how far back into the past carbon dating can take us. After about 30,000 years, the amount of carbon-14 in a sample has decayed to the point that reliable estimates of its age cannot be determined. One interesting application of carbon dating concerns the Dead Sea Scrolls. These were a group of manuscripts found in 1947 by a young boy in a cave near the ancient city of Qumran. These scrolls contained most of the books of the Old Testament, including a copy of the Book of Isaiah. Because the authenticity of some of the passages in this Book were disputed by scholars, it became important to determine the age of the manuscripts to see if certain material was added by later writers. The manuscripts had been wrapped in linen, and this substance was carbon dated. The age of the fragments was found to be about 1950 years. This age did not satisfactorily resolve the controversy, but it did involve a worthwhile application of the dating process.

SUMMARY

Nuclear stability is caused by a short-range force of attraction called the **strong nuclear force.** This force acts between protons and protons, neutrons and neutrons, and protons and neutrons.

Some isotopes are unstable, giving off particles and radiation. Such decomposition, which is called radioactive decay or **radioactivity,** may occur in a series of steps, ending when a stable isotope is produced. All radioactivity releases energy. The particles released in radioactive decay are the alpha particle, the nucleus of a helium atom, $^{4}_{2}He$; the beta particle, an electron, $_{-1}^{0}e$; or a gamma ray, a highly energetic photon of electromagnetic radiation. In all beta-decay processes, a **neutrino** is also released. The **half-life** of a radioactive isotope is the time required for half of the nuclei in a sample to decompose.

Nuclear reactions are produced when high-energy bullets are directed against a target material. In the reaction, new products are formed. In both decay and bombardment reactions, both charge and mass number are conserved. **Carbon dating** is a process for finding the age of organic relics.

KEY WORDS

Strong nuclear force	Gamma ray	Neutrino	Artificially produced
Radioactivity	Daughter	Isotope	nuclei
Alpha particle	Parent	Nuclear reactions	Carbon dating
Beta particle	Half-life		

QUESTIONS

NUCLEAR STABILITY

1. Which of the following nuclei would you expect to be stable? (a) $^{28}_{14}Si$, (b) $^{232}_{90}Th$, (c) $^{6}_{3}Li$, (d) $^{9}_{3}Li$

2. Explain why gravitational forces were not used in the explanation of how protons are held together in a nucleus. Are protons attracted to each other by gravitation?

3. Why do heavier elements require more neutrons to remain stable?

RADIOACTIVITY

4. Which of the three types of radiation, alpha, beta, or gamma, would not be deflected by the Earth's magnetic field? Defend your answer.

5. Explain the significance of the fact that the radioactivity of an element does not depend on its chemical bonding.

6. (a) A Geiger counter registered 256 counts per second (cps) near a sample of polonium-210; 276 days later the counter registers 64 cps. What is the half-life of polonium-210? What will the counter register after another 276 days? (b) Polonium-210 decays in one step to lead-206, which is not radioactive. If you were asked to give a rough estimate of the length of storage time needed to reduce the radioactivity of polonium-210 to a safe level, would you say it is a matter of months, years, decades, or centuries? Would you be concerned about any radioactive progeny that might be produced?

7. Iodine-131 is a radioactive nuclear waste product with a half-life of 8 days. How long would it take for 2000 mg of iodine-131 to decay to 125 mg? Would it be correct to say that iodine-131 is no environmental hazard because its half-life is so short? Defend your answer.

8. 24 mg of tritium ($^{3}_{1}H$) decays to 1.5 mg in 49 years. What is the half-life of tritium?

9. Strontium-90, produced in nuclear explosions and present in radioactive fallout, has a half-life of 29 years. If the activity of a "bomb-test" Sr-90 sample collected in 1957 was about 80 disintegrations per second (dps), how many years would it take to reduce the activity to the natural background count of 2.5 dps? (*Note:* The dps count is registered as clicks on a Geiger counter and is proportional to the amount of Sr-90 present in the sample.) In what year would that count be reached?

10. All the uranium from a sample of uranium ore is extracted and purified. The uranium is less radioactive than the ore from which it came. Explain.

11. Radon-222 has a half-life of 3.8 days. Radon gas is emitted from some rocks and soils. Since the half-life of radon is short, why is there any left on Earth?

12. Complete the following radioactive decay equations.

    (a) $^{238}_{92}U \longrightarrow ? + ^{4}_{2}He$

    (b) $? \longrightarrow ^{14}_{7}N + ^{0}_{-1}e$

13. Complete the following radioactive decay equations.

    (a) $^{12}_{5}Bi \longrightarrow ? + ^{0}_{-1}e$

    (b) $^{144}_{60}Nd \longrightarrow ? + ^{4}_{2}He$

14. A moving particle with a large charge is deflected more by charged plates than if it had a lesser charge. Why, then, are alpha particles bent less than beta particles?

15. If photographic film is kept in a cardboard box, alpha particles produced outside the box cannot expose the film, but beta particles can. Why?

NEUTRINOS

16. The mass-to-charge ratio of an electron can be determined by measuring its deflection in a magnetic field. Can the mass-to-charge ratio of neutrons and neutrinos be measured in the same way? Why or why not?

17. Both electrons and neutrinos are thought to be point masses. Why is a neutrino so much harder to capture than an electron?

18. Do you think it would be possible for a neutrino to pass through the nucleus of an atom? Why or why not? Would a positively charged particle the size of a neutrino pass equally well through a nucleus? Explain.

NUCLEAR REACTIONS

19. Complete the following nuclear reactions.

    (a) $^{29}_{13}Al + ^{4}_{2}He \longrightarrow ? + ^{1}_{0}n$

    (b) $^{95}_{42}Mo + ^{1}_{0}n \longrightarrow ? + ^{1}_{1}H$

20. Complete the following nuclear reactions.

    (a) $^{7}_{3}Li + ^{1}_{1}H \longrightarrow ? + ^{4}_{2}He$

    (b) $^{27}_{13}Al + ^{4}_{2}He \longrightarrow ? + ^{30}_{15}P$

21. Why is a neutron a more effective bullet for penetrating a nucleus than an alpha particle? How would an alpha particle compare to a proton bullet?

22. $^{10}_{5}$B is struck by an alpha particle. A proton and a product nucleus are released. What is the product nucleus?

23. Oxygen-18 is struck by a proton, and fluorine-18 and another particle are produced. What is the other particle?

24. Carbon dating relies on the assumption that the production of carbon-14 has been essentially constant over the last 20,000 years. If it should be determined that the production were considerably higher 20,000 years ago than it is now, would that increase the predicted life of relics or decrease it?

25. Why is the technique of carbon dating not accurate beyond about 30,000 y?

ANSWERS TO SELECTED NUMERICAL QUESTIONS

6. (a) 138 days; 16 cps

7. 32 days

8. 12.3 y

12. (a) $^{234}_{90}$Th, (b) $^{14}_{6}$C

13. (a) $^{12}_{6}$C, (b) $^{140}_{58}$Ce

19. (a) $^{32}_{15}$P, (b) $^{95}_{41}$Nb

20. (a) $^{4}_{2}$He, (b) $^{1}_{0}$n

22. $^{13}_{6}$C

23. $^{1}_{0}$n

A nuclear reactor in operation. (Courtesy Oak Ridge National Laboratory)

CHAPTER

14

Relativity and the Nuclear Age

Occasionally in our study of physics, we have encountered a concept that seems to violate common sense. The dual nature of both light and particles are two examples. Common sense says that light should be either a wave or a particle but not both; yet that is the way it is. Likewise, an electron should always behave as either a particle or a wave, but it doesn't. These are modern examples of violations of how the world looks to us, but there are also historical precedents. In the early history of astronomy, it seemed obvious to the common man that the Sun, planets, and all other heavenly bodies revolved around the Earth. This idea was promoted by Aristotle and others to the point that it became a part of religious doctrine. However, in the Sixteenth and Seventeenth Centuries, these ideas about the place of man and the Earth began to change. Even though it looks like the Sun revolves around the Earth from our vantage point, Kepler, Copernicus, Galileo, and others demonstrated in quite convincing fashion that common sense had been violated. We were relegated to the position of a quite obscure planet, far from the center of our galaxy, revolving around a quite ordinary star.

Thus, it is not uncommon in the history of mankind to find that "what seems to be" is not always true. Perhaps no other concept in the field of science demonstrates this in a more profound way than does the theory of relativity, which we shall examine briefly in this chapter. It seems that there should be certain absolutes in the world around us. Hefting a rock indicates that it has mass, and common sense tells us that nothing is likely to change the mass of that rock, especially not something as mundane as throwing it at a high speed. Likewise, the length of a piece of wood or the periodic ticking of a clock should also be sacrosanct and not subject to change just because we cause these objects to move at a high rate of speed. Yet, according to the theory of relativity, all of these things *do* change as their speed changes.

The man responsible for this change in our viewpoint of the world around us is Albert Einstein. Einstein is an unlikely personage to have wrought such changes. He was born in Germany and entered grade school in Munich, where he met with little success. At first, it was feared that he was retarded; in fact, a teacher once told him quite forthrightly that he would never amount to anything. Following his graduation from a school in Switzerland, he had trouble finding a job, but through the intervention of some friends, he found employment in a patent office in Bern. The work was undemanding, and he was able to spend several hours each day "at play" with such scientific problems as that of the photoelectric effect. His theory, which resolved the difficulties associated with this scientific dilemma and later won him a Nobel Prize, has been discussed earlier in this book. This certainly would have been sufficient to have gained him immortality in physics, but the full flower of his genius became known with his special theory of relativity. It works, it is true, and it has had consequences of immense proportions in our world, but don't expect it to be within the realm of common sense.

14.1
RELATIVITY BEFORE EINSTEIN

When you hear the word "relativity" used, you probably think of Einstein and the theory of relativity, but relative motion was of interest to scientists even as early as the days of Newton. In fact, one of the questions that these early scientists had about relative motion was intimately connected to the laws of motion. The question they raised is: Are the laws of mechanics the same for all observers? Let us examine this question a little more precisely.

When we are performing an experiment, we always choose a **reference frame** from which we make our observations. For example, if you are doing an experiment in a laboratory, the reference frame you pick is one that is at rest with respect to the laboratory. Now suppose someone passing by in a car moving at a constant velocity were to observe your experiment. Would the observations made by the observer in the reference frame of the car differ dramatically from yours?

To answer this question, let us consider a slight variation of this question, as shown in Fig-

ure 14.1a and b. Here we see two observers, one in an airplane moving at a constant velocity and another at rest on the Earth. The passenger on the airplane decides that he will perform a simple experiment of tossing a baseball into the air to see what will happen. He throws the ball straight up, and he finds that it follows a vertical path upward and falls vertically back to his hand, as shown in Figure 14.1a. This is exactly the same thing that he would observe happening if he were on Earth. The law of gravity is obeyed, and the equations of motion with constant acceleration can be used to tell the plane rider about the details of the motion. The observer on Earth, however, sees the experiment played out somewhat differently. This observer sees the path of the ball to be like that of a projectile (Figure 14.1b). While the ball was in the air, the passenger and the plane had moved to the right, so in order for the ball to be caught by the passenger, according to the earth-bound observer, the ball would have to follow the dashed path shown. The two observers thus disagree on certain aspects of the motion of the ball, but both agree that the ball obeys the law of gravity and the laws of motion. As a result of experiments such as this, we draw the following conclusion:

The laws of mechanics are the same in all reference frames moving at constant velocity with respect to one another.

A second type of question connected with relativity that was of interest to early physicists concerned absolute motion. In order to understand what is meant by absolute motion, consider the following set of circumstances. Imagine that a jet airplane is traveling at 400 mi/h, and following behind is a prop plane flying with a velocity of 100 mi/h in the same direction as the jet. The first thing that we must note about the statement of all these velocities is that they are all measured relative to the earth. In every measurement of a velocity, you have to specify the reference frame with respect to which it is measured. For example, if we took our reference frame to be that of the prop plane, that pilot would consider himself to be moving at a velocity of zero mi/h, and the jet would be moving relative to him with a velocity of

FIGURE 14.1 (a) The observer on the plane sees the ball move in a vertical path. (b) The observer on the Earth sees the ball move as a projectile.

300 mi/h. On the other hand, if the stationary reference frame were considered to be that of the jet plane, the pilot of the jet would consider herself to be at rest, and the prop plane would be moving with a velocity of 300 mi/h, but *backward*. The point to be made with these observations is that

all motion is relative to some reference frame.

Thus the question arises, When is something at rest? We have often assumed that the Earth is at rest in our problems. It obviously is not, however, because the Earth rotates on its axis and it revolves about the Sun. Is the Sun at rest? If so, we could state that an object in the Universe is also at rest if it is stationary with respect to the Sun. Again the answer is no. Our Sun moves with the Solar System in revolution about the center of our galaxy, and our galaxy moves with respect to other galaxies. This search for a reference frame that could be considered to be at absolute rest led physicists in a most unusual and interesting chase, as we shall see in the next section.

14.2
THE REST FRAME OF THE ETHER

In the early history of the study of light, several experiments, including Young's double-slit experiment, had led to the conclusion that light is a wave. The statement that light is a wave led to some serious fundamental problems for the investigators, however. The reason for their perplexity is that they knew quite a bit about several different types of wave motion such as that of water waves, sound waves, waves on strings, and so forth, and all of these types of waves had one thing in common that waves of light did not seem to share. Namely, all of these different classifications of waves have a substance through which they move. This medium of propagation is obviously water for a water wave; a solid, liquid, or gas for a sound wave; and so on. Now scientists were in the position of calling light a wave, but there did not seem to be a medium for it to travel in. For example, if you place a glowing light bulb in a container and then pump all the air out of the container, you can still see the glowing bulb inside. Likewise, you can see the Sun, stars, and other distant objects in the Universe without there apparently being any material substance in the intervening space to carry this light to Earth. Scientists were uncomfortable with the concept of a wave without a medium for its propagation, so they conceived of a medium and gave it the name, **the luminiferous ether.**

This ether would have to have some quite unusual properties. It would have to permeate all of space, because we can see distant stars as the light moves through the ether from them to us. Also, it would have to permeate matter, because light can be transmitted through a transparent object. Yet, as all-pervading as is the ether, the Earth and other heavenly objects move through it apparently unaffected by its presence. As the Earth swings around the Sun, it must move through the ether of space without being slowed by it or having its motion altered by it in any way. Thus, the ether would have to be a most unusual material.

The important attributes of the ether as far as the study of relativity is concerned were that it was an all-pervasive medium that existed throughout the Universe and, just as important, that this medium was *stationary*. As a result of these unique characteristics, the ether took on the aspect of a privileged frame of reference that could be used to determine the absolute motion of an object. We could cease worrying about having to specify the velocity of an object with respect to the Earth or anything else; instead, we would always specify its velocity with respect to the ether. Thus, if an object is at rest with respect to the ether, it is in a state of absolute rest. Also, we could avoid stating that the velocity of a baseball is 30 mi/h with respect to the Earth; instead, we would specify its velocity with respect to the ether. Thus, the search for the ether became an important research activity for scientists of about 100 years ago. In the next section, we will look at an experiment conducted by Albert Michelson and E. W. Morley in an attempt to find the ether.

14.3
THE MICHELSON-MORLEY EXPERIMENT

In 1883, Michelson and Morley devised an experiment to measure the Earth's speed through the stationary ether. They reasoned that as the Earth moves through space it must pass through the ether and that this would cause an ether "wind" to blow across the Earth. According to the accepted ideas of the time, a beam of light sent against the ether wind, as in Figure 14.2a, would travel with a speed of c − v, where c is the speed of light and v is the speed of the ether wind. The speed v is the same as the speed of the Earth in its orbit, since the ether is assumed to be in a state of absolute rest. By similar reasoning, a beam of light traveling with the ether wind, as in Figure

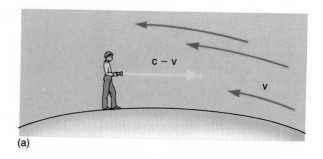

(a)

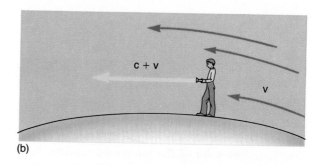

(b)

(c)

FIGURE 14.2 (a) A beam of light traveling into the ether wind was expected to have a speed c − v. (b) A beam traveling with the ether wind was expected to have a speed c + v. (c) A beam traveling perpendicular to the ether wind was expected to take longer to reach its destination than it would in the absence of the wind.

14.2b, would travel with a speed of c + v. But what about a beam of light sent on a path perpendicular to the ether wind, as in Figure 14.2c? An analysis of this trip would show that this beam would require more time to make the trip than it would if there were no ether wind.

Figure 14.3 shows the details of the Michelson-Morley experiment. Two beams of light

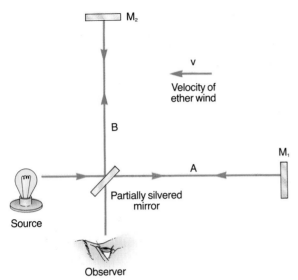

FIGURE 14.3 The light from the source is broken into two parts by the partially silvered mirror. The light that follows path A hits mirror M_1 and is reflected back to the observer. The light along path B is reflected by mirror M_2 and also returns to the observer, where an interference pattern is formed.

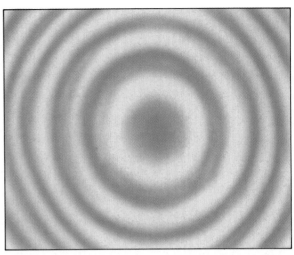

FIGURE 14.4 The beams reflected from mirrors M_1 and M_2 recombine at the eye of the observer to form an interference pattern like this one.

are sent on a race, with beam A moving parallel to the direction of the wind and beam B moving perpendicular to the wind. The beams start at the light source, break into two parts at the partially silvered mirror, and then move their respective paths to reflect from mirrors M_1 and M_2. The rays then retrace their paths and come together at the position of the observer. When the two beams recombine, an interference pattern like that shown in Figure 14.4 is formed. Suppose the experiment is started as in Figure 14.3, with path A aligned with the wind. A pattern like that shown in Figure 14.4 is observed. Michelson and Morley then rotated the device until path 2 is along the breeze, as shown in Figure 14.5. The two arms have, in effect, changed places, and a slight change in the interference pattern should be observed. (The rings should move slightly.) The amount of this movement was predicted to be quite small, but Michelson and Morley had calculated how much movement they would be able to detect, and they were convinced that the change in the pattern would be noticeable. *The end result of this investigation was that they saw no movement of the pattern at all.*

The scientific community had expected Michelson and Morley to verify the existence of the ether wind and thereby to verify the existence of the ether. The results, however, were conclusive; the ether wind did not show itself. Much time and effort were exerted by many scientists in an attempt to determine why the ether did not make its presence known. The final conclusion that was reluctantly drawn by scientists was that the ether did not affect the light at all. This means that *light travels with a speed of c regardless of the motion of the ether.* Since the primary purpose for assuming that there was an ether was that it was the medium through which light traveled, and if light's own medium was not affecting it, there was no need to assume the existence of the medium. Thus, the final result of the Michelson-Morley experiment was a negative one: *There is no ether.*

14.4
EINSTEIN'S POSTULATES

Einstein's theory of relativity is more specifically referred to as his theory of **special relativity.** The word "special" is used to mean that it applies only in a special case—that of objects that move with constant velocity with respect to one another. The more general case, in which accelerations are al-

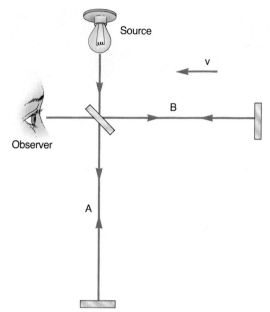

FIGURE 14.5 After rotation, path B is parallel to the ether wind and A is perpendicular to it. This should produce a change in the interference pattern.

lowed, is not included in the situations that we will discuss here. Einstein based his theory on two postulates.

1. All the laws of physics are the same for all observers moving at constant velocity with respect to one another.
2. The speed of light in a vacuum is the same for all observers regardless of the motion of the source of light or the motion of the observer.

We have already pointed out that the accepted theories of relative motion at the time of Einstein had predicted that the laws of mechanics would be the same for all observers moving at constant velocity with respect to each other. Einstein's first postulate extended this idea to cover *all* the laws of physics.

The second postulate explains why the Michelson-Morley experiment failed, although it is reasonably certain that Einstein did not know of this experiment when he made the postulate. This second postulate also deserves some elaboration because it violates our laws of common sense. To see why this second statement is so out of step

with the way we believe nature should behave, consider Figure 14.6. There we see two space travelers sending out a beam of light from their spaceships and an observer at rest on the Earth watching the action. Spaceship A is moving toward the observer, and the question is: What does the earthbound observer measure for the speed of the beam of light coming toward him? If the speed of the ship is v, common sense tells us that the speed of the ship should be added to that of the oncoming light, and the net speed measured should be c + v. Not so, says Einstein. According to him, the motion of the source or observer does not affect the speed of light. *The earthbound observer will measure the speed to be c.* Likewise, common sense tells us that the velocity measured for the beam from B emitted by the ship moving away from the observer should be c − v. Again, not so. The speed in this instance is also c. In fact, if the observers on each spaceship look at the beam from the other, they will measure these speeds to be c also.

The details of where these postulates led Einstein and the complex calculations required to move through the theory will not be examined here. Instead, we will focus on some of the strange predictions that are based on these starting points. The end results of Einstein's work profoundly altered our view of space, time, and matter. As strange as we shall find these results to be, you should keep in mind that the predictions of special relativity have been verified time and time again. There is no escaping their validity.

14.5
EINSTEIN AND TIME

The fact that the speed of light is a constant for all observers leads to some surprising conclusions concerning time intervals measured in different frames of reference. We will find that if a person at rest on Earth says his heart is beating at 70 beats per minute, an observer moving with respect to the Earth will not agree. His conclusion will be that the clock used by the person on Earth, who is in motion relative to him, is running slowly. To understand why different observers will measure different times for specific events to occur, consider the experiment illustrated in Fig-

FIGURE 14.6 Common sense says the speed of the beam of light from spaceship A should have a value of c + v as seen by the earthbound observer, and the speed of the beam from spaceship B should be c − v. Special relativity, however, says that the speed will be c in both instances.

ure 14.7. There an observer in a spaceship moving at a very high speed holds a camera with a flash attachment directly beneath a mirror. When the flash goes off, the light leaves the bulb, travels to the mirror a distance d above him and returns. Let us call the time for the round trip of the light $\Delta t'$ (read as delta t-prime). We can easily find this time interval from the definition of velocity as

$$\Delta t' = \frac{\text{distance traveled}}{\text{speed}} = \frac{2d}{c}$$

where the total distance traveled by the beam of light is 2d, and the speed of the light is $c = 3.00 \times 10^8$ m/s.

Now consider this same sequence of events as seen by an observer on the Earth. Because the spaceship is moving, he finds that the light will have to travel a different path in order to strike the mirror and return, as shown in Figure 14.8. From his point of view, the mirror will have moved to the right, from position A to position B, by the time the light from the flashbulb reaches it. Thus, according to the earthbound observer, the flashbulb will have to be held at an angle if the light from it is to hit the mirror. The light will then bounce off the mirror, obeying the law of reflection, and return to the level of the flash just as the motion of the spaceship brings the astronaut to this location (from position B to position C). Let us call the time for all this to happen Δt. Again, we can use our defining equation for speed to find the elapsed time.

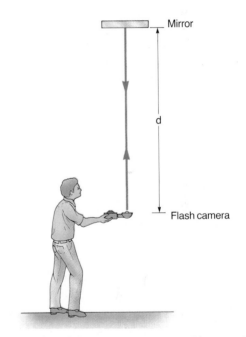

FIGURE 14.7 An observer in a spaceship sees the light from the flash camera strike the mirror and retrace its path. The distance traveled in the round trip is 2d, and the time for the trip is $\Delta t' = 2d/c$.

$$\Delta t = \frac{\text{distance traveled}}{\text{speed}}$$
$$= \frac{\text{length L1} + \text{length L2}}{c}$$

According to Einstein's postulates, the speed of

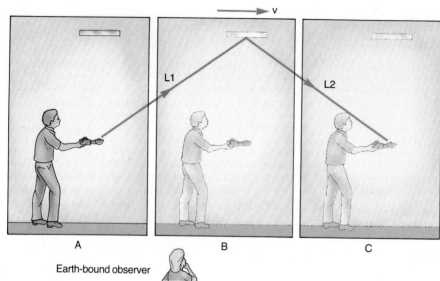

v

FIGURE 14.8 The earthbound observer sees the light from the flash camera travel a longer path because of the motion of the spaceship. The ship moves from A to B while the light moves toward the mirror, and from position B to C while the light returns.

L1   L2

A   B   C

Earth-bound observer

light remains at c, but in this case the distance traveled, length L1 plus length L2, will be greater than it was from the point of view of the person riding in the spaceship. Thus, Δt *will be greater than* $\Delta t'$. To summarize, $\Delta t'$ is the time interval measured by an observer at rest with respect to our experiment, while Δt is the time interval measured by a person who is observing the experiment while it is in motion.

We conclude that moving clocks run slowly.

We shall not repeat the details of Einstein's analysis of the exact relationship between Δt and $\Delta t'$, but he found the two are related as

$$\Delta t = \frac{\Delta t'}{\sqrt{1 - \dfrac{v^2}{c^2}}} = \gamma \Delta t' \qquad (14.1)$$

where

$$\gamma = \frac{1}{\sqrt{1 - \dfrac{v^2}{c^2}}}$$

We have always assumed in our previous study of physics that a time interval seen by one individual is the same as that seen by another. According to special relativity, this really isn't so. Moving clocks slow down. But what is a clock? It

could, of course, be an ordinary mechanical time-keeping device, but the definition of a clock can be extended to include biological processes also. Thus, the heart rate of the moving astronaut will be seen to have slowed from the point of view of the earthbound observer. In fact, all biological processes will have slowed. Thus, from the standpoint of the earthbound observer, the moving astronaut will age at a slower rate than does he. The astronaut will not have any sensation of time slowing for him. In fact, from his point of view, he will consider himself at rest while it is the observer on the Earth who is in motion. As a result, he will consider the clock of the person on Earth to be running slowly.

EXAMPLE 1.1   How time flies

(a) An astronaut in a spaceship moving at 0.9 c watches the pendulum of a grandfather clock on the spaceship swinging back and forth. He finds that the pendulum makes one complete vibration every second. How long does an earthbound observer find that it takes for the pendulum to make one complete swing?

Solution The time interval $\Delta t'$ is the time interval as measured by the observer at rest with respect to the clock. Thus, $\Delta t' = 1$ s. The time interval Δt is the time interval as seen by an observer watching the moving clock. This time interval can be found from Eq. 14.1 with

FOCUS ON . . . Atomic Bombs

An explosion develops a sudden pressure on its surroundings by the rapid production of gas and by the further expansion of the gas as the explosive energy heats it. Chemical explosives produce gases very rapidly by the decomposition of their molecules. This effect is called a blast. If a stick of dynamite explodes several hundred feet from you, the blast effect feels like a thump on your chest.

The main chemical high explosives of modern warfare have been TNT (trinitrotoluene), picric acid, and cyclonite (the explosive ingredient of "plastic explosive"). Nitroglycerin is the major explosive ingredient of dynamite, used mainly for blasting in construction and mining. The heaviest chemical bombs dropped by aircraft in World War II (the "blockbusters") contained about 1000 kg of high explosive.

The nuclear explosive in an atomic (fission) bomb is pure or highly concentrated fissile material: uranium-235 or plutonium-239. Such a material leaves only two significant fates for neutrons—fission capture or escape. The factor that determines which of these two fates will predominate is size, or mass; the minimum mass required to support a self-sustaining chain reaction is called the **critical mass.** To set off an atomic bomb, therefore, subcritical masses of uranium-235 or plutonium-239 are slammed together by precisely shaped chemical high explosives to make a supercritical mass. The chain reaction in-stantly branches, and the mass explodes.

A hydrogen (fusion) bomb derives the major portion of its energy from a nuclear fusion reaction. The most powerful bombs ever exploded have been hydrogen bombs—in tests, not warfare.

The explosive effect of a fission or fusion bomb is rated in terms of its TNT equivalent. Thus, a "1-megaton" bomb is a nuclear bomb that is equivalent to 1 megaton (10^9 kg or about 2.2 billion lb) of TNT. This equivalence, however, refers only to the blast effect. Nuclear bombs have other consequences that are not produced by chemical high explosives. These other nuclear effects include extremely high temperatures that start fires at considerable distances, prompt radiation, electromagnetic pulses that can knock out electronic systems, and climatic changes.

The question is sometimes asked, "Can a nuclear reactor explode like an atomic bomb?" Opponents of nuclear energy have complained that no one ever made such an accusation; therefore, the question just diverts attention from more credible hazards. Nonetheless, the reader should consider that the fuel in a nuclear reactor contains no concentrations that even approach the levels of bomb-grade material and that an atomic explosion therefore cannot occur.

$$\gamma = \frac{1}{\sqrt{1 - \dfrac{v^2}{c^2}}} = \frac{1}{\sqrt{1 - \dfrac{(0.9\,c)^2}{c^2}}} = 2.3$$

Thus,

$$\Delta t = \gamma \Delta t' = 5.26(1\text{ s}) = 2.3$$

This says that the earthbound observer will find that it takes 5.26 s for the pendulum to make its complete vibration. Thus, since the clock is supposed to be "ticking" once each second, we must conclude that moving clocks run slowly.

(b) If the earthbound observer has a pendulum that makes one vibration in 1 second, how long will the astronaut in the ship moving at 0.9 c say it takes to make one complete swing?

Solution In this case $\Delta t'$ is the time as measured by the observer on Earth, who is now at rest with respect to the clock. The astronaut's point of view is that he is the one at rest, while the Earth is moving backward at 0.9 c. Thus, the same calculations as done above indicate that the astronaut will say that it is the earthbound clock that runs slowly. It will take, from his viewpoint, 5.26 s to make one vibration.

14.6 LENGTH CONTRACTION

The fact that time intervals are not absolutes but instead depend on the frame of reference used is the first of the surprises that stem from Einstein's

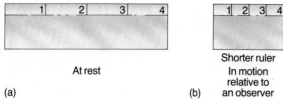

FIGURE 14.9 (a) The ruler has a certain length when it is at rest relative to an observer. (b) When the same ruler moves relative to an observer, its length is shorter.

theory of relativity. Just as surprising is the fact that lengths are not absolutes either. The results of special relativity indicate that *lengths of objects contract in the direction of motion when they are in motion with respect to an observer,* as shown in Figure 14.9. The length of an object moving with a speed v with respect to an observer is given by

$$L = L'\sqrt{1 - \frac{v^2}{c^2}} \text{ or } L = \frac{L'}{\gamma} \quad (14.2)$$

where L′ is the length as seen in a reference frame at rest with respect to the object and L is the length as seen in a reference frame in motion with respect to the object.

An additional consequence of special relativity is that length contraction takes place only along the direction of motion. *Lengths perpendicular to the direction of motion are unchanged by relative motion.* Thus, a jogger running at a speed near that of light might change from the shape shown in Figure 14.10a to that of Figure 14.10b as his speed increases.

An interesting piece of experimental evidence that confirms both the slowing down of moving clocks and length contraction involves muons, particles that can be produced high in the atmosphere in collisions between cosmic rays and atoms in the air. Muons are particles with a charge equal to that of an electron and a mass 207 times that of an electron. They are radioactive and decay with a half-life of 2.2 μs when the half-life is measured in a reference frame that is at rest with respect to them. A typical speed for these muons when they are produced in the atmosphere is about 0.97c. At this speed, a muon

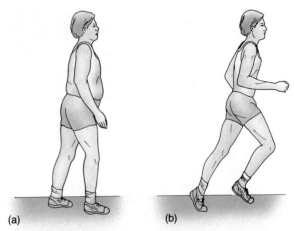

FIGURE 14.10 (a) An overweight jogger when at rest changes to a trimmer jogger (b) when he is in motion. However, his height remains the same.

would travel a distance of about 640 m during a time equal to its half-life. Thus, if 1000 muons were produced by cosmic rays at a height of 640 m above the Earth, we would expect to find approximately 500 of them reaching Earth. However, many more than this predicted number actually survive and make it to the Earth. Why?

The question posed in the last paragraph can be answered either from the point of view of Einstein's time expansion, Eq. 14.1, or from the standpoint of length contraction. Let us examine these in turn. From a frame of reference on the Earth, the half-life of a muon moving at 0.97 c will not be 2.2 μs. Instead, its half-life will be

$$\Delta t = \frac{\Delta t'}{\sqrt{1 - \frac{v^2}{c^2}}} = \frac{2.2 \ \mu s}{\sqrt{1 - \frac{(0.97 \ c)^2}{c^2}}} = 9.05 \ \mu s$$

At a speed of 0.97 c, a particle with a half-life of 9.05 μs would travel a distance of about 2630 m before decaying. Thus, based on the theory of special relativity, one would expect to find many more muons reaching the Earth, because the time before decay is longer.

An alternative explanation based on special relativity, which leads to the same result, relies on length contraction. To see how this approach works, consider an observer on Earth holding a measuring rod of length 640 m that reaches into the sky to the point where our 1000 muons are produced. However, the length of this rod is

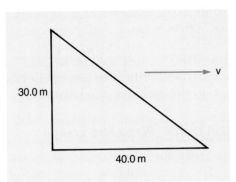

FIGURE 14.11

640 m from the point of view of the observer on the Earth. A muon considers itself to be at rest and the Earth flying up to meet it at a speed of 0.97 c. Thus, in the reference frame of the muon, the rod is not really 640 m long; its length is given by Eq. 14.2 as

$$L = L'\sqrt{1 - \frac{v^2}{c^2}}$$

$$= (640 \text{ m})\sqrt{1 - \frac{(0.97 \text{ c})^2}{c^2}} = 156 \text{ m}$$

Thus, from the point of view of length contraction, the distance that a muon has to travel before it decays is considerably shortened. As a result, many more muons than expected will reach the Earth. We have approached this problem from two different aspects of special relativity. However, in each case the results are the same. More muons reach the Earth than would be expected, and the theory of relativity achieves a major victory.

EXAMPLE 14.2   Please move to the rear of the spaceship

A spaceship is made in the shape of a triangle shown in Figure 14.11 with lengths of 40.0 m and 30.0 m when the ship is at rest. The spaceship flies by a stationary observer at 0.90 c in the direction shown. Find the shape of the spaceship as seen by the stationary observer.

Solution   The 40.0 m length along the direction of travel is contracted, and this new length is given by

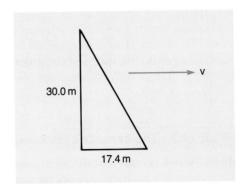

FIGURE 14.12

$$L = L'\sqrt{1 - \frac{v^2}{c^2}} = (40.0 \text{ m})\sqrt{1 - \frac{(0.90 \text{ c})^2}{c^2}}$$

$$= 17.4 \text{ m}$$

The 30 m height of the ship is not changed, because this length is perpendicular to the direction of travel. Thus, the shape of the ship from the observer's point of view is as shown in Figure 14.12.

14.7
RELATIVITY AND MASS

Time intervals and lengths have been found to be subject to the frame of reference in which they are observed. Another sacrosanct entity in physics, the mass of an object, also undergoes a change with its speed. *The mass of an object is found to increase as its speed increases* and the Einstein relationship which predicts this change is given by

$$m = \frac{m_0}{\sqrt{1 - \frac{v^2}{c^2}}} = \gamma m_0 \qquad (14.3)$$

where m_0 is the mass of an object as measured by an observer at rest with respect to it, and m is the mass of the object when in motion with respect to an observer.

This equation also points out that the greatest speed that an object can attain can never exceed the speed of light. To see that this is the case, consider what happens to the mass of an object as its speed, v, approaches c. The denominator of

Eq. 14.3 approaches zero; thus, the mass m becomes infinitely large. An infinite amount of energy would be required to accelerate an infinite mass, and, as a result, the speed of the object cannot increase to that of light. *No material object can attain the speed of light.*

EXAMPLE 14.3   The benefits of working out

In order to see if the new health club you have joined is having any positive benefits for you, you decide to play catch with a baseball having a rest mass of 0.15 kg. You find to your delight that you can now throw the ball at a speed of 0.90 c. What is the mass of the moving baseball?

Solution   The mass is given by Eq. 14.3 as

$$m = \frac{m_0}{\sqrt{1 - \dfrac{v^2}{c^2}}} = \frac{0.15 \text{ kg}}{\sqrt{1 - \dfrac{(0.90 \text{ c})^2}{c^2}}}$$

$$= 0.344 \text{ kg}$$

14.8
MASS AND ENERGY

One of the most basic principles of physics is that of the conservation of energy, which we studied in Chapter 3. There we found that energy can exist in a variety of forms, such as gravitational potential energy and kinetic energy. It may change from one of these forms to another, yet it never appears or disappears. Einstein's work with the theory of special relativity says that this idea of the conservation of energy must be modified. In particular, it says that

mass can be converted into energy and that energy can be converted into mass.

The relationship between energy E, mass m, and the speed of light c is given by the well-known equation

$$E = mc^2 \qquad (14.4)$$

We will see later in this chapter that this equation has had a tremendous impact on the course of human history in that its first application was not a peaceful one. The atomic bombs exploded over Hiroshima and Nagasaki provided the first dramatic evidence of the validity of this equation. More humane applications include the conversion of mass into energy to provide electrical power. In the next section, we shall look at another unusual verification that can be observed only within the confines of a laboratory.

EXAMPLE 14.4   Mass into energy

Suppose that the entire mass of an electron (9.11×10^{-31} kg) could be converted into energy. How many joules would be produced?

Solution   This is a direct application of Eq. 14.4. We have

$$E = mc^2$$
$$= (9.11 \times 10^{-31} \text{ kg})(3.00 \times 10^8 \text{ m/s})^2$$
$$= 8.20 \times 10^{-14} \text{ J}$$

The disappearance of a single electron does not produce a tremendous amount of energy, but a similar calculation would show that if 0.5 kg of matter were converted into energy, enough energy would be produced to keep a 100 W light bulb burning for approximately 14 million years. Thus, significant quantities of energy are produced with the disappearance of only small amounts of mass.

14.9
PAIR PRODUCTION AND ANNIHILATION

The atomic bomb and nuclear power plants provide vivid evidence of the fact that mass can be converted into energy, but does the process go in the reverse order? Can energy by converted into mass? Before we investigate the answer to this question, let us digress to examine a rather strange form of matter called **antimatter.**

In the 1920s, P. A. M. Dirac (1902–1984) developed a theory that incorporated the concepts of quantum mechanics and relativity. One of the outgrowths of the theory was the prediction that *for every type of particle, there is an antiparticle.* To investigate the properties of these particles, which are now known to exist, let us consider a special kind of decay process. Identify the missing element in the radioactive decay symbolized as

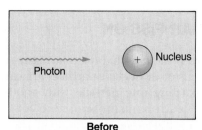

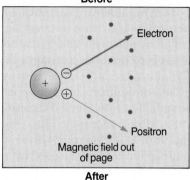

FIGURE 14.13 An incoming photon is converted to an electron-positron pair when it interacts with a nucleus. The pair can be separated by a magnetic field.

$$^{12}_{7}\text{N} \longrightarrow ^{12}_{6}\text{C} + \text{X}$$

What is X? Balancing mass numbers and atomic numbers reveals that we must have a particle symbolized as $^{0}_{1}\text{X}$. The mass number of zero reminds us of the electron, but the particle has a positive charge. In fact, this particle is indeed like the electron in all respects except for its charge. Thus, we symbolize it as $^{0}_{1}\text{e}$. This is the **positron,** and it is said to be the **antiparticle** of the electron. The positron is seen often in decay processes such as the above, but it was only discovered in 1932 by Carl Anderson, in the same year that the neutron was discovered. Positrons are often produced in high-energy cosmic ray reactions with nuclei of the atmosphere, and it was by studying such reactions that Anderson made his discovery.

One of the most common processes by which positrons are produced is through a mechanism called **pair production.** In this process, a gamma ray with sufficiently high energy collides with a nucleus; the gamma ray disappears and in its place an electron-positron pair is created. This process is pictured in Figure 14.13, where the incoming gamma ray is shown to interact with a

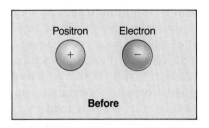

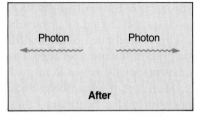

FIGURE 14.14 A positron and an electron meet (before) and annihilate into gamma rays (after).

heavy nucleus. The electron and positron can be separated by a magnetic field that causes them to bend in opposite directions. There is a lower limit on the energy that the gamma ray can have before this process can occur. The minimum energy of the gamma is $E = 2m_0c^2$, where m_0 is the mass of the positron or electron (they are the same). This equation says that the energy must be sufficient to produce a net mass equal to the sum of the masses of the positron and electron. This relationship provides a striking confirmation of the equivalence of mass and energy.

Another confirmation of Einstein's mass-energy relationship is provided by a process that is the reverse of pair production. Antiparticles cannot exist in our world for very long, because a particle such as the positron will soon encounter its counterpart in the world of ordinary particles, and the two will annihilate one another. The process is referred to as **pair annihilation.** This process is pictured in Figure 14.14. Note that two gamma rays are always produced so that momentum can be conserved, and the combined energy of the two gammas must be at least equal to $E = 2m_0c^2$.

Every ordinary particle has its antiparticle counterpart. Thus, the proton is matched with the **antiproton,** which is identical in all respects to the proton except it has a negative charge. An antiproton suffers the same fate as a positron when it encounters normal matter. The proton and the antiproton annihilate, and gamma rays having a combined energy equal to the mass-equivalent of two protons are created.

Since all particles have their antiparticles, a natural question arises: Is there, perhaps somewhere in the Universe, a world made up solely of antiparticles, as our world is made up of particles? If so, this world would consist of atoms having antiprotons and antineutrons in its negatively charged nucleus and circled by positively charged positrons. There would be no way to differentiate such a world via telescope, because it would behave in isolation exactly as would any other heavenly body. However, if this object should come into contact with ordinary matter falling into it, we might be able to detect it because of the annihilation events going on. The most distant objects in our Universe are strange creations called quasars, which are roughly the same size as a star, yet which are pouring out into space tremendous amounts of energy. For example, some quasars are emitting as much radiation as that produced in a galaxy consisting of billions of stars. What is the source of energy that drives a quasar? Several alternatives exist that might be able to explain this fantastic energy emission, but one that some astronomers have envisioned is that of matter-antimatter collisions within the quasar. (It should be noted here that other theories seem to fit the observations better then this one.)

EXAMPLE 14.5 Gone in a puff of gamma rays

A proton and an antiproton, essentially at rest, come together and annihilate. Find the energy in joules of all the gamma rays produced.

Solution The energy produced is equal to the amount of mass that disappears. The mass of the proton and antiproton are both 1.67×10^{-27} kg. Thus, the net energy of the gammas is

$$\begin{aligned} E &= 2m_0c^2 \\ &= 2(1.67 \times 10^{-27} \text{ kg})(3.00 \times 10^8 \text{ m/s})^2 \\ &= 3.01 \times 10^{-10} \text{ J} \end{aligned}$$

14.10 NUCLEAR FISSION

Recall from the last chapter that many studies of the nucleus or of nuclear particles are carried out by bombarding one particle with another. The idea that neutrons might be used to bombard and alter atomic nuclei was exciting to all the scientists who were studying nuclear reactions. The reason is that a neutron, which does not bear any charge, is not repelled by positively charged atomic nuclei and can therefore travel in a straight line until it hits one. If the neutron is absorbed by the nucleus, the ratio of neutrons to protons is changed, and so the stability of the nucleus is also changed.

In 1939, three scientists, Otto Hahn, Fritz Strassman, and Lise Meitner, discovered that when a neutron hits and is captured by a uranium nucleus, the nucleus splits into two roughly equal fragments (Fig. 14.15). This splitting is called **nuclear fission.** Further studies showed that the isotope undergoing fission is uranium-235, which makes up less than 1 percent of natural uranium. The abundant form, uranium-238, does not undergo fission.

It was also learned that extra neutrons were released in the fission reaction of uranium-235. If the reaction is *started* by neutrons and then also *releases* neutrons, a new possibility arises that is very different from anything discussed so far. It is the opportunity for a **chain reaction.** This discovery changed nuclear science from a study of purely theoretical interest to an issue of utmost importance to everyone.

A chain reaction is a series of steps that occur one after the other, in sequence, each step being added to the preceding step like the links in a chain. An example of a chemical chain reaction is a forest fire. The heat from one tree may initiate the reaction (burning) of a second tree which in turn, ignites a third, and so on. The fire will then go on at a steady rate. But if one burning tree ignites, say, two others, and each of these two ignite two more, for a total of four, and so on, the rate of burning will speed up. Such uncontrolled, runaway chain reactions are at the heart of the explosion created by a nuclear bomb (Fig. 14.16).

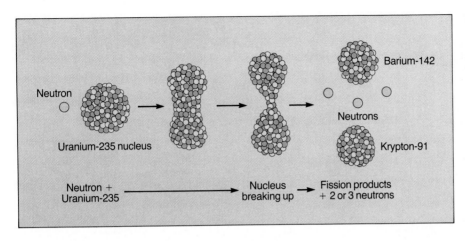

FIGURE 14.15 A neutron strikes a uranium-235 nucleus causing it to split into two nearly equal fragments (shown here as barium-142 and krypton-91) and extra neutrons.

Fission is initiated when one neutron strikes a uranium-235 nucleus and can proceed in a number of different ways. For example, as shown in Figure 14.15, uranium-235 struck by a neutron can produce barium-142 and krypton-91, while releasing three neutrons. In equation form, the fission reaction here is

$$\ce{^{1}_{0}n} + \ce{^{235}_{92}U} \longrightarrow \ce{^{142}_{56}Ba} + \ce{^{91}_{36}Kr} + 3\ce{^{1}_{0}n}$$

Note the following important points about this reaction:

1. The reaction is started by one neutron but produces three neutrons. These neutrons can, ideally, initiate three new reactions, which in turn produce more neutrons, and so forth. Thus, a chain reaction evolves.

2. The uranium-235 nucleus is split roughly in half by this reaction. The total mass of all the fission products (those on the right side of the arrow) is slightly less than the sum of the masses of the original uranium-235 atom and the incident neutron. This loss of mass is converted into energy in accordance with the equation $E = mc^2$. The energy appears mostly in the form of the kinetic energy of the fission products as they fly apart. These flying fragments then slow down as they hit other atoms, and in so doing they transfer their energy to these atoms in random patterns. This is the way in which nuclear fission releases heat. As

has been pointed out, the energies involved in nuclear transformations are much greater than those in chemical reactions. If the chain reaction continues at a very rapid rate, energy is released at an accelerating rate, and an explosion can result. (This is what happens in the detonation of a nuclear bomb.) If, on the other hand, the chain reaction is controlled, energy can be released more slowly, and the heat produced can be used to make steam, which can then be used to drive a turbine and produce electricity.

3. Fission reactions produce radioactive wastes. Barium-142 and krypton-91, the products shown in the preceding equation, are both radioactive. Furthermore, the reaction represented by this equation is only one of many that occur in nuclear fission.

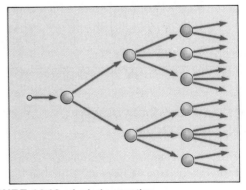

FIGURE 14.16 A chain reaction.

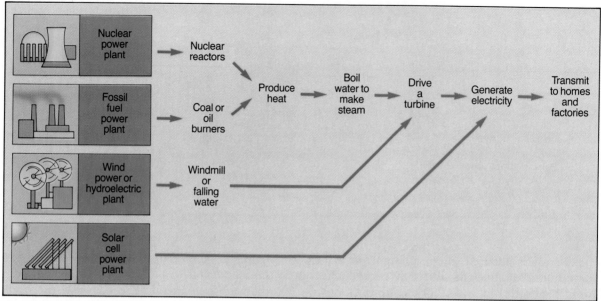

FIGURE 14.17 Different types of power plants use different systems to generate electricity.

solar power plant generates electricity directly from solar cells. In a coal-fired plant, the energy is released from the chemical combustion of the coal. In a nuclear plant, energy from nuclear fission reactions is used to heat water and to produce steam. The steam then drives a turbine to generate electricity, just as in a coal-fired plant (Fig. 14.17).

Nuclear fission reactors require fuel, and the fuel must be a substance whose nuclei can undergo fission. There are two significant nuclear fuels, uranium-235 and plutonium-239. There are a few other fissionable materials, but these are the ones on which the current nuclear energy program in the United States is based.

Uranium-235 occurs in nature, but it constitutes only 0.7 percent of natural uranium. The remaining 99.3 percent is the heavier isotope uranium-238, which does not undergo fission in a reactor.

The second fuel, plutonium-239, does not occur in nature; it is produced by bombarding uranium-238 with neutrons. Thus, the two important naturally occurring sources of fission energy are uranium-235 (which fissions but is not abundant) and uranium-238 (which does not fission, but is abundant and convertible into plutonium-239).

EXAMPLE 14.6 A fission reaction

Complete and balance the following nuclear fission reaction:

$$\,^{1}_{0}n + \,^{235}_{92}U \longrightarrow \,^{97}_{39}Y + ? + 2\,^{1}_{0}n$$

Solution The atomic numbers on the left $(92 + 0)$ must equal those on the right. Therefore, $92 + 0 = 39 + 0 + Z$, so $Z = 53$, which is the atomic number of iodine, I. For the mass numbers, note that two neutrons are produced, which count for two mass numbers. Then, $1 + 235 = 97 + 2 + N$, from which $N = 137$, and the missing isotope is therefore $^{137}_{53}I$.

14.11 NUCLEAR POWER PLANT CONSIDERATIONS

The purpose of a power plant is, of course, to generate electricity. Different kinds of power plants depend on different sources for their energy. A wind power or hydroelectric plant utilizes the mechanical energy of wind or falling water. A

Nuclear reactors require another essential ingredient besides fuel, namely neutrons. In fact, the chain reaction is initiated by neutrons. The design and operation of reactors, as well as their safety, depend on how the neutrons are managed and controlled. There are four possible events that can happen to a neutron in a reactor:

1. A neutron can be captured by a uranium-235 nucleus, which then undergoes fission. The reaction releases fast neutrons. However, slow neutrons are more readily captured by uranium-235 nuclei. The fast neutrons can be slowed down by colliding with some other particles with which they can exchange momentum and energy. The most effective particles are those with about the same mass as a neutron. As an analogy, a moving billiard ball is best slowed down by colliding with another billiard ball, not by hitting a dust particle, which will hardly affect it, or by hitting a boulder, from which it will bounce back with little loss of energy. The particles closest in mass to neutrons are hydrogen nuclei, and that is why water, H_2O, which is a convenient source of hydrogen, is a good choice. A medium that slows down neutrons is called a **neutron moderator.**

2. A neutron can be captured by a uranium-238 nucleus, producing plutonium-239. As you recall from Chapter 13, this reaction takes place in two steps:

$$^{1}_{0}n + ^{238}_{92}U \longrightarrow ^{239}_{93}Np + ^{0}_{-1}e$$

followed by the beta decay of Np as

$$^{239}_{93}Np \longrightarrow ^{239}_{94}Pu + ^{0}_{-1}e$$

Plutonium-239 can undergo fission. Thus, the production of plutonium is, in effect, a "breeding" of new fuel and is therefore attractive as a means of utilizing uranium resources more completely. The choice of whether or not to favor the breeding of plutonium determines, in large part, the design of the reactor.

3. A neutron can be captured by impurities. This causes loss of neutrons and slowing down, or "damping," of the chain reaction. It is thus necessary to have a controlled means of absorbing neutrons to regulate the reaction. The most direct method is to insert a stick of neutron-absorbing impurity. Devices used in this fashion are called **control rods;** they usually contain cadmium or boron and other elements, and they can be inserted into or withdrawn from the reactor core to regulate the neutron flow with great precision.

4. A neutron, traveling as it does in a straight line, may simply miss the other nuclei in the reactor and escape. If the reactor were very small, too many neutrons would escape and the chain reaction would not be sustained. This circumstance imposes lower limits on reactor size; there will never be a pocket-sized fission generator nor even fission engines for motorcycles. This tendency to escape also demands adequate shielding to prevent neutron leakage into the environment.

The following section discusses how all of these requirements are taken into consideration in a typical nuclear power plant in this country.

14.12
THE DESIGN OF A NUCLEAR POWER PLANT

The heart of a nuclear power plant is the reactor core, in which the essential components are (1) the nuclear fuel, (2) the moderator, (3) the coolant, and (4) the control rods (Fig. 14.18).

1. The nuclear energy source is the uranium-235 isotope, but in pure form it could serve as a nuclear explosive, not as a practical fuel. The uranium actually used is natural uranium, enriched up to 3 percent with fissionable uranium-235. Furthermore, the material used is not metallic uranium, but rather uranium dioxide. This compound is fabricated in a ceramic form that is much better than the pure metal in its ability to retain most fission products, even when overheated. The fuel is inserted in the form of pellets into long, thin tubes, called the "fuel cladding," made of stainless steel or other alloys. These "fuel rods" are then bundled into assemblies that are inserted into the reactor core (Fig. 14.19).

FOCUS ON . . . Traveling Twins

Suppose that there were a pair of twins, each 25 years old. One was a traveling astronaut, while the other stayed home. Now imagine that the astronaut boarded a rocket ship and headed out for distant stars, traveling at velocities very close to the speed of light. The twin in the rocket would exist in a different frame of reference from the earthbound twin, and time on the spaceship would slow down with respect to an observer on Earth. Upon the spaceship's return, the traveling twin would be younger than the other (see Fig.). Specifically, suppose that the rocket was traveling at 99.5 percent of the speed of light. If the clock on the spaceship wall showed that 3 years had elapsed during the journey, the traveler would have aged 3 years. The space wanderer would notice nothing unusual inside the rocket. On returning to Earth, however, the traveling twin would be only 28 years old (25 + 3). But 30 years of Earth time would have elapsed during the journey, and the other twin would be 55 years old (25 + 30)!

This seems to be something that might happen in a science fiction story, but not in reality. However, it has been demonstrated that relativity of time does occur. One experiment that added proof to its validity occurred when two extremely accurate clocks were placed one on each of two conventional airplanes. One plane flew around the world east to west, and the other made the journey west to east. Upon arrival at the starting point, the two clocks no longer read the same time. The differing motion associated with the fact that one flew with the rotation of the Earth and the other against it could be used to demonstrate that they should be out of synchronization by about one-billionth of a second relative to each other if time dilation is true. The demonstration verified exactly this. Thus, time is relative to space and velocity. There is no standard time. We each carry our own internal time with us. If you walk past me, you are not merely moving in a different portion of space but in a different time as well.

Twin sisters say goodbye as one leaves for a distant star on a very fast rocket ship.

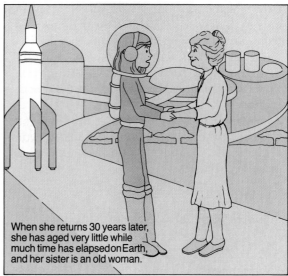

When she returns 30 years later, she has aged very little while much time has elapsed on Earth, and her sister is an old woman.

The traveler ages slower than the stay-at-home sister.

---

2. The moderator serves to slow down the neutrons. As mentioned earlier, fission is more likely to occur with slow neutrons, but the fission reaction releases fast neutrons, which must be slowed to maintain a chain reaction. Water is very convenient for this purpose be-

cause it can also serve as the coolant. In the United States, almost all commercial reactors use ordinary, or "light," water, H_2O, in which the hydrogen atoms are the ^1_1H isotope. This is cheap but not ideal, because H_2O molecules do capture neutrons to some extent. The result of

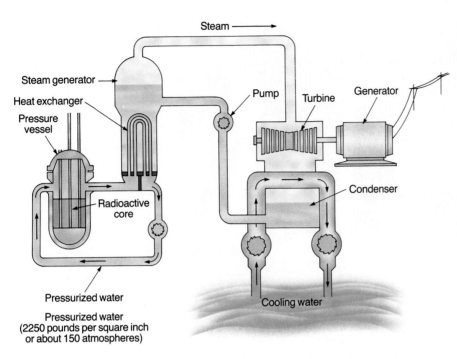

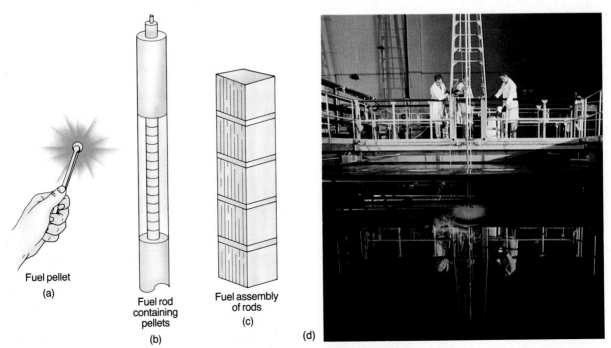

FIGURE 14.19 (a) A fuel pellet. (b) A fuel rod. (c) An assembly of fuel rods. (d) A fuel assembly being lowered into place in a reactor.

FIGURE 14.20 Control rod mechanism of a nuclear reactor.
(Courtesy U.S. Council for Energy Awareness)

such capture is that natural uranium, which contains only 0.7 percent uranium-235, cannot be used in light-water reactors. Instead, the concentration of uranium-235 must be enriched to about 3 percent to make up for the neutron loss to the coolant water.

3. When water is the moderator, it is also the coolant. In some designs, the coolant water actually boils, and the steam it produces drives the turbines. (You may think it strange for boiling water to be a coolant, but remember that heat always flows spontaneously from a higher to a lower temperature, and even boiling water is much cooler than the core of a nuclear reactor. The heat therefore flows from the reactor core to the boiling water.) In most of the commercial reactors in the United States, the water is kept under high pressure, as in a pressure cooker, and very little boiling actually occurs. This is the design shown in Figure 14.18. Note that the pressurized water flows through a heat exchanger, where it transfers its heat to a secondary loop of water that actually boils and delivers steam to the turbine.

4. Interspersed into the matrix of fuel, moderator, and coolant are the control rods (Fig. 14.20), which serve to regulate the flow of neutrons. A nuclear reactor that generates power at a constant rate must operate at a critical condition, with as many neutrons being produced as are lost by capture or escape. Since no design is that perfect, the control rods provide a means of fine tuning the operation. But the control rods serve other important functions. Recall that fission reactions produce impurities that absorb neutrons. Therefore, in the absence of any neutron regulation, the neutron flow would gradually slow down, and the fission reactions would die out before much fuel is exhausted. For this reason, there must be a provision to increase the neutron flow gradually during the life of the fuel to make up for the loss caused by impurities. The way to do this is to design for an extra large neutron flow at the outset but to limit the actual flow by control rods. As fuel is consumed, the control rods are gradually withdrawn to compensate for the accumulation of neutron-absorbing impurities. This amounts to a neat balancing of impurities to maintain steady power production. The other purpose of the control rods is to serve as an emergency shut-off system. If an emergency occurs and the fission reaction must be quenched, the rods are pushed rapidly all the way into the core.

TABLE 14–1 Isotopes of Hydrogen

Isotope	Names	Radioactive?	Natural Abundance (%)
$_1^1H$	"Ordinary" hydrogen "Light" hydrogen Hydrogen Protium	No	99.985
$_1^2H$ or $_1^2D$	"Heavy" hydrogen Deuterium	No	0.015
$_1^3H$ or $_1^3T$	Tritium	Yes (12-year half-life)	Almost none

14.13
NUCLEAR FUSION

Fission reactions occur when heavy nuclei split apart. On the other hand, *nuclear fusion* reactions *occur when nuclei of light elements are joined together.* The energy that is derived from stars in the prime of their life, such as our Sun, comes from hydrogen fusion reactions. Controlled nuclear fusion would provide abundant energy with much less environmental danger than is faced from fission reactors. However, no useful fusion reactor has yet been developed.

Unlike the fission reaction, fusion cannot be triggered by neutrons. Instead, the nuclei to be fused must be brought into contact with each other. Positive nuclei repel each other at normal interatomic distance, but if they are very close together, the strong force predominates and binds them together. In order to overcome the electrical repulsion and bring the nuclei close enough together for the strong force to take over, the nuclei must be moving very rapidly or, in other words, they must be elevated to very high temperatures. The resulting fusion is therefore called a **thermonuclear reaction.** If a large mass of hydrogen isotopes fuses in a very short time, the reaction cannot be contained, and it goes out of control; this is the explosion of the "hydrogen bomb." On the other hand, useful energy could be extracted from fusion if it were possible to devise a controlled thermonuclear reaction.

Any fusion reactor would utilize hydrogen nuclei. There are three isotopes of hydrogen: "ordinary" hydrogen, $_1^1H$; deuterium, $_1^2H$; and tritium, $_1^3H$. Some information is given about them in Table 14–1. Fusion reactions can occur between any two hydrogen isotopes. The lighter isotopes are more abundant but require much higher temperatures to initiate fusion.

For example, to start the reaction between two deuterium nuclei, the temperature would have to be raised to about 400 million degrees Celsius. The problems imposed by this requirement are so severe that it is not even being attempted. Instead, all efforts to control fusion are being directed to a cooler reaction (only about 40 million degrees Celsius), namely, the fusion of deuterium with tritium to give helium-4 plus a neutron:

$$_1^2H + _1^3H \longrightarrow {}_2^4He + _0^1n$$

This process requires a source of tritium, which is not naturally available on Earth. Tritium is produced artificially by neutron bombardment of lithium. Even though this process is expensive, it is the only feasible fusion reaction and is the one being studied at the present time.

Now think for a moment about high temperatures. An iron bar turns red hot at about 600°C, white hot around 1100°C, melts around 1500°C, and the molten iron boils at 2885°C. Other solids survive to higher temperatures—carbon and tungsten, for example, to about 3500°C—but no solids, liquids, or even any chemical bonds survive about 5000°C. At these temperatures, everything is a gas consisting of lone atoms that are much too energetic to combine chemically with each other.

At the temperatures involved in nuclear fusion reactions, measured in millions of degrees, not even atoms survive, because electrons are

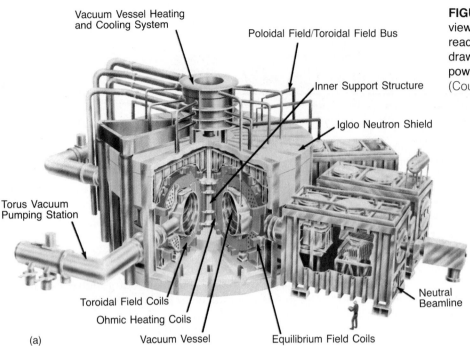

Vacuum Vessel Heating
and Cooling System

Poloidal Field/Toroidal Field Bus

Inner Support Structure

Igloo Neutron Shield

Torus Vacuum
Pumping Station

Neutral
Beamline

Toroidal Field Coils

Ohmic Heating Coils

Vacuum Vessel

Equilibrium Field Coils

(a)

FIGURE 14.21 (a) Cut-away view of a nuclear fusion reactor. (b) Schematic drawing of a thermonuclear power plant. (Courtesy Department of Energy)

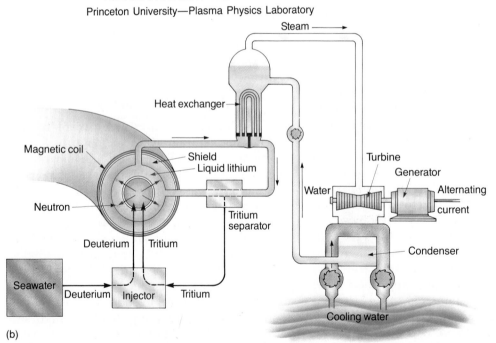

TOKAMAK FUSION TEST REACTOR
Princeton University—Plasma Physics Laboratory

Steam

Heat exchanger

Magnetic coil

Shield
Liquid lithium

Turbine

Generator

Water

Alternating
current

Neutron

Tritium
separator

Condenser

Deuterium   Tritium

Seawater

Deuterium   Injector   Tritium

Cooling water

(b)

stripped away from the nuclei. Such an extremely hot mixture of independently moving electrons and nuclei is called a **plasma.** Obviously, no container exists that can survive long enough to confine a plasma for the useful production of thermonuclear energy. Instead, what is envisaged is a

sort of "magnetic bottle," which does not consist of a physical substance at all, but rather is a magnetic field so designed that it will confine the charged particles of the plasma in which the thermonuclear reaction is going on. Research efforts have shown promise, in a modest way. On November 3, 1984, scientists at the Massachusetts Institute of Technology maintained a controlled nuclear fusion for about 50 milliseconds (about the blink of an eye), and the brief process actually yielded as much energy as was put into it.

The useful energy, once liberated, will have to be extracted in the form of the kinetic energy of the evolved neutrons. Since the neutrons carry no charge, they will pass through the magnetic field and escape from the plasma. The energy of the speeding neutrons can then be extracted by a moderator. If the moderator were water, the energy would create steam that could drive a turbine. The entire fusion reactor would be encased in a sheath or blanket in which molten lithium would be continuously circulated. The lithium would absorb the neutrons, supply the tritium, and then release its heat to water in a heat exchanger (Fig. 14.21).

In 1989, experiments in fusion took a new direction when two chemists, B. Stanley Pons of the University of Utah and Martin Fleischmann of the University of Southampton in England, announced that they had created "fusion in a bottle." At this writing, confirmation of the validity of their experiment is very much in doubt. But even if their results prove to be erroneous, the experiment could prove to have some long-range value to the scientific community by opening new avenues of investigation directed toward "cold fusion"—fusion that could take place at room temperatures.

Figure 14.22 shows the fusion bottle used by Pons and Fleischmann. They wrapped a palladium rod with a platinum wire and immersed the system in a bottle filled with "heavy water." Normal water is H_2O, while heavy water has the hydrogen replaced with deuterium. A power source connected to the palladium and platinum sends an electrical current through the water, which then dissociates into oxygen and deuterium, and the deuterium is absorbed by the palladium. The deuterium ions are then squeezed into the lattice-like structure of the palladium metal, as shown in Figure 14.23. There the claim is that the deuterium nuclei fuse together and release energy. Does this really happen, or is this only a misreading of the data by two scientists? At present, no one knows the result, but there are serious doubts that a fusion reaction has truly occurred. In fact,

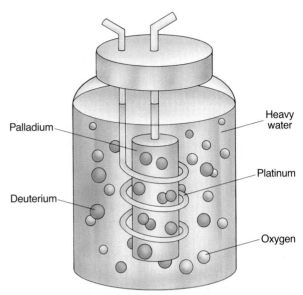

FIGURE 14.22 Cold fusion?

the probability seems overwhelming that a fusion reaction has not occurred.

If cold fusion turns out to be a reality, then the economic world order will be turned inside out. Poor countries that have inadequate supplies of energy would not have to depend on external sources. The use of fossil fuels could be redirected from burning them for fuel to using them in various manufacturing processes. Energy would become clean and cheap, because it has been estimated that there is enough deuterium in

FIGURE 14.23 The deuterium atoms are squeezed together inside the palladium.

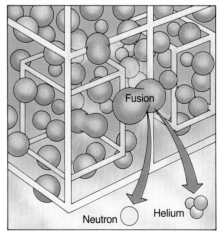

the top ten inches of one of the Great Lakes to supply this country's energy needs for 5000 years.

Could a fusion reactor get out of control and go off like a hydrogen bomb? Nuclear scientists are entirely confident that the answer is no, an explosion could not occur. The reason is that the hydrogen isotopes are continuously fed into the reactor and continuously consumed; they do not accumulate. The total quantity of fuel in the plasma of a "hot hydrogen" reactor at any one time would be very small—about 2 g—far below the critical mass required for a runaway reaction. If the temperature dropped or the plasma somehow dispersed, the reaction would stop; in effect the fusion would turn itself off.

Would there be a problem with environmental radioactivity? The answer here is yes, because both tritium and neutrons could be released. Tritium is radioactive (half-life, 12 years) and can combine with oxygen to form radioactive water. However, the beta particles emitted by tritium have so little penetrating power that it is virtually harmless to living organisms as long as its source is outside the body. The neutrons released from the reactor can be absorbed by atomic nuclei in the reactor's shielding material, and the new isotopes thus produced may be radioactive. As a result, there could be substantial quantities of radioactive matter to be disposed of. We are so far from a practical fusion reactor that we have hardly begun to study the problems of handling the wastes.

EXAMPLE 14.7 Fusion in the Sun

The fusion reactions that power the Sun are believed to occur in a three-stage process, with energy being released in each of the stages. You fill in the blanks in each of the stages outlined below.

(a) In the first stage of the cycle, two protons collide and fuse to form a positron, a neutrino, and another particle. What is the unknown particle?

Solution The reaction can be written symbolically as

$$^1_1H + ^1_1H \longrightarrow X + ^0_1e + \nu$$

Balancing mass numbers and atomic numbers, show that the unknown substance is ^2_1H, deuterium.

(b) In the second stage of the series, a deuterium nucleus produced by the reaction in part (a) fuses with a proton to produce a gamma ray and another nucleus. What is this nucleus?

Solution The reaction is

$$^2_1H + ^1_1H \longrightarrow X + \gamma$$

Balance the mass numbers and atomic numbers to show that X is ^3_2He.

(c) Finally, two of the helium nuclei produced by reactions like those of part (b) fuse to form helium-4 and two other identical particles. What are the identical particles?

Solution The reaction is

$$^3_2He + ^3_2He \longrightarrow ^4_2He + 2X$$

Solution Again, we leave it to you to balance the reaction and to show that the two particles are symbolized as ^1_1H.

(d) Often these reactions are described figuratively by saying that hydrogen is the fuel and helium-4 is the ash. Start with part (a) of the series above and count the total number of hydrogens that have been burned in one series.

Solution The reaction in part (a) had to occur twice so that we would eventually end up with two ^3_2He nuclei to fuse in part (c). Thus, part (a) of the series will consume four hydrogens. The reaction in (b) also must occur twice, so we use a total of two more hydrogens for this reaction. The grand total used is now six. However, the end result of the reaction in part (c) is that two hydrogen nuclei are left over. Thus, the total number of hydrogens used is four, so it may be said that the overall result of this series of reactions is that four hydrogen nuclei have been fused into one helium-4 nucleus. The mass of helium-4 is less than the combined mass of four hydrogen nuclei. By Einstein's mass-energy relationship, this mass is converted into energy.

SUMMARY

Ether was a material proposed as the medium through which electromagnetic waves would travel. Since it was considered to be a privileged frame of reference for the measurement of velocities, extensive research was done in an effort to detect it. This culminated with the Michelson-Morley experiment, which produced the result that the ether does not exist.

Einstein's special theory of relativity is true only for reference frames that are moving with constant velocity with respect to one another. The fantastic results of this theory indicate that moving clocks run slowly, that lengths contract for objects in motion, and that mass increases for objects in motion. Finally, it predicted that energy could be converted into mass, and vice versa. The processes of **pair production and annihilation** verify this, as well as does the use of nuclear reactors and bombs.

In **nuclear fission,** an isotope of a heavy element is split into lighter elements, releasing a large amount of energy. The important naturally occurring fissionable isotope is uranium-235. The fission is triggered by a neutron, and each atom releases two or three neutrons.

The result can be a **chain reaction,** in which a series of steps occurs. Nuclear fission can be used in an atomic bomb or in a nuclear power plant. The difference between the two applications depends on the neutrons. There are four things that can happen to a neutron:

1. Fission capture by uranium-235 to yield energy and fast neutrons. But the reaction is favored by slow neutrons. Therefore, a **moderator,** which is a substance that slows down neutrons, can be used.
2. Nonfission capture by uranium-238, thereby producing plutonium-239, which can undergo fission.
3. Nonfission capture by impurities, which causes loss of neutrons. This action can be used to control the fission process.
4. Escape. A neutron might miss everything and get lost.

In **nuclear fusion,** the nuclei of the light isotopes deuterium and tritium combine to produce helium and release energy.

EQUATIONS TO KNOW

$$\Delta t = \frac{\Delta t'}{\sqrt{1 - \frac{v^2}{c^2}}}$$ (time dilation)

$$m = \frac{m_0}{\sqrt{1 - \frac{v^2}{c^2}}}$$ (mass increase)

$$L = L'\sqrt{1 - \frac{v^2}{c^2}}$$ (length contraction)

$$E = mc^2$$ (mass-energy equivalence)

KEY WORDS

Reference frame	Length contraction	Pair annihilation	Moderators
Luminiferous ether	Mass increase	Positron	Control rods
Michelson-Morley experiment	Energy-mass equivalence	Antiproton	Nuclear fusion
Special relativity	Antimatter	Nuclear fission	Plasma
	Pair production	Chain reaction	

QUESTIONS

RELATIVITY BEFORE EINSTEIN

1. A person driving from Los Angeles to San Francisco is said to be moving northwest. But such a simple description ignores the fact that the person is rotating with the Earth, orbiting the Sun, and

simultaneously flying through intergalactic space. Would it be more accurate to define the net motion of the car as the sum of all these independent earthly celestial movements? Explain.

2. A hiker is walking northward along the shore at a rate of 2 km/h, while a canoeist is paddling south at 2 km/h. (a) What are their relative velocities with respect to each other? (b) What are their velocities relative to a stationary observer on the riverbank?

3. Rocket ship X, traveling from the Moon to Earth, is flying at 1000 km/h relative to the Earth. An astronaut in that rocket ship is floating forward at 1 km/h relative to the vessel. A second rocket, Y, is flying from the Earth to the Moon directly toward rocket X at a rate of 1500 km/h. The pilot of the second rocket is stationary relative to his rocket. Calculate the velocity of (a) rocket X with respect to the Earth, (b) rocket X with respect to rocket Y, (c) the astronaut in rocket X with respect to the Earth, (d) the astronaut in rocket X with respect to the astronaut in rocket Y.

4. Would it be possible to play a game of pool on an airplane moving at 1000 km/h? 500 km/h? Would it be possible to play a game of pool on an airplane that is accelerating? Turning a sharp corner? Explain and discuss.

5. If velocity is a relative quantity, is kinetic energy also relative to the reference frame used? Explain.

THE MICHELSON-MORLEY EXPERIMENT

6. What was the ether thought to be? How did scientists search for it? What is the significance of their findings?

7. In the situation of problem 3, a beam of light is directed at ship X from Earth. What is the speed of this light as measured by the astronaut in rocket X?

8. An extremely fast rocket is traveling toward Earth at a rate of 2×10^8 m/s. The pilot of that ship is moving forward at a rate of 1 m/s with respect to the vessel. A radio operator on Earth sends a message to this rocket. What is the speed of the radio signal with respect to (a) the Earth, (b) the moving rocket, and (c) the person in the rocket? Explain, and discuss the implications of your answer.

9. For the passage of light, the ratio of distance to time is a constant throughout the Universe. What is the value of this ratio?

10. What is the speed of light measured here on Earth from the following sources: (a) radiated directly from the Sun, (b) reflected off the Moon, and (c) originating from a distant galaxy that is flying away from us at a rate of 2×10^8 m/s?

11. A question that supposedly perplexed Einstein as a child is: What would happen if someone runs at the speed of light while carrying a mirror in his hand? What do you believe the runner would see in the mirror?

12. A passenger standing at the rear of a northbound train moving at 50 m/s throws a ball at a speed of 50 m/s southward. Where does the ball hit the ground?

13. In our discussion of refraction, we stated that the speed of light in glass is about 2×10^8 m/s. Is this a contradiction of the results of the Michelson-Morley experiment?

EINSTEIN AND TIME, LENGTH CONTRACTION, RELATIVITY AND MASS, AND MASS AND ENERGY

14. If you were traveling in a rocket ship at close to the speed of light, could you detect any contraction in length or a slow-down in time within your environment? Explain and discuss.

15. What is the reasoning behind the conclusion that mass and energy are related?

16. While working on a project in a rocketship traveling at 0.90 c, the heartbeat of an astronaut increases to 100 beats/min as measured on the ship. (a) What would an observer monitoring the heart rate from Earth measure? (b) If the observer on Earth also has a heart rate of 100 beats/min, what heart rate will the astronaut monitor for him?

17. At what speed would a rocketship have to be traveling so that its rest length of 20 m is reduced to 10 m?

18. A 0.5 kg football is kicked with a speed of 0.95 c. What is its mass at this speed?

19. (a) How fast would a proton have to travel before its mass is doubled? (b) Repeat for an electron.

20. If all the mass of a proton could be converted into energy, how many joules would be produced?

21. (a) If 1 gram of matter should be converted into energy, how many joules would be produced? (b) A 100 W light bulb utilizes energy at the rate of 100 joules/s. How many seconds would the 1 gram of matter keep this light bulb burning?

NUCLEAR FISSION AND NUCLEAR POWER PLANTS

22. Use the periodic table of the elements to find some possible products for the fission of U-235. Write your results in the form of a nuclear reaction.

23. A neutron strikes U-235 and the fission products are Sr-88 and Xe-136. How many neutrons are produced?

24. What are essential features of a nuclear fission reactor? Explain the function of each feature.
25. List the possible fates of neutrons in a fission reactor. Which of these events should be favored and which should be inhibited in order to (a) shut down a reactor, (b) breed new fissionable fuel, and (c) produce more energy?
26. Could a nuclear reactor ever be miniaturized to provide long-term power for your wristwatch? Your camera? Your pocket radio? (Assume that the proper shielding against radioactivity could be provided.) Defend your answer.
27. If no emergency ever occurred, and if a nuclear reactor always operated at steady power production, would control rods still be needed? Defend your answer.

FUSION

28. Suppose that someone claims to have found a material that can serve as a rigid container for a thermonuclear reactor. Would such a claim merit examination, or should it be ignored as a "crackpot" idea not worth the time to investigate? Defend your answer.
29. Outline the reasons why fusion reactors are expected to be far less serious sources of radioactive pollutants than fission reactors.
30. Write the following reaction in equation form. Two deuterium nuclei collide to produce tritium and an additional particle. What is the particle?
31. Write the following reaction in equation form. Deuterium and tritium collide to produce He-4 and another particle. What is the particle?

ANSWERS TO SELECTED NUMERICAL QUESTIONS

2. (a) 4 km/h, (b) 2 km/h for each
3. (a) 1000 km/h, (b) 2500 km/h, (c) 1001 km/h, (d) 2501 km/h
7. 3×10^8 m/s
16. 229 beats/min, (b) 229 beats/min
17. 0.87 c
18. 1.60 kg

19. (a) 0.87 c, (b) 0.87 c
20. 1.50×10^{-10} J
21. (a) 9×10^{13} J, (b) 9×10^{11} s (about 30,000 years)
23. 11
30. $^1_1$H
31. $^1_0$n

PART

CHEMISTRY

TWO

In some chemical reactions energy is released—as shown here—while in others energy is absorbed. (Charles D. Winters)

ROALD HOFFMANN

• INTERVIEW •

Roald Hoffmann is a remarkable individual. When he was only 44 years old, he shared the 1981 Nobel Prize in Chemistry with Kenichi Fukui of Japan for work in applied theoretical chemistry. In addition, he has received awards from the American Chemical Society in both organic chemistry and inorganic chemistry, the only person to have achieved this honor. And, in 1990, he was awarded the Priestley Medal, the highest award given by the American Chemical Society.

The numerous honors celebrating his achievements in chemistry tell only part of the story of his life. He was born to a Polish Jewish family in Zloczow, Poland, in 1937, and was named Roald after the famous Norwegian explorer Roald Amundsen. Shortly after World War II began in 1939, the Nazis first forced him and his parents into a ghetto

and later into a labor camp. His father, however, smuggled him and his mother out of the camp, and they were hidden for more than a year in the attic of a school house in the Ukraine. (His father was killed by the Nazis after trying to organize an attempt to break out of the labor camp.) After the war, Hoffmann, his mother, and his stepfather made their way west to Czechoslovakia, Austria, and then Germany. They finally emigrated to the United States, arriving in New York on Washington's birthday, 1949. That Hoffmann and his mother survived those years is our good fortune. Of the 12,000 Jews living in Zloczow in 1941 when the Nazis took over, only 80 people, three of them children, survived the Holocaust. One of those three children was Roald Hoffmann.

On arriving in New York, Hoff-

mann learned his sixth language, English. He attended public schools in New York City and went on to Stuyvesant High School, one of the city's select science schools. From there he went to Columbia University and Harvard University earning his Ph.D. at the latter in 1962. Shortly thereafter, he began the work with Professor R. B. Woodward that eventually led to the Nobel Prize. Since 1965 he has been a professor at Cornell University, where he regularly teaches first-year general chemistry.

In addition to his work in chemistry, Professor Hoffmann writes popular articles on science for the American Scientist and other magazines, and he has published two volumes of his poetry. He will be appearing in a series of 26 half-hour television programs for a chemistry course called "World of

From medicine to cement to theoretical chemistry

Visitors to Professor Hoffmann in his office at Cornell University would find mineral samples, molecular models, and Japanese art. When asked what introduced him to chemistry he replied that "I came rather late to chemistry, I was not interested in it from childhood." However, he clearly feels that one can come late to chemistry, and that it can be a very positive thing. "I am always worried about fields in which people exhibit precocity, like music and mathematics. Precocity is some sort of evidence that you have to have talent. I don't like that. I like the idea that human beings can do anything they want to. They need to be trained sometimes. They need a teacher to awaken the intelligences within them. But to be a chemist requires no special talent, I'm sorry (and glad) to say. Anyone can do it, with hard work."

He took a standard chemistry course in high school. He recalls that it was a fine course, but apparently he enjoyed biology more, because in his high school year book, "under the picture of me with a crew cut, it says 'medical research' under my name." Indeed, Hoffmann says that "medical research was a compromise between my interest in science and typical Jewish middle class family pressures to become a medical doctor."

At Columbia University, Hoffmann signed up as a pre-med student, but there were several factors that shifted him from a career in medicine. One of these was his work at the

National Bureau of Standards in Washington, D.C. for two summers, and then at Brookhaven National Laboratory in New York, for a third summer. He explains that these experiences gave him a feeling for the excitement of chemical research. Nonetheless, during his first summer at the Bureau of Standards, he "did some not very exciting work on the thermochemistry of cement." During his second summer there, he went over to the National Institute of Health to find out what medical research was about. "To my amazement, most of the people had Ph.D.'s and not M.D.'s. I just didn't know. Young people do not often know what is required for a given profession. Once I found that out, and that I did well in chemistry, it made me feel that I didn't really have to do medicine, that I could do some research in chemistry or biology. Later, the decision to pursue theoretical chemistry was a result of an excellent instructor. Had I had some really good instruction in organic chemistry, I'm pretty sure I would have become an organic chemist."

"At the very same time I was being exposed to the humanities, in part because of Columbia's core curriculum—which I think is a great idea—that had so-called contemporary civilization and humanities courses. I took advantage of the liberal arts education to the hilt, and that has remained with me all my life. The humanities professors have remained permanently fixed in my mind and have changed my ways of thinking. These were the people who really had the intellectual impact on me and helped to shape my life.

"To trace the path: I was a

latecomer to chemistry and was inspired by the research. I think *research* is the way in. It just gives you a different perception."

A love for complexity

Having discussed what introduced him to chemistry, we were interested in his view of the qualities that a student should possess if he or she wishes to pursue a career in the field. He said that "one quality needed to be a chemist is a love for complexity and richness. To some extent, this is true of biology and natural history, too. I think one of the things that is beautiful about chemistry is the great number of compounds— 10,000,000—each with different properties. What's beautiful when you make a molecule, is that you can make derivatives in which you can vary substituents, the pieces of a molecule, and these substituents give a molecule function, complexity, and richness. That's why protein or nucleic acid, with all its variety, is essential for life. That's why to me, intellectually, isomerism and stereochemistry in organic chemistry are at the heart of chemistry. I think these ideas should be taught much earlier. It requires no mathematics, only a little model building; one could do this without theory. I think it is no accident that organic chemistry drew to it the intellects of its time."

Experiment and theory

Professor Hoffmann has spent his career immersed in the theories of chemistry. However, he believes that fundamentally "chemistry is an experimental science, in spite of some of my colleagues saying otherwise.

However, the educational process certainly favors theory. It's the nature of the subject for both teacher and student to want to *understand* and to then give primacy to the soluble and the understood, at the expense of other things. We also have this reductionist philosophy of science, the idea that the social sciences derive from biology, that biology follows from chemistry, chemistry from physics, and so on. This notion gives an inordinate amount of importance to theoretical thinking: the more mathematical the better. Of course this is not true in reality, but it's an ideology; it is a religion of science."

There is, of course, a role for theory. "You can't report just the facts and nothing but the facts; by themselves they are dull. The facts have to be woven into a framework so that there is understanding. That's usually accomplished by a theory. It may not be mathematical, but a qualitative network of relationships." Indeed, Hoffmann believes that the incorporation of theory into chemistry "is what made American science better than that of many other countries. The emphasis in chemistry on theory and theoretical understanding is very important, but not nearly as important as the syntheses and reactions of molecules.

"Although I think chemists need to like conducting experiments, that doesn't mean there is no role for people like me. It turns out that I am really an experimental chemist hiding as a theoretician. I think that is the key to my success. That is, I think I can empathize with what bothers the experimentalists. In another day I could have become an experimentalist."

Major issues in chemistry and science today

Professor Hoffmann has worked, and is presently working, at the forefront of several major areas of chemistry. He is presently quite intrigued by surface science. "For instance, there is the Fischer-Tropsch process, a pretty incredible thing in detail. Carbon monoxide and hydrogen gas come onto a metal catalyst, a surface of some sort, and off come long chain hydrocarbons and alcohols. The richness of all these things happening is intriguing, and we are on the verge of understanding. We now have structural information on surfaces that's reliable, and we are just beginning to get kinetic information. Surface science is at a crossing of chemistry, physics, and engineering. The field is in some danger of being spun off on its own, but I would like to keep it in chemistry.

"Bioinorganic chemistry is another such field. In my group we are trying to understand the mechanism of oxygen production in photosynthesis, the last steps. Very little is known. There is an enzyme in photosystem II that involves 3 to 4 manganese atoms, and they are at oxidation state 3 to 4. Somehow they take oxide or hydroxide to peroxide and eventually to molecular oxygen. That's all we know. Experimentally, not theoretically, I think bioinorganic chemistry is a very interesting field."

Hoffmann remarked that "there are going to be finer and finer ways of controlling the synthesis of molecules, the most essential activity of chemists. If I were to point to a single thing that chemists do: they make molecules. Chemistry is the science of molecules and their transformations. The transformations are the essential part. I think there are exciting possibilities for chemical intervention into biological systems with an ever finer degree of control. We need not be afraid of nature. We can mimic it, and even surpass its synthetic capabilities. And find a way to cooperate with it."

Scientific literacy and democracy

Roald Hoffmann is very concerned not only about science in general and chemistry in particular, but also about our society. One of his concerns is scientific literacy because "some degree of scientific literacy is absolutely necessary today for the population at large as part of a democratic system of government. People have to make intelligent decisions about all kinds of technological issues." He recently commented on this important issue in the *New York Times*. He wrote that "What concerns me about scientific, or humanistic, illiteracy is the barrier it poses to rational democratic governance. Democracy occasionally gives in to *technocracy*—a reliance on experts on matters such as genetic engineering, nuclear waste disposal, or the cost of medical care. That is fine, but the people must be able to vote intelligently on these issues. The less we know as a nation, the more we must rely on experts, and the more likely we are to be misled by demagogues. We must know more."

Our discussion of the importance of scientific literacy led to a conversation about the broader responsibilities of scientists. "Scientists have a great obligation to speak to the public," he says. "We have an obligation as educators to train the next generation of people. We should pay as much attention to those students who are *not* going to be chemists, and sometimes need to make compromises about what is to be taught and what is the nature of our courses. I think scientists have an obligation to speak to the public broadly, and here I think they have been negligent. Society is paying scientists money to do research, and can demand an accounting in plain language. That's why I put in a lot of time on that television show [The World of Chemistry]."

This interview was conducted by John Kotz of the State University of New York at Oneonta, and appears in *Chemistry and Chemical Reactivity*, second edition, Saunders College Publishing, 1991.

A teacher of chemistry

In the Nobel Yearbook, Professor Hoffmann wrote that the technical description of his work "does not communicate what I think is my major contribution. I am a teacher, and I am proud of it. At Cornell University I have taught primarily undergraduates. I have also taught chemistry courses to non-scientists and graduate courses in bonding theory and quantum mechanics. To the chemistry community at large, and to my fellow scientists, I have tried to teach "applied theoretical chemistry": a special blend of computations stimulated by experiment and coupled to the construction of general models—frameworks for understanding." Receiving the Nobel Prize represents Dr. Hoffmann's triumph of the spirit as researcher and teacher.

FLUORITE

ROALD HOFFMANN

I was asked about my hobbies,
"Collecting minerals," I said
and stopped to think.
"Minerals in their matrix
are what I like best."

Fluorite wears a variable habit.
Colorless when pure, it is vodka
in stone. More commonly
it brandishes shades of rose to blue,
an occasional yellow. A specimen I have
tumbles in inch-long cubes,
superimposed, interpenetrating,
etched on all their faces.
The cubes have a palpable darkness,
a grainy darkness, texture
blacker than black.
Solid yet fragile when held
up to light, the darkness
deposited in this ordered
atomic form a million years ago
allows some rays through.
But only on the thin edges,
in sinister violet.

Struck with a chisel
the cubes cleave and
octahedra emerge.
I have seen it done,
but my hands tremble.
I know why it cleaves so,
but why destroy what took
centuries to grow, then
rested aeons in a cool
fissure in the rock?

Eerie crystal.
Were a Martian photograph
enlarged to reveal such polyhedral
regularity, it would be deemed
a sign of intelligence at work.
But the only work here, and it is free,
is that of entropy.

Charles D. Winters

Roald Hoffmann, the John A. Newman professor of physical science at Cornell, won the Nobel Prize in chemistry in 1981. His first book of poems, The Metamict State, *was published last year by the University Presses of Florida.*

C H A P T E R

15

Elements, Compounds, and States of Matter

15.1 CHEMICAL AND PHYSICAL CHANGES

If you cut a wooden chair into small pieces you make splinters, whereas if you burn it you produce ash and smoke. It is easy to recognize that the two changes are different, because the ash and smoke feel and smell different from the chair; the cuttings do not. We account for the similarities between the whole chair and the splinters by saying that they both consist of wood, which is a material. It is the same material whether it is in the form of a log, a chair, splinters, or sawdust. A material, also called a **substance,** is any specific kind of matter, like wood, cotton, iron, blood, air, or sugar. *Chemistry is the science that deals with the properties and transformations of materials.* A transformation that produces a new substance is called a **chemical change** or **chemical reaction.** Thus, the burning of wood to

produce ash and smoke is a chemical change. Other examples are photosynthesis, digestion, fermentation, and rusting. A transformation that does not produce a new substance is called a **physical change.** Cutting a piece of wood or stretching a rubber band are physical changes. Freezing water to make ice and boiling water to make steam are also classified as physical changes, because water, ice, and steam are considered to be different states of the same substance, water.

Properties that depend on the size or shape of a sample of matter are called **extensive properties** (sometimes also called **accidental properties**). Examples are the length of a ruler, the mass of a rock, and the volume of water in a barrel. Properties that are independent of the size or shape of a sample are **intensive properties** (also called **specific properties**). Examples are the color of emerald, the hardness of diamond, and the density of lead. Handbooks of chemistry,

349

Lavoisier, The Father of Modern Chemistry

Antoine Laurent Lavoisier (1743–1794) is often called the father of modern chemistry. Among the factors that earned him this distinction are his authorship of an important textbook, **Elementary Treatise on Chemistry,** and his being the first to use systematic names for the elements and their compounds. However, his most important contribution was that his work established chemistry as a quantitative science—primarily by way of his law of conservation of matter, which states that matter is neither lost nor gained in a chemical reaction. As an example of what he did, consider the chemical decomposition of mercury oxide (HgO) into mercury and oxygen. The chemical reaction is

$$2\text{HgO} \longrightarrow 2\text{Hg} + \text{O}_2$$

Lavoisier's careful weight measurements showed that the total weight of all the chemicals remains constant during the process of a chemical change. This means that if one finds the weight of all the matter before a chemical reaction, one will find the same weight of matter after the reaction takes place.

This was definite proof that substances can be destroyed, as is the mercury oxide above, but matter cannot. You have the same weight of mercury and oxygen atoms after the reaction as you had before. Thus, Lavoisier showed that a chemical reaction is just a recombination of atoms; it is not a process by which matter is either created or destroyed.

It is of some interest to note that all did not go well in Lavoisier's life. As a minor example, one of his primary goals as a scientist was to discover a new element, and in this he was unsuccessful. However, of even more significance was the fact that he lost his head via the guillotine during the French Revolution. The reason for his execution was that he ran a private firm that contracted with the French government to collect taxes. Individuals engaged in his practice were referred to as "tax farmers," and he was brought to judgment because of this activity only 2 months before the end of the revolution. The executioners were not impressed with the fact that he used his salary (about 100,000 francs a year) from the enterprise to support his research.

physics, and engineering list the intensive properties of substances. Catalogs that describe automobiles, furniture, or computers include, among other information, the extensive properties of these objects.

15.2 THE ELEMENTS

In Chapter 12, an element was defined as a substance made up of only one kind of atom. Obviously, then, carbon and oxygen are different elements because they consist of different kinds of atoms. But what about isotopes, such as $^{12}_{6}\text{C}$ and $^{14}_{6}\text{C}$? The atoms have different mass numbers and, what's more, carbon-12 is stable whereas carbon-14 is radioactive. Nonetheless, they are classified as the same element—carbon—because they have the same *chemical* properties, which are dependent on the atomic number. Therefore, the precise definition of an **element** is *a substance all of whose atoms have the same atomic number.*

When you think about a particular element, you may think either about its atoms or about the substance made up of those atoms. For example, if you are asked to say something about carbon, you might answer that it has an atomic number of 6, that the six nuclear protons are electrically balanced by six electrons outside the nucleus, and that the most abundant isotope has six neutrons in the nucleus, adding up to a mass number of 12. If you said those things, you would have been thinking about the carbon atom. But you might also answer, equally correctly, that carbon is a soft, black solid commonly known as graphite. Then you would have been thinking about the *substance* called carbon, which is made up of carbon atoms. Someone else, however, might offer a different answer, saying that carbon is a hard, brilliant substance, not a soft, black one. That description, too, is correct, because it refers to diamond, which is also made up of carbon atoms.

These two correct descriptions raise the question: Can one element exist in the form of different substances? The answer is yes. The definition of an element does not specify how its atoms are arranged or bonded to each other. Different arrangements or different bondings make

Chemical changes. (a) Iron rusts, (b) plants grow, (c) rocket fuel burns, (d) an egg hardens, (e) muscle action consumes sugar. [(b) and (e) courtesy Greg Perry; (c) courtesy NASA]

(a)   (b)   (c)   (d)   (e)

different substances. Such different forms of the same element are called **allotropes.** Graphite and diamond are allotropic forms of the element carbon. Different types of chemical bonding will be described in the next chapter, but a brief explanation of the allotropy of carbon is the following: The carbon atoms in graphite are arranged in layers that can easily slip past each other, making graphite soft and slippery, whereas the carbon atoms in diamond are strongly bonded in three dimensions and are difficult to separate, making a diamond a hard substance. Did you know that diamond, being a form of carbon, can burn? It's not easy, but if you heat a diamond to a very high temperature in an atmosphere of pure oxygen, it will burn with a bright glow until it is all gone. You will never get your diamond back.

The physical states of matter, such as solid, liquid, and gas, will be discussed later in this chapter. For now, note that the different physical states of an element, such as liquid mercury and solid mercury (below $-39°C$), are not considered to be allotropes. Therefore, a rigorous definition is that allotropes are elements with the same atomic number and the same physical state but with different properties. Do not confuse allotropes with isotopes. Carbon-12 and carbon-14,

for example, are isotopes, not allotropes. Carbon-12 can exist as graphite or diamond; so could carbon-14. Various other elements also exist in allotropic forms. An interesting and important example is oxygen, whose allotropes will be described in the next section.

How many elements are there?

About 108 or 109 elements occur naturally on Earth or have been synthesized in nuclear reactors and in high-energy nuclear accelerators. Since such research is still active, the total number may increase at any time. The naturally occurring elements range from atomic numbers 1 (hydrogen) to 92 (uranium), but four elements within this set are not found in any significant quantities on Earth and are therefore considered to be "missing." Thus, there are $92 - 4$, or 88 naturally occurring elements. The four missing elements, all of which have been synthesized, are technetium (atomic number 43), promethium (61), francium (87), and astatine (85). All the elements beyond atomic number 92 are synthetic.

The abundances of the elements

The most abundant element in the Universe is hydrogen, which makes up about 75 percent by

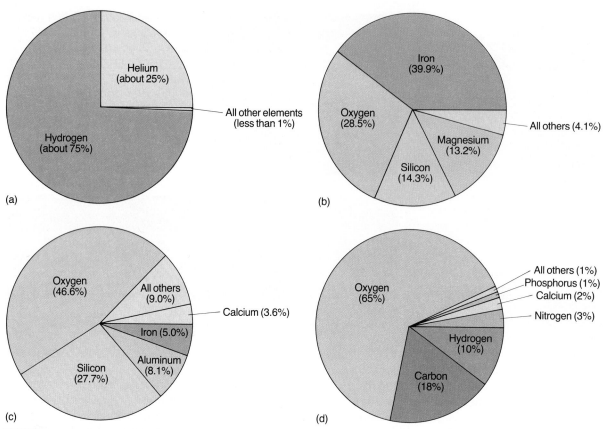

FIGURE 15.1 Abundance of the elements, in percent by mass; (a) in the Universe; (b) in the Earth; (c) in the Earth's crust; (d) in the human body.

mass of all matter known to exist. Helium is next, with about 25 percent. The remaining small proportion is made up of the rest of the elements. This universal abundance reflects, for the most part, the compositions of stars and intergalactic dust and gas, not the planets, since planetary material composes only a minute portion of the total Universe.

The planet Earth consists mostly of iron, oxygen, silicon, and magnesium, but these elements are by no means uniformly distributed. Most of the iron is in the Earth's core and therefore not available for use. The most abundant elements in the Earth's crust, which is the relatively paper-thin layer that serves as the environment for living organisms, are oxygen, silicon, aluminum, iron, and calcium. Ninety-nine percent of the mass of the human body, on the other hand, consists of oxygen, carbon, hydrogen, nitrogen, calcium, and phosphorus.

The elemental abundances are shown graphically in Figure 15.1.

Physical and chemical properties of the elements

About 75 percent of the elements are metals. Metals are, characteristically, good conductors of heat and electricity, and they reflect light well. There are about 19 or 20 nonmetallic elements and a few that are something in between. These few elements that are "on the fence" between metallic and nonmetallic are the ones that are important in the semiconductor industry for the manufacture of transistors and computer chips. Silicon and germanium are two such elements.

There are 11 elements that are gases at ordinary temperatures and pressures and two that are liquids (bromine and mercury). On a hot summer's day, two other metallic elements, cesium and gallium, would melt and join the company of liquids.

Among the metals, only a few are found in their elemental state in nature. These elements include the precious metals gold, silver, and platinum. Think of what this means. The fact that elemental gold can be found among the grains of sand in a stream bed means that it does not react chemically with the water or with the air that is dissolved in the water. Thus, gold is chemically unreactive. Most of the other metals are found only in chemically combined forms. As an example of this second category, consider calcium. Calcium is one of the more abundant elements on Earth, but you could search all the natural places on the surface of the planet and never find a speck of metallic calcium. All the calcium is chemically locked up as compounds in formations such as limestone mountains, coral reefs, animal bones, sea shells, and pearls. Metallic calcium can be prepared in the laboratory, but if it is exposed to air or water, it starts to react at once at a steady but not violent rate. Some other metals react more rapidly, others more slowly. A piece of metallic sodium or potassium, if it is dropped into water, reacts violently. The reacting materials get so hot that the evolved hydrogen catches fire. Iron, on the other hand, reacts very slowly (it rusts). Therefore, iron does not last indefinitely in natural environments, and iron is never found or mined as the natural element.

You will not be expected to memorize the names and symbols of all the elements, but it would be helpful to know at least the common ones shown in Table 15–1. (You need not memorize the atomic numbers, however.) The complete list appears inside the back cover of this book.

TABLE 15–1 Elements to Remember

Atomic Number	Name	Symbol
1	Hydrogen	H
2	Helium	He
6	Carbon	C
7	Nitrogen	N
8	Oxygen	O
9	Fluorine	F
10	Neon	Ne
11	Sodium	Na
12	Magnesium	Mg
13	Aluminum	Al
14	Silicon	Si
15	Phosphorus	P
16	Sulfur	S
17	Chlorine	Cl
18	Argon	Ar
19	Potassium	K
20	Calcium	Ca
26	Iron	Fe
29	Copper	Cu
47	Silver	Ag
78	Platinum	Pt
79	Gold	Au
80	Mercury	Hg
82	Lead	Pb
86	Radon	Rn
88	Radium	Ra
92	Uranium	U
93	Neptunium	Np
94	Plutonium	Pu

EXAMPLE 15.1

Which of the following pairs are isotopes, which are allotropes, and which are different elements? (a) "White tin," a soft metal, and "gray tin," a crumbly nonmetallic powder; (b) uranium-238 and uranium-235; (c) potassium $_{19}^{40}$K and argon $_{18}^{40}$Ar.

Solution (a) The two forms of tin have different chemical properties (one is metallic, the other is

not); they are allotropes. (b) The two uraniums have different mass numbers; they are isotopes. (c) Potassium and argon are different elements because they have different atomic numbers; the fact that they have the same mass number does not make them the same element; it just means that the sum of the nuclear protons and neutrons happens to be the same.

15.3 MOLECULES

> The atom belongs to science; the molecule belongs to chemistry.

The atoms of most elements tend to combine with other atoms. These combinations of atoms, called

FOCUS ON . . . Other Definitions of a Molecule

The word "molecule" has been used in various ways, sometimes rather vaguely. According to *Webster's Dictionary*, a molecule is "the smallest unit of matter capable of existing independently while retaining its chemical properties." That definition offers some help, but it does not cover everything. Sugar (sucrose), for example, is a white, sweet solid that melts at about 170°C. But one molecule of sugar is neither white (it is smaller than a wavelength of light, so it doesn't reflect anything), nor sweet (you can't taste one molecule), nor does it melt, because melting involves separation of molecules from each other.

The concept of a molecule is most satisfactory when it refers to gases, for in a gas each molecule does in fact have its own identity and is physically separated from other molecules except at the moment of a collision. A molecule of a gas may be a group of atoms chemically bonded to each other, such as CO_2, or it may consist of only a single atom, such as He or Ar.

In other usages, the word "molecule" simply refers to the group of atoms in the formula as it is commonly written, whether or not it is really a separate particle with its own identity. Thus, the formula NaCl for table salt, sodium chloride, is sometimes said to represent one molecule. However, as will be explained in the next chapter, salt does not consist of independent NaCl units with separate identities, so in this case the "molecule" is simply a convenient reference to the chemical formula.

Formula	Molecule
O_2	Oxygen—there are two oxygen atoms per molecule.
O_3	Ozone—there are three oxygen atoms per molecule. Oxygen and ozone are allotropes.
H_2O	Water—there are two H atoms and one O atom per molecule.
NH_3	Ammonia—there are one N atom and three H atoms per molecule.
C_3H_8	Propane (used as a fuel)—there are three C atoms and eight H atoms per molecule.
He	Helium—there is one atom per molecule.

Molecules change as the bonds between their atoms are made or broken; that's chemistry. Eat a piece of bread, and the starch molecules in it break down to sugars. The sugars react with the oxygen you breathe and eventually are converted to the carbon dioxide (CO_2) and water you exhale. These molecules in turn may be converted to parts of the molecules of a blade of grass, which are then converted to something else by the insect, cow, or deer that eats the grass. Thus, molecules are broken up, rearranged, and recombined during chemical reactions, but atoms (all but those of the radioactive isotopes) do not change. Any particular calcium atom in one of your teeth may once have been part of a dinosaur bone 100 million years ago.

EXAMPLE 15.2

Calculate the number of atoms of each element and the total number of atoms per molecule in (**a**) H_2SO_4 (sulfuric acid) and (**b**) $C_3H_5(NO_3)_3$ (nitroglycerin).

Solution Remember that the subscript follows the number of atoms it represents and includes everything inside the parentheses. The answers are: (**a**) two H, one S, and four O atoms—total 7 atoms; (**b**) three C, five H, three N, and nine O atoms—total 20 atoms.

molecules, are held together by electrical forces. In this context, the electrical forces are called chemical bonds. A **molecule** may therefore be defined as the smallest particle of a substance that is not chemically bonded to any other atoms.

A **chemical formula** represents the atomic composition of a molecule. Chemical formulas include the symbols of the atoms in the molecule and subscripts that designate the number of atoms of each element in the molecule. The subscript follows the symbol. Groups of symbols sometimes appear in parentheses; a subscript after the end parenthesis then refers to all the symbols inside. Here are some examples:

15.4 COMPOUNDS AND MIXTURES

If you buy several brands of granulated white table sugar, perhaps from different stores or at different times of the year, you will find them all to be basically alike. On a large scale level, for example, you might judge them all to be equally sweet. Additionally, they will all have the same composition corresponding to the molecular formula $C_{12}H_{22}O_{11}$. Table sugar or cane sugar, the chemical name for which is sucrose, is a **pure substance,** which is defined as a substance all of whose molecules or atoms are alike. (In practice, that means almost all. Granulated white sugar is over 99 percent sucrose.) A substance whose molecules consist of more than one kind of element is called a **compound.**

Quartz, which in the form of small particles is called sand, is the compound known as silicon dioxide, SiO_2. If you stirred some sand into your cane sugar, the result would be an impure substance, or a mixture. A **mixture** is made up of two or more substances not chemically bonded.

Now, if you spread out your mixture of sand and sugar on a smooth surface and used a magnifying glass and a very fine tweezers, and were very patient, you could separate them, because the sugar grains have different shapes than the grains of sand. Such a mixture, whose properties are not the same throughout, is said to be **heterogeneous.** A substance that is uniform, having the same properties throughout, is said to be **homogeneous.**

Now consider the question: Can an impure substance, a mixture, be homogeneous? The answer is yes, provided that the components are so intimately mixed that the material is uniform throughout. How intimate must such a mixture be? Ideally, the individual molecules of the components must be intimately mixed. Since no tweezer is fine enough to pick out individual molecules, any sample we can, in practice, extract from such a mixture will be the same as any other sample. There is nothing unusual about such intimate mixtures; in fact, they are very common. Such a uniform mixture of molecules is called a **solution.** For a liquid solution, the **solute** is the substance that is visualized as being dissolved in the other substance, which is called the **solvent.** When solids or gases are dissolved in a liquid, the liquid is considered to be the solvent. Thus, sugar (the solute) dissolves in water (the solvent). Each grain of sugar actually breaks up into about 10^{17} molecules. The sugar molecules diffuse uni-

A heterogeneous mixture of fruit.

formly among the water molecules, and there is absolutely no type of tweezer or other mechanical device that can pick them out.

Gases, too, dissolve in liquids; an example is ammonia water, which is a solution of ammonia gas in water. When liquids dissolve in each other, the choice between which is termed the solvent and which the solute doesn't matter much. Gasoline, for example, is a solution of many different compounds; therefore, like other solutions, it is a homogeneous but impure substance—a homogeneous mixture.

Finally, solids can be dissolved in each other. The most important examples are solid solutions of metals, called **alloys.** Some familiar ones are bronze (an alloy of copper and tin), brass (copper and zinc), and plumber's solder (lead and tin).

Outlined in Table 15–2 is a classification that shows the relationships among various terms that have been discussed so far in this chapter.

EXAMPLE 15.3

Which of the six substances whose formulas were listed earlier, O_2, O_3, H_2O, NH_3, C_3H_8, and He, are compounds?

Solution: H_2O, NH_3, and C_3H_8 each have more than one kind of element per molecule; they are compounds. O_2, O_3, and He are elements, not compounds.

15.5
CHEMICAL FORCES

Having introduced chemical change, elements, and molecules, and before going on to the physi-

TABLE 15–2   The Chemical Classification of Matter

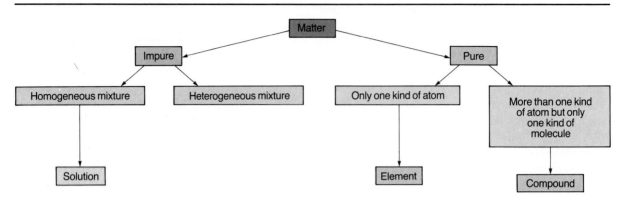

cal states of matter, it is appropriate to consider the nature of chemical forces.

Electromagnetism is the force that is involved in all chemical behavior. This includes the chemical bonds that hold the atoms together within a molecule, as well as the forces that hold molecules to each other to make liquids and solids. The electrostatic attraction or repulsion that operates between a pair of individual points of electrical charge can be calculated directly from Coulomb's law (Chapter 8), but chemical entities like atoms and molecules are not point charges. Atoms consist of positive nuclei and negative electrons, and a molecule may contain many atoms, so that the *net* forces involved are the resultants of many individual electrostatic attractions and repulsions. To make matters even more complicated, an individual entity such as an atom or a molecule is not a static object. The electronic structure of an atom can take on various energy states, and even its symmetry can be distorted by nearby charged bodies. Molecules change their shape, and therefore their distribution of electric charges, when they vibrate by stretching, bending, or twisting. These distortions are caused by collisions with neighboring molecules or by absorption of energy from photons. All of these changes can affect the electrostatic forces between atoms and molecules.

How can two electrically neutral molecules, such as HCl and HCl, attract each other electrically? The answer is that although the HCl mole-

cule as a whole is neutral, the electrical charges within the molecule are not uniformly arranged. The negative charges (the electrons) are displaced toward the chlorine atom more than the positive charges (the nuclei) are. The result is that each HCl molecule has a positive side (the H atom) and a negative side (the Cl atom), and the oppositely charged sides attract each other: (H—Cl)(H—Cl), and so forth.

Some molecules, however, do not have separated positive (+) and negative (−) sides, yet the substance can be liquefied and solidified. An example is the simple molecule H_2. Here the explanation is more subtle. The electrical symmetry of the H—H molecule is its *average* condition. Since the molecule has energy of vibration, the electrons and nuclei move away from their average pattern from instant to instant. Thus, a separation of positive and negative electrical charges does occur momentarily, and during such moments two molecules may attract each other. Of course, such attractions are quite weak, which accounts for the fact that hydrogen must be cooled to −253°C to liquefy it and to −259°C to freeze it.

The result of all these factors is that chemical forces are generally too complex to be predicted by mathematical calculations, and for that reason chemistry is, to a great extent, an empirical science. The important point for the student to remember is that chemical bonds of all sorts, as well as the forces between molecules that account for the existence of liquids and solids, are all the *net* resultants of many electrostatic attractions and repulsions.

15.6
THE STATES OF MATTER: CLASSIFICATION OF PHYSICAL STATES

Atoms bond to each other to form molecules. Molecules, too, attract each other, but the forces between molecules (intermolecular forces) are much weaker than the forces between atoms within a molecule. These relationships are summarized in Table 15–3.

The physical states of matter are generally grouped into three main categories: gases, liquids, and solids. One state can often be changed to another by the transfer of energy. For example, water can be frozen to ice by removing heat from the water; ice can be melted or water can be evaporated by adding heat to the ice or the water.

Gases are substances that have no definite shape. Instead, they disperse rapidly in space and occupy any volume available to them. All gases are transparent and most, such as hydrogen and nitrogen, are colorless. Some are colored; examples are chlorine (pale green) and nitrogen dioxide (reddish brown). The attractions between the molecules of a gas are not strong enough to bind them to each other at ordinary temperatures.

Liquids do not disperse in space, but they do change shape easily. A liquid, therefore, occupies a definite volume. Under the influence of the Earth's gravity, a liquid fills the lower part of its container. The molecules of a liquid attract each other strongly enough to prevent their dispersal in space but not enough to prevent the liquid from flowing.

Solids are rigid and maintain their own shape. Their intermolecular forces are strong enough to maintain the rigidity of the substance.

There are two other terms that refer to specific states of matter.

Liquid crystals are substances that have some properties common to solids and some that are common to ordinary liquids. Liquid crystals can maintain a definite shape but can also be made to flow with only very small inputs of mechanical or electrical energy.

Recall from our discussion of nuclear physics that a **plasma** is matter at such a high temperature that all chemical bonds are broken and even the electrons are stripped away from their atomic nuclei. A plasma is, therefore, a state of matter composed of a mixture of electrons and positive ions or nuclei. The Sun and other active stars are largely plasmas of hydrogen and helium, and the plasma state is the most common of all when the entire Universe is considered.

15.7
THE PROPERTIES OF GASES

Gases are common to our daily experience. Air is a gas and so is the "natural gas" (mostly methane, CH_4) that is used for domestic heating and cooking. The aroma of a flower and the stench of a rotten egg are gases.

If you grind a solid into a very, very fine powder, the individual particles may be small enough to be carried aloft on air currents. The same may be true of tiny liquid droplets. However, such particles or droplets are still much, much larger than gas molecules; therefore, they are not gases. Smoke is such a mixture of airborne particles; it is not a gas. Fogs and mists consist of liquid droplets and they, too, are not gases. One obvious difference is that smokes and mists are not transparent; all gases are.

TABLE 15–3 Interatomic and Intermolecular Forces

	Forces Between Atoms Within a Molecule	Forces Between Molecules
Name	Chemical bonds	Intermolecular forces
Relative strength	Stronger	Weaker
Effect	Formation of molecules	Formation of liquids and solids

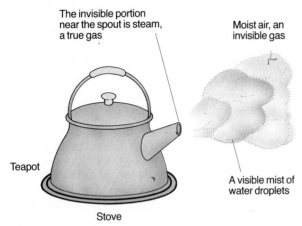

The invisible portion near the spout is steam, a true gas

Teapot

Stove

Moist air, an invisible gas

A visible mist of water droplets

FIGURE 15.2   Steam is an invisible gas.

Steam is the gaseous state of water, but the word is often misused. What you see coming out of a teapot of boiling water is not the steam but the mist of water droplets that forms after the steam cools a little; the invisible matter between the spout and the mist is the steam, as shown in Figure 15.2. The white smoke that is discharged from many industrial chimneys is mostly a water mist (with or without other waste products). A mist droplet, like a grain of sugar, contains very many molecules.

Gases show little resistance to flow compared with liquids or solids; they spread out rapidly in space and flow through very small openings, such as the tiniest pinprick in a balloon or an automobile tire. All gases can be mixed with each other in any proportions, assuming that they do not react chemically with each other. Once mixed, gases never separate from each other spontaneously, nor can they be separated by filters. For example, carbon monoxide is a gas found in all cigarette smoke; no filter can remove it.

If you squeeze an inflated balloon, its volume decreases, which means that the gas inside has been compressed. It is not easy to do the same with a rock or a piece of wood, nor can it be done with liquids. If you had a strong, sealed leather pouch completely filled with water, you would not be able to squeeze it down to a noticeably smaller volume.

A gas occupies the entire space of its container. For example, if the room you are in right now has a volume of 30 m^3, it contains 30 m^3 of air. If all that air were cooled down to about $-200°C$, it would condense to a pale blue liquid that would occupy about 0.03 m^3 (30 L), or about a tenth of 1 percent of its original volume. A reasonable conclusion from these facts is that much of a gas is empty space. Now think of any particular sample of gas, such as the hydrogen in a balloon, the air in a bicycle tire, or the helium-oxygen mixture in a diver's breathing tank. Each of these samples may be characterized by four physical properties that are related to each other. If any one of these properties is changed, there *must* be a change in at least one of the other three. These four properties are:

1. Volume of the gas
2. Pressure of the gas
3. Temperature of the gas
4. Number of molecules of the gas

These are all physical properties because a change in any of them does not produce a new chemical substance.

Now consider what may happen if the air in the bicycle tire is heated. The pressure may go up, the tire may swell so that the volume of the gas expands, some air may leak out so that the number of molecules of air in the tire decreases, or there may be some combination of these effects. No matter what, something must happen to at least one of the other three properties. The relationships among these variable properties of gases were not discovered all at once. Instead, different scientists studied these variables two at a time and came up with a set of simple laws that led to great insights into the nature of gases and the molecules that compose them.

15.8 BOYLE'S LAW

Robert Boyle (1627–1691) discovered the first of the laws relating two of the four variable properties of gases—pressure and volume. His method was to trap a volume of air in a glass tube and then to compress the air by forcing mercury into the tube, as shown in Figure 15.3. During the experiments, he measured the changes in the pressure and volume of the air. After he got all

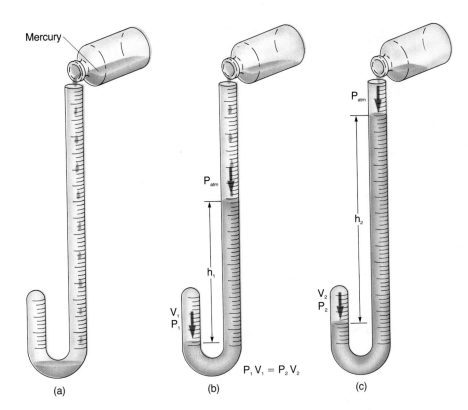

FIGURE 15.3 Boyle's experiment. The liquid is mercury. The pressure is measured in terms of the heights of the columns, plus the atmospheric pressure exerted on the open end of the tube. (a) Mercury is poured in until (b) the air in the left column is trapped at volume V_1 and pressure P_1. (c) More mercury is poured in. Now $P_2V_2 = P_1V_1$.

his results, he did something with them that set his name down forever in the history of science: He multiplied the pressure by the volume obtained in each experiment and noted that the product was always the same (within experimental error). This observation, that *the pressure times the volume of a fixed mass of gas at a given temperature is a constant*, became known as Boyle's law.

Another way of expressing the same law is:

The volume of a sample of gas (at a fixed temperature) is inversely proportional to the pressure.

(See Appendix D for a discussion of proportionality.) This means, for example, that if you double the pressure on a gas, its volume is reduced by half. Conversely, if you squeeze a gas enough to cut its volume in half, its pressure doubles. This relationship can be expressed in equation form as

$$P_1V_1 = P_2V_2 \qquad \text{(fixed mass and temperature)} \qquad (15.1)$$

where P_1 = initial pressure of the gas, P_2 = final

pressure of the gas, V_1 = initial volume of the gas, and V_2 = final volume of the gas.

The two values of pressure must be expressed in the same units, and the same is true for the two values of volume. For example, the volume units may both be expressed in liters or both in milliliters, but not one in liters and the other in milliliters. (Why not?)

EXAMPLE 15.4

A sample of neon occupies 30 L at 1.5 atm pressure. What volume will this sample occupy at 4.5 atm? Assume no change in temperature.

Solution Initial volume, V_1 = 30 L; initial pressure, P_1 = 1.5 atm; final pressure, P_2 = 4.5 atm; final volume, V_2 = ?

Solving the Boyle's law equation (Eq. 15.1) for V_2 gives

$$V_2 = V_1 \frac{P_1}{P_2} = 30 \text{ L} \frac{1.5 \text{ atm}}{4.5 \text{ atm}} = 10 \text{ L}$$

FOCUS ON . . . The Pressure of Gases

One of the properties of gases and liquids that chemists must use frequently is that of pressure. We have defined pressure in Chapter 2 as the force per unit area, P = F/A, and the units which we used were N/m^2. However, for historical reasons and for reasons of convenience, chemists often resort to other systems of units. Let us examine some of the background for these various measurement techniques.

Almost everyone is aware that the pressure increases with depth as one goes beneath the surface of a pool of water. In a lake, you can recognize this by a popping in your ears when you descend, but if you dive far beneath the surface of water, the increase of pressure can do far more than create a mildly unpleasant sensation on your eardrums. For example, in order to reach great depths under the sea, carefully designed diving bells capable of surviving tremendous pressures must be constructed. Fish living deep under water are unaware of these

great pressures because they were conceived and have lived under such conditions all their lives. Like the fish, land dwellers also live on the bottom of a great ocean—an ocean of air, and as a result, we also unknowingly exist under high pressures. As we move about here on Earth, a force is exerted on us caused by the weight of a column of air that extends to the "top" of the atmosphere. In SI units, standard atmospheric pressure at sea level is $1.013 \times 10^5 \ N/m^2$ (14.7 lb/in^2). Thus, like Atlas of mythology, who supported the world on his shoulders, each square meter of surface area on the Earth has a force of 1.013×10^5 N exerted on it by the atmosphere. An alternative way of discussing pressures of gases or liquids is in terms of a unit called an atmosphere (atm), where 1 atm = $1.013 \times 10^5 N/m^2$. Thus, a gas under 2 atm of pressure has a pressure exerted on it that is twice as great as that which air would exert on it at sea level.

15.9 CHARLES' LAW

The observation that a gas expands when it is heated is not new. It may surprise you that even rocket engines, which depend for their action on the expansion of heated gases, were known in the later ages of ancient Greece. These engines did not launch satellites, of course; they were merely pinwheel toys containing water. When the water was heated, steam escaped from opposing nozzles, making the toys spin in accordance with Newton's third law of motion, as shown in Figure 15.4.

A more precise study of the expansion of a heated gas was not done until about 1787, when Jacques Charles carried out his experiments. An easy way to show this relationship is sketched in Figure 15.5. A droplet of mercury is suspended in a thin glass tube that is closed at the bottom and open to the atmosphere at the top. Because the droplet is so small, it does not break apart and fall to the bottom of the tube. (A droplet of water in a drinking straw would behave in the same way.) The air below the mercury is therefore

trapped; it cannot get out through the glass or go through the mercury. The tube is placed in a beaker of boiling water at 100°C, as shown in Figure 15.5a, and the volume of the trapped gas below the mercury is noted. When the temperature is allowed to drop, the volume of the trapped gas decreases, and the mercury plug goes down. It is observed that the decrease in the volume of the gas as the temperature drops from 100°C to 50°C is the same as the decrease when the temperature drops from 50°C to 0°C, as shown in

FIGURE 15.4 A simple engine first developed by Hero of Alexandria.

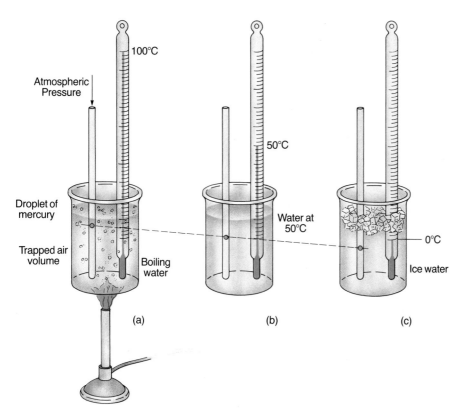

100°C

Atmospheric
Pressure

Droplet of
mercury

Trapped air
volume

Boiling
water

(a)

50°C

Water at
50°C

(b)

0°C

Ice water

(c)

FIGURE 15.5 Charles' Law. (a) A droplet of mercury traps a sample of air below it at 100°C and atmospheric pressure. (b) The same air sample at the same pressure, at 50°C. (c) Same air sample and pressure at 0°C.

Figure 15.5b and c. In fact, for every equal drop in temperature, a gas shrinks by the *same constant loss of volume.* This observation implies that if a gas were cooled sufficiently, it would shrink to nothing. Such a miracle does not occur because the molecules themselves occupy some volume, and besides, cold gases liquefy first. But if the constant shrinking were to continue without liquefaction, and if the molecules were merely points in space, zero volume would be reached at −273°C (Fig. 15.6). This temperature is therefore the lowest limit that could possibly be reached by cooling anything, because no gas, even theoretically, could have less than zero volume. This temperature is −273.15°C, but we will use the conveniently approximate value of −273°C. As we saw in Chapter 5, this lowest temperature is the basis of the *Kelvin scale of absolute temperature.* Kelvin temperature starts at 0 K, which is called *absolute zero.*

$$-273°C = 0 \text{ K}$$

and

$$0°C = 273 \text{ K}$$

Therefore, Kelvin temperature is always 273° higher than Celsius temperature, and it is always positive.

$$T_K = T_C + 273 \qquad (15.2)$$

The relationship between the volume of a gas and its Kelvin (absolute) temperature is known as **Charles' law,** which states that

the volume of a gas (at constant pressure) is directly proportional to its absolute temperature.

This law can be expressed in equation form as

$$\frac{V_1}{T_1} = \frac{V_2}{T_2} \quad \text{(fixed mass and pressure)} \qquad (15.3)$$

where T_1 = initial temperature of the gas in kelvins; T_2 = final temperature of the gas in kelvins, V_1 = initial volume of the gas, and V_2 = final volume of the gas. If the temperatures are given in °C, they must be converted to kelvins, using Eq. 15.2.

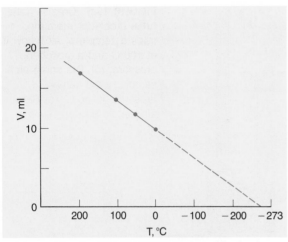

FIGURE 15.6 A Charles' Law plot (graph) of volume versus temperature of a gas shows that gases expand when heated and contract when cooled, at constant pressure. Note the dashed line indicates that the temperature approaches −273°C as the volume approaches zero.

EXAMPLE 15.5

A sample of hydrogen occupies 20 mL at 300 K. What volume will the hydrogen occupy at 750 K, assuming that the pressure remains constant?

Solution Solving the Charles' law equation (Eq. 15.3) for V_2, the final volume, gives

$$V_2 = V_1 \frac{T_2}{T_1} = 20 \text{ mL } \frac{750 \text{ K}}{300 \text{ K}} = 50 \text{ mL}$$

15.10 AVOGADRO'S LAW

If more gas is introduced into a given sample of gas and if the temperature and pressure are kept constant, the volume of the gas increases. More gas means more molecules; therefore, the volume of a gas depends on the number of molecules it contains. The quantitative expression of this relationship is known as **Avogadro's law** (after Amedeo Avogadro, 1776–1856):

Equal volumes of all gases (at the same temperature and pressure) contain the same number of molecules.

As an exercise, which of the gas laws most closely explains the reason why a hot air balloon can rise into the air?

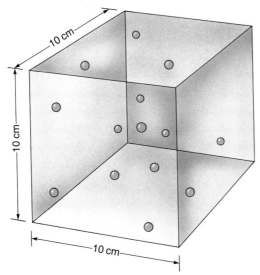

FIGURE 15.7 The kinetic theory model of a gas. A schematic representation of molecules in the space of 1 L. The volume of the gas is 1 L, but the volume of the molecules is much smaller. Therefore, most of the space in the container is empty.

EXAMPLE 15.6

Imagine that you are selling balloons to children at a county fair. You offer balloons of two sizes: 1 L or 2 L in volume. Also, you may fill the balloons with air or with the much lighter gas, helium. All the balloons are filled to a pressure of 1.1 atm. If the 1 L balloon of helium contains X molecules, how many molecules are in the 1 L balloon of air; the 2 L balloon of helium; the 2 L balloon of air?

Solution Avogadro's law applies to all gases, so the answers must be the same for the air and for the helium. Both 1 L balloons, therefore, contain X molecules and the 2 L balloons contain 2X molecules.

15.11
THE KINETIC THEORY OF GASES

Recall the properties of gases described in the previous sections: Gases are transparent, they spread out readily in space, they can flow through very small openings, and they are compressed much more easily than liquids and solids. In 1738, Daniel Bernoulli proposed a concept to account for these properties. What could Bernoulli have known in 1738? Boyle had done his experiments on pressure and volume in 1662, but modern ideas about chemical change were not to come until the work of Lavoisier in 1780 and Dalton's atomic theory in 1803. Charles' law and Avogadro's law were also not to be formulated until early in the Nineteenth Century. The idea that matter consists of small fundamental particles was fairly well appreciated, but the distinction between molecules and atoms was not so clear. Bernoulli's theory of gas behavior was based on his concept of the *motion* of molecules. This idea was quite novel and was not accepted until it was revived about a century later and gradually won out over rival theories.

Today, Bernoulli's concept is called the **kinetic theory of gases.** This theory accounts for the relationships among volume, pressure, temperature, and number of molecules of a gas. The theory makes the following assumptions:

1. Gases consist of small particles called molecules.
2. The molecules have mass and are in constant motion.
3. The volume of the molecules themselves is insignificant compared with the total space they occupy. Therefore, a gas is mostly empty space, as shown in Figure 15.7.
4. The molecules do not attract each other. Each molecule moves independently, in a straight line, until it collides with another molecule or with the walls of the container. Furthermore, the collisions are perfectly elastic, which means that the molecules do not lose any kinetic energy.

These assumptions refer to an ideal situation. Actually, there are intermolecular forces even in gases, but their effect is small compared with the effects in liquids and solids.

Now let us see how the theory accounts for the properties of gases.

Boyle's law

The pressure of a gas results from the collisions of the molecules with the walls of the container. If the volume is reduced by half, the molecules are twice as crowded and hit the walls twice as often. Twice as many hits on the wall means that the pressure doubles.

Charles' law

The temperature of a gas is a manifestation of the average energy of motion (average kinetic energy) of its molecules. As the temperature increases, the molecules speed up. As a result, the molecules hit the walls more frequently and each collision on average is more energetic. If the opposing pressure that the walls can exert remains constant (for example, by a droplet of mercury in a capillary tube or by a movable piston in a cylinder at atmospheric pressure), the walls of the "container" are pushed out. This means that the volume of the gas increases or, in other words, the gas expands. Conversely, as the temperature decreases, the gas molecules slow down, and their impacts on the walls become less frequent and less energetic. If the walls exert a constant pressure, they close in on the gas and the volume decreases.

Avogadro's law

Note that none of the statements of the kinetic theory says anything about the *kinds* of molecules in the gas. Think of two balloons of equal volume, one containing hydrogen, the other nitrogen, and both at the same temperature and pressure. Avogadro's law teaches us that the number of hydrogen molecules and the number of nitrogen molecules must be equal. If their temperatures are the same, their average kinetic energies must be equal. How can that be? A molecule of hydrogen or of nitrogen has kinetic energy because it has mass and speed. Recall that kinetic energy = $\frac{1}{2}mv^2$. Therefore, $\frac{1}{2}mv^2$ for hydrogen must be the same as $\frac{1}{2}mv^2$ for nitrogen at the same temperature. But a molecule of hydrogen has less mass than a molecule of nitrogen. The only way for its mass to be less while its kinetic energy is equal to that of the nitrogen molecules is for its

speed to be greater. Thus, lighter gases move more rapidly than heavier gases at the same temperature.

EXAMPLE 15.7

Imagine that you drive into a service station that offers a choice of gases to fill your tires. Being in an adventurous state of mind, you say, "Put helium in one, air in another, neon in a third, and hydrogen in the fourth, all at the same pressure." The attendant does as you request. Assume that all four tires have the same volume and are at the same temperature. Hydrogen is the lightest (least dense) gas. Answer true or false for each of the following statements and defend your answers: **(a)** All four tires are at the same pressure. **(b)** All four tires contain the same number of molecules. **(c)** The molecules in all four tires have the same average speeds. **(d)** If each of the tires had the same tiny leak, they would lose pressure at the same rate.

Solution **(a)** True. They were filled to the same pressure; the gauge read the same for each tire, so all the pressures are the same.

(b) True. The volume, temperature, and pressure in each tire are the same; therefore, the number of molecules in each is the same. This is the relationship expressed by Avogadro's law.

(c) False. Since all the gases are at the same temperature, the average kinetic energy of their molecules must be the same. To make up for their lower mass, the lighter molecules must have greater speeds.

(d) False. The speedier molecules reach the walls more frequently and find the hole more often and so escape faster. The tire with hydrogen therefore loses pressure most rapidly.

15.12 LIQUIDS

Electrical force, which is responsible for the attractions between molecules, does not change with temperature. What does change with the temperature of a substance is the kinetic energy of its molecules. As the temperature drops, the molecules slow down and thus lose kinetic en-

FIGURE 15.8 Liquids.

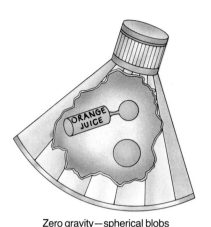

Zero gravity—spherical blobs of orange juice

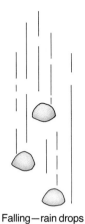

Falling—rain drops

Fallen—in the shape of the container

ergy. When the temperature is low enough, the intermolecular forces can begin to hold them together, either as a liquid or as a solid. We will consider liquids first.

Liquids like water, gasoline, and olive oil are familiar to us in our daily lives. Nevertheless, the liquid state is, in some ways, a rather strange condition of matter. Think of it this way: If the atoms or molecules have enough kinetic energy to move about independently in the space available to them, the substance is a gas. If the atoms or molecules are strongly attracted or bonded to each other, and do not have enough kinetic energy to break loose, they maintain definite positions relative to each other in space and the substance is a solid. But liquids? The atoms or molecules must be attracted to each other strongly enough not to break loose, but they are not held together strongly enough for the substance to maintain a rigid shape. Therefore, if an astronaut in a space capsule, under zero gravity, squeezes some orange juice out of a bottle, the juice does not fly apart but remains suspended as a blob. Does the blob have a characteristic shape? Yes, it is spherical; but when the capsule returns to Earth, and the round blob of juice is put into a glass or bottle, it takes the shape of its container (Fig. 15.8). Thus, the intermolecular forces in a liquid are strong enough to hold the substance together but not strong enough to resist deformation of the substance by gravity.

Some liquids, like gasoline, flow very readily. Others, like honey, flow slowly. The resistance to flow of a liquid is called its **viscosity.** The viscosity of a liquid is related to its intermolecular forces, to the actual shapes of the molecules, and to the mechanical interference they exert when they flow past each other.

Thus, we may conclude that the properties of liquids result mainly from two factors: (1) the kinetic energy of the molecules, which enables them to slide past each other, and (2) the forces of attraction between them, which hold them together. Since the cohesive forces are too weak to hold the molecules in rigid positions, the molecular arrangements in liquids shift from moment to moment. As a result, the distribution of molecules is random, or disorderly.

15.13
SOLIDS

A **crystalline solid** is a substance whose atoms or molecules are arranged in orderly or repeating patterns. These orderly arrangements are often suggested by the shapes of large particles of the substances, which are called **crystals.** The atomic or molecular patterns in such solids are based on shapes such as those shown in Figure 15.9. Imagine atoms located at the corners, sides, edges, or interiors of these imaginary shapes, and then think of a solid structure in which one such unit repeats itself many, many times in all directions. Such a structure is called a **crystal lattice,** and crystalline solids are made up of such lattice structures. The different basic repeating shapes,

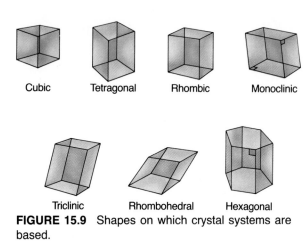

Cubic   Tetragonal   Rhombic   Monoclinic

Triclinic   Rhombohedral   Hexagonal

FIGURE 15.9 Shapes on which crystal systems are based.

and the different locations of atoms within these shapes, give a variety of crystal lattices. Three that are based on the cubic shape are shown in Figure 15.10. Furthermore, a given lattice can grow in various directions and reach different sizes, so that there are limitless possible shapes of crystals of even the same substance. This circumstance is the basis for the oft-repeated statement that of all the snowflakes that have fallen in the Earth's history, no two were ever exactly alike. One such snowflake crystal is shown in Figure 15.11; note its characteristic hexagonal pattern.

A substance made of large crystals is visually more interesting than a powder consisting of ag-

gregates of tiny crystals whose shapes are not visible to the naked eye.

The density of a crystal depends on the masses of its individual atoms and on how closely the atoms are packed together (not on how strong the bonds are nor on the sizes of the atoms). The closest packing of atoms occurs in metals, which is why the densest solids we know are metals. The densest one of all is osmium (22.5 g/cm³); a sphere of osmium the size of a grapefruit, about 15 cm in diameter, would surprise anyone who tried to pick it up, for it would have a mass of about 40 kg and so would weigh close to 90 lb. Osmium atoms are tightly packed in a lattice of the type shown in Figure 15.10c. Diamond, on the other hand, is an example of a solid whose atoms are very strongly bonded to each other but are not so tightly packed. The carbon atoms in a diamond are arranged like stacks of cubes with every other corner empty, as shown in Figure 15.12. The strong bonds are reflected in the fact that diamond remains solid even above 3500°C, which is about 800°C higher than the melting point of osmium. A research team in the Department of Geology at Cornell University reported in 1984 that a diamond surface was melted at a very high pressure with the use of a laser beam. The temperature was around 4000°C. However, the open packing of the carbon atoms gives dia-

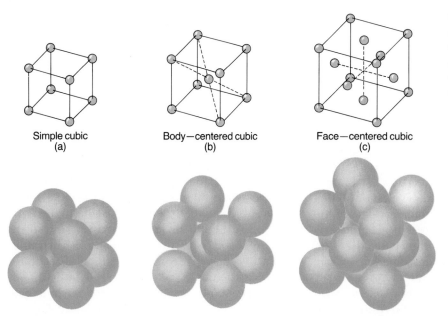

Simple cubic
(a)

Body—centered cubic
(b)

Face—centered cubic
(c)

FIGURE 15.10 Three types of crystal lattices based on the cubic system. (a) Simple cubic. (b) Body-centered cubic. (c) Face-centered cubic.

FIGURE 15.11   Snowflake crystal.

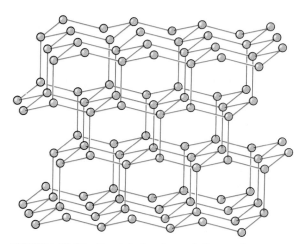

FIGURE 15.12   Crystal structure of diamond.

mond a low enough density (3.5 g/mL) for it to float easily on mercury (density 13.4 g/mL). Combinations of such factors engender an extensive variety of properties among solids. Lead bends, quartz is brittle, talc is soft, and steel is strong. All of these attributes are related in some way to the forces that hold the atoms or molecules of these solids together.

Recall that at low temperatures the molecules of a substance may slow down so much that the attractive forces among them hold them together rigidly. However, there is no guarantee that crystallization will occur. It is, to some extent, a matter of chance. Instead, the molecules may become rigid in their randomness, like frigid tar or some sugar candies. The general name for such a material is a **glass,** or **glassy solid,** which is defined as a solid in which the component atoms or molecules are randomly arranged. This category includes the common material whose specific name is glass (as in windows), which consists of uncrystallized silicate compounds. Glasses do flow very slowly; as a result, the bottoms of windowpanes found in old cathedrals are somewhat thicker than the tops. Crystals do not flow like this.

15.14
LIQUID CRYSTALS

Recall from our discussion of the physical states of matter that liquid crystals can maintain a definite shape but can also be made to flow. How can a substance be, as it were, both a solid and a liq-

uid? Must it not be one or the other? The answer is this: The atoms or molecules in a crystalline solid are arranged in orderly patterns; the regularity extends throughout the crystal in three dimensions. The molecules in a gas are random; there is no regularity at all. Between these extremes there are various degrees of partial order—less than in a true crystal but more than in a gas. The liquid state represents one such condition, but not the only one. It is possible for the molecules of a substance to be more orderly than in a liquid but less so than in a true crystalline solid. (Fig. 15.13). The forces that hold the molecules together in such a state are just barely strong enough to enable the substance to maintain a definite shape, so it is reasonable to call it a crystal. But only slight inputs of mechanical or electrical energy can disrupt these weak intermolecular forces and make the substance flow, so it is also reasonable to call it a liquid. Hence the name, liquid crystal.

One practical application of liquid crystals takes advantage of the fact that light reflected from a very thin layer of liquid, such as a slick of oil of only a few molecular layers floating on water, is colored. (Did you ever notice the rainbow effect on the surface of an oily puddle near a garage?) The reason for this effect is that some rays of light are reflected from the upper surface of the oil and some from the surface of the water on which the oil floats. When these rays recombine, they produce an interference pattern that can be either constructive or destructive, depend-

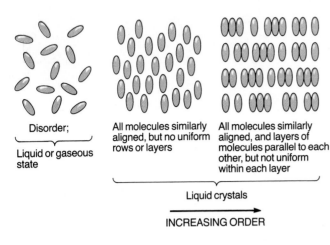

Disorder;

Liquid or gaseous state

All molecules similarly aligned, but no uniform rows or layers

All molecules similarly aligned, and layers of molecules parallel to each other, but not uniform within each layer

Molecules, rows, and layers uniform

Crystalline solid

Liquid crystals

INCREASING ORDER

FIGURE 15.13 Order and disorder in molecular arrangements.

ing on the thickness of the oil film. The particular wavelengths that undergo constructive interference determine the color of the reflected light. Because the thickness of the oil layer determines which wavelengths reinforce each other, the color depends on the thickness. In some liquid crystals, the distance between the layers is very sensitive to the temperature. This property can be used in "color-mapping" a portion of the human body. When a section of skin is coated with an appropriate liquid crystal, the warm areas over blood vessels and certain organs or diseased regions are different in color from the neighboring cooler areas, and their precise locations are thus identified.

Another application is the familiar displays on pocket calculators and wristwatches, as shown in Figure 15.14. Here the liquid crystals exist as a very thin layer between thin sheets of transparent glass or plastic that have electrical conductors embedded in them. These electrodes are arranged in patterns that can represent letters or numerals. When a voltage is applied in a particu-

FIGURE 15.14 A liquid crystal display on a laboratory instrument.
(Courtesy Bill Schulz)

lar pattern, the adjacent liquid crystals are energized and converted to a different molecular orientation—one that combines with the thin sheets around them to become opaque. These opaque areas are seen as the desired letters or numerals.

SUMMARY

A transformation that produces a new substance with different specific properties is a **chemical change;** one that does not produce a new substance is a **physical change.** Different forms of the same element are called **allotropes.** There are about 109 known elements, 88 of

which occur naturally on Earth; others have been synthesized. Most elements are metals. A **molecule** is the smallest particle of a substance that is not chemically bonded to any other atoms. The atomic composition of a substance is represented by a chemical formula. A pure substance whose molecules consist of more than one kind of element is a **compound.** A uniform, or homogeneous, mixture of different kinds of molecules is a **solution,** in which the dissolved material is the **solute** and the medium is the **solvent.**

States of matter

Gases disperse in space.

Liquids do not disperse but take the shape of their containers.

Solids maintain a rigid shape.

Liquid crystals can maintain a definite shape and can also be made to flow.

Plasmas are gases in which electrons are stripped from their nuclei.

The four properties of a sample of gas that are related to each other are pressure, volume, tempera-ture, and number of molecules. The laws that relate these properties to each other are:

Boyle's law: The volume of a given amount of gas at a constant temperature is inversely proportional to pressure.

Charles' law: The volume of a given amount of gas at a constant pressure is directly proportional to absolute (Kelvin) temperature.

Avogadro's law: Equal volumes of all gases at the same temperature and pressure contain the same number of molecules.

The **kinetic theory of gases** interprets these laws in terms of the masses and speeds of the molecules and their collisions with the walls of the container and with each other.

EQUATIONS TO KNOW

$P_1V_1 = P_2V_2$ (Boyle's law)

$\dfrac{V_1}{T_1} = \dfrac{V_2}{T_2}$ (Charles' law)

KEY WORDS

Substance	Element	Solvent	Liquid crystal
Heterogeneous	Allotropes	Solute	Plasma
Homogeneous	Molecule	Gas	Boyle's law
Chemical change	Chemical formula	Liquid	Charles' law
Physical change	Compound	Solid	Avogadro's law
Extensive property	Mixture	Crystal	Kinetic theory of gases
Intensive property	Solution	Crystal lattice	Viscosity

QUESTIONS

CHEMICAL AND PHYSICAL CHANGES

1. In each of the following sequences, which transformations are chemical changes and which are physical changes? (a) Oxygen gas is liquefied; some of this liquid oxygen is used as part of the fuel to launch a spaceship; some of it is allowed to vaporize into the passenger space; the gaseous oxygen is then used in respiration by the passengers and becomes part of CO_2 and H_2O molecules. (b) Beef fat is warmed in a pan until it melts; as it is heated further, it begins to darken and give off pungent gases; upon still further heating, it bursts into flame and burns.

THE ELEMENTS

2. White phosphorus is an extremely toxic substance that ignites spontaneously in air. Red phosphorus does not ignite spontaneously and is much less toxic than the white variety. Both substances con-

sist entirely of $^{31}_{15}$P atoms. What can account for the difference in their properties?

3. There is something wrong with each of the following definitions of an element. Explain what is wrong in each case. Supply a better definition. (a) An element is a substance all of whose atoms are exactly alike. (b) An element is a substance all of whose atoms have the same mass number. (c) An element is a substance that cannot be decomposed into simpler substances under any circumstances.

4. (a) Name two metals that are unreactive enough to survive as elements for long periods of time in contact with air and moisture. (b) Name a metal that is liquid at ordinary temperatures. (c) Name five elements that are gases at ordinary temperatures.

5. For the elements listed below, identify those that (a) occur naturally on Earth as stable isotopes, (b) occur naturally on Earth but only as unstable isotopes, (c) do not occur naturally on Earth (the atomic numbers precede the names): 92-uranium, 6-carbon, 8-oxygen, 94-plutonium, 99-einsteinium.

6. Supply examples of the following types of elements: (a) five metals; (b) three nonmetals; (c) one semimetal; (d) one noble gas; (e) one poisonous gas; (f) two elements that occur naturally in their uncombined states.

7. Which of the following pairs are isotopes, which are allotropes, and which are different elements? (a) hydrogen-2 (deuterium) and hydrogen-3 (tritium); (b) rhombic sulfur (prism-like crystals) and monoclinic sulfur (needle-like crystals); (c) radon-215 and radium-215.

MOLECULES

8. Calculate the number of atoms of each element and the total number of atoms per molecule in (a) $C_{12}H_{22}O_{11}$ (sucrose); (b) $C_7H_5(NO_2)_3$ (trinitrotoluene, TNT).

9. Recall the definition that a molecule is the smallest particle of a substance that is not chemically bonded to any other atoms. Using this definition as a guide, state whether each of the following particles can or cannot be classified as a molecule. Defend your answers. (a) An atom of neon. (Neon atoms do not enter into chemical combinations). (b) An atom of carbon in a one-carat diamond. (The carbon atoms in a diamond are all chemically bonded to each other in a continuous three-dimensional network.) (c) The entire one-carat diamond. (d) An entity consisting of 20 atoms, represented by the formula $C_6H_8O_6$, in a crystal of vitamin C. (Each such entity is held to others by intermolecular forces.)

10. The element neon is often said to be "inert." What is meant by this term?

COMPOUNDS AND MIXTURES

11. Classify each of the following substances as an element, a compound, or a mixture: (a) Dry Ice, which can be produced from combustion gases or fermentation gases, is 27.3 percent carbon and 72.7 percent oxygen by mass, and different commercial samples all have the same properties. (b) Honey has a flavor that depends in part on the bees that produce it. If honey is frozen and placed in a vacuum, water vapor escapes and sugary crystals remain behind. (c) Brass is a metallic substance composed of copper and zinc. Red brass contains 85 to 90 percent copper by mass, and yellow brass is about 67 percent copper. (d) Tantalum is another metallic substance. It has high melting and boiling points but does not separate or decompose into any other substances even at extremely high temperatures.

12. What is a solution? Can a solution be a liquid? A solid? A gas? Which of the following substances are solutions, and which are heterogeneous mixtures? (a) Cow's milk, in which the cream rises to the top; (b) antifreeze used in automobile radiators, which is a clear mixture containing water and ethylene glycol; (c) cooking oil, which is a clear liquid containing various different liquid fats; (d) silver amalgam, used for tooth fillings and consisting of silver in which mercury atoms are uniformly dispersed; (e) topsoil; (f) mud.

STATES OF MATTER

13. (a) Another way of describing the three most common states of matter is to say that state X maintains a definite volume and shape, state Y does not maintain a definite volume or shape, and state Z maintains a definite volume but not a definite shape. Which common state of matter corresponds to X? To Y? To Z? (b) Fluids are substances that flow readily. Which of the various states of matter described in the text are fluids?

14. Identify the state of matter of each of the following substances: (a) A hard lump of matter cannot be bent by hand. However, on being struck with a hammer, it shatters. Many of the resulting fragments have flat surfaces that form angles of 120 degrees with each other. (b) A rigid bar, held horizontally, sags slowly over a period of years. When the bar is smashed, the fragments seem more random than those from the previous object; no typical face angles are observed. (c) A brown transparent material is in a closed container. When the

(Question 21)

container is opened, the brown color becomes lighter, first near the top, then throughout the container. Finally, the material disappears entirely from the container. (d) A spherical object starts to fall to Earth. As its speed increases, air resistance causes it to become shaped like the top of a hamburger bun. It falls into a pore in a rock and assumes the shape of the pore.

GASES

15. (a) Name the four properties of a sample of gas that are related to each other. (b) A vendor of compressed gases receives an order for a specific volume of oxygen at a specific pressure, containing a specific number of molecules, to be delivered at a specified temperature. Explain why it is very unlikely that such an order could be filled.

16. The warning on a gas tank reads: "Caution. Contents of this tank are under pressure. Do not store in direct sunlight. Do not use or store near heat or open flame." What is the reason for this warning?

17. A novice diver asks the old-timer, "How full is this tank of compressed air?" The old-timer taps thoughtfully up and down the tank, stops about halfway, and answers jokingly, "It's down to here!" Explain what is foolish about the novice's question and how the old-timer is joking. How should the novice have asked the question to get a straight answer?

18. A sample of helium occupies 8.0 L at 10 atm pressure. (a) What volume will this sample occupy at 20 atm? (b) At 5.0 atm? (c) At what pressure will the gas occupy 2.0 L? (d) 32 L? Assume no change in temperature.

19. Another statement of Boyle's law is that the pressure multiplied by the volume of a gas is constant at constant temperature, or $PV = k$. Assume that the value of k is 12, and complete the table shown below by solving for P. Now plot your results on a graph of P (12 units on the x-axis) versus V (12 units on the y-axis). What is the shape of the curve? If you have ever pumped air into a bicycle tire by hand, can you say whether any single stroke is harder to push near the beginning or near the end of the stroke? Explain how the shape of your curve expresses this experience. Remember, you are compressing the air from a larger volume (tire + pump) into a smaller volume (tire alone).

$$PV = 12$$

V	P
1	
2	
3	
4	
6	
12	

20. (a) A sample of argon gas occupies 16 L at 100 K. What volume will the argon occupy at 200 K? At 400 K? (Assume that the pressure remains constant.) (b) A sample of air occupies 200 mL at 0°C. What volume will the air occupy at 273°C at the same pressure? (*Hint:* First convert Celsius to Kelvin temperatures.)

21. Imagine that an automobile tire is filled with cold compressed air to a certain pressure. You now drive the car rapidly on a warm road; the air in the tire warms up, and the pressure increases. To reduce the tire pressure back to its original value, you bleed out some air and catch the leaked air in a balloon (see Fig.) Would the volume of air in the balloon tell you anything about the warming of the air in the tire? Explain.

22. Look at Figure (a) on p. 372, which shows a gas thermometer. The leveling bulb can be raised or lowered so as to keep the two liquid levels equal and thus maintain atmospheric pressure inside the gas globe. Describe how you could use this device as a thermometer. Where would you place the temperature markings? Do you think that such a thermometer would be more sensitive than the ordinary liquid-in-glass thermometer (Fig. b) or less sensitive? Explain.

23. Give the name of the gas law that is illustrated by each of the following observations: (a) As a bubble of air rises from a diver's helmet to the surface of the water, it continuously expands. (b) As a rubber balloon filled with air cools during the night, it shrinks in size. (c) A 1 L volume of neon or of helium at 0°C and 1 atm pressure contains 2.7×10^{22} molecules.

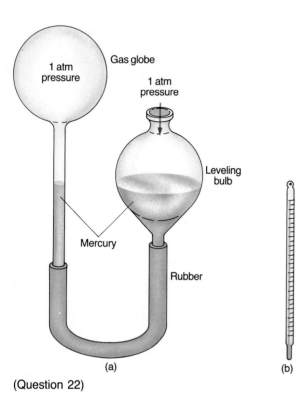

1 atm pressure

Gas globe

1 atm pressure

Leveling bulb

Mercury

Rubber

(a)

(b)

(Question 22)

24. Explain the following phenomena in terms of the theory of gases: (a) A bottle of perfume is overturned and spills onto the floor. A person standing nearby smells it right away, yet someone at the other end of the room does not smell it at once but does smell it a little while later. (b) Two equivalent gas-tight balloons are inflated with helium to the same pressure. One is kept at room temperature, while the other is stored in the freezer. The cold balloon shrinks, while the volume of the one at room temperature remains constant. (c) Two equivalent balloons made of slightly porous rubber are inflated to the same pressure, one with hydrogen, H_2, and the other with helium, He. The hydrogen leaks out more rapidly than the helium.

25. Often Boyle's law and Charles' law are combined into a single statement as PV/T = constant. Explain how this transition can be made.

26. According to the equation of question 25, does the temperature of a gas increase or decrease if the pressure is raised without changing the volume? Why? What happens to the volume if the temperature is raised without changing the pressure? Why?

27. When a closed bottle is thrown into a flame, it often shatters. Which of the gas laws best describes this situation?

LIQUIDS AND SOLIDS

28. What two factors account for the general properties of liquids? Which factor tends to make a liquid more like a gas? Which factor tends to make it more like a solid?

29. (a) The density of a solid depends on the densities of its atoms and on the geometry of their packing arrangements. Does it also depend on the sizes of the atoms? Explain. (b) A box is filled with pea-sized spheres of lead packed as closely as possible. Another box of equal size is filled with spherical grains of sand, also packed as closely as possible. Sand consists of quartz, which is less dense than lead. From this information alone, can you tell which box holds the greater mass?

30. A zinc metal rod and a glass rod are each heated in the absence of air. At a sufficiently high temperature, liquid zinc begins to drip from the zinc rod, although the rod itself remains rigid. The glass rod sags, but does not drip. Account for these phenomena.

ANSWERS TO SELECTED NUMERICAL QUESTIONS

8. (a) 12 atoms carbon, 22 atoms hydrogen, 11 atoms oxygen—45 total atoms; (b) 7 atoms of carbon, 5 atoms hydrogen, 3 atoms nitrogen, 6 atoms oxygen—21 total atoms

18. (a) 4.0 L, (b) 16 L, (c) 40 atm, (d) 2.5 atm

19.

V	P
1	12
2	6
3	4
4	3
6	2
12	1

20. (a) 32 L at 200 K, 64 L at 400 K; (b) 400 mL

Crystalline fluorite (CaF_2) is the major natural source of fluorine.
(Courtesy Leon Lewandowski)

Chemistry and physics joined hands to uncover the secrets of "The Shroud of Turin." The cloth has a faint image of the front and back of a hollow-eyed, bearded man, with what appears to be scourge marks and blood stains corresponding to the New Testament accounts of the crucified Jesus. Chemical tests and carbon-14 dating (see Chapter 13) proved the sack cloth wrappings to be about 1917 years old with an error of plus or minus 200 years. In 1988, carbon dating of the shroud indicated a date of A.D. 1320 ±60 years. (From Fine and Beall, Chemistry for Engineers and Scientists, *Saunders College Publishing)*

C H A P T E R

16

The Periodic Table and Chemical Bonds

16.1 ELECTRONIC STRUCTURES OF ATOMS

The atom was the subject of Chapter 12, and we recall from that discussion that only certain fixed energies are possible for the electrons of an atom. These fixed energy levels are designated by various quantum numbers. Here we will consider only the **principal quantum number** n, where n = 1, 2, 3, . . . , etc. According to this quantum number designation, an electron in the lowest available energy level is said to be in the n = 1 state, the next highest energy level is the n = 2 state, and so forth. There are two things to remember about electron energy levels: (1) each level can accommodate only a limited number of electrons; and (2) the electrons at a given principal energy level make up an **electron shell.**

The higher the shell number (n), the more distant are its electrons from the nucleus and the greater is the energy of the electrons. The electron that can be separated most easily from its atom is therefore an electron in the highest shell. Electrons can be pushed up to higher energy levels, or even knocked entirely away from their atoms, in a variety of ways. All chemical reactions involve shifts of electrons from one energy level to another, or even complete separation of the electrons from their atoms. The electrons in the highest shells are the ones usually involved in chemical changes; these shells are called the **valence shells** and the electrons in them are called **valence electrons.**

The larger atoms have larger clouds of electrons around them. The higher the principal quantum number, the more electrons can be accommodated within the shell. The relationship is:

$$\text{maximum number of electrons per shell} = 2n^2 \tag{16.1}$$

where n is the principal quantum number.

Table 16–1 shows the distribution of electrons in the shells of the first 18 elements. Note that the first shell is full before electrons enter the second shell, and that the second shell is full before electrons enter the third. However, it should be noted here that this pattern is not followed throughout all the shells. For example, the third and higher shells do not completely fill before some electrons go into still higher shells.

EXAMPLE 16.1

What is the maximum number of electrons in the shell for which n = 1? For the n = 2 shell? For the n = 3 shell?

Solution The maximum number is $2n^2$. Therefore, when n = 1, the number of electrons is $2(1)^2 = 2$. For n = 2, it is $2(2)^2 = 8$. For n = 3, it is $2(3)^2 = 18$.

16.2
VALENCE

Consider the following formulas of some well-known chemicals:

HCl	hydrogen chloride
NaH	sodium hydride
NaCl	sodium chloride
CaO	calcium oxide
H_2O	water
NH_3	ammonia
CH_4	methane
CaH_2	calcium hydride

In the first formula, the H and the Cl are combined in equal atomic ratios, one atom of H to one atom of Cl. They are said to have equal chemical *combining capacities*. The same is true for the atoms in the next three formulas, NaH, NaCl, and CaO. But the formulas H_2O, NH_3, CH_4, and CaH_2 clearly show that not all atoms are equal in combining capacity. In fact, these formulas show that more than one atom of hydrogen can combine with one atom of the other element. Other elements can thus have higher combining capacities than hydrogen.

The combining capacity of an atom is known as its **valence** (from Latin, *valentia*, meaning "capacity"). Valence is, of course, a chemical property. By convention, hydrogen has been assigned a valence of 1. Then *the valence of an element is defined as the number of hydrogen atoms that combine with one atom of that element.* It follows that the valences of the other elements in the above formulas must be 1 for Na (sodium) and Cl (chlorine); 2 for O (oxygen) and Ca (calcium); 3 for N (nitrogen); and 4 for C (carbon).

These assigned valences make it possible to predict other formulas. The rule is that when two elements combine, they do so in such a way that the total valences of each element are equal. For example, calcium hydride is a compound of calcium and hydrogen. What is its formula? The valence of H is 1 and the valence of Ca is 2. In the formula, the total valences of H must equal the total valences of Ca. The formula must be

CaH_2

where the valence of Ca is 2 and the total valence of H_2 is $1 + 1 = 2$. The subscripts are generally reduced to their least common denominators. Thus, for calcium oxide we write CaO, not Ca_2O_2. This reduction to the lowest common denominator is not always done. When the molecule has a definite and known structure, the formula includes all the atoms in one molecule. An example is hydrogen peroxide, H_2O_2.

The prediction of a formula from valences does not guarantee that the compound exists. For example, the valences of 4 for C and 2 for Ca predict the formula Ca_4C_2 or Ca_2C. However, there is no such compound. Conversely, many compounds whose formulas cannot be predicted from valences do exist. Therefore, valences are a useful, but not infallible, guide to the prediction of formulas of simple substances composed of two elements.

EXAMPLE 16.2

Write the formulas for the compounds formed by the combination of **(a)** Ca and Cl; **(b)** N and Cl; **(c)** C and Cl; **(d)** Ca and N; **(e)** C and O.

Solution Since Cl has a valence of 1, its subscript

TABLE 16–1 Electronic Configurations of Elements 1 to 18*

| Atomic Number | Element and Symbol | Number of Electrons in Each Shell | | | Total Number of Electrons |
		First	*Second*	*Third*	
1	Hydrogen, H	1			1
2	Helium, He	2 (full)			2
3	Lithium, Li	2	1		3
4	Beryllium, Be	2	2		4
5	Boron, B	2	3		5
6	Carbon, C	2	4		6
7	Nitrogen, N	2	5		7
8	Oxygen, O	2	6		8
9	Fluorine, F	2	7		9
10	Neon, Ne	2	8 (full)		10
11	Sodium, Na	2	8	1	11
12	Magnesium, Mg	2	8	2	12
13	Aluminum, Al	2	8	3	13
14	Silicon, Si	2	8	4	14
15	Phosphorus, P	2	8	5	15
16	Sulfur, S	2	8	6	16
17	Chlorine, Cl	2	8	7	17
18	Argon, Ar	2	8	8	18

*Among the heavier elements, the distribution of electrons becomes more complicated because of the division of shells into subshells.

in a formula will be the valence of the other element. Thus, (a) $CaCl_2$; (b) NCl_3; (c) CCl_4. (d) Ca has a valence of 2 and N has a valence of 3. In order for the total valence of each element to be equal, 3 Ca atoms must combine with 2 N atoms giving both Ca and N a total valence of 6. This gives the formula Ca_3N_2. (e) The same procedure yields a formula of C_2O_4, which reduces to CO_2. This formula actually corresponds to the composition of the molecule.

16.3
THE PERIODIC TABLE

The fact that elements can be classified by their properties into various types, such as metals and nonmetals, or into groups that are characterized by their chemical properties, interested chemists for many years. In 1864, John Newlands in England noted that when the elements were arranged in the increasing order of their atomic weights, various sets of chemical properties tended to repeat themselves. This trend was best noted in the elements from lithium to chlorine. (Helium and neon had not yet been discovered, so these elements are omitted. Modern periodic arrangements are based on atomic numbers, not weights, but the two sequences usually match each other.) We start with lithium (Table 16–2) because the first element, hydrogen, is really in a class by itself; there is nothing very much like it.

Going across the first row, from lithium to fluorine, the properties of the elements change considerably from one to the next. Lithium is a very active metal, beryllium is a less active one,

TABLE 16–2 Repeating Patterns of Chemical Properties

lithium ⟶	beryllium ⟶	boron ⟶	carbon ⟶	nitrogen ⟶	oxygen ⟶	fluorine
sodium ⟶	magnesium ⟶	aluminum ⟶	silicon ⟶	phosphorus ⟶	sulfur ⟶	chlorine

FOCUS ON . . . Poet's Corner

A representation like the periodic table of the elements is used in chemistry to convey large amounts of information, and similar representations can be used in other fields. For example, let's construct a periodic table of the poets listed below. Your table should be constructed such that reading down a column gives a history of poets in a particular country and reading across will give a view of poets living at a particular time in history. Unlike the periodic table of the elements, your poet's table will have certain blank spots, but so did that constructed by Mendeleev when he first developed his. As time went by, he was able to fill in most of these vacancies as new elements were discovered. A little library research will enable you to fill in some of the vacancies in the poet's table also.

Baudelaire	1821–1867	French	Browning	1812–1889	British
Byron	1812–1889	British	Claudel	1868–1955	French
Coleridge	1772–1834	British	Dickinson	1830–1886	British
Donne	1571–1631	British	Dryden	1631–1700	British
Eliot	1888–1965	British	Frost	1874–1964	American
Goethe	1749–1832	German	Hofmannsthal	1874–1929	German
Keats	1795–1821	British	Kipling	1865–1936	British
Longfellow	1807–1882	American	Milton	1608–1674	British
Poe	1809–1849	American	Pope	1688–1744	British
Pound	1885–1972	American	Pushkin	1799–1837	Russian
Rimbaud	1854–1891	French	Sandburg	1878–1967	American
Shelly	1792–1822	British	Tennyson	1809–1892	British
Whitman	1819–1892	American	Wordsworth	1770–1850	British

and boron is hardly metallic at all. These large changes continue from one element to the next going to the right: carbon, nitrogen, oxygen, and fluorine are all nonmetals but are very different from one another in chemical reactivity. The reactivity increases sharply all the way to fluorine, which is a pale yellow, violently reactive gas. But next comes sodium, which is a metal similar to lithium, and, progressing on through the following several elements all the way to chlorine, many of the properties of the first sequence are repeated. For example, chlorine is a pale green gas closely resembling fluorine. When Newlands suggested an analogy to the musical scale, where the frequencies establish repeating patterns, by calling these chemical relationships the "law of octaves," he was ridiculed by his contemporaries.

The classification of the elements reached the status of a serious and valuable concept in 1869, when Dmitrii Ivanovich Mendeleev in Russia and Julius Lothar Meyer in Germany published independent versions of a **periodic table of the elements.** (Mendeleev left "holes" in his chart, predicting the properties of the unknown elements that would someday fill these blank spots. Because his predictions were on target, he is given the lion's share of the credit for the development of the periodic table.) Chemists' views of the idea of such a table were transformed. Instead of looking on it as an idle curiosity, they recognized it as a broad and useful concept that could correlate a wide range of physical and chemical properties of the elements. Modern forms of the periodic table, such as that of Table 16–3, are based on the electronic configurations of the atoms. (A complete list of the elements and their atomic weights and numbers is shown on the inside back cover.)

1. Each box contains the symbol of the element and its atomic number.
2. The elements appear in the increasing order of their atomic numbers (except for two long sequences that are set below as separate rows).
3. The vertical columns, called **groups,** fall into two categories: the so-called main-group elements, where the groups are numbered from 1 to 8, and the transition elements, which lie be-

TABLE 16–3 The Periodic Table

State:
- S Solid
- L Liquid
- G Gas
- X Not found in nature

Legend:
- Metals
- Transition Metals
- Nonmetals
- Noble gases
- Lanthanide series
- Actinide series

Key:
- 92 — Atomic number
- U — Symbol
- Uranium
- 238.03 — Mass number

Group	1	2											3	4	5	6	7	8
1	1 H Hydrogen 1.01																	2 He Helium 4.00
2	3 Li Lithium 6.94	4 Be Beryllium 9.01											5 B Boron 10.81	6 C Carbon 12.01	7 N Nitrogen 14.01	8 O Oxygen 16.00	9 F Fluorine 19.00	10 Ne Neon 20.18
3	11 Na Sodium 22.99	12 Mg Magnesium 24.31											13 Al Aluminum 26.98	14 Si Silicon 28.09	15 P Phosphorus 30.97	16 S Sulfur 32.06	17 Cl Chlorine 35.45	18 Ar Argon 39.95
4	19 K Potassium 39.10	20 Ca Calcium 40.08	21 Sc Scandium 44.96	22 Ti Titanium 47.90	23 V Vanadium 50.94	24 Cr Chromium 52.00	25 Mn Manganese 54.94	26 Fe Iron 55.85	27 Co Cobalt 58.93	28 Ni Nickel 58.71	29 Cu Copper 63.55	30 Zn Zinc 65.38	31 Ga Gallium 69.72	32 Ge Germanium 72.59	33 As Arsenic 74.92	34 Se Selenium 78.96	35 Br Bromine 79.90	36 Kr Krypton 83.80
5	37 Rb Rubidium 85.47	38 Sr Strontium 87.62	39 Y Yttrium 88.91	40 Zr Zirconium 91.22	41 Nb Niobium 92.91	42 Mo Molybdenum 95.94	43 Tc Technetium 97	44 Ru Ruthenium 101.07	45 Rh Rhodium 102.91	46 Pd Palladium 106.4	47 Ag Silver 107.87	48 Cd Cadmium 112.40	49 In Indium 114.82	50 Sn Tin 118.69	51 Sb Antimony 121.75	52 Te Tellurium 127.60	53 I Iodine 126.90	54 Xe Xenon 131.30
6	55 Cs Cesium 132.91	56 Ba Barium 137.34	71 Lu Lutinium 174.97	72 Hf Hafnium 178.49	73 Ta Tantalum 180.95	74 W Tungsten 183.85	75 Re Rhenium 186.21	76 Os Osmium 190.2	77 Ir Iridium 192.22	78 Pt Platinum 195.09	79 Au Gold 196.97	80 Hg Mercury 200.59	81 Tl Thallium 204.37	82 Pb Lead 207.2	83 Bi Bismuth 208.98	84 Po Polonium 209	85 At Astatine 210	86 Rn Radon 222
7	87 Fr Francium 223	88 Ra Radium 226.03	103 Lr Lawrencium 260	104 Unq 261	105 Unp 262	106 Unh 263	107 Uns 264	108	109 Une 266									

Lanthanide series:

57 La Lanthanum 138.91	58 Ce Cerium 140.12	59 Pr Praseodymium 140.91	60 Nd Neodymium 144.24	61 Pm Promethium 145	62 Sm Samarium 150.4	63 Eu Europium 151.96	64 Gd Gadolinium 157.25	65 Tb Terbium 158.93	66 Dy Dysprosium 162.50	67 Ho Holmium 164.93	68 Er Erbium 167.26	69 Tm Thulium 168.93	70 Yb Ytterbium 173.04

Actinide series:

89 Ac Actinium 227	90 Th Thorium 232.04	91 Pa Protactinium 231.04	92 U Uranium 238.03	93 Np Neptunium 237.05	94 Pu Plutonium 244	95 Am Americium 243	96 Cm Curium 247	97 Bk Berkelium 247	98 Cf Californium 251	99 Es Einsteinium 254	100 Fm Fermium 257	101 Md Mendelevium 258	102 No Nobelium 259

tween groups 2 and 3 of the main-group elements starting with the fourth period.

The numbers of the main group correspond to the number of electrons in the highest shell of the atom, the number of valence electrons.

Thus, all the elements in Group 1, from hydrogen on down, have one electron in the highest shell of the atom; the elements of Group 2 have two electrons in the highest shell, the elements in Group 3 have three electrons in the highest shell, and so on. The transition elements are different; the important thing to remember is that there are two electrons in the highest shell for most of the transition elements.

4. *The horizontal rows, called* **periods,** *correspond to the number of occupied electron shells in the atom.*
5. The elements set off below the main table (the lanthanoids and actinoids) also usually have two electrons in their highest shells. They, like the transition elements, differ from one another in their inner electron shells.

Of course, Mendeleev did not base his periodic table on electron shells; no one knew anything about atomic structure then. The early tables were based on the chemical properties of the elements, especially valence. Some of these relationships were clear enough, such as the resemblances between fluorine and chlorine or between sodium and potassium, but others were not so clear, so there was a certain amount of confusion. Now, with the benefit of our knowledge about electron shells, we can use the modern periodic table to help us understand the chemical relationships among the elements.

The first thing to do is to consider the sets of main-group elements that belong to given groups. Remember, the elements in a given group all have the same number of electrons in their highest shells, but they differ in the number of shells they have. We will start on the right side of the table with Group 8, which consists of helium, neon, argon, krypton, xenon, and radon. These elements are all gases, and they are chemically completely inert or nearly so. They are known as the **noble gases.** The molecules of these gases consist of a single atom, so their molecular formulas are the same as their atomic symbols: He, Ne, Ar, Kr, Xe, and Rn. Except for helium,

which has only two electrons, the number of electrons in the highest shells of these Group 8 elements is eight. This fact leads to the conclusion that there is some relationship between having eight electrons in the highest shell and being chemically inert. This conclusion is a very important one that will be dealt with in more detail later in this chapter.

Next, we move on the Group 7, called the **halogens,** which consists of fluorine, chlorine, bromine, and iodine. (Astatine, the last element in Group 7, is omitted because there is so little of it on Earth, and chemists generally never deal with it.) All these elements are colored, toxic substances. Fluorine and chlorine are greenish-yellow poisonous gases. Chlorine was the first poison gas used in World War I, in 1915, on the Western Front. Bromine is a liquid that evaporates readily to a reddish-brown vapor, and iodine is a dark solid that evaporates at room temperature to give a purple vapor. These vapors, too, are toxic. The molecules of all these elements consists of two atoms; their formulas are F_2, Cl_2, Br_2, and I_2. They all show a valence of 1 in their combinations with hydrogen; the formulas are HF, HCl, HBr, and HI. So, for now, we can say that elements in the same group have similar chemical properties, including valence. Since the elements in a given group all have the same number of electrons in the outer shell, this number of outer electrons seems to be related to the properties of the elements. That statement sounds good, but a look at Group 6 shows that the situation is not so simple.

Group 6 includes oxygen, sulfur, selenium, tellurium, and polonium (look for the symbols in Table 16–3). Here the differences are more striking than the similarities. The first element, oxygen, is a nonmetal and a gas. All the other elements are solids, and the last one, polonium, is a metal. Selenium and tellurium, which are in between, have some semimetallic properties. What is evident here is a *progression* from nonmetallic to metallic character. Yet all these elements exhibit a valence of 2. For example, oxygen combines with hydrogen to form water, H_2O, and compounds with similar formulas are formed by three of the other elements of Group 6: H_2S (hydrogen sulfide, which has the odor of rotten eggs), H_2Se, and H_2Te (both of which stink much worse than

TABLE 16–4 **Comparison of Groups in the Periodic Table**

Group 4	Group 5	Group 6	
carbon, C	nitrogen, N phosphorus, P	oxygen, O sulfur, S	nonmetals
silicon, Si germanium, Ge	arsenic, As	selenium, Se tellurium, Te	semimetals
tin, Sn lead, Pb	antimony, Sb bismuth, Bi	polonium, Po	metals

H_2S). Polonium, the last element of this group, does not form a stable compound with hydrogen, but it does form $PoCl_2$.

You are not expected to memorize the periodic table, but the general trends within groups and periods illustrate important chemical principles (to be summarized later). To continue this overview, examine Table 16–4 and compare the trends in Groups 4 and 5 with those in Group 6.

Again, progression from nonmetallic to metallic character is evident as we go down the groups. The elements in each group, however, show a common typical valence. In Group 5, the typical valence is 3, as shown in the compounds NH_3, PH_3, AsH_3, SbH_3, and $BiCl_3$. In Group 4, it is 4, as in CH_4, SiH_4, GeH_4, $SnCl_4$, and $PbCl_4$.

In Group 3, all the elements but the first one, boron, are metals, and in Group 2, they are all metals, with no exceptions. Group 1 starts with hydrogen, which is really in a class by itself. Other than that, all the Group 1 elements are metals, and are called the **alkali metals.** The typical valences are 1 for Group 1, 2 for Group 2, and 3 for Group 3.

The most important property of the transition elements can be stated very simply: They are all metals. If someone asked you to name some metals, most of those you would think of would be transition elements. Try it. Six metals are mentioned in the Bible: gold, silver, copper, iron, lead, and tin. The first four of these are transition elements. (Bronze, after which an entire age of ancient history is named, is an alloy of copper and tin.) The 14 elements appearing in the two lower sections of Table 16–3 are all metals; they include all the possible nuclear fuels: uranium, plutonium, and thorium.

We may now make four simple summarizing statements about the properties of the main-group elements and their positions in the periodic table:

1. Going from left to right within a given period (the horizontal rows), the elements tend to become *less* metallic.
2. Going down within a given group (the vertical columns), the elements tend to become *more* metallic, except for Groups 7 and 8, which are all nonmetals.
3. Within a given period, the typical valences of the elements change by one unit from one group to the next.
4. The elements within the same group have the same typical valence.

Note the heavy zig-zag line in Table 16–3, which is an approximate separation between the metallic and nonmetallic elements. The nonmetals are to the right and up. The metals are to the left and down and include all the transition elements and the lanthanoids and actinoids.

EXAMPLE 16.3

State the number of electron shells and the number of electrons in the highest shell of $_7N$, $_{55}Cs$, $_{22}Ti$, $_{77}Ir$, and $_{35}Br$.

Solution $_7N$ is in the second period, so it has two electron shells. It is in Group 5, so there are five electrons in its second, or highest, shell. $_{55}Cs$ is in the sixth period, so it has six shells. It is in Group 1, so its highest (sixth) shell has one electron. $_{22}Ti$, in the fourth period, has four electron shells. But it is a transition element, for which the general rule is two electrons in the highest shell. $_{77}Ir$, in the sixth period, has six electron shells. It is a transition element, so it has two electrons in its sixth or highest shell. $_{35}Br$, in the fourth period, has four electron shells. It is in Group 7, so its highest, or fourth, shell has seven electrons.

16.4
CHEMICAL EQUATIONS AND TYPES OF CHEMICAL BONDING

Chemical changes occur all around us—in biological systems, in geological processes, in the cooking of food, in chemical manufacturing, and in fires and explosions. It would be convenient to be able to represent these changes by some kind of chemical shorthand, just as symbols represent atoms and formulas represent molecules. The key to the problem is the fact that

chemical transformations involve the making and breaking of chemical bonds but not the creation or destruction of atoms.

Therefore, the chemical formulas change, but the kinds and numbers of atoms stay the same. Chemical reactions are represented by **chemical equations,** which show the formulas of the starting materials (the reactants) and final materials (the products), as well as the relative numbers of molecules involved. The law of conservation of mass states that matter can be neither created nor destroyed. Therefore, the products must contain the same number and kinds of atoms as the reactants. An equation that shows this conservation is called a **balanced equation.**

To learn how to balance equations, let us start with a reaction expressed in words: Hydrogen reacts with chlorine to produce hydrogen chloride, or

$$hydrogen + chlorine \longrightarrow hydrogen\ chloride$$

The formulas of these substances and diagrams of their molecules are:

hydrogen, H_2, (H H)
chlorine, Cl_2, (Cl Cl) } An H atom is smaller than
hydrogen chloride, HCl (H Cl) a Cl atom.

We can now rewrite the equation with diagrams:

(H H) + (Cl Cl) $\longrightarrow$ (H Cl) (incomplete)

The above equation is incomplete because there are two H atoms and two Cl atoms on the left and only one of each on the right. The equation must be balanced by changing the numbers of molecules, never by changing the kinds of molecules:

or

(H H) + (Cl Cl) $\longrightarrow$ (H Cl) + (H Cl)

$$H_2 + Cl_2 \longrightarrow 2HCl$$
(correctly balanced)

It will not be necessary to use diagrams to balance equations. You need only adjust the numbers of molecules. In order to balance equations, it is often necessary to add a coefficient in front of one or more of the chemical formulas. For example, the number 2 in front of HCl in the equation above is the coefficient.

A chemical formula gives the composition of the molecule or of the substance that the formula represents. But the composition is given in terms of atoms, not mass. Thus, the formula H_2O tells us that there are two atoms of H and one atom of O in a molecule of water, or twice as many H atoms as O atoms in any quantity of water. Do not say that water consists of "two parts of hydrogen to one part of oxygen," because such expressions generally refer to mass, not atoms. The composition of water by mass is eight parts of oxygen to one part of hydrogen. The reason for this is related to the fact that the mass of an oxygen atom is 16 times that of a hydrogen atom. Therefore, the formula H_2O means that the oxygen-to-hydrogen composition by mass is 16 (for O) to 2 (for H), or 8 to 1. It is also important to realize that chemical formulas and equations, by themselves, do not describe the nature of the chemical bonding that is involved. Substances may be conveniently classified into four types according to their chemical bonding:

1. Ionic substances
2. Covalent substances
3. Network covalent substances
4. Metallic substances

The remainder of this chapter will deal with these four types.

EXAMPLE 16.4

Balance the following equations:

(a) $H_2 + O_2 \rightarrow H_2O$
(hydrogen + oxygen → water)

(b) $CH_4 + O_2 \rightarrow CO_2 + H_2O$
(methane + oxygen →
carbon dioxide + water)

(c) $Fe_2O_3 + CO \rightarrow Fe + CO_2$
(iron oxide + carbon monoxide →
iron + carbon dioxide)

Solution **(a)** First note that the O atoms are not balanced. To balance them, we need two H_2O molecules:

$$H_2 + O_2 \longrightarrow 2H_2O \quad \text{(still unbalanced)}$$

The H atoms now need to be balanced:

$$2H_2 + O_2 \longrightarrow 2H_2O \quad \text{(balanced)}$$

(b) This one is a bit more complex because there are three kinds of atoms, C, H, and O. The C's are already balanced. The O's occur in both of the products of the reaction, and when this occurs, one should save this for the last to be balanced. Let us balance the H atoms:

$$CH_4 + O_2 \longrightarrow CO_2 + 2H_2O$$
$$\text{(still unbalanced)}$$

Now balance the O's:

$$CH_4 + 2O_2 \longrightarrow CO_2 + 2H_2O \quad \text{(balanced)}$$

FOCUS ON . . . Name That Compound

In our study of chemistry, we encounter a variety of chemical compounds with strange sounding names. As unusual as some of these names may appear to be, there is a method to naming compounds. There are a variety of examples that one may find that violate the simple rules given below, but the technique works most of the time.

1. Compounds that contain only two elements are referred to as binary compounds. These are named by first naming the more metallic of the two elements, then giving the name of the second element followed by the suffix "ide." For example, common table salt is NaCl. Na is the more metallic of the two, so the compound is called sodium chloride.

EXERCISE

Use the information above to name the compounds **(a)** HCl (found in stomach acid) and **(b)** H_2S (the gas that gives rotten eggs their distinctive aroma).

Solution (a) HCl is called hydrogen chloride. **(b)** H_2S is called hydrogen sulfide.

2. When compounds are formed that contain polyatomic ions (see Table 16–3), the same basic procedures are used as in (1) above, except that the polyatomic ion usually retains its name. For example, NH_4Cl is called ammonium chloride.

EXERCISE

Name these compounds containing complex-ions. **(a)** $CaCO_3$, **(b)** $MgSO_4$.

Solution (a) This is calcium carbonate (limestone or marble). **(b)** This is magnesium sulfate (Epsom salt).

3. In many binary compounds, especially of nonmetals, the names are derived according to the number of atoms that occur within the molecule. The prefixes used to distinguish the number of atoms are "mono" = one, "di" = two, "tri" = three, "tetra" = four, "penta" = five, "hexa" = six, "hepta" = seven, and "octa" = eight. For example, the compound CO is called carbon monoxide, and the compound CO_2 is carbon dioxide.

EXERCISE

(a) The compound N_2O is laughing gas. Use the techniques of (3) above to give its chemical name. **(b)** P_4S_3 is the white tip on wooden matches. Name it chemically.

Solution (a) N_2O is dinitrogen monoxide. **(b)** P_4S_3 is tetraphosphorus trisulfide.

(c) First balance the Fe's, which is easy:

$$Fe_2O_3 + CO \longrightarrow 2Fe + CO_2$$
(still unbalanced)

To balance the O's, note that the CO and CO_2 molecules must be kept equal so as not to upset the balance of C atoms. Therefore, we must increase both equally to add more O atoms to the right side. The balanced equation is

$$Fe_2O_3 + 3CO \longrightarrow 2Fe + 3CO_2$$

16.5 IONIC SUBSTANCES

The preceding chapter pointed out that chemical bonds are the result of electrical forces that hold atoms together in the form of molecules. An important characteristic of each bonding type will be illustrated by the description of a simple electrical experiment carried out with familiar substances. The first substance is ordinary table salt (sodium chloride, NaCl), which will be used to illustrate the ionic bond.

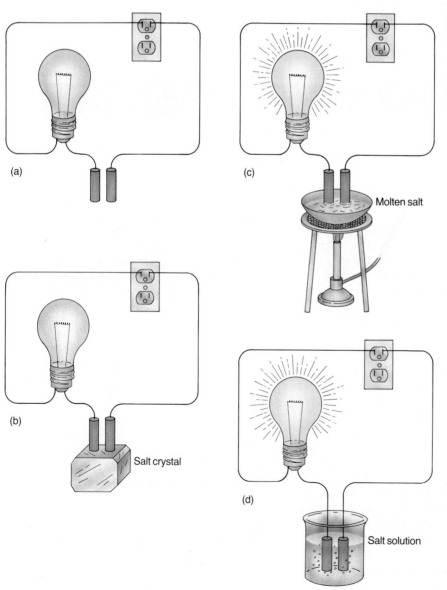

FIGURE 16.1 Experiments with electrical conductivity. (a) Open circuit. (b) Salt crystal does not conduct. (c) Molten salt does conduct. (d) Salt water does conduct.

(a)

(b) Salt crystal

(c) Molten salt

(d) Salt solution

Electrical experiment with salt

Safety Note: The electrical experiments outlined in this chapter are described only to help explain chemical bonding. They are *not* for the reader to carry out. **Do not attempt to do so.** The reason is that exposed electrodes with a potential difference of 110 volts could, if handled improperly, produce a serious electric hazard. Depending on the location of contact to the human body, the resulting electric shock could be fatal.

The experiment uses the simple electrical circuit shown in Figure 16.1a, which consists of an ordinary light bulb plugged into a household socket (110 volts). However, there is a gap in the circuit, so that no current flows and the bulb does not light. The two ends of wire at the gap are attached to sticks of graphite (carbon). These pieces of graphite are called the **electrodes.** If the gap between the electrodes is bridged by anything that conducts electricity, the bulb will light. Figure 16.1b shows the two electrodes touching a large crystal of salt. The bulb does not light, which means that the salt crystal does not conduct electricity. Next, the salt is placed in a porcelain dish and melted, which requires a temperature of about 800°C. The electrodes are now inserted into the hot liquid salt, as shown in Figure 16.1c. The bulb lights up, which means that an electric current is flowing through the molten salt. Something else also happens if direct current is used: Chlorine (Cl_2), a greenish-yellow gas, is bubbling out at the positive electrode, and sodium, a shiny metal, is being produced at the negative electrode.

Finally, some salt is dissolved in water and the electrodes are dipped into the salt water, as shown in Figure 16.1d. Again, the bulb lights up, and the action at the electrodes shows that a chemical change is taking place.

The results of this experiment are interpreted as follows:

1. An electric current is a movement of electric charge. Since sodium chloride conducts electricity when it is molten or dissolved in water, it must contain electrically charged particles that are free to move between the electrodes in response to an applied voltage. Solid sodium chloride, however, does not conduct electricity. Therefore, the electrically charged parti-cles must be locked in place in the solid crystal.

2. Since the conduction of electricity by sodium chloride is accompanied by chemical change, and chemical changes involve atoms, the charged particles in NaCl must be charged atoms. The sodium is produced at the negative electrode, so it must have a positive charge, and the chlorine is produced at the positive electrode, so it must have a negative charge.

Electrically charged atoms are called **ions.** Sodium chloride consists of positive sodium ions, Na^+, and negative chloride ions, Cl^-. The reason that the Na^+ ion has a positive charge is that one electron is missing from its outer shell. The atomic number of sodium is 11, which means that the nucleus has 11 protons and the *neutral* Na atoms has 11 electrons. The sodium *ion* thus has 11 protons and only 10 electrons, for a net charge of +1. Similarly, chlorine, atomic number 17, has 17 nuclear protons, and the neutral Cl atom has 17 electrons. The chloride ion, with one extra electron, has 17 protons and 18 electrons, for a net charge of −1.

NaCl is a stable substance because the oppositely charged ions attract each other strongly. The solid does not conduct electricity because the ions, although charged, are held rigidly in the crystal structure and do not migrate. Figure 16.2 shows a model of the NaCl crystal lattice, in which the ions are in a cubic pattern. Note that the Na^+ and Cl^- ions alternate with each other, so that oppositely charged ions, which attract each other,

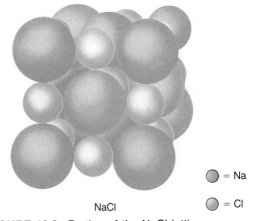

NaCl   ⬤ = Na
         ⬤ = Cl

FIGURE 16.2 Portion of the NaCl lattice.

are always close together. Such an array is very stable. When the salt melts, however, or dissolves in water, the ions are free to move. If a voltage is applied, the positive ions go one way and the negative ions go another. The liquid then conducts electricity.

The chemical changes that occur when molten salt is electrolyzed are:

At the negative electrode (cathode):

$$Na^+ + 1 \text{ electron} \longrightarrow Na \text{ (metallic sodium)}$$

At the positive electrode (anode):

$$Cl^- - 1 \text{ electron} \longrightarrow Cl \text{ (a chlorine atom)}$$

Chlorine atoms combine with each other to form molecules of Cl_2:

$$Cl + Cl \longrightarrow Cl_2$$

Not all ions have charges of $+1$ or -1. Some ions have higher charges, although rarely greater than $+3$ or -3. Furthermore, an ion need not be a single charged atom. Some groups of atoms occur together in many compounds, and the entire group may carry an electrical charge. Examples of such polyatomic ions are ammonium ion, NH_4^+, and hydroxide ion, OH^-. Table 16–5 lists the names and formulas of some common ions, both positive and negative, simple and polyatomic.

The names and formulas of common ionic compounds can be obtained directly from Table 16–5 simply by combining the names and formu-

las of the separate ions. When you write a formula, remember that the total plus and minus charges must balance each other.

Now consider the questions: How can the charge on ions be accounted for? Why are some positive and some negative? Why are some charges greater than others? To help answer these questions, refer back to the electronic structures of O, F, Ne, Na, Mg, and Al given in Table 16–1. Except for Ne, each of these elements forms an ion. The electronic structures of these ions can be calculated as follows: For O^{2-}, the shells are 2; $(6 + 2)$, or 2; 8. For Na^+, they are 2; 8; $(1 - 1)$, or 2; 8. Do the same for F^-, Mg^{2+}, and Al^{3+}, and you will see what these five ions have in common: They all have the same electronic structure as neon. They are said to be **isoelectronic** with neon and with each other. Neon itself forms no chemical compounds; it is an inert gas whose molecules are single atoms, Ne.

Similar relationships exist among the ions of elements whose atomic numbers are close to those of the other noble gases in Group 8 of the periodic table. These observations lead to the following line of reasoning:

—Electrons in atoms are involved in chemical bonds.

—The noble gases are chemically inert. They do not readily gain or lose electrons, nor do they generally enter into chemical reactions. The filled shells of the noble gases must therefore represent stable arrangements of electrons.

—Other atoms form ions by gaining or losing electrons so as to acquire the stable electronic arrangement of a noble gas.

—When atoms react by electron transfer, the total number of electrons gained and lost must be equal because the resulting ionic salt is electrically neutral.

Notice that the number of electrons lost or gained by an atom in forming an ionic bond is equal to its valence. The magnitude of the charge on the ion formed from the atom is the valence of the element. Atoms that lose electrons to form positive ions are generally the metals. Atoms that gain electrons to form negative ions are generally the nonmetals.

TABLE 16–5   Some Common Ions

Positive Ions	Negative Ions
Ammonium, NH_4^+	Bromide, Br^-
Potassium, K^+	Chlorate, ClO_3^-
Silver, Ag^+	Chloride, Cl^-
Sodium, Na^+	Fluoride, F^-
Barium, Ba^{2+}	Hydroxide, OH^-
Calcium, Ca^{2+}	Iodide, I^-
Copper, Cu^{2+}	Nitrate, NO_3^-
Lead, Pb^{2+}	Carbonate, CO_3^{2-}
Magnesium, Mg^{2+}	Oxide, O^{2-}
Zinc, Zn^{2+}	Sulfate, SO_4^{2-}
Aluminum, Al^{3+}	Sulfide, S^{2-}
	Sulfite, SO_3^{2-}
	Phosphate, PO_4^{3-}

The following paragraphs summarize these concepts about the formation of ionic compounds:

1. There are six noble gases: helium, neon, argon, krypton, xenon, and radon. They have stable electronic configurations, characterized by eight electrons in their highest shells (except for helium, which has, and can have, only two).

2. Elements whose atomic numbers are one or two (sometimes three) higher than that of a noble gas can enter chemical reactions in which they lose just enough electrons to become isoelectronic with their noble-gas neighbor. Thus, they become positive ions.

3. Elements whose atomic numbers are one or two (and sometimes three) lower than that of a noble gas can enter chemical reactions in which they gain just enough electrons to become isoelectronic with their noble-gas neighbor. Thus, they become negative ions.

4. In many cases, when elements in categories (2) and (3) are brought together, electron gains and losses occur in the same reaction, resulting in an *electron transfer* to form an ionic compound.

5. Electrons are not created or destroyed in chemical reactions, so electrons gained = electrons lost. For example,

$$Mg \longrightarrow Mg^{2+} + 2e^- \text{ (two electrons lost)}$$
$$F_2 + 2e^- \longrightarrow 2F^- \text{ (two electrons gained)}$$

The formula of the resulting ionic compound is therefore MgF_2.

EXAMPLE 16.5

Write the names and formulas of the ionic compounds formed by a combination of (a) ammonium ion with bromide ion, with chloride ion, and with sulfate ion; (b) calcium ion with hydroxide ion, with carbonate ion, and with phosphate ion.

Solution (a) Combining the names of the + and − ions gives ammonium bromide, NH_4Br; ammonium chloride, NH_4Cl; ammonium sulfate—

Compounds of the Noble Gases

For many years, the noble gases were thought to be chemically inert. In fact, the word "noble" was used to convey the idea that these elements did not associate with the ordinary ones by entering into chemical combinations with them. However, in 1962, Neil Bartlett synthesized the first xenon compound, $(XeF)(PtF_6)$. Other known xenon compounds include various fluorides and oxides. Krypton, too, is known to undergo chemical reaction, but its compounds are unstable.

There is practically no chemistry of helium, neon, or argon. Radon, which is radioactive with a short half-life, is difficult to study. Compared with other elements, these gases may still be considered inert.

here the formula cannot be NH_4SO_4 because the charges do not balance. It must be $(NH_4)_2SO_4$. Note that the procedure for writing formulas from ionic charges is the same as that which was used for writing formulas from valences.

(b) The combinations give calcium hydroxide, $Ca(OH)_2$; calcium carbonate, $CaCO_3$; and calcium phosphate, $Ca_3(PO_4)_2$, in which six + charges neutralize six − charges.

EXAMPLE 16.6

Write symbols for ions of elements number 35, 37, and 38 that are isoelectronic with krypton.

Solution Krypton is number 36, so an element just before krypton in the periodic table will need one more negative charge to become isoelectronic with it. Elements after krypton will have to lose electrons and hence become positive. The ions are Br^-, Rb^+, and Sr^{2+}.

EXAMPLE 16.7

Write formulas for the ionic compounds rubidium bromide and strontium bromide.

Solution To conserve charge, the formulas must be $RbBr$ and $SrBr_2$. In all ionic formulas, the total positive charge must equal the total negative charge.

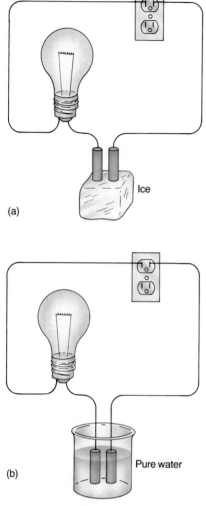

(a)

(b) Pure water

FIGURE 16.3 Neither (a) ice nor (b) pure water conducts electricity.

16.6
COVALENT SUBSTANCES
Electrical experiment with ice and water

This experiment is carried out with the same apparatus as that previously used with salt. This time, however, the electrodes are first placed on a block of ice, as shown in Figure 16.3a, and then are immersed in pure water, as shown in Figure 16.3b. The bulb does not light in either case.

The results of this experiment are interpreted to mean, simply, that water is not ionic. If it were, the ions could have been made to move in the liquid water, current would have flowed, and the bulb would have lit up. No light means no ions and no ionic bond.

Even though water is not ionic, its formula can be predicted from the valences of hydrogen (1) and oxygen (2), just as the formula for sodium oxide, Na_2O, can be predicted from the ionic charges, Na^+ and O^{2-}. Thus, the chemical bonding in H_2O must be related in some way to that in ionic compounds, even though H_2O is not ionic. To interpret these facts, we turn again to the electronic structures of these elements and to their nearest noble gas neighbors. The dots represent the electrons in the valence shells, or the valence electrons, as shown below.

Group Number

1	2	3	4	5	6	7	8
1 H ·							2 He :
					8 : O ·		10 : Ne :
11 Na ·							

The noble gas nearest to H is helium, He; the one nearest to O is neon, Ne. Hydrogen could become isoelectronic with He by gaining an electron. Oxygen could become isoelectronic with neon by gaining two electrons. These requirements cannot both be satisfied by electron transfer, because it is impossible for *both* atoms to gain electrons unless there is a loss of electrons somewhere else. However, if the atoms mutually *shared* their electrons, then each could have valence shells like that of a noble gas. Thus, the bond between H and O consists of two shared electrons, one contributed by each atom. The H now has two shared electrons, which makes it isoelectronic with He but does not convert it into an ion. The O now has seven electrons in its valence shell—the two that it shares with H and five others of its own.

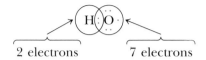

2 electrons   7 electrons

The oxygen atom is still one electron short of being isoelectronic with neon. This deficiency is satisfied by another sharing of two electrons with the second H atom, which gives us the formula

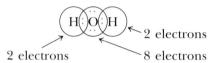

2 electrons

2 electrons   8 electrons

more simply written as H_2O. *Such as sharing of a pair of electrons is called a* **covalent bond** *and is de-picted by a pair of dots, or simply by a dash, as in the formula H—O—H. The number of covalent bonds formed by an atom is called its* **covalence.**

This description may sound very reasonable, but it doesn't explain just what holds the atoms together in a covalent bond. It is clear that in an ionic compound such as Na^+Cl^-, the attractive force is electrical. What, then, is the covalent force? Let us examine the simplest covalent sub-stance, H_2 (H—H or H:H). Remember that these atoms are not points in space; they consist of sep-arate nuclei and electrons. We must therefore consider separately the attractions between each electron and each nucleus, as well as the repul-sions between the two electrons and between the two nuclei. Then, if the sum of all the attractions exceeds the sum of all the repulsions, the mole-cule is stable. The strength of the bond results from the fact that the two electrons are concen-trated between the two nuclei to a greater degree then they would be if the atoms were unbonded, as shown in Figure 16.4. Note that whenever plus and minus electrical charges alternate,

plus minus plus minus plus minus . . .

then opposite charges, which are attracting, are closer to each other than are like charges, which repel. Hence, attractions exceed repulsions, and the array is stable.

The attractive force in a covalent bond, like that in an ionic bond, is electrical.

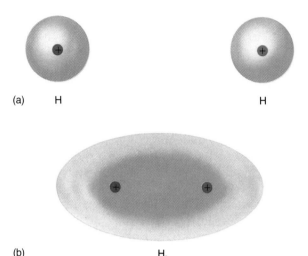

FIGURE 16.4 (a) The nuclei of two H atoms repel each other; the electrons repel each other; and the nuclei and electrons attract each other. When the two H atoms are far apart, the attractions balance the repulsions and there is no bonding. (b) At an optimal distance, the electronic charge is more concentrated between the nuclei, and the attractions exceed the repulsions. Then, a chemical bond exists and the molecule H_2 is formed.

16.7 FORMULAS OF COVALENT SUBSTANCES

Consider the formulas of some common covalent compounds of hydrogen and one other element. In these compounds, the valence of the other ele-ment is the number of H atoms in the formula.

Compound	Valence of the Element Combined with H
HCl, hydrogen chloride	Cl, 1
H_2O, water	O, 2
H_2S, hydrogen sulfide	S, 2
NH_3, ammonia	N, 3
CH_4, methane (natural gas)	C, 4

Note that the valences of these elements in their covalent compounds are the same as the electrical charges (ignoring the signs) of their ions. Thus, the ions whose charges correspond to the covalences shown above are:

Cl⁻, as in sodium chloride, NaCl
O²⁻, as in calcium oxide, CaO
S²⁻, as in sodium sulfide, Na_2S
N³⁻, as in magnesium nitride, Mg_3N_2
C⁴⁻, as in aluminum carbide, Al_4C_3

These covalences and ionic charges further illustrate the difference between covalent and ionic bonding: In ionic compounds, the atom *transfers* (gains or loses) electrons and therefore acquires an electrical charge; in covalent compounds, the atom *shares* electrons but does not gain or lose them and therefore does not become an ion. As already noted, however, the typical valence, which is the number of H or Cl atoms that the element can combine with, is the same in ionic and covalent compounds.

The covalences that you should remember are those of the nine elements in the shaded boxes shown below among the first three periods of the periodic table. Each bond is represented by a dash. The covalences equal the total number of bonds; note that they range from 1 to 4.

Except for hydrogen, which has a covalence of 1, the covalences of the other elements bear a simple relationship to the group number:

Covalence = 8 − Group Number

These covalences can be used to write **structural formulas,** which are very useful chemical expressions. In its simplest form, a structural formula illustrates the types and sequences of covalent bonds in a molecule. Recall from the previous section the structural formula for water, H—O—H, which provides more information

than the simple molecular formula H_2O. The structural formula shows that both H atoms are bonded to O. The formula H—H—O would be incorrect, because H has a valence of 1, whereas the middle H in the incorrect formula shows H with two bonds. The result is a wrong sequence of atoms.

Since structural formulas depict the arrangements of atoms in space, they can also show the angles formed by any two covalent bonds, as well as the relative lengths of the bonds. To do this properly for H_2O, it would be necessary to write

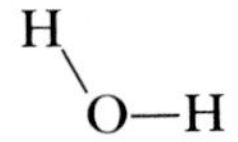

$$\begin{array}{c} H \\ \diagdown \\ O{-}H \end{array}$$

showing that the bond angle is 105° and that the two H—O bonds are of equal length.

Single bonds

With nothing more than the covalences of the nine elements shown in the preceding abbreviated periodic table, we can construct an astonishingly large number of structural formulas.

EXAMPLE 16.8

Write the molecular formula and structural formulas for the simplest covalent compound that can be formed from **(a)** H and F; **(b)** P and Cl; **(c)** C and H.

Solution **(a)** The molecular formula can only be HF, and the structural formula is H—F.

Group Number

1	*2*	*3*	*4*	*5*	*6*	*7*	*8*
H—							He
Li	Be	B	—Ċ̣—	—Ṇ̇—	—O—	F—	Ne
Na	Mg	Al	—Ṩi—	—Ṗ̇—	—S—	Cl—	Ar

(b) The covalence of P is 3, and that of Cl is 1, so the molecular formula must be PCl_3. The structural formula is

$$Cl-P-Cl$$
with Cl above P

(c) The covalence of C is 4, so the formula must be CH_4, or

H—C—H   (methane)
with H above and H below C

16.8
ORGANIC CHEMISTRY

Carbon forms a unique series of compounds of extraordinary diversity. Moreover, living organisms on Earth consist largely of carbon compounds, and the processes of life involve, in large measure, the making and breaking of bonds to carbon atoms. For these reasons, *the chemistry of carbon compounds is called* **organic chemistry.**

The uniqueness of carbon lies, in part, in the ability of carbon atoms to bond to each other in sequences of very extensive length and in continuous or branched patterns. These features can be illustrated with formulas of compounds that contain only carbon and hydrogen. Such compounds, which are called **hydrocarbons,** are the major components of natural gas, bottled gas, lighter fluid, gasoline, kerosene, diesel fuel, petroleum jelly (Vaseline), and paraffin candle wax. Illustrated below are structural formulas of propane, C_3H_8, the major component of bottled gas, and octane, C_8H_{18}, which is a component of gasoline. Recall that carbon has a valence of 4, as shown by the fact that each carbon atom in these compounds makes four bonds.

H—C—C—C—H   propane

H—C—C—C—C—C—C—C—C—H   octane

The octane pictured above is, fortunately, only a minor component of gasoline. Too much of it would cause the gasoline to burn in an uncontrolled, almost explosive, manner. The result can be heard as "engine knock," which is the sound of an inefficient and possibly damaging mode of combustion within the cylinders.

Note the preceding statement that the sequence of carbon linkages can be branched. Branched-chain hydrocarbons have much less tendency to knock in gasoline engines. A particularly good hydrocarbon (the standard for 100-octane gasoline), known as "iso-octane," is

H—C—C—C—C—C—H   (iso-octane structure)

Count the carbons and hydrogens in this formula and note that they add up to C_8H_{18}, the same formula as that for the straight-chain octane. Molecules with different structural formulas make up different compounds, even if their molecular formulas are the same. The straight-chain C_8H_{18} knocks; the branched-chain one does not. They have different properties, so they are different substances. *Such substances with the same molecular formulas but different structural formulas are called* **isomers.**

EXAMPLE 16.9

Write structural formulas for **(a)** C_4H_{10} and **(b)** C_2H_6O (two isomers of each).

Solution **(a)** The carbons must all be bonded to each other, while the hydrogens, with a covalence of 1, can be bonded only to carbon. The four carbon atoms can be linked in either a continuous or a branched chain.

$$C-C-C-C \quad \text{or} \quad C-C-C$$
with C below the middle C

If enough bonds are now added to each carbon atom to reach a total covalence of 4, and an H atom is then attached to each bond, 10 H atoms are needed for each molecule:

$$\underset{\substack{|\\H}}{H}-\overset{\substack{H\\|}}{\underset{\substack{|\\H}}{C}}-\overset{\substack{H\\|}}{\underset{\substack{|\\H}}{C}}-\overset{\substack{H\\|}}{\underset{\substack{|\\H}}{C}}-\overset{\substack{H\\|}}{\underset{\substack{|\\H}}{C}}-H \qquad H-\overset{H}{\underset{H}{C}}-\overset{H}{\underset{C}{C}}-\overset{H}{\underset{H}{C}}-H$$

These are the two C_4H_{10} isomers. The unbranched one is normal butane, and the branched one is iso-butane.

(b) There must be a chain of two C atoms and one O atom. The O atom could be in between the C atoms or at the end:

$$C-O-C \quad \text{or} \quad C-C-O$$

Bonds must now be added to the formulas to give each C atom a covalence of 4 and the O atoms a covalence of 2, and an H atom must be attached to each bond. Note that the O atom in the first chain already has two bonds, so that no more are needed:

$$H-\overset{H}{\underset{H}{C}}-O-\overset{H}{\underset{H}{C}}-H \qquad H-\overset{H}{\underset{H}{C}}-\overset{H}{\underset{H}{C}}-O-H$$

These are the two C_2H_6O isomers. The one on the left is called dimethyl ether. The one on the right is ethyl alcohol.

Double and triple bonds

Natural gas contains, among other compounds, ethylene (or ethene), whose molecular formula is C_2H_4. What is its structure? Writing

$$\underset{H}{\overset{H}{>}}C-C\underset{H}{\overset{H}{<}}$$

would satisfy the composition of ethylene and the covalences of the H atoms, but not of the carbons, which need four bonds, not three. However, by assuming that there can be more than one covalent bond between the C atoms, it is possible to write

$$\underset{H}{\overset{H}{>}}C=C\underset{H}{\overset{H}{<}}$$

and satisfy all the requirements. Now each carbon has its four bonds: one double and two single bonds. Is this merely a pencil-and-paper exercise, or does the doubly bonded formula represent some molecular reality? It is known from experiment that the bonding between the carbon atoms in ethylene is shorter and stronger than it is in an ordinary single bond between carbon atoms. This means that the attraction between the carbon nuclei and the electrons between them is greater when the carbon atoms are doubly bonded than when they are singly bonded. The double bond represents the greater concentration of negative charge between the carbon nuclei. Specifically, the double bond denotes two electron pairs (four electrons) shared by the bonded atoms. Without the double bond, each carbon atom would have a share in only seven electrons, not eight.

The concept of multiple bonding can be extended to the gas acetylene (or ethyne), C_2H_2, whose structure is represented as

$$H-C\equiv C-H$$

The triple bond is shorter and stronger than the double bond and denotes three shared electron pairs.

EXAMPLE 16.10

Write structural formulas (with one or more double bonds) for **(a)** C_2H_3Cl; **(b)** COS.

Solution **(a)** H and Cl each has a valence of 1, so they cannot have any double bonds. The double bond must therefore be between the two carbon atoms. The formula must be

$$\underset{H}{\overset{H}{>}}C=C\underset{Cl}{\overset{H}{<}}$$

This substance is vinyl chloride.

(b) With equal numbers of atoms of different elements, it is good strategy to put the element with the highest covalence in the center and to group the other elements around them. This gives S—C—O; it is then obvious where the double bonds belong, to yield the correct formula: S=C=O.

Cyclic structures

There is no reason why the ends of a chain cannot find each other and form a bond, provided that there is room for another bond on each end. The result is a cyclic structure. Note that a formula like C_3H_6 can represent either a doubly bonded or a cyclic structure:

$$H_2C=CH-CH_3 \quad \text{propylene}$$

$$\text{cyclopropane}$$

Organic compounds of oxygen and nitrogen

Carbon forms covalent bonds with many elements other than hydrogen. The most important ones in living systems are the organic compounds that contain oxygen and nitrogen. Some of the more common types of these structures are illustrated on the following:

An **alcohol** is a compound that contains the

$$-\overset{|}{\underset{|}{C}}-O-H$$

linkage, where the other three bonds from the carbon atoms must be linked to hydrogen atoms or to other carbon atoms. The simplest such compound is methyl alcohol,

$$H-\overset{H}{\underset{H}{C}}-O-H$$

sometimes also called "wood alcohol" because it can be produced by heating and decomposing wood in the absence of air. Methyl alcohol is sometimes found in adulterated alcoholic beverages; drinking it can cause blindness or death. The next compound in the series is ethyl alcohol, which is produced by fermentation:

$$H-\overset{H}{\underset{H}{C}}-\overset{H}{\underset{H}{C}}-O-H$$

The series can continue indefinitely by extending and branching the carbon chain.

Carboxylic acids are another important class of organic compounds that contain oxygen. They are characterized by the presence of a carboxyl group

$$-\overset{O}{\overset{\|}{C}}-O-H$$

where the fourth bond from the carbon atom must be linked to hydrogen or to another carbon. The first two in this series are:

$$H-\overset{O}{\overset{\|}{C}}-O-H \quad \text{and} \quad H-\overset{H}{\underset{H}{C}}-\overset{O}{\overset{\|}{C}}-O-H$$

formic acid   acetic acid

The most important class of organic nitrogen compounds is the **amines,** characterized by the linkage

$$-\overset{|}{\underset{}{C}}-\overset{|}{\underset{}{N}}-$$

where, again, the other bonds are to hydrogen or carbon atoms. The simplest such compound is therefore

$$H-\overset{H}{\underset{H}{C}}-\overset{H}{\underset{}{N}}-H \quad \text{methyl amine}$$

16.9
THE MILLER-UREY EXPERIMENT AND THE ORIGIN OF LIFE

Having seen how structural formulas are written, you may wonder whether these exercises are merely games chemists play or whether they really reflect the properties of atoms. If you cut out all the squares of a crossword puzzle and tossed them into the air, you would hardly expect the letters to fall down again in the same word

FOCUS ON . . . Bonding in DNA

DNA, or deoxyribonucleic acid, is the primary molecule in chromosomes, which are the carriers of genetic information. This molecule has a double helix structure as shown in the Figure, where the shape is maintained by bonds across the strands of the helix. In the Figure, the bonds connecting the double helix are labeled A-T and C-G. These subunits are called nucleotide base pairs and are composed of adenine (A), thymine (T), cytosine (C), and guanine (G). Bonds can form only between adenine and thymine and between cytosine and guanine. When a cell reproduces, the double helix comes apart at one end, as shown in the Figure, and separates. Each portion of the split double helix can now form a new double helix. The new helix must match the original exactly, because all the A-T and C-G bonds must match perfectly throughout the strand. Genetic information is passed from one generation of cells to the next through this matching of bonds.

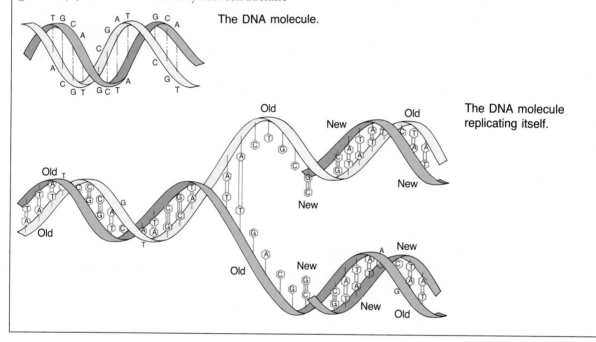

The DNA molecule.

The DNA molecule replicating itself.

pattern or indeed in any arrangement that made sense. What happens, then, when some simple covalent compounds are energized with particles or photons of high energy to break them into various molecular fragments? Do the pieces bind themselves together again? Can new products be formed? Are the normal covalences re-established?

A very interesting experiment was carried out in 1953 by Stanley Miller and Harold Urey to explore what the chemistry of the Earth's atmosphere might have been before life existed. They had speculated that the primitive Earth's atmosphere consisted largely of hydrogen (H_2), water (H_2O), methane (CH_4), and ammonia (NH_3). Miller and Urey passed an electric spark through a mixture of these gases to imitate the action of lightning, as shown in Figure 16.5. The molecules broke into a variety of fragments. These fragments then recombined spontaneously to form various more-complex molecules. The importance of this experiment lies in the fact that the spontaneously formed molecules were amino acids, which are the building blocks of proteins. The structural formulas of some of these amino acids are:

glycine

alanine

aspartic acid

Thus, the interplay of energy and probability (which molecular fragments hit each other and in what ways they collide) favors the formation of molecules of stable substances with normal valences from a chaotic mixture of molecular fragments. The atoms that make up the essential molecules of living matter, such as protein, cellulose, and DNA, are linked by the same types of bonds that exist in the simpler molecules discussed in this chapter. But this fact is not an answer to the question of how the molecules of life were formed. The Miller-Urey experiment, interesting as it was, did not provide the answer either. For one thing, the Miller-Urey hypothesis on the composition of the primitive atmosphere has been greatly modified by more recent studies and speculations. It is generally recognized that we do not have the last word on what the early Earth was like. The synthesis of amino acids, even if it did occur spontaneously, is not the same as, or even close to, the synthesis of a living organism. Living things are characterized by the ability to use an outside source of energy to form complex structures and to reproduce themselves at the expense of materials from the environment. A virus, for example, is a very complex organism, up to several hundredths of a micrometer in diameter. It is considered to be on the very borderline of life, because it can reproduce itself but cannot ingest food or grow. No such structures have ever been made in the laboratory, and there is no reliable way to estimate just how difficult such a synthesis would be or indeed whether it would ever be possible. We still do not know how life originated.

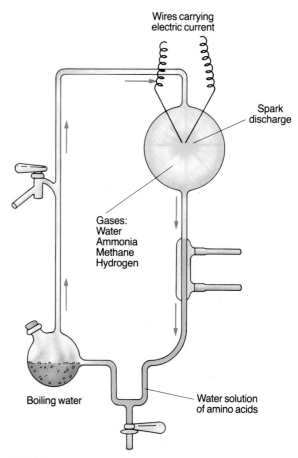

FIGURE 16.5 The Miller-Urey experiment.

16.10
NETWORK COVALENT SUBSTANCES

Electrical experiment with sand

This experiment is carried out with the same apparatus as that previously used with salt, ice, and water. When the electrodes touch the sand, as in Figure 16.6, the bulb does not light. Sand, which is particles of quartz, does not conduct electricity. Any attempt to melt the sand by heating it with an ordinary gas flame would fail because the flame is not hot enough. The melting point of sand or quartz is about 1700°C.

Quartz, like water, is a covalent compound. The molecular formula of quartz, SiO_2, is analogous to that of carbon dioxide, CO_2. (Both Si and C have a covalence of 4.) But quartz is a hard

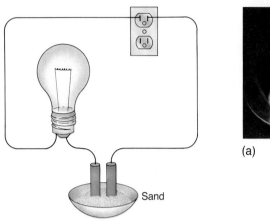

FIGURE 16.6 Sand does not conduct electricity.

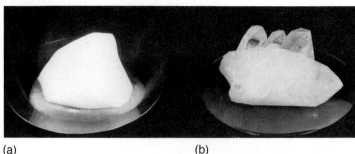

FIGURE 16.7 (a) "Dry ice" (solid CO_2); the temperature of the solid is about $-79°C$. It is rapidly turning to gaseous CO_2, and the cold fumes are generating a visible mist. The day after the photograph was taken, the dry ice was gone. (b) A quartz crystal will keep its shape indefinitely.

mineral, and carbon dioxide is a gas at ordinary temperatures, as shown in Figure 16.7. With the same valences and analogous molecular formulas, why should they be so different? The answer lies in the fact that CO_2 has double bonds, whereas SiO_2 has only single bonds. Why should this fact be so important? Recall that double bonds are shorter than single bonds. This means that doubly bonded atoms are closer to each other than are singly bonded atoms. But large atoms cannot get as close to each other as smaller atoms can, because at closer range the repulsions of the more numerous electrons in the lower shells become too great. As a result, it is more difficult for the larger atoms to form double bonds. To see how important this difference is, try to write structural formulas for the oxides CO_2 and SiO_2, using double bonds for carbon but only single bonds for silicon. Carbon dioxide presents no problem. Using double bonds, one can write O=C=O, and the structural formula is complete. But using only single bonds for Si, we start with

$$
\begin{array}{c}
\text{O} \\
| \\
\text{O—Si—O} \\
| \\
\text{O}
\end{array}
$$

This satisfies the covalence of 4 for Si, but not the covalence of 2 for O. The remedy is to add an Si to each O atom,

$$
\begin{array}{c}
\text{Si} \\
| \\
\text{O} \\
| \\
\text{Si—O—Si—O—Si} \\
| \\
\text{O} \\
| \\
\text{Si}
\end{array}
$$

and then another O to each Si atom. As we continue, the formula grows larger, but the covalences of the outermost Si and O atoms are still not satisfied. To satisfy the covalences with single bonds would require endless writing. What this means is that the covalent bonding in SiO_2 continues indefinitely; it is not limited to a small molecule. Therefore, a crystal of silicon dioxide is not composed of molecules in the ordinary sense but rather is a continuous three-dimensional network of single Si—O covalent bonds. It may be considered to be a **giant molecule.**

There have been science fiction accounts of intelligent life elsewhere in the Universe that is based on the chemistry of silicon, in contrast to the carbon-based life on Earth. Is silicon-based life plausible? Most chemists think it is not, because the Si—Si bond is much weaker than the C—C bond. Consequently, structures with Si—Si bonds would react rapidly with oxygen to form giant-molecule compounds with Si—O bonds as in SiO_2 or silicate rocks. Furthermore, Si does not tend to form double or triple bonds.

Life may exist elsewhere and its forms may be different, but we can be sure of one thing. Only carbon has the variety of bonding to furnish the multitude of compounds needed for life as we know it.

16.11
METALLIC SUBSTANCES

Electrical experiment with silver

This experiment is again carried out with the previous apparatus. The electrodes now touch a piece of silver. The bulb lights up and stays lit, as shown in Figure 16.8. No chemical action is noted on the surface of the silver. In fact, if the silver is weighed before the experiment is started, and the current is then allowed to flow through it for a few hours, or days, or months, and the silver is then removed and reweighed, no change can be detected. Nothing is lost; all is the same as before. If the silver is melted, it still conducts electricity without chemical change. In these experiments, therefore, silver does not behave like salt, water, ice, or sand.

Silver is a metal. Like other metals, silver is an excellent conductor of electricity (one of the best, in fact) in either its solid or its liquid state. Moreover, metals are distinctive enough in other properties that we can recognize them as metals after only casual observation—by looking at them, feeling them, and hefting them. (You could not tell just by looking at a white powder, or some blue crystals, whether the substance is ionic or covalent.) What is it, then, that is unique about metals, and what kind of bond holds their atoms together? Let us first look more carefully at their physical properties.

The most conspicuous attribute of metal is **luster**—the bright, highlighted appearance of a substance that reflects light well. A smooth silver surface, for example, throws back over 90 percent of the light that shines upon it. The adjective "silvery" is often used in the general sense of having luster. Two metals, gold and copper, are colored but nonetheless lustrous. Of course, many metallic objects (such as cast-iron frying pans)

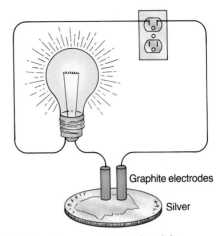

FIGURE 16.8 Silver conducts electricity.

look dark or dull, but if they are polished, the natural luster of the metal reappears, showing that the dullness was a coating of foreign, nonmetallic matter.

In the arctic winter, you don't dare place your bare hand on a piece of metal—it would freeze right onto the surface. And if one end of a metal bar is in a flame, don't touch the other end, for it is too hot. Metals are good conductors of heat.

The electrical conductivity of metals is far greater than that of molten or dissolved salts. Copper, for example, is about 60 million times better as a conductor than salt water, as shown in Table 16–6. Furthermore, when metals conduct electricity, no chemical change occurs. Thus, a silver bar or a copper wire can carry an electric current indefinitely without decomposition. Ionic substances, on the other hand, undergo chemical change when electric currents pass through them.

It was pointed out in the preceding chapter that the densest known substances are metals and that the densest one of all is osmium. Other dense metals include lead, 11 times as dense as water; mercury, almost 14 times as dense as water; and gold, 19 times as dense as water. By contrast, granite rock and concrete, which are commonly thought of as "heavy" materials, are only about two or three times as dense as water. Of course, there are "light" metals, like aluminum. How-

TABLE 16–6 Approximate Electrical Conductivities* At Room Temperature

Silver or copper	600,000
Iron	100,000
Mercury	10,000
Salt water	0.01
Pure water	0.00000006
Rubber	0.000000000000001

*The units of electrical conductivity shown here are "reciprocal ohms/cm," but it is not the purpose of this table to emphasize such absolute values. Rather, attention should be given to the relationships among the different substances. Thus, if we compare samples of mercury and salt water *of equal physical dimensions*, we find that the mercury is more conductive by a factor of 10,000 to 0.01 or 1,000,000 to 1.

ever, aluminum would sink even in the dense waters of the Great Salt Lake or the Dead Sea, and a comparison of aluminum with solid non-metallic elements of comparable or even greater atomic weight, such as sulfur or phosphorus, shows that aluminum is denser.

One of the most remarkable and unique properties of metals is their ability to maintain their crystal structure even when their shape is deformed. We say that metals are **malleable** (they can be hammered flat, as with a mallet) and **ductile** (they can be drawn into a wire). No other crystalline solids have these properties. Thus, if you strike a crystal of salt or sugar with a hammer, it will shatter, not flatten.

These sets of metallic properties are not at all consistent with our concepts of ionic or covalent bonding. The crucial difference is that the electrons involved in ionic or covalent bonds are more closely associated with specific bonded atoms. In the **metallic bond,** on the other hand, the electrons are thought to be somehow "loose," that is, they are not associated with particular atoms.

The concept of "loose" or "free" electrons in metals has led to a theory of the metallic bond. One important clue is that metals crystallize in lattice structures in which the atoms are closely packed. In such close packing, one metal atom may be equally close to as many as 12 other atoms. In these arrangements, there are not enough valence electrons to provide ordinary two-electron covalent bonds between adjacent atoms. Therefore, a given valence electron is not associated with only two atoms but rather with all the atoms in a given sample of metal. A metallic

crystal, then, has a large number of such "unattached" electrons scattered about everywhere. The energy levels of these loose valence electrons are very closely spaced; that is, the energy differences among them are very small. Taken together, the energy levels of these loose electrons make up what amounts to a continuous band, called a **conduction band.**

Now, how does this idea of the metallic state account for the known properties of metals? The key must lie in the behavior of electrons in the conduction band. This behavior is related to the properties of metals in the following ways:

Density. Electrons can establish more bonds when they are free to associate with many atoms than when they are localized between only two atoms. As a result, each atom in a metal can be closely linked to more atoms than would be possible with ionic or covalent bonding. In some metals, a given atom is closely associated with as many as 12 other atoms. (Twelve is the maximum number of spheres that can touch a given sphere of the same size.) This close packing is the reason metals are so dense.

Electrical Conductivity. The loose electrons can be promoted to slightly higher energy levels within the conduction band when an electrical potential (voltage) is applied; they are then free to move in the direction of the applied potential. Hence, metals conduct electricity. When the voltage is removed, the electrons can fall back to the lower energy levels. No nuclei migrate anywhere. When the current stops flowing, the metal is in the same condition as at the start. Hence, no chemical change occurs.

Metallic Luster. For a surface to reflect light, a photon must be absorbed and another one emitted. A photon is not like a rubber ball that literally bounces off a wall. A photon disappears when it is absorbed, and its absorption promotes an electron to a higher energy level. The loose electrons in the conduction band of metals are easily promoted to higher energies, and for most metals they fall back almost instantaneously to their original levels, emitting photons of the same frequency. That is why metals are shiny. The electrons in copper and gold, however, fall back to different levels, emitting photons of different frequencies, and therefore these metals are colored.

Conduction of Heat. The electrons in the conduction band can also absorb thermal energy readily. Recall that thermal energy can be transmitted by photons of infrared frequency. Photons in this range are less energetic than photons of visible light, but since the energy levels in the conduction band are so closely spaced, it doesn't take much to promote electrons within the band. Infrared energy will do this. That is why metals conduct heat so well.

Malleability and Ductility. Metals are ductile and malleable because under mechanical stress the atoms of the crystal can move past each other with relatively little resistance and without

breaking the metallic bonds. The bonds don't break because the valence electrons in the conduction band "belong" to all the atoms and are not localized to individuals or pairs of atoms.

EPILOGUE

Let us not forget that this chapter has dealt with chemical bonds that are common under conditions that are familiar to us on Earth. In the vast regions of outer space, matter is so sparse that a molecule such as OH or CN, with unsatisfied valences, which would react rapidly with something on Earth, might survive for a very long time without meeting any other molecule. Toward the other end of the pressure scale, say at about a million Earth atmospheres in some cold stars and in a large planet like Jupiter, atoms of nonmetallic substances are squeezed together as tightly as are atoms in metals at ordinary Earth pressures, and a substance such as hydrogen becomes a metallic conductor. At still higher pressures, atoms collapse entirely and chemistry does not exist as we know it. High temperatures, too, change chemical bonds. The temperature of a gas flame is about 2000°C. The chemistry of flames is therefore very different from that of substances at ordinary temperatures. Recall from the discussion of nuclear fusion in Chapter 14 that above about 5000°C there are no chemical bonds at all, and therefore no molecules and no chemistry.

SUMMARY

Electrons in the principal energy levels of an atom are said to be in **electron shells.** The shells corresponding to the highest energy levels include the electrons usually involved in chemical bonding and are therefore called the **valence shells.** Chemical formulas of simple compounds can often be predicted from valences.

The **periodic table** classifies elements into **periods,** which include elements with the same number of electron shells, and **groups,** which include elements with the same numbers of electrons in their valence shells. Elements within the same period differ markedly from one to the next, generally progressing from metallic to nonmetallic character as the atomic number

increases. Elements in the same group have the same typical valences but generally progress to more metallic character as the number of shells increases.

Chemical equations express the formulas and relative numbers of molecules of reactants and products in a chemical reaction. The equation is balanced when the numbers and kinds of atoms are equal on both sides of the arrow. Chemical substances are classified into four bonding types, whose properties are:

Ionic: Electrons are transferred from the metallic to the nonmetallic atom, producing ions that are isoelectronic with a noble gas. The oppositely charged ions attract each other. Ionic compounds conduct electricity only in the liquid state, either molten or dissolved in a solvent like water. The current is carried by the ions, which undergo chemical changes at the electrodes.

Covalent: Electrons are shared so that the number of valence electrons of each atom—counting the shared electrons—is the same as that in a noble gas. Covalent compounds do not conduct electricity. A single bond represents one pair of shared electrons, a double bond two pairs, and a triple bond three pairs. Carbon atoms can form covalent bonds with each other that can extend into long chains, branches, or cycles.

Network covalent: The bonds are ordinary covalent bonds but they extend indefinitely in various directions so that the molecule does not have an ordinary limited boundary. Such substances often have very high melting points and are very hard.

Metallic: The bonding electrons are not localized between specific pairs of atoms but are "loose" and are grouped in tightly spaced energy levels called conduction bands. The characteristic properties of metals result from the ease with which the loose electrons can be promoted between energy levels in the conduction bands.

KEY WORDS

Electron shell	Period (of the Periodic Table)	Chemical equation	Covalent bond
Valence shell		Ion	Covalence
Valence	Main-group elements	Chemical bond	Isomers
Group (of the Periodic Table)	Transition elements	Ionic bond	Metallic bond

QUESTIONS

ELECTRONIC STRUCTURE AND THE PERIODIC TABLE

1. (a) If the elements were arranged in the increasing order of their atomic weights rather than atomic numbers, would the sequence of elements generally be the same as in the present periodic table? (To answer this question, pick several sets of four or five consecutive elements from the periodic table, such as N, O, F, Ne, and check whether they are in the order of increasing atomic weights.) (b) Would the sequence always be the same? (Check the set Cl, Ar, K, Ca; the set Fe, Co, Ni, Cu; and the set Te, I, Xe.) (c) Explain your finding in part (b). (*Hint:* Remember isotopes?)

2. How many periods are there in the periodic table? How many groups? From this information, can you tell which is greater—maximum number of shells that atoms can have in their stable (ground) states or the maximum number of electrons that some shells can have?

3. What is the maximum number of electrons in the $n = 4$ shell?

4. Predict the number of electrons in each shell for (a) $_{13}Al$; (b) $_{18}Ar$; (c) $_4Be$.

5. Predict the number of electrons in each shell for the main-group element that appears in the (a) second period, Group 7; (b) third period, Group 1; (c) first period, Group 8.

6. If you closed your eyes and poked your finger somewhere at the periodic table, then looked and saw you had "landed" on a nonmetal, in which direction would you move your finger to reach a metal: (a) Up or down within a group? (b) Right or left across a period? (c) From main-group to transition elements or from transition to main-group elements?

7. Which of the following properties is most generally typical of all the main-group elements in a given group: (a) metallic character; (b) nonmetallic character; (c) physical properties; (d) valence?

VALENCE AND CHEMICAL EQUATIONS

8. With the aid of the periodic table, write the formulas for the compounds formed by the combination of (a) Be and F; (b) Al and Br; (c) Na and S; (d) Ca and P.

9. Given the valences of 1 for H, F, and Na; 2 for O and Ca; and 3 for N and Al, calculate the valences of the other elements in the following compounds: (a) AsH_3; (b) BaO; (c) GeO_2; (d) PF_3; (e) $GaAs$; (f) BP.

10. Balance the following equations:
    (a) $Ag + O_2 \longrightarrow Ag_2O$ (oxidation of silver)
    (b) $KClO_3 \longrightarrow KCl + O_2$ (decomposition of potassium chlorate)
    (c) $Fe_3O_4 + C \longrightarrow Fe + CO_2$ (reduction of iron ore)
    (d) $H_2S + SO_2 \longrightarrow S + H_2O$ (production of sulfur in volcanos)

11. Complete the following chemical equations by filling in the blanks.
    (a) $2O_2 + 3$ _____ $\longrightarrow Fe_3O_4$
    (b) 4 _____ $+ 3O_2 \longrightarrow 2Al_2O_3$

12. Complete the following chemical equation by filling in the blank.
    _____ $+ 2LiOH \longrightarrow Li_2CO_3 + H_2O$

13. Balance each of the following chemical equations.
    (a) $Sc + HBr \longrightarrow ScBr_3 + H_2$
    (b) $NH_3 + O_2 \longrightarrow NO + H_2O$
    (c) $HClO \longrightarrow HCl + HClO_3$

14. Balance each of the following chemical equations.
    (a) $KMnO_4 + HCl \longrightarrow MnCl_2 + Cl_2 + KCl + H_2O$
    (b) $Fe_2O_3 + S \longrightarrow Fe + SO_2$

CHEMICAL BONDS

15. What are the four major types of chemical bonds? Which type conducts electricity without chemical change? Which conducts electricity only when it is liquid? Which two are nonconductors? Which type typically has the highest melting points and the hardest solids?

16. Predict whether the structure of each of the following substances is ionic, covalent with small molecules, or covalent with a "giant" molecular network: (a) HCl, a gas at room temperature, liquefies at $-84°C$. The liquid is a poor conductor of electricity. (b) SiC. The covalency of each atom is 4, but there are no multiple bonds. (c) $BaCl_2$ a solid that does not melt on a hot plate. A solution of $BaCl_2$ in water conducts electricity.

17. Predict the type of bonding in each of the following substances: (a) BN, borazon, a substance almost as hard as diamond; (b) $SbCl_3$, antimony chloride ("butter of antimony"), which melts in hot water to form a liquid that is a poor conductor of electricity; (c) $RaBr_2$, radium bromide, a solid that will not melt in a baker's oven but that dissolves very readily in cold water. Radium bromide solution conducts electricity.

IONIC BONDS

18. Write the formulas of (a) ions of elements number 15, 16, 17, 19, and 20 that are isoelectronic with argon; (b) ions of elements number 1 and 3 that are isoelectronic with helium.

19. Using the information obtained by answering question 18, write the formulas for the following ionic compounds: (a) lithium hydride, which is a compound of lithium and hydrogen, (b) calcium chloride, (c) potassium sulfide, (d) calcium hydride, (e) lithium chloride.

20. Would a salt crystal in which the ions alternate in sets of two of each kind $(Na^+Na^+Cl^-Cl^-$ $Na^+Na^+Cl^-Cl^- \ldots)$ be (a) more stable than, (b) less stable than, or (c) as stable as the arrangement shown in Figure 16.2? Defend your answer.

21. Using the information in Table 16–5, write the names and formulas of the ionic compounds formed by a combination of (a) magnesium ion with oxide ion and with fluoride ion; (b) sulfate ion with ammonium ion, with potassium ion, and with aluminum ion.

22. Calcium ion, Ca^{2+}, has 20 protons in its nucleus. Exactly what gives it a 2+ charge?

23. What factor determines the number of electrons gained or lost by an atom in the formation of an ionic bond?

COVALENT BONDS

24. (Single bonds only) Write structural formulas, consistent with their covalences (refer to text for covalences), for each of the following: (a) H_2S; (b) SiF_4; (c) $CClF_3$; (d) PH_3; (e) C_3H_7Cl (two possibilities); (f) NH_2Cl; (g) $C_2H_3Cl_3$ (two possibilities); (h) C_2H_6S (two possibilities); (i) $C_2H_6O_2$ (four possibilities); (j) N_2H_4; (k) C_2H_7N (two possibilities).

25. (Containing one or more double bonds) Write structural formulas, consistent with their covalences (refer to text for covalences), for each of the following: (a) C_3H_6; (b) $COCl_2$; (c) $NOBr$; (d) H_2CO; (e) N_2F_2; (f) H_2CO_2; (g) H_2C_2O; (h) C_3O_2; (i) C_2H_3Cl.

26. (Containing triple bonds) Write structural formulas, consistent with their covalences (refer to text for covalences), for each of the following: (a) C_3H_4; (b) HCN; (c) C_2H_3N (two possibilities); (d) C_4H_6 (two possibilities).

27. (With single bonds and one cycle) Write structural formulas, consistent with their covalences (refer to text for covalences), for each of the following: (a) C_4H_8; (b) C_4H_8O; (c) $C_5H_{10}S$; (d) C_4H_9N; (e) $C_3H_6O_3$. (*Note:* At least two possibilities exist for each, but you should show just one.)

28. Recall the formula

$$covalence = 8 - group\ number$$

(a) Explain this relationship in terms of the electronic structures of the bonding atoms. (b) Does the formula also apply to the noble gases? Justify your answer.

METALLIC BONDS

29. (a) In your own words, summarize the physical properties of metals. (b) Iron pyrite, FeS_2, is called "fool's gold," because many inexperienced prospectors have mistaken it for the real thing. What are the probable reasons for this confusion? What step would you, as an amateur prospector, take to ensure that your find was metallic?

30. Wax can be hammered into a sheet; organic matter can occur as filaments, such as hair or fibers, resembling thin wires; molten salt conducts electricity well; cinnabar (HgS) is eight times as dense as water. None of these substances is metallic. Explain in each case why the property described does not prove that the substance is a metal.

31. In your own words, summarize the band theory of metals and state how the theory accounts for the properties of metals.

32. Lithium crystallizes in a cubic arrangement in which each Li atom may be considered to be in the center of a cube, "touching" eight other Li atoms at the corners of the cube. Explain why the theories of neither ionic nor covalent bonding could account for such a structure.

ANSWERS TO SELECTED NUMERICAL QUESTIONS

3. 32

4. (a) 2, 8, 3; (b) 2, 8, 8; (c) 2, 2

8. (a) BeF_2, (b) $AlBr_3$, (c) Na_2S, (d) Ca_3P_2

9. (a) As = 3, (b) Ba = 2, (c) Ge = 4, (d) P = 3, (e) Ga = 3, (f) B = 3

10. (a) $4Ag + O_2 \rightarrow 2Ag_2O$, (b) $2KClO_3 \rightarrow 2KCl + 3O_2$, (c) $Fe_3O_4 + 2C \rightarrow 3Fe + 2CO_2$, (d) $2H_2S + SO_2 \rightarrow 3S + 2H_2O$

11. (a) Fe, (b) Al

12. CO_2

13. (a) $2Sc + 6HBr \rightarrow 2ScBr_3 + 3H_2$, (b) $4NH_3 + 5O_2 \rightarrow 4NO + 6H_2O$, (c) $3HClO \rightarrow 2HCl + HClO_3$

14. (a) $2KMnO_4 + 16HCl \rightarrow 2MnCl_2 + 2KCl + 5Cl_2 + 8H_2O$, (b) $2Fe_2O_3 + 3S \rightarrow 4Fe + 3SO_2$

18. (a) P^{3-}, S^{2-}, Cl^-, K^+, Ca^{2+}; (b) H^-, Li^+

19. (a) LiH, (b) $CaCl_2$, (c) K_2S, (d) CaH_2, (e) LiCl

21. (a) MgO—magnesium oxide, MgF_2—magnesium fluoride, (b) $(NH_4)_2SO_4$—ammonium sulfate, K_2SO_4—potassium sulfate, $Al_2(SO_4)_3$—aluminum sulfate

24. (a) H—S—H (b) F—Si—F (with F above and F below Si)

(c) F—C—Cl (with F above and F below C), (d) H—P—H (with H above and H below P)

(e) H—C—C—C—Cl (with H, H, H above and H, H, H below), H—C—C—C—H (with H, Cl, H above and H, H, H below) (f) H—N—Cl (with H below N)

(g) H—C—C—Cl (with H, Cl above and H, Cl below), H—C—C—Cl (with Cl, Cl above and H, H below)

25. (a) H—C—C=C—H (with H, H, H above and H below) (c) O=N—Br

Three elements from Group 4: tin (Sn), silicon (Si), and lead (Pb).
(Courtesy Charles Winters)

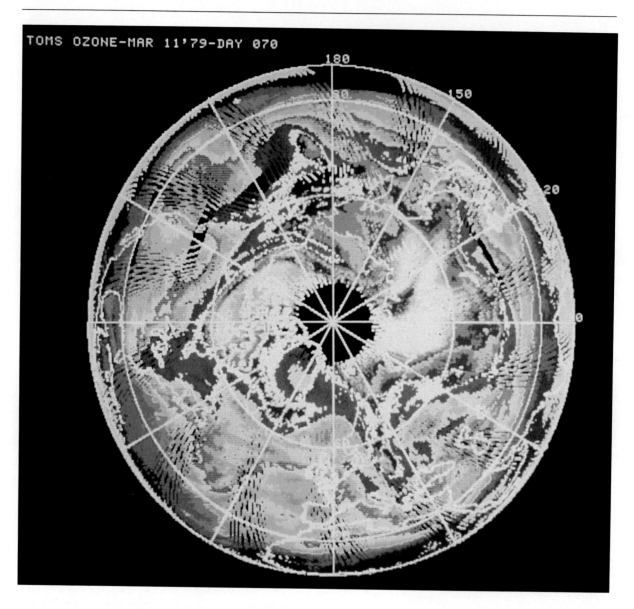

A map of the concentration of ozone in the atmosphere, as measured from space. This delicate layer may be in the process of destruction by manmade chemicals.
(Courtesy NASA)

C H A P T E R

17

Principles and Applications

17.1
ATOMIC WEIGHTS, MOLECULAR WEIGHTS, AND THE MOLE

Chemical substances are collections of small particles such as atoms, molecules, and ions. A single atom or molecule can be detected, but it cannot be handled in the way that ordinary samples of matter are dealt with in the laboratory. The smallest samples that chemists work with may be milligrams (10^{-3} g), micrograms (10^{-6} g), or even nanograms (10^{-9} g), but these small masses contain very many constituent particles. A nanogram of copper, for example, which is a billionth of a gram, contains about 10 trillion (10^{13}) copper atoms. Chemists therefore find it convenient to express the amount of a substance in terms of a standard number of its atoms or other constituent particles. Some examples in other walks of life will show that this practice is not unusual. A cook who prepares food for large numbers of people deals with eggs by the dozen, not by the kilogram

or the liter. A printer counts sheets of paper by the ream (1 ream = 500 sheets). Chemists use the **mole,** which is the amount of a substance that contains a certain number of fundamental particles. What is that certain number? The chemist answers the question in a way that may not satisfy you—the answer is this: Take *exactly* 12 g of carbon—not just any carbon, but specifically 12 g of the carbon-12 isotope. Those 12 g contain a certain number of atoms; that number is the number you are looking for, the number of fundamental particles in a mole. What is that number? The problem is that the number is not known exactly. The number is very large; it starts with 6 and is followed by 23 other digits. The best experimental value is 6.02205×10^{23}. The approximation 6.02×10^{23} is often used. Therefore, a simpler expression is that:

the mole is the amount of substance that contains 6.02×10^{23} basic particles.

The basic particles may be atoms, molecules, or

405

ions. Thus, a mole of copper, which is an element, is the amount that contains 6.02×10^{23} copper atoms, whereas a mole of water, which is a compound, is the amount that contains 6.02×10^{23} water molecules.

Neither of the above two definitions is very helpful to anyone who needs, say, a mole of nickel or a mole of water for an experiment. A chemist cannot count out the required number of atoms or molecules as a cook would count out eggs for a recipe. Fortunately, there is yet a third way of expressing what a mole is. Refer to the table of atomic weights inside the back cover. The atomic weights are not the same as mass numbers, because they take into account the distribution of isotopes as they occur on Earth. Thus, the atomic weight of carbon is not 12, but 12.011, because on Earth there is a little carbon-13 (and other isotopes) mixed in with the preponderant carbon-12 atoms. The atomic weights, then, are average values, but they are still based relative to the assignment of 12 units of mass to the carbon-12 isotope. From the first definition of the mole given above, a mole of carbon-12 weighs 12 g. Since all other atomic weights are based on carbon-12, it follows that

a mole of atoms of an element is the atomic weight of the element expressed in grams.

Thus, the atomic weight of nickel is 58.70, so a mole of nickel is 58.70 g.

What about molecules? The sum of the atomic weights in a molecular formula is the **molecular weight.** Therefore, a mole of molecules of a substance is the molecular weight of the substance expressed in grams.

It is difficult to visualize the meaning of such a very large number as 6×10^{23}, the number of atoms or molecules in a mole. One way to think about it is to recognize that atoms in any ordinary sample of matter are so numerous, and chemical changes go on so constantly all around us, that it is very likely that many of the atoms that were in the body of any ancient person you choose to name are now in your body. Or consider this: If you could identify every molecule in a cup of water and then poured that water in the ocean and stirred the water so thoroughly that your molecules were uniformly dispersed in *all* the

world's oceans, and then dipped your cup back in to refill it, you would recover over a thousand of your original molecules.

EXAMPLE 17.1

(a) Calculate the molecular weight of water, H_2O, and of carbon dioxide, CO_2. Use the following approximate atomic weights: O, 16; H, 1; C, 12.

(b) Calculate the mass of 1 mole of water and of 3 moles of carbon dioxide.

Solution **(a)** From the molecular formulas, the molecular weight of water is $1 + 1 + 16 = 18$, and for carbon dioxide it is $12 + 16 + 16 = 44$.

(b) The mass of a mole of water is its molecular weight in grams, or 18 g. For carbon dioxide, 3 moles are (3 moles)(44 g/mole) = 132 g.

17.2
ACIDS AND BASES

The original meaning of **acid** is "sour," referring to the taste of substances such as vinegar, lemon juice, unripe apples, and old milk. It has long been observed that all acidic substances have some typical properties in common. For naturally occurring **acidic solutions,** in which the solvent is always water, these properties are:

1. Acids taste sour. (But *never try to taste acids found in the laboratory; the results could be fatal.*)
2. Acids speed up the corrosion, or rusting, of metals. When the attack on a metal by an acidic solution is vigorous, hydrogen gas, H_2 is evolved in the form of visible bubbles.
3. Acidic solutions conduct electricity. As the current passes through the solution, hydrogen gas is evolved at the negative electrode (cathode).
4. Acids affect the colors of certain botanical substances, which are known as **indicators.** For example, if you add lemon juice to a cup of tea, the color of the tea becomes lighter. Lemon juice is an acid; therefore tea is an indicator that changes color when an acid is added.
5. The characteristic properties of an acid can be changed by the addition of certain other substances, known as **bases.** This chemical reaction is called **neutralization.**

(a)

(b)

(a) One mole of some common elements. Back row (left to right) bromine, aluminum, mercury, and copper. Front row (left to right) sulfur, zinc, and iron. (b) One-mole quantities of a range of compounds. The white compound is NaCl; the blue compound is $CuSO_4 \cdot 5H_2O$; the deep red compound is $CoCl_2 \cdot 6H_2O$; the green compound is $NiCl_2 \cdot 6H_2O$; and the orange compound is $K_2Cr_2O_7$.

Bases do not occur in as many common materials as do acids, but a few common basic materials are ashes, soap (especially strong dishwasher soaps), borax, lime, and lye. The characteristic properties of bases and **basic solutions** are:

1. Bases taste bitter. *But don't ever try to taste the materials referred to above or basic solutions found in the laboratory.* Two common household products, drain cleaners and oven cleaners, commonly contain lye, which is a very strong base, also called a caustic or an alkali. The chemical name is sodium hydroxide, NaOH. Products that contain this material should not be allowed near children.

2. Basic solutions in which the solvent is water conduct electricity, just as acidic solutions do. Oxygen gas, O_2, is given off at the anode (positive electrode).

3. Bases affect the colors of indicators, but the color changes are different from those pro-

duced by acids. For example, the natural indicator in purple grape juice, which is an acidic solution, turns green when enough dishwasher soap is added to make the juice basic. A common laboratory indicator is **litmus,** which is red in acid solutions and blue in basic ones. In literary parlance, a "litmus test" is used in the general sense of proving that something is definitely either this way or that—no ifs, ands, or buts.

4. Bases neutralize acids. The reaction is so rapid that it is practically instantaneous, and it gives off considerable heat. If, for example, solutions of hydrochloric acid (a very strong acid) and sodium hydroxide (a very strong base) are mixed in just the right proportions, the product is neither acidic nor basic; it is a neutral solution of ordinary salt, NaCl. Other acids and bases produce other salts. In general,

neutralization is defined as the reaction between an acid and a base to produce a salt and water.

The properties of acids and bases are interpreted as follows:

The fact that hydrogen gas, H_2, is produced when acidic solutions react with metals leads to the conclusion that acidity is related to hydrogen.

Since acidic solutions conduct electricity, they must contain charged particles. The charged particles that migrate to the cathode (negative electrode) must be positive. Since hydrogen gas is given off at the cathode, there must be some molecules or ions that contain hydrogen. The simplest positive ion is H^+, which is a proton. However, a proton cannot exist as an independent particle in water, because it becomes chemically bonded to the oxygen atom of the water molecule. The resulting ion is formulated as $H(H_2O)^+$, or simply H_3O^+, and is called the hydronium ion. It is simpler to write H^+, as many books, including this one, do; regard this as an abbreviation. The reaction at the cathode may then be written as

$$2H^+ + 2e^- \longrightarrow H_2$$

Basic solutions contain hydroxide ions, OH^-. When an electric current is passed through such a solution, the OH^- ions migrate to the anode, where oxygen gas is liberated:

$$4OH^- \longrightarrow O_2 + 2H_2O + 4e^-$$

Hydroxide ions, OH^-, can neutralize H^+ ions by reacting with them to produce water, as indicated by the equation

$$H^+ + OH^- \longrightarrow H_2O$$

Even pure water ionizes to a very slight extent. The concentration of H^+ and of OH^- in pure water at 25°C is 1.0×10^{-7} moles/L. This solution is said to be **neutral,** because the concentrations of the two ions are equal. When the hydrogen ion concentration is greater than 1.0×10^{-7} moles/L at 25°C, the solution is **acidic.** When it is less than this value, the solution is **basic.**

Hydrogen ion concentrations are usually expressed by a set of values called a **pH** scale (see next section). A neutral solution has a pH of 7. Acidic solutions have pH values below 7, and every decrease of one pH unit represents a *tenfold* increase in the concentration of H^+ ions. Basic solutions have pH values above 7, and every additional pH unit represents a tenfold *decrease* in the H^+ ion concentration (and a tenfold increase in hydroxide ion concentration).

Strong acids ionize completely in water, or nearly so. As a result, they furnish a large concentration of H^+ ions. An example in this category is nitric acid, HNO_3:

$$HNO_3 \longrightarrow H^+ + NO_3^-$$

In the case of a weak acid, such as acetic acid, only a small percentage of the molecules ionize, so that the concentration of H^+ ions remains low.

Table 17–1 gives the pH values of various common materials.

17.3 THE pH SCALE

The pH values are based on logarithms to the base 10. The relationship is

$$\text{pH} = -\log_{10} (\text{hydrogen ion concentration})$$

(17.1)

Recall that the logarithm of a number to the base 10 is the exponent to which 10 must be raised to give that number. For example, $10^2 = 100$, so log 100 = 2, and $10^{-3} = 0.001$, so log 0.001 = −3. The negative sign in the definition of pH

TABLE 17-1 Approximate pH Values of Various Substances

	pH	Substance
	14	←NaOH solution (lye), 4%
	13	←Limewater
	12	←Household ammonia
		←Washing soda (about 1%)
	11	
	10	
	9	
Basic	8	←Bicarbonate of soda (about 1%)
		←Blood -
Neutral	7	←Pure water
		←Cow's milk - - - - - - - - - - - - - - - - - -
Acidic	6	←Unpolluted rainwater
	5	←Squash, pumpkin
	4	
		←Oranges
	3	←Vinegar; soft drinks
	2	←Limes
	1	←Dilute hydrochloric acid solution, HCl

changes the sign of the exponent. Thus, if the H^+ concentration (in moles/L) is 10^{-7}, the pH is 7. If it is 10 times this value, or 10^{-6}, the pH is 6, which is an acidic solution. If it is one-tenth as concentrated, or 10^{-8}, the pH is 8, which is basic.

In summary, a solution whose pH is less than 7 is acidic. If the pH is greater than 7, it is basic. A pH of 7 is neutral.

The term pH was derived from "powers of hydrogen," where "power" refers to the exponent.

17.4
OXIDATION AND REDUCTION

Oxygen is not only the most abundant element in the Earth's crust but also one of the most versatile. In its common molecular form, O_2, it is the second most abundant constituent of the atmosphere. In its chemically combined forms, it is a major component of water, of all living organisms, and of most rocks and minerals. Oxygen forms bonds with every element except some of the noble gases. The formation of these bonds releases energy. A reaction with oxygen that releases energy rapidly enough to give off light is called **combustion.** The reacting chemical is said to *burn* in oxygen, or in a gas, such as air, that is rich enough in oxygen. You are familiar with the fact that materials such as wood and paper burn in oxygen or in air. It is not so widely known, however, that some metals, too, can burn. Magnesium, for example, is a shiny metal that burns very rapidly in oxygen. If the magnesium is dispersed in the form of a powder, all of it burns practically at the same time and the result is an explosion. The balanced equation for the reaction is:

FOCUS ON . . . Pure Water

Chemically pure water, which has a pH of 7, does not occur in nature. The reason is that water is a very good solvent, and even in an unpolluted environment it captures impurities by dissolving gases from the atmosphere and minerals from soil and rock. Sometimes these impurities are acidic, which reduces the pH below the neutral value of 7, and sometimes they are basic, yielding a pH higher than 7. When we speak of "pure drinking water" in ordinary conversation, we mean water that is wholesome, free from noxious chemicals or disease organisms, and usually containing small amounts of tasty and probably beneficial mineral matter. The preparation of chemically pure water, on the other hand, requires great care. It must be distilled in the absence of air, to avoid carbon dioxide, and without using glass, from which the water would dissolve silicates.

$$2Mg + O_2 \longrightarrow 2MgO$$

The formation of a chemical bond between oxygen and another element is called **oxidation.**

Compounds, too, can be oxidized. The familiar reactions of respiration and of the burning of natural gas (methane) are oxidations. The equation for respiration is shown as the oxidation of glucose (blood sugar):

$$C_6H_{12}O_6 + 6O_2 \longrightarrow 6CO_2 + 6H_2O$$
Glucose

The equation for the oxidation of methane is:

$$CH_4 + 2O_2 \longrightarrow CO_2 + 2H_2O$$

In general, when substances containing carbon, hydrogen, or both oxidize completely, all the carbon is oxidized to carbon dioxide, CO_2, and all the hydrogen to H_2O. Thus, the same oxidation products are formed by the complete oxidation of glucose, methane, or octane (C_8H_{18}).

We now return to magnesium: Its oxidation product, MgO, is not metallic at all; it is a white, powdery solid. Furthermore, it is an ionic compound, which can be represented as Mg^{2+} +

O^{2-}. Therefore, the equation for the oxidation of magnesium can also be written as

$$2Mg + O_2 \longrightarrow 2Mg^{2+} + 2O^{2-}$$

Magnesium also reacts with chlorine, Cl_2, to produce magnesium chloride, $MgCl_2$, as shown by the equation

$$Mg + Cl_2 \longrightarrow MgCl_2$$

But magnesium chloride, too, is ionic, so the equation could be written as

$$Mg + Cl_2 \longrightarrow Mg^{2+} + 2Cl^-$$

Now note that the Mg becomes Mg^{2+} whether it reacts with oxygen or with chlorine, so if the reaction with oxygen is oxidation, wouldn't it be reasonable to call the reaction with chlorine oxidation, even if no oxygen is involved? Consider another question: If all that happens to the magnesium is its conversion to Mg^{2+}, could such ions be produced in any other way? The answer is yes, the magnesium could simply be the anode in an electrochemical cell, as shown in Figure 17.1. Then the magnesium loses electrons, producing positive ions, which dissolve in the solution:

$$Mg \longrightarrow Mg^{2+} + 2 \text{ electrons}$$

These observations lead to a second definition of oxidation:

Oxidation is the loss of electrons.

If oxidation is either the formation of a bond to oxygen or the loss of electrons, then there must be an opposite process:

Reduction is defined as the breaking of a bond to oxygen or the gain of electrons.

The word "reduce" comes from the Latin *reducere*, to bring back or restore, but this meaning has almost entirely disappeared from ordinary usage. In chemistry, however, the original sense is preserved in that reduction of a metallic oxide brings back or restores the metal:

$$2MgO \longrightarrow 2Mg + O_2$$

In ionic form, the equation showing the reduction of the Mg^{2+} ion by the gain of electrons is

$$Mg^{2+} + 2 \text{ electrons} \longrightarrow Mg$$

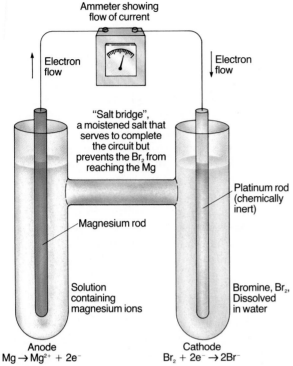

FIGURE 17.1 An electrochemical cell in which magnesium is the anode and loses electrons to become Mg^{2+}. At the cathode, bromine (Br_2) is being reduced by gaining electrons to become Br^- ions.

One way to think about oxidation and reduction is to realize that oxidation generally turns useful materials into wastes, whereas reduction does just the opposite—it converts wastes or useless products into valuable materials. In the following examples, the substances referred to are all being oxidized:

Burning of wood or coal $\rightarrow$ CO_2, H_2O, smoke, and ashes

Metabolism of food (respiration) $\rightarrow$ body wastes

Corrosion of iron $\rightarrow$ Fe_2O_3 (red iron rust)

The benefit realized from these oxidations is the release of energy.

In the following chemical changes, substances are being reduced:

CO_2 + water (photosynthesis) $\rightarrow$ sugars

Reaction of iron ore, Fe_2O_3, with carbon $\rightarrow$ iron

These reductions are not without cost; they consume energy.

One more point about oxidation and reduction: They always occur together in chemical systems, never separately. If one thing is oxidized, something else must be reduced. The reason is that the electrons that are released must go somewhere. Returning to the first reaction given in this section,

$$2Mg + O_2 \longrightarrow 2MgO$$

If the magnesium is oxidized, what is reduced? Rewriting the equation in ionic form provides the answer:

$$2Mg + O_2 \longrightarrow 2Mg^{2+} + 2O^{2-}$$

It is the oxygen that gains electrons and that therefore is reduced:

$$O_2 + 4e^- \longrightarrow 2O^{2-}$$

17.5
CASE HISTORY: ACID RAIN

Acid rain has recently been recognized as a significant environmental issue. (The more general terms are acid precipitation or acid deposition, which include acidic snow and dust as well as rain.) It is important to realize, however, that even rain that falls through an unpolluted atmosphere is slightly acidic. The reason is related to the composition of the atmosphere, which is shown in Figure 17.2. The 1 percent of "other gases" includes carbon dioxide, which reacts to a slight extent with water to produce carbonic acid, a weak acid:

$$CO_2 + H_2O \longrightarrow \quad H_2CO_3$$
$$\text{carbonic acid}$$

In addition, some nitric acid is formed during lightning storms by the oxidation of nitrogen, N_2, in the presence of water:

$$2N_2 + 5O_2 + 2H_2O \longrightarrow \quad 4HNO_3$$
$$\text{nitric acid}$$

Nitric acid is a very strong acid, but not enough of it is formed by natural processes to introduce much acidity into rainwater. The combined effects of carbonic acid and nitric acid in unpol-

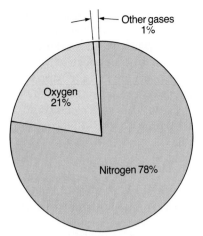

FIGURE 17.2 Approximate gaseous composition of natural dry air.

luted air make rainwater (or snow) slightly acidic, with a pH a bit below 6.

In recent years, however, rain and snow in many parts of the world have become considerably more acidic. Much of this acid precipitation has been between pH 4 and 5, but more severely acidic episodes occur from time to time. For example, a rainstorm in Baltimore in 1981 had a pH of 2.7, which is about as acidic as vinegar. Acid rain is a worldwide problem that has been estimated to cause billions of dollars worth of damage. Acids corrode exposed metallic structures, such as bridges and automobiles, and attack concrete and rock, particularly limestone and marble. Acids also cause the death of trees and fish and reduce the growth of certain agricultural crops.

Most of this excess acidity can be traced to a series of chemical reactions involving sulfur and nitrogen. The small concentrations of nitric acid that are formed by the action of lightning are augmented by the burning of fossil fuels. In the gasoline engine, for example, the electric spark generated by the spark plug serves as a substitute for lightning, and some nitrogen in the air in the cylinder is oxidized to NO,

$$N_2 + O_2 \longrightarrow 2NO$$

When the NO leaves the exhaust pipe and comes in contact with the air and moisture of the outside

atmosphere, it is further oxidized to nitric acid:

$$4NO + 3O_2 + 2H_2O \longrightarrow 4HNO_3$$

All in all, however, the total atmospheric acidity that starts with sulfur is greater than that coming from nitrogen. The reason is that sulfur, being essential to life, existed in the organisms from which fossil fuels originated. As the remains of prehistoric plants and animals gradually became transformed to coal and oil, some of the hydrogen and most of the oxygen and nitrogen escaped, but much of the sulfur stayed put and is still present in these fuels. (So is the mineral content, which also contains sulfur compounds.) Coal is especially rich in sulfur. As coal is burned and the carbon is oxidized to CO_2, the sulfur is oxidized to sulfur dioxide, SO_2, which is also a gas:

$$C + O_2 \longrightarrow CO_2$$
$$S + O_2 \longrightarrow SO_2$$

The oxidation of sulfur to SO_2 occurs directly in the flame, and therefore SO_2 is discharged to the atmosphere from the smokestack. As the SO_2 is swept along by the prevailing winds, it is slowly oxidized at ordinary temperatures to SO_3:

$$2SO_2 + O_2 \longrightarrow 2SO_3$$

SO_3 then reacts rapidly with atmospheric moisture to form sulfuric acid, which is a very strong acid:

$$SO_3 + H_2O \longrightarrow \quad H_2SO_4$$
$$\text{sulfuric acid}$$

Sulfuric acid is very soluble in water and is therefore washed out by rain. It is for these reasons that acid rain often occurs at considerable distances from the sources where the sulfur dioxide is introduced into the atmosphere. The conversion of nitrogen oxides to nitric acid is also slow enough to carry these pollutants for some distances before they come down as acid rain.

Even before the acids are formed in the atmosphere, some of the oxides of nitrogen and sulfur become attached to dust particles and adhere to them as the particles fall to earth. When they come in contact with rivers, lakes, or the moisture in soil, they react to form acidic solutions. These acidic dusts often precipitate closer

(a)

(b)

The effects of acid rain on the evergreen population atop Camel's Hump in Vermont's Green Mountain range. Part (a) was photographed 15 years earlier than part (b).
(Courtesy U.S. Environmental Protection Agency)

to the pollution sources than the acid rain and snow, as shown in Figure 17.3.

Various surveys in the Adirondack Mountains of upstate New York, as well as in Canada and elsewhere, showed that the pH of large numbers of ponds and lakes was less than 5 and that many of these had completely lost their fish populations. Plants, too, are affected. Mysterious blights that have killed increasing numbers of trees in recent years have been traced to acid rain. However, there is no simple relationship between the amount of acid and the resulting environmental damage, for some interesting chemical reasons. Here are some of the complications:

1. Most of the acid precipitation that reaches a lake does not fall directly into it but, instead, falls on the land and then runs off into the water. If the soil is rich in limestone, it can neutralize most of the acid before the rain runs off the land:

$$CaCO_3 + H_2SO_4 \longrightarrow CaSO_4 + H_2O + CO_2$$

limestone   sulfuric   calcium
            acid   sulfate

In some instances, the added sulfate even acts as an agricultural fertilizer. When the soil is poor in limestone, however, this mechanism does not operate.

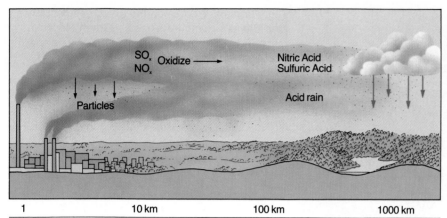

FIGURE 17.3 Acid rain. SO_x means SO_2 or SO_3. NO_x means NO, NO_2, or other oxides of nitrogen.

2. Observations in West Germany's Black Forest and elsewhere indicated that trees on mountain tops and hilltops suffered a much greater percentage of loss than those at lower levels. This curious phenomenon is related to the prevalence of fogs at these summits. Recall from Chapter 15 that a fog or mist is not a gas but a collection of tiny liquid droplets. Sulfuric acid and nitric acid are highly soluble in water, and therefore they dissolve in fog droplets, where their concentration becomes about ten times higher than their average atmospheric concentration. For this reason, fogs have been called the "vacuum cleaners of the atmosphere." It is these high acid concentrations that cause the greatest damage to trees.

3. Another unexpected effect has been the greater damage done to pine trees by nitric acid as compared to sulfuric acid, even though the atmospheric concentration of nitric acid is much less than that of sulfuric acid. Here the damage may be caused by the nutrient effect of nitrogen, which fertilizes the pine needles and causes excess growth in the late fall, making them more susceptible to winter injury.

There are several approaches to the prevention of acid precipitation or the damage it creates. One is desulfurization, the removal of sulfur from fuel before it is burned. This method utilizes a combination of chemical and mechanical processes to remove sulfur, mainly in the form of calcium sulfate, $CaSO_4$, which can then be disposed of as a nonhazardous solid waste. Another approach is the removal of the sulfur and nitrogen oxides from the combustion gases before they enter the atmosphere. This method utilizes **scrubbers,** which are devices for bringing gases and liquids into close contact with each other, as shown in Figure 17.4. Since the gases to be removed are acidic, the scrubbing liquid must contain a base to neutralize the acid. The cheapest basic material is limestone, $CaCO_3$. The base derived from limestone is calcium hydroxide, $Ca(OH)_2$, and the reactions with the sulfur oxides can be formulated as follows:

$$SO_2 + Ca(OH)_2 \longrightarrow CaSO_3 + H_2O$$
$$\text{calcium sulfite}$$

$$SO_3 + Ca(OH)_2 \longrightarrow CaSO_4 + H_2O$$
$$\text{calcium sulfate}$$

Finally, if the source cannot be controlled, the acidified area can be treated. An effective stop-gap method involves neutralizing the acid in a lake by adding the necessary quantity of lime, after which the lake can be restocked with fish.

17.6 CATALYSIS

A mixture of hydrogen and oxygen gases, ideally with twice as many hydrogen molecules as oxygen molecules, is potentially explosive. But if such a mixture is left alone at room temperature, nothing is seen to happen—no explosion, not even

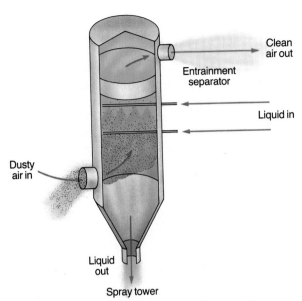

FIGURE 17.4 Schematic diagram of a scrubber.

any slow reaction. If a lighted match is introduced into the mixture, however, or even the tiniest spark, the reaction starts and spreads so rapidly that it seems to happen all at once: It explodes. The equation is

$$2H_2 + O_2 \longrightarrow 2H_2O$$

If the mixture were in a strong steel cylinder, the kind that is designed to hold gases at hundreds of atmospheres of pressure, the cylinder would burst, and fragments of steel would be blown away for distances of hundreds of meters, some perhaps for kilometers. But there is another way to set off this explosion without a match or a spark, in fact without raising the temperature of the mixture at all. All you would have to do would be to introduce a little finely divided platinum powder into the gas, and off it would go. If it were possible to gather all the products of the explosion, the platinum could be recovered and used again. The platinum has served as a **catalyst.**

A catalyst is a substance that increases the rate of a chemical reaction but is not consumed in the reaction.

The action of a catalyst may seem mysterious. How can a substance that is not used up in a reaction influence the rate of that reaction? Be-

fore addressing that question, we may ask why even a spark or a lighted match is needed to set off the hydrogen-oxygen explosion. After all, the reaction itself releases a lot of energy, so why is energy needed to start it? The process can be understood by considering the structural formulas of the substances involved: H—H for hydrogen, O—O for oxygen, and H—O—H for water, the product of the explosion. The important point is that there are no H—H or O—O bonds in a water molecule. Therefore, for the reaction to occur, the H—H bonds in hydrogen and the O—O bonds in oxygen must be broken. Some outside source of energy is needed to break those bonds. If those bonds are not broken first, nothing happens. Once the reaction does start, however, it provides more than enough energy to keep itself going. The energy released when the H—O—H bonds of water are formed serves to decompose more O_2 and H_2 molecules. There is enough energy released, in fact, to initiate a branching chain reaction, which becomes an explosion.

Now we can understand the action of the catalyst. Since the reaction cannot start before some bonds are broken, the catalyst must somehow substitute for the spark or the lighted match. What does actually happen in this case is that the hydrogen molecules form bonds to the platinum atoms, in an arrangement that may be represented as

$$\begin{array}{c} H\text{---}H \\ | \quad | \\ Pt\text{---}Pt \end{array}$$

This action weakens the bonds between the hydrogen atoms, and this weakening is just enough to get the reaction started. The start is all that is needed; the chain reaction is then on its way.

Does a catalyst participate in the reaction that it speeds up? The answer is yes. Return to the definition of a catalyst and note that it does not say that the catalyst is uninvolved, only that it is not consumed. The platinum really has to react (that is, form bonds) with the hydrogen for the catalysis to occur. When the H—H bonds break and the H atoms start to react, the platinum remains behind. It has not been consumed, which is a necessary condition for a catalyst.

Catalysis occurs in many naturally occurring chemical reactions, both in living and in nonliving systems, and is also used extensively in the chemical industry. Catalysts can operate in the gaseous, liquid, or solid states of matter, as well as in reactions that involve more than one state. The use of platinum to catalyze the reaction between hydrogen and oxygen is an example that involves more than one state: The platinum is a solid, and the hydrogen and oxygen are gases.

Humans and other animals obtain their energy by the oxidation of carbohydrates, fats, and proteins. If you were given some sugar or beef fat in the laboratory and told to oxidize it, you would have to burn it; the oxidation would then take place at the temperature of the flame, which can be somewhere between 500 and 1000°C. In air at 37°C, the reaction would be too slow to measure. But in our bodies, the oxidation does take place at 37°C, for that is normal human body temperature. Speeding up the reaction from the immeasurably slow rate in outside air to our normal metabolic rates requires very powerful catalysts. These body catalysts are called **enzymes,** which are a particular group of proteins.

EXAMPLE 17.2

The oxidation of SO_2 to SO_3 in air is slow. The equation is

$$2SO_2 + O_2 \longrightarrow 2SO_3$$

The following two reactions, however, are fast:

$$2NO + O_2 \longrightarrow 2NO_2$$

$$2NO_2 + 2SO_2 \longrightarrow 2NO + 2SO_3$$

What is the net effect of the last two reactions? What substance is the catalyst?

Solution Add the two reactions by combining all the formulas to the left of both arrows and showing that they yield all the formulas to the right of both arrows:

$$2NO + O_2 + 2NO_2 + 2SO_2 \longrightarrow$$
$$2NO_2 + 2NO + 2SO_3$$

Note that the $2NO$ and $2NO_2$ formulas appear on both sides of the equation and can therefore be

crossed out. The result will be the net reaction

$$O_2 + 2SO_2 \longrightarrow 2SO_3$$

This is the same reaction as the slow oxidation shown at the beginning of this example, but now it has taken place in two fast steps, which are faster than one slow one. The new chemical that was added was NO, but that was also recovered, not used up, so NO was the catalyst.

17.7 CASE HISTORY: OZONE IN THE ATMOSPHERE

In the stratosphere

Ozone, O_3, is very different from oxygen, O_2. Pure ozone is a blue, explosive, poisonous gas. Very few people ever see it, but most have smelled it. The pungent odor from electric sparks, such as from a worn, sparking electric motor, or the slightly pungent odor of the atmosphere after a lightning storm is the odor of ozone. Small amounts of ozone are produced naturally in the stratosphere by the action of sunlight on oxygen. (The stratosphere is the section of the atmosphere about 20 km above sea level. The structure of the atmosphere is described in Chapter 18.) Oxygen molecules, O_2, are first broken down by ultraviolet (UV) radiation to oxygen atoms. In a second step, the oxygen atoms combine with oxygen molecules to produce ozone. The equations are:

$$O_2 + UV \longrightarrow 2O$$

$$O_2 + O \longrightarrow O_3 + \text{infrared (IR) radiation}$$

The prevailing ozone concentration in the stratosphere from these processes is about 0.1 ppm.

Stratospheric ozone is involved in an interesting transition of solar radiation from a higher to a lower energy level. Specifically, a large portion of the solar ultraviolet radiation that reaches the stratosphere is converted to infrared radiation before it reaches the surface of the Earth. The conversion involves the following steps:

Step 1: Ozone absorbs UV radiation and is decomposed:

FOCUS ON . . . Expressions of Fractional Concentrations

The expressions ppm and ppb, as well as percent, are commonly used to refer to small concentrations. The meanings are:

$1\% = 1$ part per hundred $= 1/100 = 10^{-2}$
1 ppm $= 1$ part per million $= 1/1,000,000 = 10^{-6}$
1 ppb $= 1$ part per billion $= 1/1,000,000,000$
$\quad = 10^{-9}$

These expressions do not, by themselves, state what "parts" are referred to. For gases, the reference is always to parts by volume. Since the volume of a gas is proportional to the number of molecules it contains (at a definite temperature and pressure), parts by volume is the same as parts by number of molecules. Thus, the statement that the atmospheric concentration of CO_2 is 350 ppm means that there are 350 molecules of CO_2 in every million molecules

of air. A concentration of ozone of 0.1 ppm means that there is 1 molecule of ozone for every 10 million molecules of air.

For solids, the reference is usually to parts by mass, but for liquids it could be either, so it should be specified.

Concentrations expressed in parts per million or parts per billion seem quite small. In terms of atoms or molecules per unit mass or volume, however, the numbers are very large. For example, if there is 1 ppb (10^{-9}) by mass of, say, lead in water, then each gram of water contains about 3×10^{12}, or three trillion, lead atoms. It is not just the magnitude of a number—small for a concentration but large when you count atoms—that is significant. Rather, it is the effect of the components on the properties of the mixture that is important.

$$O_3 + UV \longrightarrow O_2 + O$$

Step 2: The O_2 and O recombine, releasing IR radiation (thermal energy):

$$O_2 + O \longrightarrow O_3 + IR$$

Adding steps 1 and 2 gives:

$$O_3 + UV + O_2 + O \longrightarrow$$
$$O_2 + O + O_3 + IR$$

All the chemical formulas cancel out, and the net result is

$$UV \longrightarrow IR$$

Thus, stratospheric ozone is a catalyst that provides a chemical pathway for converting some (not all) of the solar ultraviolet radiation into infrared radiation.

The chemistry of ozone in the stratosphere has attracted attention in recent years because the ozone layer is considered to be an important barrier that protects life on Earth, and some gases newly introduced into the atmosphere by human activities threaten that protective barrier. The protection offered by stratospheric ozone lies in its action in converting UV to IR radiation, as discussed above. Photons in the ultraviolet range are energetic enough to promote electrons in various

organic molecules to excited states and thus trigger chemical reactions. The UV content of sunlight can thus affect human skin in various ways, some good and some bad. The good ones include the attractive tanning of a pale skin and, more important for health, the conversion of ergosterol (a chemical naturally present in skin) to vitamin D. The bad effects may include a painful sunburn and, more seriously, skin cancer. Also, other organisms, plants as well as animals, are affected in various ways by UV radiation. The life forms that now exist on Earth have adapted to the present ranges of UV intensities. These intensities, however, differ from place to place and are about seven times greater in the tropics than in the arctic. People who migrate from high to low latitudes may be damaged by the more intense UV exposure, and it is possible that migrants in the other direction may suffer from a deficiency of UV. These complexities make it difficult to predict accurately what would happen if a depletion of the ozone layer resulted in a general increase of UV intensity on the surface of the Earth. Earlier reports predicted large increases in the incidence of skin cancer in humans, as well as retardation of the growth of some food crops. More recent estimates are less certain of what the actual damage would be. However, there is no question that UV radiation can cause cancer, that there is no "safe" level of UV, and that any increase over the present levels is therefore a potential hazard.

The possibility that human activities may reduce the natural stratospheric concentration of ozone arises from the introduction of certain atmospheric pollutants. Estimates of the seriousness of this threat are uncertain, but findings of greatly lowered stratospheric ozone in the antarctic are alarming. Two possible chemical processes are described here. One of these processes involves nitrogen oxide (NO).

$$NO + O_3 \longrightarrow NO_2 + O_2$$
$$NO_2 + O \longrightarrow NO + O_2$$
$$\text{Net equation} \quad O_3 + O \longrightarrow 2O_2$$

Note that the sum of the two equations is the destruction of ozone. Furthermore, the NO is not consumed in the process. It acts as a catalyst, and therefore small quantities of NO can destroy large quantities of ozone. These reactions have implied that a fleet of supersonic transport (SST) aircraft, flying in the stratosphere, might upset the ozone balance because the NO in the jet exhaust could initiate the ozone depletion sequences shown above.

Another possible mechanism for depletion of the ozone layer involves chlorine atoms. As shown in the following equations, chlorine atoms catalyze the ozone depletion reaction:

$$Cl + O_3 \longrightarrow ClO + O_2$$
$$ClO + O \longrightarrow Cl + O_2$$
$$\text{Net equation} \quad O_3 + O \longrightarrow 2O_2$$
(same reaction catalyzed by NO)

The important stratospheric sources of atomic chlorine are the chlorofluorocarbons, or CFCs. These compounds contain covalently bonded C, Cl, and F atoms, and they are chemically stable in the lower atmosphere. One of them, $CFCl_3$, has been used as a propellant in aerosol cans. Under compression, it provides the pressure that propels the liquid out as a fine mist. The $CFCl_3$ itself simply escapes as a gas into the atmosphere. The other important chlorofluorocarbon, CF_2Cl_2 (Freon), is the working substance that transfers energy in refrigerators and air conditioners.

When one of these appliances breaks or wears out, the CF_2Cl_2 leaks out, and it, too, escapes as a gas into the atmosphere. Since these compounds are stable, they persist long enough to diffuse into the stratosphere. There they become exposed to solar ultraviolet radiation that is energetic enough to break the C—Cl bonds and release Cl atoms. The chlorine pathway is the more effective one for the removal of atmospheric ozone. C—F bonds, on the other hand, are too strong to be broken by UV photons, so F atoms are not involved in this process.

In the lower atmosphere

The natural concentration of O_3 in the lower atmosphere is about 0.02 ppm. At one time, you could buy home "air purifiers" that were supposed to make your air fresher by producing ozone. However, ozone is a toxic gas, even though small concentrations of it in air do give a sensation of freshness. Therefore these devices did not purify the air; they polluted it.

17.8 CASE HISTORY: POLLUTION BY AUTOMOBILE EXHAUST

By the early 1900s, many industrial cities were heavily polluted. The major sources of pollution were no mystery. The burning of coal was number one. Other specific sources, such as a steel mill or a copper smelter, were readily identifiable. The major air pollutants were mixtures of soot and oxides of sulfur, together with various kinds of mineral matter that make up fly ash. When the pollution was heavy, the air was dark. Black dust collected on window sills and shirt collars, and newly fallen snow did not stay white very long.

Air pollution in Los Angeles, however, seemed to have different qualities. Especially in the years after World War II, when population boomed and automobiles became almost as numerous as people, the quality of the atmosphere began to deteriorate in a strange way. It was certainly air pollution, but it did not resemble the smog in London or Pittsburgh. The differences are summarized in Table 17-2.

TABLE 17–2 Two Types of Smog

Los Angeles Smog (Photochemical)*	London Smog (Soot and Sulfur)
Begins only during daylight	Begins mostly at night
Smells something like ozone; can also irritate the nose	Smells smoky
Looks yellow to brown	Looks gray to black
Damages certain crops such as lettuce and spinach	Damages stone buildings, especially limestone and marble
Irritates the eyes, causes blinking	Can be responsible for acute respiratory illnesses
Makes rubber crack	Causes acid rain

*The association of the type of smog with the city is historical, and does not necessarily refer to typical conditions now or in the future. Major efforts at improving air quality have been made in various areas that were once polluted and, conversely, air quality has deteriorated in other areas that were once pristine.

Scientists looked for possible sources of the mysterious air pollution. There was relatively little heavy industry in Los Angeles in those years, the warm climate minimized the need for home heating, and people did not burn coal. However, domestic garbage was commonly burned in open back-yard incinerators, which were inevitably smoky, so these were banned. Nevertheless, the pollution persisted. Finally, A. J. Haagen-Smit, a chemist, turned his attention to automobiles, and in 1951 he reported on his crucial experiments with them. He piped automobile exhaust into a sealed room equipped with ultraviolet lamps, as shown in Figure 17.5. The room contained various green plants and pieces of rubber. The room was also provided with little mask-like windows that permitted people to stick their faces in and smell the inside air. His results are summarized in Table 17–3. There were the crucial experiments that identified the source of Los Angeles smog and eventually caused its name to be changed to **photochemical smog.** (It isn't just in Los Angeles.) Haagen-Smit's results showed that the combination of auto exhaust and sunlight is responsible. Solar UV radiation promotes the formation of ozone, which must be involved in the production of the smog, as shown by the results of Experiment 3. Haagen-Smit and other chemists then searched for reactions that caused this pollution.

Gasoline is a very complex mixture of chemicals, mostly hydrocarbons containing seven or eight carbon atoms per molecule. A typical formula is C_8H_{18}, which represents a set of isomers known as "octanes." In an ideal combustion, the octanes would all burn completely to CO_2 and water, as shown by the equation

$$2C_8H_{18} + 25O_2 \longrightarrow 16CO_2 + 18H_2O$$

The molecular structure of the hydrocarbon includes various C—C and C—H bonds, as illustrated by the structural formula for "iso-octane" given in Chapter 16.

However, there are no C—C bonds in CO_2. Therefore, all the bonds in the octane must be broken for the reaction to go to completion. This decomposition does not occur instantaneously; it takes a little time. When an automobile engine is running, the piston moves so rapidly (typically 1500 to 3000 strokes per minute) that the gasoline vapor spends very little time in the cylinder.

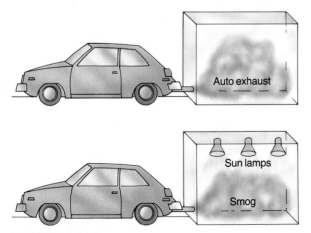

FIGURE 17.5 Smog is produced when automobile exhaust is exposed to sunlight.

TABLE 17–3 Results of Automobile Exhaust Experiments

Experiment	Condition	Results
1	Auto exhaust piped into room; UV lamps turned off.	Smelled like auto exhaust but not like smog; smog effects not evident.
2	Auto exhaust piped into room; UV lamps turned on.	Smog! Plants were damaged and rubber was cracked; if you stuck your head in the window, your eyes became irritated.
3	Auto exhaust piped into room; UV lamps turned off; and ozone added to room.	Smog again, as in Experiment 2.

This is not enough time, in fact, for the octanes to break down completely. The incompletely decomposed fragments, however, do oxidize to some extent, and these partly oxidized fragments are released through the exhaust pipe into the atmosphere. Here they react with ozone or other chemically active forms of oxygen to produce various new reactive molecules such as

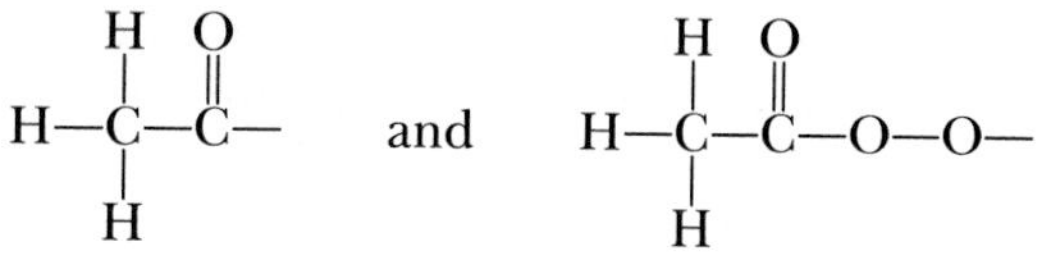

and

EXAMPLE 17.3

Recalling the normal covalences H = 1, O = 2, N = 3, and C = 4, identify the atoms in the above formulas that have abnormal covalences.

Solution In the first formula, the C bonded to O has only 3 covalences—a single bond to the other C atom and a double bond to O. In the second formula, the last O atom has an abnormal covalence because it has only one bond to another O atom. These unsatisfied valences are indicated by the incomplete bonds.

Molecules with abnormal covalences are often very reactive. The typical reactions are those in which the atom with an abnormally low covalence forms a new bond to bring its covalence up to the normal value. A particularly important reaction is the one in which NO_2 combines with a reactive molecule to produce a compound called peroxyacetyl nitrate (PAN), which is known to be one of the major irritants in photochemical smog, as well as a toxicant to plant life:

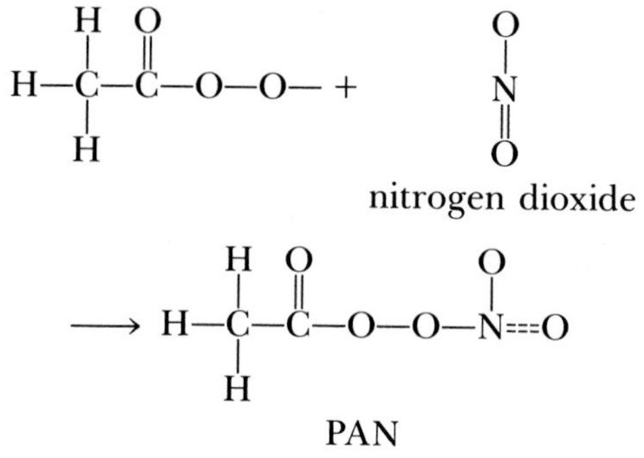

nitrogen dioxide

PAN

EXAMPLE 17.4

(a) One of the reactive products of incomplete combustion of gasoline formulated earlier is

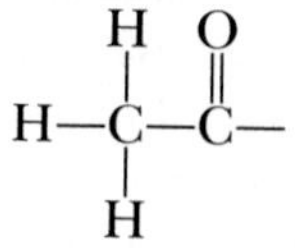

Write the structural formula of the product of the reaction of this molecule with oxygen. (Write the formula for oxygen as O—O. This is not unreasonable because the valences in the O_2 molecule are in fact not fully satisfied.) Is this product still chemically reactive?

(b) Write the structural formula for the product of the reaction between

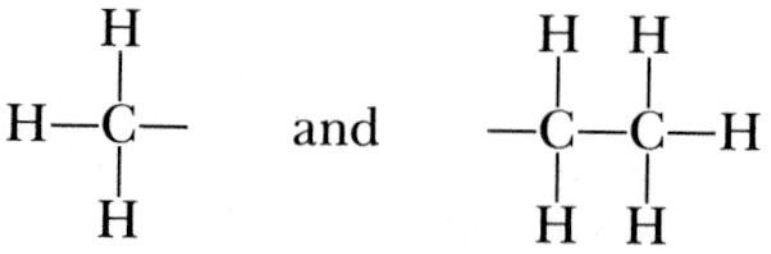

and

Is this product still chemically reactive?

Solution **(a)** Note that the carbon bonded to oxygen has only three completed bonds, so its

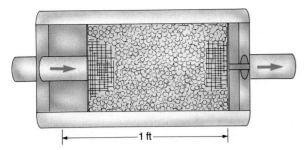

FIGURE 17.6 Cutaway view of catalytic converter, showing catalyst pellets.

covalence is low and it will bond with the O_2 molecule to form

$$\underset{\underset{H}{|}}{\overset{\overset{H}{|}}{H-C}}-\overset{\overset{O}{\|}}{C}-O-O-$$

This molecule is still reactive because the last O atom has only one completed bond, not two.

(b) One carbon atom of each molecule has only three completed bonds, so these atoms will bond to each other to produce

$$\underset{\underset{H}{|}\ \underset{H}{|}\ \underset{H}{|}}{\overset{\overset{H}{|}\ \overset{H}{|}\ \overset{H}{|}}{H-C-C-C-H}}$$

Now all the covalences are normal, so this molecule will be less reactive. It is propane.

The detailed chemistry of photochemical smog is devastatingly complex. Many separate chemical steps are involved. The rates of these various reactions change during the course of the day as the intensity of the sunlight changes. In fact, the whole story—all the reactions and their rates—is not yet known. One thing, however, is known. If hydrocarbons and other organic compounds were not introduced into the atmosphere, the polluting process would not take place. The best remedy would be to burn the gasoline completely within the cylinder to CO_2 and

H_2O. This objective can be approached, but not fully realized, by improving the design of the engine. Any such modification also enhances engine efficiency and results in substantial fuel savings.

Another approach is to oxidize unburned fuel after it leaves the cylinder but before it is exhausted into the air. This objective is achieved by using the catalytic converter, as shown in Figure 17.6. The best catalysts for speeding up the oxidation of organic molecules to CO_2 and H_2O are certain heavy precious metals, especially platinum and palladium. (Recall how platinum catalyzes the oxidation of hydrogen.) Such catalytic converters are now required by law in the United States for gasoline engines.

However, two major problems arise. For many years, a lead compound, tetraethyl lead, $Pb(C_2H_5)_4$, was added to gasoline to improve engine performance. But the lead poisons the catalyst, destroying its effectiveness. It is for this reason that automobiles equipped with catalytic converters must use unleaded gasoline. The second problem is that the catalytic oxidation of the gasoline hydrocarbons generates heat within the catalytic converter that promotes other environmentally unfavorable oxidations. Probably the most harmful of these is the increased conversion of N_2 to NO and NO_2. The production of oxides of nitrogen is just what we want to avoid, since these compounds are also involved in the photochemical smog sequence. To this extent, the catalyst can have a detrimental effect on the atmosphere.

The technical development of antipollution systems for automobiles is far from complete, and the next step may well be directed to the objective of minimizing the production of oxides of nitrogen. Our experience with the problems of atmospheric pollution from automobiles shows that the environmental aspects of this single process are extremely complex. Of course, decisions of public policy are also involved; pollution from the use of gasoline can be effectively reduced by using less gasoline—by driving fewer miles in smaller cars with more efficient gasoline consumption.

SUMMARY

A **mole** is the amount of substance that contains 6.02×10^{23} basic particles; it is also the atomic weight of an element or the molecular weight of a compound expressed in grams.

Acids taste sour and introduce H^+ ions into an aqueous solution. **Bases** taste bitter and introduce OH^- ions into an aqueous solution. Acids and bases change the colors of indicators and neutralize each other to form a salt and water. The acidity or basicity of a solution is expressed by pH values. A neutral solution has equal concentrations of H^+ and OH^- ions and a pH of 7. The more basic the solution, the higher is the pH above 7; the more acidic the solution, the lower is the pH below 7.

Oxidation refers to the formation of bonds to oxygen or, more generally, to the loss of electrons. **Reduction** refers to the breaking of bonds to oxygen or, more generally, to the gain of electrons. Oxidations typically turn useful materials into wastes, whereas reductions typically do the opposite, converting wastes to useful materials. A **catalyst** speeds up a chemical reaction but it not consumed in the reaction.

Acid rain refers to the excess acidification resulting from the oxidation of nitrogen, ultimately to nitric acid, HNO_3, and of sulfur, ultimately to sulfuric acid, H_2SO_4. Acids can be controlled by neutralizing them with a base such as calcium hydroxide, $Ca(OH)_2$, which converts them to calcium salts. **Ozone, O_3**, in the stratosphere acts as a catalyst for the conversion of solar ultraviolet (UV) radiation to the less energetic infrared (IR) radiation. Nitrogen oxides and chlorofluorocarbons can act as catalysts to convert ozone back to oxygen, O_2, thereby increasing the amount of UV radiation that reaches the Earth's surface. Automobile exhaust contains oxides of nitrogen and unburned, as well as partly burned, hydrocarbons. This mixture reacts in the presence of sunlight to produce **photochemical smog.** The effect can be prevented by the use of a catalytic converter that helps to complete the oxidation of the partly burned hydrocarbons.

EQUATIONS TO KNOW

pH = $-\log_{10}$(hydrogen ion concentration)

KEY WORDS

Mole	Neutralization	Acid rain (also, acid	Enzyme
Acid	pH	precipitation or acid	Ozone
Base	Oxidation	deposition)	Photochemical smog
Indicator	Reduction	Catalyst	

QUESTIONS

MOLECULAR WEIGHTS AND THE MOLE

1. (a) Calculate the molecular weight of (i) ozone, O_3; and (ii) sulfur dioxide, SO_2. Use the atomic weights from the table in the inside back cover of the book, and carry the answers to one decimal place. (b) Which has more mass, 1 mole of O_3 or one mole of SO_2? (c) Which contains more moles, 1 kg of O_3 or 1 kg of SO_2? (d) Which contains more molecules, 1 kg of O_3 or 1 kg of SO_2? (e) Which contains more molecules, 1 mole of O_3 or 1 mole of SO_2?

2. What is the mass of a mole of (a) argon, (b) carbon dioxide, CO_2?

3. Sulfur boils at 445°C to form a vapor consisting of S_8 molecules. At still higher temperatures, this vapor decomposes to form S_4 molecules:

$$S_8 \longrightarrow 2S_4$$

As this change occurs, which of the following quantities remain constant, which increase, and which decrease? (a) mass; (b) number of molecules;

(c) number of atoms; (d) number of moles; (e) atomic weight of sulfur; (f) molecular weight of the sulfur vapor.

4. Container X holds 1 mole of carbon monoxide gas, CO, at 100°C and at a pressure of 2 atmospheres. Container Y holds 1 mole of carbon dioxide gas, CO_2, at the same temperature and pressure. Which container, if either, has (a) the greater number of molecules; (b) the greater mass of gas; (c) the greater volume of gas?

ACIDS AND BASES

5. (a) Summarize the properties of acidic and basic solutions. (b) Imagine that you are hiking in the wilderness in summer and come upon a small pool of water and want to determine whether the water is strongly acidic, strongly basic, or fairly close to neutral. You have no chemical indicator papers, but your pack does include lemonade and soap. There are various berry bushes nearby, but the varieties are not familiar to you. You may handle the unknown water, but assume that you are not willing to taste it. Describe how you could test the water.

6. From their names or other evidence, tell whether each of the following substances are acidic, basic, or neutral: (a) caustic potash; (b) sour salt; (c) pickling liquor (used to etch metals); (d) Alka-Seltzer; (e) a solution of hydrogen iodide, HI, which ionizes in water; (f) a solution of sugar, which does not ionize.

7. Characterize each of the following solutions, from their given pH values, as weakly acidic, strongly acidic, weakly basic, strongly basic, or neutral; (a) 7.1; (b) 1.1; (c) 7.0; (d) 6.85; (e) 13.7.

8. What is the pH of a solution in which the H^+ concentration in moles per liter is (a) 10^{-5}; (b) 10^{-2}; (c) 10^{-9}; (d) 10^{-12}?

9. (a) Could an acidic solution be so concentrated that its pH is 0? (b) Could the pH be a negative number? (c) What is the pH of a solution whose H^+ concentration is 1 mol/L? (d) 10 mol/L?

OXIDATION AND REDUCTION

10. Give two definitions of oxidation and two of reduction.

11. In each of the following oxidation-reduction reactions, state which substance is being oxidized and which is being reduced:
(a) $H_2O_2 + Fe \longrightarrow FeO + H_2O$ (reaction of hydrogen peroxide with iron)
(b) $Zn + F_2 \longrightarrow Zn^{2+} + 2F^-$ (zinc + fluorine $\longrightarrow$ zinc fluoride)

(c) $2Na^+ + 2Cl^- \longrightarrow 2Na + Cl_2$ (electrolysis of molten salt)
(d) $2H_2 + O_2 \longrightarrow 2H_2O$ (burning or explosion of hydrogen in oxygen)

12. In each of the following cases, state whether the material referred to is being oxidized or reduced: (a) A fallen tree slowly rots. (b) Copper ore is processed to produce the pure metal. (c) Bamboo is eaten and metabolized by a panda. (d) Propane gas leaks out to the atmosphere through a faulty valve, comes in contact with a spark, and explodes.

13. Can an oxidation occur without a reduction? If your answer is yes, show how. If it is no, explain why not.

ACID RAIN

14. What are the sources of the slight acidity of rainwater in unpolluted atmospheres?

15. How can automobile exhaust contribute to the acidity of rainwater? (Assume that the automobile uses sulfur-free gasoline.) What happens in the cylinders? What happens in the outside atmosphere?

16. How can sulfur in coal contribute to the acidity of rainwater? What happens in the flame when coal is burned? What happens in the outside atmosphere?

17. Sulfur dioxide emitted from a stack is responsible for acid dusts and acid rain. Which is more likely to fall to earth closer to the stack? Explain.

18. What are the two basic approaches to the prevention of acid precipitation from sulfur compounds?

CATALYSIS

19. What is wrong with defining a catalyst as a substance that speeds up a reaction without entering into the reaction? Suggest a better definition.

20. What is the catalyst in the following transformation? Unsaturated oil + nickel + hydrogen → nickel + saturated fat.

21. The catalytic oxidation of vanadium from V^{3+} to V^{4+} occurs as follows:

$$V^{3+} + Cu^{2+} \longrightarrow V^{4+} + Cu^+$$
$$Cu^+ + Fe^{3+} \longrightarrow Cu^{2+} + Fe^{2+}$$

What is the net effect of these two reactions? What is the catalyst?

OZONE

22. Stratospheric ozone protects us from excessive ultraviolet irradiation. If you were trying to get a suntan by using ultraviolet sunlamps, do you think it would be a good idea to use an ozone-producing device in your room to protect you against a burn

from excessive exposure? Defend your answer.

23. The law of conservation of energy tells us that energy cannot be created or destroyed. Ultraviolet radiation is a form of energy. How, then, can stratospheric ozone reduce the solar UV radiation that reaches the surface of the Earth? What happens to it?

24. Explain how a supersonic airplane or a can of aerosol spray might reduce the level of stratospheric ozone.

25. Which of the following compounds could threaten the ozone layer if they reached the stratosphere? (a) Carbon tetrachloride, CCl_4, formerly used as a cleaning solvent but now banned because of its high toxicity; (b) benzene, C_6H_6, another toxic solvent; (c) carbon tetrafluoride, CF_4, a refrigerant; (d) methyl bromide, CH_3Br, an agricultural fumigant. (C—Br bonds are weaker than C—Cl bonds. C—F bonds are stronger than C—Cl bonds.)

SMOG

26. Write the balanced equation for each of the following "smog" reactions: (a) the formation of nitrogen dioxide and atomic oxygen from NO and O_2; (b) the formation of nitrogen dioxide from NO and ozone; (c) the decomposition of one molecule of NO_2 to form atomic oxygen and another product; (d) the formation of ozone from two other forms of oxygen.

27. Gasoline vapor and UV lamps do not produce the same smog symptoms that auto exhaust and UV lamps do. What do you think is missing from gasoline vapor that helps to produce smog?

28. Why is it illegal, as well as harmful, to use leaded gasoline in a modern automobile?

ANSWERS TO SELECTED NUMERICAL QUESTIONS

1. (a) 48.0, 64.1, (b) SO_2, (c) O_3, (d) O_3, (e) same
2. (a) 39.9 g, (b) 44.0 g
4. (a) both have same number, (b) container Y, (c) both have same volume
7. (a) weakly basic, (b) strongly acidic, (c) neutral, (d) weakly acidic, (e) strongly basic

8. (a) 5, (b) 2, (c) 9, (d) 12
9. (a) yes, (b) yes, (c) 0, (d) -1
11. (a) Fe oxidized, H_2O_2 reduced, (b) Zn oxidized, F_2 reduced, (c) Cl^- oxidized, Na^+ reduced, (d) H_2 oxidized, O_2 reduced
26. (a) $NO + O_2 \rightarrow NO_2 + O$, (b) $3NO + O_3 \rightarrow 3NO_2$, (c) $NO_2 \rightarrow O + NO$, (d) $O_2 + O \rightarrow O_3$

White light incident upon soap bubbles gives rise to the highly colored "interference" patterns we see.
(Peter Aprahamian)

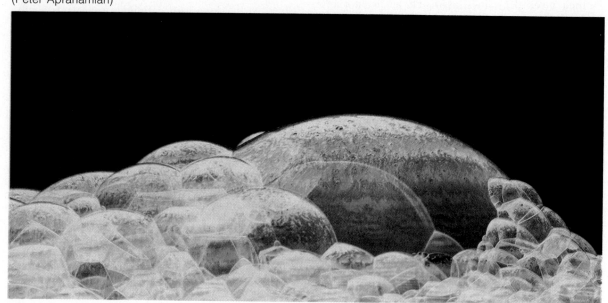

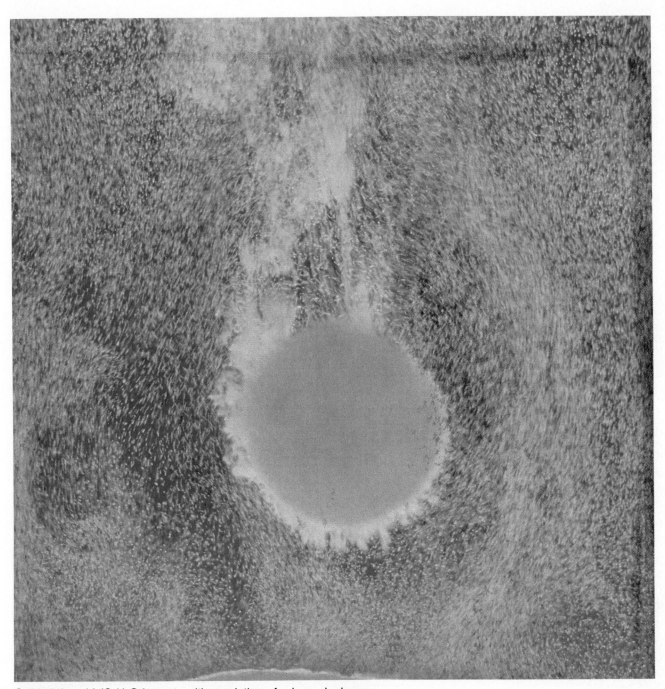

Solid citric acid ($C_6H_8O_7$) reacts with a solution of a base, hydrogen carbonate ion (HCO_3^-), to liberate carbon.
(Courtesy Charles D. Winters)

PART

GEOLOGY

THREE

The exploration of space has also enabled geologists to study the dynamic features of our planet.

JOHN R. HORNER

• INTERVIEW •

John R. Horner was born in 1946 in Shelby, Montana, where his father operated a large gravel quarry and his mother ran the home. He developed an early interest in fossils and geology during youthful forays on the High Plains around Shelby, where the remains of marine animals, dinosaurs, and Ice Age mammals are plentiful. His study of geology at the University of Montana was interrupted by service in Vietnam with the U.S. Marine Corps. from 1966 to 1968. He returned to the University of Montana in 1968. He was Research Assistant in the Department of Geological Sciences at Princeton

University from 1975 to 1982 and also was Museum Scientist at the American Museum of Natural History in New York City from 1978 to 1982. While at Princeton and the American Museum, he returned occasionally to the area near his home in Montana during summer field seasons to explore and recover the fossils of large animals that he had noticed as a child. Many of his discoveries are now prized specimens in the Princeton and American Museum of Natural History collections. During this time, he began to concentrate on dinosaurs, and he realized that the rocks near Shelby contain abundant

and well-preserved dinosaur remains. In 1982 he was appointed Curator of Paleontology at the Museum of the Rockies in Bozeman, Montana, where he has continued his research on the nature and living habits of dinosaurs. He was awarded an Honorary Doctorate of Science from the University of Montana in 1986.

He is author of many publications, most dealing with dinosaurs and their social habits, and of two popular books about dinosaurs. He has received several large National Science Foundation grants in support of his research. In 1986 he was awarded the MacAr-

thur Foundation Fellowship, commonly known as the "Genius Award." It is a large cash grant given to unusually creative and productive individuals. He has been featured in many national news and science magazines, including Readers' Digest, Natural History, Omni, Life, National Geographic, People, *and* U.S. News and World Report.

Where were you born and raised? What initiated your interest in geology and paleontology?

I was born and raised in Shelby, Montana. I really don't know how my interest in geology and paleontology started. My father owned a gravel quarry, and he says that when I was very young, I was always sorting the rocks out of the big gravel pile into what I thought were groups of different kinds of rocks.

My father would take me to places that he had ridden horseback when he was a rancher, and he would show me areas where he had found what he thought were dinosaur bones. I was eight years old when he first took me to one of these places, and I collected a couple of bone fragments. Then we went back to the same area when I was in high school, and I collected two partial dinosaur skeletons.

Through high school, I was really interested in science. I spent most of my time working on science projects. In my senior year I did a project on fossils. I was trying to figure out why the dinosaur remains found in the Judith River Formation in Montana were different from the ones found in the Judith River Formation in Canada.

On the Canadian side there are articulated dinosaurs and lots of duck-billed bones. But even at that time I knew that most of my dinosaur bones from Montana were flat-headed kinds and were strewn all over. I never could find an articulated one. I saw this difference as a problem, but I was unable to resolve it then. In fact, I didn't resolve it until I published a paper just two years ago on it. Now we know that the dinosaur bone beds in Alberta are a slightly different type from those in Montana, so they are stratigraphically not the same. They also had different environments of deposition.

You are now one of the most highly visible and best known paleontologists/geologists in the country, yet in high school and as a university student you got low grades. How do you reconcile your success as a scientist with your grades as a student?

My academic record at the university was even worse than my high school grades. After my first year and a half at the University of Montana, my cumulative grade point average was a 0.06. If I went to college now, I would still have the same problem. I have a learning disability called dyslexia. It's a problem that didn't stop me from wanting to learn, but it did stop me from being able to. It made it almost impossible for me to absorb information that was assigned to me. I just couldn't assimilate the material fast enough. It got to the point where I didn't care what the grade was. I would learn as much as I could, and if at the end of the course that was D work for the professor, then I got a D. If I felt that the course seemed really interesting and if it was something that I thought I could get more out of, I'd just take it again. If I thought I knew enough for what I wanted to do, then I wouldn't take it again regardless of my grade. I didn't know what the problem was at the time, but I did know that I really wanted to learn.

I spent a year and a half at the university, and then in 1966 I got drafted into the Marine Corps. Special Forces in Vietnam. I got out of the Marine Corps. in February, 1968 and went back to the University of Montana. My grades were a lot better when I came back, and I started taking zoology and geology courses. But my grades, were still lousy by university standards, and that's when they started throwing me out of school every quarter. Each time I could demonstrate that I was bringing my grades up so they had to let me back in. I took every geology course that was offered and all the zoology courses that looked appropriate, as well as botany, physical geography, and a few anthropology courses. I never finished my undergraduate degree, but I did eventually get a doctorate from the University of Montana.

What did you do after college? What was your first job?

After I took all the courses I was interested in, I went back to Shelby where my brother and I bought the gravel company from my father, who was retiring. I had been there for about a year and a half when I began sending letters to all the museums in the English-speaking world to apply for jobs from janitor up to curator. I didn't really care what it was, I would have taken anything. I got three responses:

from the Los Angeles County Museum for the position of Chief Preparator, from the Royal Ontario Museum in Toronto for Assistant Curator, and from Princeton for a Preparator and Research Assistant. I took the job at Princeton and worked there for seven years.

In 1979, I was talking to the Director of the Museum of the Rockies and he told me that he had heard about all the dinosaur eggs we were finding in Montana around Choteau and Shelby. He asked me if we would donate some to the Museum of the Rockies, so I donated a clutch of eggs. In 1981, I saw the Director again, and I told him I was from Montana and that I really wanted to work there. He offered me the Curator of Paleontology position at the Museum of the Rockies and I came home.

Tell us about your work at the Museum of the Rockies.

The paleontology crew at the museum includes one curator (me), four full-time preparers, a thin-section histology technician, an illustrator, a collection manager, a computer illustrator who does all of the mathematical simulation, six graduate students, and a staff of about 15 part-time people.

My research is primarily on dinosaur behavior, ecology, and evolution. My graduate students are all geologically oriented to do field studies in stratigraphy and sedimentology. One of the students is doing comparative studies of different kinds of bone beds we have found. A bone bed is one geologic horizon on which lots of specimens occur together. For example, one at Choteau appears to be a volcanic ash kill, and our evidence shows that over 10,000 animals are buried there.

All of the bone beds we work on cover at least one square mile.

What would large herds of dinosaurs eat? How could so many of them live in a square mile? What was their environment like?

Well, one of the interesting things is that dinosaurs that lived in large herds, hadrosaurs and ceratopsians, did not evolve until Late Cretaceous time. That coincides with the evolution of the angiosperms, deciduous plants that can be stripped one season and still grow back the next. So, I don't think the large groups or herds existed before angiosperms appeared. But once these plants existed as a food source, then all that's required of these big groups is for them to migrate with the seasons. About 75 to 80 million years ago, the western part of North America was a linear continent extending north to south. The Rocky Mountains were young and were actively building up. So there was a mountain barrier on the west and an ocean to the east with a coastal plain in between. The dinosaurs would migrate north to south on that coastal plain with the seasons. From their size and stride lengths, it appears that they could have easily walked 1500 to 2000 miles each year following seasonal shifts in temperature and food supply.

The only constraint on the migrating dinosaurs was the nesting period. The dinosaurs had to wait at least a month somewhere for the incubation of their eggs. Most of the incubation, or growth of the fetuses, may have occurred in the mothers' bodies prior to egg laying. After the eggs were laid, it was probably a relatively short

time (possibly only two or three weeks) before the eggs hatched. Another three or four weeks at the most would have been needed for the nesting period. During this short time the young would grow large enough to walk with the adults. We're guessing from preliminary information that duck-billed dinosaurs hatched out of their eggs at about 18 to 20 inches long and grew to about 45 to 50 inches long, possibly more, by the time they left the nest. But at that same growth rate, they would have grown to 9 to 12 feet the first year. The maximum size of an adult is about 35 feet. This suggests that growth rate was a little faster than an ostrich, but a lot slower than most birds.

Much of your work has changed the traditional views of dinosaurs, how they behaved, and even what they were. Tell us what the traditional view is and how your work has affected this view.

Originally, scientists decided, on the basis of certain cranial features, that the dinosaurs should be classified as reptiles. Once they had been placed in this group, there were certain characteristics, certain little labels, that were automatically assigned to dinosaurs simply because they were called reptiles. For example, modern reptiles are cold-blooded, therefore dinosaurs were considered cold-blooded. Modern reptiles are slow moving, so dinosaurs were probably slow. Modern reptiles drag their tails, so dinosaurs must have dragged their tails.

You have to realize that in the early days of dinosaur discoveries, people didn't really study dinosaurs; they collected their remains for museums. It wasn't until the last 15 years or so that people actually started to consider how dinosaurs lived. In those 15 years we've come to find out that the original classification is probably wrong. Dinosaurs don't belong to the reptile group. They are much more like birds, and it is likely that modern day birds evolved from dinosaurs.

Most reptiles dig a hole in the ground, lay their eggs in it, cover it up, and then leave. So it was assumed dinosaurs did the same thing. But we now know that dinosaurs, like modern birds, put a lot of time and care into building a nest, laying the eggs, and guarding the eggs. Our evidence shows that dinosaurs even guarded their young after they hatched by herding or flocking in large groups. These large groups of dinosaurs were very similar to what we observe in modern herding animals or flocking birds, which are not just aggregations of animals, but actually structured groups with certain individuals in charge.

The animals that we find in large groups, such as the duck-billed dinosaurs and the horned dinosaurs, all have some type of cranial display features such as horns on their heads. We know that the horns of modern mammals are a primary adaptation for determining hierarchies within a society. Modern horned animals, such as elk, live in big groups, and generally the males use their horns for male-to-male combat to determine the hierarchy individual. We see a similar thing with the horned dinosaurs.

What do you think about the traditional view of dinosaurs being green and scaly as opposed to more recent suggestions that they were furry, hairy, or covered with feathers?

I think the babies had some kind of downy cover. All our evidence suggests that dinosaurs were warm-blooded just like we are. In fact, they were probably warmer blooded than we are. Their fast growth rate suggests a very high metabolism, suggesting a relatively warm internal body temperature. So, the babies had to have some kind of insulatory mechanism to keep the heat in. But our bone histology shows that the rapid growth ended at about 20 feet in length. Then we see rest lines in the bones suggesting that the metabolism had slowed way down. At this point, the dinosaurs were still creating heat, but much less than before.

So this means that their rate of food consumption decreased when they became large. This would address those who argue that dinosaurs could not be warm-blooded because there wouldn't be enough food for many animals of that size. If there are hundreds of dinosaurs or possibly more nesting in one area, there would have to be a large food supply for the young, and the adults would be eating, too. Well, I don't think the adults were eating at all. It's like when birds are feeding their young—they don't eat. They just haul in food for the babies. So all that was needed was a supply large enough for the young, and it really makes little difference how far the adults had to walk to get food. For example, penguins often go a hundred miles to get food for their babies.

What good is a study of geology to someone who doesn't plan to become a geologist?

I think our environment is going to be the next century's biggest topic. And you cannot have a good handle on the environment unless you understand both biology and geology. So as the environment becomes more and more of an issue, there are going to be more and more jobs in the field of environmental geology. I think we are going to see the pendulum swing toward fields that figure out how to save our world. We're going to have to understand animals as animals, how rivers work, and similar concepts. A strong understanding of geology and biology is what most people are going to need to address environmental concerns.

This interview was conducted by Graham Thompson, University of Montana, and appears in *Modern Physical Geology*, Saunders College Publishing, 1991.

Sky Bridge, Kentucky. (Courtesy Bill Schulz)

CHAPTER

18

The Atmosphere and Meteorology

In this chapter we will be concerned with our atmosphere. Why is it like it is, and what factors have caused it to evolve into its present form? We will also be concerned with both climate and weather. **Climate** is a description of long-term meteorological conditions. For example, New York City has a temperate climate with moderate rainfall; one can always expect warm summers and cold winters there. **Weather,** on the other hand, is a description of relatively unpredictable short-term conditions. Storms, a heat wave, or a cold spell are all part of the weather in a given area. But why does a particular region have the climate that it does? What causes rain, clouds, and so forth, and how can one predict the weather? Understanding and predicting the weather fall under the heading of **meteorology.**

18.1
EVOLUTION OF THE ATMOSPHERE

During the past few decades, there has been an intense period of exploration of the Solar System.

By the end of 1989, spacecraft had landed on or flown past all of the outermost planets except Pluto, and many of the planetary moons had been studied as well. Many of these missions will be discussed in more detail in Chapter 22. At this point, two important conclusions are relevant. First, many of the planets and moons have some sort of atmosphere. Second, no other celestial body has an atmosphere that is similar to our own.

Most scientists agree that the Earth's atmosphere today is quite different from its original, primordial atmosphere. At first, the Earth was a loosely collected mass of dust and gases, mostly hydrogen. Over the millennia, the dust coalesced into a solid sphere, most of the hydrogen escaped into space, and a secondary atmosphere was formed. This secondary atmosphere was produced by a variety of means, including volcanic ejection of gases that had been trapped within the planet as it formed and accretion of gases from comets and meteorites that bombarded the Earth. There is some uncertainty as to the exact composition of this atmosphere. By comparing the Earth with its two nearest neighbors, Venus and Mars, and by studying the composition of very old rocks on the Earth's surface, scientists now believe that the original gases consisted mainly of carbon dioxide (CO_2), nitrogen (N_2), and water vapor (H_2O), with smaller concentrations of methane (CH_4), ammonia (NH_3), hydrogen (H_2), and carbon monoxide (CO). The best evidence indicates that oxygen was present in trace quantities only, although this conclusion has not been entirely substantiated.

Living creatures that exist today would not survive in a carbon dioxide, nitrogen, and water atmosphere. How was the modern atmosphere formed? Although some theorists believe that geological processes altered atmospheric composition, most scientists believe that living organisms and a favorable atmospheric environment evolved hand in hand. In the beginning, when there was little free gaseous molecular oxygen (O_2), there would be no ozone (O_3). Without a protective ozone layer, intense ultraviolet radiation from the Sun could reach the Earth. Ironically, these rays, which could harm or destroy life today, may have been responsible for the formation of the first organic compounds and living cells. One theory suggests that the synthesis of simple organic molecules was initiated by the action of energetic ultraviolet light on the molecules of the primitive atmosphere. Once the first organic molecules were formed, they presumably combined to form both proteins and complex molecules that carry hereditary information. Then these large molecules joined together to form simple living organisms.

Not all scientists agree that life evolved from reactions involving atmospheric gases. One alternative theory suggests that perhaps the first living organisms were formed in the vicinity of underwater volcanic vents, where carbon, nitrogen, and sulfur compounds were more concentrated. Yet another theory postulates that the earliest forms of life evolved on the surface of certain clay minerals.

In any case, it is fairly certain that bacteria, the first living creatures preserved in the fossil records, must have lived in water. Since there was little oxygen, they could not have metabolized their food as most organisms do today but must have lived by some **anaerobic** (without oxygen) process. Up until this point, one theory contends, there was little free oxygen in the atmosphere, and the organisms required none. The next evolutionary step was a crucial one. Blue-green algae evolved, able to synthesize their own complex organic molecules by combining simple organic molecules in the presence of sunlight, as discussed in the previous chapter. Recall that during this process, carbon dioxide and water are combined in the presence of sunlight to form glucose (a sugar) and oxygen.

Most scientists believe that the excess oxygen released by these first plants accumulated slowly over the millennia until its concentration reached about 0.6 percent of the atmosphere. Most multicellular organisms require oxygen to survive and could have evolved only at this point. The emergence of various multicelled organisms about 1 billion years ago triggered an accelerated biological production of oxygen. The present oxygen level of 21 percent of the atmosphere was reached about 450 million years ago. While the concentration has not been precisely constant since that time, an overall oxygen balance has always been maintained. This scenario is illustrated schematically in Figure 18.1.

If the oxygen concentration in the atmosphere were to increase by even a few percent, fires would burn uncontrollably across the planet; if the carbon dioxide concentration were to rise by a small amount, plant production would increase appreciably and the climate of the Earth would probably change. Since these apocalyptic events have not occurred, the atmospheric oxygen must have been balanced to the needs of the biosphere during the long span of life on Earth. By what mechanism has this gaseous atmospheric balance been maintained? Some scientists believe that it is maintained by the living systems themselves, as shown in Figure 18.2. According to this theory, not only is the delicate oxygen–carbon dioxide balance biologically maintained, but the very presence of oxygen in our atmosphere can be explained only by biological activity. If all life on Earth were to cease and the chemistry of our planet were to depend solely on inorganic processes, oxygen would become a trace gas and the

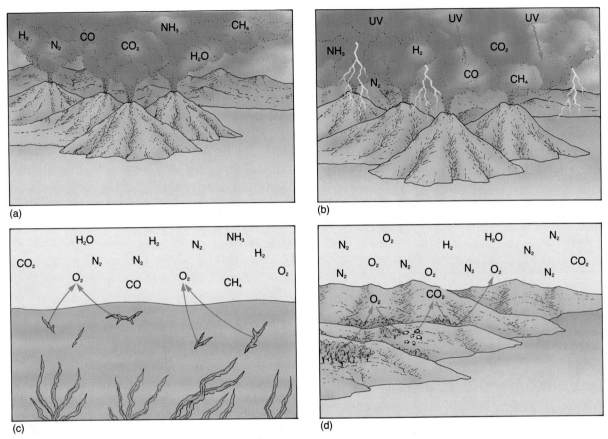

FIGURE 18.1 Evolution of the atmosphere. (a) The primitive atmosphere contained gases released from volcanic eruptions. (b) The action of lightning and ultraviolet light on this atmosphere initiated a chain of reactions that led to the evolution of simple organisms. These microbes lived in the oceans. (c) As plants evolved, the composition of the atmosphere began to change. Oxygen released during photosynthesis began to accumulate. (d) The modern atmosphere is composed mainly of nitrogen and oxygen, with smaller concentrations of carbon dioxide, water, and other gases. The ratio of oxygen to carbon dioxide is maintained by dynamic exchange among plants and animals.

atmosphere would revert to its primitive condition and be poisonous to any complex plants and animals that were reintroduced.

If it is true that the required atmospheric oxygen concentration of about 20 or 21 percent is maintained by biological processes, then the Earth's atmosphere is not in danger of disastrous changes so long as living species survive. An alternative theory claims that our physical environment has evolved through a series of inorganic reactions and that biological evolution and physi-

cal evolution were independent. The difference between these two theories is not trivial. If the biological theory is correct, then a large ecological catastrophe, such as the death of the oceans or the destruction of the rain forests in the Amazon Basin, could cause reverberations throughout our physical world that might create an inhospitable environment for life on Earth. Alternatively, if the physical world did evolve independently of the biological and is now controlled by inorganic processes, such a doomsday prediction concern-

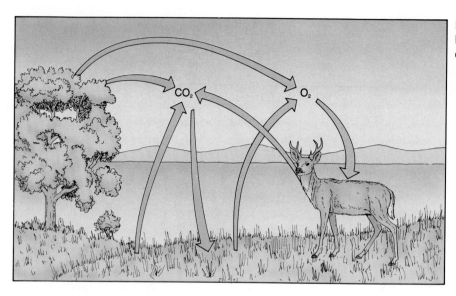

FIGURE 18.2 Simplified biological oxygen–carbon cycle.

ing oxygen balance might be considered unnecessarily alarming.

18.2
THE ATMOSPHERE TODAY

Did you ever wonder what keeps the atmosphere up there? Consider, for example, a molecule of oxygen at the level of your nose. It is being pulled downward by the force of gravity, so why doesn't it fall—like Newton's apple—and land with a clunk on the floor? The reason is that molecules are always moving and colliding with each other, and the hotter the gas is, the faster they go. At 0°C, oxygen molecules travel at an average speed of about 425 m/s, although at any one instant some go much faster while others move more slowly. They travel randomly in all directions, but the average distance between collisions is very short. Thus, each molecule may be considered to have at least two kinds of motion: the thermal motion influenced by collisions with the other gas molecules, and a downward acceleration caused by gravity at 9.8 m/s². If a box of gas were floating about in free space, where there are no gravitational forces, the gas would be distributed uniformly throughout the box. But if this container were placed on Earth, the gravitational acceleration would draw molecules downward, and there would be more gas molecules in a given volume at the floor level than at the ceiling (Fig. 18.3).

On Earth, the atmosphere grows less and less dense with increasing altitude because gravitational forces are weaker at greater distances. Anyone who has ever climbed a high mountain has experienced the effects of this atmospheric thinning with height. At about 3000 m (about 10,000 ft), even a person in good physical condition readily notices that exertion is more difficult than it is at sea level. At 4500 m, a person's actions are slowed considerably, and above 6000 m, climbers find that they move surprisingly slowly and lose their breath quite quickly.

The blanket of air resting above the Earth has a great deal of mass, and, when pulled downward by the force of gravity, this mass is heavy. How can this column of air be weighed? If a two-pan balance is resting on a table, it reads zero, not because the air is weightless but because the air pushes down equally on both pans, as shown in Figure 18.4. But if one end of an open glass tube is placed in a dish of a liquid, such as mercury, and the air is evacuated from the other end, the liquid rises in the tube. Why? The reason is that the air column pushes down on the liquid in the dish while there is no equivalent force from within the evacuated tube, as shown in Figure 18.5. This device is called a **barometer.** The height to which the liquid rises is a measure of the downward force exerted by the air. At sea level, the mercury will rise approximately 76 cm, or 760 mm, into the tube. The effect of the column

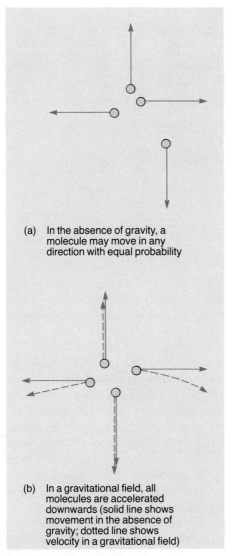

(a) In the absence of gravity, a molecule may move in any direction with equal probability

(b) In a gravitational field, all molecules are accelerated downwards (solid line shows movement in the absence of gravity; dotted line shows velocity in a gravitational field)

FIGURE 18.3 Gas molecules under the influence of gravity (in the absence of wind and air currents).

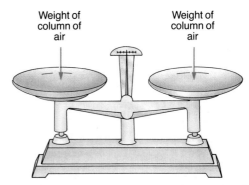

Weight of column of air   Weight of column of air

FIGURE 18.4 You cannot measure the weight of the atmosphere with a two-pan balance because the weight of the air is equal on both pans.

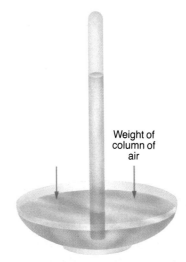

Weight of column of air

FIGURE 18.5 A barometer. The weight of the column of air on the mercury in the dish is not balanced by any air pressure from within the tube, because the upper region is evacuated. Instead, the air pressure is balanced by the weight of the mercury in the tube. Equilibrium is attained, and the height of the mercury in the column is a measure of the outside air pressure.

of air is generally expressed in pressure units (pressure = force/area) and is referred to as **barometric pressure,** as explained in Chapter 2. The pressure exerted by the Earth's atmosphere at sea level fluctuates with climate conditions. On the average, this pressure supports a column of mercury 76 cm high, as noted above. One standard atmosphere is defined as the atmospheric pressure that would cause a column of mercury to stand at a height of 76 cm. In SI units, pressures are measured in N/m^2, and these units are further defined such that 1 N/m^2 is called a **pascal**

(Pa). Another unit accepted by the SI is the **bar,** which is 10^5 pascals. The bar is a convenient unit because it is very close to a standard atmosphere:

1 std atm = 1.01325 bar = 1013.25 millibar

Sometimes other pressure units such as lb/in^2 (abbreviated psi for pounds per square inch) are used to express atmospheric pressures.

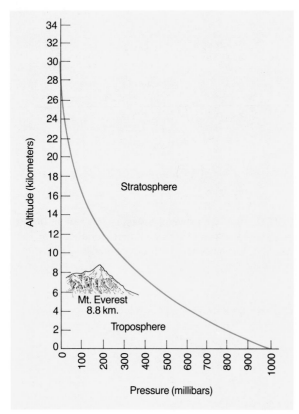

FIGURE 18.6 Decrease of atmospheric pressure with altitude.

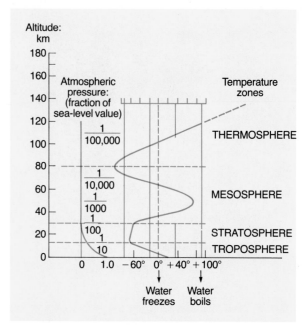

FIGURE 18.7 Atmospheric temperature change with altitude.

thus loses energy, which goes back out to space. If there were no atmosphere above us, the loss of energy by radiation would be extremely rapid; at night, when this loss is not compensated by thermal energy received from the Sun, the surface would cool drastically. In our world, nighttime temperatures are maintained at a fairly high level because the atmosphere absorbs and retains a great deal of thermal energy before it can escape to outer space. But at higher and higher elevations in the troposphere, the atmosphere becomes thinner, its insulation properties decrease, and the average temperature also decreases, as shown in Figure 18.7. Thus mountaintops are generally colder than valley floors, and pilots flying at high altitudes must keep their cabins well heated.

The steady decline of temperature with increasing altitude ceases abruptly about 12.5 km above the Earth. If we ascend farther, a gradual warming trend is observed, and above that level the temperature increases rapidly with altitude. The layer of air of fairly constant temperature is called the **stratosphere,** and above that lies the **mesosphere.** Throughout the stratosphere and the lower portion of the mesosphere, where the air is too thin to support life, high-energy ultraviolet rays are absorbed by ozone. It is this absorp-

Figure 18.6 is a graph of the atmospheric pressure as a function of altitude. If the change of temperature with altitude is studied, no such smooth curve is observed. Rather, as shown in Figure 18.7, the temperature profile alternates. If a rocket with a recording thermometer mounted on it rose straight upward from the surface of the Earth, it would register first a cooling trend, then a gradual warming, a rapid warming, another cooling trend, and finally a pronounced warming. Let us see why these variations are observed.

The layer of air closest to the Earth is known as the **troposphere.** This is where we live, where our weather occurs, and here the air is close to the land and oceans. When sunlight strikes the Earth, the energy is readily absorbed by soil, rock, water, and living organisms, and thus the surface of the planet becomes warm. But just as a piece of rock or soil can absorb thermal energy, it also emits radiation (at infrared wavelengths) and

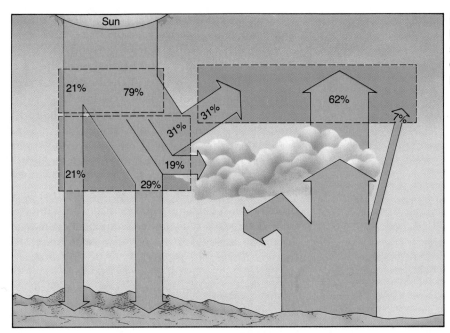

FIGURE 18.8 Energy balance of the Earth. The sets of numbers in the dashed areas total 100 percent.

tion of UV radiation and reradiation as infrared, or heat rays, that is largely responsible for the temperature increase at these altitudes. In the upper regions of the mesosphere, little radiation is absorbed and the thin air is extremely cold. Starting at about 80 km above the Earth, the temperature again starts to rise rapidly in a region known as the **thermosphere.** Here high-energy X-rays and ultraviolet radiation from the Sun are absorbed by atoms and molecules in the atmosphere. High-energy reactions result, which strip electrons from atoms and molecules to produce ions. Again, subsequent reradiation of infrared produces a warming effect.

18.3 THE ENERGY BALANCE OF THE EARTH

In its daily and seasonal variations, the climate of any given region of the Earth follows a reasonably definite, recurring pattern from year to year. This pattern results from a balance of opposing processes. On the one hand, the Earth receives a continuous influx of energy from the Sun. If there were no energy losses, the Earth would get hotter and hotter until rocks melted and vaporized. Since these catastrophes have not occurred

for billions of years, we know that energy must also be radiated away from the Earth at a constant rate. This opposition of inflow and outflow is called the **energy balance of the Earth.**

Figure 18.8 shows how the balance is distributed. Only 21 percent of the incident solar radiation strikes the Earth directly. The other 79 percent is intercepted by the atmosphere—the clouds, gases, and small particles. For example, much of the ultraviolet light is absorbed by ozone in the atmosphere. Some of this intercepted radiation is reflected back to space (31 percent), some is absorbed as heat (19 percent), and some is rescattered down to Earth (29 percent). On a global average, just about half of the energy received from the Sun reaches the surface of the Earth. Some of the energy that reaches the surface is reflected back into the atmosphere, while some is absorbed, thereby warming rock, soil, and the water of the oceans. Energy is then carried back into the atmosphere by several processes, including conduction, convection, and reradiation. The warm atmosphere radiates some thermal energy back to the Earth and some to outer space. Thus, a complex series of interactions occurs, in which radiation bounces back and forth between the surface and the atmosphere until it is ultimately lost to space.

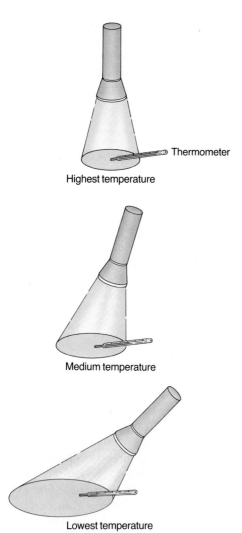

Highest temperature

Thermometer

Medium temperature

Lowest temperature

FIGURE 18.9 Intensity of light varies with angle of light source.

The temperature of the Earth's surface and its surrounding atmosphere depends on the total rate of absorption and reflection. Thus, the reflectivity of the Earth and its atmospheric components is an important factor controlling temperature. Of course, some surfaces are better reflectors than others. The **albedo** is a measure of the reflectivity of a surface. Clouds, snowfields, and sparkling glaciers reflect sunlight efficiently and are said to have a high albedo. On the other hand, city smog and dark, rough surfaces do not reflect light well and have a low albedo. If the albedo of the Earth were to increase, as for exam-

ple if the ice cover grew, the surface of our planet would cool; alternatively, a decrease in albedo would cause a gradual warming trend.

This discussion of energy balance does not explain why some regions of the Earth are warm and others are generally much colder. Thus, temperate zones experience distinct summer and winter seasons, and the polar regions are always cooler than the equatorial ones. To understand the reasons behind these temperature differences, first consider what happens if a flashlight is shined onto a flat board. If the light is held directly overhead and the beam is shined vertically downward, the light illuminates a smaller area than it would if the beam were shined onto the board from an angle (Fig. 18.9). Of the three positions shown in the figure, the one aimed at the shallowest angle illuminates the greatest area. Thus, the more the flashlight is tilted, the less concentrated is the light on the board, and the lower is the temperature of any point in the illuminated area. (Of course, the same results would be observed if the board were tilted instead of the flashlight.)

With this in mind, let us consider what happens when light is beamed onto a spherical surface such as the Earth. First, imagine that the globe were held in a fixed vertical position. The top-most and bottom-most points of the globe are the poles, the imaginary line that encircles the middle is the Equator, and the imaginary straight line through the Earth from pole to pole is the Earth's axis. If a light is shined perpendicular to the Equator, all the other surfaces receive light at an angle. Specifically, the polar surfaces are angled farther and farther from the perpendicular until at the pole itself the surface of the sphere is actually parallel to the light, as shown in Figure 18.10. Thus, the light intensity per unit area *decreases* from the equator to the poles. At the poles, the light is parallel to the surface, and no direct radiation strikes the area at all. A uniform light source shining perpendicular to the center of the globe delivers the most radiant energy to the Equator and the least to the poles.

It is now obvious why the hottest climates on Earth are found near the Equator and why climates become progressively cooler north or south of that line. This explanation accounts for gen-

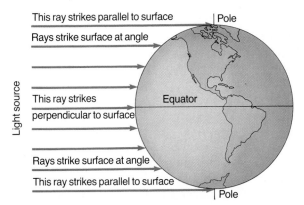

FIGURE 18.10 If the Earth were not tilted, the Sun's rays would always strike perpendicularly at the Equator and parallel at the poles. Therefore, the equatorial regions would receive the most intense solar radiation, and the polar regions would receive hardly any at all.

eral climatic regions but does not tell us why summer and winter seasons occur in the higher latitudes. To understand these effects better, it is necessary to consider the Earth's rotation around an imaginary axis running through the North and South Poles. If this axis were perpendicular to the plane of the Earth's orbit, the Sun would always be directly above the Equator, and there would not be any seasonal temperature variations. In reality, the Earth's axis is tilted at an angle of 23.5° with respect to a line drawn perpendicular to the plane of its orbit. As a result of this tilt, the angle of incidence of the Sun's rays on the Earth changes as the planet moves around in its orbit.

On June 21, the Earth is located so that the North Pole leans the full 23.5° toward the Sun, as shown in Figure 18.11. This condition is called

FIGURE 18.11 A schematic view of the Earth's orbit showing the progression of the seasons.

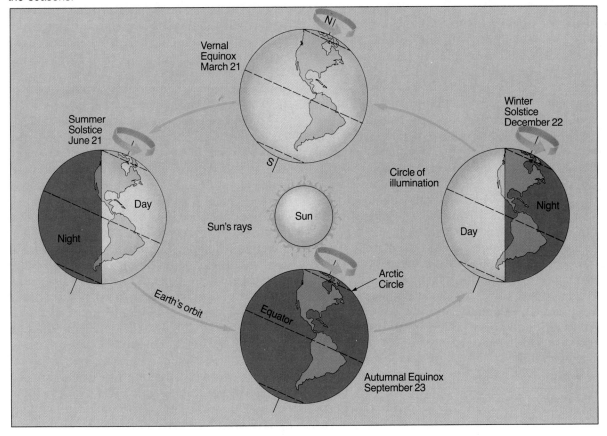

the **summer solstice.*** The Sun strikes the Earth directly overhead at a location 23.5° north of the Equator and not at the Equator itself. The northern latitudes receive more direct sunlight than the southern ones, and the North Pole, tilted toward the Sun, receives a continuous 24 hours of daylight. Polar regions are often called "lands of the midnight Sun" because the Sun never sets in the summertime. While it is summer in the Northern Hemisphere, the South Pole is tilted away from the Sun and lies in continuous darkness. June 21 marks the first day of winter in the Southern Hemisphere. Six months later, on December 22, the situation is reversed: The North Pole is tilted away from the Sun and lies in continuous darkness, whereas the South Pole is bathed in constant light. It is summer in the Southern Hemisphere and winter in the Northern.

What happens midway between these two extremes, on March 21 and September 23? Although the Earth is still tilted at 23.5°, the tilt aims the North Pole neither toward nor away from the Sun but rather at right angles to it. Since no part of the Earth is angled toward or away from the Sun, the most direct light shines at the middle, that is, at the Equator, and the North and South Poles each receive equal periods of night and day. In fact, when the Earth is in either of these two positions, which are called spring and autumn **equinoxes,** respectively, every portion of the globe receives 12 hours of daylight and 12 hours of darkness. This does not mean that all areas of the globe receive equal quantities of solar energy. Locations on the Equator receive the most while the polar regions receive hardly any direct radiant energy at all; it is just that all areas are in daylight for the same length of time, 12 hours.

It is interesting to note that all areas of the globe receive the same total number of hours of sunlight every year. The North and South Poles receive their sunlight in dramatic opposition—six months of continuous light and six months of continuous darkness—whereas at the equator each day and night is close to 12 hours long throughout the year. However, once again, while the poles receive the same number of hours of sunlight as do the equatorial regions, the sunlight reaches the poles from a very acute (shallow) angle and therefore delivers much less total radiant energy.

18.4 WIND SYSTEMS

A steamship chugging across the ocean derives its power from the thermal energy released when coal or oil is burned to heat the boilers. Where does the energy come from that powers a sailboat across the ocean? From the wind, of course. But wind isn't an energy source like coal or oil; it is more closely analogous to a working substance like the steam that drives a turbine. So the question remains: What energy source drives the winds?

The wind systems of our globe represent the functioning of a great, natural heat engine that is analogous in many respects to a mechanical heat engine. In the engine room of a steamship, fuel is used to boil water and to heat the resultant steam. The moving steam then forces the blades of the turbine to rotate.

The power source of our natural wind systems is the Sun. When the Sun heats one part of the atmosphere more than another, this warm air expands. The expansion makes the heated air less dense than the surrounding air, and the warm air rises. Thus, heat is converted to motion. As the less dense air rises, colder, denser air moves along the surface to replace it. This surface movement is **wind.** As you can see, wind systems operate much like convection currents in a room. Recall from our discussion of convection currents in Chapter 5 that if a heater is placed in one corner of a room, it heats the air adjacent to it. The heated air expands, becoming less dense. This light air rises and is replaced by denser cool air moving along the floor, and an air current is thus established. Wind systems are basically large convection currents that operate on local, continental, or global scales. However, this picture of wind systems as a straightforward and predictable convection cycle is, of course, greatly simplified. Wind systems are affected by a great many other forces, as we shall see.

---

*In some years the solstices and equinoxes occur one day earlier or later than mentioned here. Thus, the summer solstice sometimes occurs on June 22.

FOCUS ON . . . The Effects of the Speed of the Wind— The Wind Chill Factor

Wind can affect our comfort and safety in a variety of ways, but the two that we shall consider in this section are those produced by the speed of wind. One obvious effect that wind speed can have is that high wind velocities can cause damage to structures. How fast does a wind have to be to produce certain kinds of damage? The question is best answered by referring to the table below, which shows the effect that winds of different speeds can have on our environment.

Wind speed also has an effect on our comfort. For example, on a hot day, a slight breeze may bring a welcome relief. The reason for this is that a wind blowing across the body enhances the evaporation of perspiration, thus cooling the body. (See the box on relative humidity and comfort.) However, a strong wind during a cold day may quickly cause injury to exposed body tissue. The wind speed and the temperature are related through the **wind chill index.** The wind chill index is best illustrated by way of a table, which shows air temperatures along the horizontal axis and wind speeds along the vertical. Let us examine what this table tells us about a day when the air temperature is 5°F and the wind speed is 10 mph. Find these values in the table; the point

Speed (mph)	Effect
0	Smoke rises vertically from a chimney
1–3	Smoke is deflected in the direction of the wind
4–7	Slight rustling of leaves on trees
8–12	Leaves move perceptibly
13–18	Dust is kicked up and small branches are set into motion
19–24	Small trees sway
25–31	Large branches on trees move
32–38	Large trees sway
39–46	Walking is difficult
47–54	Roofs may be damaged
55–63	Trees are uprooted
64–74	Structures are damaged considerably
74+	Storm conditions

where the row and column intersect is the wind chill index, or the effective temperature of this day. We find this value to be −15°F, a bitterly cold day. Also, notice that if the wind picks up only slightly to 15 mph at the same temperature, the effective temperature is −25°F, an extremely cold temperature.

MPH	Equivalent Temperature* of Wind Chill Index (°F)																	
Calm	35	30	25	20	15	10	5	0	−5	−10	−15	−20	−25	−30	−35	−40	−45
5	33	27	21	16	12	7	1	−6	−11	−15	−20	−26	−31	−35	−41	−47	−54
10	21	16	9	2	−2	−9	−15	−22	−27	−31	−38	−45	−52	−58	−64	−70	−77
15	16	11	1	−6	−11	−18	−25	−33	−40	−45	−51	−60	−65	−70	−78	−85	−90
20	12	3	−4	−9	−17	−24	−32	−40	−46	−52	−60	−68	−76	−81	−88	−96	−103
25	7	0	−7	−15	−22	−29	−37	−45	−52	−58	−67	−75	−83	−89	−96	−104	−112
30	5	−2	−11	−18	−26	−33	−41	−49	−56	−63	−70	−78	−87	−94	−101	−109	−117
35	3	−4	−13	−20	−27	−35	−43	−52	−60	−67	−72	−83	−90	−98	−105	−113	−123
40	1	−4	−15	−22	−29	−36	−45	−54	−62	−69	−76	−87	−94	−101	−107	−116	−128
45	1	−6	−17	−24	−31	−38	−46	−54	−63	−70	−78	−87	−94	−101	−108	−118	−128
50	0	−7	−17	−24	−31	−38	−47	−56	−63	−70	−79	−88	−96	−103	−110	−120	−128

Cold / Very cold / Bitterly cold / Extremely cold

*Wind speeds greater than 40 mph have little additional chilling effect.   The Wind Chill Index   (ESSA, Washington, D.C.)

Pressure gradients and winds

The pressure at the Earth's surface is not constant from place to place or from time to time. Daily weather maps show the positions of areas of high and low pressure. These variations in pressure within the atmosphere act as the primary mechanism to drive winds. The actual variations at any one time in a particular region are best discussed in terms of a pressure map like that of Figure 18.12.

To produce one of these maps, all pressures are reduced to the value they would have if the

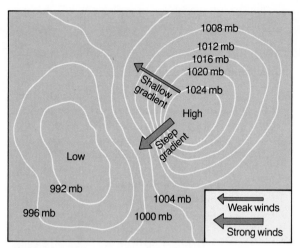

FIGURE 18.12 Along a steep pressure gradient, winds are strong. Likewise, they are weak along a shallow gradient.

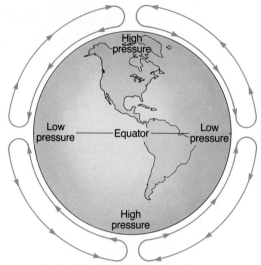

Surface winds would come from the north in the Northern Hemisphere and from the south in the Southern Hemisphere

FIGURE 18.13 Wind systems on an imaginary nonrotating Earth.

altitude were sea level. This is necessary because pressure also declines with increasing altitude, and this decline is far greater than that which would be produced by regional differences in pressure. Since it is the regional differences that are important in determining local winds, it is necessary to mask out any effect on pressure caused by altitude. Once these atmospheric pressures are determined, a map like that of Figure 18.12 is plotted. These maps have contour lines, called **isobars**, drawn through all locations that have equal pressures. At regions where these lines are close together, large changes in pressure occur as one moves between these locations. These regions are said to exhibit large **pressure gradients.** If the lines are spaced far apart, the changes in pressure from one location to another within this region are not great. Such regions have low pressure gradients. Notice on these maps that the contour lines form a closed pattern. The central part of this region is referred to as either a **high pressure** or a **low pressure region.** The importance of these maps is that wind tends to move from high pressure regions to low pressure regions, and the winds are fastest and strongest along steep gradients in pressure, as shown in Figure 18.12.

At first thought, one might be led to believe that winds move directly from a high pressure region along a straight-line path to a low pressure region. Such is not the case, however. The direction of wind flow is affected by a force called the Coriolis force, as we shall see in the next section.

The Coriolis effect

Sunlight reaches the polar regions at such an acute angle, even in summertime, that the quantity of radiation received is quite low. If the Sun were the only source of heat, these regions would be cold in summer and near absolute zero ($-273°C$) in winter, when no sunlight at all is received. Throughout the year, polar regions are considerably warmer than one would expect solely on the basis of the amount of sunlight received. On the other hand, measurements and calculations show that the equatorial regions of our planet receive so much radiant energy that they should be much hotter than they actually are. Thus, heat must be carried in some way from the tropics to the poles. Winds and ocean currents are responsible for the transfer.

To understand global wind systems, consider for a moment what would happen if the Earth did not rotate about its axis. The intense sunlight at the Equator would heat the air there, causing it to rise. As the air rose, cooler air would move in from the polar regions, and a set of convection

train and that you rolled your bowling ball directly down the center of the alley on a perfect path toward the strike zone. But just as the ball left your hand, the train reached a curve and started to turn toward the west (left). The ball continued to travel due north, but the target was moving away toward the left, so the ball appeared to veer to the right, thereby missing the pins. The observed path, shown in Figure 18.14, results from the motion of the ball moving straight along an alley that is moving to the left. The bowling ball situation is an example of relative motion. To an observer on the train, the ball has moved off to the right, regardless of its motion relative to some other frame of reference such as a compass.

What does this have to do with the wind? The distance around the Earth, measured parallel to the Equator, is greatest at the Equator and decreases toward the poles. But all parts of the planet make one complete revolution every day. Thus, since a point on the Equator must travel farther than any other point on the Earth in 24 hours, the equatorial region moves faster. At the equator, all objects move eastward with a speed of about 1600 km/h; at the poles, there is no eastward movement at all, and the speed is 0 km/h. Now imagine a parcel of air located at the Equator. It is traveling eastward at 1600 km/h and then gets heated and moves upward and starts to travel poleward. Let us say that this parcel moves north. As it starts, there are two components to the velocity, an eastward component and a northward component. At any distance north of the Equator, it is traveling eastward *faster* than the Earth beneath it. Thus, the motion of the air relative to Earth is curved toward the east, or the right, as shown in Figure 18.15a.

Let us imagine what would happen if the situation were reversed and a parcel of air moved southward from the North Pole to the Equator. This air is moving more slowly than air in tropical regions; therefore, it lags behind the ground below. Since the Earth moves in an easterly direction, our parcel of air veers toward the west, as shown in Figure 18.15b.

As a rule of thumb, when one looks along the direction of motion of the wind, all winds veer to the right in the Northern Hemisphere. In the Southern Hemisphere, the opposite effect is observed, and winds veer toward the left. The de-

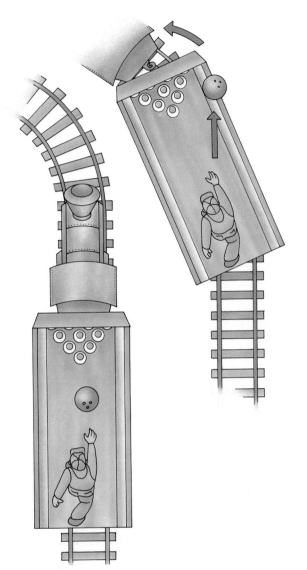

FIGURE 18.14 Observed path of bowling ball thrown in a straight line while train turns to the left.

currents would be established, as shown in Figure 18.13. According to this model, in the Northern Hemisphere the predominant surface winds would blow southward, while in the upper atmosphere, warm air would move poleward. In the Southern Hemisphere, the situation would be reversed.

Of course, our static model is too simple, because the Earth does rotate. To visualize the effect of this rotation, imagine what it would be like if there were a bowling alley on a moving train. Suppose that you were riding due north on the

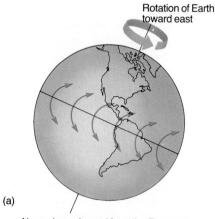

Rotation of Earth
toward east

(a)

Air moving poleward from the Equator is
traveling east faster than the land beneath
it and veers to the east (turns right in the
Northern Hemisphere and left in the
Southern Hemisphere)

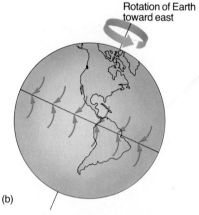

Rotation of Earth
toward east

(b)

Air moving toward the Equator is
traveling east slower than the land
beneath it and veers to the west
(turns right in the Northern Hemisphere
and left in the Southern Hemisphere)

FIGURE 18.15 The Coriolis
effect. Air moving north or
south will be deflected by the
rotation of the Earth.

flection of airflow caused by the rotation of the
Earth is called the **Coriolis effect.**

Figure 18.16 shows how the Coriolis effect
affects the wind moving away from a high-pressure region. For simplicity, we have shown the
isobars as concentric circles about a high pressure
region in Figure 18.16a. As the wind moves down
the pressure gradient, it is deflected toward the
right in the Northern Hemisphere. The end result is that the wind moves in a clockwise spiral
about a high-pressure region. As the wind moves
toward the center of the low shown in Figure
18.16b, the deflection caused by the Coriolis effect produces a counterclockwise spiral about the
center of the low. These flow patterns are reversed in the Southern Hemisphere. That is, the
winds move in a counterclockwise spiral about a
high and in a clockwise spiral about a low.

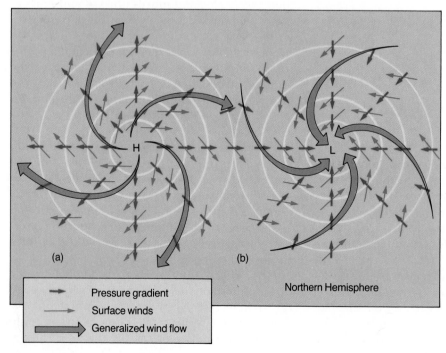

(a)   (b)

Northern Hemisphere

→ Pressure gradient
→ Surface winds
⇒ Generalized wind flow

FIGURE 18.16 (a) The
Coriolis effect causes winds
to circle clockwise about a
high-pressure region and
(b) counterclockwise about a
low.

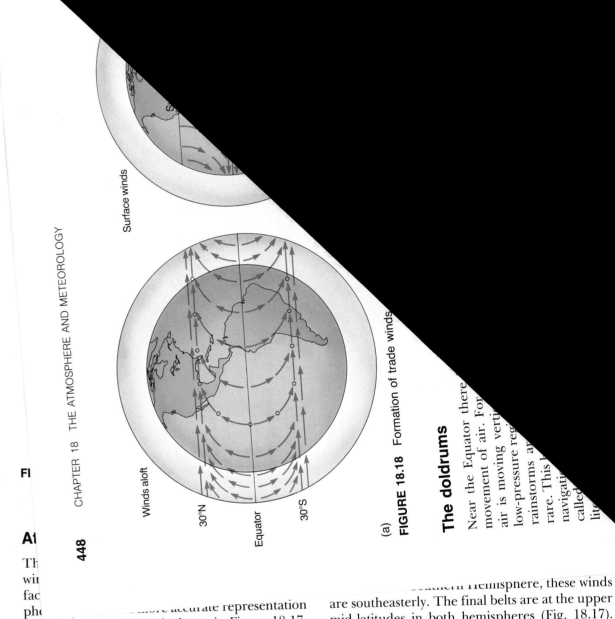

Surface winds

Winds aloft

30°N

Equator

30°S

(a)

FIGURE 18.18   Formation of trade winds

The doldrums

Near the Equator there⋯
movement of air. For⋯
air is moving verti⋯
low-pressure reg⋯
rainstorms a⋯
rare. This ⋯
navigati⋯
called ⋯
lite⋯

A⋯

Th⋯
win⋯
fac⋯
ph⋯ ⋯ more accurate representation
of global wind patterns is shown in Figure 18.17.
As shown, the Earth is divided into six basic wind
belts. Two of these belts, one in each hemisphere,
are produced by air moving from the high pressure regions at the poles toward the Equator. The
Coriolis effect causes those winds from the North
Pole to be deflected toward the right, or from east
to west. Thus, these winds are called the **polar
easterlies.** You use the Coriolis effect to show that
the winds from the South Pole should also be referred to as polar easterlies.

Atmospheric factors cause the winds flowing
from the low-pressure zone at the Equator to sink

⋯outhern Hemisphere, these winds
are southeasterly. The final belts are at the upper
mid-latitudes in both hemispheres (Fig. 18.17).
The Coriolis effect causes these winds to move
from the west in both hemispheres, hence the
name **westerlies.** Although this map of global
wind patterns is more correct than the one of Figure 18.13, it is still an idealized situation. One of
the reasons for this is that our map assumes that
the Sun is shining down directly on the Equator.
This is obviously not the case in the winter and
summer seasons. Thus there are seasonal variations in the strength of the winds in each of these
belts. Let us now take a closer look at conditions
in these wind belts.

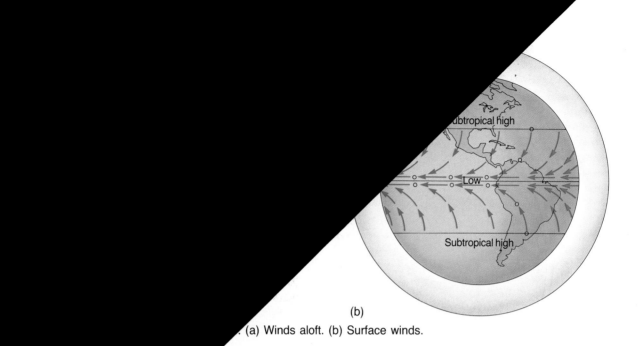

(b)

. (a) Winds aloft. (b) Surface winds.

...is very little horizontal... ...the most part, hot humid ...cally upward, forming a vast ...ion. Therefore, local squalls and ...e common, but steady winds are ...ot, still region was a serious barrier to ...on in the age of sailing ships. Mariners ...the equatorial region the **doldrums,** and ...rature is alive with descriptions of the despair of being unable to move across the vast, windless seas. Over the land areas, the frequent rains near the equatorial low-pressure regions give rise to the tropical rain forests that abound at these latitudes.

The horse latitudes and trade winds

The air that rises at the Equator splits and travels north and south at high altitudes. These high-altitude winds are not felt at the surface of the Earth. They move high in the troposphere, thereby affecting airplane travel. In both Northern and Southern Hemispheres, this air veers toward the east until it has turned so much that it is moving due east, as shown in Figure 18.18. These eastward movements are realized at about 30° north and south latitudes. This high-altitude

air, having lost its poleward speed, has time to cool and drop to the surface. Thus, a generally high-pressure area exists in these regions. Since the air is dropping vertically and not moving horizontally, few steady surface winds are experienced; and since the region is one of predominantly high pressure, there is little rain. These regions have sometimes been called **horse latitudes.** Sea stories tell us that the name was given because sailing ships were often becalmed in this region for long periods of time, and horses transported as cargo often died of thirst and hunger. The warm, dry, descending air in these high-pressure regions has helped to form the world's great deserts, like those found in the Sahara region of North Africa.

The air that descends at the horse latitudes then travels toward the Equator to complete the convection cycle initiated when the warm air ascended. This convection current is so steady and predictable that surface winds almost always blow between the horse latitudes and the doldrums. As the air moves toward the Equator, it is deflected toward the right (west) by the Coriolis effect, and thus these surface winds veer so that they blow from the northeast in the Northern Hemisphere. You show that these winds should be from the southeast in the Southern Hemisphere. Sailors have traditionally depended on these reliable winds and hence called them the **trade winds.**

The westerlies

As we have seen, the predominant surface winds in the temperate regions come from the southwest in the Northern Hemisphere and from the northwest in the Southern Hemisphere. Thus, heat from the Equator is carried toward the poles. Of course, the poleward movement of air must be balanced by air moving toward the Equator from higher latitudes. Although the prevailing winds in the Northern Hemisphere, for example, move from the southwest, polar air masses frequently move southward. These winds have an important effect on local weather conditions. In the Southern Hemisphere, there is almost no land between 50° and 65° south latitude to interrupt the flow of air, so the winds build up and cause giant waves to roll along with them. The sailing ships of old followed these ferocious winds along what were called the "roaring forties," the "howling fifties," and the "screaming sixties." Sea stories are filled with tales of the legendary high winds and huge waves. In the Northern Hemisphere, the continents and mountain ranges interrupt the steady flow of air, and weather is less predictable. As arctic air flows southward into the path of the westerlies, it grows turbulent, bringing rain and generally favorable conditions for agriculture. The great wheat belts of the United States, Canada, and Russia all lie above 30° north latitude.

Polar winds

The polar regions are cold for most of the year. Dramatic temperature gradients, interactions between sea and land, and other local conditions cause the winds in these regions to be quite unpredictable.

Upper air winds

The winds that we have discussed thus far have been those near the surface of the Earth. Of equal importance, and in some instances of even greater importance, are winds in the atmosphere above 5000 m, and the most important of these are fast moving "rivers of air" called the **jet streams.** The presence of these winds became apparent during World War II when high-alti-

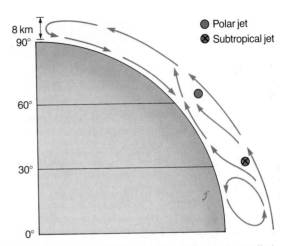

FIGURE 18.19 The jet streams and a more realistic representation of the circulation of air on the Earth.

tude bombers found that they could travel to their destination more quickly when flying toward the east than they could when flying toward the west.

The jet streams are located at roughly 30° and 60° latitude in both hemispheres, at or near the boundaries between the wind belts. These streams are shown in Figure 18.19. Also shown in this figure is a more accurate representation of actual global air circulation. At these boundaries, there are strong pressure gradients, which can lead to intense winds. The **polar jet stream** at 60° latitude ranges from 60 to 160 km in width and up to 2 or 3 km in depth. The jet stream at 30° latitude is the **subtropical jet stream.** Both of these jets are best developed in winter when hemispherical temperatures have their steepest gradient. During the summer, these jet streams weaken in intensity, and the subtropical jet often disappears completely. For reasons that have not been fully explained, the upper air goes into oscillations. One possible triggering mechanism is the variation of the surface temperature of the oceans. For example, if the water temperature in the oceans near the Equator or at high latitudes becomes either colder or warmer than usual, oscillations in the upper atmosphere are initiated that continue until the ocean temperatures return to normal. Whatever the mechanism, severe disturbances in the jet streams may significantly change the climate at certain regions on the

FOCUS ON . . . Relative Humidity and Comfort

Many southwestern states advertise that their high temperatures are more comfortable than the same temperature found, say, in a midwestern state. They base this statement on the fact that their arid climate lends itself to an atmosphere with a low relative humidity. Is this assertion true or false? The answer is that it is indeed true. Let's see why. One of the primary cooling mechanisms of the human body is perspiration, and relative humidity affects our comfort through its relationship to the rate of evaporation. As perspiration evaporates from the skin, the skin is cooled, because as we saw in Chapter 5, evaporation is a cooling process. This occurs because the heat required to change the droplets of sweat to a vapor is subtracted from the skin. Obviously, if the relative humidity is high, perspiration evaporates slowly from the skin, and you are left with a damp, sticky feeling. But just how important is relative humidity to our comfort? The answer can best be demonstrated with reference to the table. To see how this table is used, first consider a state like Arizona, which might have a temperature of 90°F but a relative humidity of only about 10 percent. Find 90°F along the horizontal and move down the column beneath it until you are directly across from the 10 percent relative humidity reading. The intersection of this row and column shows you that 90°F feels like 85°F when the relative humidity is only 10 percent. However, consider a midwestern state on a day when the temperature is 90°F. On such a day, a relative humidity of 70 percent is not uncommon. Finding the value at the intersection of the row and column for these values shows that this particular 90°F day would actually feel more like 106°F. In this latter case, the air is already almost saturated, which makes it far more difficult for evaporation to take place.

Effective air temperatures as a function of relative humidity

	Air Temperature (°F)						
	70	75	80	85	90	95	100
Relative Humidity	*Effective Temperature*						
0%	64	69	73	78	83	87	91
10%	65	70	75	80	85	90	95
20%	66	72	77	82	87	93	99
30%	67	73	78	84	90	96	104
40%	68	74	79	86	93	101	110
50%	69	75	81	88	96	107	120
60%	70	76	82	90	100	114	132
70%	70	77	85	93	106	124	144
80%	71	78	86	97	113	136	
90%	71	79	88	102	122		
100%	72	80	91	108			

Earth. For example, droughts may occur, winters may be colder than normal, and so forth.

Jet streams are also significant in that they can carry pollutants for large distances and at very high speeds. For example, the polar jet stream carried ash from Mt. St. Helens eastward into Idaho and Montana.

18.5 CONDENSATION AND PRECIPITATION

If you boil water on a stove, you can see a steamy mist above the kettle, and then higher still the mist seems to disappear into the air. Of course, the water molecules have not been lost. In the

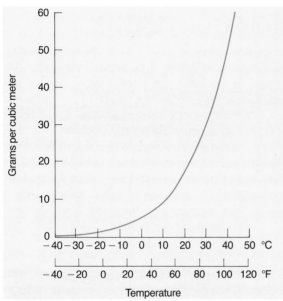

FIGURE 18.20 Maximum water content of air as a function of temperature.

pan, water is in the liquid phase, and in the mist above the kettle the water exists as tiny droplets. These droplets then evaporate, and the water vapor mixes with air and becomes invisible. Air generally contains some water vapor. **Humidity** is a measure of the amount of water vapor in air. **Absolute humidity** is defined as the mass of water vapor contained in a given volume of air and is generally expressed in units such as g/m^3. But air cannot hold an unlimited quantity of water vapor. If you poured liquid water into a container of dry air, the water would start to evaporate; that is, some water molecules would leave the liquid and mix with the air molecules in the form of a gas. Of course, some molecules would go the other way, from gas to liquid. When the two opposing rates become equal, net evaporation stops, and the air is said to be **saturated** with moisture.

At the saturation point, the air can hold no more moisture. The saturation quantity varies with temperature. Figure 18.20 shows that warm air can hold more water vapor than can cold air. For example, a cubic meter of air can hold 23 g of water vapor at 25°C, but if the air is cooled to 12°C, it can hold only half that quantity, 11.5 g/m^3. The **relative humidity** is a measure of the

amount of water vapor in the air compared with the saturation quantity at a given temperature.

Relative humidity (%)   (18.1)
$$= \frac{\text{actual quantity of water per unit of air}}{\text{saturation quantity at the same temperature}} 100\%$$

Suppose that there are 11.5 g/m^3 of water vapor in a parcel of air at 25°C; that is, the absolute humidity is 11.5 g/m^3. Since air at that temperature can hold 23 g/m^3 when it is saturated, it is carrying half of the saturation quantity, and the relative humidity is

$$\text{Rel humidity} = \frac{11.5 \text{ g}}{23 \text{ g}} \, 100\% = 50\%$$

As an experiment, let us take some of this air and cool it without adding or removing any water vapor. Since cold air can hold less water vapor than warm air, the relative humidity rises even though the absolute humidity remains constant at 11.5 g/m^3. If the air is cooled to about 12°C, the relative humidity reaches 100 percent because air at that temperature can hold only 11.5 g/m^3 and that amount is already present. Any further cooling causes the water to condense and to form droplets.

You can observe how water condenses upon cooling by performing a very simple experiment. Heat some water on a stove until it is boiling rapidly. The clear air that lies just above the steamy mist will be hot and full of water vapor. Hold a drinking glass in this space and you will observe droplets of water condensing on the surface. Because the surface of the glass is cool, it absorbs heat from the warm, moist air that comes in contact with it. As the air cools, some of the water vapor condenses into liquid droplets. The same effect can be observed if you are inside a house on a cold day. If you breathe onto a window, droplets of water or crystals of ice appear as your moist breath cools on the glass.

With this background, we can easily understand how **dew** is formed. On a typical summer evening in a moist temperate zone, the air is likely to be warm and laden with water vapor. After the sun has set, the surfaces of various objects such as plants, houses, and windows begin to lose heat by radiation. In the early hours of the morning,

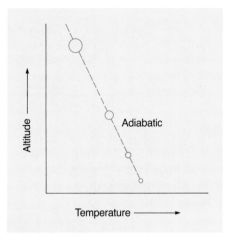

FIGURE 18.21   Adiabatic cooling.

when these surfaces are quite cool, water condenses from the warm moist air just as your breath condenses onto a window on a cold day. This condensation is called dew. The **dew point** is the temperature to which a sample of air must be cooled to become saturated with moisture.

If the temperature of the air is at or below freezing, the moisture in the air goes directly from the vapor state to the solid state upon condensation. This is the process that occurs in the formation of **frost.**

18.6 CLOUDS

Clouds are formed when warm, moisture-laden air is cooled below its dew point. However, condensation occurs most readily when there are small particles present in the air. These particles, called **condensation nuclei,** serve as surfaces upon which water molecules can collect. One type of cloud formation is **fog.** When warm moist air from an ocean blows inland to a cold coast, the air cools, water condenses, and low cloud formations known as fog are formed. Thus, San Francisco (California), Seattle (Washington), and Vancouver (British Columbia) all experience foggy winters accompanied by rain and drizzle.

But cloud formation can occur in other ways as well. When a body of air rises, it ascends into a region of lower atmospheric pressure. As it moves into the zone of lower pressure, it expands, just as a balloon would expand if you placed it in a partial vacuum. As a result, the thermal energy spreads out over a larger volume, and the air cools. This effect is called **adiabatic cooling,** as shown in Figure 18.21.

Now imagine that a moist mass of air in a given region is moving toward the Equator. If the temperature of the air could remain constant, the water would remain in a vapor form indefinitely.

FIGURE 18.22   Cirrus clouds are wisps of fine ice crystals. (Courtesy of Kenneth R. Martin Collection)

FIGURE 18.23 Stratus clouds are formed of horizontal sheets of water droplets. (Courtesy of S. Brazier)

However, as this air moves into the tropics, it is heated, becomes less dense, and therefore starts to expand. It then rises, expands further, and begins to cool adiabatically. If conditions are favorable, the air cools so much that water vapor condenses out and clouds are formed. Thus, cloud formation occurs most readily when moist air rises.

There are three basic types of clouds: cirrus, stratus, and cumulus. In addition, under each of these three categories are various subtypes. Let us take a look at each of these basic categories in turn.

Cirrus

The name *cirrus* comes from the Latin meaning lock or wisp of hair. These clouds are the type often painted by artists because they can be applied to canvas as feathered patches of white. Figure 18.22 shows that this is an accurate rendering of this type of cloud. Cirrus clouds are usually found at very high altitudes, normally between 6,000 and 11,000 m. Because of the low temperature at this altitude, these clouds are actually formed of ice crystals rather than water vapor.

Stratus

The name *stratus* is derived from the Latin for layer. As Figure 18.23 indicates, these clouds are characterized by their sheet-like or layered ap-

pearance. These clouds can be formed at any height ranging from near the surface of the Earth to about 6000 m, and they consist of water droplets. It is stratus clouds that are present during the winter months when a drab gray cloud cover blankets the entire sky. These clouds have an almost constant thickness throughout their extent. Fog can be considered a low-lying stratus cloud. During those times when this thickness is not particularly great, the Moon or Sun can be seen through the overcast. Often during these times, a halo, or corona, can be seen surrounding the Moon or Sun.

Cumulus

The name *cumulus* comes from the Latin for heap or pile. In contrast to stratus clouds, cumulus clouds have a vertical rather than a horizontal development, as shown in Figure 18.24. They generally have a flat base with a towering, majestic pile mushrooming up from the top. These clouds can have their base at a height of anywhere from about 500 m to about 12,000 m. Cumulus clouds are formed by an upward movement of air. The base of the cloud is formed at the height at which condensation begins. These clouds are often seen on bright summer days.

Nimbus

Nimbus is a word from the Latin that means precipitation and is often used as a prefix or a suffix to describe clouds. For example, a nimbostratus cloud is a stratus cloud from which a drizzle is falling. The familiar thunderhead seen during summer rains is an example of a cumulonimbus cloud. The dark appearance of this latter type occurs as condensation builds up in the cloud to the point at which sunlight is completely blocked. Figure 18.25 is an example of a cumulonimbus cloud.

18.7 PRECIPITATION

We can now understand a few simple relationships between atmospheric conditions and weather. When warm air rises, it expands and cools. The expansion makes it less dense, which means that it exerts less pressure and the barome-

FIGURE 18.24 Cumulus clouds form over locations where there are rising currents of air. The base of these clouds are located at the height where condensation begins. (Courtesy of Kenneth R. Martin Collection)

ter reading in that region is lower. But, as already noted, these same conditions also generally lead to cloud formation and, therefore, to the possibility of rain or snow. Thus, a falling barometer is a good indication that precipitation may soon follow. Alternatively, when cool air falls to the Earth, the barometric pressure rises. Air is compressed and heats up as it falls. Since warm air can hold more moisture than cold air, clouds generally do not form under high-pressure conditions. Instead, the warm air with low relative humidity tends to absorb moisture from the Earth's surface. Thus, a rising barometer generally predicts fair weather.

Of course, an understanding of this general relationship between barometric pressure and cloud formation does not instantly transform anyone into a seasoned weather forecaster. Often the barometric pressure drops but no rain falls, or, conversely, the pressure may rise amid cloudy skies. The temperature, rainfall, and wind patterns in a region are a result of many complex factors. Some of these are caused by local disturbances and others by global atmospheric patterns.

When clouds form, the droplets usually are light enough that they do not fall to the Earth. As a result, they remain suspended in near-equilibrium under the action of gravity (downward) and the movement of air (upward). However, if the mass of the droplets becomes too large, they will fall out of the sky as either rain, snow, hail, or sleet. Let us look at each of these in turn.

Rain

The exact process that produces rain is not fully understood, but the method seems to depend upon the type of cloud structure present. Raindrops falling from a cumulus cloud may have

FIGURE 18.25 A cumulonimbus cloud. (NOAA)

originated in a different way than those falling from a stratus cloud. Let us examine these two different processes.

The essential elements for the formation of rain within a cumulus cloud are (1) ice crystals, (2) a supersaturated vapor, and (3) a mixing process within the cloud, as shown in Figure 18.26. The first of these, the ice crystals, are present at the top of a cumulus cloud because the cloud extends high enough into the atmosphere for the crystallization process to occur. Supersaturated vapor is present at or near the base of the cloud, but we must first examine what supersaturated vapor really is. We have stated that when the temperature drops below the dew point, condensation occurs. However, it is possible for the temperature to drop below the dew point without condensation taking place, and when this occurs, the vapor is said to be supersaturated. In the presence of a mixing agent, supplied perhaps by air currents, the ice crystals and the supersaturated vapor can come into contact. The supersaturated vapor then condenses on the ice crystals, forming droplets, which in turn fuse with other droplets to form a drop of rain. The final result is a typical raindrop of about 2.5 to 6 mm in size and heavy enough to fall from the cloud as rain.

It can be easily demonstrated that the process described above is an important one in the formation of rain, because rainmakers often hasten or help along the processes described above. During a drought, the essential element often missing in the prescription for rain is the presence of ice crystals in the clouds. As a result, modern-day rainmakers seed the clouds, via airplane, with silver iodide or carbon dioxide pellets. The structure of silver iodide crystals is similar enough to ice that they become a direct substitute for ice crystals. The carbon dioxide crystals, on the other hand, are at a temperature of about $-80°C$, and as they go directly into the vapor state, they produce a cooling action within the cloud, which initiates the formation of ice crystals.

The rainfall from a stratus cloud is produced by a different and more direct method. In this case, the water droplets are formed by the coalescing of smaller droplets until a drop large enough to fall is formed. Because of this method

FIGURE 18.26 Ice crystals, a supersaturated vapor, and mixing are necessary to produce rain from a cumulus cloud.

of formation, the droplets formed are often very small, of the order of 0.5 mm, and the rain may fall as a mist or drizzle.

Snow

After rain, the most common form of precipitation is snow. Snow is produced when the dew point is below 0°C, the freezing point of water. Under this condition, the water vapor passes directly from the vapor state to that of minute ice crystals. These crystals may fall as individual crystals, or, if the air temperature is high enough, several crystals may melt slightly and become stuck together to form a rather large snowflake.

Sleet

Sleet is formed when rain falls through a layer of cold air and freezes. The result is a small globule of pure ice.

Hail

Hail is a form of precipitation that is less likely to occur than any of the others discussed. It usually occurs during the summer months, and the cloud from which it falls is often in the cumulonimbus family. Hail is produced when there are strong updrafts within a cloud formation. Ice crystals fall through the cloud and grow as supersaturated vapor coalesces on them. However, because of the updrafts, one of these ice particles may be lifted upward again and again to pass through the supersaturated vapor. As a result, layers of ice form around the original crystal. When the hailstone produced finally exits the cloud, it may range in size from 5 mm up to the size of a baseball. The world record is a hailstone 30 cm in diameter that fell in Australia.

18.8
THE OCEANS AND WORLD CLIMATE

The concepts discussed in the preceding sections enable us to predict wind and weather from season to season in a general way over vast areas of the Earth. But climates are influenced by local factors as well. Continents, oceans, prairies, and mountain ranges and the many dynamic interactions among them continuously influence winds, rains, and cloud formations. The ocean, in particular, affects climate significantly, in large measure because water absorbs and stores heat at a rate different than does the solid earth.

Recall from our discussion of heat in Chapter 5 that if equal masses of water and ethyl alcohol are both heated on equal hot plates, the *temperature* of the alcohol rises nearly twice as fast as that of the water. Also, we saw that if water and sand absorb equal amounts of thermal energy, the temperature of the sand rises about five times as fast. Similarly, if equal masses of hot water and hot sand, both starting at the same temperature, are placed in a refrigerator, the sand cools faster. This means that more thermal energy is stored in 1 g of hot water than in 1 g of hot sand at the same temperature, which is to say that the specific heat of water is greater than that of sand. As a matter of fact, water has an unusually high specific heat—about five times that of sand, granite rock, or dry clay. Consider, then, what happens near the boundary between an ocean and a continental land mass. Suppose for a moment that both the water and the land are at the same temperature at some time in the spring. As summer approaches, both land and sea receive equal amounts of solar energy. But the land warms up faster, just as the ethyl alcohol or the sand gets

FIGURE 18.27   Continental areas experience greater temperature extremes than do coastal areas. St. Louis: average January temperature = 0°C; average July temperature = 26°C. San Francisco: average January temperature = 9°C; average July temperature = 17°C.

hotter than the water even though they both absorb equal amounts of thermal energy.

There is yet another factor that complements the difference in specific heats and makes the land warm up faster than the sea. The ocean is turbulent, and waves and currents carry warm surface water downward and bring the cold deep water up to the surface. On land there is no comparable mixing process, and much of the thermal energy is concentrated in a shallow surface layer of soil or rock. Therefore, on land a small surface layer becomes quite hot, so that at times you may burn your feet walking across dry sand or concrete. Yet a meter below the surface, the soil is 10° to 15°C cooler. In the sea, the temperature is much more uniform throughout. The same amount of energy that warms a large mass of water slightly causes a small layer of soil to become much hotter.

For these reasons, summertime temperatures are generally hotter in the central areas of large continents than near the seacoasts. In the winter, the opposite effect is observed, and the land tends to cool more than the sea. Thus, inland areas are generally colder in winter and warmer in summer than the coastal regions. For example, the coldest temperatures recorded in the Northern Hemisphere are found in central Siberia and not at the North Pole, because Siberia is landlocked, whereas the North Pole is located in the middle of the Arctic Ocean. However, in the summer, Siberia is considerably warmer than the North Pole. Or, as another example, let us compare the climates of two large cities in the United States, San Francisco and St. Louis, that are both located at approximately 38° north latitude. As shown in Figure 18.27, St. Louis experiences much greater temperature extremes than San Francisco.

Coastal climate is also influenced by several other factors. The oceans are not like a huge bathtub full of water that sloshes about randomly. Water travels across oceans in well-defined and predictable **currents,** much like the prevailing wind patterns in the atmosphere. These currents can be thought of as rivers that flow within the ocean.

Surface currents are caused partially by winds. When the trade winds blow steadily across the sea, they push along the upper layers of water, and waves are formed. But just as the air pushes the surface water, so the moving surface waves drag deeper waters along with them. Thus, the steady winds cause water to flow in predictable patterns across the oceans. Of course, there is considerable slipping between wind and water, and therefore the water moves much more slowly than the air; most midocean currents move at a rate of only 3 to 5 km/h. Ocean currents, like wind systems, are deflected by the spin of the Earth. They are forced to veer clockwise in the northern oceans and counterclockwise south of the Equator. As the moving waters strike continental land masses, they are deflected even further, so that they rotate in circular motions called **gyres.**

Currents may carry either warm water toward the poles or cold water toward the Equator. Either way, the moving water profoundly affects climates on land. Portland, Maine, in the United States is at about the same latitude, 44° north, as the north coast of Spain. However, the Spanish coast is a warm resort area where people swim and sunbathe nearly year-round, whereas the Maine coast is noted for its hard, snowy winters and cool summers. Much of the reason for this climatic difference lies in the fact that the cold Labrador Current moves down from the Arctic Ocean to the coast of Maine, whereas the warm Gulf Stream tempers the climate of Spain (and the entire western coast of Europe). Figure 18.28 shows the global system of currents.

Sea breezes

Anyone who has lived near an ocean or large lake has undoubtedly become acquainted with recurring winds that blow from water to land and from land to water. Local sea breezes are caused by uneven heating and cooling of the land and the ocean, as shown in Figure 18.29. We have already learned that the land heats up faster than the sea in summertime and cools more quickly in winter. The same effect occurs on a daily time scale as well. If land and sea exist at close to the same temperature on a summer morning, the land and the air above it will become hotter by noon. Hot air then rises over the land, a local convection current is established, and cooler air from the sea flows toward the land. Thus, on a hot sunny day at the beach, winds generally blow from the sea inward. At night, the reverse process occurs. The land cools faster than the sea until it becomes colder by nighttime. Then the winds reverse, and breezes blow from the shore out toward the sea.

Monsoons

Sea breezes also occur on a larger, continental scale. Seasonally reversing winds caused by uneven heating of land and sea dominate the weather over most large land masses. The **monsoons** that are common in Asia and Africa are an example of such seasonal winds. In the summertime, the continents heat up faster and become warmer than the sea. The warm earth heats the air above it, and this warm air mass rises, creating a large low-pressure area. This rising air draws moisture-laden sea breezes inland. When the wet air rises, clouds form and heavy rains fall. In the winter the process is reversed. The land cools below the temperature of the sea, and air rises over the ocean to descend over the land, producing a continental high-pressure zone. The surface winds blow from the land out to sea, and this dry high-pressure air produces very little rain. More than half of the inhabitants of the Earth depend on the monsoons for their survival, because the predictable heavy summer rains bring water to the prairies and grain fields of Africa and Asia. If the monsoons fail to arrive, crops cannot grow and people have nothing to eat.

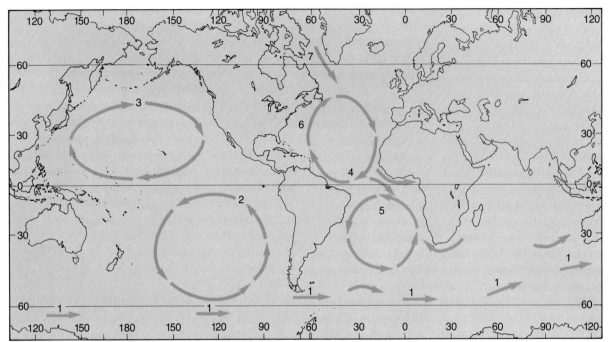

FIGURE 18.28 Surface currents of the world in July. Key: 1, West Wind Drift; 2, Pacific South Equatorial Current; 3, North Pacific Current; 4, Atlantic North Equatorial Current; 5, Atlantic South Equatorial Current; 6, Gulf Stream; 7, Labrador Current.
(After Naval Oceanographic Office, SP-68)

FIGURE 18.29 Sea breezes caused by uneven heating of land and water. (a) During the day, the land is warmer than the water, warm air rises over the land, and a breeze blows from the sea to the land. (b) During the evening, the sea is warmer than the land because it doesn't cool as quickly, and wind blows from the land to the sea.

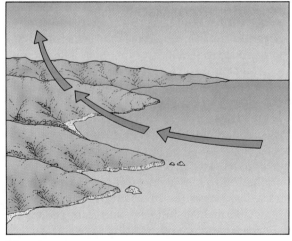

(a)

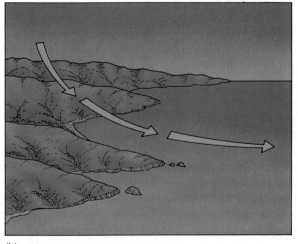

(b)

FIGURE 18.30 Rainfall patterns in the state of California. Note how the rain shadow deserts lie east of the mountain ranges. Rainfall is reported in centimeters per year.

18.9
WEATHER PATTERNS

In any particular location on Earth, seasonal changes are fairly predictable. Thus, ranchers in Colorado bring in the hay and round up scattered herds of cattle every fall in expectation of the winter snows they know will come. Skiers tune their skis in late November or early December. By the end of March, sports shops display tennis racquets, and the ranchers repair broken machinery and prepare for spring plowing. Yet no one can predict accurately when the first snow will fall or exactly when warm weather will return. **Weather** is a description of the day-to-day changes in temperature, wind, and precipitation.

Recall from an earlier discussion in this chapter that precipitation is likely to occur when moist air rises. The rising air cools, causing the moisture to condense and fall as rain or snow. There are several ways in which moisture-laden air can

rise. As mentioned previously, if the Sun heats the air in one region more than in another, the warmer air expands and travels upward. This type of pattern is responsible for the frequent rain squalls near the Equator. The same effect may occur on a local scale. Visualize a large area of plowed fields adjacent to a vast forest. Bare soil absorbs heat more efficiently than green leaves, and the air above the farms will be heated relative to adjacent air masses over the woods. This air may rise, and if conditions are favorable, local summer thunderstorms will occur.

Mountains also affect local weather. An air mass traveling across a mountain range must necessarily rise to flow over the mountains. If this rising air is laden with moisture, conditions will be favorable for condensation. Rain or snow will then be likely. Conversely, the air must fall on the downward side of the mountains. As it descends, it is compressed and thereby heated. Since this warm, high-pressure air is unlikely to discharge moisture, a belt of dry climate, called a rain shadow, often exists on the lee side of major mountain ranges. Death Valley in California is a rain shadow desert (Fig. 18.30).

18.10
FRONTAL WEATHER SYSTEMS

If you had an insulated container of cold, dry air and blew a breath of warm, moist air into it, the two parcels of air would eventually mix until the temperature and humidity in all parts of the container were the same. These changes are predicted by the second law of thermodynamics, which states that any undisturbed system will spontaneously tend toward maximum disorder, or sameness. Similar considerations apply whether the sample of air is a small volume in a balloon or the entire Earth's atmosphere. If the atmosphere were insulated and undisturbed, it would drift toward sameness, and eventually its temperature would be identical at the poles, the Equator, and everywhere else. But the atmosphere is dynamic, not static. It is warmed unevenly by the Sun, and it loses energy by radiation out into space.

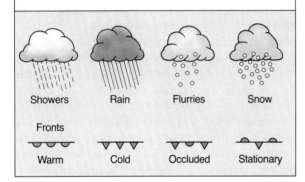
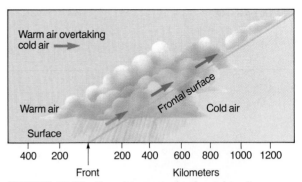

FIGURE 18.31 Development of a frontal system. When a warm front overtakes a cold front, the warm air rises and forms a wide zone of low pressure and precipitation.

turn leads to a zone of low barometric pressure, which frequently gives rise to cloud formation and precipitation. However, the character of the low pressure zone and the weather that results depend on the relative velocities of the two parcels of air.

Warm fronts

If a warm air mass overtakes a mass of cold air, the warm air rises, as explained above. But as it rises, it continues to move forward. The net result is that it flows up and over the cold air, pushing the cold air mass into a wedge. Figure 18.31 shows the shape of the two air masses at a range of elevations, from ground level to an altitude high in the troposphere. You can see that the warm air extends over the cold air for a distance of several hundred kilometers. This rising air leads to a storm system that is often 500 to 600 km wide and is characterized by low barometric pressure, cloud formation, and precipitation.

Cold fronts

A cold front, like a warm front, occurs along a zone where warm air and cold air masses come in contact. In this case, however, it is the cold air that is overtaking and displacing the warm air. During this collision, the faster moving cold air cannot slide over the warm air, because it is more

Air that lies over the polar icecaps is frequently dry and cold, whereas air lying over tropical oceans is usually warm and moist. The term **air mass** refers to a large body of air that has approximately the same temperature and humidity throughout.

As the Earth turns and winds blow, different air masses move about and collide. When one air mass collides with another, at first the two behave as separate entities, each with its own specific properties, like a drop of ink in a glass of water. If no other interactions occurred, they would, in time, mix to become one homogeneous parcel of air. But the Earth's atmosphere is not at all static. The Sun shines, winds blow, other air masses move in, and changes occur in a continuous and often unpredictable manner.

When two air masses collide, the zone where they meet is called a **front.** The movement of air along a weather front is fundamental to our understanding of weather. Whenever warm air comes in contact with cold air, the warm air, being less dense than the cold air, rises. This in

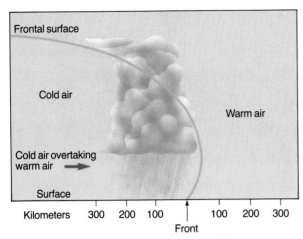

FIGURE 18.32 When a cold front overtakes a warm front, the zone of rising air is steep and narrow, leading to a narrow zone of turbulent weather.

dense. Therefore, the cold air is distorted into a blunt wedge or bulge as it pushes under the warmer air. A typical frontal profile is shown in Figure 18.32. Note that the zone of rapidly rising air and precipitation is generally quite narrow, as opposed to the broader zone (500 to 600 km) observed for a warm front. However, the steep zone of contact between the two air masses causes the warm air to rise quite rapidly. The net result is a narrow band of rather violent weather. In a typical situation, a line of storms may be only 50 to 150 km wide, but within this zone downpours and thunderstorms are prevalent.

Figure 18.33 shows a sequence of four weather maps of the western United States. A satellite photograph of the cloud cover accompanies each map. You can see from the map that on November 19 a small cold front was developing in the Northwest, but the central mountain district was experiencing a period of high pressure. Extensive clouds covered Washington and Oregon, but the cloud cover over the mountains was lighter and broken. The weather map for the following day shows that within the 24-hour period, the cold front had moved toward the south and east. A low-pressure zone formed along the front, and the photograph shows the development of a thick layer of clouds that brought snow to the region. This storm continued throughout the day

of November 21, as shown. By November 22, conditions had changed. Cold air at higher elevations began to sink, creating a high pressure in the northern Rockies, and sunny skies were predominant. At the same time, a low-pressure system was developing in the Southwest. The photograph shows the development of a cyclonic cloud system characteristic of many low-pressure storm systems.

Occluded fronts

A third type of system occurs when two fronts (and three air masses) collide. Figure 18.34 illustrates the case when a cold front overtakes a warm front. In this situation, a parcel of warm air is trapped between two parcels of cold air and is lifted completely off the ground. Such a system combines both the narrow zone of violent winds and precipitation of the cold front with the wider and more gentle precipitation zone of the warm front. The net result is a large zone of inclement weather.

Once again, it is important to understand that weather prediction is complex. An air parcel may rise and then, for any of a wide variety of reasons, precipitation may not occur. Storm fronts sometimes seem to build and then suddenly dissipate. Weather forecasters use sophisticated ground instruments, computer analysis, and extensive data from specially designed weather satellites, yet they still achieve only about 80 percent accuracy in their predictions.

18.11
THE EARTH'S CHANGING CLIMATE

After studying the orderly processes that control global wind systems, one might imagine that the average worldwide climate must be constant from year to year. But extreme changes occur over long spans of time. Thus, what is now the temperate zone in North America has experienced both tropical warmth and the cold of the Ice Ages in the distant past. One hundred fifty million years ago, giant dinosaurs wallowed in hot, humid swamps, while only 25,000 years ago woolly mammoths roamed the edges of giant glaciers that covered much of the continent. During the past

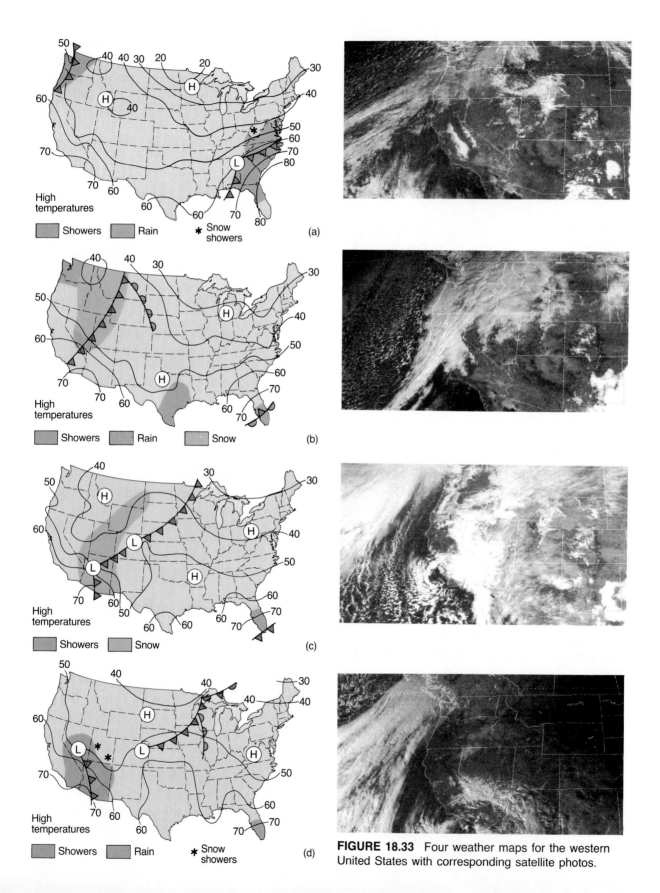

FIGURE 18.33 Four weather maps for the western United States with corresponding satellite photos.

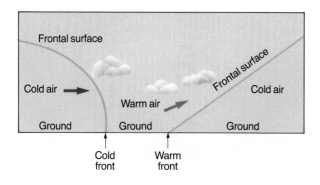

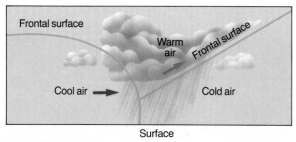

FIGURE 18.34 An occluded front occurs when two fronts collide. In this example, a cold front overtakes a warm front. The warm air mass is lifted right off the ground, and a wide zone of turbulent weather results.

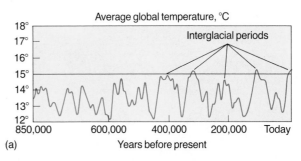

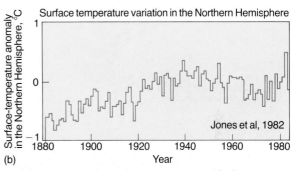

FIGURE 18.35 Past fluctuations in world climate. (Reprinted with permission from P. Jones, et al.: "Variations in Surface Air Temperatures," *Monthly Weather Review,* 110:59–70, 1982.)

100 years, climate has continued to fluctuate on a smaller scale. Figure 18.35 shows past temperature changes in the Northern Hemisphere. As you can see from Figure 18.35b, the period from 1930 to 1960 was relatively warm, and a slight cooling trend followed. Temperatures started to rise again during the decade between 1975 and 1985. Notice, however, that the temperature fluctuations have been small, less than 1°C. You may ask: What difference does it make if the average temperature rises or falls 1°C? Why can't one put on another sweater or turn the thermostat up a touch? The answer to these questions is that small changes are not particularly important as far as human comfort is concerned, but they can be vital to agriculture. For example, if the temperature drops an average of 1°C in the northern United States or southern Canada, this usually means that the last killing frost occurs a week later in the spring and a week earlier in the fall than it did previously. Thus, a seemingly insignificant change in average temperature becomes quite significant when it reduces the growing season by 2 weeks. Even frost-resistant plants grow

much more slowly if spring and fall temperatures are cold. In recent years, a 1°C cooling trend was blamed for a 25 percent decline in hay production in Iceland.

A warming trend could also have adverse effects. If the climate grows warmer than it is now, then the zone of dry descending air, the horse latitudes, would tend to move toward higher latitudes. Approximately 5000 to 8000 years ago, the Northern Hemisphere was much warmer than it is now. During that time, grass grew in the Sahara, but little rain fell in parts of the Great Plains in North America and in the steppes of eastern Russia, as shown in Figure 18.36. Sand dunes blew across the now fertile fields and ranches of eastern Colorado and western Nebraska. Today, the farms in North American and Eurasia produce a significant portion of the grain needed to feed a hungry world. No one can say for sure that a future warming trend would affect rainfall in the same manner, but many scientists believe that it would.

During the period between 1900 and the present, human development has accelerated

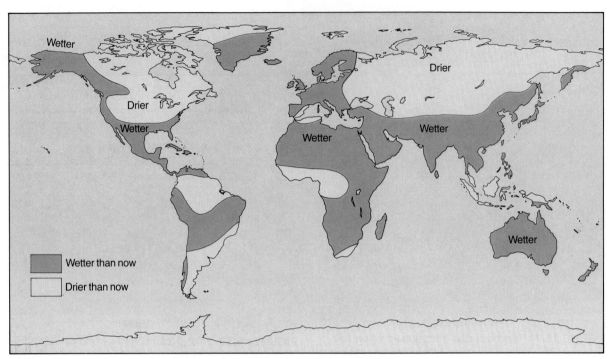

FIGURE 18.36   This map reconstructs climate conditions 5000 to 8000 years ago when the Earth was warmer than it is now.

rapidly. This era has witnessed an abnormally favorable global climate that has led to high agricultural productivity. Even with all the agricultural development that has occurred, millions of people are undernourished or even literally starving. The problem that causes great concern is this: Food production and population have been pushed to the practical limit during an unusual period of particularly favorable climate. World population has increased exponentially since 1900 and shows no sign of slowing down. If the climate should change appreciably, farms in many regions of the world could fail. In fact, many climatologists believe that *either* a cooling or a warming trend might disrupt agriculture significantly.

18.12 NATURAL FACTORS AND CHANGE OF CLIMATE

Climate has fluctuated throughout the history of our planet, and therefore climatic change is a natural phenomenon. Some of the factors that may

cause temperature and rainfall patterns to change are listed below.

Climate and atmospheric composition

If there were no atmosphere, the view from the Earth would be much like that which the astronauts saw from the Moon—a terrain where starkly bright surfaces contrast with deep shadows and a black sky from which the Sun glares and the stars shine but do not twinkle. The atmosphere protects us by serving as a light-scattering and heat-mediating blanket. As shown in Figure 18.8, about half of the incident radiation from the Sun passes through the atmosphere to the Earth; the rest is reflected or absorbed in the atmosphere. The energy absorbed by the Earth is eventually re-emitted to the atmosphere as infrared radiation. A large portion of this infrared energy is reabsorbed by the atmosphere and is, in effect, conserved, with the result that the surface of the Earth is warmer than it would otherwise be.

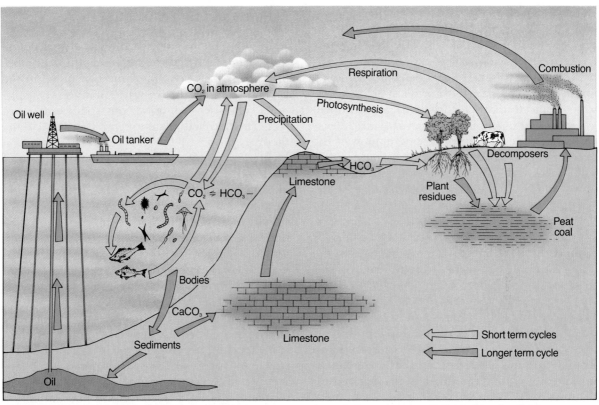

FIGURE 18.37 A simplified carbon cycle. Carbon passes through the processes indicated by the lighter-colored arrows much more rapidly than through those indicated by the darker-colored arrows.

Some molecules in the atmosphere absorb infrared radiation, and others do not. Oxygen and nitrogen, which together compose almost 99 percent of the total composition of dry air at ground level, do not absorb infrared. On the other hand, molecules of water, carbon dioxide, methane, and ozone do absorb infrared and thereby warm the atmosphere. Water plays the major role in absorbing infrared, because it is so abundant. Carbon dioxide is also important, because it cycles in both natural and industrial systems, as shown in Figure 18.37. The natural cycles are quite varied. Only a small fraction of the total carbon near the Earth's surface is present in the atmosphere at any one time. In addition to the available pool of atmospheric carbon, large quantities of this element are (1) dissolved in seawater, (2) combined with other elements to form certain types of rocks, and (3) combined with other elements to form plant and animal tissue.

The chemistry of carbon compounds in natural systems is extremely complex. For simplicity, three generalized reaction sequences are summarized below:

$$\text{atmospheric } CO_2 \longleftrightarrow CO_2 \text{ dissolved in seawater}$$
$$\text{atmospheric } CO_2 \longleftrightarrow \text{carbonate rocks}$$
$$\text{atmospheric } CO_2 \longleftrightarrow \text{plant and animal tissue}$$

Each of the equations is written with arrows going both ways, for all of these reactions can proceed in either direction. Atmospheric carbon dioxide can dissolve in seawater; conversely, dissolved carbon dioxide can escape into the atmosphere. Similarly, gaseous carbon dioxide can react to form carbonate rocks, and carbonate rocks can react to form carbon dioxide gas. During photosynthesis, carbon dioxide gas is converted into plant tissue, and during respiration, plant and animal tissue react to form carbon dioxide and other products. Various physical factors such as pressure, temperature, and atmospheric composition affect all of these reactions. In addition, biological factors such as the vitality of ecosystems are important. The net result is that it is extremely difficult to predict how atmospheric carbon dioxide levels are likely to change. Some scientists speculate that the Earth's systems are so well balanced that perturbations are compensated for. For example, if human activities release carbon dioxide into the atmosphere, most of the excess dissolves in the oceans or is converted to solid forms such as limestone. Other scientists disagree and argue that the entire balance of the Earth may be disrupted by a relatively small initial change.

One possible mechanism for such a change is outlined here: Suppose that for some reason the temperature of our planet were to rise by a few degrees. In the absence of other factors, this small increase in temperature would cause some of the carbon dioxide that is now dissolved in the oceans to be released to the atmosphere. But atmospheric carbon dioxide absorbs infrared and would warm the Earth further. Carbon dioxide is not the only compound that absorbs infrared radiation. Water is another important component of the climate cycle. If the Earth becomes warmer, there will be increased evaporation from the oceans, thereby increasing the quantity of water vapor in the atmosphere. Since water vapor also absorbs infrared, this factor would add still further to a global warming. Thus the original temperature rise, which was caused by some totally unrelated factor, would be amplified by the increase in the concentrations of various atmospheric gases. A spiral could be envisioned whereby increased warming leads to more IR-absorbing gases emitted into the atmosphere, which then leads to even a warmer climate. Scientists believe that a spiral of this sort has actually occurred on the planet Venus. Venus is similar to the Earth in many ways, but it is a little closer to the Sun. As a result, it would naturally receive more solar energy and be a little warmer than the Earth. This relatively small amount of additional heat has caused carbon dioxide and other compounds present on Venus to be released as gases. In turn, these gases absorbed infrared, causing the temperature to rise even more. Today, the average temperature on the surface of Venus is approximately 500°C, hot enough to melt lead.

Climate change and dust

About 65 million years ago, fully two-thirds of all known animal species and an uncounted number of plant species became extinct. Included in the list are the dinosaurs, as well as many less conspicuous creatures such as various species of clams, fish, single-celled plankton, and insects. Many theories proposed for this period of extinction focused on the demise of the dinosaurs alone and made no attempt to explain why numerous types of smaller creatures with vastly different food supplies, habitats, and reproductive mechanisms should accompany the larger reptiles into oblivion. In recent years, a plausible answer has been developed for the entire ecological catastrophe. This theory states that the extinctions were caused by a rapid and catastrophic climate change. In turn, the climate change was caused by the injection of large quantities of dust into the atmosphere. The dust blocked so much sunlight from reaching the Earth that plants died, rivers froze, and many entire species perished. There are two different opinions about the original source of the dust. According to one opinion, a giant meteorite or a rain of comets struck the Earth with such powerful impact that rock exploded, shooting dust and debris into the air. This dust cloud was carried by high-altitude winds and dispersed throughout the upper atmosphere. Other scientists claim that the best evidence indicates that the dust was produced during a period of intense volcanic activity. A single volcano can eject large quantities of dust into the upper atmosphere. If many eruptions occurred in a geologically short period of time, the accumulated dust would be sufficient to cause the catastrophic cooling that is believed to have occurred.

Evidence for one or the other of these scenarios is quite compelling. The key clue to the mystery centered on the discovery that rocks that were formed just about the time of this extinction contain 100 times higher concentrations of the element iridium than do rocks that were formed just prior to or just after the extinction. Iridium is extremely rare on the surface of the Earth, but it is much more abundant both in stony meteorites and in certain types of volcanic ash. Thus, the geochemical evidence indicates that some sort of

FOCUS ON . . . Volcanoes and Weather

In our own times, two explosive volcanic eruptions have occurred. One was Mount St. Helens, which erupted in the state of Washington in May of 1980. This eruption released mostly relatively large particles of dust and ash. These settled back to Earth quickly, and no significant effect on climate was recorded. A second eruption occurred from the Mexican volcano El Chichon in the spring of 1982. Although the eruption was not particularly large, significant quantities of sulfur dioxide were ejected straight up into the stratosphere. As explained in Chapter 17, with reference to acid rain, the sulfur dioxide reacts with water in the air to form sulfuric acid mists. These mists stay in the stratosphere for long periods of time and reflect some of the solar energy that reaches the Earth. Scientists have calculated that if no other atmospheric changes were occurring at the same time, the eruption of El Chichon would have led to a global cooling of 0.3°C to 0.5°C. The problem is this: Many other factors were occurring at the same time, and a half a degree is a small change. Did El Chichon cause global cooling or not? This question is hard to answer. The year 1981 was the warmest in several decades in the Northern Hemisphere. Temperatures had begun to decline in 1982 several months before the eruption. The temperature decline continued after the eruption. No one knows whether the continued decline in temperature was a result of the volcano or was caused by some other factor or group of factors.

geological change occurred simultaneously with the biological changes, and a cause-and-effect relationship is assumed.

The conclusion that interests us here is that high-altitude dust can cause severe climatic change. In modern times, there have been no catastrophic meteorite impacts, but volcanic eruptions also inject large quantities of dust into the atmosphere. In 1815, a volcano erupted from the crater of Mount Tambora in Indonesia. It was the largest eruption in modern times; approximately 100 km^3 of ash and rock were ejected into the atmosphere, and the top 1300 m of the mountain was completely blown away. The booming noise was so intense that on Java, nearly 500 km away, naval officers believed that pirates were shelling nearby coastal towns and sent warships to offer relief. Large quantities of dust from the volcano collected in the stratosphere and began to circle the Earth. One year later, in 1816, spring came late to the Northern Hemisphere. May was unusually cold, snow fell in northern New England in June, and frosts in July and August virtually wiped out crops of corn and vegetables. Thousands starved and the year was commonly called "eighteen hundred and froze to death." In several other instances, abnormally cold seasons have followed major volcanic eruptions. Many scientists now believe that gases and small particles of the high-level atmospheric dust that are ejected during volcanic eruptions reflect significant quantities of sunlight back into space, thereby reducing temperatures on Earth.

Change of climate and the Earth's orbit

Many climatic variations occur in regular, periodic intervals, although the time period for these cycles differs widely. For example, the temperature of the Earth is known to oscillate on 23,000-, 41,000-, and 100,000-year cycles. During the cooler portions of these cycles, huge glaciers have advanced across the continents. Many different theories have been proposed to explain the advance and retreat of Ice-Age glaciers. Some scientists believe that there is a relationship between variations in the Earth's orbital path around the Sun and the Ice Ages. Figure 18.38 shows a possible correlation between the shape of the Earth's orbit and global temperature. If this theory is used to extrapolate future trends, a glacial period could begin in the next few thousand years.

Change of climate and the Sun

Astronomers have noted that the energy output of some stars varies with time, and there is increasing evidence that our Sun's output may vary as well. If this is true, then some long-term climatic variations may be explained quite simply. When the Sun emits more energy, the Earth's climate becomes warmer; when the Sun cools, so does our planet. Unfortunately, not enough data are available to prove or to disprove this theory.

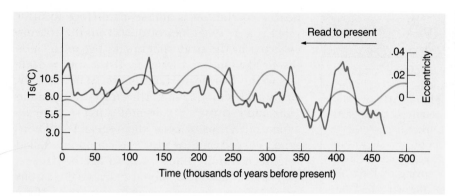

FIGURE 18.38 Relationship between observed temperature (red line) and temperature changes predicted by calculating changes in the Earth's orbit (blue line).

18.13
HUMAN ACTIVITIES AND GLOBAL CLIMATE CHANGE

The first human activities to affect the environment significantly were related to food gathering and agriculture. Thus, while industrial air pollution is only about 100 years old, the burning and cutting of forests and the development of agricultural systems have been going on for many centuries.

A natural temperate forest system affects climate in many ways. The shade provided by the trees maintains a cool environment at ground level. In addition, a temperate forest floor is a thick bed of partially decayed leaves, needles, and rotting wood. This spongy mass retains moisture, absorbing excess water during the wet seasons and releasing it slowly through evaporation and runoff during drier months. Think of the difference between a parking lot at one extreme and a woodland at the other. During even a mild rainstorm, puddles of water sit in depressions of the parking lot while tiny rivers and streams pour off high places. When the Sun comes out, this surface water disappears quickly, and usually the area is dry and hot within a few hours. On the other hand, fresh rainwater penetrates gently into a forest floor and remains absorbed in the soil and organic matter for long periods of time. Thus, a forest maintains a cool, moist environment. When a forest is cut and converted to a field of grain, many of these natural control mechanisms are disturbed. The shade, of course, is reduced. When soil is plowed and in some instances left bare for long periods of time, much of

the organic matter is lost; poorly tended fields can be more similar to a parking lot than to the original forest system. Thus, as compared with a forest, a field tends to be hotter in the summertime and more vulnerable to cycles of flood and drought.

However, when continental-size regions of forest are cut, opposing effects are observed. Bare soil and snow-covered fields reflect light better (have a higher albedo) than dark, shady forests. This effect would tend to associate cooling with loss of forest land. Thus, when a forest is cut, the loss of shade and moisture retention tends to cause a warming of the environment. At the same time, the change in albedo would tend to lead to a cooling effect. Do the two opposing factors cancel each other out? That question is hard to answer. On a local scale, deforestation has sometimes led to the loss of groundwater, with the subsequent formation of deserts or unproductive steppelands. This problem is particularly severe with respect to the extensive destruction of tropical rain forests that is occurring today. On a global scale, it is often difficult to sort out cause and effect. Over the past few thousand years, the temperature of the Earth has alternately risen and fallen, but not all these changes can be attributed solely to deforestation. Many other factors are operating at the same time, and it is impossible to predict what would have happened if the forests had been left alone.

The greenhouse effect

Recall from the last section that carbon dioxide in the atmosphere absorbs infrared radiation from the Earth and tends to warm the lower atmosphere. Carbon dioxide is cycled not only by many natural processes but by many human ones as well.

Carbon dioxide is released into the air whenever any organic matter is burned or is consumed and digested by an animal. Four representative transformations are listed below:

1. Carbon (as found in coal) $+ O_2 \xrightarrow{\text{burns}} CO_2$

2. Wood $+ O_2 \xrightarrow{\text{burns}} CO_2 + H_2O$

3. Sugar $+ O_2 \xrightarrow{\text{respiration}} CO_2 + H_2O$

4. Cellulose $+ O_2 \xrightarrow{\text{respiration}} CO_2 + H_2O$

Today people are introducing significant quantities of carbon dioxide into the atmosphere in two different ways. The first is by the destruction of forests. When forests are cut and converted to farmland, wood, branches, and leaves are often burned, and large quantities of carbon dioxide gas are released to the atmosphere. In addition, a forest floor consists of a thick, spongy layer of partly rotted leaves, needles, twigs, and other organic debris. In a healthy, undisturbed system, some of this material rots every year, while at the same time more falls from the trees, so that the total quantity of organic matter in the soil remains fairly constant. However, if the timber is cut, organic material in the soil rots faster than it is replenished and the balance is disrupted. Large quantities of organic matter decompose, and even more carbon dioxide gas is released.

In recent years, the carbon dioxide gas released from the burning of fossil fuels has also become significant on a global scale. The most accurate measurements indicate that the carbon dioxide concentration in the atmosphere has increased from about 290 ppm (parts per million) in 1870 to 340 ppm in 1985 and that the rate of increase is accelerating. Since carbon dioxide absorbs infrared, increased quantities of this gas could lead to global warming. Some scientists estimate that the carbon dioxide concentration could increase enough by the year 2040 to warm the Earth by as much as 2° to 3°C. Even if such a

temperature rise were to occur, it is impossible to predict the effects on world climate. Growing seasons in the Northern Hemisphere may be prolonged by as much as 2 or 3 weeks, but, as mentioned previously, a warming trend may lead to drought conditions in many temperate zones. A significant global warming might also melt the polar icecaps sufficiently to raise the ocean level and flood coastal cities.

Dust

As mentioned previously, gases and small particles of dust injected into the stratosphere by volcanic activity tend to reflect solar radiation and cause a cooling effect. What, then, of the air pollutants injected into the atmosphere by human activity? These materials are derived from a variety of sources, including dust from agriculture and smoke from the incomplete combustion of coal, oil, wood, and garbage. Of course, much of the larger particulate matter falls to the ground or is washed down by rain. However, there is some persistent introduction of small particles into the upper atmosphere.

To date, this reflection of radiation by industrial air pollution is very small compared with the reflection by volcanic material. On the other hand, the effect of pollutant particles in the lower atmosphere, where most of them are concentrated, is more complicated. Particles can absorb as well as reflect radiation, and it is not at all easy to determine whether the net effect of industrial and agricultural dusts near the ground is one of heating or cooling. There is therefore no convincing evidence that pollutant dusts in the lower atmosphere have any important effect on the Earth's temperature. However, the dust that settles to the ground is another matter. Most of us have seen how snow in the city can become dirty after a few days. If you live in a rural area where snowfall is common, spread a thin layer of ashes on a one-square-meter section of snow on a warmish, sunny, winter day. By evening, you will notice that the snow under the ashes has melted faster than the snow nearby. The dark ash absorbs sunlight (lowers the snow's albedo) and causes the snow to melt. If increasingly larger quantities of dust from industry and agriculture were to settle on snow packs over wide areas, a change in global climate might possibly occur.

Chemical pollutants and ozone

In Chapter 17, we learned that various industrial pollutants serve as catalysts in the destruction of atmospheric ozone. In turn, if the ozone layer is destroyed, abnormal amounts of solar ultraviolet radiation penetrate the lower atmosphere. It is uncertain how this affects climate, although it is known that ultraviolet light can promote skin cancer and reduce the growth of certain crops.

18.14 NUCLEAR WARFARE AND CLIMATE CHANGE

Earlier in this chapter, it was pointed out that scientists believe that a cloud of dust ejected into the upper atmosphere by volcanic eruptions or an exploding meteorite led to the extinction of about two-thirds of all the species of animals alive at that time. What, then, about the dust that would be raised by multiple explosions in the event of a nuclear war? In October 1983, an international conference of 100 atmospheric physicists and biologists from the United States, Western Europe, and the Soviet Union studied the problem. This group focused on the atmospheric and ecological consequences of such a war and not on the direct effects of the blasts and the ionizing radiation generated. The conclusions of the study are summarized below:

1. Multiple ground-level nuclear blasts would lift phenomenal amounts of very finely pulverized soil particles into the atmosphere. This soil would be accompanied by soot from fires initiated by the blasts. The soil and soot would be concentrated enough to block out 95 percent of the normal solar radiation. Temperatures in the Northern Hemisphere would plummet to −25°C (−13°F) even if the war occurred during summertime. Crops and natural ecosystems would die, and billions of humans as well as an uncountable number of animals would starve to death.

2. The intense heat of the blasts would vaporize a large variety of different materials, which would then be carried aloft by dust clouds. Many of these would include industrial chemicals of all sorts. Some industrial chemicals such as pesticides, certain solvents, and a great many other compounds are already poisonous. Others are benign in their present composition but would be converted to poisons if heated. For example, polyvinyl chloride (PVC) is a clear plastic used for a variety of purposes, including the construction of shatterproof bottles for shampoo, cleansers, and other consumer products. PVC is an inert plastic, suitable as a general packaging material; but if it is heated or burned, toxic fumes containing hydrochloric acid are released. Today, very large quantities of different chemical compounds and products are integrated throughout our society. In the event of a nuclear war, many of these would be blasted into the atmosphere as vapors or aerosol particles and then dispersed over the entire globe in the form of deadly acid and chemical rains.

3. The heat of the blasts would convert large quantities of atmospheric nitrogen to nitrogen oxides. In turn, these compounds would destroy much of the ozone layer, exposing survivors and ecosystems to high levels of ultraviolet radiation. (See discussion of the ozone layer in Chapter 17.) In conclusion, an article published in *Science* in 1983 and co-authored by 20 prominent scientists states:

*The extinction of a large fraction of the Earth's animals, plants, and microorganisms seems possible. The population size of Homo sapiens conceivably could be reduced to prehistoric levels or below, and extinction of the human species itself cannot be excluded.**

---

*See R. P. Turco, O. B. Toon, T. P. Ackerman, J. B. Pollack, and C. Sagan: "Nuclear Winter: Global Consequences of Multiple Nuclear Explosions." *Science*, Vol. 222, Dec. 23, 1983, p. 1283ff. Also see P. Ehrlich, et al: "Long Term Biological Consequences of Nuclear War." *Science*, Vol. 222, Dec. 23, 1983, p. 1293ff.

SUMMARY

One theory states that the composition of the Earth's atmosphere is maintained by living organisms. **Atmospheric pressure** is the weight of the atmosphere per unit area. Pressure decreases smoothly with altitude. The layers of the atmosphere, starting with the one closest to the Earth, are the **troposphere,** the **stratosphere,** the **mesosphere,** and the **thermosphere.**

The energy balance of the Earth is achieved by an opposition of energy inflow and outflow to and from ground level and all layers of the atmosphere. The general temperature gradient from the Equator to the poles and the change of seasons are caused by the spherical nature of the Earth and the tilt of the Earth on its axis.

Wind is a convection current powered by uneven heating of the Earth's surface. There are six major wind belts on the Earth. The movement of wind is affected primarily by convection and the **Coriolis effect** that causes objects to be deflected toward the right, as seen along the direction of motion, when moving in the Northern Hemisphere. The basic wind systems are the **polar easterlies,** the **trade winds,** and the **westerlies.** The **doldrums** is a region around the Equator where heated air is rising. This air falls at the **horse latitudes.**

Between the two, the **trade winds** blow in a steady and predictable pattern. The three basic types of clouds are **cirrus, stratus,** and **cumulus. Condensation** occurs when moist air is cooled below its **dew point.** Condensation can lead to the formation of clouds, and when clouds form, there is a chance for precipitation to fall in the form of rain, snow, sleet, or hail.

Many wind systems arise because the temperature of the oceans changes slowly in response to a change in solar radiation, whereas the temperature of the land surfaces changes much more quickly.

Stormy weather generally occurs when warm and cold air masses collide. When two air masses collide, the warm air often rises, leading to a low pressure, cloud formation, and often precipitation.

Climate change is affected by change in atmospheric composition, dust, changes in the Earth's orbit, and changes in radiation output of the Sun. Human activities such as cutting forests, changing the carbon dioxide concentration, or introducing dust and pollutants, can alter climate. Nuclear war could potentially destroy life on Earth by altering climate catastrophically.

EQUATIONS TO KNOW

$$\text{Relative Humidity (\%)} = \frac{\text{actual quantity of water per unit of air}}{\text{saturation quantity at the same temperature}}\ 100\%$$

KEY WORDS

Barometric pressure	Equinox	Humidity	Cumulus
Barometer	Coriolis effect	Absolute humidity	Nimbus
Troposphere	Polar easterlies	Relative humidity	Climate
Stratosphere	Trade winds	Dew	Weather
Mesosphere	Horse latitudes	Dew point	Sea breeze
Thermosphere	Doldrums	Adiabatic cooling	Monsoon
Albedo	Westerlies	Cirrus	Air mass
Solstice	Jet Stream	Stratus	Front

QUESTIONS

PRIMITIVE ATMOSPHERE

1. How did the primitive atmosphere differ from our atmosphere today?
2. Humans could not survive in the Earth's primitive atmosphere, yet life as we know it could not have evolved in the present one. Explain and discuss.
3. Explain why the compositions of the rocks on Earth, Venus, and Mars are all much more similar than the compositions of the atmospheres on the three different planets.

ATMOSPHERE

4. Explain how plants help to maintain an atmosphere that can support animal life.
5. List the four primary layers of the atmosphere. Discuss the physical properties of each.
6. Imagine that enough matter vanished from the Earth's core that the mass of the Earth were reduced to half its present value. In what ways do you think the atmosphere would change? Would the normal pressure at sea level be affected? Would the thickness of the atmosphere change? Would more molecules be lost to outer space? Explain.
7. An astronaut out on a spacewalk must wear protective clothing as a shield against the Sun's rays, but the same person is likely to relax in a bathing suit in the sunlight down on Earth. Explain.
8. Climbers on high mountains must wear dark glasses to protect their eyes; also they get sunburns even when the temperature is below freezing. Explain.
9. The temperatures are quite high in the thermosphere, but little thermal energy is stored there. Explain.

ENERGY BALANCE OF THE EARTH

10. As the winter ends, the snow generally starts to melt around trees, twigs, and rocks. The line of melting radiates outward from these objects. The snow in open areas melts last. Explain.
11. Refer to Figure 18.8a. What percent of the incident solar energy is received by the Earth? (b) Is the Earth growing warmer or colder, or is the global temperature fairly constant? Explain.

CLIMATE AND WEATHER

12. If we lived on the surface of a flat Earth, would different regions experience different climates or similar ones? In answering, assume that the flat Earth is tilted 23.5° with respect to the plane of its orbit.
13. Explain how the tilt of the Earth affects climates in the temperate and polar regions.
14. If the North Pole receives the same number of hours of sunlight per year as do the equatorial regions, why is it so much colder at the North Pole?

WIND SYSTEMS AND CONDENSATION

15. Explain why frost forms on the inside of a refrigerator, assuming it is an old-fashioned one and not a modern frost-free unit. Would more frost tend to form in (a) summer or winter? (b) In a dry desert region or a humid region? Explain.
16. Which of the following conditions will produce frost? Which will produce dew? Explain. (a) A constant temperature throughout the day. (b) A warm summer day followed by a cool night. (c) A cool fall afternoon followed by a freezing temperature at night.
17. Explain why a falling barometer is a good indication that cloudy weather may soon follow.
18. In what way are the winds analogous to a heat engine? What is the energy source that powers the wind?
19. If the wind is blowing southward in the Northern Hemisphere, will the Earth's spin cause it to veer east or west? If the wind is moving south in the Southern Hemisphere, which way will it veer? Justify your answer.
20. What is a trade wind? Why are trade winds so predictable?
21. Why is the doldrum region relatively calm and rainy? Why are the horse latitudes calm and dry?
22. Would the exact location of the doldrum low-pressure area be likely to change from month to month? From year to year? Explain.
23. Sailors traveling in the Northern Hemisphere expect to incur predictable winds from the northeast between about 5 and 30° north latitudes. Should airplane pilots expect northeast trade winds while flying at high altitudes in the same region? Explain.
24. Why doesn't the air that is heated at the Equator continue to rise indefinitely?
25. If you were firing a long-range rocket and aimed it due north at a target due north of your launching pad, would you score a hit or a miss? What would

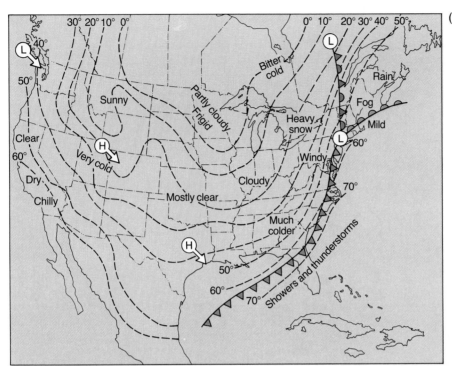

(Question 34)

happen if you fired due west at a target located due west? Explain.

WATER AND CLIMATE

26. Discuss the effect of the oceans on world climate. Would it be correct to say that coastal regions are always warmer than inland areas? Explain.

27. Would a large inland lake be likely to affect the climate of the land surrounding it? Deep lakes seldom freeze completely in winter, whereas shallow ones do. Would a deep lake have a greater or a lesser effect on weather than a shallow one? Explain.

28. Would sea breezes be more likely to be strong on an overcast day or on a bright sunny one? Explain.

29. What is a monsoon? How are they formed? At what time of the year do monsoons generally bring rain?

30. Explain how arctic air moving south can cause the monsoon rainfall patterns to change.

WEATHER AND CLIMATE

31. Describe the difference between weather and climate. Which is more predictable?

32. Describe four different weather conditions that

may cause air to rise, resulting in precipitation.

33. What is an air mass? Describe what would be likely to happen if a polar air mass collided with a humid subtropical air mass.

34. Study the weather map shown in the Figure and predict the weather in Salt Lake City, Chicago, and New York City two days after this map was drawn. Defend your prediction.

CLIMATE CHANGE

35. Discuss the problems inherent in trying to determine whether world climate is changing.

36. Explain why a 2°C drop in global temperatures would be alarming.

37. Name six factors that may cause climate to change. How many of these factors are at least partially controlled by people? Discuss.

HUMAN ACTIVITIES AND CLIMATE

38. Discuss four ways in which human activities may be changing world climate. In each instance, explain what types of activities are causing the potential disruption and by what mechanism the climate may be changed.

A lava flow from a volcano. (Courtesy U.S. Geological Survey)

CHAPTER

19

Our Geological Environment— Earth

19.1 INTRODUCTION

Imagine yourself on a rocky beach, walking toward the surf. You can see, hear, and feel the wind and the water. But you do not see the cliffs move, and the Earth doesn't shake under your feet. The solid Earth seems to be a firm base beneath the blowing winds and the breaking waves. However, this apparent rigidity is deceptive— Earth's crust is actually dynamic, not static. Continents move, mountains rise and fall, and rocks flow or are pushed from place to place. These movements escape most casual observations because they are generally slow, although every year volcanic eruptions, earthquakes, and other types of rapid movement do occur somewhere on our planet.

There are two types of movement of solid material that affect our environment. Mountain-building, continental migration, erosion, and other movements of large masses of materials are powered by natural energy sources of far greater magnitude than any that humans can harness. These phenomena will be discussed in this chapter. The second type of movement, to be discussed in the next chapter, is initiated by human beings and involves comparatively tiny amounts of energy and relatively insignificant masses of material. We refer here mainly to mining and farming. Yet these activities, which are insignificant on a scale of global energy, are vitally important on a human scale. Farmers generally dig up less than 1 m of a planet whose radius is 6,400,000 m, and miners probe only a few kilometers downward at most, yet the impoverishment of soil or the depletion of mineral reserves has major technological, political, environmental, and economic consequences. Therefore, we will study transformations that occur on widely different energy scales, from the movement of continents to the displacement of relatively small quantities of minerals and topsoils.

19.2
THE STRUCTURE OF EARTH

In our study of the atom, we learned how scientists are able to understand the structures of objects that are too small to see. Now we ask how it is possible to study the interior of the Earth, thousands of kilometers below the deepest well. Again an analogy will be helpful. Have you ever gone to the store to buy a watermelon? If you have, you know that there is always a concern that you pick a really juicy, perfectly ripe melon. The problem is that you have to make the choice without looking inside. One trick is to tap the melon gently with your knuckle. If you can hear a "sharp," "clean" reverberation traveling through the core of the melon, it is probably ripe; a dull thud indicates that it may be overripe and mushy. Of course, the words "sharp" and "clean," as used here, are not scientifically precise, but if you tap many different melons, you can hear differences. The point is that sound waves are affected differently by different types of liquid or solid media.

The same general technique is used to study the structure of the interior of the Earth. If you could somehow give the Earth a sharp tap, sound waves would travel through the rock and other material that lies beneath the crust. Recall from Chapter 7 that if a vibration is established in a medium, the speed of the resulting wave will be characteristic of the medium. Thus, if a bell is rung in air, a sound wave will travel outward with a speed of about 340 m/s. However, sound travels much faster through solids and liquids: It travels at 1500 m/s through water, 3810 m/s through marble, and 5200 m/s through iron. Thus, a measurement of the speed of a wave provides a clue to the composition of any medium through which it travels.

Additional information is available as well. For example, sound travels at 1500 m/s through water. Now what would happen if one person rang a bell under water while an observer listened for the sound on shore? The sound wave would travel through the water at 1500 m/s and then slow down as it entered the air. At the water-air boundary, the wave would also refract, just as light waves refract when passing from one medium to another. The same technique can be

FOCUS ON . . . "Revolutions" of the Earth

"During the past few thousand years, there have been three great revolutions in our understanding of the Earth. One was the realization that the planet is round, demonstrated by the ancient Greeks, particularly Eratosthenes, but not widely accepted until after voyages of the great navigators. Another, likewise anticipated by the Greeks 22 centuries ago but not generally recognized until much later, was that the Earth circles the Sun, instead of being central to the Universe. The third such revolution is reaching its climax as evidence accumulates that the continents of today are not venerable land masses but amalgams of other lands repeatedly broken up, juggled, rotated, scattered far and wide, then crunched together into new configurations like ice floes swept along the shore of a swift-flowing stream." (From Walter Sullivan: "Geologists add more pieces to a global jigsaw puzzle," *Smithsonian,* January, 1985.)

used to study the Earth. If refraction patterns of waves are studied, information can be obtained about the boundaries between different layers that exist within our planet. Other studies of interference patterns and changes in wave forms are also useful in analyzing the chemical and physical properties of an unknown substance.

The next question is: How do you tap the Earth? One way would be to set off explosions near the surface. This technique is a valuable one and is used to study rocks that lie close to the surface. However, conventional explosions are not nearly energetic enough to transmit waves that can travel through the center of the Earth and back to the surface and still be detected and analyzed. However, earthquakes release tremendous amounts of stored energy. When an earthquake occurs near the surface of the Earth, built-up stress is released as huge segments of rock suddenly slip past each other. This movement initiates waves in the rock, just as a clanging bell initiates waves in water or air. By studying the speed of these waves through Earth, as well as their refraction and reflection at boundaries where the density changes, scientists can deduce much about the structure of the Earth's interior, as shown in Figure 19.1.

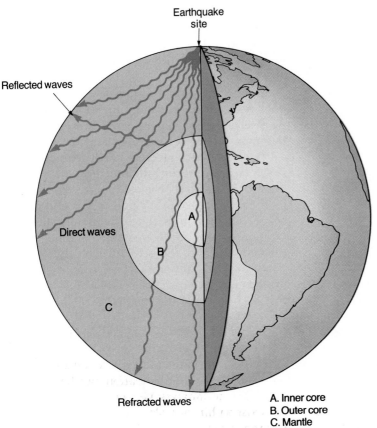

Earthquake site

Reflected waves

Direct waves

A

B

C

Refracted waves

A. Inner core
B. Outer core
C. Mantle

FIGURE 19.1 Seismic waves can be used to study the Earth's interior. Note that the waves refract (bend) as they pass through different layers.

The science of measuring and recording the shock waves of earthquakes is called **seismology.** Earthquakes are commonly referred to as seismic events or seismicity. Studies of earthquake waves have shown that Earth is composed of several distinct layers of solid and liquid matter, as represented in Figure 19.2. The **core** is predominantly iron with some nickel. The inner core is solid, and is surrounded by an outer core of molten iron and nickel. A large solid layer called the **mantle** surrounds the core. The chemical composition of the entire mantle is fairly homogeneous. However, temperature and pressure generally increase with depth, and these factors affect the chemical properties of different portions of the mantle.

Seismic studies show that the mantle is divided into discrete layers. The **upper mantle** extends from the base of the **crust,** the rocky surface on which we live, downward to about 670 km beneath the surface. The uppermost portion of the mantle and the crust above it are close enough to the surface that temperature and pressure are relatively low. Therefore, the rock in this region is rigid and brittle. Although the composition of the crust and the mantle is different, the zone including crust and uppermost mantle is distinguished by its physical properties. Thus, the crust and the uppermost mantle together are called the **lithosphere** (Greek for "rock layer"). The lithosphere is about 100 km thick in most places. This lithosphere, broken into plates, rides on the semifluid layers beneath it, and, as we shall see in later sections, the semifluids flow slowly, causing the lithosphere to shift and to be distorted.

Starting at a depth of about 100 km, there is a sharp and distinct boundary between the lithosphere and the layer beneath it, called the **asthenosphere.** The asthenosphere extends from the base of the lithosphere to a depth of about 350 km. The asthenosphere is unique in that

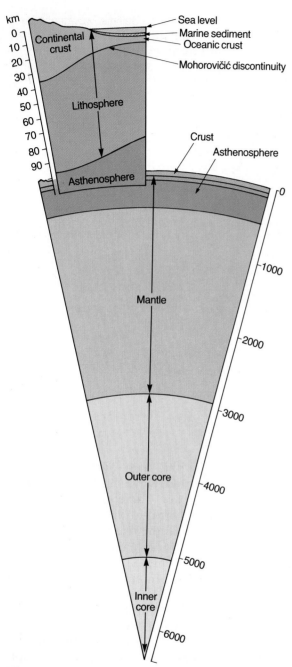

FIGURE 19.2 Cutaway view of the interior structure of the Earth.

overwhelms the effect of increasing temperature, and the mantle becomes wholly solid and less plastic. At a depth of about 670 km, pressure is great enough to cause some minerals to collapse and to form new minerals that are more dense. Thus, the chemical composition of the mantle does not change, but the mineral structures do. This change in mineral structure marks the boundary between the upper and lower mantle.

How was this layered Earth formed in the first place? According to the most widely held theory, the Earth is believed to have been formed by the coalescence of a cloud of gas and dust in space, left over after the formation of the Sun. Originally, this mass was fairly homogeneous throughout, and the structure of the modern Earth evolved only after this primordial planet was heated. Two basic processes were responsible for the heating of the Earth:

1. *Gravitational Coalescence.* As particles of dust and gas were attracted to each other by their mutual gravitation, they began to accelerate toward each other. The kinetic energy attained by this acceleration was converted to thermal energy as particles collided with each other. Thus, the Earth warmed as it was being formed.

   After the Earth coalesced, additional matter was attracted by gravity. Intense meteorite bombardment heated the surface.

2. *Radiogenic Energy.* Some of the elements of the Earth are naturally radioactive. Many of these have long half-lives, so that the radioactivity has persisted throughout the geological history of the Earth and is still evident today. When a radioactive atom breaks apart, energy is released as heat. If this heat is released near the surface of the globe, it is quickly radiated off into space. On the other hand, because of the low thermal conductivity of rock, heat energy produced deep within the interior of our planet cannot escape so easily and remains trapped. Even though the quantity of heat released per day is small, geological time spans are large. Hundreds of millions of years after the Earth first became solid, this radiogenic heat had gradually accumulated until the interior of the planet became so hot that rocks began to melt.

most of the rock is solid, although a few percent is melted. Because of the higher temperature, the solid rock is plastic and flows more readily than the rock either above it or below it.

Below the base of the asthenosphere, at a depth of about 350 km, the increasing pressure

Geophysicists believe that as a result of these two processes much of the planet's interior melted sometime after its original formation. Now, if you have a mass of molten rock, the denser material settles to the center and less dense materials float to the top. This is exactly what happened in the Earth. The dense elements, such as iron and nickel, gravitated toward the center. These were surrounded by a mantle of lighter silicon- and oxygen-rich (silicate) minerals and a surface crust that was generally composed of comparatively low-density silicates. This gravitational overturning produced more heat.

As the Solar System aged, many of the small free particles were swept up by the planets. As a result, bombardment from space was reduced to a tiny fraction of what it once was. Within the molten Earth, most of the radioactive elements reacted to form compounds with a comparatively low density that rose toward the surface with the lighter rock. Rock is such a good insulator, however, that the Earth would have continued to heat up, but the heating caused a softening and decrease in density. This resulted in convectional overturning of mantle and crust, releasing heat to the atmosphere and thereby cooling the Earth. The Earth's surface layer (lithosphere) cooled, becoming solid and brittle, but convection continues in the underlying plastic asthenosphere, releasing heat and driving the process of plate tectonics described below.

19.3 THE ROCK CYCLE

Pick up a rock that you find in your neighborhood and look at it carefully. Chances are that you will see particles of different kinds of materials. You may see a speck of pink matter, another of black, and a third that is white. A **rock** is a natural solid made up of one or more minerals or other natural solids. A **mineral** is a naturally occurring inorganic solid that has a definite chemical composition and crystal structure. Quartz is a

FIGURE 19.3 Photo of course-grained granite. (Courtesy Geoffrey Sutton. From Thompson/Turk *Modern Physical Geology,* Saunders College Publishing, 1991.)

mineral; a piece of quartz can be a single crystal of silicon dioxide. On the other hand, granite is a rock, made up of several minerals, and there is no single chemical formula that can be used to describe it.

In the asthenosphere there are pockets of molten rock, called **magma,** which have gases such as water vapor dissolved in them. When magma flows quickly to the surface of the Earth, a volcano is formed. The outpouring magma is called **lava.** If magma protrudes up into the crust slowly but does not travel all the way to the surface, it will cool and crystallize slowly deep within the crust. Slow cooling produces coarse-grained rock such as **granite,** as shown in Figure 19.3. Rock formed directly from cooling magma is called **igneous rock.** There are many types of igneous rocks with various chemical compositions. Silicon, oxygen, aluminum, calcium, magnesium, iron, potassium, sodium, or any of several other elements are present in various combinations. In addition to chemical differences, there are also variations in texture. Texture is generally determined by the rate of cooling and solidification, which in turn depends on the rock's history. Thus granite, which is formed when magma cools and crystallizes slowly, is coarse-grained. **Obsidian,**

FIGURE 19.4 Obsidian is an igneous rock formed from lava that cools so quickly that crystals do not have a chance to form, so the rock becomes a type of glass. This material can be chipped relatively easily and formed into sharp objects such as knives and arrowheads. Obsidian was considered very valuable to the Indians before Europeans brought metal tools and weapons to the North American continent.

shown in Figure 19.4, is a smooth volcanic glass that forms when lava cools too rapidly to crystallize. **Pumice,** a volcanic rock formed from lava that is frothing and bubbling, is light and porous because voids are left as the hot gases escape from the cooling liquid.

When igneous rocks reach the surface, they are exposed to surface forces. Wind, rain, freezing water, the movement of streams and glaciers, wave action, and biological processes serve to chip off tiny pieces or to dissolve the rocks. These small particles are carried downslope by streams and rivers, where they collect in valleys or where the river enters a lake or ocean. As they are deposited, the particles that collect on top of each other in layers are called **sediments.** Sedimentary

deposits are easy to find all around us. If there is a river or a creek near your home, notice that at a bend in the stream the water travels fastest on the outside of the turn and slowest on the inside, as shown in Figure 19.5. Go to where the water is slow, and dig up a section of a river bottom. Most probably it contains numerous fine particles of different colors. Some are shiny, others dull, some reddish, some black, and so on. The sediment in your sample probably contains pieces of rock from different locations, mixed with organic debris. Thick layers of sediment are formed at the edge of oceans or along the bottoms of the slopes of high mountains or plateaus.

If sediments are buried, compressed for long periods of time, and heated by the hot mantle beneath them, the pieces cement together and coalesce to a solid known as **sedimentary rock,** such as that shown in Figure 19.6. We learned earlier that the term "igneous" refers to a process of formation, not a chemical or physical composition. Similarly, the term "sedimentary" refers to a generalized process, and sedimentary rocks can be quite different from one another. **Limestone,** which is a sedimentary rock, is nearly pure calcium carbonate. **Sandstone** is a mixture of many small grains. Generally, most of the grains are quartz, but they may also include feldspar or in rare instances gold or any of a number of other types of rocks or minerals. **Shale** is nearly pure clay, microscopic grains produced during the chemical weathering of other minerals.

Sediments can build up to impressive thicknesses. In the Grand Canyon, these layers, or strata, were exposed 1.5 km deep as the Colorado River gradually cut its way into a raised sedimentary plateau, and the strata are clearly visible today. Other less visible sedimentary formations are even thicker yet. For example, a sedimentary layer that lies beneath parts of western Montana is 20 km thick!

Rock and sediments may also undergo geologic change—**metamorphism**—if they are heated and/or compressed for long periods of time. For example, clay, a naturally occurring type of soil, is soft and pliable; but if it is heated, it can be converted to a hard, strong solid. Thus, clay dishes and sculptures are molded when the material is soft and are then hardened in a type of

FIGURE 19.5 Sedimentary deposits are beginning to form on the inside curve of this creek.

oven called a kiln. A similar process may occur naturally. Suppose, for example, that hot magma came in close contact with a natural clay deposit. The heat from the magma would harden the clay. Many other types of heat-hardening processes occur in the Earth. If limestone is subjected to heat and pressure, it is converted to **marble.** Granite can be metamorphosed as well. During metamorphism, the minerals in granite often react to form new minerals, and textures change. One type of rock that can be formed in this way is called **gneiss,** which is often characterized by parallel layers of different compositions, such as is shown in Figure 19.7. Note that in the examples given here, the minerals aren't melted. The heat and pressure alter the texture or chemical composition (or both) and form new minerals in the affected rock and generally increase its hardness. The resulting product is called **metamorphic rock.**

Mineral matter is slowly but continuously being changed from one form to another. Igneous rocks are broken apart to form sediments and then, depending on conditions, may be converted to sedimentary and then to metamorphic rock. Similarly, metamorphic rock may erode and be deposited as sediments. In addition, under certain conditions surface rocks are slowly forced downward by geological action where they are heated and melted. Once molten, the liquid

FIGURE 19.6 Sedimentary rock in Garfield County, Utah. Note how the rock has been formed in horizontal layers.
(Courtesy of U.S. Geological Survey; photographer, C. B. Hunt)

FIGURE 19.7 Metamorphic rock on the Grand Teton in Wyoming. Note the random streaks characteristic of metamorphic material.

magma mixes with other minerals, and undergoes physical and chemical changes. This material may return to the surface millions or hundreds of millions of years later as newly formed igneous rocks. Thus, crustal material is formed, altered, and removed in a slow but continuous cycle called the **rock cycle,** as shown in Figure 19.8.

19.4 GEOLOGIC TIME

When we think about the movement of rocks and the cycling of crustal material, we must contemplate very large spans of time. Earth is some 4.6 billion years old, and parts of the solid crust today are more than 3 billion years old. Time spans of billions of years are truly difficult to comprehend. To grasp the concept of a billion, think first of a single coin, say a penny. A penny is about 2 mm thick. If a billion pennies were stacked in a pile, the pile would be 2000 kilometers high.

How can scientists measure geologic time? How do they know when certain events in the dis-

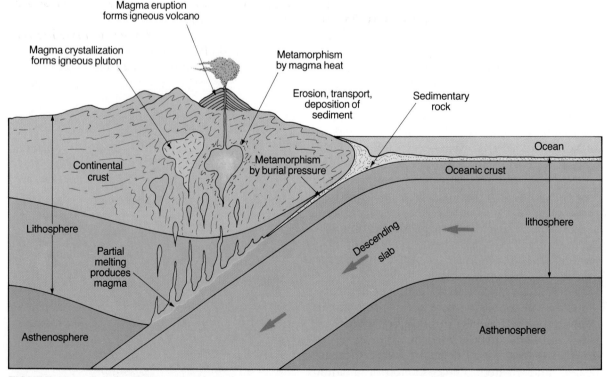

FIGURE 19.8 Schematic view of rock cycle.

FOCUS ON . . . Potassium-Argon Dating of Rocks

A commonly used method for dating rocks, in addition to that of uranium-238 mentioned in the text, uses radioactive potassium-40. Potassium-40 has two decay schemes: One is a beta-minus decay that leads to the daughter product calcium; the second, and the one important here, is a positron decay leading to the daughter argon-40. The potassium-to-argon decays occur in about 11 percent of the potassium-40 nuclei. Thus, in a sample of rocks that originally contained potassium-40, over a period of millions of years, argon accumulates within the rocks. One advantage to this method is that argon is inert—it does not combine chemically with other elements. Thus, when one finds argon entrapped in a rock along with a remnant of the original potassium, it is likely that the argon originated by potassium-40 decay.

The half-life of potassium-40 is 1251 million years. Thus, if one tests a sample of rock and finds that the ratio of entrapped argon to potassium-40 is 1 to 1, the age of the sample is 1251 million years. If the ratio is 1 to 3, the sample is 2502 million years.

As an exercise for you, what would the ratio of potassium to argon have to be for the age of the sample to be three half-lives, 3753 million years, or four half-lives, 5004 million years. (The answers are, respectively, 1 to 7 and 1 to 15.) For the sake of simplicity, we have not considered the fact that only 11% of the potassium-40 decays by positron emission. If we did so, we would have to adjust our ratios.

The potassium-argon method is important in determining the age of many geologic samples. However, it should be understood that there are many factors to be considered in determining whether or not this method has yielded a valid age. For example, it is obvious that this method will fail if any of the argon has leaked out. This is likely to occur if the temperature of the sample ever reaches 125°C. Thus, the age determined by this method may actually be a determination of the period since the last heating of the sample took place.

tant past occurred? The first studies of geologic time were conducted on a relative basis. For example, imagine a series of layers of sedimentary rocks. Since the sediments were laid down in sequence one on top of another, one could presume that the deepest layer is the oldest and that the surface layer is the youngest. If an entire sedimentary formation lies on top of a bed of metamorphic material, it logically follows that the metamorphic rock is older than any of the sediments above it. The concept that rocks are deposited in sequential order is known as the **principle of superposition.** Although this principle is useful in a great many situations, the age of rocks is not *always* directly related to their position. For example, molten magma could push upward into the center of an older sedimentary layer, and in this situation the youngest rocks would lie beneath older ones.

A relative time scale establishes a chronological sequence but does not specify absolute dating. To measure the age of a rock in terms of years, one must search for some sort of internal clock within the rock itself. One very useful natural clock measures time by the process of radioactive

decay. Consider a radioactive isotope such as uranium-238. As discussed in Chapter 13, a given radioactive isotope decomposes at a specific rate known as its half-life. Additionally, the decay process produces a known set of products. For example, uranium-238 decays to form other radioactive isotopes, which decay relatively rapidly in turn to produce lead-206. The half-life for uranium-238 is 4.5 billion years. Therefore, if a sample of rock contains equal quantities of uranium-238 and lead-206, that rock is 4.5 billion years old. The absolute age of rocks is determined by studying the radioactive decay patterns of uranium-238 and a variety of other isotopes with long half-lives, especially potassium-40.

The geological time scale is shown in Table 19–1. To put our own existence into perspective, imagine that the entire history of the Earth were recorded on a linear scale 1 meter long. The early history, largely devoid of anything but one-celled life, would occupy 86 percent, or 86 cm, of that scale. The great coal-producing swamps and marshes would appear 5.5 cm from the end and the extinction of the dinosaurs at the 1.4 cm mark; the Ice Ages would be found in the last

TABLE 19–1 Geologic Time

Time Units of the Geologic Time Scale				Distinctive Plants and Animals	
Eon	Era	Period	Epoch		
Phanerozoic Eon (*Phaneros* = "evident"; *Zoon* = "life")	Cenozoic Era	Quaternary	Recent or Holocene	"Age of Mammals"	Humans
			Pleistocene		
		Tertiary — Neogene	Pliocene —2— —5—		Mammals develop and become dominant
			Miocene —24—		
		Tertiary — Paleogene	Oligocene —37—		
			Eocene —58—		
			Paleocene —66—		Extinction of dinosaurs and many other species
	Mesozoic Era	Cretaceous —144—		"Age of Reptiles"	First flowering plants, greatest development of dinosaurs
		Jurassic —208—			First birds and mammals, abundant dinosaurs
		Triassic —245—			First dinosaurs
	Paleozoic Era	Permian —286—		"Age of Amphibians"	Extinction of trilobites and many other marine animals
		Carboniferous — Pennsylvanian —320—			Great coal forests; abundant insects, first reptiles
		Carboniferous — Mississippian —360—			Large primitive trees
		Devonian —408—		"Age of Fishes"	First amphibians
		Silurian —438—			First land plant fossils
		Ordovician —505—		"Age of Marine Invertebrates"	First fish
		Cambrian —570—			First organisms with shells, trilobites dominant
Proterozoic		Sometimes collectively called Precambrian —2500—			First multicelled organisms
Archean		—3800—			First one-celled organisms
Hadean		—4600 ±—			Approximate age of oldest rocks
					Origin of the Earth

0.03 cm; and recorded history would be squeezed into the final 0.003 cm, a zone thinner than the diameter of the period at the end of this sentence.

19.5
CONTINENTS AND CONTINENTAL MOVEMENT—THE IDEA IS BORN

The beginning of the Twentieth Century was an exciting period in the history of scientific discovery. Recall that J. J. Thomson had discovered the electron in 1897, and Rutherford elucidated the nuclear structure of the atom in 1909. During the same time period, other scientists were exploring the remote regions of the Earth. In 1908, a British expedition guided by Sir Ernst Shakleton was traveling in a desolate, frozen region of Antarctica when its members made a startling discovery. They found a large deposit of coal on the frozen continent where virtually no plants live and the mean annual temperature is far below freezing. A few years later, another expedition was organized to study the coal deposits in more detail. Geologists found fossilized bones of many different types of animals embedded in the layers of coal. This discovery, by itself, was not particularly surprising. After all, coal is known to be formed from rotting vegetation, so if forests once thrived in Antarctica, one would naturally assume that animals lived there as well. But the next discovery *was* surprising. Many of the fossil bones of various reptiles and amphibians were virtually identical to bones found in Africa and South America. It is conceivable that plant seeds could have floated from one continent to another, but it is preposterous to imagine that several species of slow, wallowing, freshwater swamp-dwellers had swum thousands of kilometers across the tempestuous southern ocean to land and thrive in Antarctica.

Geologists speculated that several hundred million years ago the entire Earth was warmer than it is now. During that time, the planet was so warm that even the polar regions were tropical. The theory continued that there must have been thin strips of land connecting Africa, Antarctica, and South America. Animals migrated across these land bridges, populating all three diverse regions with the same species. In time, climates changed and the land bridges were eroded and washed away by the sea.

However, this explanation was not entirely satisfactory. The gap between Africa and Antarctica is about 4000 km, and there are no geological remnants of the phenomenally long land bridges that the theory required.

In 1912, a German scientist, Alfred Wegener, proposed an alternative theory. He suggested that many millions of years ago, all the land surfaces on the Earth were fused into one single continent. No oceans separated one region from another, and animals could migrate freely without swimming across oceans or negotiating narrow land bridges. Then this giant land mass cracked and broke into various segments. In time, the segments gradually drifted apart to their present locations.

This theory was quite radical then. Most geologists naturally assumed that the Earth's crust is sold, permanent, and immobile, and no one had really thought that the continents were migrating over the surface of the planet. But Wegener was a careful man and did not base his theory solely on a single line of reasoning. Although the fossil evidence was compelling, several additional observations supported the same conclusion. Two of these are explained below.

1. Look at a map of the world as shown in Figure 19.9a. If the continents are cut out and repositioned like the pieces of a jigsaw puzzle, they fit together amazingly well, as shown in Figure 19.9b. From this evidence alone, Wegener deduced that perhaps there once existed one, or perhaps two, large supercontinents. The supercontinent(s) then broke apart, and the pieces slowly drifted away from each other to their present positions.

2. Many mineral deposits on the Earth are concentrated in geographical strips, or belts, on the various continents. For example, belt-like formations of low-grade tin deposits are shown in Figure 19.10. As seen on the map, many of these belts appear to end abruptly at the ocean. But if the continents are fitted together as in a jigsaw puzzle, many of the tin belts line up. Belts of coal seams, salt deposits, and gypsum concentrations also seem to disappear into the ocean but align with each other when conti-

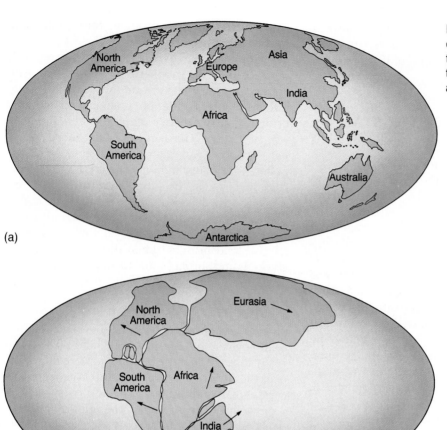

FIGURE 19.9 Continental drift. (a) The world as it is today. (b) The continents as they were 250 million years ago.

nents are pieced together. Wegener theorized that belts of this sort were formed many millions of years ago on a primordial supercontinent by individual processes, such as sedimentation, or by the movement of large masses of ore-bearing magma through massive fissures in the Earth's crust. When the supercontinent broke up, the belts separated into the patterns that are observed today.

This new idea, called the **theory of continental drift,** was developed in an elegant and convincing manner. It was supported by three separate and seemingly unrelated observations. In addition, there were no facts that seemed to contradict the theory. However, it was not well received at the time.

The Earth's crust is a continuous and com-

plete shell. People asked; How can one part of this shell move with respect to another? The most plausible answer at the time seemed to be that the continents must plow through the ocean floor as a rigid boat would slowly push through heavy mud. However, when physicists calculated the relative strengths of the rocks that made up the continents and the rocks that made up the ocean floor, they learned that the continents could not possibly plow through the ocean floor. The continents were not strong enough and would break apart first. It would be like trying to push a matchstick boat through heavy tar; the boat would break apart before it could move very far.

In short, Wegener's theory elegantly explained a variety of different and otherwise perplexing observations that could not be satisfactorily explained in any other way. On the other

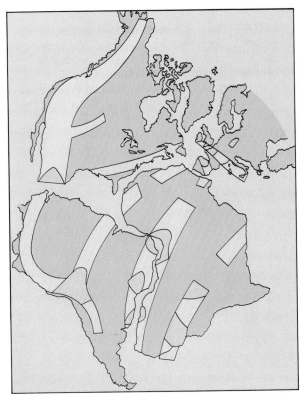

FIGURE 19.10 Low-grade tin belts on a reconstructed map as the continents were positioned 250 million years ago.

FOCUS ON . . . Lasers and Continental Drift

In recent years, the theory of plate tectonics has been substantiated by a fascinating experiment. A laser beam has been transmitted from one continent, reflected off the Moon, and then received on another continent. By carefully controlling conditions, the distance between the two experiment stations can be measured with extreme accuracy. When the experiment was repeated at various time intervals, the measurements showed that the continents are, in fact, drifting apart at a rate of a few centimeters per year.

Movement of a few centimeters per year may seem slow but that depends on the time scale used as a reference. At a rate of 5 cm per year, a continent will move 2000 km in 40 million years, which is not long by geological reckoning.

hand, critics asked for a mechanism; if the continents moved, they wanted to know how. When no one could offer a satisfactory mechanism, the theory was rejected. Today, with the perfect vision of hindsight, it is easy to accuse the critics of muddled thinking. If the theory accounts for many observed facts, but the proposed mechanism is unsatisfactory, one might look for a better mechanism, not throw away the theory. But in this case the theory itself was seemingly so outrageous that it was discarded.

19.6 CONTINENTS AND CONTINENTAL MOVEMENT—THE MODERN THEORY

Wegener died in 1930 in a blizzard while traveling across Greenland by dogsled on a scientific expedition, and for 40 years his ideas were largely forgotten. However, several decades after Wegener's death, additional lines of evidence were found to support his conclusion that the continents must move.

During the 1950s, several groups of geologists were studying magnetism in rocks. It is common knowledge that a compass needle aligns itself with the Earth's magnetic field and points toward the magnetic North Pole. Certain minerals containing iron and other elements are naturally magnetic. Now, imagine that some of these minerals crystallize from a magma that is cooling on or near the surface of the Earth. As the magma cools below around 500°C, particles become magnetized and are aligned parallel to the Earth's field. The same type of magnetism can also develop during the formation of sedimentary rocks. Imagine that a tiny magnetic crystal is carried downstream by a river along with other, nonmagnetic particles of silt and sediment. This sediment may then be deposited in a calm pool along the way and settle to the bottom. During the settling process, the still-floating magnetic particles align themselves north to south. Then they fall to the bottom and eventually are trapped by layers of sediment falling on top of them. Gradually, as these sediments coalesce to form rock, millions of tiny magnetic fingers pointing north are cemented into a fixed position.

Because of this process scientists can study ancient rocks and determine where the magnetic North Pole was in relation to these rocks at the time of formation. Now let us say that there is a

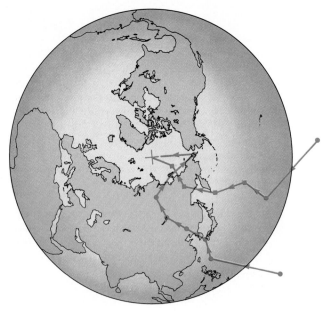

FIGURE 19.11   Apparent wandering for the magnetic North Pole for the past 500 million years, as deduced from European and North American rocks. + = present North Pole, ● = positions of North Pole deduced from study of North American rocks. ◆ = positions of North Pole deduced from European rocks.

thick deposit of sedimentary rock. The bottom layer is the oldest and may have been formed tens or even hundreds of millions of years before the uppermost sediments were laid down. If the magnetic North Pole and the continents had both been stationary, all the magnetic needles of all the rock layers would point in the same direction. But in fact the magnetic pointers shift direction from age to age, indicating that either the magnetic North Pole or the continents, or both, must have moved at some time in the past. How can we tell which event occurred?

Geologists have studied rock magnetism all over the world. Rocks found in Europe that are about 200 million years old show that the magnetic North Pole lay somewhat east of its present location. However, North American rocks of the same age point to a different location for magnetic north, as shown in Figure 19.11. This cannot be, of course—the magnetic North Pole can be in only one place at a time. However, if the continents were once joined together, the magnetic pointers from both regions would aim toward the same point.

When geologists searched for an explanation of these data, they returned to Wegener's old hypothesis. If the continents were once all joined together as Wegener had proposed, then the magnetic pointers from both North America and Europe would aim toward the same point. No other theory could explain the magnetic behavior of the rocks. When this analysis was added to the already existing information, the evidence became compelling. Wegener's ideas were incorporated into the modern theory of **plate tectonics.** The independent sections of the Earth's lithosphere that move about are called **tectonic plates.** The word "tectonic" is derived from the Greek word *tektonikos*, which is an adjective meaning "pertaining to construction." This, in turn, comes from the root *tekton*, which means a carpenter. In modern geology, tectonic refers to the building, development, or "construction" of Earth's crust.

The modern theory states that about 320 million years ago, the precursors to today's continents converged. They fused into a single continent, which then broke apart about 180 million years ago. The segments then gradually moved apart to form the Earth as we know it today, as shown in Figure 19.12. At present, there are seven major plates and several smaller fragments. Most probably, North and South America are now moving away from Europe, Asia, and Africa at about 2 cm per year, and other plates are believed to be moving at speeds varying between 1 and 15 cm per year.

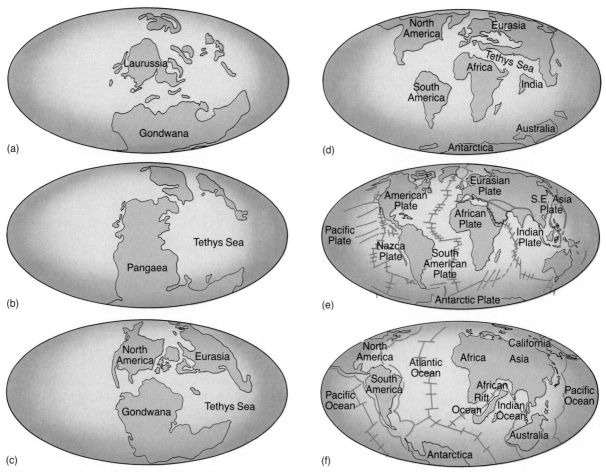

FIGURE 19.12 (a) About 320 million years ago, the precursors of today's continents converged. (b) About 250 million years ago, most of the land surface on Earth was fused into one single continent, called Pangaea, meaning "all lands." (c) About 180 million years ago, this supercontinent started to break up, and the individual pieces started to drift apart. This map shows the world as it was 135 million years ago. (d) The Earth 45 million years ago. India had separated from Eurasia and was traveling northward, and Australia had recently separated from Antarctica. (e) The Earth as it is today, showing the plates. (f) If plate movement continues as predicted, the map of the Earth will look like this 100 million years from now. Note that the Atlantic Ocean will be larger than it is today, and the Pacific will be smaller. Africa is believed to be splitting apart today, and in the future, it will be separated by a new seaway. At this same time, the bulk of the African continent is plowing into Eurasia, and new mountain ranges will be formed.

Although plate tectonic theory was becoming accepted, no satisfactory mechanism for plate movement had been proposed, so the old question of how the continents move remained. Shortly after World War II, oceanographers found that a distinct and sharp mountain range extends the whole length of the floor of the Atlantic Ocean from north to south. This mountain range is now called the **Mid-Atlantic Ridge.** Studies of this ridge showed that it is different from any mountain range found on dry land. The differences fall into five major categories.

1. There is a deep and very narrow split, called a **rift,** running right down the center of the range. This split is narrower, sharper, and longer than any extensive series of valleys or gorges found in any mountain range on land.

2. The heat flow (heat escaping from the Earth's interior) on either side of the rift was measured and found to be greater than that of other areas on the sea floor.

3. The Mid-Atlantic Ridge is composed almost entirely of igneous rock. In contrast, most continental areas and mountain ranges contain a mixture of igneous, sedimentary, and metamorphic rock.

4. All of the rocks immediately adjacent to the rift are quite young. The igneous rocks progressively farther from the ridge, located either toward the east or toward the west, were found to be progressively older. Thus, for example, Ascension Island, which lies quite near the ridge, is composed of rocks that are about 1 million years old. St. Helena Island, a few hundred kilometers to the east, is 20 million years old, and Sao Tome Island, still farther east, is 120 million years old, as shown in Figure 19.13.

5. Igneous rocks on either side of the ridge are covered by a thin blanket of deep sea sediments. This blanket gets progressively thicker, and sediment at the bottom of the blanket gets progressively older, with distance from the ridge. These sediments are always slightly younger than the underlying igneous rock.

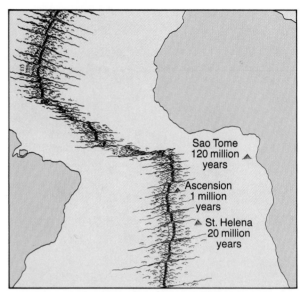

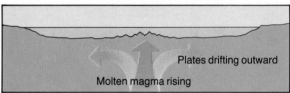

FIGURE 19.13   A cross section of the Atlantic Ocean floor, showing the age of the rocks on three islands located at various distances from the Mid-Atlantic Ridge. Each island was originally formed at the ridge and has since migrated outward as the plates moved. Note that the island farthest from the ridge is the oldest; the one closest is the youngest. This progression is a marker for the formation of rock on the ocean floor and the movement of the plates.

How could these five facts be explained? Geologists have reasoned that the Mid-Atlantic Ridge marks the boundary between tectonic plates. The plates on either side of the boundary are moving away from each other. As they spread apart, magma from deep within the Earth oozes upward to form new crustal material. Thus, 200 million years ago, when North and South America were fused to Africa and Europe, there was no Atlantic Ocean floor at all. The material that has by now become the Atlantic Ocean floor was, at that time, molten magma lying in the asthenosphere deep beneath the crust. As the continents cracked and moved away from each other, magma rose to the surface to form new rock. This newly formed rock then drifted outward away from the ridge with the drifting plates, as shown

in Figure 19.13, and was slowly buried under a blanket of deep sea sediment. This theory, called the **theory of sea floor spreading,** explains these five observations about the geology of the sea floor.

When researchers mapped and studied the floors of other oceans, they found ocean ridges in other regions as well. Each of these ridges marks the boundary between separating plates.

At this point you may ask, if the sea floor is spreading along the mid-ocean ridges, is the Earth's crust growing, and is our planet slowly expanding like a marshmallow that is being roasted in a fire? The answer is most certainly no. Again, evidence comes from a study of the floor

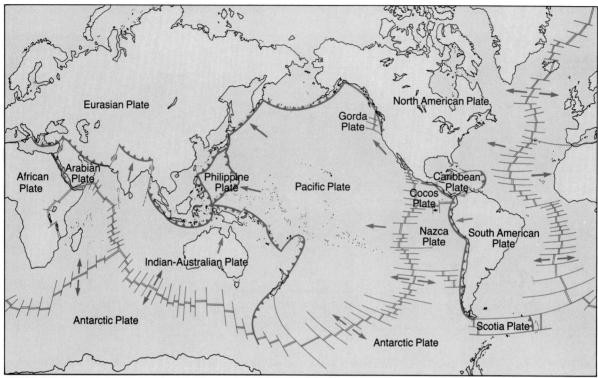

FIGURE 19.14 The major tectonic plates. The arrows indicate the direction of their movement.

of the ocean. In several regions, notably on the west coast of South America and the east coast of Asia, there are deep trenches quite near the continents, as shown in Figure 19.14. How are these trenches formed, and why haven't they filled up with sediment over the millennia? The answer is that the ocean trenches occur along boundaries where two tectonic plates collide. Remember that plates move away from each other along the mid-ocean ridges, and new plate material is formed from rising magma. The deep ocean trenches mark regions where plates collide. The relatively thin but dense oceanic lithosphere is unable to crack the continents apart or rise up over these lower density, thicker plates. Because the oceanic lithosphere is cool and dense, it sinks under the continental lithosphere. This downward movement of the lithosphere is called **subduction.** As the ocean crust sinks farther and farther beneath the surface, it eventually descends into the hot asthenosphere below. Some of the rock partially melts, becoming magma once again, and the remaining solid is incorporated into the astheno-

sphere. In a sense, the cycle is completed, as shown in Figure 19.15.

Sea floor spreading and subduction provide the mechanism that the geologists of Wegener's time searched for and could not find. Wegener's critics were correct in calculating that the continents could not plow through the ocean floor as a boat moves through the sea or through thick mud. Nevertheless, the theory has been substantiated. The searched-for mechanism is as novel as the original idea itself. Ocean crust is being continuously cycled; it is formed along the ocean ridges and absorbed and remelted along the ocean trenches. Continents don't have to push through the ocean floor because the ocean floor is moving as well.

19.7
CONTINENTAL AND OCEAN CRUSTS

Most geologists believe that the Earth first formed about 4.6 billion years ago. As discussed earlier, the primordial planet was heated, par-

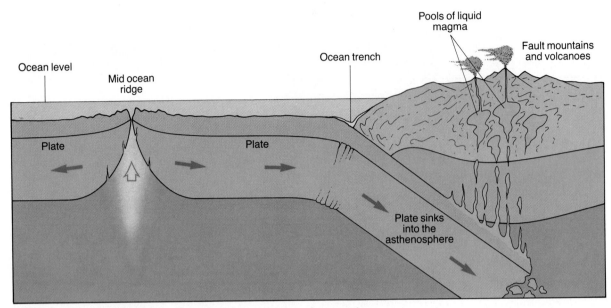

FIGURE 19.15 Formation of mid-ocean ridge and rise of mountains.

tially melted, and then cooled somewhat. The best estimate places the formation of the solid crust at about 4 billion years ago. This original crust is believed to have been more uniform than the crust is today, and there may have been less separation into ocean basins and continents.

After the crust cooled, the asthenosphere beneath was still hot and active. (In fact, it was probably hotter and more active than it is today.) Volcanic eruptions pushed through the surface, spewing forth lava and building mountains and mountain ranges. The mountains eroded over a period of millions or hundreds of millions of years, and sediments accumulated along the edges of the ranges. Some of these sediments were then buried and heated by close proximity to the hot mantle. As rocks were mixed, melted, and recycled over long periods of time, their chemical compositions and physical properties changed. Sections of the initial, dense crust were replaced by less dense materials, and these remained on the surface, forming the beginnings of continents. Over the millennia, the process of sedimentation, reheating, and remixing continued, causing the continents to grow.

Today, the continental crust is significantly different from the ocean crust. Continental crust is predominately granitic, while the ocean crust is composed of basalt. Granite is less dense than basalt. Additionally, the continents are much thicker than the ocean crust. Both the continents and the ocean crust are believed to be floating on the denser lithosphere beneath them. But because the continents are less dense than the ocean crust, they are more buoyant and float higher into the air. At the same time, since the ocean crust is nearly as dense as the mantle, it is relatively easy for this material to sink along the subduction zones.

The concept of floating continents is an important one. Granitic rock is not a particularly strong structural material. If an engineer were hypothetically asked to use granite to build a freestanding mountain range the size of the Himalayas on some imaginary solid foundation, the mountains would not be strong enough to stand up for very long. Don't think of mountains and continents as structures that are built up on some rigid base, but rather think of them as iceberg-like objects that are floating in a denser semifluid material beneath. Thus, mountain ranges have roots that sink beneath the surface, just as most of the material in an iceberg lies beneath the surface of the water. This concept that the continents and mountains are floating is called the principle of **isostasy.**

A second important difference between the continents and the ocean crust relates to their relative ages. In general, continental rocks are much older than rocks in the ocean crust. Recall that the original crust of the Earth formed about 4 billion years ago. All of this original material has since been remelted and recycled. The oldest rocks that have been found on Earth are about 3.96 billion years old. These are rare, but many others have been found that range in age from 1 to 3 billion years old. All of these old rocks are found on continents. In contrast, the oldest sections of any ocean crust are of the order of 200 million years old. This observation fits with tectonic plate theory. According to plate tectonics, new ocean crust is formed along the mid-ocean ridges. This material then migrates outward as the sea floor spreads. The ocean crust is denser but thinner than the continents. Therefore, when the two collide, the ocean floor sinks and is remelted, as discussed previously. The continents remain floating on the surface. Thus, the ocean crust is continuously being formed and destroyed, whereas the continents float about and grow older.

In one sense, the theory of sea floor spreading provides a mechanism for Wegener's original theory of continental drift. But, in another sense, the concept of sea floor spreading is an incomplete explanation, for it doesn't answer the question of what actually causes the plates to move. The lithosphere is less dense than the asthenosphere; thus, the tectonic plates may be visualized as "floating" along the surface of the globe. One theory states that the asthenosphere may be heated unevenly, causing large convection currents to be established, as shown in Figure 19.16. The lithosphere may then ride along these convection currents much as icebergs are carried long distances as they float on ocean currents.

19.8 PLATE TECTONICS AND MOUNTAIN BUILDING

Mountains form some of the most spectacular landforms on Earth and provide a constant re-

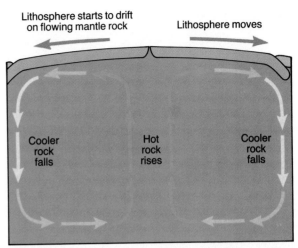

FIGURE 19.16 One theory suggests that convection currents exist in the asthenosphere and that lithospheric plates ride on these currents.

minder of the tremendous forces that exist within our dynamic planet. Several different types of mountain ranges can be distinguished.

The Andes

The Andes Mountains lie in a long, thin line along the west coast of South America. This region marks the boundary between the South American continental plate and an ocean plate to the west. Geologists believe that the two plates are colliding along this boundary. A deep trench exists in the ocean floor along the collision zone. As explained earlier, the ocean plate is descending, forming the trench. As this ocean plate dives under the edge of the continent, the rock above it moves upward. However, the rocks do not slide past each other smoothly. The continental rock catches on the moving rock beneath it and is compressed, crumpled, and distorted. Think of trying to slip two pieces of sandpaper past each other. They will, in fact, move, but one or the other will crumple, rise, and fall in the process. As portions of the subsiding ocean plate melt, some of the molten material moves upward through the cracks in the stressed continental plate to form both volcanoes and large bodies of granite. This combination of distorted, uplifted crust and volcanic action has formed the Andes

(a)

(b)

FIGURE 19.17   (a) Mountain in the Peruvian Andes formed by crustal uplifting. (b) Mt. Sajema in the Bolivian Andes, a volcanic cone.

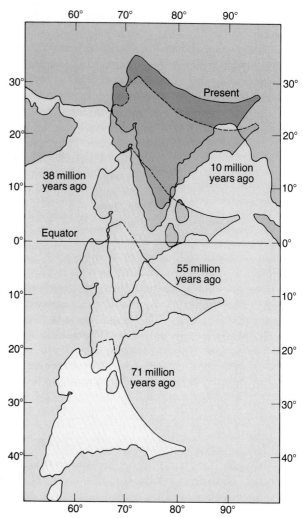

FIGURE 19.18   India's northward movement. In geologically recent times, the Indian subcontinent has collided with the Asian continent, creating the Himalayan mountain range.

mountains, the second highest mountain range on Earth, as shown in Figure 19.17. The core of the range is composed mainly of granitic rock. Interspersed here and there, the volcanic cones stand as visual reminders of the forces at work beneath the surface.

The Himalayas

Whereas the Andes were formed from the collision of a continental plate with an oceanic plate, the Himalayas were formed from the collision of two continents. Figures 19.18 and 19.19 show how the Indian subcontinent has collided with Eurasia. Measurements indicate that the two continents collided some 50 million years ago, and since the initial impact, India has continued to push northward. Neither continental plate has been forced downward. Instead, they both have buckled upward and formed the Himalayan mountain range.

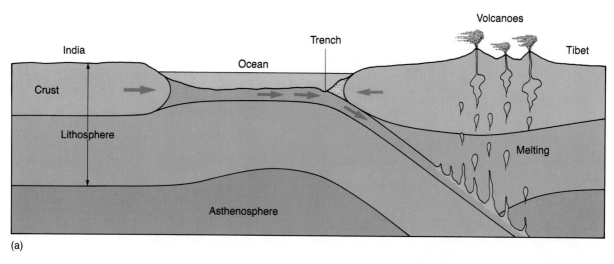

(a)

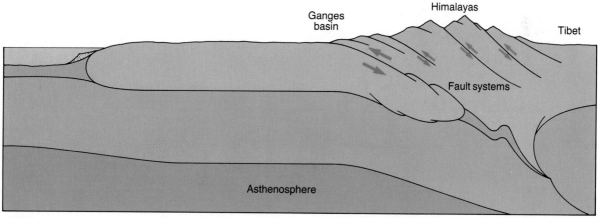

(b)

FIGURE 19.19 Formation of the Himalayas. (a) Geological condition as the Indian subcontinent approached Asia (about 50 million years ago). (b) As India collided with Asia, land masses buckled upward to form the Himalayan mountains.

The Rockies

The Rocky Mountain region of North America has experienced several periods of mountain building. About 2 to 3 billion years ago, an ancient mountain range existed where the modern Rockies are found today. These eroded over the millennia, and for many eons much of the region was relatively flat. In some places, inland seas covered the landscape, and dinosaurs roamed across swampy marshes. About 100 to 120 million years ago, the west coast of North America was shaped in such a way that what is now part of the inland mountains was then relatively close to the coast, as shown in Figure 19.20a. At that time, the Pacific Ocean plate rubbed against North America, and a compressional mountain range, similar to the Andes, was formed. Notice from Figure 19.20a that the ocean plate was moving outward from the East Pacific Rise, a mid-ocean ridge system similar to the Mid-Atlantic Ridge, with offsets along transform faults. As the two plates con-

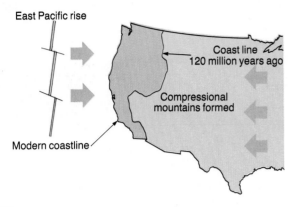

East Pacific rise

Coast line
120 million years ago

Compressional
mountains formed

Modern coastline

(a)

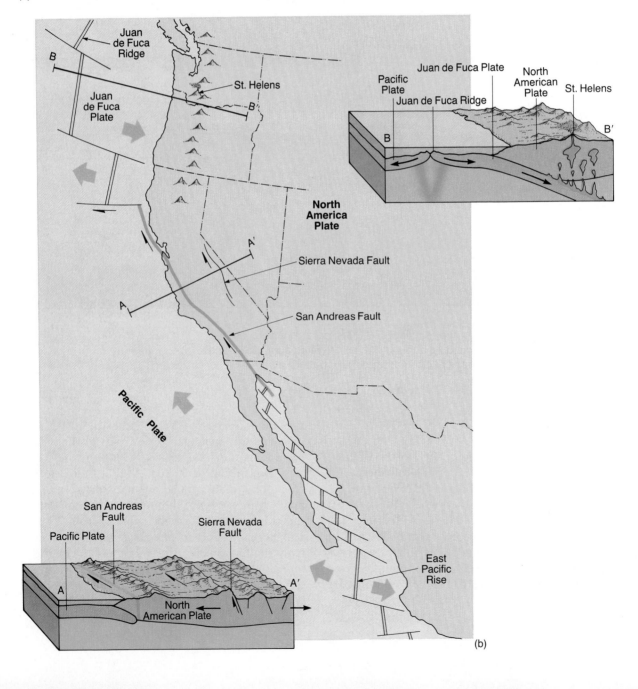

Juan
de Fuca
Ridge

B

Juan
de Fuca
Plate

St. Helens

B'

North
America
Plate

Sierra Nevada Fault

San Andreas Fault

Pacific Plate

San Andreas
Fault

Pacific Plate

Sierra Nevada
Fault

A

North
American Plate

A'

Pacific
Plate

Juan de Fuca Plate

Juan de Fuca Ridge

North
American
Plate

St. Helens

B

B'

East
Pacific
Rise

(b)

tinued to converge, gradually the East Pacific Rise came in contact with the continent itself.

Today, this ridge system runs through the Gulf of Baja, where it is offset to the northwest across the state of California as the San Andreas Fault. It then proceeds outward in a zigzag fashion into the North Pacific Ocean, as shown in Figure 19.20b. About 50 to 35 million years ago, the relative motion along this fault initiated a lateral distortion of the rock. In some places, segments of the crust were stretched apart. At the same time, hot magma from below began to push upward. This combination of stretching and upward flow produced mountain formation.

We, as humans, live in one very tiny period of geological history. Change is still occurring all around us. The continents continue to move. Many hot springs that exist throughout the Rocky Mountains remind us that hot magma is still rising close to the surface, and geologists would not be surprised if a volcano erupted in Yellowstone National Park or some other active region.

19.9
EARTHQUAKES AND VOLCANOES

In most parts of the Earth, the crust is whole and solid and is held together firmly. In some places, especially at the intersections of tectonic plates, there are weak points in the rock structure. Try to imagine the forces that operate on the boundaries between plates. On one hand, a large segment of rock—perhaps a whole continent—is being pushed in a given direction. But other forces oppose this movement. Another continent may be pushing in the opposite direction, or alternatively, the frictional forces between the rock

◄ **FIGURE 19.20** Formation of the Rockies. (a) This map shows the coast of the western United States as it was 120 to 100 million years ago compared with the present outline. At that time, a compressional collision between the plates caused mountain formation in the Rocky Mountain region. (b) Today part of the East Pacific Rise has migrated under part of western California, where it is called the San Andreas fault. At present, plate movement is causing lateral distortion and expansion of the continental crust.

FOCUS ON . . . The Richter Scale

Perhaps the most widely recognized method used to measure the magnitude of an earthquake is the Richter scale, devised by C. F. Richter. This method specifies the magnitude of the earthquake based on the movement of the earth at a distance of 100 km from the source of the quake. The scale is a logarithmic scale, like the decibel scale studied in Chapter 7. On this scale, an increase of one whole number corresponds to an earthquake wave with an amplitude ten times longer than the next lower number. Thus, an earthquake of magnitude 6 produces ten times more ground motion than one of magnitude 5. Likewise, a magnitude 6 earthquake produces one hundred times the ground motion of a magnitude 4 earthquake. The table below specifies the severity of damage associated with several magnitudes on the Richter scale.

Richter Magnitude	Effects
less than 2.5	Generally goes unnoticed, but can be recorded
2.5–5.4	Occasionally felt. Produces only minor damage
5.5–6.0	Slight damage to structures
6.1–6.9	Destructive in populated regions
7.0–8.0	A major earthquake producing serious damage
greater than 8.0	A great earthquake, producing total destruction to adjacent regions

---

of adjacent tectonic plates may hold the continent stationary. When this happens, tremendous stresses build up. Years and years may elapse without motion, until suddenly the stress becomes so great that it overcomes the friction, and rocks on either side of the fault slip past each other. This slippage is called an **earthquake.** A place or zone in the rocks along which relative motion has occurred is known as a **fault.**

From our knowledge of plate tectonics, we can predict that earthquakes are not likely to occur randomly but are most frequent at the intersections of plates. Thus, there are distinct regions of earthquake activity on the face of the

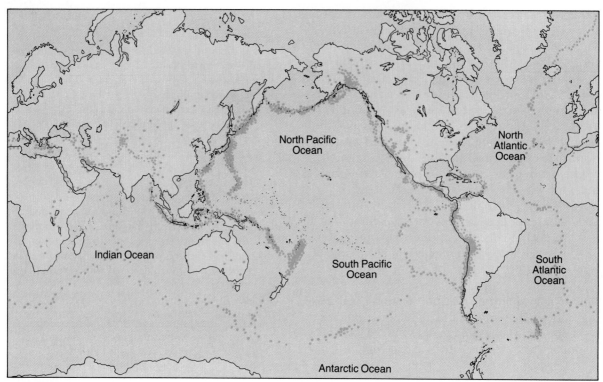

FIGURE 19.21   Map showing major centers of earthquake activity.

FIGURE 19.22   Earthquakes. (a) Near vertical motion. (b) Horizontal motion. ▶

globe, as shown on the map of Figure 19.21, and these correlate precisely with the plate boundaries shown in Figure 19.14.

During some earthquakes, the land on one side of the fault may slide up and over, while the other side drops under as tectonic plates converge. In other cases, the plates may move in opposite horizontal directions. Both patterns are shown in Figure 19.22. The **San Andreas fault** in California is an example of an earthquake zone of primarily horizontal movement. This major fault system stretches from the Gulf of Baja northward beyond San Francisco and out to sea, as shown in Figures 19.20 and 19.23. It is thought that this fault is caused by the collision of part of the Pacific Ocean floor with the North American continent.

Sometimes horizontal movement can be slow, gradual, and relatively nonviolent. Such motion is known as **fault creep.** In several locations in California, buildings have been uninten-

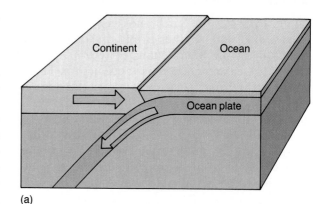

(a)

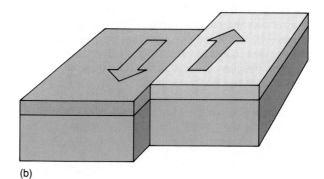

(b)

Oblique aerial view southeast along the San Andreas (fault) zone. A linear valley has been eroded along the main trace of the fault. Black line at the right is not a fault but a fence line against which tumbleweed has collected.
(Courtesy U.S. Geological Survey, R. E. Wallace)

A fault bed near Arches National Park in eastern Utah. The valley running down the middle of the photograph, with the road in the bottom, is a fault line. Many years ago, the rock on the right side of the valley and the rock on the left were connected and were part of a single bed of sedimentary rock. At this time, the surface of the land was a level plain. During the era of mountain building, the sedimentary bed cracked and the rock on the right tilted and sank into the fault zone, producing the structure you see at the present time.

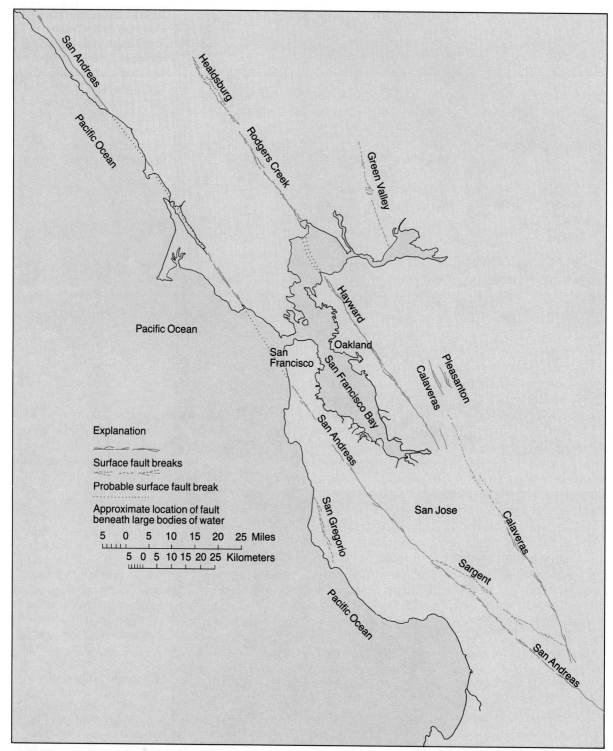

FIGURE 19.23 Historically and prehistorically active faults in San Francisco Bay region.

FIGURE 19.24 California earthquake, 1906. The wrecked Hibernia Bank building in San Francisco. (Photo by W. C. Mendenhall, U.S. Department of the Interior Geological Survey)

tionally located so that they straddle the great San Andreas fault right where creep is occurring. Slowly, the two sections of earth slide past each other. Many tiny shakes and quakes occur, but none is severe enough to topple the buildings. However, as the earth on each side of the fault moves, it cracks the foundation until walls fracture and break apart.

Such gradual slippage can destroy a few buildings along the fault, but since there is no serious sudden movement, nearby buildings are not harmed. But if rock is stretched or compressed too severely without continual release by many small earthquakes, large sections of the Earth may move suddenly along a fault, resulting in a disastrous earthquake. For example, if part of a tectonic plate moves at a rate of 5 cm a year, and another part is held rigid by frictional forces, the rock may stretch like a giant rubber band. Tremendous forces are involved. The slow stretch may accumulate for 100 years or more; then suddenly rock surfaces break loose and a very large earthquake occurs. A 5-cm stretch per year for 100 years would result in a total displacement of 5 m.

In 1906, sections of rock near San Francisco jumped 4.5 to 6 m in a matter of seconds and then abruptly came to a halt, generating huge shock waves. Buildings were split and separated. In the city of San Francisco itself, disaster struck,

as shown in Figure 19.24. The rapidly shifting rock caused the soil above it to move and settle. Many buildings whose foundations were anchored in soil toppled immediately. As buildings fell and underground gas lines were cut by the moving Earth, great fires were started. The fires spread throughout the city, causing widespread destruction that actually was more devastating than the effects of the quake itself.

Earthquakes are some of the most devastating of all natural phenomena. They generally strike without forewarning, destroying homes, apartment buildings, and sometimes even entire cities. In many regions of the world, lumber is scarce and people build their houses out of dried mud bricks called adobe. The roofs of these building are generally constructed of red clay tile. Adobe tile buildings are easily toppled by earthquakes, and millions of people have died in recent years when the heavy mud blocks have collapsed. If scientists could predict earthquake activity, people could be evacuated from unsafe buildings just before a potential disaster. Chinese geophysicists initiated a massive research program directed toward earthquake prediction. In late January 1975, these scientists noted unusual motions of the subsurface rock layers near the city of Haichang and ordered a massive evacuation of the region. The evacuation was completed on the morning of February 4, and in the early evening of the same day a large earthquake occurred. Houses, apartments, and factories were destroyed, but there were very few deaths.

Immediately after that success, many geologists optimistically hoped that a new era of quake prediction had begun. But, in 1976, Chinese geophysicists failed to predict an earthquake in Tangshan that killed approximately 650,000 people, and in many other regions of the world scientists have been unable to make accurate forecasts. Likewise, the devastating earthquakes that struck San Francisco and Beijing, China, in 1989 were not predicted.

Some of the general principles of earthquake prediction are relatively well understood. If segments of the Earth's crust have moved significantly along one region of a large fault system but other regions along the fault have remained stationary, then the rock that has not moved becomes strained and elastic potential energy builds up. These strained regions may crack, fracture, and move at any time. For example, earthquake activity has caused rock to shift all along the coast of Alaska, with the exception of the gap regions shown in Figure 19.25. Geologists concluded that the stress must be high in this gap region and that an earthquake was likely. On February 28, 1979, a large earthquake struck the eastern border of the gap located near the city of Valdez. Many scientists feel that another quake is now likely in the Valdez area. There is significant environmental concern because Valdez is the southern terminal of the Alaska pipeline. Huge storage tanks hold great quantities of crude oil, loaded supertankers are sailing southward out of the port, and, of course, oil is flowing through the pipeline itself. When will the quake occur? In the spring of 1979, geologist William McCann, who has studied the region, stated, "These patterns indicate that a quake may occur in about 20 years. But, in fact, it may be much longer. It could be another 20 to 40 years. But we would not be at all surprised if it were tomorrow."

After an earthquake has occurred along an active earthquake zone, tectonic stresses start to build up again. Several decades or even a century later, another quake is likely to occur in the same region. According to this reckoning, southern California, central Japan, central Chile, Taiwan,

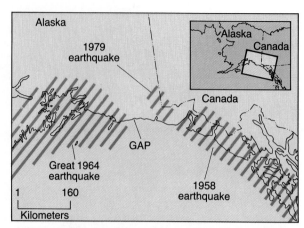

FIGURE 19.25 Recent earthquake activity on the south coast of Alaska.

the west coast of Sumatra, parts of the west coast of Mexico and, once again, the San Francisco Bay area, are all threatened by major disasters in the near future. In fact, the Memphis–St. Louis area (along the New Madrid Fault) is perhaps the most threatened area in this country. But what is meant by "near"? A week? A month? Ten years? Another century? No one knows for sure. Of course, it would be impractical to evacuate Los Angeles in anticipation of a quake that may or may not strike for 25 years!

Many scientists have theorized that a large quake is preceded by many small ones. As an analogy, if you try to break a stick by bending it slowly, you may hear a few small cracking sounds just before the final snap. However, an analysis of data shows that only about half of the major earthquakes in recent years were preceded by small shocks. Other scientists have concluded that fault creep is a natural means of relieving stress. They say that if rock surfaces slide by each other gradually, severe stress will not build up, and a major earthquake will not occur. Therefore, a quake is not likely to occur in active creep zones, and the danger points are those that lie along known faults but are currently immobile. However, others disagree and argue that fault creep, like small quakes, is simply a signal that should warn us of impending disaster. Until scientists are able to answer these and other complex questions, earthquake prediction will not be reliable.

FOCUS ON . . . Two Hundred Years of Earthquake Activity in the San Francisco Bay Area

In the early 1800s, San Francisco consisted of little more than a fort, a small port, and some scattered dwellings. When gold was discovered in northern California in 1848, the city boomed as a transportation, supply, and distribution center. By 1900 it had more than 340,000 inhabitants.

The city of San Francisco sits almost directly atop a portion of the San Andreas fault. To the east and across the bay, the Hayward fault runs past San Jose and northward through the modern city of Oakland. These faults are currently active, and numerous earthquakes have occurred over the past 200 years. The most devastating was a magnitude 8.3 quake that struck the city on April 18, 1906. The maximum movement along the fault was horizontal, and exceeded 6 m just north of the city. Some vertical displacement was observed as well, but it was less than 1 m. Although the ground motion toppled many buildings, the major damage was caused by fire. As buildings fell and underground gas lines were ripped apart by the moving earth, flames spread throughout the city, totally destroying the downtown area.

More recently, a magnitude 6.9 earthquake on October 17, 1989, left 55 dead in the city and the surrounding areas. The 1989 quake was centered in Loma Prieta, just east of Santa Cruz, about 125 km south of San Francisco. During this event, a large segment of the downtown business district in Santa Cruz was destroyed, motorists were killed when a segment of interstate highway 880 collapsed, and damage was heavy in specific areas such as San Francisco's marina district, which had been built on an old landfill. The total damage was estimated at about $4 billion.

Historical data suggest that the 1989 quake may be followed by a more devastating one. In the 1800s there were two significant pairs of earthquakes. A quake on the eastern, Hayward fault in 1836 was followed by one on the San Francisco Peninsula two years later. Then, a second quake on the peninsula in 1865 was followed three years later by a quake to the east. If this pattern repeats itself, an earthquake might occur near Oakland in the near future.

Predictions based on historical patterns are not entirely reliable and no one knows when and where the next quake might strike. Yet it is important to ask: How much damage would be expected from another major earthquake in modern times? The Loma Prieta quake was devastating, but not catastrophic. Although structural repairs take months or years, within a week to a month after the event, conditions returned to near normalcy for most of the residents of the area. Two factors reduced the damage and loss of life. (1) The epicenter was located in a sparsely populated mountainous district in Loma Prieta, not in one of the densely populated bay area cities. (2) While a magnitude 6.9 quake is certainly significant, the 8.3 magnitude quake in 1906 released 60 times as much energy.

City officials, engineers, and architects are well aware of the hazard of a catastrophic metropolitan quake. As a result, buildings, pipelines, roadways, and bridges have all been designed to be earthquake resistant. However, the recent disaster on highway 880 serves as a grim reminder that some precautionary measures are not adequate. Two scenarios have been outlined. Many homes in the San Francisco Bay areas are made of wood, and as explained above, wood-frame structures resist structural damage. Therefore, if a quake were to strike at night or during the weekend, experts predict a low mortality, perhaps 1000 out of a total population of 2 million in the Bay area. However, if the quake struck late in a summer afternoon, when bridges and subways were jammed with commuters and the streets were still packed with late shoppers, the situation could be much worse. Some structures would almost certainly collapse, and the death toll would be high.

Volcanoes

Perhaps the most spectacular and rapid geological event occurring on Earth is a volcanic eruption, which occurs when hot, molten magma moves upward through fissures in the rocks and escapes to the surface as fiery lava or hot ash, as shown in Figure 19.26. The eruption is generally accompanied by the release of large quantities of steam and other gases. Sometimes magma is ejected in relatively calm lava flows, while in other instances violently explosive eruptions occur. Mauna Loa, a towering mountain that rises above the sea to form the second highest peak on the

FIGURE 19.26   This unique photo from a weather satellite shows Mt. St. Helens moments after its eruption. Notice the circular "smoke ring" billowing outward from the mountain. The ring is caused by the shock wave from the explosion. (Courtesy NASA)

island of Hawaii, has erupted frequently in modern times, but the nonviolence of these events has enabled people to work and farm only kilometers from the volcanic crater. When the mountain becomes active, masses of molten rock flow smoothly out of a central crater and down the sides of the mountain until they cool and solidify. These frequent gentle lava flows have gradually accumulated to form the mountain, which rises 4500 m from the ocean floor to the surface of the sea and another 4200 m above that. From ocean floor to summit, Mauna Loa is about as tall as Mt. Everest.

Not all volcanoes are so gentle, however. In the early 1800s, the island of Krakatoa in Indonesia was a landmark for clipper ships that carried tea and other freight from India. The mountain on the center of the island was conical, covered with trees, and rose nearly 800 m above sea level. On August 26, 1883, a huge volcanic explosion rocked the island. The crew of a ship sailing offshore witnessed an immense cloud of dust, ash, and steam that darkened the horizon. Lightning storms and intense squalls developed as the sailors headed out to sea to escape the violence. On the next day, four more great explosions rocked the island; and when the dust had cleared away,

the island of Krakatoa and its 800 m mountain had disappeared. A few tiny islets remained on what had been the rim of the former island, but the rest was gone. It is believed that approximately 20 km^3 of volcanic material shot skyward. As the exploding lava shot into the air, a huge hole appeared in the center of the island, and when the eruption subsided, the mountain had disappeared. A very similar eruption about 7000 years ago formed Crater Lake, Oregon, by collapse of much of the mountain into the partially emptied magma chamber, as represented in Figure 19.27.

All volcanic eruptions occur when molten magma forms in the interior of the Earth and moves toward the surface through fissures in the rock. The violence of the eruption is controlled by factors such as the chemical composition of the magma, its temperature, the shape and size of the fissures, and the quantity of gas in the fluid mixture. If the lava is thick and viscous and if large amounts of gas are trapped in the molten rock, eruptions are likely to be violent. The viscous lava does not flow easily and is prevented from moving upward until the gaseous pressure causes it to explode violently. On the other hand, the magma of a gentle volcano such as Mauna Loa is more

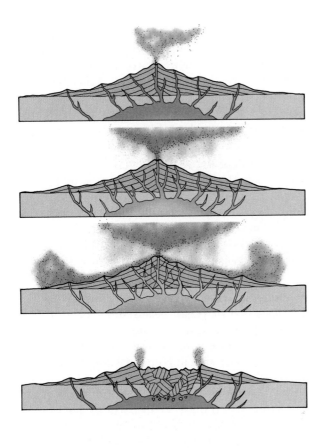

FIGURE 19.27 The collapse of a mountain during a cataclysmic volcanic eruption.

fluid and contains comparatively less steam. The fluid lava moves upward and out of rock fissures easily, and, as there is little steam, extreme pressures and violent explosions never develop.

Volcanoes are common in any region where magma is rising upward through fissures in the Earth's crust. The boundaries between tectonic plates are particularly active volcanic zones. Thus, frequent eruptions occur along the mid-ocean ridges where plates are separating. They also occur along the contact zones where ocean plates are colliding with continents. As discussed earlier, portions of the subsiding ocean plate melt, and some of this molten material may rise

upward to form volcanoes. Mount St. Helens was formed in this manner. Volcanoes are also observed in certain portions of the interior of plates where magma is moving upward. These regions, called hot spots, are as yet not completely understood. For example, there are many upwellings in the central Rockies. Hot springs are common, and there are several extinct volcanic craters in the area.

Volcanoes have played a vital role in the evolution of planetary atmospheres and of life itself. When solid planets such as Earth, Venus, and Mars were originally formed, various gases were trapped in the rocky interiors. Many of these gases—compounds of hydrogen, carbon, nitrogen, oxygen, and sulfur—were released during volcanic eruptions. As discussed in the previous chapter, the volcanic gases accumulated to form an atmosphere on each planet; and on Earth some of these gases probably combined to form amino acids that served as the building blocks for proteins and finally for living organisms.

19.10
WEATHERING AND EROSION

Millions or tens of millions of years are required for rock to form, distort, and rise upward to become a mountain range. During this time, the rock is exposed to many different kinds of surface forces. These forces, acting collectively, remove small pieces of rock and carry them downslope. The wearing away and removal of material from the Earth's surface occurs in two stages. First, the rock is broken into small fragments by various chemical and mechanical processes. The deterioration of rock into small pieces is called **weathering.** The small bits of rock and soil are then carried away by the action of running water, glacial ice, winds, or waves. This movement of material is called **erosion.**

Chemical weathering

Air and water, especially when carrying impurities, may be corrosive and therefore can react with many types of rocks and minerals. For example, pure iron is a hard, strong metal. But there are very few natural deposits of pure iron near the surface of the Earth. As Earth's crust was being formed, any iron in contact with oxygen reacted to form iron oxides. The oxide known as rust (Fe_2O_3) usually becomes loose and flaky. The conversion of a hard, abrasion-resistant material (iron) to a softer, flaky one (rust) is an example of chemical weathering.

Water is another chemically active substance. Water never exists in its purest state in nature. Many minerals dissolve to some extent in pure water; a few, such as sodium chloride (used commonly as table salt), are highly soluble. Even distilled water contains dissolved air, which includes two important reactive components—oxygen (O_2) and carbon dioxide (CO_2). Dissolved oxygen is an oxidizing agent and contributes to the rusting of iron under water. Dissolved carbon dioxide reacts with water to form an acidic solution. Thus, rainwater is slightly acidic and therefore slightly corrosive. This solution is capable of dissolving many types of rocks and therefore carries dissolved mineral matter to the ocean. Other corrosive impurities may enter water systems from a large variety of sources. (Refer back to the discussion of acid rain in Chapter 17.) For example, sulfur compounds present in polluted air or in certain natural rock formations dissolve in water to form strongly acidic solutions. In many industrial regions of the world, atmospheric pollutants have mixed with airborne water droplets to such an extent that the rainfall is acidic enough to kill fish and forests and to corrode statues and buildings. Salt from ocean spray, oxides of nitrogen from combustion or lightning, and many other substances may all dissolve in water and enhance its corrosiveness.

Mechanical weathering

Temperature Changes. Rocks can also be broken apart by purely mechanical processes such as expansion and abrasion. For example, most liquids contract when cooled and shrink even more when they freeze. Water is anomalous: It expands when it freezes. Thus, if water drips into a crack in a rock and then freezes, the resultant expansion acts to push the rock apart. The ice holds the rock from falling, but when the ice

melts, sections of rock crumble apart. If you ever climb in a high mountain range in the spring or early summer, when water freezes at night and thaws during the day, you will find that the mountains come alive with falling rocks. You can stand in a narrow valley and listen as the debris tumbles off the high cliffs.

Biological Processes. Plant roots also can crack rocks by expansion, as shown in Figure 19.28a. If a little bit of soil collects in a fissure in solid rock, a seed that falls there may start to grow. The roots then work their way down into the rock. As the plant grows and the roots expand, they push the rock apart just as ice does.

Moving Water. Have you ever walked along a stream bed or ocean beach and looked at the rocks lying in or near the water? If you have, you may have noticed that many are rounded and smooth, as indicated in Figure 19.28b. Pure water by itself has little abrasive power, but when water is moving rapidly, it picks up bits and pieces of silt and sand. When these small particles are hurled against the rocks, the solid material is gradually ground away. During storms and floods, fist-sized stones or even large boulders are pushed by the violent water, and as they tumble along and rub against each other, small bits are broken off. Over long periods of time, the weathering and erosive action of streams and rivers can reshape huge land masses. In Utah and Arizona, the Colorado River has dug tremendous trenches below the level of uplifted plains to form the Grand Canyon and its tributaries, as shown in Figure 19.29. Ocean waves are also abrasive. They can carve away significant portions of a sandstone cliff in a single winter storm by rolling rocks against them in the surf.

Wind. If purely gaseous air blows against a rocky mountainside, it has little effect. But if small pieces of silt or sand are suspended in the air and are blown against the rocks by wind action, they chip away at the solid material, as shown in Figure 19.30. Thus, the mechanism of weathering by wind is similar to the action by water.

(a)

(b)

FIGURE 19.28 (a) The roots of this tree have cracked this large boulder apart. Root action is a powerful weathering agent. (Photo by Grant Lashbrook) (b) Wave action has produced both depositional features (the beach in the foreground), and erosional features (the island-like stacks beyond the beach) along this section of the Oregon coast. (Courtesy of L. E. Davis)

FIGURE 19.29 The Grand Canyon was formed as an uplifted plateau and was weathered and eroded.

FIGURE 19.30 Winds and water carve spectacular rock formations like this one at Natural Bridge in Kentucky.
(Courtesy Bill Schulz)

(a)

(b)

FIGURE 19.31 (a) Aerial view of the Sherman Glacier, Alaska. (b) Granite Creek with an alpine glacier in the Eastern Chugach Mountains in Alaska.
(Courtesy of U.S. Geological Survey)

Glaciers. In high mountains and near polar regions, the snow that falls in winter never melts completely during the summer months and therefore accumulates from year to year. As the snow is melted and refrozen and compressed by the layers of snow above it, it gradually becomes dense enough to be called glacial ice. When the ice builds to considerable thickness, it becomes quite massive and moves slowly downslope. Such a formation is known as a **glacier,** as shown in Figure 19.31. When ice near the bottom of a glacier is subjected to the weight of thousands of tons of ice above, it takes on unusual properties and flows slowly, like a semifluid plastic. For this reason, glaciers slide downslope as a unit, oozing along on a layer of cold, fluid material.

Huge glacial deposits of ice exist in many parts of the Earth. In Greenland, the ice layer is 3000 m thick, while in Antarctica the ice cap is 4000 m deep in certain locations. Mountain glaciers are much smaller, sometimes being as thin as 60 to 90 m. Glaciers are particularly powerful and grind the rocks beneath them. As the glacier flows downhill, it picks up pieces of rock and soil, small stones, and even huge boulders weighing many tons. When this solid material is dragged seaward, grinding against bedrock, it carves huge valleys and shapes the topography of mountains or even continents.

Erosion

Once rock has been weathered, small pieces of it can be carried away by some moving substance. *Erosion is the process whereby weathered material is removed.* Running water, moving wind, and flowing ice, which are active weathering agents, cause erosion as well. Bare soil is particularly vulnerable to erosion. This movement of material with the subsequent loss of valuable farmland is a serious environmental problem that will be discussed in the next chapter.

Weathering and erosion do not act equally upon all types of rock. Some rocks are hard and resistant to weathering, whereas others are physically soft or chemically reactive. Therefore, some regions wear away faster than others. As a result, valleys may be cut between bastions of harder rock, and sharp cliffs are shaped or rounded in

many different ways. Mountain building, weathering, and erosion act together to shape the Earth's crust. As subterranean forces push rock upward, surface forces scour the uplifting mass. Thus, mountains and other landforms that we see today were created by a series of opposing forces.

Sometimes it is relatively straightforward to read the history of a landform by studying its present structure. Consider the following two examples:

1. Figure 19.32 is a photograph of rocks in the Utah desert. Even a cursory examination reveals that there are horizontal beds of sedimentary rock. Sedimentary rock is formed in ocean basins or at the bottom of valleys, lakes, or rivers, not on the tops of mountains. Therefore, the structures shown here were once in a low region and have since been uplifted. If you draw imaginary lines from one pillar of rock to another, you see that the horizontal layers line up. From this evidence, we can deduce that all of the rock pillars were once part of a single deposit, which was uplifted and eroded.

2. Figure 19.33 is a cross-sectional map of a mountainous region, showing the different types of rock present in the cores of the mountains. Such a map tells us a great deal about the history of the region. To interpret the data, first draw imaginary lines to reconstruct the landform as if it were not sculptured by erosion, as shown in Figure 19.34. The reconstructed structure shows a sequence of sedimentary beds overlying an older igneous granite. This sequence was folded upward. At some later time, a fracture developed and hot magma pushed upward to cool and form a dike of hard, young rock cutting through the older sequence.

19.11
FORMATION OF MINERAL DEPOSITS

If you collected a few samples of rock from your neighborhood and had them analyzed very carefully, chances are good that you would find traces of iron, copper, silver, gold and a variety of other

FIGURE 19.32 A desert rock formation in Canyon Lands, Utah.

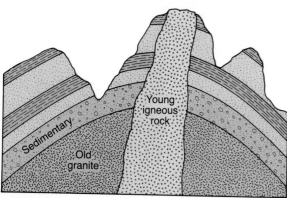

FIGURE 19.33 Cross section of a hypothetical mountain.

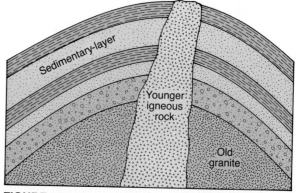

FIGURE 19.34 The mountain reconstructed as if no weathering or erosion had occurred.

valuable metals. However, the concentration of these metals would be so low that it would be impractical to start a mine in the area. On the other hand, there are a variety of natural processes that have concentrated certain minerals in distinct regions. Thus, if you explored the Earth's surface, you would find that the chemical composition of the rock and soil varies. These changes

FIGURE 19.35 Veins of different minerals embedded in older bedrock.

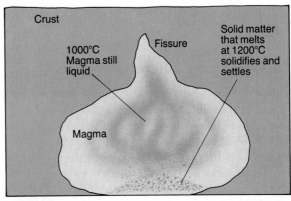

FIGURE 19.36 Minerals can separate by gravity from magma that cools very slowly.

often occur sharply, especially in mountainous regions. For example, outside of Boulder, Colorado, in the foothills of the Rocky Mountains, a series of uplifted sedimentary rock marks the landscape. Just a few kilometers away, however, many of the imposing cliffs of Boulder Canyon are granite. Rock chemistry may change even more abruptly, and two or more distinct rock types may appear in a small area. Figure 19.35 shows small segments of one type of rock, called a **vein,** embedded in another type of rock. One important aspect of geology is to try to understand the processes that cause local concentrations of one mineral or another.

Hot magma lying beneath the Earth's surface is not the same from place to place. Therefore, the chemical composition of the lava coming from Mauna Loa in Hawaii is likely to be different from that coming from Mt. St. Helens in Washington. But only a small fraction of the Earth's continental crust is composed of lava that shot rapidly out of a volcanic opening. Much of the igneous rock oozed up slowly through cracks in the crust, cooling gradually during its travel. If a layer of older rock is cracked apart and invaded by some upflowing magma, the final formation will contain veins of foreign rock.

Weathering and erosion also play an important role in developing heterogeneity. If several different rock types lie in the same region, the softer rock erodes away more quickly, leaving exposed layers of hard rock rising over a valley or plain of sedimentary material.

Many other chemical and physical processes are responsible for concentrating specific types of rocks and minerals. In reviewing some of these, we will emphasize the development of economically significant deposits of ore, fuel, and fertilizer.

Separation of gravity

Suppose that two minerals crystallize at the same time from magma. If the magma is agitated and pushed upward to the surface by a volcanic eruption, it then hardens quickly, and the minerals are more or less uniformly dispersed in the newly formed rock. But suppose, instead, that the magma had started to move up slowly through a fissure in the crust and cooled gradually while still kilometers below the surface. Suppose, for example, that one of the minerals solidifies when cooled to 1200°C, and the other solidifies at 1000°C. As the total mixture cooled, the mineral with the 1200°C freezing point could start to solidify while the rest of the magma remained liquid. But remember that the cooling process occurs slowly, sometimes requiring many thousands of years. During this time, if the crystals are denser than the remaining magma, they settle downward through the lighter liquid melt until there exists a concentration of one mineral at the bottom of the magma body, as shown in Figure 19.36. The magma finishes crystallizing, and this deposit may remain deep in the crust, unreachable by modern mining techniques, or it may be uplifted by tectonic processes and exposed by erosion.

Separation by differential solubility

Minerals may also become concentrated by differential solubility. To understand how this process works, take a little bit of salt and mix it up with a lot of sand. It would be physically tedious to pick out the grains of salt from the grains of sand. But suppose you put the entire mixture in a glass of water and stirred it up. The salt would dissolve into the water, and the sand would settle to the bottom. You could then pour off the water and collect the sand. The salt, which remains in the water, could be retrieved by evaporating the water and collecting the residue that remains.

Similar processes can occur at the Earth's surface. Many valuable mineral deposits in Utah were formed by differential solution of rock in water. Rainwater traveling across rock and soil dissolves minerals such as salt (NaCl) and potash (K_2CO_3), whereas other minerals such as quartz (SiO_2) are largely unaffected. Most of these dissolved salts are either absorbed by plant roots or carried into the ocean. But occasionally, in arid regions, large land-locked lakes develop that have no outlet into open sea. Water flows into these lakes through streams and rivers but can escape only by evaporation. However, only the water evaporates; the salt remains behind. As time passes and water is lost, the mineral concentration of land-locked lakes increases. There was once a large land-locked sea covering much of northwest Utah. The salinity of this sea gradually increased. Then a major change of climate occurred. The region got hotter and drier, and streams flowing into this sea slowed down or dried up. The result was that more water escaped through evaporation than entered via streams, and the lake shrank to its present size. As the waters receded, mineral deposits were left on dry land, and these are now mined commercially.

Placer deposits

There has always been a great deal of legend and intrigue about mining, especially in the pioneer-

ing days when Europeans colonized new continents of the Americas, Africa, and Australia and searched for mineral wealth. Men hunted the streams and hillsides for all kinds of metals, but most of all it was the search for gold that led them to face long periods of loneliness, frigid cold, and the dangers of unknown lands. Occasionally, miners discovered veins of nearly solid gold, but much of the gold was taken out of stream-deposited sediment. This source of gold is called a **placer deposit.** Gold is a particularly dense metal; if you swirl a mixture of gold dust, sand, and soil in a glass of water and allow the solids to settle out, the gold falls to the bottom first. This action allows gold to be collected by **panning.** A miner who suspects that gold may be found in a certain area dumps a small shovelful of sand and gravel in a pan, adds water, and swirls the mixture. Then gradually and carefully the particles on the surface of the swirling liquid are spilled out, water is added, and the process is continued. If there is any gold in the sediment, it remains in the bottom of the pan to the end.

Differential settling may occur by natural processes. Suppose, for example, that small amounts of gold from a mountainside are carried downstream by a river. Chances are that the concentrations of gold in the river water and surface sediment is too small to mine economically. But now imagine that the river is partially blocked downstream by a beaver dam. As the water reaches the pond behind the dam, it slows down but does not stop and continues over the dam and on toward the sea. When rains fall and snows melt in the early spring, the stream may carry a lot of fine sediment with an occasional speck of gold. When the water reaches the beaver pond and slows down, the sediment settles. But what settles first? Gold, of course! If the stream flows at just the right speed, the water carries most of the lighter sediments downstream. As the years go by, more and more gold specks collect in one location, until a small concentration, called a placer deposit, is formed.

Minerals under the sea

Many mineral deposits have formed on the ocean floor. Some of these were uplifted during the col-

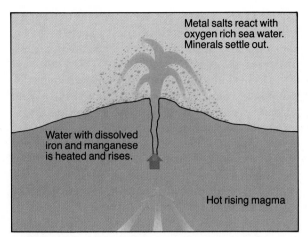

FIGURE 19.37 Formation of undersea mineral deposits.

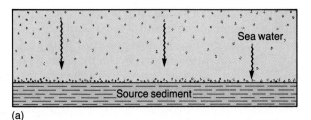

(a)

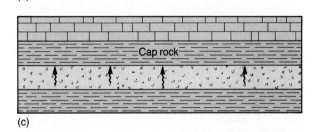

(b)

(c)

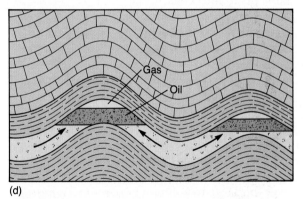

(d)

FIGURE 19.38 Formation of petroleum. (a) Organic rich marine sediments fall to the floor of the ocean. (b) This sedimentary layer accumulates. (c) The sediments are covered by a layer of impervious rock. Oil and residual water are trapped by this rock. (d) The rock is deformed. Gas and oil slowly rise and are concentrated in dome-shaped structures.

lisions of tectonic plates and became easily accessible. The rich iron, copper, lead, and zinc deposits that contributed greatly to the wealth and power of the early Roman empire were formed in this manner. Other valuable deposits lie relatively inaccessible under several kilometers of sea water.

Rapid mineralization occurs near the mid-ocean ridges. Recall that tectonic plates separate at the mid-ocean ridges, and as they travel away from each other, magma from the asthenosphere is constantly flowing upward and solidifying. Huge, deep cracks and crevices have been observed along these boundaries. Sea water seeps downward through the cracks and comes in contact with fresh, rising magma. Valuable metals, including iron, manganese, copper, lead, and zinc, may be dissolved or suspended in this water. As the water rises back up to the ocean floor, it mixes with the oxygen-rich waters above, and many metallic elements can react with the oxygen to form metal oxides. These metal oxides are insoluble minerals, and they fall to the sea floor near the ocean ridges in rich layers, as shown in Figure 19.37.

Fossil fuel deposits

So far we have discussed how inorganic mineral deposits were formed. Three organic deposits found in the Earth are coal, natural gas, and pe-

troleum, the three fossil fuels. These materials are called fossil fuels because they were formed from the remains of decayed plants and animals that lived millions of years ago. Coal is composed primarily of carbon, and gas and petroleum are composed primarily of carbon and hydrogen. Living organisms are composed mainly of carbon, hydrogen, and oxygen. In order to convert organic tissue into fossil fuels, the tissue must be entrapped in an oxygen-poor environment. If tissue decays in an oxygen-rich environment, it burns or rots and is converted to carbon dioxide and water.

Coal was formed primarily from plant matter. In any forest or swamp, leaves, twigs, branches, and entire trees fall to the ground, where they slowly decompose. However, decomposition is not always complete. If newly fallen litter or sediment covers the old, partially rotted debris before it decomposes fully, some of the organic matter is preserved. If conditions are favorable, the surface layers prevent oxygen from penetrating the deeper sediments. Over time, heat and pressure gradually build up as this preserved matter is further compressed by accumulating debris and soil. The combined heat and pressure from accumulating layers initiate a series of transformations that change the buried plant tissue into **coal.**

Go to a swamp or marsh and dig up a shovelful of the muck on the bottom. If you examine it closely, you will find that there is very little mineral content in your sample; it is mostly decayed plant matter. If such a swamp bottom is covered with inorganic sediment and compressed for hundreds of thousands of years, a small coal deposit develops. However, most modern swamps are poor coal producers. It is estimated that a layer of compressed organic debris 12 m thick is required to produce a 1 m layer of coal. For a layer this large to accumulate, conditions must be favorable and stable in a region for a great many years. Coal deposits probably are being formed today in many areas, notably in the Ganges River delta in India, but the process is extremely slow—much, much slower than the exploitation of existing reserves. Therefore, since we cannot expect formation of new deposits to keep pace with use, there is all the more reason to conserve our present reserves.

Oil and natural gas are also organic deposits, but these fuels were formed from tiny marine microorganisms rather than from the debris of large plants. As microscopic sea creatures settled to the bottom of the ocean and were later covered with mineral sediment, partial decomposition in the absence of oxygen occurred. This decomposition of marine organisms produced droplets of oil, which then accumulated in natural pores of the rock, as illustrated in Figure 19.38a to c on p. 513. However, such a dispersed accumulation of tiny droplets of oil in a porous rock would hardly make a commercial oil well. In order for commercial concentrations of oil to accumulate, the oil droplets from a fairly wide area must migrate to a single location. This has occurred by a variety of mechanisms.

In one possible sequence, imagine that some seawater was trapped in the rock along with the oil. Imagine further that a layer of impervious clay settled on top. This clay would gradually turn to a type of rock called shale, which, in turn, would serve as a cap to prevent oil or trapped water from escaping. Over the millennia, as the oil was being formed, the sediments might have been folded as shown in Figure 19.38d. In this ideal sequence, the drops of oil would naturally float toward the surface of the entrapped water. However, none of the liquids can escape through the shale. Therefore, the oil collects in concentrated regions at the tops, or domes, of the folds. It is important to understand that although this oil concentration would be a commercial deposit, it is not an underground pool, or lake, of oil. Rather, it is a porous rock formation filled with oil, more like a sponge soaked with oil than a glass full of oil.

SUMMARY

The inner core of the Earth is solid iron and nickel, surrounded by an outer core of molten iron and nickel. A large solid **mantle** surrounds the core. The **asthenosphere** is in an upper section of the mantle and is semi-fluid and plastic. The **lithosphere** includes the surface crust and the uppermost portion of the mantle. The

Earth was originally heated by the kinetic energy of the collapsing primordial cloud, radioactive decay, and bombardment from outer space. **Igneous** rocks are formed directly from molten magma, or lava, and **sedimentary** rocks are formed from sediments that are compressed or cemented together. **Metamorphic** rocks are formed when other types of rock are heated (but not melted) and compressed for long periods of time.

Plate tectonic theory states that the Earth's surface is broken up into a number of distinct plates that move relative to each other. Some lines of experimental evidence for this theory include the fossil record, the shape of continents, the deposition of ore deposits, and the orientation of magnetism in rocks. Plates separate along mid-ocean ridges. If an ocean plate collides with a continental plate, the ocean plate subsides and the continental plate buckles and rises to form mountains. The concept of **isostasy** states that the continents can be viewed as masses of less dense rock floating on denser material beneath them. The Andes were formed along a subduction zone, the Himalayas were formed when two continents collided, and the modern Rockies were formed by compression followed by expansion. Earthquakes generally occur where plates slip past each other, where one is subducted, or at mid-ocean ridges where plates are separating. Volcanoes occur when magma rises rapidly to the surface.

The deterioration of rock into small pieces is called **weathering.** Weathering can occur by chemical action or mechanical processes, including temperature changes, biological processes, moving water, wind, and glaciers. **Erosion** is the movement of small bits of rock and soil. Mineral deposits are formed by many processes, including separation by gravity, separation by differential solubility, formation of placer deposits in streams and lakes, and chemical deposition of minerals in seawater. Fossil fuel deposits are partially decomposed organic debris.

KEY WORDS

Seismology	Lava	Subduction	Volcano
Core (of the Earth)	Igneous rock	Mid-Atlantic Ridge	Weathering
Mantle	Sediments	Rift	Erosion
Asthenosphere	Sedimentary rock	Subduction	Glacier
Lithosphere	Metamorphism	Isostasy	Vein
Crust	Metamorphic rock	Earthquake	
Rock	Continental drift	Fault	
Mineral	Tectonic plates	Fault creep	
Magma			

QUESTIONS

STRUCTURE OF THE EARTH

1. Briefly outline the interior structure of the Earth. What regions are solid and brittle? Plastic? Molten?
2. An intense rain of debris from outer space probably fell to the Earth over 4 billion years ago. Explain why there are no craters on the Earth from this bombardment.
3. Explain how radioactivity can heat the Earth.

ROCK CYCLE

4. Trace possible geological processes for each of the following transformations: (a) metamorphic rock changing to sedimentary rock, (b) sedimentary rock changing to metamorphic rock, (c) sedimentary rock changing to igneous rock.

5. What is magma? How does the rate of cooling of magma affect the texture of the resultant rocks? What do we call magma when it travels quickly and violently to the surface? What kind of rock is formed when magma flows slowly upward but cools and solidifies before it reaches the surface?

PLATE TECTONICS

6. Explain why a geologist searching for mineral or fuel deposits would be wise to study plate tectonics.
7. Did the discovery of coal in Antarctica prove the theory of continental drift? Why or why not?
8. The climate of many regions of the globe is changing. Could these climatic changes be caused by continental drift? Defend your answer.

9. When continental drift theory was first proposed, it was believed that the continents must somehow plow through mantle rock as a ship plows through the water. This analogy has since been discarded. Explain how the continents move and why this movement cannot be compared to a ship traveling across the sea.

10. The deepest ocean trenches in the world lie on the western edge of the Pacific Ocean. What types of geological activity would you expect to find in the region?

MOUNTAIN FORMATION

11. Why are there deep ocean trenches adjacent to some large mountain ranges such as the Andes?

12. The Himalayas contain extensive regions of sedimentary rock, and marine fossils have been found at high elevations. Explain these facts in terms of the theory for the Himalayan formation as described in the text.

13. Geologists believe that at the present time Africa is moving toward Eurasia. When the two collide, new mountains will be formed. Will this mountain building be most similar to the formation of the Andes, the Himalayas, or the Rockies? Explain.

14. Explain why the cores of some of the mountains in the Andes are granitic, whereas others are volcanic cones.

EARTHQUAKES

15. Explain why some regions of the globe are more likely than others to experience earthquakes.

16. It has been suggested that engineers should inject large quantities of liquids into locked portions of the San Andreas fault. Proponents of the plan believe that these liquids will reduce friction by lubricating the sides of the fault, and thus the rock will slide along slowly, pressure will be relieved, and a major earthquake will be averted. If you were the Mayor of San Francisco, would you encourage or discourage the injection of fluids into the fault structure? Defend your decision.

17. Compare Figure 19.21, which shows centers of earthquake activity, with Figure 19.12e, which shows tectonic plate structure. Comment on any similarities as well as any differences.

VOLCANOES

18. Are the Hawaiian Islands composed primarily of igneous, sedimentary, or metamorphic rock? Explain.

19. Two volcanoes that have erupted in recent time are Mt. St. Helens in Washington and Mauna Loa in Hawaii. Mt. St. Helens erupted violently, and Mauna Loa has erupted in gentle lava flows. What types of eruption would you predict for each mountain in the future? Defend your answer.

20. Do you think that volcanic activity might be likely to occur near earthquake zones, or would you guess that no correlation would be observed? Defend your answer.

PLANETARY GEOLOGY

21. Today the surface of Mars is drier than any desert on Earth. Yet deep canyons are observed on the surface. Speculate on the nature of these canyons. If you were in charge of a space program and could send an unmanned rocket to Mars, what information would you seek to test your hypothesis?

22. Astronauts have established several permanent seismographic recording stations on the Moon. (A seismograph is an instrument that measures earthquake waves.) What can be learned about the structure of the Moon from these recorders? If no moonquakes occur, would you consider the experiment to be a failure? Explain.

23. Volcanic eruptions have been observed on one of Jupiter's moons, Io, but not on the Earth's moon. Which body would be more likely to experience quakes? Why?

WEATHERING AND EROSION

24. Baffin Island is located in the arctic region of Canada. The southern and eastern parts of the island are mountainous and are composed mostly of granite and other igneous rocks. The climate is extremely cold, with bitter winters, short summers, and consequently very little plant growth. There is some precipitation year round, but the region is generally dry. Predict what types of erosion and weathering predominate.

25. The western plains of North America are relatively flat. Before farmers came to the region, the natural prairie vegetation held the soil very efficiently, and only small amounts of material were washed into streams and rivers. In fact, the depth of the soil increased with time. Would you say that weathering was insignificant, or that erosion was insignificant, or both, or neither? Defend your answer.

FORMATION OF MINERAL DEPOSITS

26. It is common for a single mine to contain fairly high concentrations of two or more minerals. Discuss how geological processes might favor the deposition of two similar minerals in a single location.

27. If one compound is to be separated from a com-

plex mixture, it must somehow be transported away from the rest of the material. Explain how ores are moved out of a mixture in each of the following processes: (a) separation by gravity, (b) separation by differential solubility, (c) formation of placer deposits.

28. If you were searching for petroleum, would you search primarily for sedimentary rocks, metamorphic rocks, or igneous rocks? Which of these three would be the least likely to contain petroleum? Explain.

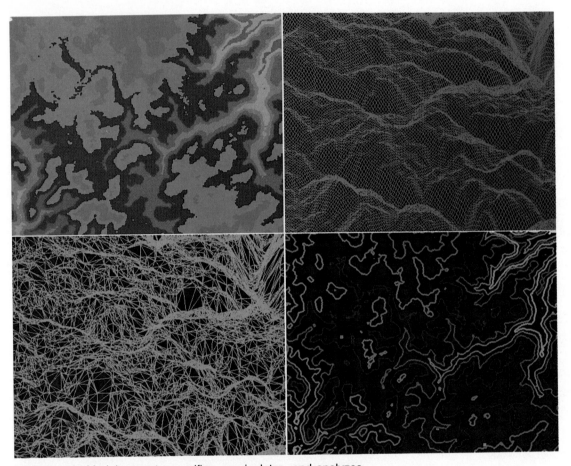

MGE Terrain Modeler creates, verifies, manipulates, and analyzes terrain models represented as triangulated irregular networks (TIN), elevation matrices (GRID), or contour areas. Upper left image shows color-filled contours; upper right, GRID model; lower left, TIN model; and lower right, contours color-coded by elevation.
(Courtesy of Intergraph Corporation, Huntsville, Alabama)

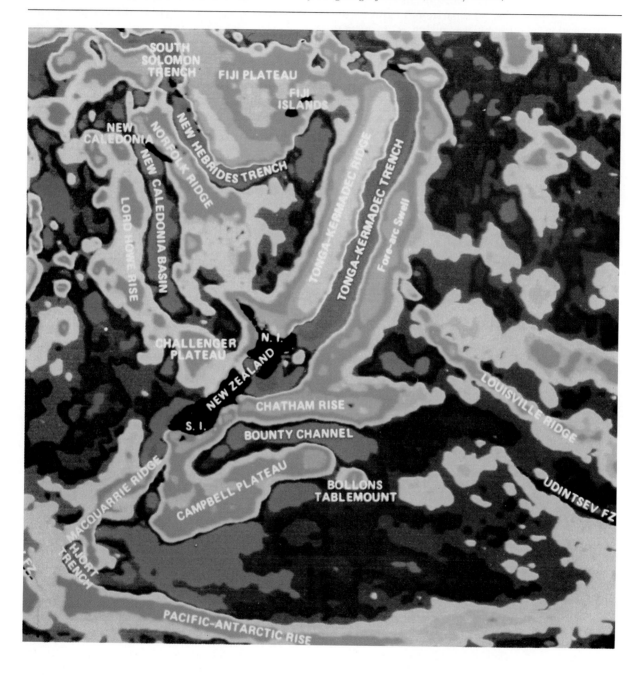

Computer map of Pacific shows sea floor geologic features. (Courtesy NASA)

C H A P T E R

20

Environmental Geology

20.1
INTRODUCTION

If you go to an urban elementary school and show the class a picture of a field of corn or a stalk of wheat, many of the children will not be able to identify the plants. Yet, wheat and corn are two of the main food staples of our civilization. Living in our technological world, it is too easy for even educated adults to assume that bread comes in plastic bags from "bread factories" and corn flakes come in cardboard boxes and are also somehow manufactured. This emotional separation of people from their source of sustenance can lead to the attitude that technology has somehow liberated humans from their dependence on the natural world. Nothing could be further from the truth. In this chapter, we will study soil, water, and minerals, three of the basic necessities of civilization and of life itself. We will also see how certain kinds of human activities can endanger these resources.

SOIL

20.2
SOIL COMPOSITION

Rock is a poor medium to support the growth of plants. Plants need a continuous supply of minerals, water, and sunlight. Although an organism growing on rock may be exposed to adequate sunlight, the minerals are generally bound chemically into the structure of the rock and are not readily accessible, and water runs off the surface and is gone soon after a rainstorm has passed. When rock is cracked and broken into tiny pieces by various weathering processes, the resultant material becomes gravel, sand, silt, or clay, depending on the size of the various particles. These inorganic mixtures are also unsatisfactory for maintaining a healthy and diverse plant community. True soil, which does support vigorous plant growth, is an intimate mixture of pulverized rock and organic debris, as shown in Figure

519

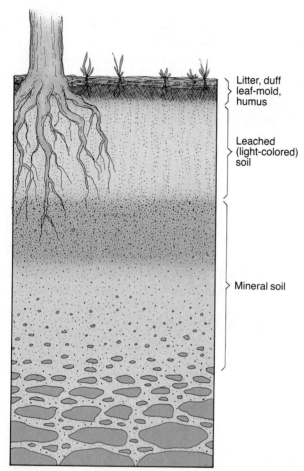

Litter, duff
leaf-mold,
humus

Leached
(light-colored)
soil

Mineral soil

FIGURE 20.1   Cross section of forest soil.

20.1. The organic matter contains many nutrients necessary for plant growth, For example, plants need an adequate supply of nitrogen for the synthesis of protein. Although nitrogen is plentiful in the air, it exists as the gaseous element N_2, and most plants cannot utilize nitrogen in this form. Instead, plants need certain nitrogen compounds, such as ammonia or nitrates, found in the soil or dissolved in water. Pure silica sand, SiO_2, contains no nitrogen and thus cannot support the growth of plants. However, fertile soil contains available nitrogen compounds as part of the organic component. These materials are introduced into soil by a variety of processes. Some bacteria, including those that live on the roots of

peas, beans, and alfalfa, have the ability to utilize elemental nitrogen and convert it to usable nitrogen compounds. These organisms, therefore, represent an important asset to a plant community. Once nitrogen is incorporated into proteins and introduced into the soil, it can be recycled many times, as shown in Figure 20.2. For example, rotting leaves and animal urine contain ammonia, NH_3. The nitrogen in ammonia is bound in a chemical form that is readily assimilated by almost all plants for the synthesis of protein.

In addition to being a rich source of vital nutrients, soil plays an important role in the regulation of essential plant processes. Plants need soil with properly regulated nutrients, density, moisture, salinity, and acidity. Soil conditions are regulated by the **humus,** a very complex mixture of compounds resulting from the decomposition of plant tissue. A given piece of tissue, such as a leaf or a stalk of grass, is considered to be humus rather than debris when it has decomposed sufficiently in the soil system so that it is no longer recognizable. Compared with inorganic soil, humic soil is less dense, holds moisture better, and is more effectively buffered against rapid changes of acidity. Additionally, certain chemicals present in humus aid the transfer and retention of nutrients. For example, calcium ions, Ca^{2+}, can exist in water solutions from which they are available to plants. However, atmospheric carbon dioxide reacts with water to form carbonate ion, CO_3^{2-}, which reacts with calcium ions in water to form the sparingly soluble compound, calcium carbonate, $CaCO_3$, the major component of limestone.

$$Ca^{2+} \; + \; CO_3^{2-} \; \longrightarrow \; CaCO_3$$
$$\text{calcium} + \text{carbonate} \longrightarrow \text{calcium}$$
$$\text{ion} \qquad \text{ion} \qquad \text{carbonate}$$

Plant roots cannot absorb calcium efficiently from calcium carbonate. However, calcium ions also react with certain organic compounds, called **chelates,** in the humus to form organic calcium complexes, which we will represent as calcium $^{2+}$ (chelate). Plant roots can readily absorb the calcium from these metal-organic complexes. Thus, two possible fates of calcium in the soil are

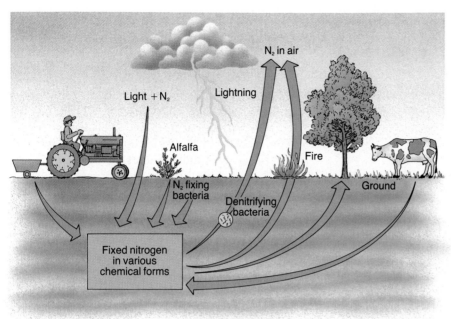

FIGURE 20.2 The nitrogen cycle.

calcium ions + carbonate ions $\longrightarrow$
calcium carbonate
(relatively inaccessible
to plants)

calcium ions + chelate $\longrightarrow$
calcium^{2+} (chelate)
(relatively accessible
to plants)

The organic components of the soil tend to react with various ions and retain them in a form that is easily absorbed by plant roots and utilized in the growing systems, as shown in Figure 20.3.

20.3 SOIL EROSION

In a natural temperate ecosystem, new soil is formed by the decomposition of rock at a rate varying between 2 and 11 tonnes per hectare per year (1 tonne = 10^3 kg; 1 hectare = 10,000 m^2), depending on the region. At the same time, some of the soil is carried away by wind and water. In most temperate ecosystems, the rate of production is equal to or greater than the rate of re-moval, so soil depth and fertility increase slowly with time. When natural forests and prairies are cut and plowed, soil is exposed for at least part of the year and is susceptible to erosion.

In 1977, the United States Soil and Water Resources Conservation Act called for a detailed survey of soils in the United States. According to the report, approximately one-third of the natural topsoil in North America has already been lost since the start of agriculture on the continent. At the time of the study, erosion was continuing in the United States at an average rate of about 10.5 tonnes per hectare per year. However, in some regions the figure was considerably higher, as shown in Figure 20.4. Average yearly losses of 31 tonnes per hectare were reported in parts of Tennessee, 25 tonnes per hectare in New Jersey, 33 tonnes per hectare in Texas, and 22 tonnes per hectare in Iowa. A little arithmetic puts these numbers in perspective.

A 1 cm layer of soil weighs approximately 200 tonnes per hectare.

In many agricultural areas, topsoil is about 30 cm deep.

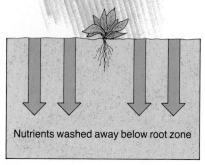

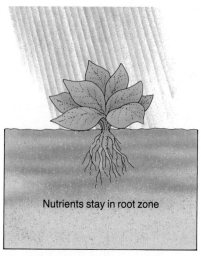

(a)   Soil with low humus content

(b)   Soil with high humus content

FIGURE 20.3 (a) In soil without humus, soil nutrients are often washed away below the root zone. (b) Humus retains nutrients and makes them available to plants.

At an erosion rate of 30 tonnes per hectare per year, the net annual loss is only about 25 tonnes per hectare because on the average 5 tonnes are replaced by natural processes. This loss amounts to approximately 1/8 cm of soil every year, which would lead to complete eradication of the topsoil in about 240 years.

On a national average, soil erosion in the United States destroys an equivalent of 1.2 million hectares (3 million acres) of prime farmland every year. (This number was derived by calculating the partial soil loss over all the cropland in the country and mathematically concentrating it to a total loss over a smaller area.) This loss amounts to the destruction of the agricultural capacity of an area the size of the state of Connecticut every year. Another study, conducted in 1982, showed that soil loss was greater than soil formation on 44 percent of the cropland in the United States.

In many other regions of the world, erosion rates are unacceptably high as well. For example, in India erosion rates are greater than the rate of soil formation on 60 percent of the farmland, and the average quantity of soil lost per hectare per year is twice the U.S. average. In the Soviet Union, an estimated half million hectares of cropland are abandoned each year because they have been eroded so severely that they are no longer profitable to farm. One Soviet journalist reported that "two-thirds of the plowed land in the Soviet Union has been subjected to the influence of vari-

FIGURE 20.4 Fertile topsoil being washed away in an untended field in eastern Missouri.
(From *Contemporary Physical Geology*, third edition, Levin, Saunders College Publishing)

A view of the ramparts along the McKenzie River. The high, white cliff is made of limestone. The thin dark line beneath the trees is the soil that supports the forest above it. Note how thin the soil zone is. Yet, if such zones are destroyed, life on Earth as we know it could not survive.

ous forms of erosion." Erosion is a serious problem in China, too. Rivers are becoming silted with soil that once was part of fertile fields. Wind erosion poses yet another danger. Scientists in Hawaii can tell when plowing has begun in China, because the concentration of certain characteristic soil particles in the atmosphere increases dramatically. Most other nations in the world are also threatened by alarming rates of soil erosion.

People living in urban centers do not see the immediate effect of soil erosion, and the problem is much less obvious than other environmental disruptions such as air and water pollution. But the soil is the basic resource needed to produce food.

Soil erosion can be prevented by simple measures. First, land should not be left unplanted for long, since plant roots hold the soil in place. Terracing and contour plowing prevent the washing of soil down slopes, and planting windbreaks checks erosion by high winds, as shown in Figure 20.5. Also, since organic debris can bind a great deal of water, soil with a high content of organic matter is less likely to be washed away; thus, the use of manure instead of chemical fertilizers is a considerable deterrent to erosion.

The soil erosion problem is largely economic. An individual farmer can profit more in a single year by planting crops instead of rows of windbreak trees, by using inorganic rather than organic fertilizers, or even by leaving land without plant cover for a period. On the other hand, the effects of soil erosion may not be felt for several years or even until the next generation. Often, it is tempting to respond to short-term economic

During the 1920s and 1930s, windblown dust had a devastating effect on agriculture in the United States. This photograph, entitled "Buried machinery," was taken in South Dakota on May 13, 1936.
(Reprinted with permission of the National Archives)

(a)

(b)

FIGURE 20.5 Wind erosion—the Dust Bowl—and part of the cure. (a) When the wind stopped, this road in Idaho was covered with soil, which in places was 65 cm deep. (b) Windbreaks of willow prevent soil erosion on this farm in Michigan.
(Courtesy of USDA Soil Conservation Service)

needs rather than to conserve for the future. Although governments in many countries are waking up to the disastrous effects of continued soil erosion, they face the difficult task of designing economic incentives that will encourage farmers to conserve soil.

20.4
LOSS OF FARMLAND TO URBANIZATION

A person living in a rural community can raise enough vegetables on 0.07 hectare of land to support a family of four. In contrast, the average urban American requires almost twice as much land per person (0.13 hectare) for nonagricultural purposes—for homes, lawns, roadways, parking lots, shopping centers, and factories. Much of this urban development has occurred in regions that were once prime farming areas. The loss of farmland to urban sprawl has become a serious concern in recent years.

In the United States, a total of 1.2 million hectares of land is paved with concrete and asphalt or replanted with ornamental lawns every year. This is equivalent to an area nearly the size of the state of Connecticut. Unfortunately, the loss is concentrated in many of the most productive agricultural areas. If current trends continue, large quantities of fertile farmland will be urbanized by the year 2000. For example, unless our priorities change, most of Florida's citrus groves, 16 percent of the vegetable-producing regions of Southern California, and 24 percent of the prime agricultural land in Virginia will be removed from production by that time.

The same trend is occurring in the less developed countries, although at a less rapid rate. Although many people are urbanized in these countries, they use less space per person because many live in crowded slums, and elaborate roadways and shopping centers are practically nonexistent. Yet estimates show that if present trends continue, the cropland that will be urbanized in these areas between 1980 and the year 2000, represents the agricultural capacity to feed 84 million people.

Is there any way this trend can be reversed? In a well-planned society, the rate of change could certainly be reduced. Two examples illustrate the point.

Suburban subdivisions are characterized by single story, one-family dwellings on small plots of land. Suburban shopping centers are often single story, sprawling stores and malls. If these were replaced with multistory apartments and department stores, a considerable area of land could be saved.

Transportation systems are similarly inefficient. A two-track local subway uses a roadbed 11 m wide and can carry 80,000 passengers per hour. On the other hand, an eight-lane superhighway is 38 m wide and carries only 20,000 people per hour under normal traffic conditions. A superhighway capable of carrying 80,000 people per hour would have to be 152 m wide (approximately 1.5 times as wide as the *length* of a football field). Therefore, a shift to mass transit would conserve land surfaces.

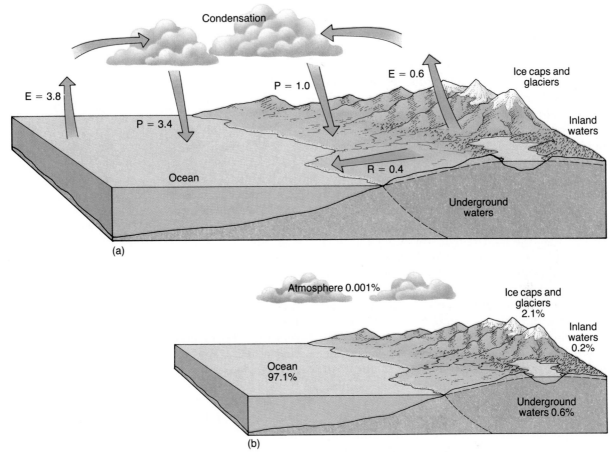

FIGURE 20.6 (a) The natural water cycle. E = evaporation; P = precipitation; R = runoff. Numbers are in units of 10^{20} grams. 10^{20} grams is about 100,000 cubic kilometers of water. (b) Percentages of total global water in different portions of the Earth.

WATER

20.5
THE HYDROLOGIC CYCLE (WATER CYCLE)

If you travel widely on land, by sea, and in the air and look at the Earth's waters, three observations become apparent. First, much of the water is *stored* in places that look rather permanent. The largest quantities, of course, are in the oceans. But there are also the Greenland and Antarctic Ice Caps, as well as many lakes and glaciers. Second, much of the Earth's water is in motion: Snow and rain fall, clouds drift, and rivers flow toward the sea. Third, the water on land is very unevenly distributed. As you wander through tropical jungles, everything is wet, and water often drips on you throughout the day. But you had better not try to trek across Australia, Libya, or even southern California without taking all your water with you. These regions get very little rainfall; most of the year they get none.

The movement of water on Earth is called the **hydrologic cycle.** It is convenient to break down the different methods of water transport into three simple categories—evaporation, precipitation, and runoff, as shown in Figure 20.6a.

Evaporation, or **vaporization,** is the transformation of liquid water to water vapor. Dissolved minerals remain behind when water evaporates. Most water vapor is produced by evaporation of liquid water from the surface of the oceans. Water can also vaporize *through* the tissues of plants, especially from leaf surfaces. This process is called **transpiration. Precipitation** means falling from a height. Referring to water, precipitation includes all forms in which atmospheric moisture descends to earth: rain, snow, hail, and sleet. The water that enters the atmosphere by vaporization must first condense into liquid (clouds and rain) or solid (snow, hail, and sleet) before it can fall. **Runoff** is the flow back to the oceans of the precipitation that falls on land. In this way, the land returns the water that was carried to it by clouds that drifted in from the ocean. Runoff occurs both from the land surface (rivers) and from underground water.

Now refer to Figure 20.6b, which is the same diagram with different numbers. Here the numbers tell you where the waters of the Earth *are* at any given time, not where they are going. The numbers are given in percentages of the total quantity of global water, which is about 1.35 billion km^3.

Note that of this vast amount, only 0.8 percent is in the form of inland and underground waters, and that most of this quantity (0.6 percent) is underground. The remaining 0.2 percent is the inland surface water such as lakes and streams. The least amount of water is in the atmosphere (0.001 percent). Since all this water is in motion, it follows that all the waters of the Earth renew themselves; that is, they move from place to place.

An important question is: How long does it take for water in a given part of the Earth to renew itself? Consider, for example, two flows of pure water, one that goes into a small basin and the other into a large one, as shown in Figure 20.7. Both basins are well stirred, so that the water in each is always uniform throughout. If the water in both basins is polluted, the time it takes for the fresh water to rinse out the pollutant depends on the flow rate of the water and the volume of the basin. The greater the flow rate and the smaller the basin, the faster is the rinsing action. The *average* time that a water molecule spends in the basin is called the **residence time.**

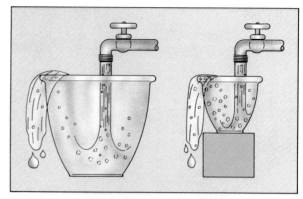

FIGURE 20.7 The same rate of water flow rinses pollutants out of a small basin faster than out of a large one.

Some molecules, however, remain for a longer or shorter time than the average. Any particular water molecule in a natural basin, such as Lake Erie, may be lost by evaporation or runoff the very next day or may still be there 1 year or 100 years later. Table 20–1 gives average residence times for water in various parts of the hydrologic cycle.

Note that water spends the least time in the atmosphere and the longest time in the deepest ocean layers. Changes in global energy patterns can therefore readily affect atmospheric moisture and hence rainfall and agricultural productivity. The fresh water available for human use is the runoff from rivers and underground sources. Rivers renew themselves rapidly (in weeks), but groundwater takes much longer (hundreds to thousands of years). Pollution of waters with long residence times is not easily reversed.

20.6 HUMAN USE OF WATER

Water is used in the home or office, in industry, in agriculture, and for recreation. Both the quantities used and the water quality needed vary widely, depending on the application. In the home, for example, the amount of water used in one toilet flush would satisfy the drinking requirements of an adult (1 liter per day) for about 3 weeks; the water used for one load of laundry in a clothes washer would be enough for drinking for almost 6 months. The amounts used in indus-

TABLE 20–1 Average Residence Times of Water Resources

Location	Average Residence Time
Atmosphere	9–10 days
Ocean	
Shallow layers	100–150 years
Deepest layers	30,000–40,000 years
World ocean average	3000 years
Continents	
Rivers	2–3 weeks
Lakes	10–100 years
Ice caps and glaciers	10,000–15,000 years
Shallow groundwater	up to 100s of years
Deep groundwater	up to 1000s of years

try and agriculture are far greater than those needed for any personal use. For example, the water used in industry to refine a tonne of petroleum would be enough to do about 200 loads in a clothes washer. When crops are irrigated, it takes much more water to grow a tonne of grain than it does to manufacture a tonne of most industrial materials such as metals or plastics.

The strictest requirements for quality apply to drinking water for humans. The least strict requirements probably apply to water used for cooling, where the prime concern is its temperature. Seawater is therefore adequate. For some industrial applications, the most important consideration is whether the water will corrode the equipment; control of acidity is often the only requirement in such cases. For most human needs, however, including the large amounts used in agriculture and industry, water must be fresh, not salty. Figure 20.8 shows the pattern of consumption of fresh water in the United States. Note that the highest water consumption does not occur in the most densely populated areas. In general, rates of water consumption reflect the needs of agriculture much more than those of the home, of commerce, or of industry. One conclusion from this difference is that efforts to conserve water in the home, while locally helpful, cannot make a significant contribution to the demands of agriculture. A comparison of Figure 20.9 with Figure 20.8 shows that some of the areas where the surface water supply is often depleted are also the areas where the largest

quantities of fresh water are needed. Therefore, if the only fresh water available were the water on the Earth's surface in lakes and rivers, it would hardly be enough. Even in years of average rainfall, much of the Midwest and Southwest of the United States depletes most of its surface waters. Therefore, people transport water large distances or use groundwater.

20.7
WATER DIVERSION PROBLEMS

One obvious solution to the problem of local water shortages is to divert water from an abundant source to a dry region where the water is needed. Southern California offers examples of this type of water diversion. The All American Canal channels water from the Colorado River to the farms and cattle ranches of the Imperial Valley, east of San Diego. The great Los Angeles Aqueduct, completed in 1913, brings water south from Owens Valley to Los Angeles. More recently, several large diversion projects have been considered. Perhaps the most ambitious of these is a proposal to divert water from the southern portions of Hudson's Bay, in northeast Canada, to the midwestern regions of the United States. Such large water diversion projects can create a number of problems at both ends—at the source and at the area to which the water is supplied.

1. Encouragement of Waste. Planners often underestimate the ability and willingness of people to conserve when it is necessary to do so. (The unexpected decline in energy use in the early 1980s is one such example.) Most people use water wastefully when they know that it is abundant, but when it is scarce, conservation is not seen as a serious burden. Thus, it is convenient to leave the water running while you are washing your hands or brushing your teeth. But if you must, you can use only about one-tenth as much water in a stoppered sink to accomplish the same purpose in about the same time. There are many other ways to conserve water in the home, on the farm, and in water-distribution systems. In many cases, these measures would be an adequate or at least a partial substitute for water diversion projects.

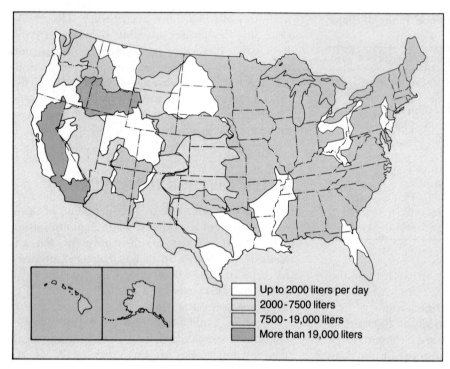

FIGURE 20.8 Per capita fresh water consumption in the United States. (1 gallon = 3.785 L) This includes all types of consumption, household, industrial, and agricultural.

Up to 2000 liters per day
2000-7500 liters
7500-19,000 liters
More than 19,000 liters

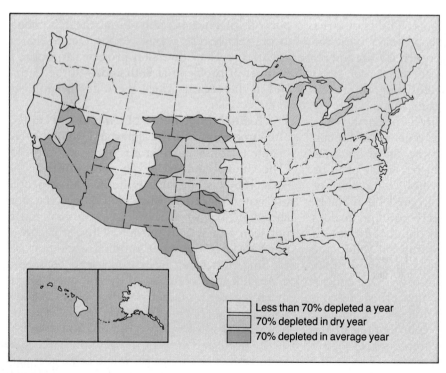

FIGURE 20.9 Fresh water consumption compared with water supply in the United States.

Less than 70% depleted a year
70% depleted in dry year
70% depleted in average year

2. Salinity. Irrigation is almost as old as agriculture itself. The ancient Egyptians, Babylonians, Chinese, and Incas all brought water from nearby rivers to increase the yields of their crops. Today, a large portion of the world's crops of vegetables and some grains depends on irrigation. With imported water, marginal farmland has become more productive, and even former deserts are being farmed. Despite these successes, irrigation leads to some environmental problems.

When rainwater falls on mountainsides, it collects in small streams and in groundwater. As it flows downward, it filters over, under, and through rock formations. The water dissolves various mineral salts present in the rock and soil. Therefore, river water is slightly salty. In most cases, you can't taste the salt, but it is there. If this water is used for irrigation, farmers are bringing slightly salty water to their fields. When water evaporates, the salt remains. Thus, over the years, the salt content of the soil increases slowly. Because most plants cannot grow in salty soil, the fertility of the land decreases. In Pakistan, an increase in salinity (saltiness) decreased soil fertility alarmingly after 100 years of irrigation. In parts of what is now the Syrian desert, archaeologists have uncovered ruins of rich farming cultures. However, the land lying near the ancient irrigation canals is now too salty to support plant growth. California farms now produce about 40 percent of the vegetables consumed in the United States. Here, too, salinity is threatening productivity. As a result, elaborate drainage systems have been built to draw off salty water. These measures alleviate the problem but do not eliminate it entirely.

3. Energy Consumption. The great aqueducts of ancient Rome are sloped downward; there were no pumps. Ancient Egyptian farmers pumped the waters of the Nile a meter or so up to their farms by human or animal power; many do the same to this day, as shown in Figure 20.10. Modern water diversion projects, however, carrying water over hilly terrain, use electrically driven pumps. Diversion projects involve *large* quantities of water (hundreds of cubic meters per second) and therefore need much energy. One estimate for an expanded California State Water Project

FIGURE 20.10 Waters from the Nile even today are pumped up to farmland by human power, using a hand pump.

foresees a use of 10 billion kWh of electricity in the year 2000—about as much energy as is used in 2 million homes.

20.8 THE POLLUTION OF INLAND WATER BY NUTRIENTS

All animals, even those that live under water, require oxygen to survive. On the surface of the Earth, oxygen is readily available. Oxygen is also dissolved in most bodies of water, but in general the supply under water is much less plentiful than it is on land.

In a natural system, the growth of all organisms is controlled by the quantity of the nutrients available. The entire system is delicately balanced. If the supply of any essential ingredients is increased or decreased, the entire system may be upset. Imagine that there is a clean, cool, flowing river that supports a healthy population of trout and salmon. The fish share the stream with populations of microorganisms, larger plants, small aquatic worms, and many other types of organisms. Now suppose that someone dumps some sewage into this stream. The sewage provides

FIGURE 20.11   The effect of phosphorus on the productivity of a lake. This lake in Manitoba was divided in two by plastic sheeting across the narrow neck in the middle of the photograph. Phosphorus was added to the half of the lake in the upper part of the photograph. Several weeks later, the phosphorus-fertilized half of the lake was opaque as a result of massive plankton bloom; the lower part of the lake was as clear as it was before the experiment. (Photograph courtesy of David Schindler)

nutrients for plants and animals. Microorganisms and algae grow faster, as shown in Figure 20.11. The fish that eat these organisms are nourished as well. Yet the introduction of sewage may lead to severe ecological disruptions. In general, populations of smaller animals and plants grow and reproduce faster than populations of larger animals. All these animals consume large quantities of oxygen. If the growth is rapid enough, most of the oxygen in the water is used up, and fish may eventually suffocate and die.

It is important to understand that the sewage, by itself, does not kill fish. In fact, it nourishes them. But the sewage supports the growth of other forms of life that consume the oxygen. It is the lack of oxygen that kills fish. When so many

nutrients have been added to a body of water that the fish die, the resulting condition is called **eutrophication.** Many lakes, rivers, and bays throughout the world are polluted in this manner.

Sewage is not the only material that can fertilize lakes and rivers and lead to eutrophication. Sometimes fertilizers that have been spread on farmers' fields wash into waterways. Fertilizers promote plant growth on land and in the water. If they are applied to a cabbage field, the cabbage grows well. If they dissolve in rainwater and flow into aquatic systems, algae and other aquatic plants and the organisms that feed on them grow well, too. These growing organisms consume, and may eventually deplete, the available oxygen. Then, as mentioned above, fish die.

In recent years, there has been considerable public discussion of laundry detergents and their role in water pollution. Many modern detergents contain phosphates, which are an essential component of agricultural fertilizers. More importantly, phosphates are frequently in short supply in natural waters. Without this nutrient, aquatic plants cannot grow in abundance. When phosphate detergents are discharged into waterways, they supply a needed nutrient for the growth of plants. Sometimes, the results far exceed unsophisticated expectations. In many areas of the world, aquatic weeds have multiplied explosively, as shown in Figure 20.12. They have interfered with fishing, navigation, irrigation, and the production of hydroelectric power. They have brought disease and starvation to communities that depended on these bodies of water. Water hyacinth in the Congo, Nile, and Mississippi rivers, the water fern in southern Africa, and water lettuce in Ghana are a few examples of such catastrophic infestations. People have always loved the water's edge. To destroy the quality of these limited areas of the Earth is to detract from our humanity as well as from the resources that sustain us.

20.9
INDUSTRIAL WASTES IN WATER

Many large American cities obtain their drinking water from nearby rivers. A problem arises be-

(a)

(b)

FIGURE 20.12 The choking of waters by weeds. (a) The dam on the White Nile at Jebel Aulia near Khartoum, Sudan. The area was clean when photographed in October 1958. (b) The same area in October 1965, showing the accumulation of water hyacinth above the dam.
(From Holm: "Aquatic Weeds." *Science,* 166:699–709, November 7, 1969. © 1969 by the American Association for the Advancement of Science.)

cause industries find it convenient to dump their wastes in the same waterways. In the United States and most other developed countries, there are legal requirements for cleanup of such wastes before they leave the factory boundaries. Most large factories and chemical plants include some sort of water purification system as part of their manufacturing operations. But the law permits the discharge of some wastes, and the purification process is never 100 percent complete. Therefore, "purified" industrial wastewater cannot be assumed to be fit to drink. As a result, the major rivers of the industrialized countries, such as the Mississippi, the Rhine, and the Volga, contain small concentrations of thousands of different industrial chemicals. Cities and towns along the rivers draw this water and purify it again before it is piped to homes for drinking. This purification process, however, is also incomplete; sometimes the major objective is only to kill microorganisms and to make the water look clear. Trace concentrations of heavy metals, pesticides, and industrial organic chemicals—some of them suspected carcinogens—are therefore found in the drinking water of many cities that are located near major rivers.

Certainly, goods can be manufactured more cheaply if industrial wastes are discharged directly into the environment. But then everyone suffers from the effects of unhealthy water or unclean air. Therefore, it is a legitimate role of government to regulate waste disposal practices; the arguments arise when trying to decide *how much* regulation is desirable. Pollution control is not a yes or no, on or off affair. Imagine that a chemical factory with no pollution control devices releases a certain quantity of wastes into the water every month. Equipment can be designed to remove any portion of the pollutant, from a minor amount to practically all of it. In general, the more pollution that is removed, the more expensive the process becomes. Limited pollution control can be relatively inexpensive, but an essentially pollution-free environment is very costly.

How much pollution control should we pay for? Some people suggest that pollution control measures should be applied only when it can be shown that there is a positive economic return on

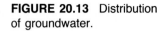

FIGURE 20.13 Distribution of groundwater.

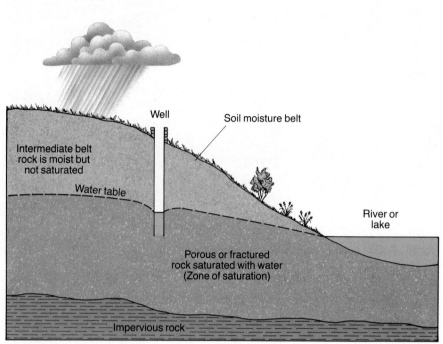

the investment. Opponents of this argument claim that money is not the only measurement of the quality of life. How can you place a dollar value on the annoyance of a vile odor or of unclean water? What about recreational opportunities? How much is it worth to be able to float quietly down a river and fly-fish for trout? Going beyond annoyance, how can you measure the dollar value of human suffering and misery caused by illness, or the value of a human life that ends too soon? Many people feel that such costs are beyond our right to judge.

20.10
GROUNDWATER

When rain falls on the land, two forces act on it. One is the Earth's gravity, which pulls the water downward through any open path. The second force is the attraction between individual molecules of water and the molecules of other materials to which water tends to stick. Thus, drops of rainwater stick to the vertical side of a window without falling. Water can also be pulled into tiny

holes. If a corner of a paper towel is placed in a dish of water, the liquid travels upward, against the force of gravity. In this case, the electrical force attracting water molecules to paper molecules is stronger than the gravitational force pulling the water downward. The movement of water upward through small holes is called **capillary action.**

Now, consider what happens when rain falls on dry soil. The first raindrops simply wet the soil; they do not flow down or away. As the rain continues after all the land surface is wet, gravity pulls the excess water down through the spaces between the soil particles to the rock, sand, or gravel that lies beneath the soil. Eventually, the downward flow is stopped when the water meets rock that is impermeable. Since the water can go no farther, it backs up, filling all the pores in the rock above the barrier. This completely wet section is called the **zone of saturation,** as shown in Figure 20.13. The upper boundary of the zone of saturation is called the **water table.** Below the water table, the ground is saturated; above it, the ground may be moist but is not saturated.

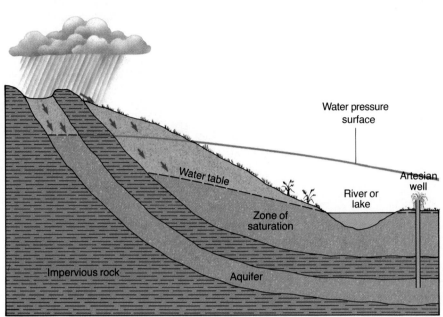

FIGURE 20.14 An aquifer is a layer of water-bearing rock. An artesian aquifer forms when permeable rock is sandwiched between two layers of impervious rock and the rock beds are sloped as shown in this drawing. Water will flow to the surface of the well without being pumped. This is called an artesian well. Water that is trapped for long periods of time in an aquifer is called "fossil water."

Once underground, the water moves at widely varying rates, depending on a variety of geological conditions. At one extreme, moisture can flow fairly rapidly through subsurface voids or caverns, much like an underground river. However, underground caverns are rare. By far the largest proportion of underground water moves slowly through pores in rock. The total quantity of slow-moving groundwater is very large. Note, however, from Table 20–1 that these waters are replaced *very* slowly (in up to thousands of years). Much of the water in some underground reservoirs was accumulated many centuries ago in wetter climates than the present one. Under such conditions, deep groundwater may be considered to be, for all practical purposes, nonrenewable. Just as coal and petroleum are called fossil fuels, so is deep groundwater sometimes called "fossil" water. The removal of deep groundwater is therefore analogous to mining.

An **aquifer** is a body of rock that is porous and permeable enough to yield economically significant quantities of water. If an aquifer is trapped between two layers of impervious rock, as shown in Figure 20.14, then an artesian aqui-

fer forms. Water in an aquifer can move horizontally, but its vertical movement is limited.

The Ogallala aquifer of the American Midwest is one of the world's largest reservoirs of fresh groundwater. In recent years, farmers in the Midwest have found that it is profitable to "mine" this water and use it to irrigate field crops. As a result, more water is taken out of the Ogallala aquifer than is replaced by rainfall. It is estimated that some of the Ogallala groundwater levels are being lowered at rates of several centimeters to about half a meter a year. At such rates, serious depletion of the aquifers can occur early in the next century, as shown in Figure 20.15. It is important to understand that the viability of an agricultural-industrial-urbanized society that depends on fossil water can be at risk when that source is "seriously depleted."

Two other problems besides depletion can arise as a result of the excessive removal of groundwater. One of these problems is **subsidence,** or settling, of the ground as deep groundwater is removed. This removal allows the rock particles to shift somewhat closer to each other, filling some of the space left by the departed water. As a result, the volume of the entire rock

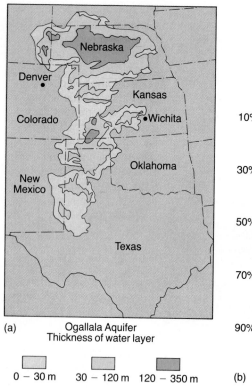

(a) Ogallala Aquifer
Thickness of water layer

0 – 30 m   30 – 120 m   120 – 350 m

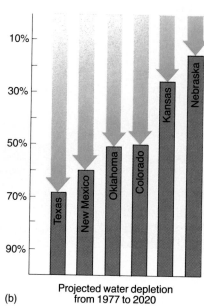

(b) Projected water depletion
from 1977 to 2020

FIGURE 20.15 (a) Location of the Ogallala Aquifer. (b) Projected water depletion in various sections of the Ogallala Aquifer, from 1977 to 2020. (Numbers are given in percent of original supply.)

layer decreases, and the surface of the ground subsides, as shown in Figure 20.16. (Removal of oil from oil wells has the same effect.) Subsidence rates can reach 5 to 10 cm per year, depending on the rate of water removal. These effects have been observed in such areas as the San Joaquin Valley of California, Houston (Texas), and Mexico City. Unfortunately, subsidence is not a readily reversible process. In many cases, the pores in the underground rock are squeezed shut by the weight of surface rock and soil. As a result, the water-holding capacity of a depleted aquifer may be permanently reduced so that it cannot be completely recharged even when water becomes abundant again.

The other problem is **saltwater intrusion,** as shown in Figure 20.17. As groundwater is removed from a coastal area, the zone of fresh water saturation is reduced both from above and below. From above, the water table declines. From below, saltwater seeps in. As a result, saltwater may be drawn into wells, making the water unfit for drinking.

20.11 POLLUTION OF GROUNDWATER

In recent years, the pollution of groundwater in some areas has become a serious problem. Groundwater reserves can be polluted either by sewage residues or by industrial chemicals. The problem is particularly serious for three reasons: (1) Once polluted water has accumulated in deep layers of rock, there is little oxygen for decomposition. Thus, natural decay and cleansing do not operate efficiently. (2) It is very difficult to assess the full extent of the pollution of underground supplies or to purify the system artificially. (3) Many groundwater systems move very slowly, so the removal of pollutants by flushing out to sea could take decades or even centuries.

One short case history illustrates the type of problem that can affect a community. Long Island is a flat, sandy island located just east of New York City. Although the region was once predominantly rural and agricultural, heavy industrialization has occurred in recent years, espe-

FIGURE 20.16 An extreme example of subsidence. In some regions of Florida, large quantities of water were withdrawn from local aquifers during the late 1970s and early 1980s. Certain underground rock structures collapsed, drawing houses, cars, and commercial buildings into the gaping holes. (Courtesy of Wide World Photos)

cially on the western end. By 1970, groundwater reserves on the western side of the island were polluted, whereas supplies on the more rural eastern side were pure. Local city governments in these eastern regions fought to maintain their rural environment and hence—they thought—the high quality of the water. However, in 1979, traces of a pesticide named Temik were found in wells in eastern Long Island. Temik had been used since 1973 to combat various pests of potato crops, but the chemical filtered through the soil and collected in the groundwater. A study showed that more than 100 wells in the region were contaminated. One of the geologists working on the project reported, "Any pesticide they put down will get into the water supply. People

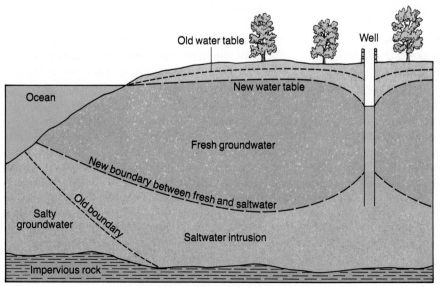

FIGURE 20.17 Saltwater intrusion.

FIGURE 20.18 Beaches along the area shown were polluted with large amounts of oil from the wrecked tanker *Amoco Cadiz*.

must realize that using chemicals on the surface of Long Island is a game of high risk in which you are playing with your community's drinking water."

20.12 POLLUTION OF THE OCEANS

For many years, people were relatively unconcerned about the pollution of the oceans, because the great mass of the sea can dilute a huge volume of foreign matter to the point that it has little effect. In recent years, this attitude has changed for several reasons. For one, pollutants are often added in such relatively high concentrations that environmental disruption does occur in a local area before the pollutants have a chance to disperse. In addition, the quantity of pollutants dumped into the ocean has grown so large that some scientists fear that global effects may be significant.

Oil

On March 16, 1978, the oil tanker *Amoco Cadiz* lost steering control off the coast of Brittany in France. High winds blew the ship ashore, and within the next few days it broke apart, spilling approximately 1.6 million barrels (220,000 tonnes) of crude oil onto the beaches and into the water, as shown in Figure 20.18. Brittany had long been a vacation area, and tourists avoided the area for years after the spill. Also, over a million sea birds were killed within a few days of the disaster. The oil clogged their feathers and respiratory tracts so that they drowned or died from inhaling the oil. A huge oyster fishery was destroyed. Plankton, fish, and other sea animals were also killed.

Between 1969 and 1985, there were more than 500 tanker accidents that involved oil spills. Altogether, more than 1 million tonnes of oil were released. In 1976 alone, five major accidents off the coast of the United States combined to spill 35,000 tonnes of oil. But accidents are not the only source of oil in the sea. Tanker captains often clean the oily holds of their ships by washing them out with seawater, despite the fact that this is strictly illegal in many parts of the world. It has been estimated that up to 90 percent of the oil in the oceans comes from these small discharges.

On March 24, 1989, the worst oil spill in U.S. history occurred when the 987 ft tanker *Exxon Valdez* struck a reef in Alaska's Prince William Sound. More than 250,000 barrels of crude oil were spilled, resulting in devastation along 14 miles of shoreline. Marine life, such as otters and sea birds, suffered high losses. The tragedy was compounded when, one week later, the *Valdez* left an 18-mile-long oil slick off the coast of San Diego while being towed in for repairs. Because of the enormity of this disaster, let us briefly recount the details, as they are currently known.

At 9 PM, the fully loaded *Valdez* set sail for California, one hour ahead of schedule. At 11:15 the captain radioed the Coast Guard that he was moving the vessel from the outbound shipping lane to the inbound lane in order to avoid ice. At this time, control of the ship was turned over to the second mate, who was told by the captain to move back into the outbound lane after the ship passed a navigational point near Busby Island, an island near the mouth of Prince William Sound. This island is three miles north of Bligh Reef, where the accident occurred. At 11:55, the ship's log indicates that the turning point had been

Oil-soaked gannet (a gull-like bird), Jones Beach, Long Island.
(Komorowski, from National Audubon Society)

reached, but the ship's course recorder indicates that the turn did not actually begin until 7 minutes later. Shortly thereafter, a lookout reported to the pilothouse that a flashing red buoy near Bligh Reef had been spotted on the right side of the ship when it should have been on the left. For a reason yet to be determined, the ship was not responding well to the turn. The captain was called back to the bridge with the information that the ship was in trouble. Moments later the ship hit the rocks, and the damage that was produced was devastating.

Offshore drilling operations also contribute their share of oil to the sea. The largest spill from this source started on June 3, 1979, at an offshore oil well owned by Pemex, the Mexican national oil company. Mud, rapidly followed by oil and gas, started to gush through an unsealed drip pipe. The fumes ignited on contact with the pump motors, the drilling tower collapsed, and the spill was out of control. It took 9 months to stop it, by which time 3.1 million barrels (440,000 tonnes) of oil had been spilled into the Gulf of Mexico.

A large quantity of oil finds its way into the sea every year. It has been estimated that about 1 million tonnes of oil are spilled into the ocean each year from ships and oil drilling operations alone. But there are also many "mini-spills," such as sludges from automobile crankcases that are dumped into sewers, routine oil-handling losses at seaports, leaks from pipes, and the like. Some oil aerosols also settle into the sea from the atmosphere. The grand total from all these sources is difficult to estimate, but it could well reach 10 million tonnes or more per year.

Other chemical wastes

There is no known inexpensive and guaranteed safe method of disposing of highly poisonous chemical wastes, such as by-products from chemical manufacturing, chemical warfare agents, and pesticide residues. It is cheap and therefore tempting to seal such material in a drum and dump it into the sea. But drums rust, and outbound freighters do not always wait to unload until they reach the waters above the sea's depths. As a result, many such drums are found in the fisheries on continental shelves or are even washed ashore. It is estimated that tens of thousands of such drums have been dropped into the sea.

Of course, all the river pollutants enter the same sink, the world ocean. The organic nutrients are recycled in the aqueous food web, but the chemical wastes from factories and the seepages from mines are carried by the streams and rivers of the world into the sea. And where do the air pollutants go? Airborne lead and other metals from automobile exhaust, mercury vapor, and agricultural sprays dissolve in rainwater or become attached to dust particles and ride the winds. Many of these are eventually deposited into the oceans.

Is there an overall threat to life in the sea?

In ocean regions near large cities such an New York, pollution has killed most marine life in wide areas. These places have come to be known as "dead seas." Is it possible that the entire ocean may die?

These concerns, expressed in 1972 at the United Nations Conference on the Human Environment in Stockholm, led to a study called "The Health of the Oceans" that was released in October 1982. The findings of the study were optimistic. The world's oceans actually seemed healthier in 1982 than they were in 1972. Some of this improvement is a result of environmental laws that now restrict the production and distribution of many toxic substances, such as pesticides and harmful metals, in the most industrialized countries. In addition, various natural biological and chemical processes serve to assimilate or degrade oil spills and other toxic materials enough to render them harmless. These conclusions apply specifically to the open oceans, hundreds of kilometers from any shoreline. Along the coasts,

many bays and coastlines are already severely polluted, aquatic systems have been disrupted, and populations of fish have declined drastically. This problem is particularly severe because most of the richest fisheries have been close to the shore or to river mouths. Thus, areas like Chesapeake Bay, on the east coast of the United States, were extremely productive only a few decades ago, but today their waters are so severely polluted that the populations of fish have declined dramatically.

MINERALS

20.13 NONRENEWABLE MINERAL RESOURCES

Living organisms use a source of energy (ultimately, the Sun) to convert materials from the environment (nutrients) into body tissues. The time scale of these conversions is the time scale of the spans of life—months for plant fibers such as cotton, years for animal material such as bone and hide, and decades for wood. On the time scale of human lives, therefore, these materials are **renewable;** if they are not consumed faster than they are produced, they need never be exhausted. Geological processes, like those of life, can also organize and concentrate materials, but here the time spans extend to millions and billions of years. Since humans cannot wait that long, mineral resources are said to be **nonrenewable.**

An **ore** is considered to be a rock mixture that contains enough valuable minerals to be mined profitably with currently available technology. The **mineral reserves** of a region are defined as the estimated supply of ore in the ground. Reserves are depleted when they are dug up, but our reserve supply may be increased by either of two circumstances. First, new reserves may be discovered. Second, the value of a known deposit may change. For example, many known deposits are not being mined because it would not be profitable to do so under the current economic climate. If technology improves so that the materials can be refined cheaply, or if the market price of the ore increases, the deposit will suddenly become an ore reserve.

Many of the very high grade, concentrated, and easily accessible ores, such as the 50 percent iron deposit of the Mesabi Range in Minnesota, are being used up rapidly and either have been or will be depleted in the near future. These mines are essentially nonrenewable. Once they are gone, our civilization will have suffered an irreplaceable loss. But our technological life will not end with the exhaustion of these rich reserves, because less concentrated deposits are still available. In some situations, the less concentrated ores are more plentiful than the concentrated ones are. Returning to our example of iron, in 1966 it was estimated* that the global resource reserve of iron was about 5 billion tonnes. At that time, the global annual consumption rate was about 280 million tonnes. If these figures were correct, and if consumption continued at a constant rate, then the iron reserve of the Earth would have been consumed in 18 years (5 billion/280 million), bringing the end of iron reserves to 1966 + 18 years, or 1984. There must be something wrong with such calculations. Thus, we see again that reserves are not a constant factor but can change markedly with exploration and with the development of methods suitable for processing ores of lower grade. In the case of iron, the big change has been the development of improved methods of processing taconite rock for its iron content.

Would it be reasonable to expect similar dramatic improvements in the technology of mining other materials? If so, continued progress would make it profitable to mine less and less concentrated sources, and then we need never run out of anything. Such an approach ignores some serious difficulties. As discussed in more detail in the paragraphs below, as lower-grade ores are mined, more energy is needed to produce a tonne of product. Imagine for a moment that you were to mine iron, for example, in your back yard. Iron could undoubtedly be found in the rock and soil around your house. The problem is one of concentration. For many metals, the concentrations in ordinary rock may be as low as one ten-thousandth of those in commercial ores. The

energy consumption required to mine this rock and the pollution resulting from such a mine would be prohibitive.

The question remains: Are we, or are we not, in danger of running out of mineral resources? There is no clear-cut answer to this question; instead, there are two opposing opinions. One view holds that as rich deposits are depleted and people must mine increasingly lower-grade reserves, the cost and the total environmental consequences of mining will become so great that many minerals will become prohibitively expensive.

Others disagree. They claim that as the richest mines become depleted, three factors will act together to ensure that acute shortages do not develop. (1) Mining technology will continue to improve and new reserves will be found. (2) As minerals become more scarce, recycling will automatically become attractive, thereby extending the life of present reserves. (3) When certain minerals do become expensive, satisfactory substitutes will be found. Therefore, our technological existence will continue uninterrupted.

The argument: mineral reserves will be depleted in the near future

The ancient Greeks and Romans found copper ore under 10 cm or so of soil. Today, the depths of mines are measured in kilometers. It is impossible to go back to the days of picks and shovels. In fact, as increasingly lower-grade ores are sought, the technological problems inherent in all aspects of mining and refining rise sharply (Figure 20.19). Dependence on technology to solve all problems may lead to grave disappointment. In addition, the future availability of metals depends on many factors besides the quantity of ore in the ground and the state of refining technology. Some of these are discussed in the following paragraphs.

Availability of Energy. To extract metal from ore, the dirt and rock must be dug up and crushed, the ore itself must be separated and chemically reduced to the metal, and the metal must finally be refined to purify it. Each step, especially the chemical reduction, requires energy. Moreover, low-grade ores require much more energy to process than do high-grade ores.

---

*B. Mason: *Principles of Geochemistry.* 3rd ed. New York, John Wiley & Sons, 1966, Appendix III.

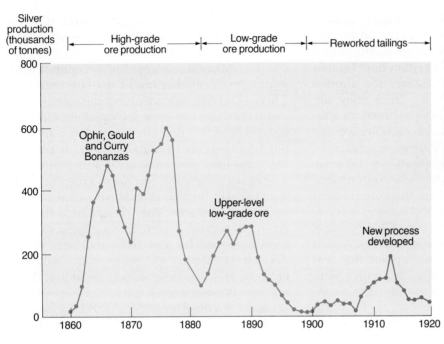

FIGURE 20.19 The depletion of the Comstock silver lode in Nevada. The first discoveries were very rich in silver. Later periods of mining followed, but the quantity of silver produced in each period was less than in the previous one. This pattern is followed in many other mining districts. The graph reminds us that mineral depletion is a real problem and that once rich mines are exhausted, a real loss has occurred.

Some low-grade ores differ from high-grade ores not only in concentration but in chemical composition as well. Some chemicals are easier to purify than others. Chemists are certainly able to purify these low-grade ores, but in many cases more energy is needed. Thus, the price and availability of many ores is linked to the price and availability of energy.

Pollution and Land Use. Most mining processes cause significant pollution of land, water, and air. For example, sulfur is found in large quantities in many ore deposits. This sulfur, chemically bound to metals in the Earth's crust, is brought to the surface when minerals are mined. Sulfur reacts with water in the presence of air to produce sulfuric acid, which runs off into the

Copper deposit being mined at Butte, Montana. Butte was once the copper capital of the West, but now all the mines have shut down because the quality of the remaining ore is too low to mine profitably.

streams below the mine. This pollution, known as **acid mine drainage,** kills fish and disrupts normal life cycles in streams, rivers, and lakes. When sulfur combines with other chemicals during the refining processes, it is often converted to gaseous air pollutants such as hydrogen sulfide and sulfur dioxide. These compounds, in turn, react in the atmosphere and fall to the ground as acid rain. Sulfur, of course, is not the only polluting chemical from mining operations. Many other mine pollutants cause serious air and water pollution.

Just as more energy is required to handle low-grade ores than high-grade ones, more pollution generally results from processing these impure materials. The pollution can be controlled with highly specialized pollution-abatement equipment, but such measures are expensive and add to the total cost of refining ore.

The world is running short of food, energy, and recreational areas, as well as high-grade mineral deposits. What should our policy be if a valuable ore or fuel lies under fertile farmland or a beautiful mountain? Which resource takes precedence? At present, this question is being raised principally with respect to exploitation of fuel reserves, for vast coal seams lie under the fertile wheat fields of North America. If large areas of low-grade ore must be exploited, the problem will extend to metal reserves as well.

The argument: mineral reserves will last for generations to come

Those who believe that our mineral reserves are not likely to be depleted in the foreseeable future point to past successes. They argue that pessimists predicted iron shortages by the early 1980s when, in fact, no such shortages occurred at that time. Moreover, this line of reasoning continues, there are several options in addition to mining conventional, land-based reserves.

Nonconventional Reserves. As conventional mines become depleted, new ones can be explored. What about mining the sea floor, or the Moon, or the planets? Various explorations have indicated that the mineral deposits on the sea floor are vast. Much of this material is concentrated in the form of round, flat, or odd-shaped pieces, typically weighing about a kilogram or so, that are rich in manganese. They are called **manganese nodules,** but they also contain copper, iron, nickel, aluminum, cobalt, and about 30 or 40 other metals. It is estimated that there are a trillion or more tonnes of these nodules on the sea floor. The sea floor does not have to be drilled or blasted, and explorations can be done with undersea television cameras. Furthermore, no one "owns" the sea floor, but the question of just how the mining rights are to be allocated is still unclear. Nonetheless, the technology of collection is complicated, and costs are expected to be high. The sea is not the easiest environment in which to operate complex machinery. Possible methods of collection include scoops, dredges, and vacuum devices, as shown in Figure 20.20. Various groups of corporations are already involved in exploration and in planning undersea mining operations.

In the early years of the space age, some enthusiasts wrote optimistic speculations about minerals from the Moon and the planets. These voices are hardly ever heard anymore. The expected costs would simply be enormous.

Conservation and Recycling. An alternative approach is to use less rather than to mine more. Consumption can be reduced by either conservation or recycling. Figure 20.21 shows the extent to which recycling in the United States could increase the life of iron reserves. The depletion of reserves is accounted for by the increase in stock plus waste, but the waste accounts for over 80 percent (210 kg/260 kg per year per person). This means that if all the discarded iron and steel were recycled, reserves could last four times as long.

Such objectives could also be realized by manufacturing products that are smaller and last longer or by taking better care of them. (It is much easier to keep on using a device that still works than it is to recycle it after it stops working.) Perhaps best of all is an improvement in technology or methods that results in extensive conservation of materials. A good example is the modern pocket calculator replacing the old slow, mechanical, office calculating machine, which weighed about 20 or 25 kg.

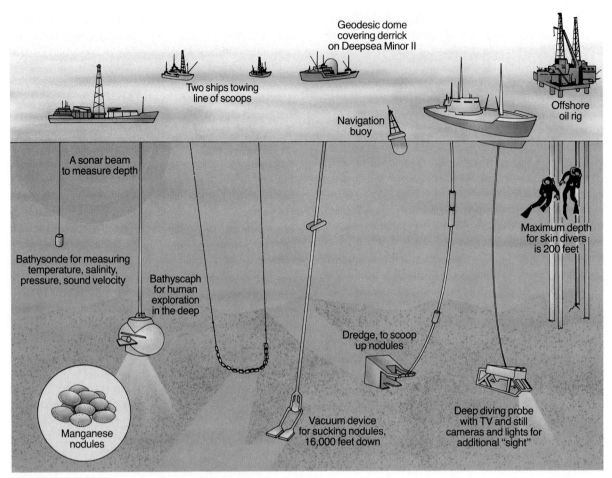

FIGURE 20.20 Mining the sea floor for manganese nodules.

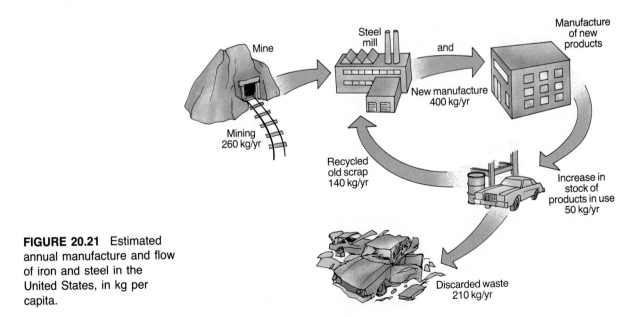

FIGURE 20.21 Estimated annual manufacture and flow of iron and steel in the United States, in kg per capita.

FIGURE 20.22 Concrete is often used in place of steel in the construction of highway bridges.

Substitutes. Finally, what about substitutes for at least some nonrenewable mineral resources? Why not go back to the simpler styles of using wood, stone, and fiber, or go forward to using plastics that can be synthesized from coal, air, and water? To some extent, both paths are being followed. The use of sand and gravel (for concrete) and of stone is increasing, as shown in Figure 20.22. Wood, too, is still a very desirable construction material. The use of many varieties of plastics is increasing much more rapidly than the use of metals. Improvements in the chemistry, as well as in the mechanical make-up, of plastics have made them competitive with metals in many applications, even where strength is an important factor.

OVERVIEW

Seven different factors must be estimated in order to predict the future availability of metals.

These are (1) the quantities, concentrations, and locations of minerals in the Earth's crust; (2) the availability of energy; (3) the effects of pollution from mining and refining; (4) land-use conflicts between mining and agriculture or urbanization; (5) future demands for metals; (6) future population levels; and (7) future rates of recycling. Since there are large differences in predictions of any *one* of these factors, it becomes nearly hopeless to predict how all seven will operate in concert. However, we can be sure of two things: The reserves of many important minerals are not inexhaustible, and once they are dispersed and discarded in old dumps where they are mixed with many other waste materials, they will be very expensive to recover.

SUMMARY

Soil is an intimate mixture of pulverized rock and organic debris. Soil erosion and loss of soil to urbanization are significant environmental problems.

Water leaves the atmosphere as rain or snow, moves on the Earth in the form of ice or liquid water, and returns to the atmosphere by evaporation. At any one time, only a small fraction of the Earth's water is in the atmosphere; most of it is in the oceans, which serve as the ultimate sink for water-borne impurities. Humans use the least amounts of water for drinking, but this water must be of the highest quality. The largest amounts are used for industry and agriculture. **Water-diversion projects** are used to bring water to regions where it is needed. However, these systems are expensive. In addition, they encourage waste, they may cause salinity, and some require large consumption of energy for pumping. Nutrients pollute water by promoting the rapid growth of organisms that then deplete the dissolved oxygen. Many water pollutants, such as heavy metals and organic chemicals, come from industrial sources. Some groundwater reserves are "fossil" and are not replaced quickly. **Groundwater** can be polluted by contamination from a variety of sources. Natural purification of groundwater is very slow because it is not readily diluted and does not have access to air. A major source of pollution in the oceans is petroleum from tankers. The ultimate effect of pollution on aquatic life is uncertain, although the immediate and local effects can be devastating.

An **ore** is a rock mixture that contains enough valuable mineral matter to be mined profitably. **Mineral reserves** are the estimated supplies of ore in the ground. This estimate can change with the discovery of new reserves, with improvements in extraction and refining, with new prices for minerals and for energy, and with increased pollution associated with the mining and processing of ores. Future prospects for metals cannot be predicted accurately, because they depend on changes in reserves, on conservation and recycling, on changes in technology, and on the use of substitute materials. However, once the prime ores are depleted, the poorer sources will be more expensive to mine, will require much more energy, and will cause much more pollution. One possible source of new mineral resources is the store of manganese nodules on the sea floor.

KEY WORDS

Soil	Transpiration	Capillary action	Salt water intrusion
Humus	Precipitation	Zone of saturation	Ore
Chelates	Runoff	Water table	Mineral reserve
Hydrologic cycle	Residence time	Aquifer	Manganese nodules
Evaporation	Eutrophication	Subsidence	Acid mine drainage
Vaporization	Groundwater		

QUESTIONS

SOIL

1. How does fertile soil differ from beach sand?
2. Would humus help prevent excess soil salinity caused by irrigation? Why or why not?
3. One of the advantages of chemical fertilizers is that they are so concentrated that only a small amount of material is needed to add the chemicals necessary for plant growth. Manure contains fewer nutrients per ton, so that fertilization with manure imposes much higher handling costs. What other arguments are important in deciding whether to use manure or chemical fertilizers on a farm? Discuss.
4. If a given land area is farmed for many years, does its productivity necessarily decrease? Justify your answer.

THE HYDROLOGIC CYCLE

5. Describe the various ways in which (a) a rise and (b) a fall in the average global temperature could affect the hydrologic cycle.
6. Many urban areas draw their drinking water from a nearby river. Much of this water is then returned to the river in the form of treated sewage. How does this practice affect the hydrologic cycle of the region? How does it affect the water quality?

HUMAN USES OF WATER

7. Assume that you had water available from the following five sources: (1) rainwater drained from your roof; (2) good well water or tap water; (3) water from the wash cycle of your dishwasher or clothes washer; (4) water from the rinse cycle of your dishwasher or clothes washer; (5) water drained from your bath or shower. List all the applications in your home, in your garden, for your pets, or for other purposes for which each of these water supplies could be used.

8. List as many uses as you can think of for water in industry, in agriculture, and for recreation.

DIVERSION PROJECTS

9. Imagine that you live in an area with abundant water, and it is proposed that some of the excess water be diverted to another region that needs it. List the questions that people on both sides of the pipeline should consider before construction is started.

10. Explain how irrigation can, in some instances, be economically favorable in the short term but harmful over the long range.

GROUNDWATER

11. (a) Describe what happens to rainwater that starts to fall onto a very dry area and then continues heavily for several days. (b) Describe what happens to this water when there is no more rain for a month.

12. Why is pollution of groundwater potentially more serious than pollution of surface waters?

13. Recently, a team of engineers proposed to dig a deep well to supply irrigation water for a village in the desert in Egypt. The town elders asked whether the water from the well would be like water from a river or like oil from an oil well. What do you think they meant by the question?

14. Explain why land subsides when groundwater is depleted. If the removal of groundwater is stopped, will the land necessarily rise again to its original level? Defend your answer.

WATER POLLUTION

15. In its article on "Sewerage," the eleventh edition of the *Encyclopedia Brittanica,* published in 1910, states, "Nearly every town upon the coast turns its sewage into the sea. That the sea has a purifying effect is obvious. . . . It has been urged by competent authorities that this system is not wasteful, since the organic matter forms the food of lower organisms, which in turn are devoured by fish. Thus the sea is richer, if the land is the poorer, by the adoption of this cleanly method of disposal." Was this statement wrong when it was made? Defend your answer. Comment on its appropriateness today.

16. Explain how a nontoxic organic substance, such as chicken soup, can be a water pollutant.

17. What is eutrophication? Explain how it occurs and why it is hastened by the addition of inorganic matter such as phosphates.

18. List an activity that would pollute (a) inland surface waters, (b) groundwater, (c) the oceans. Discuss the effects of this pollution on humans and on plants and other animals.

19. Pesticides are poisonous to many species of plants and animals, whereas phosphate is a form of fertilizer. Yet both are considered to be water pollutants. Explain the differences.

MINERAL RESERVES

20. Explain why energy is an important factor in estimating future mineral reserves.

21. Petroleum is generally burned as a fuel, but in many applications the chemical compounds in the oil are used for the manufacture of plastics and other materials. Explain why petroleum cannot be economically recycled after it is burned, but if it is used for the synthesis of plastics, recycling may be possible.

22. A noted environmental scientist reported that the world tin reserves may be depleted in 2000. In making this prediction, he assumed that (a) mining technology and world economic activity will remain constant, (b) consumption levels and population will remain constant, and (c) no new deposits will be discovered. Do you think that these assumptions are reasonable? If so, defend your position. If not, suggest more likely assumptions.

RECYCLING

23. Sand and bauxite, which are the raw materials for glass and aluminum, respectively, are plentiful in the Earth's crust. If we are in no danger of depleting these resources in the near future, why should we concern ourselves with recycling glass bottles and aluminum cans?

24. When animals are slaughtered for human consumption, various waste products are produced. These include fat and bones from cattle, chicken feathers and entrails, blood, and unused fish parts. These wastes, when discarded, are a large source of water pollution. Some factories, called rendering plants, recycle these materials by converting them to tallow (which is used to make soap) and to animal feed products. However, some rendering plants discharge odorous pollutants into the atmosphere. Discuss the overall environmental impact of these rendering plants.

25. Referring to a discussion of the difficulty of recovering ores from the soil and rock in a back yard, Professor Peter Frank of the Department of Biology of the University of Oregon wrote, "The second law of thermodynamics comes in with a vengeance." Explain what he meant.

PART

ASTRONOMY

FOUR

The only object in the Universe, other than Earth, on which a human has set foot is the Moon—yet as we shall see in our study of astronomy, much is known about the heavens. Let us follow Astronaut Aldrin as we step forward on our voyage of discovery of astronomy. (Courtesy NASA)

SIDNEY C. WOLFF

• INTERVIEW •

Sidney C. Wolff received her Ph.D. from the University of California at Berkeley. She served as a research associate at the Lick Observatory for one year before joining the Institute for Astronomy at the University of Hawaii. During the seventeen years that Dr. Wolff spent in Hawaii, the Institute for Astronomy developed Mauna Kea into the world's premier observatory. Dr. Wolff earned international recognition for her research, particularly on stellar atmospheres—the evolution, formation and composition of stars. In 1984 she was named Director of the Kitt Peak National Observatory, Tucson, Arizona and in 1987 became Director of the National Optical Astronomy Observatories (NOAO). She is the first woman to head a major observatory in this country. As Director of NOAO, she directs a staff of 460 and facilities used by more than 100 visiting scientists

annually, while continuing her own work in astronomy.

Q: What was your earliest exposure to astronomy and how did you get involved in the field?

A: I had a spelling lesson in the third grade and it included astronomical words such as planet and telescope. I started reading about astronomy and was just fascinated by it. In high school I wrote to a professor at a college and asked him to suggest some books and astronomy magazines to read. There are many magazines now, but the big one then was *Sky and Telescope,* so I kept reading. I was never much of an amateur astronomer; I never had a telescope to look

through. I just read. Then I picked a college that offered an astronomy major. I didn't think I would ever be a professional astronomer because I didn't think I could make a living that way. But I thought at least I could take some of the courses for fun. And I discovered that I could do both: make a living and do something I enjoy.

Q: It seems you decided to focus on astronomy very early in your life. Did you have any ideas what you could do with astronomy as a career?

A: Most of the things I have done have been related to astronomy in some way. When I went to graduate school I expected to do research and to teach. And I did that for a while. I didn't expect to manage a large scientific organization, which I do now. I didn't expect to write textbooks. I didn't expect to spend a lot of time in Washington interacting with government funding agencies that provide support for astronomy. There are a lot of things that you do that you don't necessarily expect. I went to a small liberal arts college, and I'm glad I had that broad grounding in English, history, and all of the subjects that you learn in a liberal arts school.

Q: Are there other jobs where a background in astronomy would be helpful, or perhaps give an advantage?

A: I think so. The kind of technical background you acquire when you major in astronomy is useful for many jobs. It can be used for technical writing or science writing, if you want to go into journalism. The general technical background can help you in business because more and more businesses make extensive use of computers and communications links. So I think there are a lot of directions you can go in. I was just on a visiting committee for some liberal arts colleges in Massachusetts and I found out that many people are majoring in astronomy just the way you would major in history or English—because it is fun and interesting. They do not necessarily intend to be professional astronomers.

Q: You are the first woman to head a major observatory in this country. Tell us about the role of women in the field of astronomy, what they used to do, and what they are doing now.

A: About twelve percent of professional astronomers are women, and that is high for physical science—much higher than, say, physics, but still it is obviously a very small number. Historically, the number has been about eight percent, so it's not a very big change. What has changed is the kind of things that women are allowed to do. Some of the best known women astronomers, at the beginning of this century, were not given faculty positions. They worked mainly as research assistants. But they were not given much opportunity to choose their own research programs. I think that women are now full partners in this effort. They are appointed to the best faculties and they do choose their own research directions and they do participate in the committees that set priorities for the subject, and so on. So while the numbers haven't changed a lot, I think the role has.

Q: I have heard that a lot of astronomers don't look through a telescope anymore because the information is gathered from computers. Is that true? Do you look through a telescope?

A: Not very often. We use much the same kind of sensors that are in a TV camera, so one rarely looks through a telescope. Telescopes like the largest one on Kitt Peak, the observatory I operate, you *can't* look through. I mean, there is simply no provision, no eyepiece, nowhere you could possibly look through it. It is all done with TV cameras and computers.

The last time I looked through a telescope was when I took a group of people from the National Science Foundation to one of our small telescopes. We walked around, showed them all of our sophisticated equipment and big telescopes, and they were very interested. But then they viewed Saturn for real through a small telescope—*that* is when they *really* got excited!

Q: If a student is interested in astronomy, and wanted to learn more, what do you think would be the next step to take?

A: There are a large number of groups of amateur astronomers around the country. Almost any city of reasonable size has a group of amateur astronomers. And there are people who are enthusiastic about astronomy. They often will arrange programs where people will come in and lecture about recent advances in astronomy. They have star parties and they build telescopes. They will give you an opportunity to look at the stars. These clubs are really good for anyone interested in astronomy. They usually have a wide range of ages from children through adults, who are all very enthusiastic. Your college may be able to recommend programs or special summer sessions, or even provide jobs for people who want to gain some research experience. There are a lot of ways to get involved in astronomy. You just need to take the initiative.

Q: Do you think astronomy, as a science, is receiving enough emphasis in education compared to other subjects? It seems there is a lot of information available, it is just up to the individual to pursue it.

A: Yes. I think that astronomy, more than most sciences, has a special role to play in attracting people to science in the first place. There aren't too many sciences for which there are three or four popular magazines published, for example. I think that is because we answer the kinds of questions that everybody is interested in—where do we all come from? What is the ultimate fate of the universe? Is there life elsewhere in the universe? Everybody is interested in those questions. I know a lot of scientists who started out with an interest in astronomy and then pursued some other science.

Q: Can studying astronomy help in other areas of a student's education?

A: Absolutely. Astronomy could stimulate motivation in sticking with some of the harder subjects, such as math, because you realize that you must know some math in order to understand astronomy. I think astronomy motivates students. They may, eventually, go into other fields, but at least they take the basic courses in order to make that choice.

We desperately need more scientists and engineers in this country, and this is a major issue. When I go to Washington, D.C., one of the topics frequently discussed is how we encourage more students from the junior high level to become interested in science. But then how do we keep them interested in science? I think astronomy is having less trouble in retaining that interest than most areas of science, but the overall attrition in most sciences is very high as the people go through high school and college. One of the areas we are looking at is the way we teach science. You see, it is the

way we teach science that makes people drop out, and this is a major concern. Toward the end of this decade a large fraction of the faculty at the colleges and universities will be retiring and it is difficult to see how we will even replace, let alone provide, the people for business and industry.

Q: Is there a movement to change the way science is taught in our schools?

A: I was talking to the president at Hampshire College, Massachusetts, a few weeks ago. It is one of the few colleges where they graduate more science majors than those who begin as science majors, because of their particular approach. As students enter they complete a survey and are asked if they are interested in science as a major. It turns out that they actually have more people interested by the end of the program than at the beginning. And this is very rare.

Q: What are they doing to achieve this response?

A: For freshmen they have what in most colleges would be a senior seminar, where they attempt to understand current research papers. This is difficult, of course, because the freshmen don't have the background, but they do get a good idea of scientific principles, research, and applications. When students see what can be done in the sciences, they are strongly motivated to learn the basics, to take introductory courses, and to study mathematics, because they can see what they can accomplish ultimately.

I think it's important to introduce modern research very

A solar-powered satellite.

early in undergraduate education, and I believe astronomy is especially good at that. We discuss black holes and other modern concepts in the elementary textbooks used in the introductory college courses. That's one of the reasons that astronomy is so attractive—it introduces, at the earliest possible level, the problems that scientists are working on now.

Q: Can you tell us what areas of astronomical research will be the most interesting in the near future?

A: Over the next decade, one of the most interesting areas of research, that is one of the areas where we are most likely to make advances, is the formation of planets around other stars and the possibility of life in the universe other than here.

We are about to have a new generation of telescopes in space and on the ground that will allow us to study the formation of planets around other stars. I think we will be able to understand whether planets are common in some ways. This is a fundamental question that you have to answer before you can answer whether there is other life in the universe, which is a very vexing question. But there can't be life unless there is the right kind of habitat for it, and astronomy examines habitats other than Earth's. We should be able to answer in the next decade what fraction of stars have planets and whether those planets are the right distance from the stars in order to have the kinds of temperatures that are conducive to development of life.

Q: What kinds of answers do you expect the recently-launched Hubble Space Telescope to eventually provide?

A: It is going to have a big impact in exploring the origin of the universe and the formation of galaxies. When we look at very distant objects we are seeing them as they were a long time ago. We can find galaxies that are so far away that it takes 10 billion light-years to get from there to here. So that means we are seeing them the way they were 10 billion years ago. We think the universe is only 13–15 billion years old, so we are already looking toward the beginning of time, and together with the Hubble Space Telescope and some new ground-based very large telescopes that we hope to build, we should be able to probe even closer to the beginning. So we will really have a kind of time machine in the Hubble Space Telescope where we can look

back into the past and see how the universe used to be and get an idea of how it was in the beginning, how galaxies originated, how stars first formed.

Q: What are some of the challenges facing scientists today?

A: One of the important challenges for a scientist is to not simply accept conventional wisdom, but to keep asking "Is this really right?" I asked one of my colleagues what he is researching and he said he is attempting to prove that the universe is actually expanding. This has been conventional wisdom since 1929, and it is quite likely to be true, but he's devising a new test that would distinguish between that possibility and some of the more exotic possibilities. You always have to question the conventional wisdom and think up new ways to look at it.

Q: If there were thoughts you would like to impart to students to keep in mind as they continue their study, what would they be?

A: I would emphasize the need for a broad education as an undergraduate, and not to specialize too early because you can't be sure how your career will develop. I would say that even if you are committed to a research career in science, you can't really know what direction that career will take you. It may involve research, you may end up managing large scientific organizations, you may end up

writing textbooks, or you may become a popular science writer. There are so many directions that you can go. The broader your education while you are in college, the better.

It is particularly important that you learn how to write well and to express yourself clearly, because part of science is telling other people what you have discovered. If you only discover it and keep it to yourself, that's not science. You have to be able to express your scientific ideas very clearly to other people so that the ideas will have more impact.

The other thing I would say about science as a career is that for most of the scientists that I know, science is not only their vocation, but their *avocation*. They really like what they are doing. I think that's a more important factor to consider in choosing a career than how much money you will make or how famous you will be. The question is, when you get up in the morning, do you look forward to going to work? I think that most of the scientists I know do.

Q: Good! And do you?

A: Yes!

If you are interested in astronomy, and would like to become more involved, consult Appendix F for a list of amateur astronomer clubs and write to the one nearest you for more information.

This interview was conducted by Kate Pachuta for Saunders College Publishing.

CHAPTER

21

Motion in the Heavens

Astronomy, the study of the heavens, holds the unique position of being both the oldest science and the youngest. It is the oldest in the sense that probably the first intelligent creatures to walk this planet looked toward the heavens and asked: Why? Every civilization has had its legends concerning the stars, the Milky Way, and the wonders of its known universe. Astronomy was also the first of the sciences to be investigated on a systematic basis. Many of the reasons for these studies were based on practical concerns such as using the stars for navigational purposes. Other reasons were far from practical, even though they were thought to be practical at the time, because they arose from a belief in astrology among the common people as well as the royalty of the time. Astrologers held that events on Earth could be predicted by the stars; in fact, they believed that the characteristics of an individual's personality were influenced by the position of the stars and planets at the time of his birth. The detailed ob-

servations made by astrologers were futile in predicting the course of human history, but they did have some value in that many of the observations of the heavens made by them helped eventually to form that portion of physics referred to as mechanics.

Astronomy is also a new science because items from the popular press frequently discuss new discoveries made by astronomers that cause us to alter our perception of our place in the scheme of the heavens. Thus, astronomy is evolving rapidly, even today, and this is the characteristic of a new science. Whatever the reasons, the fascination of astronomy was strong in the past and continues to be just as compelling today.

Our study of astronomy will be from the ground up. We will look first, in this chapter, at motion in the heavens, as seen from Earth, and at some of the central figures from the past who have contributed to our present knowledge of astronomy. Chapter 22 will deal with the Solar

System, and then in Chapter 23 we shall focus our attention on the stars. Finally, in Chapter 24, we will look at galaxies and other unusual objects that populate our Universe, such as quasars, black holes, and pulsars.

21.1
THE CONSTELLATIONS

The fixed stars that dot the night sky form the backdrop for observing motion in the heavens. These stars are called fixed stars because day in and day out, year in and year out, these stars seem to maintain their same position relative to one another. Everyone at some time has gazed at clouds drifting overhead and observed familiar shapes formed by them. Likewise, ancient observers gazed heavenward and found that their imagination could associate an image with the arrangement of a particular grouping of stars. The Chinese, the Egyptians, and finally the Greeks did this, and these groupings of stars are called **constellations.** Modern-day astronomers still make use of these constellations because they provide a road map in the sky. For example, if you want to tell someone the general location of an object in the sky, to say it is in or near the constellation Orion provides a simple method for doing so.

There are 88 of these recognized constellations today, and most of them were first identified by the Greeks. Many of these have names that are Latin translations of the names given originally by the Greeks. Don't be concerned if you look at a constellation and see no apparent relationship at all with its namesake. For example, the best-known of all the constellations is Ursa Major, the great bear. However, finding a bear in the shape of this grouping of stars is not easy. In fact, part of this constellation is more commonly known in the United States as the Big Dipper, a shape that is more easily discernible. Be aware that the stars that make up a constellation are seldom related to one another in any way except for the fact that they are seen from Earth in roughly the same part of the sky. One star in the constellation may be relatively close to Earth, whereas another may be trillions of miles farther out in space.

The star charts at the end of this chapter show the primary constellations as seen from the Northern Hemisphere during the four seasons. Because there is something satisfying about being able to find your way around the sky, we encourage you to spend a few evenings under the stars seeing if you can find these constellations. We don't intend to give you an exhaustive discussion of these star groupings, but let us take a brief look at a few of them.

Turn to the star chart for the month of December and locate the constellation Orion, the Hunter. An easy way to locate Orion in the sky is to find the three stars that are aligned to form the belt of the hunter. This constellation is one of the most easily found and prettiest in the sky. It contains several bright stars, including Rigel and Betelgeuse. Rigel is the brightest star in Orion and has a brilliant blue color when observed through a telescope. It locates the left kneecap of Orion. The star Betelgeuse (from the Arabic for armpit) is the second brightest star in Orion and has a distinctive red color. It locates the right shoulder of Orion. Now refer to the star charts and note how the constellations appear to move in the sky. In December, Orion has just risen in the east at 9 P.M. Turning to the star chart for March, we find that Orion has moved across the sky toward the western horizon at the same time of night. Thus, the general motion of the constellations through the seasons is from east to west. Finally, in June, Orion has moved completely out of the sky at 9 P.M.

Let us look at another constellation on the charts, Ursa Major, the Big Bear. In June, Ursa Major is found in the northern part of the sky and slightly toward the western horizon. Moving to the sky chart for September, we see that Ursa Major has moved more toward the western horizon and has also moved lower down on the northern horizon. In December, the constellation is still low on the northern horizon, but it has now shifted slightly toward the eastern portion of the sky. Finally, in March, it has moved higher in the northern sky and is still located slightly toward the east. The geographical latitude of these charts is 34°N, and at this latitude the constellation Ursa Major is never out of the sky. Any constellation that is always above the horizon at a particular

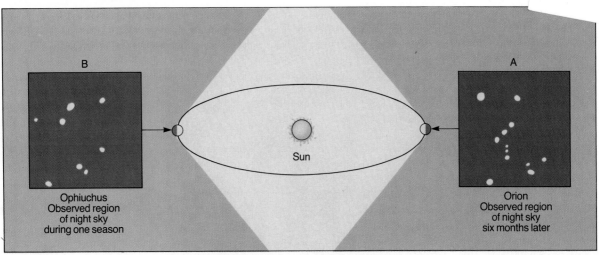

FIGURE 21.1 The night sky changes with the seasons because the Earth is continuously changing position as it orbits the Sun.

location on Earth is called a **circumpolar constellation.** Before we leave the constellation Ursa Major, look at the star chart for March. Note that the two stars forming the outer edge of the bowl of the dipper can be used as pointer stars to find the location of the star Polaris, the North Star. (The dashed lines in the figure point the way.) Polaris is not a particularly bright star, but is famous because of its seemingly constant location in the northern sky. If one were to extend a line outward into space through the axis of rotation of Earth, Polaris would be almost on this line.

EXAMPLE 21.1 Staying out all night

We have noted that the motion of the constellations during the course of a few weeks is such that they tend to move toward the west. But what about their movement during the course of a single night? We will give you the answer below, but a question like this can be answered a little more satisfactorily if you go outside and observe for yourself. On a clear night, find a particular constellation and periodically check its position over the course of a few hours. You can gauge the movement of the stars by comparing their position to an object on the Earth's horizon. How do they move?

Solution This kind of motion is called daily motion, because it can be observed during the course of a single day. You should have noted that the daily motion of the stars also appears to be from east to west.

21.2
MOTION IN THE HEAVENS

In the last section, we noted two different types of motion that the stars seem to undergo—one during the course of several months and another during the course of a single day. Let us now investigate why these motions occur.

Figure 21.1 shows why the stars shift position during the course of a year. This shift occurs because the Earth orbits the Sun, and as a result different portions of the heavens are visible in the night sky at different times of the year. During the winter season, the stars within region A are visible at, say, 9 P.M., whereas 6 months later an entirely different vista, B, is visible at the same hour of the night. The motion of the Earth in its orbit causes an individual star to rise approximately 4 minutes earlier each night.

The daily motion of the stars, as discussed in the exercise above, takes place because of the rotation of the Earth on its axis. The stars appear to

FIGURE 21.2 A time exposure showing circular arcs of stars around the North Pole. (Courtesy Yerkes Observatory)

rise in the east and set in the west, but this is only an apparent motion caused by the rotation of the Earth from west to east. The axis of rotation of the Earth points toward the star Polaris, and the motion of the Earth on its axis can be seen by a time-exposure photograph, as shown in Figure 21.2. This time-exposure shows star trails as they circle about Polaris, which remains essentially motionless because it is within 1° of the axis of rotation of the Earth. If you were to live at the North Pole directly below the star Polaris, you would observe that night in and night out, season in and season out, the stars would always circle above you in the sky with the North Star as the center of their circular path. As a result, you would never see the stars sink below the horizon.

This means that if you lived at the North Pole, all the stars you could see would be circumpolar. An observer at the Earth's Equator would see Polaris on the northern horizon, and all the stars circling about it would rise and set. In fact, all the stars seen would rise straight up on the eastern horizon and set straight down on the western horizon. Thus, at the Equator there are no circumpolar stars.

So far in this section we have glibly spoken of the motion of the Earth in its orbit about the Sun and of the rotation of the Earth on its axis. However, as we shall see before the end of this chapter, such notions were not easily arrived at by early astronomers. The Sun appears to rise each morning in the east and set each evening in the

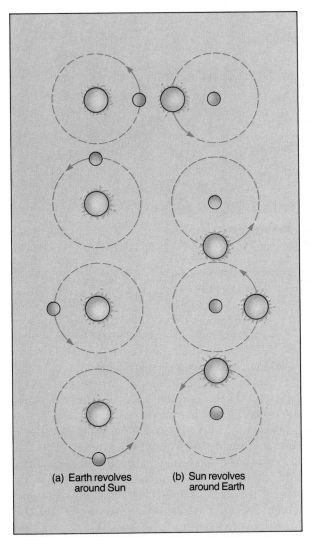

(a) Earth revolves around Sun

(b) Sun revolves around Earth

FIGURE 21.3 Series (a) shows the Earth revolving around the Sun, and series (b) shows the Sun revolving around the Earth. Now lay some thin paper over series (a) and trace the outlines of the Sun and Earth, but not the arrows or orbits. Lay this tracing over Series (b) and note that they match exactly, after you shift the paper for each sketch to make sure the Sun and the Earth superimpose. *Conclusion:* There is no difference between the Earth revolving around the Sun or vice versa, provided you do not refer to anything else, such as the outline of these diagrams or another star.

western sky. But such an apparent motion could result either if the Earth rotated once every 24 hours on its axis or if the Sun were orbiting around our planet. Until the Sixteenth Century,

most astronomers believed that the Earth was the center of the Universe and the Sun and the stars revolved around it. Students now have a tendency to laugh at the foolishness of these ancient scientists, because everyone knows that the Earth revolves around the Sun. However, if there were only two objects in the Universe, the Earth and the Sun, there would be no way to tell which one was moving around which, and it would be meaningful to say only that they were orbiting relative to each other. The exercise in Figure 21.3 illustrates the dilemma.

To understand how astronomers eventually proved that the Earth revolves around the Sun, think of the following situation. Suppose that you were on a raft drifting on an ocean. If there were no landmarks anywhere, it would be impossible to tell whether you were moving or stationary. But now suppose that there were two islands in sight, one close to your boat and the other farther away. If you were stationary, the two islands would remain in the same relative position to each other. However, if you moved, they would appear to shift positions. This apparent shift in position of objects against a background, due actually to the motion of the observer, is known as **parallax.** This shift is illustrated by the two photographs of the columns of a building shown in Figure 21.4.

The ancients understood the concept of parallax, and they correctly reasoned that if the Earth did move around the Sun, the stars should change position relative to one another, as shown in Figure 21.5. Since no such shift was observed, they concluded that the Earth must be stationary. This was the only logical conclusion possible from the data at hand. The mistake arose not out of faulty reasoning but out of inaccurate measurements. The diameter of the Earth's orbit is small in comparison with the distance to even the nearest stars. Therefore, the parallax angle changes very little in 6 months' time and was not measurable until more accurate measuring instruments became available. Only then could astronomers prove that stellar positions change relative to one another and therefore that our planet is in motion.

We now know that the Sun is the center of our Solar System and that nine planets revolve

(a) (b)

FIGURE 21.4 The concept of parallax is illustrated by two photographs of the columns of a building, taken from different positions. (a) The photographer is in line with the row of columns. (b) When the photographer moved, the relative positions of the columns with respect to each other appear to have shifted. The same effect has been observed in astronomical studies. As the Earth revolves about the Sun, the relative positions of the stars with respect to each other appear to shift.

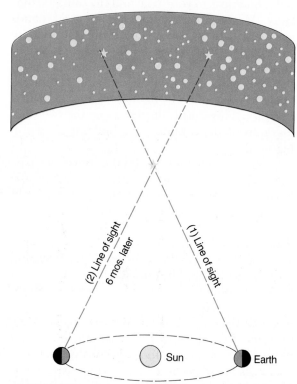

FIGURE 21.5 *(Left)* A nearby star appears to change position with respect to the distant stars as the Earth orbits around the Sun. This drawing is greatly exaggerated; in reality, the distance to the nearest star is so much greater than the diameter of the Earth's orbit that the parallax angle is only a small fraction of a degree.

around the Sun in elliptical orbits. In addition to revolving, the planets simultaneously spin on their axes. The Earth spins approximately 365 times for each complete orbit around the Sun. Each complete rotation of the Earth represents one day. As an important vocabulary item, note that the words "revolve" and "rotate" are frequently used interchangeably in everyday life, but scientists distinguish between the two. The word **"revolve"** is used for those situations in which *one object moves around another*. Thus, the Earth revolves around the Sun. The word **"rotate"** is reserved to designate an object *turning on its axis*. Thus, the Earth rotates on its axis once each day.

The story of how it was determined that objects move in the heavens reads, at times, more like a detective story with a strange cast of characters than like a study of pure science. Later in this chapter, we will look at the contributions made to our eventual understanding of motion in the heavens by these early observers of the sky.

21.3
THE MOTION OF THE MOON
Phases of the Moon

Even the most casual observer of the Moon has seen that it appears to change shape on a regular basis. If on one evening the Moon is circular and bright, it is said to be full. A few evenings later, part of the disk is darkened. As the days progress, the dark portion grows until only a tiny sliver of moon is visible. This phase of the Moon is called a crescent. Finally, about 15 days after the Moon was full, it becomes invisible. This day is called the time of the new moon. Shortly after the new moon, the thin crescent reappears. As the nights go by, the visible portion of the Moon grows larger and larger until the Moon is full again after a total cycle of about 29.5 Earth days. Further observation reveals that when the Moon is full, it rises approximately at sunset and on each successive evening it rises, on the average, about 53 minutes later, so that in 7 days it rises in the middle of the night. In about a month, the cycle is complete and the next full moon rises on schedule in the early evening (Figs. 21.6 and 21.7).

To understand the phases of the Moon, we must first realize that the Moon does not emit its own light but simply reflects light from the Sun. Thus, the amount of moonlight received depends on the relative positions of the Sun, the Moon, and an observer on the Earth. The Earth revolves around the Sun in a flat orbit, as if the two were positioned on an imaginary table top. As shown in Figure 21.8, the Moon's orbit around the Earth is tilted with respect to this plane, so that generally the Moon, the Sun, and the Earth do not lie in a straight line. As the Moon orbits, the half of the sphere facing the Sun is continuously illuminated, as shown in Figure 21.7. If the Moon is positioned directly behind the Earth away from the Sun, the entire sunlight area is visible, and the Moon appears to be full. However, if the Moon is between the Earth and the Sun, the sunlit side is facing away from the Earth, and the unlit surface that faces the Earth cannot be seen. The Moon is then virtually invisible to us, and it is said to be new. (The word "virtually" is used in the last sentence because there is a slight illumination of the Moon caused by light reflected from the Earth to the Moon.) In Figure 21.7, the position of the Moon is shown in its different phases. Midway between a new moon and a full moon, half of the side of the Moon facing Earth is illuminated. Because the Moon at this time is one-quarter of the

FIGURE 21.6 The phases of the Moon. (Courtesy of Hale Observatories)

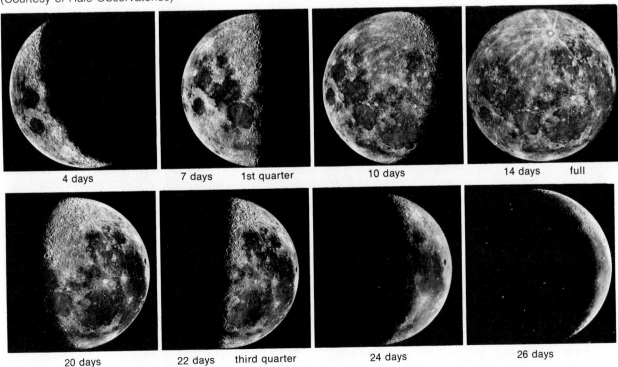

4 days   7 days 1st quarter   10 days   14 days full

20 days   22 days third quarter   24 days   26 days

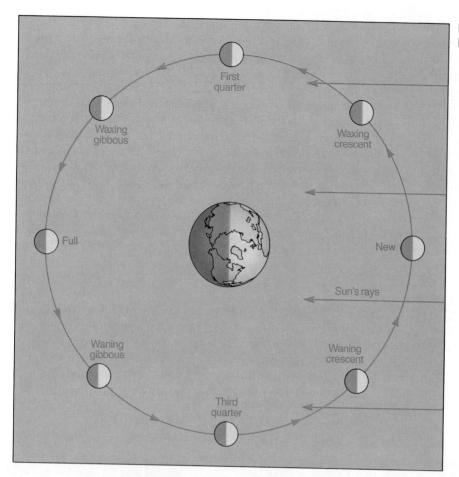

FIGURE 21.7 Phases of the Moon.

way through its complete cycle, starting with the new moon as the beginning of the cycle, this moon is called a first-quarter moon. Note that during the interval between a new moon and a full moon, the Moon appears to grow as seen from the Earth. Those phases of the Moon in which we see a little more of the Moon on one night than we did the night before are said to be **waxing phases.** Thus, a few days after a new moon, we have a waxing crescent phase and a few

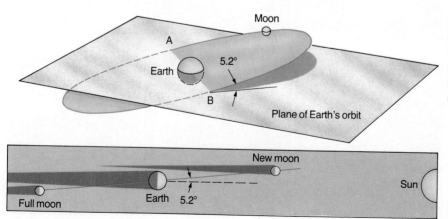

FIGURE 21.8 The Sun and the Earth lie in one plane, while the Moon's orbit around the Earth lies in another. (Scales are exaggerated for emphasis.)

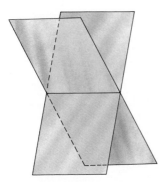

FIGURE 21.9 Two planes intersect in a straight line.

days after a first-quarter moon, we have a waxing gibbous moon. As the Moon moves from the full phase back to the new phase, we see a little less of the Moon each night than we did the night before, and these phases are referred to as the **waning phases.** Thus, a few nights after a full moon, we have a waning gibbous phase, which eventually becomes a waning crescent phase and then a new moon once again. Midway between a full moon and a new moon, half of the face of the Moon visible from Earth is again illuminated, and we have a third-quarter moon.

Eclipses of the Sun and Moon

Recall that the Moon's orbit is tilted with respect to the plane of the Earth's orbit about the Sun. But two planes tilted with respect to each other must necessarily intersect, as shown in Figure 21.9. Therefore, as the Moon orbits, it must pass through the Earth-Sun plane twice during each revolution. These points are called the **nodes** of the orbit. If a line is drawn between the Sun and the Earth, normally the nodes of the Moon's orbit do not fall on that line. Instead, the alignment is as shown in Figure 21.10b or c. But the nodes of the Moon's orbit do line up with the Sun on occasion. Suppose that there is a full Moon at one of these special times, as shown in Figure 21.10f. The Moon is positioned behind the Earth, but now instead of lying above or below its orbital plane, it is directly behind the Earth, in its shadow. When this happens, the full

Moon becomes temporarily invisible from Earth. This phenomenon is called a **lunar eclipse.** Lunar eclipses last a few hours from beginning to end and occur at periodic intervals.

What happens if there is a new moon at the time when the lunar orbital nodes line up with the Sun? The Moon's shadow falls onto the Earth, thereby blocking out, or eclipsing, the Sun, as shown in Figure 21.10e. When a total solar eclipse occurs, an unnatural darkness descends, and the Earth becomes still and quiet. Birds seem to become confused, return to their nests, and cease their singing. While the eclipse is total, the Sun itself is hidden; but the outer solar atmosphere (called the corona), normally invisible because of the Sun's brilliance, appears as a halo around the dark Moon.

21.4 THE MOTION OF THE PLANETS AND THE SUN

The word "planets" is derived from the Greek word "*planetes*," which is translated as *wanderers*. They were so named because the planets change their position relative to the fixed, or background, stars. Early observers of the heavens noted that the Sun and Moon did not stay fixed in position relative to the backdrop of the stars. Instead, these two objects drift eastward with respect to the stars behind them. Most of the time, the planets also obey this eastward drift. This means that if you observe a planet on a particular night and pay attention to a particular star close to it in the sky, then in a couple of weeks the position of the planet has changed such that it is farther east than is the background star.

A problem that confronted early astronomers was that the Sun and Moon always drift eastward, but at times the planets do not. Sometimes the planets move westward. Why do they wander around in the way that they do? To show how this works for a planet, consider Figure 21.11. On the first day of observation, the planet is at position 1 on the figure, and on day 2 it has moved to position 2. Note that the planet is drifting eastward in the sky relative to the background star S. This gradual eastward drift continues until we reach position 4. It then begins to back up in

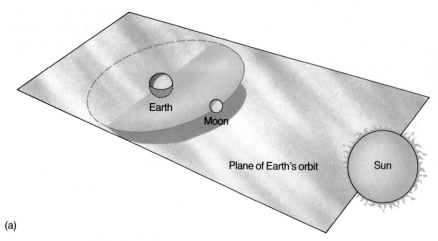

(a)

FIGURE 21.10 Eclipses of the Sun and Moon. The Sun and the Earth lie in one plane, while the Moon's orbit around the Earth lies in another. (Scales are exaggerated for emphasis.) (a) Normally the Moon lies out of the plane of the Earth-Sun orbit . . . (b) so at new Moon the Moon's shadow misses the Earth, (c) and at full Moon the Earth's shadow misses the Moon.

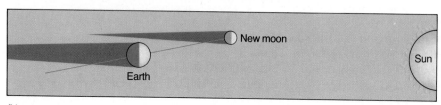

(b)

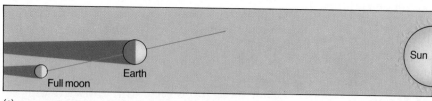

(c)

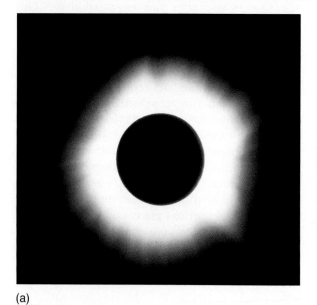

(a)

(b)

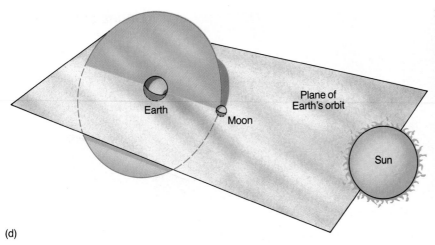

(d)

(d) However, if the Moon passes through the Earth-Sun plane when the three bodies are aligned properly, then an eclipse will occur. (e) An eclipse of the Sun occurs when the Moon is directly between the Sun and the Earth, and the Moon's shadow is cast on the Earth. (f) An eclipse of the Moon occurs when the Earth's shadow is cast on the Moon.

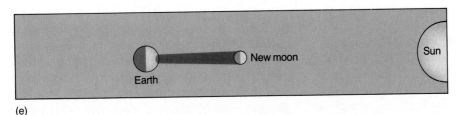

(e)

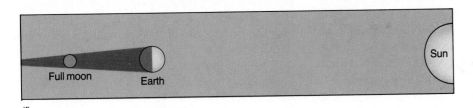

(f)

the sky, drifting westward with respect to the star until it reaches position 7. The planet then reverts to its original eastward drift as shown by positions 8 through 10. When a planet is backing up in the sky, drifting westward, the planet is said to be undergoing **retrograde motion.** What causes this? In a later section, we shall see the incorrect explanation of early astronomers and then discuss what actually is happening.

Let us review our discussion, first presented in Chapter 18, of one more feature of the motion of the Earth that affects our life on this planet.

◄ (a) The solar corona, or atmosphere, is seen during the March 7, 1970, total eclipse of the Sun. (b) A small spot of light, looking much like a diamond ring, is seen in the Apollo spacecraft as the Sun goes into eclipse from their viewpoint.
((a) Courtesy NOAO, (b) courtesy NASA)

The seasonal changes we observe on Earth are due to a relationship that exists between the revolution of the planet and its axis of rotation. It is a common misconception that the Northern Hemisphere of our planet is cold in the winter because it is then that Earth is farthest from the Sun, and the hotter temperatures of the summer are because it is closer to the Sun. As we shall elaborate later in this chapter, the exact opposite of this is true. The seasons are not caused by the changing Earth-Sun distance: They are produced because the axis of rotation of the Earth is tipped at an angle of approximately 23.5° away from the perpendicular, as shown in Figure 21.12. During the summer season, position S in the figure, the Northern Hemisphere is tilted toward the Sun, and at position W, winter, the Northern Hemisphere is tilted away from the Sun. The summer

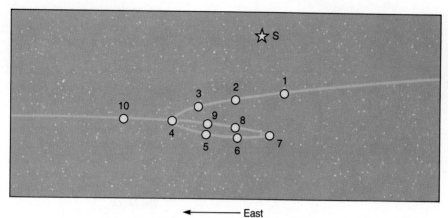

FIGURE 21.11 Retrograde motion of a planet. The planet drifts eastward with respect to the background star S until position 4. It then moves westward with respect to S until it reaches position 7. It then resumes its eastward march.

← East

tilt means that the incoming sunlight hits that part of the Earth more nearly head-on (Figure 21.13a) than it does in the winter, when the sunlight arrives at a more glancing angle (Figure 21.13b). When the sunlight hits head-on, the incoming energy is spread out over a smaller surface area than it is when the light hits at a glancing angle. The Southern Hemisphere is warm when it is winter and cold when it is summer in the Northern Hemisphere. In the spring, position SP, and the fall, position F, as shown in Figure 21.12, the Northern and Southern Hemispheres are both at the same angle with respect to the incoming sunlight and therefore have close to the same temperature.

A large sphere with its center concentric with the Earth, as shown in Figure 21.14, is called the **celestial sphere.** The projection of the Earth's Equator on this sphere is called the **celestial equator.** The Sun appears to move around the sky, changing its position with respect to the background stars by about 1° each day. This path of the Sun is called the **ecliptic.** Because of the tilt of the Earth's axis of rotation, the celestial equator and the ecliptic are also tilted with respect to one another by 23.5°. Note that the celestial equator and the ecliptic intersect at two points called the equinoxes (from the Latin for equal nights). At these times, there are 12 hours of daylight in both the Northern and Southern Hemispheres. When

FIGURE 21.12 The seasons are caused by the tilt of the axis of rotation of the Earth with respect to a line drawn perpendicular to its orbit.

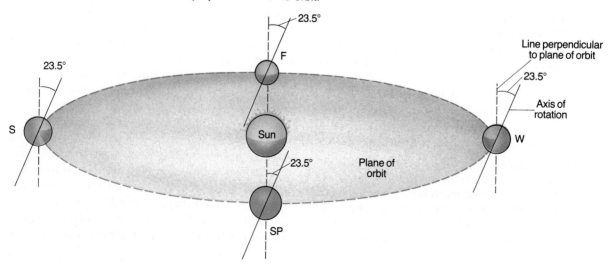

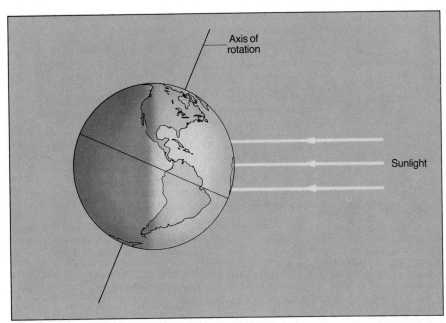

Axis of
rotation

Sunlight

(a)

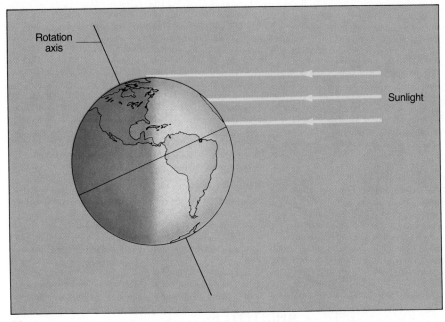

Rotation
axis

Sunlight

(b)

FIGURE 21.13 (a) In the Northern Hemisphere during the summer season, the sunlight hits more nearly head-on to the Earth than it does (b) in the winter. Just the opposite is true for the Southern Hemisphere.

the Sun passes the point indicated as the **vernal equinox,** spring begins in the Northern Hemisphere. This occurs each year about March 21. The sun continues its northward movement in the sky until it reaches a point called the **summer solstice** on about June 21. At this time, the Sun stands as high in the sky as it is going to get. As the Sun continues its motion, it reaches a point called the **autumnal equinox** on about September 22, and fall begins for the Northern Hemisphere. Finally, on December 21, the winter season begins when the Sun reaches its lowest point in the sky at the **winter solstice.**

As the Sun makes its yearly journey around the heavens, it passes through 12 constellations that are called the constellations of the **zodiac.**

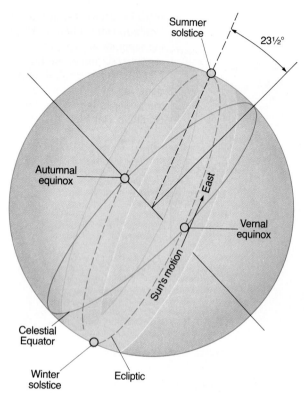

FIGURE 21.14 The celestial sphere. The celestial equator is the extension of the Earth's equator into space. The ecliptic is the apparent path of the Sun around the celestial sphere during the course of a year. The Moon and planets are always close to the ecliptic, and those constellations close to the ecliptic are called the constellations of the zodiac. The ecliptic and the celestial equator intersect at the vernal equinox and the autumnal equinox. The Sun is highest in the sky, in the Northern Hemisphere, at the summer solstice, and lowest at the winter solstice.

These constellations lie within a belt that extends about 9° on each side of the ecliptic. Some of these constellations are so dim that they are hardly visible to the naked eye, yet more people have heard of these than of any other constellations in the sky, because they form the basis of astrological horoscope tables. Virtually everyone knows his or her "sign." But what does it mean if you are a Libra, for example. Astrologists would have you believe that a person born under this sign has a tendency toward certain behaviors that are governed by the stars. Nothing could be fur-

ther from the truth. There is absolutely no scientific way in which objects as remote as the stars could affect a person's life simply because he was born under a particular grouping of them. If you are a Libra, this means that if your father or mother had taken you outside on the day you were born and let you look at the sky, the constellation Libra would have been in the sky directly behind the Sun. You would, obviously, not have been able to see it, because it would be hidden in the Sun's glow, but you would have been a Libra nonetheless. To cast one more aspersion on the beliefs of astrology before we move forward, you actually would not be a Libra in today's world. When the beliefs in astrology were first being formed, the constellation Libra *was* behind the Sun in October, the time for Libras. However, since that time, the position of the Sun as seen from Earth has shifted slightly, and now if you are born in October, the constellation that is behind the Sun is Virgo. Libra has moved to a position such that it would be known as a November sign if astrologers were developing their charts in the present day.

Finally, let us consider one more feature of the motion of the planets. If one were to go to the North Pole of the Earth and go directly upward in space, after having traveled far out into space, one would see a view of the Solar System as shown in Figure 21.15. All the planets would be seen to be revolving counterclockwise about the Sun, and most of them would be seen to be rotating counterclockwise. As we shall see, there is one notable exception to this; Venus revolves counterclockwise, but it rotates clockwise. Any theory of the formation of the Solar System must account for these features of planetary motion.

21.5
ARISTOTLE

We encountered Aristotle when we examined some of the early ideas about motion in our study of physics. Aristotle wrote on a wide variety of subjects ranging from politics to medicine, and among his fields of philosophical thought was astronomy. Let us examine a few of his viewpoints concerning the heavens. His philosophy held that

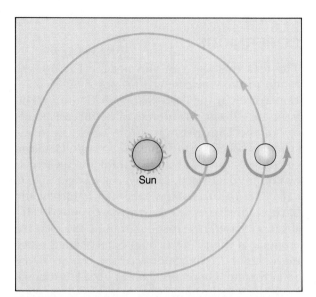

FIGURE 21.15 From above the North Pole of the Earth, all the planets revolve counterclockwise about the Sun and most also rotate counterclockwise.

the Earth was the center of the Universe about which all the planets, the Sun, the Moon, and the stars revolved in perfectly circular orbits. This idea came, in time, to be called the **geocentric** (Earth-centered) model of the Universe. The circle was selected as the shape of the orbits of these objects because the ancients tended to fit their concepts in with a spiritual view of the heavens. Heavenly objects seemed to behave differently than earth-bound objects. A moving object on Earth always stops, but the planets and other objects in the sky never stop their motion. There was just something different about the heavens and the Earth. The circle seemed to fit in with this view of heavenly perfection. The circle is symmetrical, and it has no beginning and no end. It seems a shape destined for the heavens and for the presumed perfection that resides there. Another concept that Aristotle held that fit in with this perfection of the heavens was that the Moon was perfectly spherical and polished like a looking glass so that it could reflect the light from the Sun and reveal Man in all his glory. Granted the Moon can be observed to have some blemishes. Even with the naked eye, light and dark regions are obvious. However, in the spiritual realm, a few imperfections should be expected because the Moon is the heavenly body that is closest to the corrupt and imperfect Earth.

We have already discussed the fact that most of these Aristotelian ideas of the way the Universe works are wrong, and we shall see shortly that, in fact, *all* of the ideas mentioned above are incorrect. What is so important about the fact that Aristotle was wrong in his view of the Universe? Many scientists have been wrong about many topics throughout the course of human history, and scientists are often incorrect or incomplete about the results of their research even today. Yet these scientists are not mentioned in a textbook. The reason that Aristotle is so important in the progress of astronomy lies with the fact that his beliefs became inculcated into religious beliefs in the Thirteenth Century by Saint Thomas Aquinas. Aquinas believed that there should not be a conflict between religious thought and beliefs concerning how the Universe works and how it is laid out. The end result of this expression was that Aristotle's concept of the heavens came to hold second place only to the revelations of the Bible. Aristotle's teachings became so strongly ingrained in Christian thought that it became difficult to question the ideas of the ancients in any way. Thus, an idea that the Earth might not be at the center of the Universe was not simply viewed with skepticism; the originator of such an heretical idea could be subject to severe religious persecution.

It should be noted here that not all of Aristotle's ideas were incorrect. For example, he correctly concluded that the Earth was round. He did so by observing that the shadow cast on the Moon by the Earth during an eclipse is always round. The only shape that the Earth could have and still always produce a round shadow is a sphere. If the Earth were disk-shaped, it would occasionally cast a shadow that was a line. Additionally, he noted that a traveler moving north was able to observe stars that had not been visible before. This bore witness to the curvature of the Earth.

The geocentric model was brought to its greatest fruition by Claudius Ptolemy about A.D. 140. He developed a model of the Universe that was able to predict the positions and motions

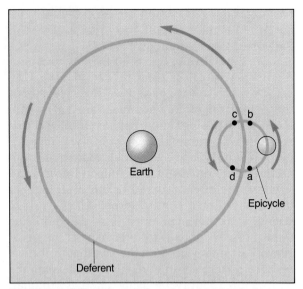

FIGURE 21.16 The deferent and an epicycle for a planet revolving about the Earth according to the geocentric theory of the Universe.

of the planets with great accuracy. Figure 21.16 shows the basic approach that he used. He assumed that the motion of a planet about the Earth followed a circular path that he called the **deferent.** Thus, if the planets always moved eastward against the background stars as they followed their deferents, all would have been well. However, as we have seen, the planets often back up in their orbits and drift westward in retrograde motion. To explain this aspect of the motion of a planet, Ptolemy had to modify his basic model by adding **epicycles,** which were small orbits that the planet followed as it moved on the deferent, as shown in Figure 21.16. Thus, we have circles rolling on circles. When the planet is moving from a to b on the epicycle of Figure 21.16, it is moving eastward, but when the planet moves from c to d, it is moving against the direction of its basic motion along the deferent, so it drifts westward. This scheme solves the problem of retrograde motion, and by adding epicycles on epicycles, and by choosing the distances and speeds of the planets with care, it is possible to explain most of the basics of planetary motion. However, the scheme is cumbersome, and the way was open for a better idea.

21.6 COPERNICUS

The idea that the Sun and not the Earth is the center of the Universe was proposed by Nicolaus Copernicus in the early Sixteenth Century. This basic idea of a Sun-centered, or **heliocentric,** theory of the Universe was not in accord with the religious beliefs of the time. As a result, Copernicus did not publish his ideas until 1543 when he was on his death bed. Copernicus was not able to detect the parallax shift of the stars, which would have verified that the Earth was in motion. Instead, he based his reasoning on the fact that the motions of the planets could be explained much more simply if the Sun were the center of the Solar System. He retained many of the traditional ideas, such as the belief that the orbits of the planets were circular. In his book, *De Revolutionibus,* he set forth his postulates, but the book was not widely read. It might be said that his book and his ideas suffered from the lack of a public relations person to hype it for him. Such a man, Galileo, was to arrive on the scene later.

We have seen that much of the work done by Ptolemy with his deferents and epicycles was an attempt to explain features of planetary motion such as retrograde motion. He was able to do so, but the scheme relied on complex geometrical manipulations. On the other hand, the explanation based on the heliocentric idea is considerably easier to comprehend. Figure 21.17 shows how retrograde motion can be explained for the planet Mars in a Sun-centered Solar System. We first find the Earth in the figure at position 1 while Mars is in its orbit more distant from the Sun at its position 1. The line of sight toward Mars of an observer on Earth is indicated by the dashed line which passes through both points 1 in the figure. This observer will see certain stars behind the planet at this time. As the weeks pass, the Earth moves to position 2 in its orbit and Mars also moves to a new position 2. However, note that the Earth is catching up with Mars because it is moving faster in its smaller orbit. The line of sight of the observer on Earth is shown by the dashed line passing through these points labeled as 2. Different stars are now seen behind Mars by the observer, but note that the motion of the two

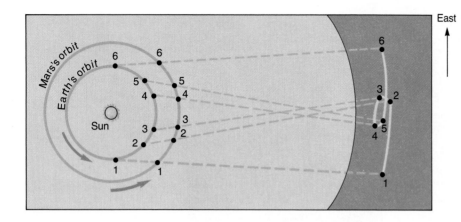

East

FIGURE 21.17 Retrograde motion from the point of view of the heliocentric theory.

planets has been such that Mars has appeared to drift eastward against the backdrop of stars. Note now what happens between those positions labeled as 3 and 4 for Earth and Mars. During this interval, the Earth has caught up with Mars and passed it. The dashed lines in the figure indicate that during this interval Mars has appeared to drift westward. Thus, retrograde motion has been occurring. You can continue this procedure through points 5 and 6 to see that this retrograde motion soon stops and the normal eastward drift of the planet resumes.

Thus, Mars never really moves backward. It just moves forward more slowly than Earth and seems to move backward when the Earth catches up to it and passes it.

21.7
BRAHE AND KEPLER

Tycho Brahe (1546–1601) was a Danish astronomer who was born into a family of nobility. His privileged youth allowed him to become adept in astronomy. As a result of his reputation, he gained the patronage of Frederick II and was able to establish an excellent observatory on the island of Hveen. Brahe's work was done prior to the discovery of the telescope, but he employed all of the best available instruments of the time, including some that he developed himself, in an effort to chart the course of the Sun, Moon, and

planets. He is regarded by many as the most systematic observer of the heavens in the history of astronomy. He compiled enormous amounts of data on the position of all the known bodies in the Solar System. This information enabled him to note that the positions of the planets varied from predictions that were based on the work of Ptolemy.

If you had lived at that time and for some reason wanted to know the position of the planet Venus in the sky on the night of June 15, 1585, Brahe would be the source to whom you would direct your question. Thus, Brahe was an observer of considerable note, but other predilections of his may have been responsible for his never developing any laws or theories based on these observations. As an example of the kind of outside interests Brahe had, consider the fact that he had his nose cut off in a duel as a young man. To cover this deformity, he had false noses fashioned out of gold and silver. Brahe also had a strong inclination for late-night drinking parties. In fact, this led to his death because at one party he drank to such excess that his bladder ruptured.

When Frederick II died in 1597, Brahe was forced to leave Denmark by the enemies he had accumulated in abundance while he served as royal astronomer. He moved to Prague where he undertook the same duties for Emperor Rudolph of Bohemia. While there, Brahe made a great contribution to astronomy when he hired a young assistant named Johannes Kepler.

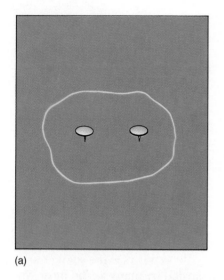

(a)

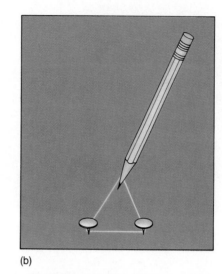

(b)

FIGURE 21.18 (a) Drape a loop of string over two thumbtacks. (b) Draw the string taut with a pencil. (c) Keep the string taut and trace out a curve, and (d) you end up with an ellipse.

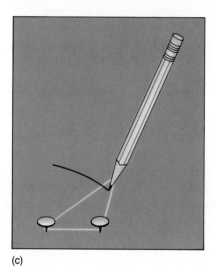

(c)

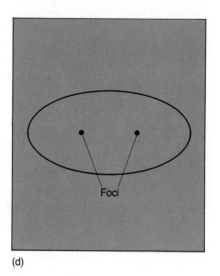

Foci

(d)

21.8 KEPLER

Johannes Kepler (1571–1630) fled his Catholic homeland of Germany as a protestant refugee to work with Brahe. The relationship was not a pleasant one, however, because Brahe was jealous of Kepler's brilliance. As a result, Brahe kept much of his observational data from Kepler. However, when Brahe died, this wealth of information came into Kepler's hands and he studied it for approximately 20 years. The first of Kepler's works was published in his book *The New As-*

tronomy in 1609. Today the work of Kepler is summarized in his three laws of planetary motion. It should be emphasized here that Kepler's laws are representative of a class of laws called empirical laws. This means that there was at the time no method by which they could be derived based on the known laws of physics. He simply observed that they were true and did not attempt to show that they should be valid based on other, more basic, principles of science. The status of these laws as empirical laws came to an end when Newton formulated his law of universal gravitation. Based on his law, Newton was able to show that

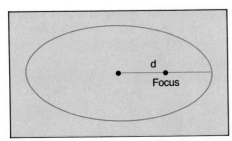

FIGURE 21.19 The distance d from the center of an ellipse through a focus is called the semimajor axis of the ellipse.

Kepler's laws were a necessary consequence of any two objects bound to one another by an inverse-square attractive force. Let us take a look at each of Kepler's laws in turn.

Kepler's first law

In order to understand Kepler's first law of planetary motion, we must first understand a geometrical curve called an ellipse. Figure 21.18 shows how to draw an ellipse. First, two thumbtacks are pressed into a table, as in Figure 21.18a; the location of the tacks will be referred to as the *foci* (singular *focus*) of the ellipse. Second, a loop of string is placed over the tacks, as indicated in Figure 21.18a. In part (b) of the figure, the string has been drawn taut by a pencil pressed against it. A curve is then drawn by tracing out those points that the pencil can reach while the string is kept taut. This is indicated in Figure 21.18c. The result is a curve called an ellipse, like that shown in Figure 21.18d. The distance from the center of the ellipse, through one of the foci, and out to the ellipse is called the **semimajor axis.** This distance is d in Figure 21.19.

Kepler's first law states that:

The orbit of a planet is an ellipse with the Sun at one focus. The other focus is empty.

It should be pointed out here that although Kepler formulated his law based on observations of the planets, the basic form of the law is found to hold for any object that orbits another. Thus, the Moon moves about the Earth in obedience to Kepler's first law, as do double star systems orbiting one another and as does a man-made satellite orbiting the Earth.

Kepler's second law

Kepler's second law is often referred to as the law of equal areas. To see why it is given this designation, consider Figure 21.20, which shows the orbit of a planet moving about the Sun. First, we draw a line from the Sun to the planet when the planet is at position 1 in its orbit; then we wait a certain period of time, say 1 month. In that month's time, the planet will have moved to a new position in its orbit, position 2 in the figure. After this time interval, we draw a second line from the Sun to the planet. These two lines form a pie-shaped segment of the ellipse, as indicated in the figure by area A. Now, let us wait awhile until the planet is at some new position in its orbit, say position 3 in the figure. We now draw a line from the Sun out to position 3 and wait for the *same* time interval as when we were sketching out our previous area, 1 month. The planet has again moved in its orbit during this month and is now located at position 4. Again, a pie-shaped area, B, has been swept out by the line during this interval of time. Kepler found that segments A and B had the same area. He stated his second law as:

The line from the Sun to a planet sweeps out equal areas in equal intervals of time.

As an outgrowth of this law, it can be seen that the planets do not move in their orbits with a constant speed. To see this, consider Figure 21.20 once again. If the areas A and B are to be equal, then the planet would have to move faster in its orbit when close to the Sun as it swept out area B than it would when farther from the Sun sweeping out area A. Thus, the Earth moves faster in its orbit during the winter season when it is close to the Sun than it does in the summer when it is farther from the Sun.

Kepler's third law

Kepler's third law is a mathematical relationship between the period of a planet and its average distance from the Sun. The period is the time it takes a planet to make one complete revolution

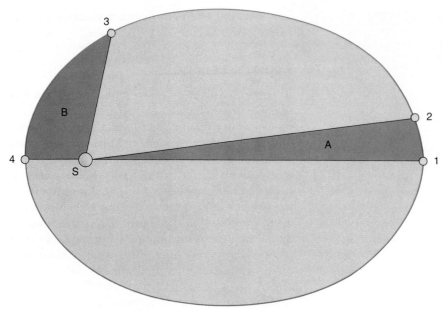

FIGURE 21.20 Kepler's second law, the law of equal areas.

in the orbit, and in Kepler's formulation of the law, this period is expressed in Earth-years. *The average distance from the Sun is the length of the semi-major axis of its elliptical path, and this distance must be expressed in* **astronomical units, A.U.** The average distance of the Earth from the Sun is about 93,000,000 miles, and this distance is defined to be the astronomical unit. Table 21–1 gives the period of the planets that were known in Kepler's time and their average distance from the Sun in A.U., the units appropriate to Kepler's third law.

Kepler's third law is expressed in terms of a mathematical relationship between the period P and the average distance from the Sun d. The equation is

$$P^2 = d^3$$

In words, the law is stated as:

The squares of the periods (in years) of the planets are equal to the cubes of their average distance (in A.U.) from the Sun.

EXAMPLE 21.2   Checking out the third law

Consider the information given in Table 21–1 and show that Kepler's third law is satisfied by the motion of Venus.

Solution   From the table, we find that the period of Venus is 0.61 years. Let us square this number to find

$$P^2 = (0.61)^2 = 0.37$$

Likewise, we find from the table that the average distance of Venus from the Sun is 0.72 A.U. Let us cube this number.

$$d^3 = (0.72)^3 = 0.37$$

Thus, we find that $P^2 = d^3$. You try this for a few of the other planets. Does the law work as well for Saturn, a distant planet, as it does for Mercury, which is very close to the Sun?

TABLE 21–1

Planet	Period (Years)	Distance (A.U.)
Mercury	0.24	0.39
Venus	0.61	0.72
Earth	1.00	1.00
Mars	1.88	1.52
Jupiter	11.86	5.20
Saturn	29.46	9.54

EXAMPLE 21.3 An invisible planet

Let us suppose that an astronomer discovers a planet 4 A.U. from the Sun. (No such planet really exists.) What would be the period of this planet in years?

Solution The cube of this planet's average distance from the Sun is

$$d^3 = (4)^3 = 64$$

Thus, we find

$$P^2 = 64$$

from which, P = 8 years.

21.9 GALILEO

We have discussed in our study of physics some of the discoveries and methods used by Galileo in his investigation of mechanics, and we found that his adherence to experimental observations became a basic part of science. Let us now turn our attention to his contribution to astronomy, along with the trials and tribulations that he encountered along the way. Galileo was born about a century after Copernicus, and during his life the teachings of Copernicus were still banned by religious authorities. However, Galileo came to support the idea of a heliocentric solar system based on his observations with a telescope. Galileo did not invent the refracting telescope, but he has gained distinction as the first person to use it in a systematic way for observing the heavens. Let us consider a few of the things that Galileo saw through his telescope that caused him to accept the Copernican viewpoint.

1. Galileo turned his telescope to the hazy pathway across the sky that we call the Milky Way and saw that it was not simply a misty blur, but instead was a vast collection of individual stars too faint to be distinguished individually by the naked eye. This drew him into controversy with contemporary thought, because the fundamental outlook on science at the time was that if it were true that the Milky Way was a vast collection of stars, the ancients would have already mentioned it. There was no room for new information about the heavens. This observation had nothing to say about the validity of the heliocentric theory, but it caused Galileo to ponder the idea that if the ancients could be wrong about this, perhaps they could also be wrong about the geocentric theory.

2. When he turned his telescope toward Jupiter, he found that it was circled by four moons. (We now know that Jupiter has many more moons than this, but the best telescope that Galileo ever owned had a magnification of only about 30, and as a result, he was not able to see these smaller, dimmer moons.) This was an important observation for Galileo, and a blow to his detractors because they had held that *everything* orbited the Earth. Now Jupiter was found to have moons orbiting it. The idea that the Earth was the center of all motion was in serious trouble.

3. Galileo's telescope revealed that the Moon had craters and tall mountains. This was in contradiction to the idea of the time that heavenly objects were more nearly perfect than is the Earth. That is, they were smoother and more symmetrical. Yet, here was an object that had about the same kind of blemishes on it as could be found on the "corruptible" Earth. Thus, the Earth could also find its place among the heavenly objects. Again, this observation did not prove the Copernican system to be correct, but it once more raised the issue that if one idea can be wrong, others can be incorrect also.

4. The deciding factor that convinced Galileo of the truth of the Copernican concept was his observations of the planet Venus. He observed that Venus goes through phases just like the Moon. At some observation times, Galileo would find Venus in a crescent phase; at others he would find it to be in a gibbous phase, and so forth. The only way that Venus could ever be seen in a gibbous phase is if it is circling the Sun. Figure 21.21 shows why Venus undergoes these phase changes.

Based on his telescopic observations, Galileo became a staunch defender of the heliocentric theory of the Solar System. But old ideas die hard. In 1633, Galileo was taken before a high court of the Church and asked to recant his belief that the Sun and not the Earth was the center of the Solar System. Galileo bowed to the pressure of the Church rather than face imprisonment

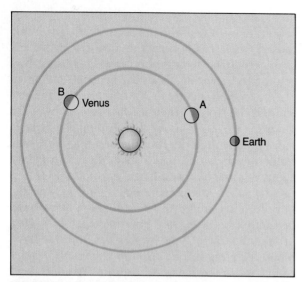

FIGURE 21.21 At position A, Venus is in a crescent phase; at B, it is in a gibbous phase.

and torture. The main issue of the controversy concerning Galileo centered on his book *Dialogue Concerning Two Chief World Systems*. The book was written in a language understandable by the common man and consisted of a series of discussions, or dialogues, among three characters, Salviati, Sagredo, and Simplicio. The name Salviati implies Savior, in the sense that he is an adherent of the heliocentric theory. Sagredo is neutral, but his name implies wisdom or intelligence. The third character is a defender of the geocentric theory, and his name Simplicio, or simpleton, needs no explanation. The end result of the dialogue is that Salviati convinces Sagredo of the wisdom of the heliocentric system.

Following his inquisition by the Catholic Church, Galileo's book was published in protestant Holland, where his ideas met with widespread acceptance. Thus, the death knell had been struck for the geocentric view of the Solar System.

21.10 NEWTON

Galileo was to spend the rest of his life under house arrest, but his work was to serve as the inspiration for a sickly English lad named Isaac Newton. It is of interest to note that Newton was born on Christmas Day of 1642, the same year as Galileo's death. As close as they were to being contemporaries, the world in which they worked was considerably different. Galileo suffered from persecutions because of his scientific teachings, but Newton was to gain great fame and wealth during his lifetime because of his contributions to the endeavor of science.

The first great contribution to astronomy that Newton made was his invention of the reflecting telescope, discussed in Chapter 11. However, it was his three laws of motion and the law of universal gravitation that united the heavens with the Earth. It had been held that the heavens operated under a different set of laws and rules than did the Earth. For example, it was Aristotle's belief that the natural tendency of moving objects on the Earth was that they should slow down. This did not, of course, apply to the heavens, since they obeyed a different set of rules. The motion of the stars, the Sun, the Moon, and the planets never ceased. Newton, on the other hand, showed that it was the natural tendency of objects to continue to move endlessly unless a force, such as friction, acted to retard their motion. This brought the heavens closer to Earth. With this law of gravitation, Newton was able to show that Kepler's laws of planetary motion were natural consequences of the fact that they were held in their orbits by a force of mutual attraction described by his expression for the gravitation force. So, Newton moved Kepler's laws from the realm of empirical equations to equations that could be derived from more fundamental principles. Thus, it may be said that Newton's laws unified the Universe. The laws of nature apply equally to falling apples and moving comets.

SUMMARY

Constellations are groupings of stars that are imagined to have certain shapes from Earth. There are 88 of these now recognized, and they are still of importance to astronomy, because they serve as guides to the sky. Those constellations that never rise or set at a particular location on Earth are said to be **circumpolar constellations** for that location.

Astronomers proved that the Earth revolves around the Sun by showing that the stars change position relative to each other during the course of a year. The phases of the Moon are caused by the revolution of the Moon around the Earth in a plane slightly tilted with respect to the plane of Earth's orbit. **Eclipses** occur when the Sun, Moon, and Earth all lie in a straight line and the Moon is in the plane of the Earth's orbit.

The general motion of the planets is toward the east against the background of fixed stars. However, at certain times the motion shifts to a westward drift. This backing up of a planet in its orbit is called **retrograde**

motion and can be explained from both the geocentric and heliocentric theories of the Universe.

The path followed by the Sun as seen from Earth is called the **ecliptic.** The time when the Sun is highest in the Northern Hemisphere is called the **summer solstice,** and the Sun is lowest in the sky at the **winter solstice.** The ecliptic and the **celestial equator** intersect at the **autumnal equinox** and the **vernal equinox.**

Aristotle and Ptolemy are primarily responsible for the geocentric theory of the Universe. Copernicus is primarily responsible for the heliocentric theory. Brahe is famed for his careful observations of the heavens. Kepler is known for his three laws of planetary motion. Galileo is sometimes called the father of modern science, primarily because he introduced into science the method of experimentation as the basis for deciding between conflicting theories. Newton, by means of his law of universal gravitation, showed that motion in the heavens obeys the same physical laws as on Earth.

EQUATIONS TO KNOW

$P^2 = d^3$ (Kepler's third law)

KEY WORDS

Constellations	Waxing	Ecliptic	Deferent
Circumpolar	Waning	Vernal equinox	Epicycle
constellations	Eclipses	Summer solstice	Heliocentric theory
Parallax	Planets	Autumnal equinox	Kepler's laws
Revolve	Retrograde motion	Winter solstice	Astronomical Unit
Rotate	Celestial sphere	Zodiac	
Phases of the Moon	Celestial equator	Geocentric theory	

QUESTIONS

THE CONSTELLATIONS AND MOTION OF THE EARTH

1. Using the star charts at the end of the chapter, go outside and find several constellations now visible and several of the brighter stars now in the sky.
2. Use the star charts at the end of the chapter to

identify several constellations that are circumpolar and several that are not.
3. Explain why all constellations are circumpolar at the North Pole but none are at the Equator.

4. Are the same stars visible in the sky on a winter night as a summer night? If you were in a balloon high in the atmosphere, could you see the summer stars in winter? Explain.

5. The motion of the Earth relative to the center of the galaxy is composed of at least four different independent movements. Describe each one.

6. Suppose that you are driving a car and the speedometer reads 80 km/h. A person sitting next to you and looking at the speedometer at the same time may read a different value, perhaps 76 km/h. Explain how two people can look at the same instrument at the same time and read different values.

7. With rigidly fixed modern telescopes, astronomers can determine how far away some stars are by observing the parallactic shift as the Earth travels around the Sun. Using this technique, would it be easier to estimate distance to nearby stars or to distant stars, or would it be equally difficult for all stars? Defend your answer.

THE MOTION OF THE MOON

8. How is the Moon positioned with respect to the Earth and the Sun when it is (a) full, (b) new, (c) crescent?

9. Suppose the Sun, Earth, and Moon lay in a single plane, but all other motions were the same as now. Describe the lunar cycle in this hypothetical situation.

10. Many societies use a lunar calendar instead of a solar one. In a lunar calendar, each month represents one full cycle of the Moon. How many days are there in a lunar month? If there are 12 months to a lunar year, will a lunar year be longer or shorter than a solar year? Explain.

11. Why does the Moon rise approximately 50 minutes later each day than it did the previous day?

12. Draw a picture of the Sun, Earth, and Moon as they will be during an eclipse of the Sun. Draw a picture of the Sun, Earth, and Moon as they will appear during an eclipse of the Moon. Explain how these positions produce eclipses.

THE MOTION OF THE PLANETS AND THE SUN

13. If the tilt of the Earth's axis of rotation with respect to its orbital plane were (a) greater than it is now, how would seasons on the Earth be affected? (b) Repeat assuming the tilt is less than now.

14. Explain how the tilt of the Earth's axis of rotation produces the seasons.

15. (a) Is the Sun farther from the ecliptic in the summer or winter? (Be wary of trick questions.) (b) In what season is the Sun farthest from the celestial equator? (c) When is the Sun closest to the celestial equator? (d) Do the ecliptic and the celestial equator ever intersect?

16. Explain why it is cold during the winter in the Northern Hemisphere while it is warm at the same time in the Southern Hemisphere.

17. How would our view of the heavens be changed (all other features being the same) if (a) Earth revolved clockwise rather than counterclockwise, (b) rotated clockwise rather than counterclockwise?

ANCIENT ASTRONOMERS

18. Present the main arguments for and against the geocentric theory and the heliocentric theory.

19. List some of the beliefs that Aristotle held about astronomy that are now known to be incorrect.

20. Explain retrograde motion by means of both the geocentric theory and the heliocentric theory.

21. In the heliocentric theory, we state the order of the planets from the Sun outward. Thus, the order is Mercury, Venus, Earth, Mars, Jupiter, and Saturn. How would this ordering change from the point of view of the geocentric theory?

22. It was once hypothesized that a planet similar to the Earth circled the Sun such that it was always behind the Sun as seen from Earth. What distance would this planet have to be from the Sun?

23. Prove that Kepler's third law is valid for Jupiter.

24. If the Earth moved in a perfectly circular orbit rather than its elliptical orbit, would our seasons be affected?

25. Use Kepler's second law to verify that the Earth moves fastest in its orbit during the winter season.

26. List the observations made by Galileo through his telescope and show how they tended to disprove the geocentric ideas of Aristotle.

27. Show the position of Venus when it is in a full phase. Why could this position never be seen from Earth?

28. Show the position of Venus for its various phases from the point of view of the (a) geocentric theory, and (b) the heliocentric theory.

29. Explain the statement that Newton changed our outlook on the heavens from that of a duoverse to that of a universe.

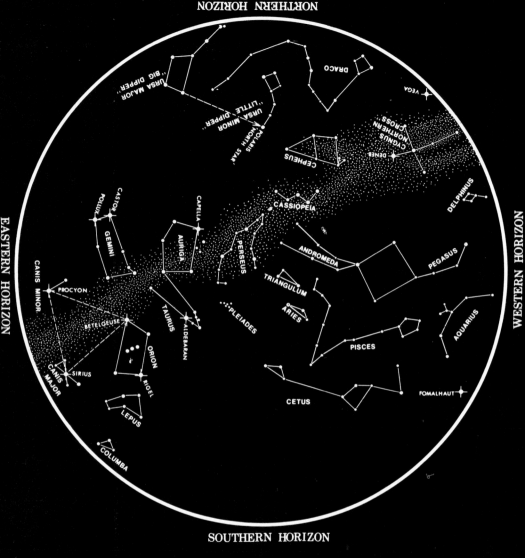

THE NIGHT SKY IN DECEMBER

Latitude of chart is 34°N, but it is practical throughout the continental United States.

Chart time (Local Standard):

10 p.m. First of month
9 p.m. Middle of month
8 p.m. Last of month

Star Chart from GRIFFITH OBSERVER, Griffith Observatory, Los Angeles

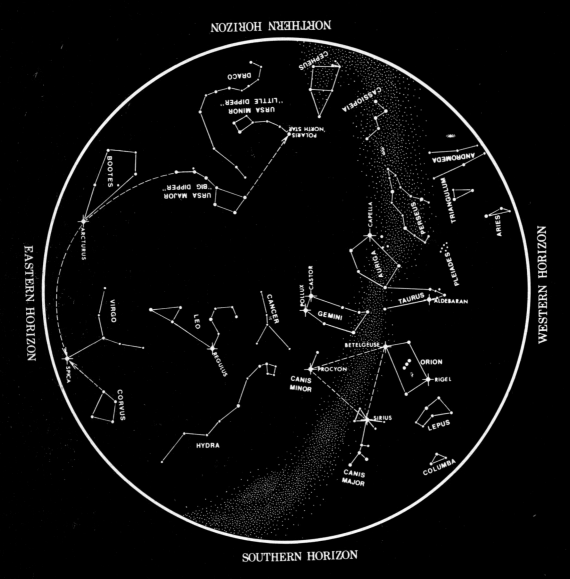

NORTHERN HORIZON

EASTERN HORIZON

WESTERN HORIZON

NORTHERN HORIZON

DRACO

URSA MINOR "LITTLE DIPPER"

CEPHEUS

CASSIOPEIA

ANDROMEDA

POLARIS "NORTH STAR"

BOOTES

URSA MAJOR "BIG DIPPER"

PERSEUS

TRIANGULUM

ARIES

ARCTURUS

CAPELLA

AURIGA

PLEIADES

VIRGO

CANCER

POLLUX
CASTOR

TAURUS

ALDEBARAN

LEO

GEMINI

REGULUS

SPICA

BETELGEUSE

ORION

CORVUS

PROCYON

RIGEL

CANIS MINOR

HYDRA

SIRIUS

LEPUS

COLUMBA

CANIS MAJOR

SOUTHERN HORIZON

THE NIGHT SKY IN MARCH

Latitude of chart is 34°N, but it is
practical throughout the continental
United States.

Chart time (Local Standard):

10 p.m. First of month

9 p.m. Middle of month

8 p.m. Last of month

Star Chart from *GRIFFITH OBSERVER*, Griffith Observatory, Los Angeles

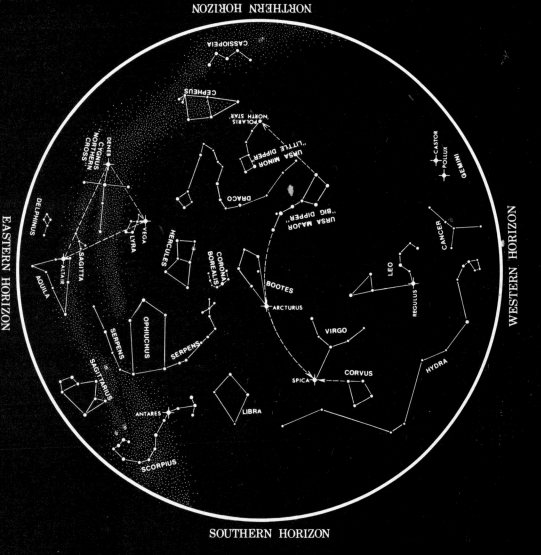

NORTHERN HORIZON

EASTERN HORIZON

WESTERN HORIZON

SOUTHERN HORIZON

THE NIGHT SKY IN JUNE

Latitude of chart is 34°N, but it is
practical throughout the continental
United States.

Chart time (Local Standard):
  10 p.m. First of month
  9 p.m. Middle of month
  8 p.m. Last of month

Star Chart from GRIFFITH OBSERVER, Griffith Observatory, Los Angeles

579

THE NIGHT SKY IN SEPTEMBER

Latitude of chart is 34°N, but it is practical throughout the continental United States.

Chart time (Local Standard):
10 p.m. First of month
9 p.m. Middle of month
8 p.m. Last of month

Star Chart from *GRIFFITH OBSERVER*, Griffith Observatory, Los Angeles

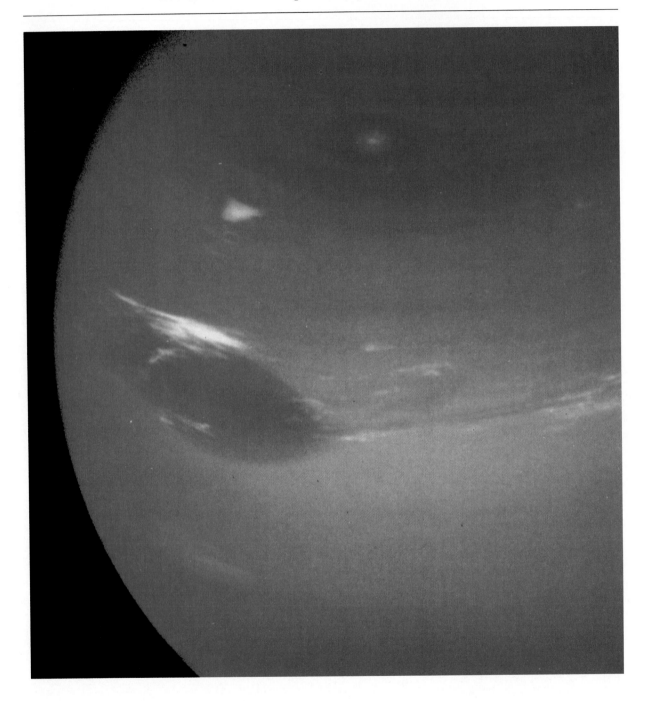

The last planet to have been studied by Voyager 2 is Neptune. The spacecrafts Voyager 1 and Voyager 2 have added enormous amounts of information to aid astronomers in understanding the Solar System. (Courtesy NASA)

C H A P T E R

The Solar System

Our journey through the Universe actually starts with this chapter. Here we focus our attention on our nearest neighbors in space: the planets and their moons, and our Sun. Our progression will be from the Sun outward. We will find that much is known about these objects but that there is still much to be learned: Only in the past decade has a spacecraft passed close enough to some of our outermost planets to provide us with surprising details about them and their moons. As a result of voyages such as these, many statements reported as facts in astronomy textbooks of a decade ago are now known to be incorrect. Perhaps no other science has changed as rapidly in the last ten years as has astronomy. Let's begin our voyage.

22.1
THE FORMATION AND STRUCTURE OF THE SOLAR SYSTEM

There are several conflicting scientific theories that attempt to explain the origin of our Solar System and the planet Earth. According to the most widely accepted of these ideas, the **nebular hypothesis,** our Solar System evolved from a cloud of gas and frozen dust. This cloud was nothing like the clouds you can see in our atmosphere. For one thing, it was extremely diffuse and would be considered to be a good vacuum by terrestrial standards. Second, it was composed mainly of light elements and is believed to have been approximately 79 percent hydrogen, 20 percent helium, and 1 percent other elements.

Approximately 5 billion years ago, billions of years after the galaxy itself had taken form, the cloud that was to become our Solar System began to contract. This diffuse mass was originally rotating quite slowly. The material in the cloud was quite cold, with temperatures perhaps as low as 3 K. At these extremely low temperatures, particles move about so slowly that even a very slight force will be able to affect them appreciably. Some scientists believe that the slight gravitational attractions among the dust and gas particles themselves caused the cloud to condense slowly into a spherical ball, as shown in Figure

FIGURE 22.1 Formation of the Solar System. (a) The Solar System was originally a diffuse cloud of dust and gas. (b) This dust and gas began to coalesce under its internal gravitation. (c) The shrinking mass began to rotate and was distorted. (d) The mass broke up into a discrete protosun orbited by large planets composed primarily of hydrogen and helium. (e) The Sun heated up until fusion temperatures were reached. The heat from the Sun then drove most of the hydrogen and helium away from the closest planets, leaving small, solid cores behind. The massive outer planets remain mostly composed of hydrogen and helium.

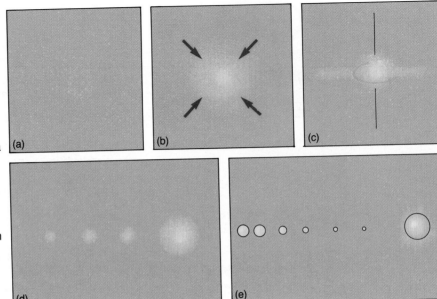

22.1a and b. Alternatively, perhaps a star exploded in a nearby region of space and the shock wave triggered the condensation. As the condensation continued, more and more matter gravitated toward the center of the newly formed sphere. Therefore, the density and pressure began to increase within the cloud. As atoms were pulled inward, they accelerated under the influence of the gravitational field. Some of the energy from the collapse was converted into heat, and the temperature of the gas and dust began to rise. This shrinking ball began to rotate more and more rapidly as it contracted.

As the rotation accelerated, the matter at the outer fringes of the ball started to move quite rapidly, thereby orbiting the contracting core. This motion distorted the ball to form a body shaped something like that shown in Figure 22.1c. The center of this cloud then coalesced into a large mass that is called the **protosun,** meaning the "earliest form of the Sun." Within the protosun, the gravitational attraction was very great, and the gases were pulled inward rapidly until intense pressures and temperatures were reached. The highest temperature was found in the very center of the protosun, where atoms were accelerating inward at the fastest rate. As the contraction progressed, the temperature continued to rise. Eventually, when the critical temperature of 10 million K was reached, collisions between hydrogen nuclei were so forceful that nuclear fusion reactions were started. Hydrogen fused into helium, great quantities of energy were released, and the Sun was born. Our Sun and all the other stars in the sky live out their lives under a delicate balance between forces tending to collapse them and forces tending to make them expand, as shown in Figure 22.2. The force of gravity tends to make the star fall inward on itself, while thermal pressure created by the intense heat and radiation present at the core of the star tries to make it expand. At some particular radius, these two forces are equal and an equilibrium is achieved that is maintained until the star begins to die. We will see what happens then in the next chapter.

Not all of the material present in the original cloud was drawn inward to form the Sun, and some of the dust and gas remained in the orbiting disk. As fusion reactions were initiated in the core

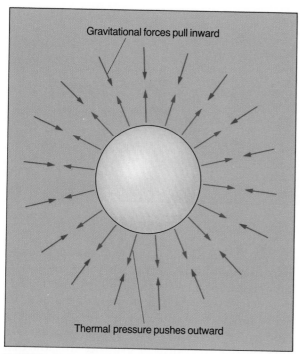

Gravitational forces pull inward

Thermal pressure pushes outward

FIGURE 22.2 Equilibrium in a star.

of the Sun, the temperature of the Solar System began to rise. In the inner regions of the system, light elements such as oxygen and nitrogen existed in the gaseous state, while the heavier elements remained in the solid state as particles of dust. In time, many of these dust particles came together to form small chunks of rock, which in turn coalesced to form miniplanets, called **planetesimals.** This aggregation continued, and the planetesimals collided to form larger masses that eventually became the four inner planets, Mercury, Venus, Earth, and Mars (see Fig. 22.1d and e). As these planets grew in size, many changes occurred. Heat was generated when particles and planetesimals fell together. This heat melted certain solids, vaporized some of the liquids, and boiled many of the gases off into space. Later, the planets cooled, and over the course of time additional solid particles and gases were drawn inward until the present structure and composition were achieved. Today, these inner planets are made up mainly of heavy elements in their solid or liquid state, although two—Earth and Venus—have substantial gaseous atmospheres.

There is some disagreement about the evolution of the giant outer planets—Jupiter, Saturn, Uranus, and Neptune. These four have a composition similar to that of the Sun and consist mainly of hydrogen and helium. According to one idea, the outer giants evolved directly from the condensation of the original cloud of dust and gas, much as the Sun evolved. The major difference was that the masses of the outer planets were not great enough to pull the elements inward with sufficient speed to raise their temperature to the point that fusion reactions could occur. Another theory states that the formation of the outer planets was similar to that of the inner ones, except that at the extremely cold temperatures of the outer Solar System, more of the lighter elements were retained.

Most of the mass in the Solar System is concentrated in the Sun. Nine major planets orbit the Sun: Mercury, Venus, Earth, Mars, Jupiter, Saturn, Uranus, Neptune, and Pluto. Most of the planets have one or more satellites of their own orbiting about them. For example, one moon orbits Earth, there are two moons orbiting Mars, and 16 (at last count) revolve around Jupiter. In addition to the planetary system, other bodies such as asteroids, meteoroids, and comets orbit about the Sun. Figure 22.3 is a pictorial representation of the entire system. The four planets closest to the Sun—Mercury, Venus, Earth, and Mars—are all relatively small, dense, and rocky. They are called the **terrestrial,** or Earth-like, planets. The next four—Jupiter, Saturn, Uranus, and Neptune—are much larger, more massive, and less dense, being composed mainly of hydrogen and helium. They are called the **Jovian** (for Jupiter) planets. Pluto, the farthest known planet, has a density approximately equal to that of the four gas giants, but it is much smaller, about the size of our Moon.

The theory for the formation of the Solar System, outlined earlier, was originally based mainly on calculations of how scientists believe that matter would behave under certain conditions. However, in recent years, these theoretical models have been supported by a series of observations that have provided direct evidence for (1) star formation, (2) the formation of a solar system, and (3) the existence of at least one solid planet circling one other star.

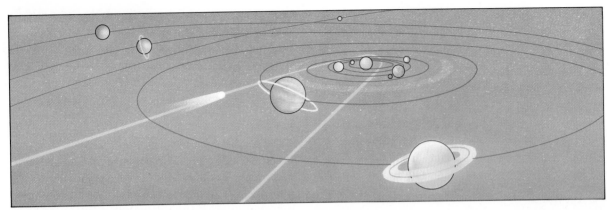

FIGURE 22.3 A schematic view of the Solar System.

Several clouds exist in outer space which are believed to be condensing to form new stars. But most of these are in too early a stage of evolution to determine whether they will condense into a single star or break up into smaller fragments that will become planets. However, in 1984, a satellite recorded unusual heat emissions from a star, Beta Pictoris, about 50 light-years from Earth. Further observations from ground-based instruments showed a halo of light around the star, as shown in Figure 22.4. If this light is being reflected off many small particles, as astronomers believe, the photograph might be a record of a solar system in formation. Other astronomers have observed that at least one nearby star is wobbling just a tiny bit around its normal position. This wobble could be explained by the existence of a planet in orbit around the star. This is the first evidence of a solar system besides our own any place in the Universe. The odds seems tremendously in favor of there being many other such solar systems circling the myriad of stars that dot our galaxy and other galaxies in space.

22.2 THE SUN

Ancient Greek scientists and philosophers believed that the Sun was a perfect, symmetrical, homogeneous sphere, unblemished in any way. This belief was first questioned by Galileo in the Seventeenth Century. While studying the Sun with the aid of a telescope, he noted occasional dark spots appearing on its surface. Many critics

disagreed with Galileo, although they refused to look through his telescope. They argued instead on philosophical grounds, claiming that if Galileo questioned the perfect symmetry of the Sun, he was simultaneously questioning the perfection of God, and this was heresy of the highest order. We now have a great deal of evidence to show that the Sun is, in fact, a complex heterogeneous sphere. Some of the evidence has been obtained visually, from frequent observations of dark spots and granular structures on the surface of the Sun

FIGURE 22.4 This photograph may be an image of another Solar System in the process of formation. The thin disk around the central star, Beta Pictoris, is composed of bits of dust, which are believed to be similar to the material that condensed to form the planets of our own Solar System. (Courtesy of University of Arizona and Jet Propulsion Laboratories.)

and from huge flares of hot gas that occasionally shoot outward from its surface. Yet our visual information represents only a small part of the solar data that have been collected. Astronomers also study many other forms of electromagnetic emissions: radio, infrared, ultraviolet, X-rays, and gamma rays. Additionally, much of what is now deduced about the unseen internal structure of the Sun has been inferred by calculating how matter must behave under the conditions of the solar environment. Of course, such inferences are uncertain, and many solar phenomena remain poorly understood.

Although nearly 70 elements have been detected in the Sun's atmosphere, most exist only in trace quantities. The main ingredient is hydrogen, which alone accounts for nearly 75 percent of the total matter. Helium, second in abundance, makes up close to 25 percent of the total, and the sum of all the other elements accounts for about 1 percent.

The central **core** of the Sun is extremely hot and dense, reaching temperatures of over 15,000,000 K at a pressure 1 billion times that of atmospheric pressure on Earth. What happens to the hydrogen and helium under these conditions? Individual gaseous atoms cannot remain electrically neutral at these extreme conditions. Interatomic collisions are so intense and so frequent that the electrons are simply knocked out of their atoms. The result is a homogeneous mixture of hydrogen and helium nuclei surrounded by a rapidly moving sea of electrons. This is a fourth state of matter called a **plasma.** Suppose that, within this plasma, hydrogen nuclei approach each other on a collision course. If their kinetic energies are great enough to overcome their mutual electrical repulsions, they fuse together to form helium. This nuclear fusion releases energy. Eventually, most of the Sun's hydrogen will be converted to helium; when that happens, some 5 billion years from now, the Sun will change dramatically, as we will see in the next chapter.

A great deal of energy is released from the core during these fusion reactions. This energy is carried radially outward by energetic photons. However, radiation cannot pass from the core directly through the body of the Sun to the sur-face. The photons are absorbed by many particles along their route, re-emitted, and finally absorbed again in a region just under the surface. Energy is carried outward from there by convection. The structure of the Sun is shown in Figure 22.5.

At the visible surface of the Sun, called the **photosphere,** energy is radiated out into space in the form of photons. Typically, the temperature of the photosphere may be 6000 K with a pressure only one-tenth that of air at the Earth's surface. The part of the Sun that we see is an extremely diffuse region of glowing, gaseous hydrogen and helium. Close examination reveals that it has a granular structure, as shown in Figure 22.6. The bright spots are regions where hot gases are rising upward, and the dark spots are cooler areas of descending gases. Thus, there is direct visual evidence that convection carries energy to the surface of the Sun.

Large dark spots, called **sunspots,** also appear regularly on the surface of the Sun, as shown in Figure 22.7. These are the same phenomena that were observed by Galileo over 350 years ago. A single sunspot may be small and last for only a few days, or it may be as large as 150,000 km in diameter and remain visible for several months. From spectral analysis, astronomers now know that sunspots are simply regions of the Sun that are 1000 K cooler than the gases around them. Because they are cooler, they radiate less energy and hence appear dark by comparison with the rest of the photosphere. No one is able to predict exactly when or where a particular sunspot will appear or how long it will last, but it is known that sunspot activity becomes more frequent on a 22-year cycle.

Since heat normally flows from a hot body to a cooler one, it would seem logical to suppose that a large, cool region of gas would be heated quickly and disappear. Yet relatively long-lived sunspots have been observed. Sunspot formation is associated with the presence of intense magnetic fields on the Sun. Astronomers believe that these strong magnetic fields restrict the solar turbulence and somehow inhibit the transfer of heat from hot regions into the nearby sunspots.

Large flares of hot gas sometimes explode from amid a group of sunspots. Gases accelerate

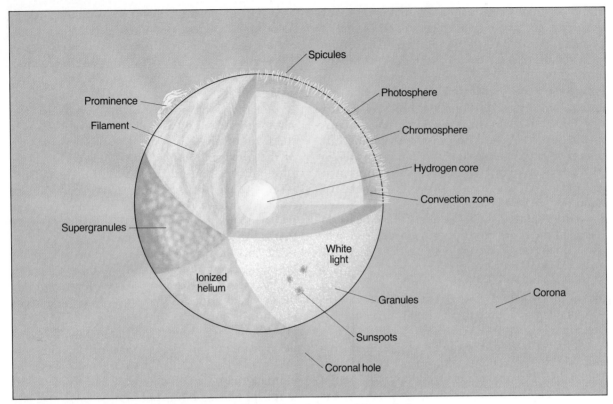

FIGURE 22.5 The structure of the Sun. The pie-shaped views of the exterior show what the photosphere looks like when viewed from cameras sensitive to different frequencies.

FIGURE 22.6 A view of the Sun showing its granulated structure. (Courtesy NASA)

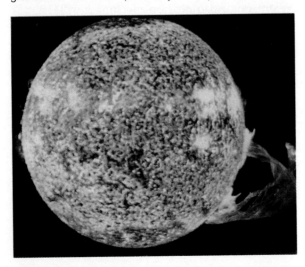

upward and outward, attaining speeds often in excess of 900 km/h, and shock waves smash through the solar atmosphere. Truly the Sun is not a static, homogeneous, symmetrical sphere such as the Greek philosophers envisioned. Sometimes these flares are powerful enough to have direct effects here on Earth; they initiate reactions that can interrupt radio communication and are associated with auroral displays in our polar regions. Also, such solar activity may affect terrestrial climate, but no definite relationship has been proved.

The Sun's outer layer

The Sun is not bounded by a sharply defined surface. Rather, its gaseous regions extend far out into space. Above the photosphere lies a turbu-

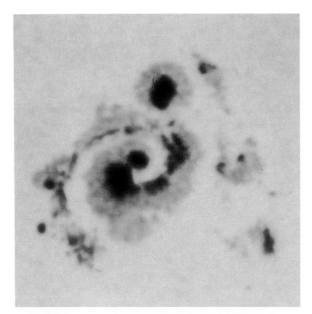

FIGURE 22.7 This unprecedented spiral sunspot was photographed on February 19, 1982. Sunspots are normally seen as irregularly shaped dark holes. This spiral sunspot had a diameter of approximately 50,000 miles—about six times the diameter of the Earth—and held its shape for about two days before it broke up and changed its form.

lent, diffuse, gaseous layer called the **chromosphere.** The chromosphere consists of a series of spikes, or spicules, that look something like flames from a burning log, as shown in Figure 22.8. A representative spike is about 700 km across and 7000 km high and lasts about 5 to 15 minutes. These spikes are composed of hot gases that are shot upward from the turbulent photosphere below.

Farther out, beyond the chromosphere, there exists an even more diffuse region called the **corona.** The corona can be observed as a beautiful halo around the Sun during a full solar eclipse. The corona is extremely hot, about 2,000,000 K. However, even though the temperature is extremely high, there is comparatively little thermal energy in the region. Recall that the thermal energy in a sample of matter is related both to the temperature and the mass of the substance. The corona has a density equal to one-billionth of the density of the atmosphere at the surface of the Earth. In a physics laboratory, such a density would be considered a good vacuum.

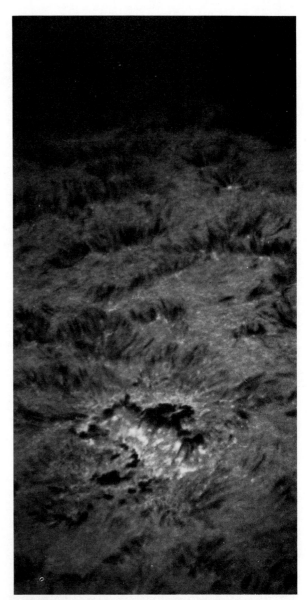

FIGURE 22.8 Numerous spicules in the chromosphere of the Sun.
(Courtesy NOAO)

How does the photosphere, which is at a temperature of 6000 K, heat the corona to 2,000,000 K? That is one of those questions with no satisfactory answer as yet. Different theories have been offered and disproven. At the present time, astronomers believe that perhaps twisting magnetic fields accelerate charged particles to high speeds, which are equivalent to very high temperatures.

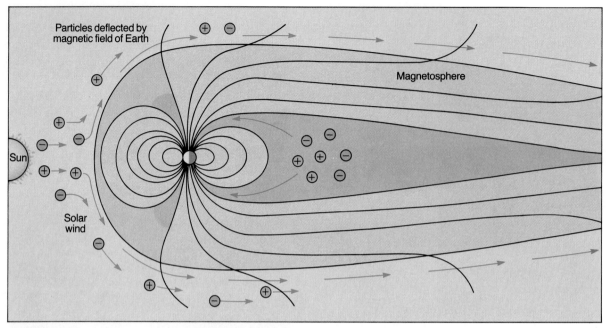

FIGURE 22.9 The Earth and its magnetosphere, showing how aurora are produced.

The solar wind and auroras

Matter behaves very differently at one-billionth of an atmosphere and 2 million degrees than it does in any terrestrial system. Electrons are stripped off their atoms. Hydrogen and helium are reduced to bare nuclei in a sea of electrons, and many of these particles fly off into space. This stream of nuclei and electrons is called the **solar wind.** It surrounds the Earth and extends

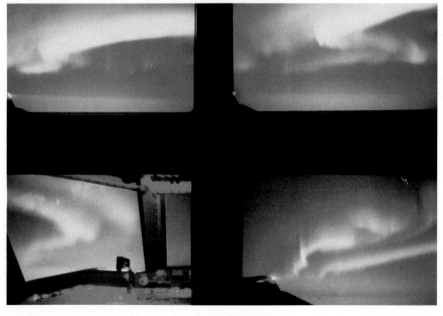

An aurora as seen from high above the Earth.
(Courtesy NASA)

This outstanding view of a full Moon was photographed from the Apollo 11 spacecraft from a distance of 10,000 nautical miles. Note the cratered regions and the flat plains, which are erroneously called seas or maria.
(Courtesy NASA)

outward toward the far reaches of the Solar System. When you think about the solar wind, don't think of a gentle breeze blowing against your face. The atmosphere on Earth contains approximately 10^{19} particles/cm^3, and you can feel the effect of these particles striking your cheek. However, the solar wind contains only 5 particles/cm^3 and certainly cannot be felt.

Despite this incredibly low density, the solar wind affects the Earth in many ways. These effects arise out of interactions between the charged particles of the solar wind and the magnetic field of the Earth. Recall from the discussion of magnetism in Chapter 9 that two different types of interactions can occur when a charged particle moves into a magnetic field. One possibility is that the particle will be deflected. The other possibility is that the charged particle will be trapped by the magnetic field, as we will discuss below. Both types of interactions occur between the solar wind and the magnetic field of our

planet. When high-speed nuclei and electrons from the Sun intersect the Earth's magnetic field, most of them are deflected and move around the Earth, leaving a comet-shaped area around our planet called the **magnetosphere,** as shown in Figure 22.9. Thus, the magnetic field of the Earth is a protective shield that prevents most of these high-energy particles from reaching the surface.

Other particles, those that come in almost parallel to the magnetic field lines of the Earth, may become trapped in this field and spiral from pole to pole circling about a field line. The number of particles trapped in this way builds up to the point that some of them are dumped out. These smash into the upper atmosphere, ionizing some of the atoms and molecules in the region and exciting others to higher energy states. When these molecules or atoms return to their original condition by acquiring an electron or by losing energy, they emit radiation of characteristic wavelengths. The visible wavelengths produce the auroras that are commonly seen from high latitudes on Earth.

22.3
THE STUDY OF THE MOON AND THE PLANETS

Before the exploration of space in the 1960s, the study of the Moon and nearby planets had been a great deal more difficult than the study of the Sun. Scientists are able to study the Sun and other stars through the interpretation of spectral data. But the Moon and the planets do not emit their own light; they reflect the rays of the Sun. Therefore, the spectrum of moonlight is nearly identical to that of sunlight, and it teaches us very little about the Moon. The nature of some regions of planetary atmospheres can be deduced by noting which frequencies of sunlight are absorbed, but this information tells us nothing about the nature of the surfaces of the planets. Unfortunately, all earth-bound studies of the surface detail of planets and moons must always be limited. Let us suppose, for example, that it were possible to build an optically perfect telescope as large as we wanted. No matter how efficient the telescope, it

FOCUS ON . . . How Is the Geological History of the Moon and the Planets Deduced?

Considerable information about the geological history of a moon or planet can be deduced by studying the number and type of meteorite craters. If there are two regions, and one is marked by a series of meteorite craters while the other is smooth and level, we can deduce that both regions were once covered with craters, for external bombardment would affect the entire planet equally. At a later date, some of these craters must have been obliterated by volcanic or tectonic activity or through extensive erosion. Thus, the cratered region is geologically older, and the smooth region would have been formed by geological processes. If a few scattered craters appear in the smooth areas, and if smaller craters lie inside larger ones, it seems reasonable to believe that a second, later era of meteorite bombardment followed the first.

This chronology serves as a rough guide to the sequence of events but does not date the various time periods. In order to establish lunar chronology, astronauts have collected samples of rock from various representative areas on the Moon and brought them back to laboratories on Earth. These have been dated by studying patterns of radioactive decay, and the oldest date back 3.9 billion years.

No one has yet retrieved samples of rock and soil from any other celestial bodies for careful analysis here on Earth. However, the lunar chronology may be used to establish a general geological time scale for the Solar System. First, we assume that any significant period of meteorite bombardment must have affected the Moon and nearby planets at the same time. Then it is a simple matter to compare crater patterns on planets with those on the Moon. If the densities, sizes, and general shapes of the craters on a planet are similar to those on the Moon, we may deduce the age of the planetary rock from our knowledge of the Moon. This type of deductive reasoning is not foolproof, but at the moment it is the best we have.

---

would not enable us to see the details of the surface of another planet. The problem arises because the Earth's atmosphere is turbulent and heterogeneous. Light entering this atmosphere is refracted unevenly. This effect produces an inherent blurring in any observation of a distant object, thus limiting the resolution of any terrestrial telescope. This problem was unsolvable until it became possible to build rockets to carry telescopes, cameras, and other scientific instruments above our own atmosphere.

These instruments have provided pictures and recorded data about planetary temperatures, the strengths of magnetic and gravitational fields, the chemical composition of planetary atmospheres, the nature of atomic particles in interplanetary space, and much more. Astronauts have landed on the Moon, carried out a variety of experiments, and returned to Earth with valuable samples of lunar rock and soil. As a result of all this exploration, our knowledge of the Solar System has expanded at an extremely rapid rate.

22.4 THE MOON

The Moon is close enough that its gross surface features are detectable by telescope. Its mountain ranges, smooth flat plains (called **maria** from the Latin for seas), and thousands of craters have been well known since the time Galileo first observed them in the 1600s. Before the Apollo space program, however, astronomers really knew very little about the geology of the Moon. The two most significant questions asked during the space exploration program were: First, how was the Moon formed, and how did it begin to orbit the Earth? Second, what is the geological history of the Moon: Was it once hot and molten like the Earth, and if so, does it still have a molten core?

The answer to the question of how the Moon was formed has been and continues to be the subject of considerable debate. There are several theories. Let us examine these in turn.

The surface of the Moon showing numerous craters.
(Courtesy NASA)

Theory: The Moon split off from the Earth soon after the formation of the planet. According to this theory, the Earth was spinning so rapidly that a large chunk of molten matter spun off, much as water spins out of wet clothes on the spin cycle of a washing machine.

This concept is largely rejected because no one can explain how the proto-Earth came to spin so rapidly, and if it did rotate at this rate, no one can explain how the rate of rotation of the Earth-Moon system has decreased to its present value.

Theory: The Moon was formed in some remote region of the Solar System and then was captured by the gravitational field of the Earth at some later time.

This event is highly improbable. Imagine that a Moon-sized object were flying randomly through space. The probability of passing close enough to a planet to be captured is very small. Even if it did pass closely enough to a planet, its trajectory and energy would have to be within very precise limits for capture to occur. Such an event might happen once, but most of the planets in our solar system have moons and some have many. It is difficult to believe that captures would have occurred repeatedly.

Theory: The Earth and Moon were formed simultaneously from a single cloud of dust and gas. As this cloud condensed, a high concentration of heavy elements gravitated toward the center of this sphere to form the Earth, while the outer ring, which became the Moon, was made up of minerals of lower density.

This theory has been popular for many years but has one major drawback: The chemical compositions of the Earth and Moon appear to be more different from one another than can be satisfactorily explained by this theory.

Theory: Shortly after the Earth was formed, a giant object, perhaps a miniplanet, smashed into the Earth. Parts of this object and segments of the Earth's crust were ejected into space by the force of the collisions, and these fragments began to orbit the central planet. Eventually, the fragments coalesced into a single satellite, the Moon.

Both calculations and comparative observations of the other planet-satellite systems support this theory as the most plausible hypothesis. However, questions still remain, and any theory, by its very nature, is open to question and dispute.

The second question, posed earlier, is: What is the geological history of the Moon, and how did it evolve in the eons after its original formation? One of the most significant discoveries of the Apollo program was that the entire surface of the Moon is covered with different types of igneous rock. Since igneous rock is formed only when lava cools, it is clear that at least the surface was once hot and liquid. The next logical question arose: How was it heated? Our planet was heated both by radioactive decay from within its interior and by intense meteorite bombardment from outer space. But what about the Moon? Theoretical geologists have calculated that it would take

Earthrise, as seen from the Moon.
(Courtesy NASA)

roughly 1 billion years for radioactive decay inside the Moon to build up enough heat to melt the rock. Analysis of lunar rock showed that the oldest igneous rocks were crystallized from molten lava when the Moon was a mere 400 million years old. Thus, since there was insufficient time for the rock to have been melted by the heat from radioactive decay, at least some of the melting of the Moon was caused by meteorite bombardment. Further study of lunar geology has given us a partial picture of the rest of its history, as outlined below.

1. Astronomers believe that the Moon was formed shortly after the original condensation of the Earth, some 5 billion years ago. A few hundred million years after its formation, much or all of the lunar material grew hot enough to melt. Possibly some of the heat required to melt the Moon was generated during the condensation of the protomoon, while most of the rest was supplied by intense meteorite bombardment. Within the molten interior, the denser elements such as iron and nickel gravitated toward the center, while the less dense minerals floated upward toward the surface. Thus, the Moon, like the Earth, has a small metallic core surrounded by a shell of rocks of lower density. Later, the Moon cooled and solidified so that this structure was maintained.

2. During the time period from about 4.2 billion years to 3.9 billion years ago, a second gigantic series of meteor showers rained down upon the Moon. Billions of meteors, some small and others as large in diameter as the state of Rhode Island, smashed into the surface and gouged out most of the craters that can be seen today.

3. At the same time that the meteor shower was marking the surface, the lunar interior was being heated by radioactive decay. Huge lava beds were formed deep beneath the surface. About 3.8 billion years ago, some of this molten material flowed upward through the surface crust to form many active volcanoes. Smooth lava flows covered vast regions of the Moon, forming what is now called the lunar maria, or seas. (The use of the word "seas" is a misnomer; on the contrary, these regions are dry, barren, flat expanses of rock. Early observers of the Moon believed that these dark regions actually were seas, and thus the term arose.) This volcanic activity lasted approximately 700 million years and ended about 3 billion years ago.

4. Because the Moon is small, the heat produced by radioactive decay was quickly dissipated into space, so that the Moon soon cooled considerably and now lies geologically quiet and inactive. Seismographs left on the lunar surface by Apollo astronauts indicate that the energy released by moonquakes is only one-billionth to one-trillionth as much as is released by earthquakes here on our own planet. Seismic data indicate that the core of the Moon is probably molten, or if it is not, it is at least hot enough to be soft and plastic. However, the cool, solid upper mantle and crustal layers are thick enough to inhibit appreciable seismic activity. Meteor bombardment of the Moon has continued throughout this history of geologic dormancy, but there has never again been an intense rain of meteors such as the one that occurred about 4 billion years ago.

Close-up of lunar plain with mountain range in background. The vehicle on the plain is the Lunar Rover.
(Courtesy NASA)

The lunar experiments tell us a great deal about the history of our own planet. If the Moon was subject to intense meteor bombardment 4 billion years ago, the Earth, being larger and exerting a greater gravitational force, must have attracted more debris and thus must also have been melted by the same process at the same time. We can imagine that the Earth also cooled at the end of this epoch, only to be reheated by its own internal radioactivity. Since the Earth is so much larger than the Moon, it has not cooled as rapidly; as a result, part of the core continues to be molten and the mantle is hot and geologically active to this day. Most of the original rocks of the Earth's crust have long since been pushed down into the mantle by tectonic activity, to be remelted and reformed in a continuous and dynamic process, and other ancient landforms have been altered or destroyed by erosion. Thus, evidence of the first

1.5 billion years of the Earth's history has been largely lost. But the Moon—cold, lifeless, and not subject to erosion from wind and water—has preserved a record of its history and has allowed us to probe more deeply into the origins of the Earth.

22.5 MERCURY

Our studies of lunar geology have taught us much about the state of our Solar System several billion years ago. But to complete the picture and to really understand what conditions were like throughout our local region of space, we must also consider the planets.

Mercury has a radius less than four-tenths that of the Earth. It is also the closest planet to the Sun, and therefore it orbits the Sun faster than any other planet, in obedience to Kepler's third law. Each Mercurial year is only 88 days. Mercury rotates on its axis rather slowly, so that it rotates only three times for each two complete revolutions around the Sun. Note here that the period of revolution and the period of rotation for this planet are almost the same. At first thought this might be considered a coincidence, but it actually isn't. When the period of revolution and rotation are the same, the smaller object is said to be in **synchronous** rotation with the larger. Thus, Mercury is almost in synchronous rotation with the Sun. Synchronous rotation occurs when the orbiting body is slightly more massive on one side than it is on the other. In this case, the gravitational force exerted on it by the object about which it revolves is slightly off-center, and this uneven gravitational force will cause the smaller one to move into a synchronous rotation after several million revolutions. Such synchronous rotation seems to be almost the rule rather than the exception in the Solar System. For example, the Moon is in synchronous rotation around the larger Earth, as are the moons of Mars. Figure 22.10 shows an interesting side-effect that occurs with synchronous rotation. Consider an observer at A in the figure. At the beginning of the revolution he is facing the Sun, and as the revolution progresses to B, one-fourth of the way through, the rotation carries the observer also one-fourth

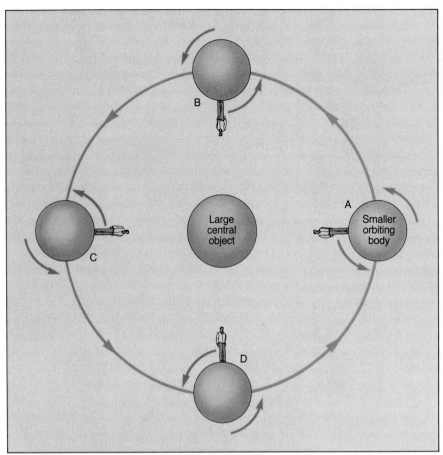

FIGURE 22.10 The smaller orbiting body is in synchronous rotation about the larger central object. Note that the same face of the orbiting body is always seen from the central object.

of the way through a complete turn. As we follow the motion through C and D, we see that when the rotation and revolution rates are the same, the observer is always facing the central body.

When the rotation is synchronous, the same side of the smaller object always faces the central object.

The above is true for our Moon. Until the Apollo astronauts flew by the back side of the Moon, we had never seen any of the features of that side of its surface.

Mercury is close to the Sun, and because it keeps one face pointing toward the Sun for a long period of time, its sunny side becomes extremely hot. Studies of infrared and radio emissions indicate a daily high of about 450°C (hot enough to melt tin or lead). On the other hand, the temperature of the side of the planet facing away from

the Sun drops to lows of −175°C (nearly cold enough to liquefy oxygen). These extremes of temperature are enhanced by the fact that there is virtually no atmosphere on Mercury, so there can be no wind to carry heat from one region to another and no cloud cover to retain heat on the dark side. The absence of an atmosphere can be explained by the fact that the temperatures on the planet are so high that air molecules would achieve a high enough speed to escape the small gravitational pull of a planet as low in mass as Mercury.

Little was known about the surface of Mercury before the spring of 1974. At that time, a sophisticated spacecraft called Mariner 10 passed close to Mercury and began relaying information back to Earth. The first Mariner 10 photographs revealed a cratered surface, similar in many respects to that of the Moon, as shown in Figure

FIGURE 22.11 A close-up view of Mercury. The photograph shows an area 550 km from side to side.
(Courtesy NASA)

22.11. The similarities in surface contours indicate that the geological histories of the two bodies must be similar. Thus, although many questions remain unanswered, most scientists agree that, after its initial formation, Mercury was subject to a period of intense meteor bombardment, similar to the bombardment that occurred on the Moon. Perhaps the interior of the planet was once hot, but no one is sure, for there are no vast lava plains comparable to the lunar maria. The ancient meteor craters stand out sharply, unmarked by extensive erosion or tectonic leveling, for there has been little geological activity during the past few billion years.

Perhaps the most striking discovery of the Mariner probe was that Mercury has a small but distinctly measurable magnetic field. Scientists believe that the Earth's magnetic field results from the effect of the relatively rapid rotation of its iron core. The discovery of a magnetic field on Mercury was a surprise because this planet rotates slowly. This anomaly is best explained by assuming that the metallic core of Mercury is rel-

atively larger than the core of the Earth. In turn, this assumption can also be justified. Mercury is closer to the Sun than the Earth, so during its formation a greater percentage of the lighter elements would have vaporized into space, leaving behind a proportionally larger core of dense metals.

Since the surface of Mercury is so hot, and the gravitational field of the planet is so small, we would expect that any atmospheric gases would have vaporized into space long ago. Surprisingly, an atmosphere was detected on Mercury, and although it is only one-billionth as dense as that on Earth, even this small amount has puzzled scientists.

22.6
VENUS

Of all the planets in the Solar System, Venus most closely resembles Earth with respect to size, density, and mean distance from the Sun. From this information alone, astronomers once believed

In 1975, a Soviet spacecraft landed on the surface of Venus. The intense heat and pressure rapidly destroyed the instruments aboard, but before the radio transmitter failed, a single photograph was sent back to Earth. This picture shows sharply angular rock, as can be seen in Figure 22.14a. A second Soviet craft soon landed 2000 km away and returned a photograph of a smooth landscape interspersed with sections of cooled lava or highly weathered rock (Figure 22.14b). What a wealth of information is contained in just two photographs! The angular rocks must be geologically young, for they would be expected to erode rapidly in the harsh conditions of the Venusian atmosphere. The smooth landscape and weathered rock of the second landing site would necessarily be older. Thus, scientists were able to deduce from these data that there has been recent tectonic activity on Venus.

that environmental conditions on Venus might be expected to be similar to those on Earth and thus that some form of life might be found there. However, until recently it was impossible to study the surface of Venus, for it is wrapped in a thick, dense atmosphere with an opaque cloud cover, as shown in Figure 22.12. But now it can be said with certainty that no life exists on Venus. The environment there is so harsh that it is often referred to as the "Hell Planet." Temperatures on the planet are extremely high; the surface actually has a slightly higher average temperature, about 460°C, than that of Mercury. As noted in the last chapter, most of the planets and moons in the solar system revolve and rotate counterclockwise as seen from north of the Solar System. Venus revolves in this counterclockwise fashion, but its rotation is retrograde—it rotates clockwise as shown in Figure 22.13.

The atmosphere of Venus

Venus is slightly smaller than the Earth with a radius equal to 0.95 that of Earth, and its gravitational force is also less. Nonetheless, the atmospheric density on Venus is 90 times greater than that of our planet. Thus, the pressure acting on an object on the surface of Venus is equal to the pressure on an object that is 1000 m beneath the surface of the ocean on Earth. The Venusian atmosphere is composed of 96 percent carbon dioxide and 3.5 percent nitrogen, with the remainder being helium, neon, and other gases. The clouds we see contain a high percentage of sulfuric acid. Thus, a rainfall on the planet would bring down the same acidic solution that is in a car battery.

One is immediately made to wonder why conditions on Venus should be so different from those on Earth. The most widely accepted answer to this question centers on the greenhouse effect, discussed briefly in Chapter 5. Recall that both carbon dioxide gas and water vapor absorb infrared radiation and warm the environment. Carbon dioxide can exist as a gas in the atmosphere, can be dissolved in water, or can be chemically bonded with other substances to form rock. Water commonly exists in its vapor form, or as a liquid, or as ice, a solid. According to the most widely accepted theory, at one time Venus was much cooler than it is today. If this premise is correct, rivers would have flowed over its surface, and oceans would have filled the central basins. Since Venus is slightly closer to the Sun than the Earth, Venus naturally receives more sunlight if all other conditions are equal. Therefore, the rate of evaporation on Venus must have been just a little greater than it was on Earth. But water vapor absorbs infrared, so a greenhouse warming occurred. In turn, as the temperature increased, more water evaporated. Conditions spiraled. As more and more water evaporated, the carbon dioxide that was dissolved in the oceans was released as well. Thus, Venus became hotter and hotter until eventually the liquid water all boiled away, and then the carbonate rocks decomposed, releasing even more carbon dioxide into the at-

(a)

FIGURE 22.12 Two views of Venus. Note that the solid surface of the planet is obscured by a turbulent cloud cover.
(Courtesy NASA)

(b)

mosphere. Eventually, an equilibrium temperature of about 460°C was reached.

Despite the incredible harshness of the Venusian environment, the atmosphere at the surface is not as turbulent and changeable as that on Earth. Venus rotates quite slowly on its axis, so slowly in fact that one day on this planet is actually longer than a year. Astronomers believe that the slow rotation is one fact that is responsible for the relative stability of the lower atmosphere. In contrast, the upper atmosphere is quite windy, with speeds up to 300 km/h. An ongoing problem in planetary astronomy is to explain these weather conditions on Venus. Such knowledge might help us to better understand wind circulation patterns on our own planet.

Another important difference between Venus and the Earth is that Venus has no magnetic field. In the absence of a protective magnetic shield, the particles of the solar wind penetrate easily into the upper atmosphere. In turn, these particles ionize many of the atmospheric gases, and the electrical turbulence generates continual lightning storms that travel back and forth across the upper layers of the clouds.

The surface of Venus

Unlike the surfaces of the Moon and Mercury, the surface of Venus shows many signs of relatively recent tectonic activity, as shown in Figure 22.14. Radar sensors mounted on spacecraft have

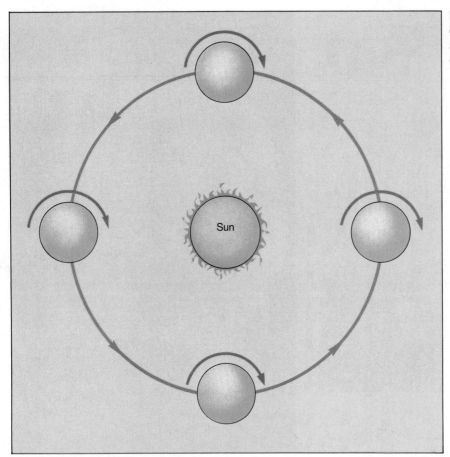

FIGURE 22.13 Venus has a counterclockwise revolution about the Sun but a clockwise rotation on its axis.

detected rolling plains, high mountains and deep valleys. Most of the terrain of Venus consists of plains, with only about 10 percent of the surface considered highland regions comparable to continents on Earth. There are two of these large highland regions on the planet, as shown in Figure 22.15. The larger, Aphrodite, is along the equator of the planet, stretching almost halfway around the globe. The northern highland region, Ishtar, is about the same size as Australia. This "continent" contains the highest mountains on the planet, the Maxwell Mountains, which reach a height of about 11 km above the lowland regions.

22.7
MARS

Mars has captured the imagination of scientists and lay people alike, since early observations seemed to indicate that if extraterrestrial life were to be found in our Solar System, it would be found on that planet. A part of the reason for this began with some observations of Mars made in 1863. An Italian astronomer, Giovanni Schiaparelli, drew a map of the planet that contained some lines he called *canali*. The best translation of this word to English would be channels; however, the translation to English became canals. This left the impression in the mind of most lay people that these were canals laid out by intelligent beings on Mars. In fact, in 1894, a noted American astronomer, Percival Lowell, drew a map of the planet based on his own observations that showed details of 500 canals.

An additional observation of Mars that led to the idea that it might have a hospitable environment for life was the existence of polar caps. Unlike Venus, Mars is covered by a thin, nearly cloudless atmosphere that enables astronomers to easily view its surface with telescopes. Several

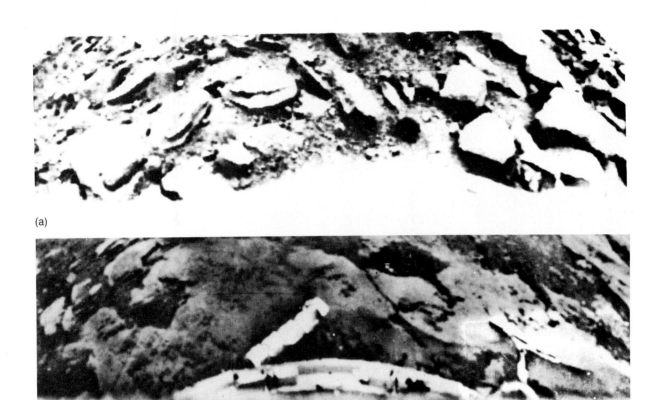

(a)

(b)

FIGURE 22.14 (a) The surface of Venus photographed from Venera 9. Notice the sharply angled rocks. (b) The surface of Venus 2000 km from the Venera 9 landing site, photographed by Venera 10. Note how much smoother the surface is.

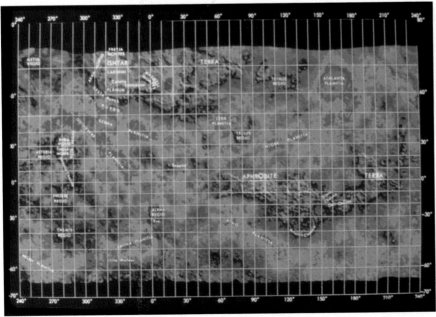

FIGURE 22.15 The surface of Venus as based on the Pioneer Venus radar map. Two continents exist on Venus: Aphrodite Terra, which is comparable in scale with Africa, and Ishtar Terra, which is comparable in scale with the continental United States or Australia.

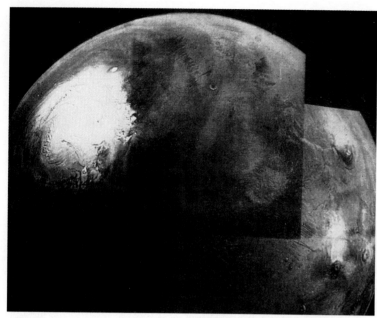

(a)

FIGURE 22.16 (a) A view of the north polar cap of Mars as seen by the Mariner spacecraft. (b) A close-up view of this same cap.
(Courtesy NASA)

(b)

hundred years ago, it was noted that the Martian polar regions are white and that the white ground cover shrinks in summer and expands in winter, as shown in Figure 22.16. These changes strongly suggest that the white regions are ice caps. If this speculation were correct, and if the ice caps melt significantly in summer, water must be available. This led to the idea that the purpose of the canals was to bring water from the polar ice caps to irrigate desert regions near the equator. Astronomers also observed that, each spring, large regions of the globe near the equatorial region darken, only to become light in winter. Many people thought that these color changes were caused by annual blooms of vast areas of vegetation. The axis of rotation of Mars is tipped at an angle of approximately 24° with respect to the plane of its orbit. As you recall, it is such a tilt that causes the Earth, and hence Mars, to have seasons. Thus, seasonal activity in the form of the growth of plant life might be expected, and this idea helped to solidify the theory of life on Mars.

In the 11 years between 1965 and 1976, a total of 12 United States and Soviet spacecraft were sent to Mars, including two spectacularly successful Viking vehicles that landed on the surface of the planet, where they collected and analyzed samples of Martian soil. The data from the spacecraft have drastically changed our picture of the Martian environment and have led to the conclusion that probably no life, not even that of microorganisms, now exists, or ever has existed on the planet. The postulated vast forests and canal systems have never been photographed, and we can be sure they do not exist. The seasonal changes of color are actually caused by great, dry dust storms powered by seasonal winds. When the large global winds subside, bright particles of dust are thought to settle in certain areas, causing these regions to be light in color. In the spring, local winds stir up the dust into suspension in the atmosphere. This sweeping action reveals the darker underlying surface of these regions.

You should not think of these dust storms as being of the type one might observe in a desert sandstorm. The atmosphere of Mars is extremely rarefied, with the total amount of atmospheric gases being about 1 percent of that on Earth and consisting largely of carbon dioxide. Thus, very little material is moved about in one of these storms. In fact, cameras located on the surface of the planet have barely detected any loss of visibility at all, even during an intense storm.

The winter ice caps, once thought to be the source of spring floods, are largely composed of frozen carbon dioxide, commonly called dry ice. Considerable quantities of water ice are present as well, but polar temperatures remain below the freezing point of water throughout the entire year and this water ice never melts. Thus, the shrinking of the polar caps during the summer months is caused by evaporation of the dry ice, leaving behind a permanently frozen water ice cap. At some time in the distant past, however, liquid water did exist on Mars. Rivers must have

FIGURE 22.17 Close-up of Mars taken by Mariner 9 spacecraft, showing a deep canyon. This channel is thought to have been formed by running water in Mars' geologic past.
(Courtesy NASA)

flowed across the surface, gouging out stream beds and deep canyons, for these features can be observed today, as shown in Figure 22.17. However, no rain has fallen for millions or hundreds of millions of years, and now there is no liquid water any where on the surface of the planet. This barren, rocky land is also quite cold. Near the equator, midday temperatures may reach as high as 20°C during the summer season, but at night these same locations experience a temperature drop to as low as −140°C. These extremes of temperature during a single day are explained by the thin atmosphere that is too rarefied to hold in the heat during the cloudless nights.

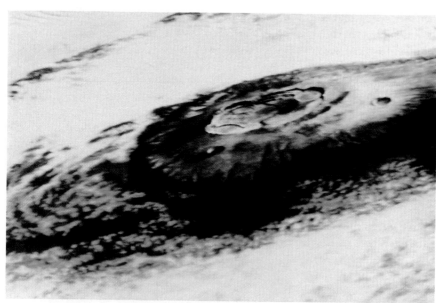

FIGURE 22.18 The largest volcano on Mars, and probably in the Solar System, is Olympus Mons.

When the spacecraft Mariner 9 was sent to Mars, one of the dust storms mentioned earlier arose, and there was considerable fear that the batteries on board would be drained before the storm subsided to the point where any surface details could be seen. For a while only three features could be seen near the equatorial belt of the planet. They were dark spots that could be explained only by assuming that they were mountains high enough to be seen above the enveloping dust storm. When the storm subsided, the spots were, indeed, found to be gigantic extinct volcanoes. The largest of these has been named Olympus Mons (Fig. 22.18). Its height is about 24 km, and its base would reach across the state of California. Thus, a getaway vacation to Mars would have to take you to this natural wonder. While there, you should also visit a huge canyon, named Valles Marineris in honor of the Mariner spacecraft, that stretches away from the base of Olympus Mons. This canyon makes our own Grand Canyon look like a small gully. On Earth, Marineris would stretch from New York City to Los Angeles, and in some places would be as much as four miles deep.

The moons of Mars

When seen, even by the naked eye, Mars has a distinct reddish coloration. This color is due to a complex series of interactions that arise basically because of the thin atmosphere of the planet. On Earth, very little ultraviolet radiation reaches the surface of the planet because it is absorbed by our relatively thick atmosphere. However, on Mars this protective blanket is not present, and the ultraviolet penetrates easily. One of the effects that this has on the planet is that any water vapor that was present in the atmosphere has been broken down into hydrogen and oxygen. The oxygen forms compounds with other elements present, notably iron. In general, iron oxides are distinguished by their red color. The chemical origin of the color of Mars was not known by ancient observers, but its color was known. As a result, the distinctive color of the red planet may have brought to mind the color of blood and had some influence on its having been named for Mars, the god of war. Students of mythology may recall that the Mars of myth had two colleagues who helped him with his dastardly deeds. They were aptly named Phobos (fear) and Deimos (panic). Thus, it seems appropriate that the two Martian moons should also bear these names.

Both Phobos and Deimos are irregularly shaped moons, as shown in Figure 22.19. Phobos is about 29 km at its longest point and about 16 km across at its narrowest point. Deimos is even smaller, with an approximate size of about 14 by 11 km. Phobos has the distinction of being

(a)

(b)

FIGURE 22.19 (a) The Martian satellite Phobos, and (b) the satellite Deimos. (Courtesy NASA)

the moon that is the closest to its planet of any in the Solar System. Its revolution rate is about 7 h and 10 min. Both moons are in synchronous rotation about Mars; the surfaces of both are heavily cratered; they are very dark in color, and they also have densities of about twice that of water. The dark coloration and the low densities are characteristics of asteroids. Thus, there is much speculation that these moons are asteroids that have been captured by the gravitational pull of Mars.

22.8 JUPITER

Mercury, Venus, Earth, and Mars constitute a foursome called the terrestrial planets. They are all relatively small, have a solid mineral crust, orbit close to the Sun, and have rotation rates that are relatively slow. On the other hand, the giant outer planets—Jupiter, Saturn, Uranus, and Neptune—are considerably different from the terrestrial group. Visualize once again the primordial dust cloud that was eventually to condense and become the Solar System. As the cloud shrank and broke apart, the protosun and all the protoplanets were originally composed mainly of hydrogen and helium. The protosun's gravitation

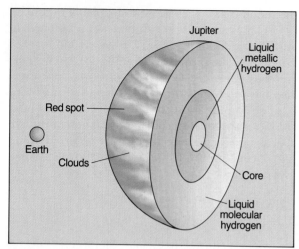

FIGURE 22.20 The current model of the interior of Jupiter.

FIGURE 22.21 Voyager 1 took this photo of Jupiter at a range of 32.7 million km. Notice the different colors in the clouds which form a band structure around the planet. Also, note the Great Red Spot near the equator.
(Courtesy NASA)

was so great that it pulled its gases inward with enough force to initiate fusion reactions. On the other hand, the gravitational fields of the terrestrial planets were so weak that most of their light gases escaped and boiled off into space or were blown away by the solar wind. Jupiter is very much larger than any of the terrestrial planets, yet very much smaller than the Sun. Therefore, it is physically similar to neither. Jupiter is the largest of the planets. In fact, its size is such that approximately 1300 Earths could fit inside it. The axis of rotation of Jupiter is tipped only about 3° with respect to the plane of its orbit, so it does not have seasons like the Earth.

As Jupiter was being formed, the inward condensation provided enough energy to heat the dust appreciably, but fusion temperatures were never reached. Yet the internal mass was sufficient for the gravitational forces to retain most of the original hydrogen and helium. Therefore, the chemical composition of Jupiter is much like that of the Sun, but its internal temperature and structure are very different. As shown in Figure 22.20, there is no hard, solid, rocky surface where an astronaut could land or walk about. The surface of the planet and more than half of the volume of its interior is a vast sea of cold, liquid hydrogen. An inner layer, composed of hydrogen in a different form lies beneath this hydrogen ocean. At the extreme pressures and

temperatures found there, the electrons become separated from the nucleus. In this state, the substance is referred to as **liquid metallic hydrogen.** The term "metallic" is used to indicate that it is a good conductor of electricity. Jupiter's core is believed to be a solid, rocky sphere about 20 times as massive as the Earth and probably composed of iron, nickel, and other metals and minerals.

The Jovian atmosphere contains a mixture of gases, liquid droplets, and crystalline particles consisting of hydrogen, helium, ammonia, methane, water, hydrogen sulfide, and other substances. This atmosphere is indeed a turbulent region, as even a casual glance at Figure 22.21 reveals. It is heated from above by the Sun and from below, to an even greater extent, by the interior of the planet. Moreover, the giant planet

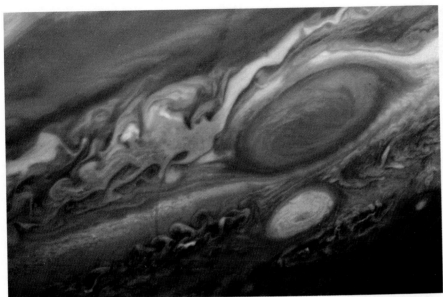

FIGURE 22.22 Two close-up views of the Great Red Spot, a hurricane that has been swirling for over 300 years.
(Courtesy NASA)

(a)

(b)

spins quite rapidly, rotating once approximately every ten hours. (A rapid rotation rate is a characteristic of all the gas-giants.) All of these effects combine to generate turbulent wind systems, great storms, and changing weather patterns on the surface. Most of the recognizable storm systems appear to form, distort, and move on within a few hours or days, but some of them are surprisingly stable over long periods of time. Over

300 years ago, two European astronomers reported seeing a **Great Red Spot** on the surface of Jupiter; although its shape and color have changed noticeably from year to year, the spot remains intact to this day, as shown in Figure 22.22. Data from space probes of the planet indicate that the spot is a giant hurricane-like storm. If the entire Earth's crust were peeled off like a giant orange rind and laid flat, it would fit en-

FIGURE 22.23 A volcanic eruption, on the horizon, as photographed by Voyager 1. Solid material is being thrown up to an altitude of about 160 km. (Courtesy NASA)

tirely within the Great Red Spot. Astronomers have devoted considerable effort toward attempting to explain how the Great Red Spot has maintained itself for over 300 years. Certainly, our theoretical understanding of the Great Red Spot has improved in the last decade, but no one can explain for certain how it has remained so stable for so long or predict how long it will last. Hurricanes on Earth dissipate after a week or so, yet on Jupiter a storm has been raging for centuries.

Another obvious feature of Jupiter that can be noted from Figure 22.21 is the dark and light colored bands that circle the planet. The light areas are regions where gas is rising from the interior of the planet, and the parallel darker regions are locations where gases are descending toward the interior.

Apparently, another characteristic of the Jovian planets is that they have rings around them. The rings of Saturn are well known by elementary school children, and until recently it was thought that this was a feature only of Saturn. However, the Voyager 1 spacecraft returned photographs of a thin ring circling Jupiter as well. The particles in this ring are extremely small, about the size of those found in cigarette smoke. As a result, these rings could not have been around since the formation of the Solar System; the solar wind would have driven them off into space. This means that they are constantly being replenished by some source, perhaps by particles from Jupiter itself.

22.9 JUPITER'S MOONS

Recall from the last chapter that the discovery of Jupiter's moons played an important role in the development of our present understanding of the Solar System. In 1610, Galileo discovered four tiny specks of light close to Jupiter. He noted that they distinctly orbited Jupiter and correctly reasoned that they were satellites of the giant planet. This direct visual evidence that at least some objects did not orbit the Earth was the first concrete proof that the Earth was not the center of all motion in the heavens.

A total of 16 moons (at last count) revolve around Jupiter. Of these, the four that were originally discovered by Galileo have been the most widely studied. The innermost of the so-called Galilean moons, Io, is small, dense, and rocky, while the outermost, Callisto, is significantly less dense and is believed to consist largely of water ice. Thus, the Jovian moon system is reminiscent of the Solar System itself, for, as we have seen, the

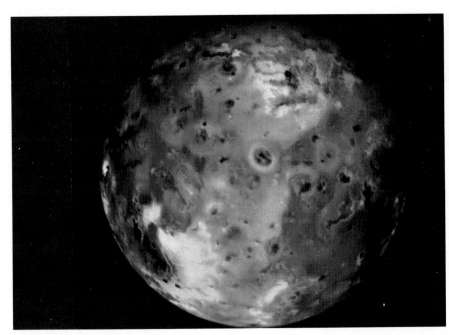

FIGURE 22.24 The colorful, volcano-dotted face of Io is unlike any other planet or satellite.
(Courtesy NASA)

inner four planets are much denser than the five outer ones.

These similarities imply that perhaps Jupiter and its moons were formed simultaneously, much as the Solar System itself was formed. Just as the Sun and planets condensed out of a single nebula, it is possible that Jupiter and its moons condensed as sort of a mini–solar system of their own. As Jupiter coalesced, huge quantities of heat must have been generated by gravitational forces. This heat was sufficient to boil most of the lighter elements off the surface of Io, leaving behind a relatively dense sphere. The outer two Galilean moons, Ganymede and Callisto, retained more of their lighter elements and are less dense than Io.

Io

Io is a small satellite, only two-thirds the size of the Earth's Moon and just slightly more dense. Since it is too small to have retained the heat released by radioactive decay in its interior, some observers expected to see a cold, lifeless, cratered, lunar-like surface. Nothing could be further from the truth. Spectacular photographs transmitted from Voyager spacecraft showed clear images of volcanoes erupting on the surface of Io, as shown in Figure 22.23. The pictures provided the first evidence of currently active extraterrestrial volcanism in the Solar System. Huge masses of gas and rock were seen to be ejected higher than 200 km above the surface. This material is not lava, steam, or carbon dioxide, which are the normal components of the material ejected from terrestrial volcanoes. Instead, the material ejected from the volcanoes of Io consists of sulfur and sulfur dioxide. The plumes from the volcanoes cool as they rise, condense, and drift downward as "snowfall" that covers wide areas of the surface. Sulfur and sulfur compounds generally have colors that range from white through all the hues and shades of orange and red. As a result, a photograph of Io looks like a huge pizza because of its unusual coloration, as shown in Figure 22.24. The average daytime temperature on Io is about 130 K, and the ejected gases give it a very rarefied atmosphere.

A question that scientists had to grapple with was: Why is Io geologically active? The answer seems to be that Io is affected by a complex interplay of gravitational forces between Jupiter and two other moons of Jupiter, Europa and Ganymede. Jupiter is about 300 times more massive than the Earth, yet Io is about the same distance

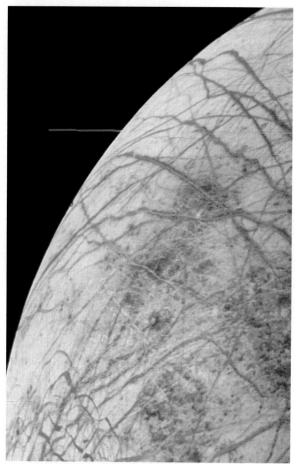

FIGURE 22.25 Europa has a surface of water ice, crossed by complex cracks.
(Courtesy NASA)

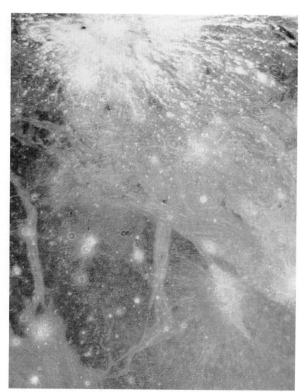

FIGURE 22.26 A close-up view of the surface of Jupiter's largest satellite, Ganymede.
(Courtesy NASA)

from that planet as is our own Moon from Earth. Thus, Jupiter exerts enormous gravitational forces on Io. In addition, Europa and Ganymede also exert gravitational forces, but in the direction opposite to that of Jupiter. The opposition of these forces causes Io to exhibit tidal flexing, and this flexing causes enough friction within the interior of Io to generate the heat required for nearly continuous volcanic activity. The frequent lava flows have obliterated all ancient land forms, giving Io a smooth and nearly craterless surface.

Europa

The next moon out from Jupiter is Europa, as shown in Figure 22.25. Because of its greater dis-

tance from Jupiter, Europa is not subjected to enormous tidal forces, and as a result, it has no active volcanoes. However, the surface is smooth and relatively craterless. Since meteor bombardment probably occurred sometime in Europa's history, the smooth surface seen today indicates that some kind of geologic activity has produced a degree of self-renewal on its surface.

Europa basically has a rocky composition, but its surface is covered with ice. Large streaks were observed in this ice surface, which might be caused by cracks in the ice that have opened and then been refrozen.

Ganymede

Ganymede is the largest moon in the Solar System, with a diameter of 5270 km, larger than the planet Mercury. It is a large sphere composed of a mixture of rock and water ice, as shown in Figure 22.26. Two distinctly different types of terrain are observed on this moon; one is heavily

FIGURE 22.27 Callisto photographed from Voyager 1. The bullseye is an impact crater about 2600 km across.

cratered, while the other is grooved and contains fewer craters. Astronomers believe that the heavily cratered regions are more than 4 billion years old, whereas the smooth regions are much newer and have been formed by recent internal activity.

Callisto

Callisto, the outermost Galilean moon, is heavily cratered and shows no grooved or smooth terrain, as shown in Figure 22.27. These data indicate that its surface crust is very old. This moon has a diameter of 4820 km, about the same as the planet Mercury. However, the density of Callisto is about one-third that of Mercury, which tells us that it is an icy body. The temperature of Callisto is about 120 K at the equator. Its most significant surface feature is a huge bull's-eye impact crater. The meteor that caused the bull's-eye did such extensive damage to the moon that it almost shattered it.

22.10 SATURN

Saturn, the second largest planet, is similar in many respects to Jupiter. It has the lowest density of all the planets, so low in fact that the entire planet could float on water if there were a bathtub large enough to hold it, as shown in Figure 22.28. This low density implies that it, too, must be composed primarily of hydrogen and helium. The atmosphere of the planet is similar to that of Jupiter. Dense clouds cover the planet, and several distinct storm systems have been photographed, although none as notable or persistent as the Great Red Spot of Jupiter. Additionally, the atmosphere of Saturn has the same banded structure as does Jupiter, as shown in Figure 22.29, but the variation in the colors of these bands is not quite as distinct as on Jupiter. Like Jupiter, and all the gas giants, Saturn has a rapid rotation rate, turning on its axis in about 10 hours and 40 minutes. Because of its high rate of rotation and low density, Saturn is the most oblate of all the planets. Its diameter at the equator is about ten percent greater than its diameter as measured pole to pole. Its axis of rotation is tipped about 27° with respect to a line drawn perpendicular to its orbital plane, so it does have seasons. The mean temperature of Saturn is lower than that of her sister planet, Jupiter; thus some substances that are gases on Jupiter are frozen solid on Saturn. In addition, the colder temperatures have favored the formation of certain organic molecules such as ethane, C_2H_6.

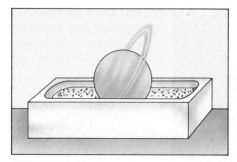

FIGURE 22.28 Saturn's density is lower than that of water.

FIGURE 22.29 Voyager view of the cloud structure of Saturn, which is much less spectacular than that on Jupiter.

Certainly, the most distinctive feature of Saturn is its spectacular rings that are readily visible from Earth, even through a small telescope, as shown in Figure 22.30. In addition to the rings, there are at least 17 satellites orbiting the planet. Before the space program, astronomers had observed three distinct rings with dark gaps separating them. However, observations from the Voyager space probe showed that the rings are incredibly more complex than was previously expected. A total of seven major rings was recorded, and each of these is further differentiated into thousands of smaller ringlets, as shown in Figure 22.31. This ring system is extremely

FIGURE 22.30 The ring system of Saturn as it can be seen by a good telescope on Earth.
(Courtesy NASA)

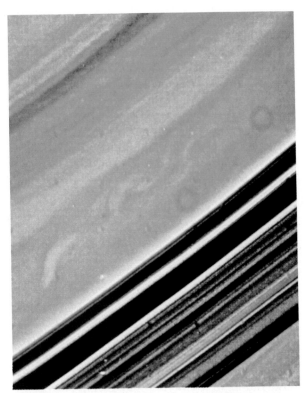

FIGURE 22.31 A close-up of view of Saturn's rings. (Courtesy NASA)

thin. Estimates vary, but observations indicate that the rings are only 10 to 25 m thick, considerably thinner than the length of a football field. However, they have a large diameter and cover a distance of some 425,000 km from the inner edge to the outer edge. If you were to make a scale model of the ring system and you chose a plastic disk the thickness of a phonograph record, your model would have to be 30 km in diameter.

The rings are not rigid but are composed of many separate particles of dust, rock, and ice. The particles in the outer rings are only a few ten-thousandth of a centimeter in diameter, but the innermost ones are made up of larger chunks a few meters across. Each piece orbits the planet independently, and some are moving faster, some slower, in a continuous jumbled parade. Two major questions have been asked about the rings: First, How were they formed? Second, Why are they so intricate and complex?

According to current theory, these rings are believed to be the scattered remnants of a moon that was never formed or that was formed and then ripped apart by the gravitational field of Saturn. Small objects, such as a rock or an ice cube, are held together by electrical attractions between the atoms and molecules. But large objects, such as the Sun, the planets, and their satellites, are held together mainly by gravitational forces. Now imagine what happens to a small satellite orbiting a larger central planet. The surface of the moon closest to the planet is attracted more strongly than the region farther away, as shown in Figure 22.32. A strong force on one side of the moon and a weaker force on the other tend to elongate the moon and break it into pieces. Such forces are called **tidal forces.** Tidal forces between the Earth and Moon cause ocean movements on the Earth and seismic rumblings on the Moon, whereas forces between Io and Jupiter are comparatively greater, causing the internal layers of rock in Io to move and heat up. If a satellite is too close to its planet, then the tidal forces can be greater than the internal forces that hold the moon together, and the moon is pulled apart or never formed in the first place.

The explanation of the fine structure of the rings involves three separate mechanisms. Each of these mechanisms originates independently of the others, yet the three combine to create a single system.

1. From fundamental laws of physics, any rotating, flattened disk that is made up of independent particles and is orbiting a central massive object will generate spirals like the grooves of a phonograph record. This type of structure is clearly seen when billions of stars orbit around a massive galactic core. The same effect occurs with the small pieces of debris that orbit Saturn.

2. The particles in the rings of Saturn are attracted not only by the planet but also by the 17 moons that orbit it. Over the years, the gravitational pull of the satellites has altered the orbits of the individual particles, spreading out the ring system and creating gaps within it. In some instances, two satellites work together to "shepherd" particles into precise orbital zones,

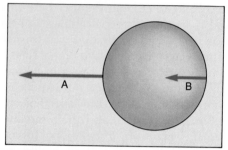

FIGURE 22.32 The force on side A is greater than on side B. This tends to stretch the moon and to create internal stress.

much as two sheep dogs work together to channel a flock of sheep into a thin column that can pass through a narrow gate.

3. The third mechanism is not gravitational but electromagnetic. Saturn has a strong magnetic field. It follows that the field would trap charged particles in belts, as does the Earth. Scientists believe that the solid particles of the rings become charged and complex attractions and repulsions between these particles then alter their orbital paths.

Although it is difficult to analyze each of these independent factors mathematically and then integrate them into a single, unified picture, the overall effect of these factors has produced the enormous, beautiful ring pattern that makes Saturn one of the most awe-inspiring sights in our Solar System.

22.11
TITAN

Of the 17 moons of Saturn, Titan is the largest. It is also one of the most unique of the moons in our Solar System, because it is the only satellite with an appreciable atmosphere. This atmosphere is largely methane, CH_4, a material that is commonly used on Earth as a fuel and is the major component of natural gas. The mean temperature on the surface of Titan is 90 K ($-183°C$), and the atmospheric pressure is 50 percent greater than the atmospheric pressure on the surface of the Earth. These conditions are close to the point where methane can exist in the solid, liquid, or vapor form. Therefore, small changes in atmospheric conditions on Titan cause meth-

ane to freeze, melt, vaporize, or condense. This situation is analogous to the environment on Earth, where water exists in each of its three phases and transfers back and forth readily between them. At present, astronomers believe that the surface of Titan is covered by a methane ocean.

At one time, astronomers held out the hope that conditions on Titan might be suitable for life. It was thought that the thick atmosphere might, via the greenhouse effect, have produced high enough surface temperatures so that some microbial life might have formed. However, the conditions as presented above by way of data from the Voyager probes are so harsh that we cannot visualize any such life forms.

22.12
URANUS AND NEPTUNE

Uranus and Neptune, both invisible to the naked eye from Earth, were unknown to the ancients. They are so distant that even today ground-based observation of them is limited. They are both quite large in size and low in density. The best evidence indicates that they are similar in structure to Jupiter and Saturn, being composed of a dense atmosphere, a liquid surface, and a solid mineral core. In 1977, scientists discovered rings around Uranus. In 1989, astronomical observations from the Voyager 2 spacecraft found five rings around Neptune.

Data from the Voyager 2 satellite indicate that, as expected, Uranus is enveloped in a thick atmosphere, composed primarily of hydrogen and helium with smaller amounts of compounds of carbon, nitrogen, and oxygen. The surface is believed to be a sea of methane, ammonia, and possibly water, and the core is probably rocky and about the size of the Earth. Figure 22.33 is a view of Uranus one week before the closest approach of Voyager 2.

Uranus is the only planet in the Solar System that is "tipped over." This means that its axis of rotation is in the same plane as its plane of revolution, as shown in Figure 22.34. Thus, at times the polar regions are aimed toward the Sun. To make matters even more confusing, the magnetic pole is tilted 55° off the planet's rotational axis. (Earth's magnetic North Pole is tilted 11.7° from

FOCUS ON . . . Extraterrestrial Life

For many years, people have asked whether or not the Earth is unique in its ability to support life. In the beginning, the search for extraterrestrial life was concentrated on the nearby planets, Venus and Mars. But, as we have seen, Venus is too hot to support life, and Mars, although potentially more hospitable, has been found, thus far, to be completely devoid of any living organisms. Today, astronomers are expanding their search for life toward other regions of our galaxy.

As for life in other solar systems, one can only make some sort of guess. As we will learn in the next chapter, there are roughly 100 billion stars in an average galaxy and billions of galaxies in the Universe. Thus, there are so many stars in existence that it is reasonable to believe that solar systems have formed around other stars. Distances are so large in the Universe that the only reasonable way to search for life is via radiotelescopes. Such a telescope, pictured in the Figure, detects radio waves from space instead of visible light. In 1960, a radiotelescope followed two nearby stars looking for any tell-tale signs of radio emissions different from what would be expected normally from a heavenly object. Later in 1970, about 600 nearby stars were scanned in the same manner, but thus far there is no indication of extraterrestrials. In 1971, a joint project between NASA and the American Society of Engineering Education suggested that a giant array of 100 radiotelescopes be constructed for the sole purpose of searching for signals from extraterrestrial life. Thus far, this project has not been funded.

The acronym used to describe searches such as those described above is SETI, for **s**earch for **e**xtra**t**errestrial **i**ntelligence. There is another side to the

The 43 m radio telescope at Green Bank. Note the size of the astronomer standing under the telescope.
(Courtesy of Marc Kutner)

coin, called CETI, for **c**ommunication with **e**xtra**t**errestrial **i**ntelligence. By this it is meant that we should also use the radiotelescopes mentioned above as transmitters of information in an effort to let others "find us." You should be aware that if such an effort ever meets with success, the dialogue will be considerably different from that normally found in radio transmissions here on Earth. The distance to even the nearest stars is such that if we say "hello," the message would not reach a nearby star for about 4 years, and it would take another 4 years minimum to receive a reply. For more distant stars, the answer to a question could require several lifetimes.

---

its geographic pole.) Scientists speculate that maybe a collision with a giant object knocked the planet over, giving it its strange tilt. However, the anomaly of the magnetic field has not yet been explained.

A complex ring system and a total of at least 15 moons have been discovered orbiting Uranus. Several of these moons are small and irregular, indicating that they may be debris from the postulated collision that tipped the planet. The largest moons show complex surfaces indicative of a varied geological history. For example, ten differ-

ent kinds of terrain have been identified on the moon Miranda.

In contrast to the symmetrical rings of Saturn, those around Uranus are warped, tilted, bizarrely elliptical, and varied in width. Some of these features can be explained by the proximity of "shepherd moons," but further analysis of the data is needed before a full explanation is available.

Neptune is often referred to as a triumph of gravitational theory. This description occurs because its presence and position were predicted

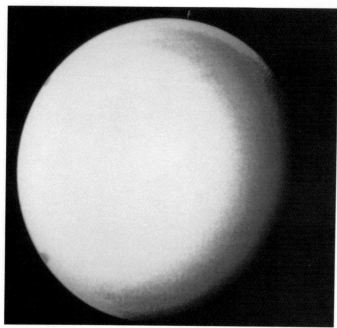

FIGURE 22.33 The cloud structure of Uranus is much less spectacular than that on either Jupiter or Saturn.
(Courtesy NASA)

FIGURE 22.34 The axis of rotation of Uranus is tipped over.

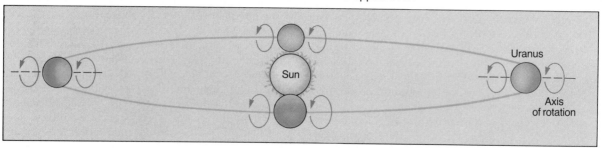

before it was found by telescope. The story began as astronomers attempted to explain some unusual variations in the orbit of Uranus. The only way to explain these deviations was to postulate the presence of another planet that was producing them. Thus, two mathematicians, Adams and Leverrier, predicted the location of the planet and are given credit for its discovery even though it was first actually seen by the astronomer Galle. The Voyager 2 spacecraft passed by the planet in 1989, and at this time the evaluation of the data obtained is still incomplete. However, several significant pieces of information have come forward. Photographs reveal a Great Dark Spot near the equatorial region that seems to be a storm like that seen in the Great Red Spot of Jupiter (Fig.

22.35). Six new moons have been found to give a total of eight at present, but more are likely to be found as analysis of the photographs continues. It has also been observed that the planet has bands around the polar caps.

It is also hoped that the Voyager 2 data will help to unravel the history of the Neptunian moons, including Triton and Nereid. Triton is only slightly smaller than our own Moon. One of its unusual features is that it has a retrograde revolution in that it revolves about Neptune clockwise. Triton is the only moon in the Solar System to exhibit this behavior. The moon Nereid is also unusual in that it has the most eccentric orbit of any moon in the Solar System. (The more eccentric the orbit of an object, the more squashed and

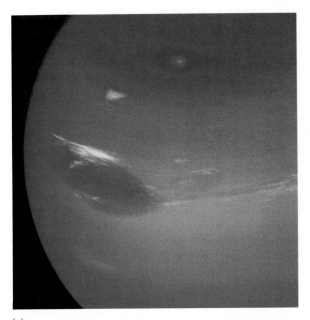

(a)

FIGURE 22.35   (a) A view of Neptune as seen from Voyager 2, and (b) a close-up of the Great Dark Spot.

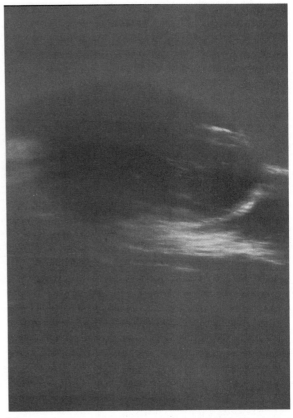

(b)

elongated is the orbit. For example, a perfectly circular orbit is said to have zero eccentricity.) These unusual features of Triton and Nereid have led many astronomers to believe that some catastrophe happened early in the history of the formation of the Solar System. One of the most common speculations is that a near-collision occurred between these moons and some larger object. The near-miss sent Triton into its retrograde orbit and caused Nereid to deviate into its strange orbit.

22.13
PLUTO

Pluto, the outermost of the known planets, is quite small, roughly about the size of Earth's Moon. Spectroscopic studies of Pluto's surface indicate that methane ice exists on the surface. Therefore, temperatures must be extremely low, about 40 K (−230°C). For many years, astronomers believed that Pluto was dense and rocky,

like the terrestrial planets. However, our understanding was altered drastically in 1978 when a satellite, now named **Charon,** was discovered orbiting this distant planet, as shown in Figure 22.36. This moon has the distinction of being the largest moon relative to the size of its planet of any in the Solar System. Using Kepler's third law, it is possible to calculate the relative masses of the planet and satellite when the radius of the satellite's orbit and its period are known. Using the best available data, the density of Pluto has been estimated to be only slightly greater than that of Saturn, indicating that it is composed of light elements.

22.14
VAGABONDS OF THE SOLAR SYSTEM

The imposing parts of our solar system include the Sun, the planets, and the moons, but now we

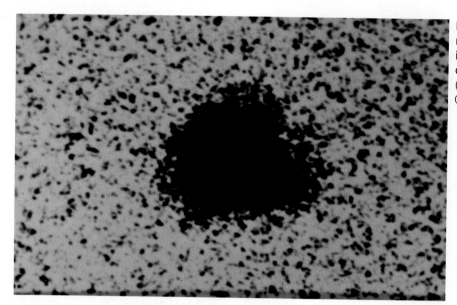

FIGURE 22.36 Highly magnified image of Pluto and its satellite (the small bump on the image). (Courtesy U.S. Naval Observatory)

shall turn our attention to some objects of lesser significance—meteoroids, asteroids, and comets. As we shall see, some of these objects produce brief flickers in our night sky, others have caused tremendous devastation on our planet in the past, whereas others can lead to dramatic displays sufficient to cause people to ask the age-old question: Why? Let us look at these vagabonds of the Solar System in turn.

Meteoroids

The terrestrial planets are no longer subject to the intense bombardment from outer space that once generated enough heat to melt large volumes of crustal rock. But if you sit outside on almost any clear night, watching the sky for a few hours, you may see a fiery streak called a **meteor,** or colloquially a **shooting star,** descend toward Earth, as shown in Figure 22.37. Shooting stars appear when small bits of interplanetary solid matter called **meteoroids** are caught by Earth's gravity and accelerated through the atmosphere. Friction between the meteoroid and the atmosphere produces enough heat to raise the temperature of the meteoroid to a high level. The fiery streak across the sky, seen by an observer on Earth, is heated air and vapor from the meteor. Most meteoroids are barely larger than a grain of

FIGURE 22.37 A meteor, or shooting Star. (Courtesy of Smithsonian Astrophysical Observatory)

sand when they enter the atmosphere and are completely vaporized before they reach the Earth. If they are larger, say the size of a basketball, some of the original material falls to the ground. A fallen meteoroid is called a **meteorite.** Examination of fallen debris indicates that most meteorites are roughly as old as the Solar System, approximately 4.6 billion years old. Some meteorites have never been subjected to planetary heating and remelting, and scientists believe that these fragments represent the kind of primordial material that originally condensed out of the interspacial dust to form the terrestrial planets.

One fascinating observation is that some meteorites contain fairly complex organic compounds, including amino acids and other molecules that are vital components of living organisms. Thus, complex organic molecules are not unique to the Earth but have been formed in other regions of the Solar System. This fact by itself does not mean that life exists elsewhere in our Solar System, but it does tell us that the molecules of life have been synthesized by inorganic (nonbiological) reactions occurring in outer space.

Asteroids

Eighteenth-Century astronomers noted that the dimensions of the planetary orbits increased in a regular pattern, starting with Mercury's orbit, the smallest, and going on to those of Venus, Earth, and Mars. There seemed to be a gap before Jupiter. On the basis of this interruption in the pattern, it was predicted that a planet might be found in the "open space" between Mars and Jupiter. Instead of a full-sized planet, however, observers have found tens of thousands of smaller bodies orbiting in a wide ring. These bodies are called **asteroids.** The largest asteroid, Ceres, has a diameter of 770 km. Three others are about half that size, and most are far smaller. The orbit of a given asteroid is not permanently fixed like that of a planet. If an asteroid passes too close to a nearby planet, it will be pulled toward it and fall onto the planet's surface. However, if an asteroid passes by a planet without getting too close, the gravitational force of the planet will pull the asteroid out of its current orbit and deflect it into a new orbit about the Sun. Thus, a given asteroid may change its orbit frequently in an erratic manner. Frequently, small fragments and pieces of dust are deflected helter-skelter in widely divergent directions. Some of these fragments cross the orbit of the Earth and are at-

tracted by our gravitational field. These then fall through the atmosphere and are visible as meteors. Asteroids are not the only source of meteors. When a comet passes too near the Sun, it can be broken up by tidal forces. When this occurs, the debris drifts through space, covering a large volume of the Solar System. When the Earth passes through this trail, intense meteor showers are visible. In fact, most meteors are remnants of comets, not asteroids.

What is the probability that a large fragment, or even an entire asteroid, will strike the Earth some day? Geological evidence indicates that such events have occurred in the past. For example, there is a large crater near Winslow, Arizona, approximately 1.5 km in diameter, that was formed by a falling meteorite about the size of a large semi-truck, as shown in Figure 22.38. This meteorite is believed to have landed in recent geological history, and the crater is perhaps no more than 50,000 years old. Other curious circular basins exist on the surface of the Earth that may well be eroded remnants of large meteorite craters.

Comets

Occasionally, a fuzzy object appears in the sky, travels slowly around the Sun in an elongated elliptical orbit, and then disappears again out into space, as shown in Figure 22.39. These objects are called **comets** after the Greek word for "longhaired," and have been considered to be powerful astrological omens. Despite their fiery appearance, comets are quite cold, and most of the light that we see is reflected sunlight. When a comet is far out in space, millions of kilometers from the Sun, it has no tail at all, but is simply a cohesive ball. This ball is believed to be a tenuous collection of rock and metallic particles coated with water, ammonia, methane, and carbon dioxide, all frozen as solids. As the comet approaches the Sun, solar radiation vaporizes its surface, and the force of the solar wind blows some of the lighter particles away from the head to form a long tail. As the comet orbits the Sun, the solar wind constantly blows the tail so that it always points away from the Sun, as shown in Figure 22.40. Comet tails have been observed to be over 90 million

FIGURE 22.38 Aerial view of the meteor crater near Flagstaff, Arizona.

A meteorite about the size of a basketball.
(Courtesy NASA)

miles long (almost as long as the distance from the Earth to the Sun). There is very little matter in a comet tail. By terrestrial standards, this region would represent a good, cold laboratory vacuum, yet viewed from a celestial perspective, such an object looks like a hot, dense, fiery arrow. Some comets travel in very long, cigar-shaped orbits that carry them well out into space, beyond even the orbit of Pluto.

Approximately 12 comets per year pass by the Earth on their orbits around the Sun, but most of these are too faint to be seen by the naked eye. Very few comets are visible enough to attract the attention of nonscientists. The most famous of all is Halley's Comet, which made an unspectacular pass by the Earth in late 1985 and early 1986.

FOCUS ON . . . Halley's Comet

"When beggars die, there are no comets seen;
The heavens themselves blaze forth the death
of princes."

Shakespeare, *Julius Caesar*

Actually, no comet "blazed forth" the death of Julius Caesar, but one was seen suspended over Rome just before the general and statesman Marcus Agrippa died in 11 B.C. This was the fourth recorded return of what is now called Halley's Comet.

In 1705, the British scientist Edmund Halley noted that the orbits of bright comets that had appeared in 1531, 1607, and 1682 were about the same. This similarity, coupled with the regularity of return, suggested that the three appearances were of a single orbiting comet. Halley predicted that it would return again in 1758, which turned out to be 16 years after Halley's death. Since then Halley's Comet, as it is now called, has reappeared three times, in 1835, 1910, and most recently in late 1985 and early 1986. The 1910 sighting was particularly spectacular. The comet was so close that it was observed to stretch across a 100° to 120° arc in the night sky. Some people predicted that poisonous gases within the comet's tail would pass through the Earth's atmosphere and destroy civilization, but responsible scientists understood that a comet's tail is too diffuse to cause any significant impact on terrestrial life.

The 1985 to 1986 return was disappointing compared with earlier encounters. The comet was considerably fainter than anticipated, and it was not an impressive naked-eye object. Nonetheless, a battery of five spacecraft studied the comet, one from the European Space Agency, two from Japan, and two from the Soviet Union.

The United States dispatched a mission called ICE (International Cometary Explorer) to study another comet, Giacobini-Zinner, several months before Halley's close approach. The instruments on this spacecraft recorded energetic electrons and lobes of opposing magnetic polarity. Thus, scientists learned that a comet is more turbulent, more complex, and considerably more energetic than had been anticipated.

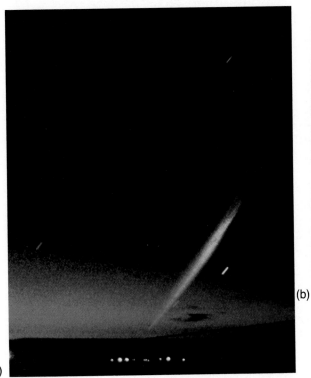

(a)

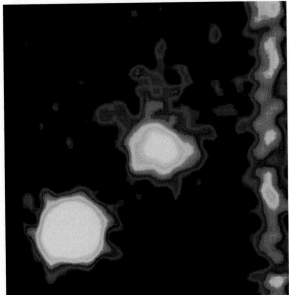

(b)

FIGURE 22.39 (a) The 1978 appearance of comet Ikeya-Seki, (Courtesy NASA) (b) Comet Halley when first rediscovered in 1984 is shown in computer-enhanced false colors to emphasize its features. (Courtesy NOAO)

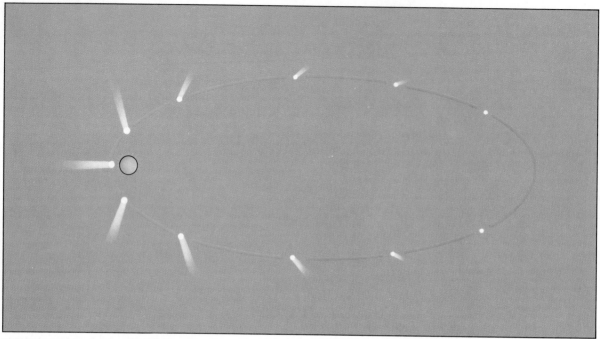

FIGURE 22.40 A comet's tail always points away from the Sun.

SUMMARY

The Solar System was formed from a mass of dust and gas that was rotating slowly in space. Within the center, the gravitational attraction was so great that the gases were pulled inward with enough speed that fusion temperatures were reached. The planets were also formed from coalescing clouds of matter, but fusion temperatures were not achieved.

The **central core** of the Sun is hot and dense. Hydrogen fusion occurs in this region. The visible surface of the Sun, called the **photosphere,** is appreciably cooler and has a pressure of 1/10 of an atmosphere. **Sunspots** are magnetic storms on the surface of the Sun. The outer layers, called the **chromosphere** and the **corona,** are turbulent and diffuse. The **solar wind** is a stream of particles coming from the Sun. **Auroras** occur when particles from the solar wind interact with the magnetic field of the Earth.

A recent theory states that the Moon was formed from the debris of a collision between a giant object and the Earth. The Moon was heated by the energy released during condensation, by radioactive decay, and by meteorite bombardment, but it is cold and inactive today.

Mercury is hot on the sunny side and cold on the shaded portion. Its topography is similar to that of the Moon. Venus has a hot, dense atmosphere and a surface that shows signs of tectonic activity. Mars is a dry, cold planet with a thin atmosphere, but the surface bears signs of ancient periods of erosion and tectonic activity.

Jupiter, Saturn, Uranus, and Neptune are all large planets with a low density. All are believed to have an appreciable atmosphere, a surface region composed largely of hydrogen, an inner zone of liquid metallic hydrogen, and a solid inner mineral core. The largest moons of Jupiter are Io, which is heated by gravitational forces; Europa, which is smooth and ice covered; and Ganymede and Callisto, which are large spheres made up of rock and ice. The rings of Saturn are made up of many small particles of dust, rock, and ice. They were formed from a moon that was pulled apart (or not allowed to form) by gravitational forces. Titan, the largest moon of Saturn, has an atmosphere and may be tectonically active. Pluto has a low density and is quite small.

A **meteorite** is a fallen **meteoroid,** a piece of matter from interplanetary space. **Asteroids** are small planet-like bodies. **Comets** are diffuse collections of solid mineral particles coated with frozen films of various compounds.

KEY WORDS

Protosun	Maria	The Great Red Spot	The Great Dark Spot
Planetesimal	Mercury	Europa	Pluto
Plasma	Synchronous rotation	Ganymede	Charon
Photosphere	Venus	Callisto	Meteor
Sunspot	Mars	Io	Shooting star
Solar flare	Phobos	Saturn	Meteoroid
Chromosphere	Deimos	Titan	Meteorite
Corona	Jupiter	Uranus	Asteroid
Solar wind	Terrestrial planets	Neptune	Comet
Magnetosphere	Liquid metallic hydrogen		

QUESTIONS

THE FORMATION AND STRUCTURE OF THE SOLAR SYSTEM

1. Explain why the protosun gradually became warmer as it shrank. Did the protoplanets also become warmer as they coalesced? If so, why are they so much cooler today?

2. Astronomers have observed that all the planets of our Solar System revolve around the Sun in the same direction and in nearly circular orbits. Is the theory of the origin of the Solar System offered in this text consistent with this ordered revolution? Defend your answer.

3. Explain why a star would appear to wobble as seen from Earth if it is being circled by a planet.

4. Venus rotates slowly clockwise while all the other planets rotate counterclockwise. What does this suggest about the history of Venus?

THE SUN

5. What are the most abundant elements present on the Sun? How does this composition differ from that of the Earth? Explain how this difference evolved.

6. Draw a diagram of the Sun, labeling the core, the photosphere, and the corona. Label the temperatures and the relative densities of each region.

7. What is the fundamental source of energy within the Sun? Is the Sun's chemical composition constant, or is it continuously changing? Explain.

8. How does energy travel from the core of the Sun to the surface? How does it travel from the surface of the Sun to the Earth?

9. Compare and contrast the core of the Sun with its outer surface.

10. Describe the formation of sunspots. Why do sunspots appear black to an observer here on Earth?

11. What did the first observations of sunspots have to say about earlier theories of the heavens?

12. Auroras are observed on Earth. On which of the following would you expect to observe auroras: the Moon, Mercury, Venus, Jupiter, Saturn? Defend your answer in each case.

13. Briefly outline some changes that would occur if the Earth's magnetic poles shifted so that they lay on the Equator.

14. Compare and contrast the solar wind with a breeze that blows on the surface of the Earth.

THE MOON

15. What leads us to believe that the Moon was hot at one time in its history? According to modern theory, how was the Moon heated?

16. Suppose the oldest igneous rocks on the Moon had been formed when the Moon was 1 billion years old. What conclusions would we then draw about the geological history of the Moon? Could we positively answer the question: Was the Moon heated by internal radioactivity, or by external bombardment? Defend your answer.

17. Explain how we can learn about Earth's geology by studying the Moon.

MERCURY

18. Give a brief description of the planet Mercury. Include its atmosphere, magnetic field, surface temperature, type of terrain, and speed of revolution around the Sun.

19. Why is the presence of a magnetic field around Mercury a curious phenomenon?

20. If Mercury rotated once every 24 hours as the Earth does, would you expect daytime tempera-

tures on that planet to be higher or lower than they are today? Defend your answer.

21. Why was it a surprise to astronomers to discover an atmosphere (even a very thin one) on Mercury?

VENUS

22. At one time, Venus and Earth probably had similar environments, except that Venus was about 20° warmer. If you could somehow cool the surface of Venus by 20°, would conditions on that planet be likely to become similar to those on Earth? Explain.

23. Why are there very few meteorite craters visible on Venus?

24. Venus is often referred to as the sister planet to Earth. Why?

25. If you were to take a vacation on Venus, list several features of the planet that you would like to see.

MARS

26. Discuss the evidence that indicates that the Martian atmosphere was once considerably different than it is today. Explain why this atmosphere could not have been similar to that of the Earth for long periods of time.

27. Refer to Figure 22.17. Imagine that you knew nothing about Mars except that it is a planet and this photograph was taken of its surface. What information could you deduce from this picture? Defend your conclusions.

28. In the late 1980s, President George Bush called for a manned mission to Mars. Do you believe this should be undertaken? Defend your answer.

29. Discuss the early arguments that led people to believe that there was life on Mars.

30. At the surface of Phobos, the acceleration due to gravity is 0.001 that of Earth. If a baseball is released from the top of a tall building on that moon, what speed will it have after falling for 3 seconds? (*Hint:* Refer to the equations for a freely falling object given in Chapter 2.)

31. It is said that a baseball player standing on Phobos could throw a baseball such that it would go into orbit around that moon. Based on the acceleration due to gravity on Phobos as given in question 30, do you believe this is likely?

32. Explain how the relative masses and distance from the Sun of Venus, Earth, and Mars led to markedly different environments on each of these three planets.

JUPITER

33. Describe the composition of the planet Jupiter. How does it differ from that of Earth?

34. Explain why the mass of Jupiter was an important factor in determining its present composition and structure.

35. About 4 billion years from now, the Sun will probably grow significantly larger and hotter. How will this change affect the composition and structure of Jupiter?

36. Compare and contrast the four Galilean moons of Jupiter. Describe some similarities between the Galilean moon system and the Solar System as a whole.

SATURN

37. Compare and contrast Saturn with Jupiter.

38. Compare and contrast Titan with the Earth.

39. Would you expect to find gases in the ring system of Saturn? Why or why not?

40. A science fiction story describes a visit by a spaceship to a planet with an enormous gravitational field. Before the ship could land, however, it was ripped apart by tidal forces. Explain how tidal forces could have done this.

41. Give arguments for and against the possibility of life on Titan.

URANUS AND NEPTUNE

42. Why is the discovery of Neptune often called a triumph of gravitational theory?

43. Draw a sketch of Uranus at several positions in its orbit showing its axis of rotation with respect to the plane of its orbit.

PLUTO

44. At one time, it was thought that Pluto might once have been a moon of Neptune. However, the discovery of the moon Charon has caused some astronomers to doubt this scenario. Why should Charon play a role in this debate?

45. Sometimes Pluto is the eighth planet and Neptune is the ninth. Sketch possible orbits for these two planets which would allow for this possibility.

METEOROIDS AND COMETS

46. Astronomers once thought that perhaps meteoroids were fragments from a planet that formed and then exploded or was destroyed in a collision with another planet. Do you think that such an origin is likely? Defend your answer.

47. Is a comet really hot, dense, and fiery? If so, what is the energy source? If not, why do comets look as though they consist of burning masses of gas?

48. When meteor showers occur, they are most easily visible after midnight. Why should the time of day be a factor?

ANSWERS TO SELECTED NUMERICAL QUESTIONS

30. 0.03 m/s

The Hubble Space Telescope, still in the grasp of Discovery's remote manipulator system, is backdropped over the intersection of the Caribbean Sea and the Atlantic Ocean. In this scene, the telescope has yet to have deployment of its solar array panels and its high gain antennae.
(Courtesy NASA)

The Rosette Nebula in Monoceros is a huge cloud of dust and gas, a region where star formation is possible.

C H A P T E R

23

The Life and Death of Stars

Throughout most of this book we have tried to explain physical phenomena in terms of concepts that can be related to our everyday experiences. But in some instances this is quite difficult to do. For example, in our discussion of relativity, such effects as the increase in mass with speed, the decrease in length, and the slowing down of high-speed clocks do not fall into the realm of common observation. Likewise in this chapter, we will find some happenings in space that stagger the imagination. Large objects and great distances are difficult to comprehend, and when the mysteries of distant galaxies and the vast regions of intergalactic space are considered, we simply cannot use common "house and garden" analogies. Such comparisons are simply inadequate. For example, some distant galaxies are over 8 billion light-years from Earth. That means that light moving at 3.00×10^8 m/s must travel for 8 billion years to reach us here. Think about it: The light we see now left that galaxy 3.5 billion years before our Solar System was even formed. It has been travel-ing all this time at 3.00×10^8 m every second (186,000 miles/s).

Not only are distances in space incomprehensibly large, but forces are similarly unimaginable. For example, some dead and dying stars become compressed so severely that protons and electrons are squeezed together to form neutrons. Other stars, compressed even more vigorously, become so dense that nothing, not even light, can escape from their gravitational tug. If we are to comprehend these and other related concepts, we must release our imagination and let our thoughts fly beyond terrestrial standards of force, size, mass, distance, and time.

23.1 STUDYING THE STARS

The nearest star (Proxima Centauri), other than our Sun, to Earth is about 4.2 light-years distant from us, and most are much farther. Yet, as distant as are these stars, astronomers have been

As we have seen, astronomical distances are incredibly large, even within the neighborhood of our own Solar System. For example, in the last chapter we found it convenient to speak of distances in terms of astronomical units, where 1 A.U. is equal to the average distance from the Earth to the Sun, 1 A.U. = 93,000,000 miles. Likewise, when speaking of distances beyond our Solar System, it is convenient to measure distances in terms of light-years, where, as we have seen, one light-year is the distance that light travels in one year, 1 light-year = 9.5×10^{12} km. In these units, the closest star to our Solar System is about 4.2 light-years distant. A light ray leaving that star for Earth must travel for over 4 years before it reaches us. If our focus of attention is expanded farther out into space, even a distance of 4 light-years can seem small. The Andromeda Galaxy is 2.25 light-years from Earth, and the most distant galaxies are about 10 billion light-years.

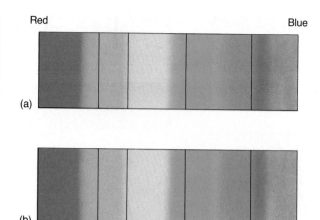

Red   Blue

(a)

(b)

FIGURE 23.1 (a) Absorption spectrum from a star. (b) Absorption spectrum of hydrogen made from a light source on Earth.

able to piece together much information about them. But how do we know these facts? No one has ever sampled even a fragment of a star. Yet we know their temperatures, their masses, their chemical compositions, their velocities through space, and many other pieces of information. The answer to how we found out most of the information lies with the study of the spectra of the stars. We have discussed the various kinds of spectra previously, but in this section we shall review them briefly and then make use of our findings to see how information can be gleaned from starlight.

Spectra and chemical composition of stars

As we have seen, one of the most useful methods available to astronomers is the study of atomic and molecular spectra. As you recall from Chapter 12, an excited atom emits light when an electron returns to a lower energy state from a higher state. The light, however, is not emitted in a random pattern. Instead, only certain frequencies

and wavelengths are observed. Each element has a characteristic **emission spectrum** and can be identified by it. If the spectrum of an unknown element matches that of neon, for example, the unknown element must be neon.

Light is emitted deep inside a star as a **continuous spectrum** over a wide range of frequencies. As this light passes through the outer layers, some of it is selectively absorbed by the various atoms in the cooler gas layers surrounding the star. Therefore, an observer on Earth sees a spectrum showing lines of darkness crossing the continuous band of colors. This is called an **absorption spectrum.** To see how an absorption spectrum can be used to determine the chemical composition of a star, consider Figure 23.1a. There we see an imaginary spectrum from a similarly imaginary star. There are only a few absorption lines missing from the spectra, and the task is to find out what element did the absorbing. Figure 23.1b shows an absorption spectrum produced here on Earth by allowing a continuous spectrum to pass through hydrogen gas. Note that the lines in (a) and (b) match up perfectly. Thus, we can conclude correctly that this star contains hydrogen. In actual practice, one would never find an absorption spectrum from a star looking quite as clean as the one in Figure 23.1. Because a star contains many different gases, a

FIGURE 23.2 The absorption spectrum of the Sun; similar spectra are observed from certain stars. (Courtesy NOAO)

stellar spectrum contains the superimposed images of many individual absorption spectra. As an example, the absorption spectrum of the Sun is shown in Figure 23.2. By a careful matching up of lines with spectra taken in earthbound laboratories, astronomers can identify which of these lines belong to what elements, and thus they can determine the chemical composition of the star.

The temperature of a star

Chemical composition is not the only information that can be obtained from absorption spectra. The amount of absorption caused by a given element depends on the temperature of the gas that does the absorbing. Thus, absorption spectra can give quite precise estimates of the temperature of a star. However, there is a more direct method for determining rough information about a star's temperature that you can actually try on your own. To see how this can be done, let us consider what happens to a piece of metal when you heat it. As the temperature of the metal increases, it eventually begins to glow with a dull red color. Then, as the temperature continues to increase, the color changes to orange, yellow, and finally to white and then to blue. Similarly, the overall color

of starlight changes with temperature, so by studying the color of a star, astronomers can obtain an idea of stellar temperatures. The color of some very bright stars can be detected with the naked eye. Examples are the red star Betelgeuse on the right shoulder of Orion the Hunter and the blue star Rigel at his knee. A telescope provides even better visual observation. If you have access to a telescope, slowly scan across the sky and pay attention to the various hues and shades that you find. Each color has its own story to tell about the temperature of that star.

Spectra and the speed of stars

Stars are all traveling through space, and scientists can determine their speed relative to Earth by studying stellar spectra. To recall how this is done, return to Chapter 7 and review the discussion of the Doppler effect. We found in our discussion of the Doppler effect that objects moving toward us have their sound frequencies shifted toward a higher value than when the object is stationary. Likewise, the sound frequency is shifted toward smaller values if the object is in motion away from us. The frequency of light waves also changes with relative motion. Spectral lines

Red   Blue

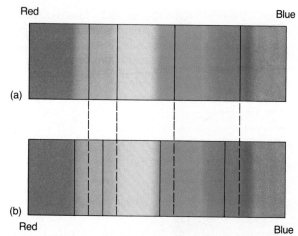

(a)

(b)

Red   Blue

FIGURE 23.3 (a) Hydrogen spectrum on Earth, and (b) hydrogen spectrum from a star. Note the shift of the lines from the star toward the red.

reaching us from a star that is flying rapidly away from Earth appear at a lower frequency (closer to the red end of the spectrum) than would be expected if the star were stationary with respect to Earth. For example, Figure 23.3 shows a spectrum of hydrogen taken in a laboratory on Earth, and a spectrum from a star that contains hydrogen. Note that the spectral lines from the star are not quite where they should be. Instead, they are shifted slightly toward the red end of the spectrum. We say that this light has been **red-shifted.** Such a red shift is indicative of the fact that the frequency of the light from the star has been decreased because of its motion away from Earth. Similarly, light from stars traveling toward us appears at higher frequencies and is said to be **blue-shifted.**

The distance to the star

As discussed in the Focus box on measuring distance, the distance to nearby stars can be found by use of parallax. However, this method breaks down for more distant objects because it is impossible to measure such small angles precisely enough. Yet we have been talking about finding the distance to galaxies as far away as 8 billion light-years, and the distance to the farthest objects in our Universe is now thought to be between 18 and 20 billion light-years. How do we know these distances?

FOCUS ON . . . Measuring the Distance to Nearby Stars by Parallax

The distance to the nearest stars can be measured by use of a technique called the method of parallax. To understand how this works, let's take an example from ordinary life. Imagine that you and a friend are driving down a straight highway in two separate cars. You spot a tree off in the distance, and you decide that you would like to know how far the tree is away from the highway. The technique of parallax can provide the answer, as shown in the Figure. The two of you stop your cars along the side of the road and measure the distance between the two. This distance indicated is called the baseline. Each of you then measures the angles shown as angle A and B. We will not go through the details of how to find the distance to the tree, indicated by the distance D in the Figure. It suffices here to say that it is a simple technique from trigonometry to find D if you know the length of the baseline and the angles

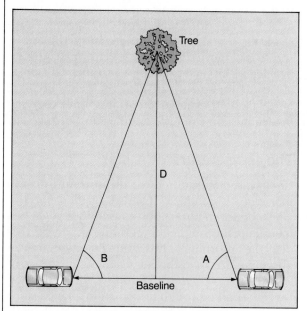

The distance between two cars is the baseline for measuring the distance to the tree by parallax.

In the 1920s, an astronomer named Edwin Hubble noted that light from almost every galaxy outside our own local group of galaxies is red-shifted. Thus, all the galaxies are flying away

A and B. Now, let's move into space and see how this same procedure can be used to find the distance to some stars.

First, we must set up a baseline, and the next Figure shows how this is done. We observe a nearby star and measure the angle A to it at a particular time of year. We then wait until the Earth has moved to the other side of its orbit and again measure the angle to the star, indicated as B. Our baseline is now equal to twice the distance from the Earth to the Sun, 2 × 93,000,000 miles, and we know the angles A and B. As before, simple trigonometry enables us to find the distance to the star D.

To see why this technique works only for nearby stars, consider the third Figure. Note what is happening to the angles A and B as the distance to a star becomes larger and larger. Even in the best of circumstances, the angles A and B are very close to 90°.

(They are shown here as being considerably smaller than 90° for convenience.) However, as the distance increases, the angles become so close to 90° that the measuring instruments simply cannot find the exact value of the angle. Thus, the distance to only about 600 nearby stars can be found by this method.

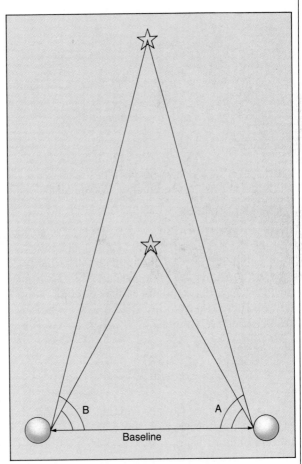

Using parallax to measure the distance to a star.

The angles A and B become impossible to measure accurately for very distant stars.

from each other; the Universe is expanding. Moreover, he observed that the more distant the galaxy, the more the light is red-shifted. Thus, these galaxies are moving outward at great

speeds, while the closer ones are receding more slowly. This relationship has been quantified and is known as Hubble's law, as shown in Figure 23.4. This diagram is a plot of the speed away

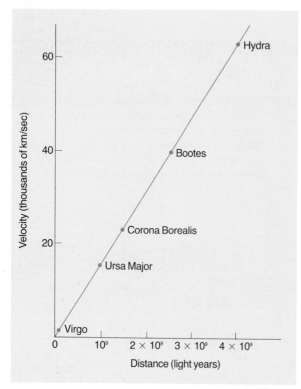

FIGURE 23.4 The Hubble diagram for the galaxies Virgo, Ursa Major, Corona Borealis, Bootes, and Hydra.

from Earth for several galaxies versus the distance to these galaxies in light-years. The straight line indicates that there is a direct relationship between these two quantities. The exact relationship is given by **Hubble's law,** which is stated as

$$v = Hd \qquad (23.1)$$

where v is the speed of the galaxy or the star in km/s, d is the distance of the galaxy or star in millions of light-years (Mly), and H is Hubble's constant given by

$$H = 15 \frac{km/s}{Mly}$$

EXAMPLE 23.1   How far away is that star?

The red shift in the light from a distant object indicates that it is receding from the Earth at a

speed of 30,000 km/s. What is the distance to this object?

Solution   This is a direct application of Hubble's law, which we can solve for the distance d as

$$d = \frac{v}{H} = \frac{30000 \text{ km/s}}{15 \dfrac{km/s}{Mly}} = 2000 \text{ Mly}$$

23.2
THE LIFE OF A STAR

An active star is a large, spherical mass of gases that are so hot that nuclear fusion reactions are occurring within the central regions. As described in the last chapter, tremendous gravitational pressures pull these gases inward, yet the star does not collapse. In fact, the outer layers of most stars are so diffuse and tenuous that they are less dense than Earth's atmosphere at sea level. Because stars do not collapse, some pressure must push outward against the force of gravity. This outward pressure is created by energetic photons and fast-moving subatomic particles released by nuclear reactions at the core. These particles and photons push against the outer layers of gas and prevent the gases from falling inward. The resultant density of a star is determined by two opposing forces: the gravitational force pulling in and thermal forces pushing out (Fig. 23.5). These forces leave a star like the Sun in a condition of equilibrium that can last for billions of years.

Over a period of many millions or billions of years, stars evolve, change, and die, and new ones are born. Throughout the life of a star, the gravitational and thermal forces oppose each other until, as we shall see, the star dies and undergoes radical change.

When a large, diffuse cloud of cold dust and gas is pulled together by gravitation, it forms the early stages of a star, known as a **protostar.** As the particles are pulled inward even faster, they eventually reach the extremely high temperatures required for nuclear fusion. At the onset of fusion, additional energy is released, and the gases move even more rapidly in all directions. These extremely energetic particles push outward

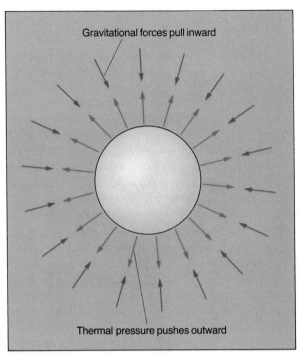

FIGURE 23.5 Equilibrium in a star is a balance between the gravitational forces pulling matter inward and the radiation pressure pushing outward.

(a)

(b)

Gigantic clouds of dust and gas like (a) the Horsehead Nebula, and (b) the Dumbbell Nebula in Vulpecula are regions where star formation may be possible.
((a) Courtesy Hale Observatory; (b) Courtesy California Institute of Technology and Carnegie Institution of Washington.)

against the gravitational force. Therefore, the gravitational force pulling inward is balanced by the thermal and radiation forces pushing outward. At equilibrium, an average-sized star has a dense core surrounded by a less dense envelope.

The outer region of a star is considerably cooler and less dense than the core. The surface temperature of our Sun, for example, is only about 6000 K, whereas the temperature in the core soars to 15,000,000 K. At the relatively cool outer temperatures, hydrogen nuclei do not collide forcefully enough to fuse, so no energy is produced in these regions. Within the dense core, however, temperatures are so high that hydrogen nuclei fuse together to form helium, with the release of large amounts of energy (Fig. 23.6a).

The process described above happens to all stars regardless of their mass while they are living out their normal lives. Take particular note of the result once again. The fuel being consumed by the nuclear reactions is hydrogen, while the resultant ash is helium. Eventually, as in all normal processes, the fuel will all be consumed at the core. At this point, conditions change drastically, and the star enters the first stage of its death throes.

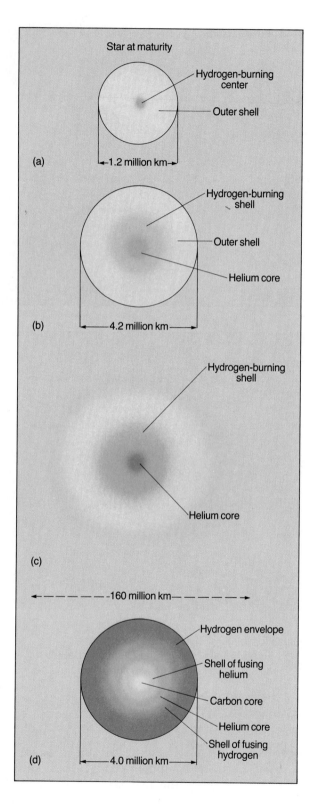

FIGURE 23.6 Aging of a star the size of our Sun. (In all cases the core region has been drawn larger than scale to show detail.)

23.3
THE DEATH OF A STAR LIKE OUR SUN

As a star grows older, increasing quantities of the ash, helium, accumulate at the core of the star, and the fuel, hydrogen, is depleted in the core. The helium nuclei do not fuse with each other at the temperature of an average mature star like our Sun. Recall that for fusion to occur two nuclei must be pushed very close together. A hydrogen nucleus contains only one proton, so to fuse two of them, the electrostatic repulsion of one proton for another must be overcome. However, each helium nucleus contains two protons, so in order for helium fusion to occur, a much greater repulsion must be overcome. As a result, the helium nuclei cannot fuse at this time. A star the size of our Sun has enough hydrogen fuel to last for about 10 billion years. After that time, the outer shell still contains large quantities of hydrogen, but the central core is mostly hot helium (Fig. 23.6b). The star now begins to behave quite differently than it did earlier in its life. Since little nuclear energy is produced in the central core, the core cools. When the core temperature decreases, the outward pressure that kept the star from falling inward also decreases, and the central regions start to shrink under gravitational forces. This gravitational contraction causes the core to stop cooling and begin to grow hotter. It seems a paradox that when the nuclear fire starts to diminish, the core should get hotter, but that is exactly what happens. To repeat,

when hydrogen fusion ends within the core, the equilibrium of the star is upset so that the central core is compressed by gravity and its temperature rises.

The hot shell of hydrogen around the helium core then fuses more rapidly. At this point, the star is releasing hundreds of times as much energy as it did when it was mature. The situation

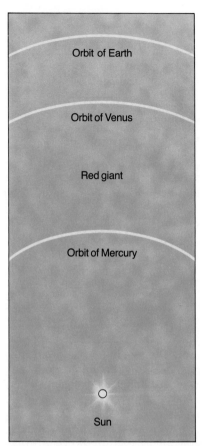

FIGURE 23.7 The Sun, shown in its original position, will expand in the red giant stage, indicated by the red haze, to a size that it will engulf the Earth.

is analogous to a flash fire of hydrogen which is rapidly burning just outside the helium-rich core. The intense energy output now causes the outer parts of the star to expand (Fig. 23.6c). As this outer shell grows in size, it cools. In fact, its temperature rapidly reaches the stage at which the star appears red in color. When our Sun goes through this phase of its death throes, its outer shell will expand so that its surface will be somewhere between Earth and Mars (Fig. 23.7). To recap, the star has grown cool as it has expanded, and it has become enormous compared to its original size. From these two facts, a star in this stage of its life is given its name: a **red giant.**

Meanwhile, the inner core continues to contract under the influence of its gravitation. Because of this contraction, the core gets hotter and

hotter until the critical temperature of 100 million degrees is reached. At this temperature, a new nuclear fusion reaction becomes possible. Helium nuclei begin to fuse together to form carbon, according to the following reaction sequence:

$$\tfrac{4}{2}\text{He} + \tfrac{4}{2}\text{He} + \text{energy} \longrightarrow \tfrac{8}{4}\text{Be}$$

This first reaction indicates that two helium nuclei have fused to form beryllium. (Note that an input of energy is required to make this reaction occur.) The beryllium nuclei produced then undergo a fusion reaction with another helium nucleus as

$$\tfrac{4}{2}\text{He} + \tfrac{8}{4}\text{Be} \longrightarrow \tfrac{12}{6}\text{C} + \text{energy}$$

Thus, the end result of this chain of events is that helium is being burned and the ash that is now accumulating at the core of the star is carbon, $\tfrac{12}{6}\text{C}$. The onset of the helium fusion reactions initiates several drastic and rapid changes. After a few hundred thousand years of instability, the star shrinks and then enters a new and stable phase. The newly structured core is now composed of fusing helium. Gradually, as increasing quantities of helium fuse to form carbon, the carbon starts to accumulate in the core just as helium had done during the early life of the star (Fig. 23.6d). As the helium fuel is consumed, the nuclear fire diminishes once again and the carbon core contracts. This gravitational contraction causes the core to become hotter once again. At this point, the history of a star depends on its initial mass. If it is about the mass of our Sun, the temperature will not rise enough to initiate fusion of the carbon nuclei. The gravitational collapse provides enough thermal energy to blow a shell of gas away from the star. This expanding shell is called a planetary nebula (Fig. 23.8). Meanwhile, the core continues to contract. A star as massive as our Sun will eventually shrink so that its diameter will be approximately that of Earth. Such a shrunken star no longer produces energy of its own and glows solely from the residual heat produced during past eras. The star is no longer kept from collapsing by thermal pressure. Rather, it has become so small that particles are squeezed

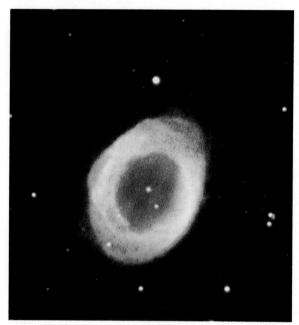

FIGURE 23.8 The Ring Nebula.
(Courtesy NOAO)

FIGURE 23.9 The size of a white dwarf is not very different from the size of the Earth. However, a white dwarf contains about 300,000 times more mass than does the Earth.

together. Finally, a point is reached at which the particles resist further squeezing and further contraction stops. At this stage, the star is still hot; in fact, it is white hot, and it is also very small. Hence, such a star is called a **white dwarf** star. The size of a white dwarf is not very different from the size of Earth (Fig. 23.9). However, a white dwarf contains about 300,000 times more mass than does Earth. Thus, the star is extremely dense; in fact, if one were to pick up a teaspoon of it, one would find that it would weigh about a ton!

White dwarf stars are extremely common in the Universe. Our own galaxy, the Milky Way, is estimated to contain several billion. Such a star will cool slowly over the course of tens of billions of years, but it will never change size again. No nuclear fire will cause it to expand, and the gravitational force isn't strong enough to compress it further. Such a star will cool until it reaches its final stage as a **black dwarf** star. Eventually, the Universe will be populated by trillions of black dwarf corpses, but none are seen now. Many astronomers believe that even if the sequence of events for the death of a star had occurred

shortly after the formation of the Universe, there has not been sufficient time for a white dwarf to have cooled to the black dwarf stage.

23.4
THE DEATH OF MASSIVE STARS

Some stars do not die as gently as will our Sun. If a star has a mass greater than about four times the mass of the Sun, the slowdown of helium fusion and the contraction of the resulting carbon core does not terminate in a white dwarf. Instead, as the core contracts slightly, the fusion reactions accelerate, and the outer regions expand into a second red giant phase.

After some time, the gravitational contraction of the carbon core will be strong enough to heat the core to such extreme temperatures that new fusion reactions start to occur. As an indication of what can happen, consider the following processes. (1) When the temperature of the core reaches 600 million K, carbon burning begins,

(a)

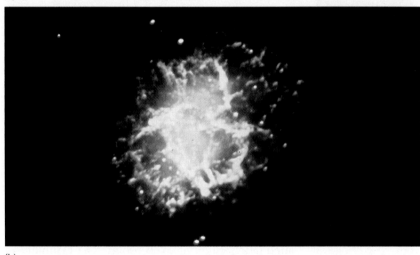

(b)

FIGURE 23.10 (a) Direct photographs of a region of space before and during a supernova explosion. (b) The Crab Nebula in Taurus is the remains of a supernova of AD 1054.
[(a) Courtesy of Lick Observatory; (b) Courtesy of NOAO]

and the ash produced consists primarily of magnesium and neon. (2) As the core continues to collapse, the temperature can soar to 1 billion K. At this temperature, neon burning begins, and the ash is largely oxygen and magnesium. (3) At a temperature of 1.5 billion K, oxygen burning begins, and the principal by-products of this reaction are sulfur, silicon, and phosphorus. (4) For a very massive star, a final gravitational collapse can cause the temperature to reach 3 billion K. Here, silicon can burn, and the primary ash is a crucial one in the life of the star, iron. Iron is important to the future life of the star because when fusion occurs with iron, energy is *absorbed*. Thus, the fusion of iron nuclei *cools* the core. When this happens, the thermal pressure that pushed the stellar gases outward is reduced, and the star begins to collapse under its own gravitation. This collapse releases large amounts of heat. In turn, the intense heat initiates a complex series of nuclear reactions that quickly lead to cataclys-

mic changes. Within a few seconds—a fantastically short period of time in the life of a star that is measured in terms of billions of years—the temperature reaches trillions of degrees, and the star explodes, hurling matter outward into space. This exploding star is called a **supernova** (Fig. 23.10). The word *nova* is translated to the English equivalent of *new*. Thus, the word implies that a super, or very bright, new star has appeared in the sky. This is an accurate way of describing such an explosion. Imagine the following scenario. You look at a particular region of the sky, and you see no star at all. There *is* a star there, but it is so dim that you can't distinguish it with your naked eye. However, the next night when you come out and look at that particular region of space, there is a star there which is brighter than any near it. What has happened, of course, is that the star at that location in space has undergone a supernova explosion, and you now are able to see it. To the ancients, however, it was as though a

Supernova in 1987A in the Large Magellanic Cloud appears as a very bright object near the lower right area of this photograph.
(Courtesy NOAO)

new star had miraculously appeared. This new star will not retain its brilliance for long. It will fade and disappear to the naked eye within a few weeks or months.

A supernova explosion is truly fantastic. For a brief period of time, a single star shines as brightly as hundreds of billions of stars and can emit as much energy as an entire galaxy. Four of these events have been seen in our galaxy in the last 1000 years, and one occurred in 1987 in a nearby galaxy, the Large Magellanic Cloud. A supernova explosion is violent enough to fragment many atomic nuclei, thereby shooting subatomic particles about in all directions. Shock waves—giant sonic booms—race through the atmosphere of the star. Under these conditions, many of the nuclear particles collide with sufficient energy both to fuse and to split apart. These processes form all of the known elements heavier than iron.

Thus, in studying the evolution of stars, scientists have learned how the natural elements were formed. We now believe that in the beginning of time, when the Universe was first formed, hydrogen was the predominant element. There was some helium, but there were no elements heavier than boron present. Within the cores of stars, hydrogen was converted to helium, helium to carbon, carbon to magnesium, and so forth, until many elements, including iron, became abundant. Then, in giant supernova explosions, the rest of the elements were formed by fission and fusion of atomic nuclei and nuclear fragments.

The death of a giant star is, in a way, only a beginning. A great number of nuclei are shot out into space. As countless supernova explosions occur in the course of time, the concentration of heavy elements slowly accumulates. Then, when conditions are favorable, the heavy elements, mix with quantities of hydrogen and helium, condense into new stars and new solar systems. Our own Sun and the planet we live on are condensates of the remnants of supernova explosions that occurred billions of years ago. In fact, every object in the Universe that is composed of elements heavier than the lightest ones originated in some supernova explosion. Since you are made of such heavy elements, you too are the remnant of some long-dead stars. Thus, the next time someone looks into your eye and refers to you lovingly as star-stuff, they are absolutely correct. Life itself rises out of the debris of dying stars.

23.5 NEUTRON STARS, PULSARS, AND BLACK HOLES

In a supernova explosion, not all the original matter is shot out into space. Some of it (perhaps half) is left behind with the core, compressed into a tight sphere. In the 1930s, scientists tried to imagine the physical characteristics of the residual matter. If the sphere were more than 1.4 times as massive as the Sun, the gravitational forces would be extremely intense, so intense, in fact, that the star could not resist further compression in the same manner as a white dwarf. Instead, the theory said that in every atom the electrons would be squeezed into the nucleus where they would join together with protons to form neutrons as

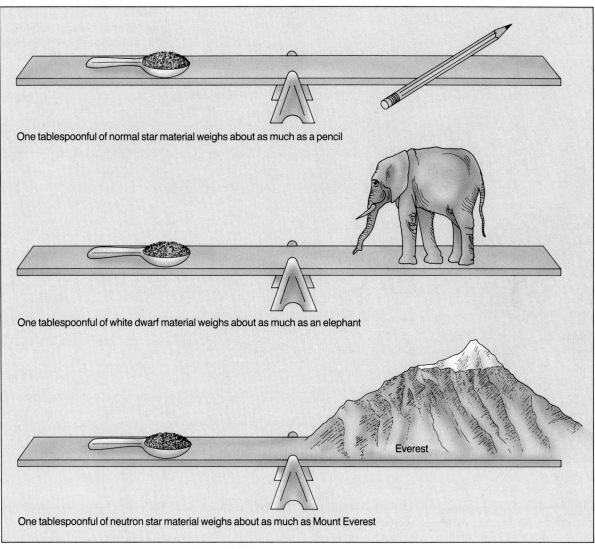

One tablespoonful of normal star material weighs about as much as a pencil

One tablespoonful of white dwarf material weighs about as much as an elephant

Everest

One tablespoonful of neutron star material weighs about as much as Mount Everest

FIGURE 23.11 Comparison of the densities of some stellar bodies.

$$\text{electrons} + \text{protons} \longrightarrow \text{neutrons}$$

or

$$_{-1}^{0}e + _{1}^{1}H \longrightarrow _{0}^{1}n$$

The neutrons would then resist further compression and remain tightly packed. This ball of compressed neutrons is called a **neutron star.** A neutron star would be so dense that a typical star would become about the size of a medium-sized city. It would have a radius of about 10 miles. If one could pick up a teaspoon of this star, it would have about the same weight as Mount Everest

(Fig. 23.11). Theory predicts that such stars should exist, but because of their small size, they would be very dim. Thus, astronomers had never seen one and had very little hope of finding one in space until an accidental discovery in 1967.

In 1967, Jocelyn Bell (now Burnell) was a graduate student in astronomy at Cambridge University in England. Her doctoral dissertation was to be concerned with radio emissions from distant galaxies, and as a result, she constructed a telescope to detect frequencies within the radio range. In one part of the sky, she detected a radio signal that pulsed with a frequency of about one

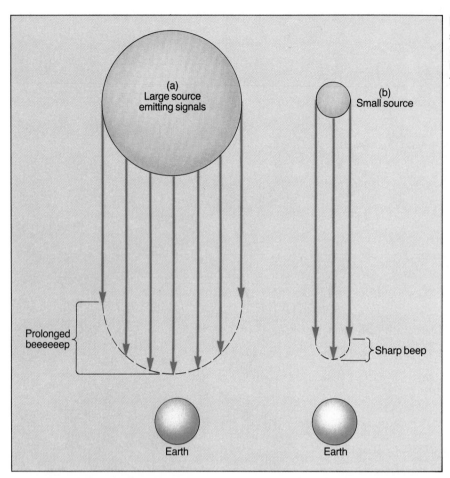

FIGURE 23.12 A sharp signal from a larger sphere (a) arrives over a longer time interval than a sharp signal from a smaller sphere (b).

pulse every 1.33 seconds. If such a signal could be amplified and fed into the speaker of a conventional radio, you would hear a beep, beep, beep evenly spaced at one beep every 1.33 seconds. There are many radio emissions arriving at Earth from outer space, but what made these particularly unusual was that they were (a) sharp, (b) regular, and (c) spaced only a little over one second apart. At first, astronomers seriously considered that they might represent a signal from intelligent life, so for a short while they called the pulse signals LGM, for "little green men." But when Burnell found a second similar pulsating source in a different region of the sky, scientists ruled out the possibility of two widely divergent civilizations, each sending out bizarre signals in almost identical fashion and in a way that was quite wasteful of energy. Once it was established that the signals were of natural origin, the unknown

pulsing sources were called **pulsars.** But naming the objects didn't help to explain them. What a puzzle! How is it possible to study an object that is many trillions of kilometers away and that emits no visible light, only radio signals?

The first step was to estimate the size of the mysterious object. Not all parts of an object in space are equidistant from Earth (Fig. 23.12). If a large sphere emits a sharp burst of radio signals from over its entire surface, some of the photons start off on their journey significantly closer to Earth than others and therefore arrive sooner. A person on Earth listening to the radio noise hears not a sharp beep, but a more pronounced beeeeep, because it takes a while for all the photons to arrive. Alternatively, a signal from a small sphere is much sharper, for the difference in distance is not nearly so great. The pulsar signals were unusually sharp, indicating that the source

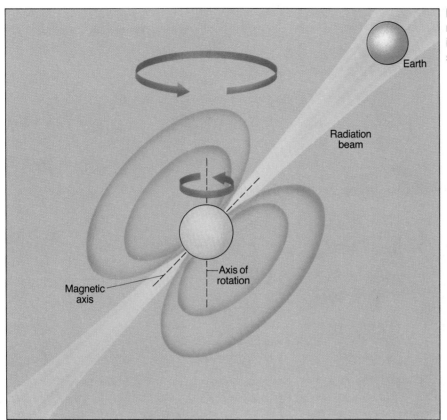

FIGURE 23.13 The radiation beam of the pulsar is observed only when it sweeps across the Earth.

Earth

Radiation beam

Axis of rotation

Magnetic axis

must be extremely small, perhaps 30 km in diameter. Was the source a star-like object only 30 km across? The smallest star previously recorded was a white dwarf 16,000 km in diameter, and a white dwarf, in any case, could not pulse so fast. Pulsation speed varies with the density of the object, and white dwarfs are not sufficiently dense. Scientists then reasoned that the pulsar might be the long-searched-for neutron star.

The question remains, however: Why does a neutron star beep periodically. The following theory, sometimes called the **lighthouse theory,** explains how this occurs. Imagine yourself lost at sea in a fog. Safety comes when you see a lighthouse beam blinking on and off in the distance. In the absence of the fog, you could actually see the lighthouse beam sweeping around and coming toward you, but with the presence of the fog, you will see the light only when it is pointing directly toward you. Thus, the conclusion that astronomers drew is that *a pulsar is a rotating neutron star*. Stars, including the Sun, have magnetic

fields, and if the Sun were shrunk to the size of a neutron star, the magnetic field would become very intense. As we have seen, the magnetic field of the Earth traps charged particles in our atmosphere, and they make their presence known when they spill out over the poles to produce auroras. A somewhat similar occurrence takes place on a pulsar, except the radiation is produced by accelerating and decelerating charges. Recall that X-rays are produced when electrons are decelerated upon collision with a block of metal. Any charged particle radiates energy when it is accelerated or decelerated, and such changes in velocity occur for particles falling into a pulsar. Additionally, the beam from the star is concentrated in a very narrow beam away from the star's poles. The north and south magnetic poles of the Earth are not located directly at the geographic north and south axis of rotation of the Earth, and neither are these locations the same on a pulsar. As shown in Figure 23.13, we see the radiation from the rotating neutron star only when the beam is

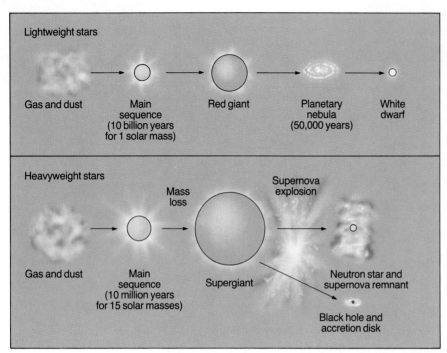

FIGURE 23.14 A summary of the stages of evolution for stars of different masses.

directed toward us on Earth. Thus, there may be many pulsars in space which we will never see because their lighthouse beam never sweeps across our planet.

The crucial test for the pulsar as rotating-neutron-star theory came when radio astronomers looked at the Crab Nebula. The Crab is the remnant of a supernova explosion that occurred in the constellation of Taurus (the bull) in A.D. 1054 (see Fig. 23.10). The event was recorded by Chinese astronomers, who called it a "guest star." According to their records, the star was visible in the daytime for 23 days and was bright enough to read by at night. The radio astronomers found a pulsar in the center of this debris. Thus, a pulsar was found precisely where a neutron star should be. This made it even more convincing that pulsars must be neutron stars that are emitting radio-frequency energy.

Figure 23.14 summarizes the death cycles of lightweight stars like our Sun and of heavyweight stars. The end result for the light star is a white dwarf and for the heavier, it is a supernova remnant with a neutron star at the center. However, also notice that there is an alternative pathway for even heavier stars, leading to a remnant that bor-

ders on science fiction, the **black hole.** Let us investigate these fantastic creations of the Universe.

What happens when a *very* large star dies? Astronomers believe that if the central core remaining after a supernova explosion is greater than three to five times the mass of the Sun, the neutrons are not able to resist the inward gravitational force. Then the star shrinks to a size much smaller than a neutron star and becomes a black hole. Such a collapse is impossible to imagine in earthly terms. A tremendous mass, perhaps a trillion, trillion, trillion kilograms of matter, shrinks smaller and smaller. If known laws of physics are obeyed, this faint star will contract to the size of a pinhead and then continue to shrink to the size of an atom, and then even smaller. Eventually, it will collapse to a point of infinite density, called a singularity. Incredible! Can this really happen?

A black hole is so small and massive that it creates an extremely intense gravitational field. According to Einstein's theory of general relativity, a photon is affected by a gravitational field as though it had mass. As proof of this, light from nearby stars has been observed to be bent by the gravitational field of the Sun. If a star becomes dense enough, its gravitational field becomes so

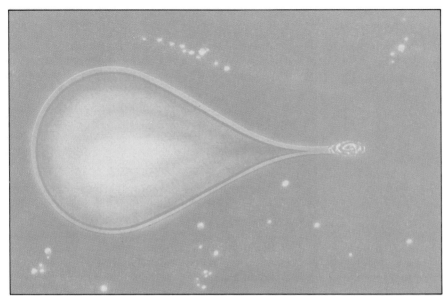

FIGURE 23.15 An artist's conception of the disk of swirling gas that would develop around a black hole (right) as its gravity pulled matter off the companion supergiant star (left). The X-ray radiation would arise in the disk.

intense that photons cannot escape. If you throw a baseball up into the air here on Earth, it will return to you, as you well know. However, if you throw it such that when it leaves your hand, it is traveling at a speed of 7 miles/s, it will have achieved escape velocity, and it will travel upward never to return. On a black hole, the escape velocity exceeds 3×10^8 m/s, the speed of light. Since no light can escape such an object, it must always be invisible; hence the name "black hole." If you were to shine a flashlight beam, or a radar beam, or any kind of radiation at a black hole, the photons would simply be absorbed. The beam could never be reflected back to your eyes; therefore, you would never see the light again. It would be as if the beam just vanished into space. Similarly, if a spaceship flew too close to a black hole, it would be sucked in forever. No rocket engine could possibly be powerful enough to accelerate the ship back out, for no object can travel faster than the speed of light.

The search for a black hole has been even more difficult than the search for a neutron star. How do you find an object that is invisible, emitting no energy in any form? In short, how do you find a hole in space? Although it is theoretically impossible ever to see a black hole itself, it is possible to observe the effects of its gravitational field. Many of the stars in the Universe exist in pairs or small clusters. If two stars are close together, they will orbit about each other. What if one of the pair were a black hole? The visible one would appear to be orbiting around an invisible partner. Astronomers have studied several stars that appear to vary in this unusual manner. In at least one case, the invisible member of the pair has been shown to have more than five times the mass of the Sun. Since a normal star this massive would be visible at that distance, the invisible partner might be a black hole. However, observation of such movement is not complete proof that a black hole exists. The unseen partner could be some other object, such as a neutron star, although theory indicates that a neutron star could not be so massive.

Astronomers have calculated that if a supergiant star were mutually orbiting with a black hole, great masses of gas from the supergiant would be sucked into the black hole, to disappear forever from view (Fig. 23.15). As this matter started to fall into the hole, it would naturally accelerate, just as a meteor accelerates as it falls toward Earth. The gravitational field of a black hole is so intense that the acceleration of the falling mass would be very great. As we saw in our discussion of X-rays in Chapter 12, when charged matter, such as a stellar plasma, is accelerated, radiation is produced. For accelerations as large as those that would be produced by a black hole, the photons produced would be in the X-ray range. These X-rays might then be detected here on Earth because they are produced far enough from the black hole that they can escape its gravitational tug. Just such a scenario seems to be being played out in the constellation of Cygnus (the swan). There an intense X-ray source, called Cygnus X-1, is pouring X-rays out into space. At present, this seems to be the best candidate for a black hole, but several others have also been nominated.

SUMMARY

Atomic spectra are used to determine various physical characteristics of an object in space. The relative velocity of unknown objects can be determined by measuring the **Doppler shift.** The distance to a nearby star can be found by parallax measurements and to more distant ones by Hubble's law.

Stars are formed from condensing clouds of dust and gas. During maturity, hydrogen nuclei in the core of a star fuse to form helium, with the release of large amounts of energy. When the hydrogen fuel in the core is exhausted, and hydrogen fusion ends within the core, the central mass is compressed by gravity, and the temperature rises. Fusion in the outer shell accelerates, and the star expands to become a **red giant.** In the following stage, helium nuclei fuse to form carbon. In an average-size star, after helium fusion ends, the star sends off a **planetary nebula** and then shrinks to become a **white dwarf.** A larger star continues to undergo a sequence of fusion steps, then explodes to become a **supernova.** The remnant can contract to become a **neutron star** or, if the mass is great enough, a **black hole.**

EQUATIONS TO KNOW

v = Hd (Hubble's law)

KEY WORDS

Spectrum	Hubble's law	White dwarf	Pulsar
Absorption spectrum	Protostar	Black dwarf	Lighthouse theory
Emission spectrum	Red giant	Supernova	Black hole
Doppler effect	Planetary nebula	Neutron star	

QUESTIONS

STUDYING THE STARS AND THE DISTANCE TO A STAR

1. Explain how the chemical composition, the temperature, the velocity of a distant star, and the distance to a star can be determined.
2. In recent years, astronomers have learned a great deal about our Universe by studying radio, X-ray, and other invisible radiation. Do you think that more information could be gained by building large microphones to detect the sounds of giant explosions in space? Defend your answer.
3. What happens to the (a) wavelength of light from a star moving away from us, (b) the frequency of the light from that star, and (c) the speed of the light from that star?

4. A particular star is observed to be light blue in color, and upon careful observations, the hydrogen and helium lines predominate and these lines are red-shifted. Based upon these pieces of information, what can you determine about that star?

5. As a review of the Doppler shift, consider dipping one tine of a vibrating tuning fork into a pool of water and then moving the fork through the water. If you were in front of the moving tuning fork, would you observe more waves, fewer waves, or the same number of waves per unit time than if you were behind the fork? Explain.

6. Many of the stars in our galaxy exhibit spectra that are blue-shifted. Does this invalidate Hubble's law? Explain.

7. In example 23.1, we calculated the distance to a star based on the value of Hubble's constant of $15 \frac{\text{km/s}}{\text{Mly}}$. However, in the last few years disagreements have arisen over the exact value of this constant. Some astronomers believe that a more accurate value is $25 \frac{\text{km/s}}{\text{Mly}}$. If this number should prove to be correct, what is the distance to this star?

8. Some galaxies and other strange objects have been detected as far away as 15 billion light-years. According to Hubble's law, what is the speed of recession of these objects?

THE LIFE AND DEATH OF STARS

9. Hydrogen burns in air according to the following equation:

$$2H_2 + O_2 \longrightarrow 2H_2O$$

Is this chemical combustion of hydrogen an important process within a star? Why or why not? Explain.

10. Explain why the density of the gases near the surface of the Sun is less than the density of the gases near the surface of the Earth, even though the gravitational force of the Sun is much greater.

11. Explain why the core of a star becomes hotter after the nuclear fusion reactions diminish.

12. What could you tell about the past history of a star if you knew that its core was composed primarily of carbon? Explain.

13. Compare and contrast a white dwarf with a red giant. Can a single star ever be both a red giant and a white dwarf during its lifetime? Explain.

14. Would you be likely to find life on planets orbiting around a star in which the primary fusion process is that of the "burning" of carbon? Explain.

15. Certain stars that lie above and below the plane of the Milky Way contain fewer heavy elements than our own Sun has. From this information alone, what can you tell about the history of these stars?

16. What is a supernova? Do all stars eventually explode as supernovae? Explain. What is a neutron star? Do all supernovae lead to the formation of neutron stars?

PULSARS AND BLACK HOLES

17. Explain why the pulsar signals detected by Jocelyn Bell Burnell could not have originated from (a) a star, (b) an unknown planet in our Solar System, (c) a distant galaxy, or (d) a large magnetic storm on a nearby star.

18. Do you think that a black hole could be hidden in our Solar System? Explain.

19. Some astronomers believe that there are tiny black holes flying through space. Would such objects represent a hazard to a rocket ship traveling to distant stars? Could the crew of such a rocket detect a black hole well in advance and avoid an encounter? Explain.

20. Do you think that a very heavy concentration of black holes in intergalactic space could be detected? What about a low concentration? Discuss.

21. Why would parallax measurements work better for a planet than a star?

22. Trace the steps in the death of a star in which the remnant is (a) a white dwarf, (b) a neutron star, (c) a black hole.

23. Use the lighthouse theory of pulsars to explain why there are probably many more of these in our Galaxy than we have detected or can detect.

24. It is estimated that neutron stars may have some structure such as one-inch-high mountains. It is also estimated that to climb such a mountain would require the energy expenditure of a full lifetime. Explain why so much energy would be required.

25. Why is the Doppler effect of no value for determining the speed of a star moving perpendicular to the line of sight from the Earth to that star?

C H A P T E R

24

Galaxies and Time

This chapter begins with a look at collections of stars, called galaxies, and moves from there to objects, called quasars, that are the most distant objects in the Universe. Finally, we conclude with an attempt to answer some questions that have been asked by all civilizations since time immemorial. They are: Where did we come from? and Where are we going?

24.1
THE MILKY WAY GALAXY

The structure of our galaxy

Because of light pollution in our environment caused by street lights, lighted parking lots, and so forth, modern people do not have the appreciation of the sky that their ancestors had. You should make an effort to get away from the city lights sometime to take a good look at the sky from a dark country meadow. From such a van-

tage point, you will note that most of space contains a diffuse scattering of stars that are spaced so far apart that you can see large expanses of black space between these points of light. However, in one region of the sky, a nearly continuous band of light, called the **Milky Way,** stretches from the northern to the southern horizon, as shown in Figure 24.1. In 1610, Galileo turned his telescope toward the Milky Way and found that this haze of light is produced by multitudes of stars. Approximately 150 years later, the astronomer Thomas Wright hypothesized (correctly) that the Earth is located in a disk-shaped group of stars. When we look toward the band of light, we are looking into the plane of the disk. What you are seeing are the billions of billions of stars, some relatively close to Earth and some farther away, that make up the disk of the Milky Way. A view perpendicular to this disk shows only a few scattered stars, as noted above. This grouping of stars is now known to be only one of many such group-

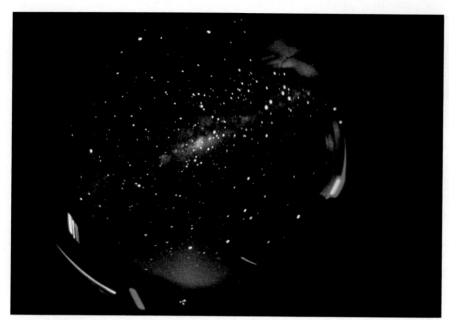

FIGURE 24.1 This fish-eye lens view shows the arch of the Milky Way over Cerro Tololo Inter-American Observatory in Chile. (Courtesy NOAO)

ings throughout the Universe, and these bundles of stars are referred to as **galaxies.**

A galaxy is a large volume of space containing many billions of stars all held together by their mutual gravitational attraction.

In 1785, the idea of the geocentric universe returned in a slightly different form, and again it was incorrect. In that year, the noted astronomer William Herschel did a telescopic star count in all directions from the Earth. As you might expect, he found that there were many more stars in the plane of the Milky Way than there were in directions away from this plane. However, his count revealed that the number of stars were about the same in all directions when looking in the plane of the galaxy. Based on this observation, he concluded that our Solar System lies near the center of the Milky Way galaxy. Thus, the old idea that there is something special about the Earth that puts it at the center of the Universe reared its head again. Much later it was discovered that the star count was inaccurate because of large clouds of interstellar dust that block our view of space. Herschel was actually counting only stars that were within about 2000 light-years of Earth, because the light from stars more distant than this is blocked, in some directions, by obscuring clouds of dust and gas.

The first accurate estimate of the size of our galaxy and of our position within it came in 1917 from work done by Harlow Shapley, who at that time was studying **globular clusters.** Globular clusters are collections of hundreds of thousands of stars bundled together by gravity. Figure 24.2 shows a typical globular cluster. At the center of one of these clusters, the stars are packed together such that their average distance of separation is only about 0.5 light-year. It is of interest to note that if you were born in a solar system located in one of these clusters, you would not have to worry about studying astronomy, because very little would be known about the subject. You would find that in whatever direction you looked there would be a star. Thus, you would live on a world of endless daytime.

Shapley's primary interest in these clusters was to find the distance to them and to plot their relative distribution in space. There were 93 known globular clusters at the time of Shapley's work, and they were easy to study since most of them did not lie in the plane of the galaxy where they would be obscured by the intervening dust clouds. He found that the clusters were grouped into a roughly spherical shape whose center was approximately 25,000 to 30,000 light-years away and in the direction of the constellation Sagittar-

FIGURE 24.2 A globular cluster in the constellation Hercules.
(Palomar Observatory, California Institute of Technology)

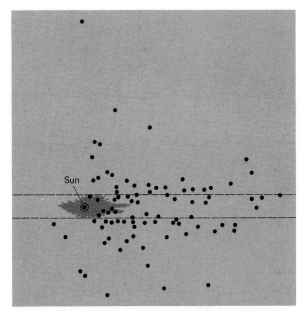

FIGURE 24.3 A copy of a diagram by Shapley, showing the distribution of globular clusters in a plane perpendicular to the Milky Way and containing the Sun and the center of the Galaxy.

ius. Figure 24.3 shows a two-dimensional plot of his observations, along with the location of the Sun. Based on these observations, Shapley proposed that the center of these globular clusters was actually the center of our galaxy. Many pieces of information since that time have confirmed Shapley's theory.

Additional information about the shape of our galaxy began in a negative way with telescopic observations made by Charles Messier in 1781. Messier was an inveterate comet hunter who did not want the distractions of any other heavenly objects to veer him from his single-minded path. As he scanned the night sky in his relentless search for new comets, he would occasionally happen upon a nebulous looking object in space that was so hazy and indistinct that he could not tell whether or not it was a comet. After a few nights of observation, he would find that the object did not move and thus could not be a comet. Since there were others interested in comet hunting, he made a detailed catalog of these so-called nebulae in which he listed 109 of them under such names as M41 (for Messier object number 41). His goal with this catalog was to help others

by telling them, "Don't waste time looking here; this is not a comet." Fortunately, other astronomers in later years did not take Messier's advice, because these objects he rejected as not being worth the time for observation turned out to be of extreme astronomical importance. Many of these objects were later found to be star clusters or clouds of dust and gas in our own galaxy. However, some stood out as having a distinct spiral structure like that shown in Figure 24.4. The idea gradually grew that some of these groupings of stars might be "island universes" far removed from our own Milky Way galaxy. Others held that this idea was ridiculous. Shapley was one of these. He believed that they were simply unusually shaped groupings of stars that were a part of our galaxy, just as were the globular clusters that he had been studying. The controversy was finally resolved by a young Kentucky lawyer named Edwin Hubble, who abandoned his law practice in favor of astronomy. He was able to find the distance to one of these Messier objects, M31, in 1924 and found it to be 2.25 million light-years from Earth. Thus, for the first time, we began to appreciate the enormous size and complexity of

FIGURE 24.4 The Whirlpool Galaxy. (Courtesy Naval Observatory)

FIGURE 24.5 The Andromeda Galaxy. (Courtesy NOAO)

the Universe. M31 is a large spiral-shaped galaxy now called the Andromeda galaxy (Fig. 24.5).

The observation that many of these island universes are spiral in shape gave astronomers the clue to what shape our own Milky Way might have. We now know that the Milky Way is a spiral galaxy shaped as shown in Figure 24.6. The disk of our galaxy is approximately 5000 light-years thick and nearly 100,000 light-years in diameter. As seen from the edge, the Milky Way looks like two fried eggs placed back to back. The central bulge is called the nucleus of the galaxy, and it is

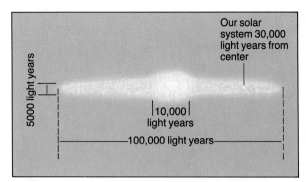

FIGURE 24.6 An edgewise view of the Milky Way galaxy.

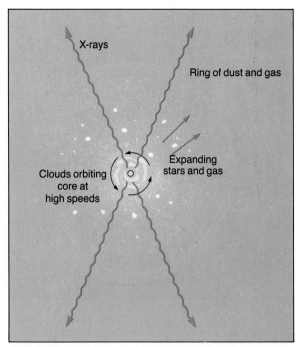

FIGURE 24.7 The core of our galaxy is invisible to us, but we can detect several signs of violent activity within the core.

about 10,000 light-years thick. Extending outward from this central bulge are relatively diffuse spiral arms. Our Sun lies in one of the spiral arms of the Milky Way, some 30,000 light-years from the center. The concentration of stars in the galactic core is perhaps one million times greater than is found in the outer disk. This core is like the globular clusters in the sense that if you could visit a planet orbiting one of these stars, you would never experience night, for the accumulated starlight would provide ample light all the time. However, stable solar systems are not likely to exist at all in this region, for collisions or near collisions between stars would rip planets out of their orbits fairly frequently, as judged by the expanses of astronomical time.

The core of the Milky Way

The core of the Milky Way is not only much denser than the disk but is different in many other ways as well. For one thing, almost all the stars within the core are relatively cool. In addition, these stars are chemically different from the stars in our own region of space. Recall that most stars are composed primarily of hydrogen and helium. However, a star such as our Sun was formed from the remnants of supernova explosions and contains 1 percent heavier elements. In contrast, stars in the core contain a much smaller percentage of heavier elements. The chemical composition and temperature, taken together, imply that these stars are quite old and were formed as the galaxy itself was forming, before heavy elements were synthesized during stellar evolution.

While it is true that the stars near the galactic center are old and cool, this region is not at all dead and static, as shown in Figure 24.7. Radio and infrared observations have detected violent motion in this region. Close to the core of the galaxy, interstellar clouds are orbiting some unseen object located at the very center. The orbital speeds are quite high, indicating that this central object is quite massive. Farther out, one group of stars and clouds of hydrogen are expanding outward, moving at 100 km/s. This motion suggests that some sort of massive explosion has expelled material with great force. Even farther out in space a gigantic ring surrounds the center. This ring also seems to have been formed by some sort of giant explosion. Finally, energetic X-rays are streaming out of the very small, unseen object in the center of all this activity. Only highly energetic interactions can generate X-ray emissions.

The most widely accepted explanation of these findings is both fantastic and conjectural. According to this theory, the evolution of the gal-

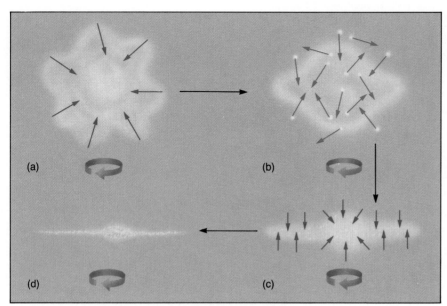

FIGURE 24.8 The evolution of the Milky Way galaxy.

axy is thought to be roughly analogous to the evolution of the Solar System, as discussed in Chapter 22. Originally, a pregalactic cloud rotated slowly in space. This cloud began to condense under the influence of gravity. As it condensed, it began to rotate faster, and this motion flattened the cloud into a disk, as shown in Figure 24.8. As described previously, the Solar System broke up into discrete planets orbiting a central Sun. The protogalaxy, on the other hand, separated into discrete stars, with the highest concentration toward the center. But what is the core itself? Calculations vary considerably, but the core is believed to be between 1 million to 1 billion times as massive as the Sun. Despite the great mass, measurements indicate that it is considerably smaller than our own Solar System. Try to imagine the forces involved when such a tremendous amount of material came together under the influence of its own gravitation! Many astronomers now believe that some sort of super supernova explosion occurred around 5 billion years ago, sending matter and shock waves into space and leaving behind an extraordinarily massive black hole as the nucleus of our galaxy. Today, this black hole is sucking matter into its void, releasing energetic X-rays and other forms of energetic radiation.

In our region of the galaxy, we are quite oblivious of the turmoil at the center. The primary galactic motion that we undergo is one of a slow orbit about the nucleus. The Sun is believed to have made about 15 to 20 complete revolutions within its lifetime. Most of the stars in this outer region contain a high percentage of heavy elements as compared with the stars near the core. As explained earlier, these characteristics indicate that the stars are formed from the remnants of stellar explosions that were energetic enough to fuse lighter elements into heavier ones. In addition, many of these stars are young and hot, which gives further evidence that they were formed in a later stage of galactic evolution.

24.2 GALAXIES AND CLUSTERS OF GALAXIES

As mentioned earlier, one of the pioneers who provided us with much of our present information about the galaxies was Edwin Hubble. He found that galaxies occur in four basic shapes: spiral, barred spiral, elliptical, and irregular.

The Milky Way is an example of a **spiral galaxy,** as is the Andromeda galaxy, which is the closest large galaxy to us. Spirals all have the basic shape shown in Figure 24.9, but they differ in the size of the central bulge and the winding of the spiral arms. Spirals range between the extremes

(a)

(b)

FIGURE 24.9 Various types of spiral galaxies. (Courtesy NASA and NOAO)

of those having a large nucleus with tightly wound spiral arms to those with a small nucleus and loosely wound spiral arms.

An example of a **barred spiral** is shown in Figure 24.10. These are characterized by a bar that runs through the nucleus and from the ends of which the spiral arms emerge. These galaxies range over the same gamut as do the spirals, from

those with a large nucleus and closely wound arms to a small nucleus with loosely wound arms.

An **elliptical galaxy** is shown in Figure 24.11. These are named for the shape, and as the figure shows, they have no spiral arms. These ellipticals are found in shapes ranging from the almost spherical to the highly flattened elliptical. These galaxies have almost no interstellar dust

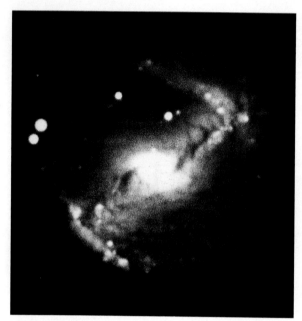

FIGURE 24.10 A barred spiral in the constellation Canes Venatici. (Courtesy NOAO)

FIGURE 24.11 A giant elliptical galaxy in the constellation Virgo. (Courtesy NOAO)

and gas, and perhaps because dust clouds are the maternity wards in space for star formation, there are no young stars found in ellipticals. The largest galaxies found in the Universe are elliptical in shape, as are the smallest. Giant ellipticals are relatively rare, but the smaller, dwarf ellipticals, containing only a few million stars, are numerous.

Irregular galaxies, like those shown in Figure 24.12, have no obvious geometrical shape. The two galaxies closest to us are irregulars. They are the Large Magellanic Cloud and the Small Magellanic Cloud. As their name implies, they were discovered by an astronomer who traveled with Magellan on his around-the-world voyage from 1519 to 1522. These galaxies cannot be seen from the Northern Hemisphere.

It would be satisfying at this point to conclude our discussion of the various kinds of galaxies that inhabit space with a discussion of the evolution of galaxies. Do galaxies begin as irregulars, evolve into spirals, and finally, after many revolutions, wind their arms in on themselves to become ellipticals? Or does the progression move in some other direction, such as from irregular to elliptical, which then throws off spiral arms. Unfortu-

nately, this discussion cannot be completed. No one knows how galaxies evolve. Perhaps there is no evolutionary track. Maybe they are born as ellipticals or spirals or irregulars and remain that way throughout their life.

Approximately 1 billion galaxies can be observed from Earth. When the distribution of galaxies was mapped, astronomers learned that they exist in **clusters.** Thus, just as there are galaxies of stars, there are also galaxies of galaxies. For example, the Milky Way, Andromeda, the Magellanic Clouds, and about another two dozen small ellipticals are bound into a cluster that is called the **Local Group.** These clusters range in size and number of galaxies from relatively sparse ones, like ours, to giant superclusters that contain more than 100,000 galaxies. A typical group of galaxies may occupy a volume of 3 million light-years in diameter.

24.3 ENERGETIC GALAXIES AND QUASARS

Recall that the cores of many spiral galaxies exhibit evidence of explosions or unusually power-

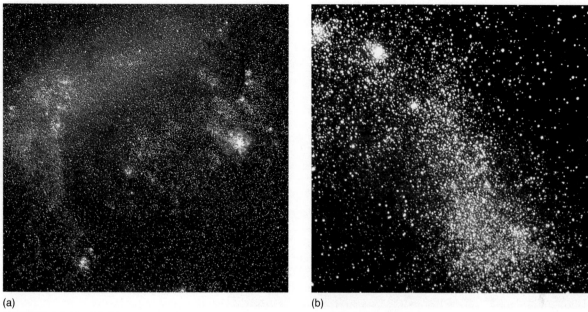

(a)   (b)

FIGURE 24.12 (a) The Large Magellanic Cloud, and (b) the Small Magellanic Cloud.
(Courtesy NOAO)

ful sources of energy. Even more energetic reactions have been recorded in other galaxy-sized objects. In 1940, astronomers discovered a distant object that glowed only dimly in the visible region of the spectrum but was an intense emitter of energy at radio frequencies. More than three times as much radio-frequency energy was radiated from this source as is emitted by the entire Milky Way galaxy at all frequencies. This unusual object appeared to be a galaxy-sized collection of stars and thus was called a radio galaxy. Since that time, many other similar objects have been detected. Pictures of some radio galaxies show large lobes of gas that seem to be flying away from the central core, as shown in Figure 24.13. Do these lobes represent material ejected from the center of the galaxy by a giant explosion? Perhaps. If so, the intense radio signals are being emitted by electrons that were accelerated by shock waves or by intense magnetic and electric fields generated by the explosion.

As the study of the Universe has progressed during the past few decades, the list of unusual and highly energetic objects has grown. The most energetic objects discovered so far are called **quasars,** short for quasi-stellar radio sources. Qua-

FIGURE 24.13 This galaxy is a source of radio emissions. The jet of gas may be a huge mass of material that has been ejected by a giant explosion in the galactic core.
(Courtesy of Lick Observatory)

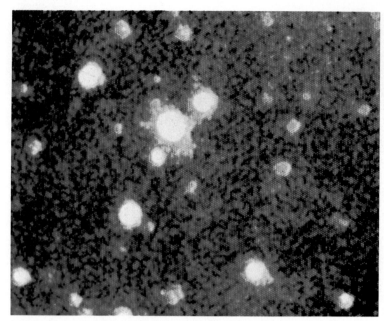

FIGURE 24.14 Quasar 3C 275.1, the first to be found at the center of a galactic cluster, appears as the brightest object near the center of this pseudocolor image. The quasar nucleus is surrounded by a rotating elliptical gas cloud. The quasar is believed to be some 7 billion light-years away, and the light that formed this image left the quasar more than 2 billion years before our Solar System was formed. (Courtesy NOAO)

sars exhibit tremendous red shifts, far greater than any other objects yet observed in the Universe. This indicates that they are moving away from us at very great speeds. Indeed, the fastest quasar is traveling at 92 percent of the speed of light! According to Hubble's law, these tremendous speeds mean that they must be very far away, on the order of several billion light-years from Earth. Quasars emit tremendous quantities of energy, perhaps 100 times as much as the largest galaxies known. Despite their energy output, quasars are quite small; most are less than 1 light-year in diameter (recall that the Milky Way galaxy is 100,000 light-years in diameter), and one has been found that is as small as 1 light-week in diameter.

Often when astronomers look at objects in space, they say that they are looking backward in time. We have touched on the reason for this earlier, but let us reconsider it for a moment and see what it has to say about quasars. Light from the Sun takes about 8 minutes to reach the Earth. Thus, when we look at the Sun, we see it not as it is right this moment, but as it was 8 minutes ago. Now, consider the Andromeda Galaxy, which is 2.25 million light-years from us. When we focus a telescope on this galaxy, we see it not as it is now, but as it was 2.25 million years ago. Thus, when

you think about quasars, it is important to remember that the light we see from them is billions of years old. In fact, the most distant quasar is about 14 billion light-years from Earth. This distance is one piece of evidence which is used by astronomers to determine the age of the Universe. As a result, the Universe is now determined to be between 14 to 20 billion years old.

Quasars are a window to the past, a view of what the Universe was like long ago. Extending this line of reasoning, quasars might be young galaxies in formation and might provide a picture of the Milky Way as it was long, long ago. Many astronomers believe that a quasar is the core of an evolving galaxy. To support this hypothesis, the best photographs of quasars indicate that at least a few of them actually lie in the center of a galaxy, as shown in Figure 24.14. If stars surround other quasars, they are hard to see because of their great distance.

One theory states that the core of a quasar is a massive black hole, and the emissions are coming from material that is accelerated by this intense source of gravitational energy. If this idea is correct, it could provide both an explanation for the energy in a quasar and a model for the formation of our own galaxy, but at present the theories are still highly speculative and our view of the

Universe may change as more data become available.

24.4 COSMOLOGY—A STUDY OF THE BEGINNING AND THE END OF TIME

During our brief look at astronomy, we have studied many of the celestial bodies and some of the physics of the events that occur within our Solar System and the regions beyond. The knowledge that we have gained does not detract from the mystery and wonder, for one can hardly contemplate the objects in space and the distances between them without being filled with awe. Perhaps the ultimate mystery, the great question that never ceases to boggle the mind, is: When did time and space begin, what was here before it, and when, if ever, will it end? **Cosmology** is the study of the origin, structure, and fate of the Universe. To contemplate cosmological questions, you must stretch your imagination even farther than you have done previously in this chapter and attempt to contemplate new dimensions of space, time, and nothingness.

The Big Bang Theory

As part of our search for an ultimate beginning, we must search for signs that the Universe is progressively aging. If the state of the Universe doesn't change with time, one might be led to conclude that perhaps there was no evolution, no beginning, no start—ever. There are, in fact, indications that the Universe is growing older. One piece of evidence is based on the quantity of elements heavier than hydrogen and helium in the stars and galaxies. Within the core of the Sun and other mature stars, hydrogen is being converted to helium, and helium is being converted to heavier elements. As stars fade, change, and die amid giant supernova explosions, the heavier elements are gradually formed in increasing quantities. This conversion of hydrogen into other elements is essentially irreversible. No known process now occurring within the Universe breaks down large quantities of these heavier elements into hydrogen. Thus, the chemical composition of the Universe is a type of evolutionary clock.

A second marker of the passage of time is found in the movement of the galaxies. Recall that all the galaxies are flying away from each other. If they are all flying away from each other, it seems reasonable to assume that they were all in the same location at some time in the past. Following this logic still further, scientists now believe that the Universe was once compressed into a single ball of pure energy. Since an observer on Earth sees everything moving outward, does this mean that our own Milky Way is, by chance, the center of the Universe? No, not at all. There is no center. Think of a baker making a loaf of raisin bread, as shown in Figure 24.15. Imagine that each raisin is a galaxy and that the dough represents the empty space between them. As the dough rises, each raisin moves farther away from every other one. Any observer on any raisin always sees all the other raisins moving away, and thus no one can truly be considered to be the center. This conclusion is somehow disquieting, and you may ask, "If everything, all matter, was once in one spot, then that spot had to be the center." But, to repeat, that type of reasoning is not correct; there is no center. The formation of the Universe may have been the creation of space itself; before the beginning there was nothing at all—not even emptiness—not even time.

Let us return once again to the original ball of infinite density that is believed to have contained all the matter of the Universe. It is meaningless to try to describe this collection of all matter compressed into no space, because there is no frame of reference to start with. If you think about this point sphere, you cannot help but feel its mystery. Such feelings may be expressed in religious or other personal ways, but they are beyond science. Science can deal only with events that started after the initial formation of the Universe.

According to the most accepted theory, at the beginning of time, this ball of pure energy exploded. This cataclysmic event, the **Big Bang,** marked the beginning of the Universe and the start of time. Since scientists know how far apart the galaxies are at present and approximately how fast they are moving, it is possible to calculate backward through time and estimate how long ago it all started. Most current estimates place the

FIGURE 24.15 From every raisin in a raisin cake, every other raisin seems to be moving away from you at a speed that depends on its distance from you. This leads to a relation like the Hubble Law between the speed and the distance. Note also that each raisin would be at the center of the expansion measured from its own position, yet the cake is expanding uniformly. For a better analogy with the Universe, consider an infinite cake; clearly, there is no center to its expansion.

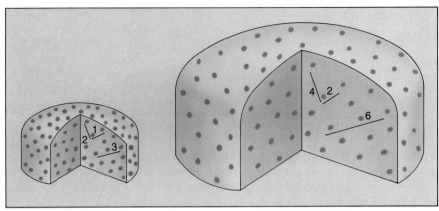

Big Bang between 18 and 20 billion years ago. Matter and energy, in the sense in which we think of them, came into existence with this explosion. This was no ordinary event such as one might observe if a supergiant firecracker or a massive star exploded. Ordinary explosions, even thermonuclear ones, start in a definite center and spread outward to engulf more and more matter. No! This must have been an explosion that started simultaneously everywhere (and nowhere because there was no space before the explosion), instantly creating the Universe.

At first, there were no atoms, or even protons and neutrons. The Universe was entirely filled with radiation and then elementary subatomic particles. Theoretical physicists believe that about 1/100th of a second after the start of time, the temperature of the Universe was about 100 billion K. Certainly, no molecules or even atomic nuclei could exist at these temperatures, and the Universe consisted of a uniform sea of photons and atomic particles, such as electrons, positrons, and neutrinos. However, the temperature started to drop rapidly. Approximately 1 second after the Big Bang, the Universe cooled to 10 billion degrees K, and particles joined together to form protons and neutrons. Within a few minutes, temperatures dropped sufficiently so that fusion reactions could occur, and the simplest atomic nuclei were formed. About 25 per-

cent of the mass of the Universe combined to form helium, as shown in Figure 24.16. Nearly all the rest remained as hydrogen, which even to this day composes about 75 percent of all known matter. After about 1 million years, the temperature was low enough that electrons and simple nuclei could condense to form atoms. At this point, photons left over from the Big Bang itself were no longer absorbed efficiently, for remember that atoms can absorb light only at certain specific frequencies. These primordial photons dispersed unhindered into the void. About 15 billion years later, in 1965, two astronomers at Bell Laboratories recorded a faint, low-energy radio signal that was uniformly distributed throughout all of space. Most scientists believe this radiation could only have been left over from the Big Bang itself. This discovery provides one of the most significant pieces of information in support of the Big Bang theory.

As matter expanded farther, it separated into galaxy-sized agglomerations held together by mutual gravitation. These galaxies continued to fly apart, as part of the expanding Universe. Meanwhile, within each galaxy, matter collected into stars, and the stars slowly aged. Supernova explosions ripped through dying stars to produce heavier elements, and, in at least one case, a solar system was formed and life evolved. This evolution of the Universe is pictured in Figure 24.17.

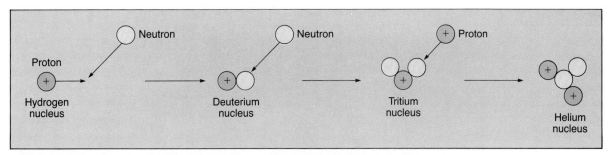

FIGURE 24.16 The formation of helium nuclei in the primordial Universe. This process predominated when the Universe was of the order of several minutes old. Approximately 25 percent (by mass) of the protons and neutrons of the Universe combined to form helium.

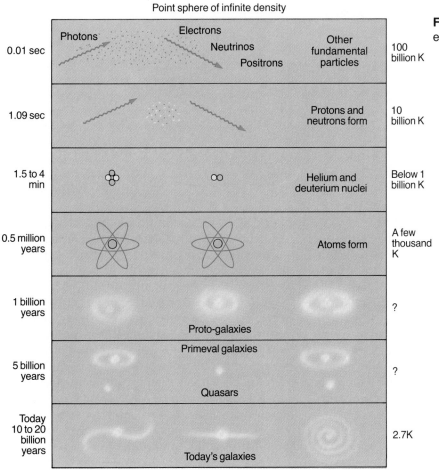

FIGURE 24.17 The evolution of the Universe.

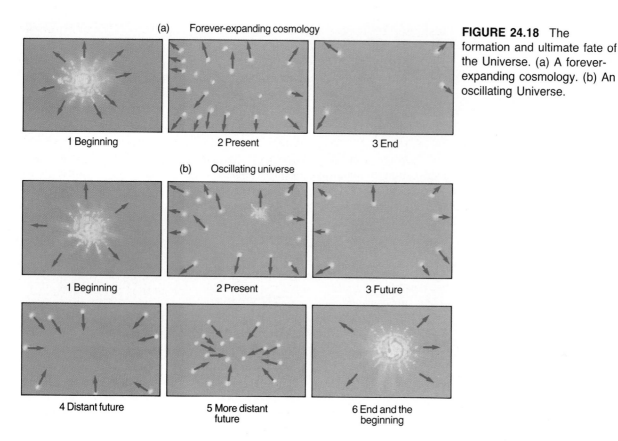

FIGURE 24.18 The formation and ultimate fate of the Universe. (a) A forever-expanding cosmology. (b) An oscillating Universe.

The end of the Universe

Although scientists feel fairly confident that they know something of what happened in the beginning of time, they feel less sure about what will happen in the future. One possibility is that the Universe will expand forever. Galaxies will continue to disperse, and the space between them will gradually grow larger and emptier, as shown in Figure 24.18a. Simultaneously, fuel for stellar fusion reactions will be consumed, so the energy output within each galaxy will slowly diminish. Eventually, space will become emptier and emptier, and only small bits of cold matter in a frozen void will remain.

Another possibility is that the expansion will eventually come to a halt. Although the galaxies are flying away from each other, they are also being pulled inward by mutual gravitation. If the gravitational attraction is sufficiently large, it will cause the galaxies to decelerate gradually to a standstill, then accelerate back inward, falling closer and closer until they all join together. All the galaxies, all the stars, all matter, and all space will be unified once again. This giant agglomeration of matter containing all the mass in the Universe would be exactly identical to the mass that was present at the beginning of our current Universe. This dense matter could then "explode" again and begin expanding to form a new Universe. This scenario is called the **oscillating Universe,** which alternately expands and contracts indefinitely. In this version, the death of one universe would simultaneously represent the birth of a new one. Thus, the oscillating universe cosmology predicts innumerable beginnings and endings in an infinite continuum, as shown in Figure 24.18b.

Will our Universe slowly expand into space and gradually fade into oblivion, or will it be pulled back together by its own gravitation so that all matter becomes reunified into a dense volume? Astronomers trying to answer this question have attempted to measure the density of the entire Universe. They reason that if they knew the density of the system, they could calculate its internal gravitation. Then, knowing the momentum of the receding galaxies, it would be possible to calculate whether they would be slowed down and drawn in again or whether they would continue to move outward forever. If space contains approximately one hydrogen atom in every cubic meter, that is enough mass to stop the expansion and cause the Universe to collapse.

It is very difficult to calculate the amount of mass in the Universe, and there are a great many uncertainties in the attempts that have been made. To understand the nature of the problem, let us start by focusing on something that we can observe today rather than on an event that may or may not occur sometime in the distant future. Throughout the Universe, there are many large groups, or clusters, of galaxies. Each cluster is held together by the mutual gravitation of the component galaxies. However, if astronomers calculate the mass of each galaxy as measured by the estimated number of stars in it, they come to a disturbing conclusion. The galactic masses calculated in this fashion account for only about 10 percent of the mass required to hold the galaxies together. Ninety percent of the mass is unaccounted for. The same conclusion is reached from the study of the motion of stars within a single galaxy. Stars revolve around the galactic core, and this motion obeys Newton's laws. However, the galactic revolutions do not follow the known laws of physics unless the galaxies are much more massive than we measure them to be. In observation after observation, the same conclusion has been reached: Over 90 percent of the Universe is not visibly accounted for. Scientists have added the masses of interstellar and intergalactic dust and stray hydrogen atoms, but still there is not nearly enough mass to cause the Universe to behave the way it does.

What about black holes? Could they provide the missing mass? The answer appears to be no. To understand this conclusion, let us return to our study of the origin of the Universe. Recall that when the Universe was very young—on the order of several minutes old—there were large quantities of deuterium nuclei floating about in a sea of protons and neutrons. If deuterium is struck first by a neutron and then by a proton, it will react to form helium. The rate of the reaction is partially dependent on the density of the various particles. If the density of deuterium, protons, and neutrons were very high, all the deuterium would have reacted to form helium; if the density were very low, very little helium would have been formed. By measuring the ratio of helium to deuterium observed today, it is possible to calculate the density of the nuclear particles in the primordial Universe. This calculation indicates that there was not enough of these particles to form enough black holes to provide the missing mass.

Another consideration is neutrinos. Physicists estimate that throughout the Universe there are 10 billion neutrinos for every atom. If each neutrino had even a tiny mass, then all of them together would add up to quite a lot of mass. But recall from Chapter 13 that scientists are not sure whether or not neutrinos have any mass at all.

The search for the missing mass, often called the **"dark matter,"** continues. Some physicists postulate that there are other subatomic particles that may exist but have never been detected that may make up some of the missing mass. But this line of reasoning is highly conjectural, to say the least.

In short, no one really knows how massive the Universe is and whether it will expand forever or oscillate. Perhaps the Hubble Space Telescope, launched by shuttle in April 1990, will provide more information bearing on this question.

EPILOGUE

In a search for the origin and conclusion of the Universe, we have thought about events that must have occurred many billions of years ago or that will occur many billions of years from now. We have tried to extend our vision across expanses of space, and in our imagination we have traveled to the very edge of what is known and beyond.

Let us ask the final question: Have we gone beyond the bounds of the merely unknown and perhaps encroached on the unknowable? The Big Bang cosmologies assume a beginning, a small, dense, hot proto-universe. Can science ever address itself to the question of how that Universe originally appeared or where it came from? If there were nothingness before, what was nothingness like or even what is meant by "before"? At the present, there is no experiment, not even a thought experiment, that begins to bear on this question. It is inherently unanswerable. If you think about it very long, it is difficult not to wonder: If our Universe is a unique "one-shot" affair, why are we so special? On the other hand, if we live in an oscillating Universe, the search for a beginning may be meaningless, for perhaps there was none; perhaps matter has been expanding and contracting for all of past time and will continue to oscillate forever.

SUMMARY

A **galaxy** is a large volume of space containing many billions of stars, all held together by their mutual gravitational attraction. Our Sun lies in the plane of the disk of the **Milky Way galaxy.** There are large, diffuse clouds of dust and gas between the stars in our galaxy. There is evidence that there has been a gigantic explosion at the core of the galaxy. One theory is that the central core exploded in a colossal supernova-type explosion, and a giant black hole was left behind. There are four basic types of galaxies. They are **spiral** galaxies like our Milky Way, **elliptical** galaxies, **barred spirals,** and **irregulars.** Galaxies also form collections held together by gravitational attraction. The Milky Way is part of the **Local Group,** an assemblage of about 28 galaxies.

Scientists believe that **quasars** are objects that are very far away, emit perhaps 100 times as much energy as an entire galaxy, and are quite small compared with the size of a normal galaxy. Quasars may be young galaxies in the process of evolution. (Remember, the light we see from them is several billion years old.)

The Big Bang theory states that about 18 to 20 billion years ago, the Universe was confined to a small globule of pure energy. This exploded to form the Universe itself. The original cloud of matter eventually cooled and separated into galaxies, stars, and planets, and in at least one place, living creatures came into being. Our Universe is expanding, and it is not known whether this expansion will continue or whether the galaxies will eventually fall back together. If so, another Big Bang could occur, creating another universe. This latter concept is called the **Oscillating Universe** theory.

KEY WORDS

Galaxy	Elliptical galaxy	Local Group	Big Bang theory
Globular clusters	Barred spiral galaxies	Radio galaxy	Oscillating Universe
Milky Way	Irregular galaxies	Quasar	theory
Spiral galaxy	Clusters of galaxies	Cosmology	Dark matter

QUESTIONS

THE MILKY WAY GALAXY

1. In what way were our early perceptions of our place in the galaxy similar to those of the geocentric theory of the Solar System?
2. What evidence indicates that the center of our galaxy was once the scene of a violent explosion?
3. Explain how globular clusters were important in determining the position of our Sun in the galaxy.
4. If a gigantic explosion should occur at the center of our galaxy at this moment, should we be concerned about the effects that this might have on us

during our lifetime? How long would an explosion with its front traveling at the speed of light take to reach us?

5. It is often said that if you lived on a planet circling a star embedded in a globular cluster, your knowledge of astronomy would be very limited. Explain this statement.

GALAXIES AND CLUSTERS OF GALAXIES

6. What is the closest spiral galaxy to the Milky Way? What is the closest irregular galaxy?

7. Why did we not give a more comprehensive treatment of the evolution of galaxies?

ENERGETIC GALAXIES AND QUASARS

8. Would you consider a quasar to be similar to a star, a galaxy, or neither? Discuss.

9. Quasars are all very far away. Using the current theory of quasars, explain why they are not found (a) in our Solar System, (b) in our galaxy, (c) approximately the same distance from Earth as the closest galaxies.

10. Based on your knowledge of physics, devise an explanation for how the enormous energy of a quasar could be produced.

11. Why is it often said that quasars are "windows on time"?

COSMOLOGY

12. Arrange the following different environments in the order of increasing densities: (a) intergalactic space, (b) the core of the Sun, (c) the corona of the Sun, (d) the region of space between planets in our Solar System, (e) galactic space, (f) the Earth's atmosphere at sea level.

13. According to the Big Bang theory, the Universe was once a homogeneous mass of photons and atomic particles. Trace the evolution of hydrogen, helium, and the heavier elements.

14. What evidence supports the statement that our Universe is very old?

15. Some Big Bang cosmologists predict ours is an ever-expanding universe, while others predict an oscillating universe. Discuss the similarities and differences between the two ideas. What information is needed to show which one is correct?

16. If we live in an oscillating universe, would there be any way to find out about what happened in previous universes?

17. What evidence supports the contention that the Universe is evolving?

18. What evidence supports the Big Bang theory?

19. The contention is often made that space and time began at the instant of the Big Bang. Defend this statement.

NASA's Infrared Astronomical Satellite (IRAS) has discovered a shell or ring of large particles surrounding Vega, the third brightest star in the sky. The material could be a solar system at a different stage of development than our own. (Courtesy NASA)

A P P E N D I C E S

A
HANDY CONVERSION FACTORS

To Convert From	To	Multiply By	
Centimeters	Feet	0.0328 ft/cm	
	Inches	0.394 in/cm	
	Meters	0.01 m/cm	(exactly)
	Micrometers	10,000 μm/cm	(″)
	Miles (statute)	6.214×10^{-6} mi/cm	
	Millimeters	10 mm/cm	(exactly)
Feet	Centimeters	30.48 cm/ft	(exactly)
	Inches	12 in/ft	(″)
	Meters	0.3048 m/ft	(″)
	Micrometers	304800 μm/ft	(″)
	Miles (statute)	0.000189 mi/ft	
	Yards	0.3333 yd/ft	
Gallons (U.S., liq.)	Cu. centimeters	3785 cm^3/gal	
	Cu. feet	0.1337 ft^3/gal	
	Cu. inches	231 in^3/gal	
	Cu. meters	0.003785 m^3/gal	
	Cu. yards	0.004951 yd^3/gal	
	Liters	3.785 L/gal	
	Quarts (U.S., liq.)	4 qt/gal	(exactly)
Grams	Kilograms	0.001 kg/g	(exactly)
	Micrograms	1×10^6 μg/g	(″)
	Ounces (avdp.)	0.03527 oz/g	
	Pounds (avdp.)	0.002205 lb/g	
Inches	Centimeters	2.54 cm/in	(exactly)
	Feet	0.0833 ft/in	
	Meters	0.0254 m/in	(exactly)
	Yards	0.0278 yd/in	
Kilograms	Ounces (avdp.)	35.27 oz/kg	
	Pounds (avdp.)	2.205 lb/kg	
Liters	Cu. centimeters	1000 cm^3/L	(exactly)
	Cu. feet	0.0352 ft^3/L	
	Cu. inches	61.02 in^3/L	
	Cu. meters	0.001 m^3/L	(exactly)
	Cu. yards	0.001308 yd^3/L	
	Gallons (U.S., liq.)	0.264 gal/L	
	Ounces (U.S., fluid)	33.81 oz/L	
	Quarts (U.S., liq.)	1.0567 qt/L	

continued next page

To Convert From	To	Multiply By	
Meters	Centimeters	100 cm/m	(exactly)
	Feet	3.2808 ft/m	
	Inches	39.37 in/m	
	Kilometers	0.001 km/m	(exactly)
	Miles (statute)	0.0006214 mi/m	
	Millimeters	1000 mm/m	(exactly)
	Yards	1.0936 yd/m	
Miles (statute)	Centimeters	160934 cm/mi	
	Feet	5280 ft/mi	(exactly)
	Inches	63360 in/mi	(exactly)
	Kilometers	1.609 km/mi	
	Meters	1609 m/mi	
	Yards	1760 yd/mi	(exactly)
Ounces (avdp.)	Grams	28.35 g/oz	
	Pounds (avdp.)	0.0625 lb/oz	(exactly)
Pounds (avdp.)	Grams	453.6 g/lb	
	Kilograms	0.454 kg/lb	
	Ounces (avdp.)	16 oz/lb	(exactly)
Ounces (troy)*	Grams	31.1 g/oz (troy)	
	Ounces (avdp.)	1.097 oz (troy)/oz (avdp.)	
Pounds (troy)	Grams	373 g/lb (troy)	
	Ounces (troy)	12 oz (troy)/lb (troy) (exactly)	

*The price of gold and other precious metals is commonly quoted in troy weight. Note that a troy ounce (31.1 g) is heavier than an avdp. ounce (28.35 g), but a troy pound (12 troy oz) is lighter than an avdp. pound (16 avdp. oz).

B
SIGNIFICANT FIGURES

The discussions in this text have not dealt with the significance of measurement figures, but if the reader wishes to do so, this section will serve as a guide.

Suppose you weigh an object four times on a balance that provides readings to the hundredth of a gram, and you get the following values:

5.14 g
5.13 g
5.12 g
5.13 g

You assume that the mass is actually constant; why, then, are there different values? Perhaps the balance is influenced by uneven air currents, or maybe you don't eye the scale the same way each time. These variations may change the reading over a range of one or two hundredths of a gram, so it would be reasonable to state that the mass of the object is between 5.12 and 5.14 g, or 5.13 ± 0.01 g. More commonly, the value would be written simply as 5.13 g, with the understanding that there may be some uncertainty in the information provided by the last figure.

Now suppose that the same object is weighed on a more sensitive balance, with the following results:

5.12904 g
5.12903 g
5.12904 g

The mass would now be expressed as 5.12904 ± 0.00001 g, or simply as 5.12904 g, again with the understanding that there may be some uncertainty in the last figure. With the more sensitive balance you are now sure of the 5.1290, but the last place is still uncertain.

In the value of 5.13 g there are three digits that provide significant information; in the value 5.12904 g there are six such digits. A **significant figure,** therefore, is defined as a digit that is believed to be correct, or nearly so. The value 5.13 g has three significant figures, and 5.12904 g has six. Note that even though the last number is uncertain, it is still counted as a significant figure, because its uncertainty is limited, usually to ± 1 or 2.

Decimal points have nothing to do with significant figures. If the value of 5.12904 g is expressed in mg, it becomes 5129.04 mg, which still has six significant figures.

The concept of significant figures does not apply to all kinds of numbers; it applies only to numbers that express measurements or computations derived from measurements. Thus, the concept does not apply to the numbers in the relationships 1 m = 100 cm, or 1 ft = 12 in, because these are exact definitions, and the numbers are not measured ones.

The following rules govern the use of significant figures:

Rule 1

To count the significant figures in a measured number, read the number from left to right starting with the *first digit that is not zero*. Count that first number and all the numbers that follow, including all the later zeros. The position of the decimal point, if any, is irrelevant. Thus, the value 0.10 mg has two significant figures. If the same mass is expressed as 0.00010 g, it still has two significant figures. There are instances, however, where this rule is not followed, and your common sense will recognize such exceptions. For example, if you read a statement that Mt. Everest is "about 9000 meters high," you would probably not conclude that the statement means 9000 ± 1 m, implying 4 significant figures. First, you may doubt that the measurement could be so precise; second, you may think it unlikely that the height would be a whole number of thousands of meters; and third, the word "about" implies a rough approximation. The three zeros are in fact "spacers," or place-holders, to define the magnitude of the first digit. The ambiguity could be

avoided by using exponential notation, such as $9 \cdot 10^3$ m, which shows only 1 significant figure, or $9.0 \cdot 10^3$ m, which shows 2 significant figures, as explained in Rule 2 below.

EXAMPLE 1

How many significant figures are there in each of the following measured quantities? **(a)** 0.00406 mm; **(b)** 31.020 L; **(c)** 0.020 s; **(d)** $6.00 \cdot 10^8$ kg; **(e)** 50,000,000 years.

Answer: (a) 3; **(b)** 5; **(c)** 2; **(d)** 3; **(e)** ambiguous, but probably only 1.

Rule 2

If the number of digits needed to express the magnitude of a measurement exceeds the permissible number of significant figures, exponential notation should be used. For example, assume that a length is measured to be 5.2 m. This value contains two significant figures. What is that length expressed in mm? Multiplying 5.2 mm by 1000 gives 5200 mm, but that appears to have four significant figures, which are too many. The answer should therefore be expressed as $5.2 \cdot 10^3$ mm, which expresses the correct magnitude and retains the correct number of significant figures.

Rule 3

In addition or subtraction, the value with the *fewest decimal places* will determine how many significant figures are used in the answer:

308.7810 g	(4 decimal places; 7 significant figures)
0.00034 g	(5 decimal places; 2 significant figures)
10.31 g	(2 decimal places; the fewest; 4 significant figures)
Sum: 319.09 g	(2 decimal places; 5 significant figures)

Rule 4

In multiplication and division, the value with the *fewest significant figures* will determine how far the significant figures should be carried in the answer:

$$\frac{3.0 \cdot 4297}{0.0721} = 1.8 \cdot 10^5$$

Note that the number with the fewest significant figures is 3.0 (two significant figures), and therefore the answer must also have two significant figures.

Rule 5

When a number is "rounded off" (that is, nonsignificant figures are discarded), the last significant figure is increased by 1 if the next figure is 5 or more and is unchanged if the next figure is less than 5:

4.6349, rounded off to four significant
figures $\longrightarrow$ 4.635
4.6349, rounded off to three significant
figures $\longrightarrow$ 4.63
2.815, rounded off to three significant
figures $\longrightarrow$ 2.82

C
LOGARITHMS TO THE BASE 10 (THREE PLACES)

N	0	1	2	3	4	5	6	7	8	9
1	000	041	079	114	146	176	204	230	255	279
2	301	322	342	362	380	398	415	431	447	462
3	477	491	505	519	532	544	556	568	580	591
4	602	613	623	634	644	653	663	672	681	690
5	699	708	716	724	732	740	748	756	763	771
6	778	785	792	799	806	813	820	826	833	839
7	845	851	857	863	869	875	881	887	892	898
8	903	909	914	919	924	929	935	941	945	949
9	954	959	964	969	973	978	982	987	991	996

D
DIRECT AND INVERSE PROPORTIONALITY

Two variable properties are said to be directly proportional to each other when a change in one of the variables produces a proportionate change in the other. If one variable is doubled, the other doubles; if one triples, the other triples. An example is Charles' law (Chapter 15) which states, "The volume of a gas is directly proportional to its absolute temperature." The expression is

$$V \propto T$$

(The symbol $\propto$ means "is proportional to.") For example, if a sample of gas at 100 K occupies 2 L, if we double the absolute temperature to 200 K, the gas volume also doubles, to 4 L.

In an inverse proportionality, when one variable changes by a given factor, the other variable changes by the *reciprocal* of that factor. For example, Boyle's law (Chapter 15) states that "the volume of a gas is inversely proportional to its pressure." That is,

$$V \propto \frac{1}{P}$$

Thus, if the pressure of a sample of gas is doubled, multiplied by 2, the volume of the gas is halved, multiplied by 1/2.

E
ALGEBRA

When algebraic operations are performed, the laws of arithmetic apply. Symbols such as x, y, and z are usually used to represent quantities that are not specified or have unknown values.

First, consider the equation

$$4x = 12$$

If we wish to solve for x, we can divide (or multiply) each side of the equation by the same factor without affecting the validity of the equation. In this case, let us divide both sides by 4. We have

$$\frac{4x}{4} = \frac{12}{4}$$

or

$$x = 3.$$

Next, consider the equation

$$x + 5 = 9$$

In this type of expression, we can add or subtract the same quantity from each side without affecting the validity of the equation. If we subtract 5 from each side, we get

$$x + 5 - 5 = 9 - 5$$

or

$$x = 4$$

Now, let us put these facts together to solve a slightly more complex equation. Consider the expression

$$\frac{2x}{5} + 3 = 13$$

The overall objective in any algebraic expression such as this is to isolate x by itself on one side of the equation. We can begin this task by getting rid of the 3 on the left side of the equation. We do so by subtracting 3 from each side of the equation, as

$$\frac{2x}{5} + 3 - 3 = 13 - 3$$

which reduces to

$$\frac{2x}{5} = 10$$

Now, let us get rid of the 5 on the left side of the equation. This is done by multiplying both sides of the equation by 5.

$$\frac{2x}{5} 5 = 10 \times 5$$

or

$$2x = 50.$$

Finally, we get rid of the 2 on the left side of the equation by dividing both sides of the equation by 2. We find

$$\frac{2x}{2} = \frac{50}{2}$$

We are left with

$$x = 25.$$

Regardless of how complicated an equation in this text may look, it can be solved by carefully following the rules specified above.

F
AMATEUR ASTRONOMY ORGANIZATIONS

ALABAMA

Birmingham Astronomical Society
P.O. Box 36311
Birmingham, AL 35236
(205) 979-9343

Mobile Astronomical Society
P.O. Box 190042
Mobile, AL 36619
(205) 973-1325

Von Braun Astronomical Society, Inc.
Box 1142
Huntsville, AL 35807
(205) 881-0793

ARIZONA

Phoenix Astronomical Society
6945 E. Gary Road
Scottsdale, AZ 85254
(602) 996-3617

Tucson Amateur Astronomy Assn.
7222 E. Brooks Dr.
Tucson, AZ 85730
(602) 790-5053

ARKANSAS

Arkansas-Oklahoma Astronomical Society
P.O. Box 31
Fort Smith, AR 72902
(501) 452-4614

CALIFORNIA

Central Coast Astronomical Society
P.O. Box 1415
San Luis Obispo, CA 93406
(805) 528-6682

Central Valley Astronomers, Inc.
5790 E. Tarpey Dr.
Fresno, CA 93727
(209) 291-7879

China Lake Astronomical Society
P.O. Box 1783
Ridgecrest, CA 93555
(619) 375-5681

Eastbay Astronomical Society
4917 Mountain Boulevard
Oakland, CA 94619
(415) 533-2394

Idyll-Gazers Astronomy Club
P.O. Box 1245
Idyllwild, CA 92349
(714) 659-3562

Los Angeles Astronomical Society
2800 E. Observatory Road
Los Angeles, CA 90027
(213) 926-4071

Oceanside Photo and Telescope Astronomical Society
929 Buena Rosa Ct.
Fallbrook, CA 92028
(619) 723-0684

Orange County Astronomers
2215 Martha Ave.
Orange, CA 92667
(714) 639-8446

Peninsula Astronomical Society
P.O. Box 4542
Mountainview, CA 94040
(415) 566-3116

Polaris Astronomical Society
22018 Ybarra Road
Woodland Hills, CA 91364
(818) 347-8922

Pomona Valley Amateur Astronomers
546 Prospectors Road
Diamond Bar, CA 91765
(714) 860-5373

Riverside Astronomical Society
P.O. Box 7213
Riverside, CA 92503
(714) 689-6893

Sacramento Valley Astronomical Society
P.O. Box 575
Rocklin, CA 95677
(916) 624-3333

San Bernardino Valley Amateur Astronomers
1345 Garner Ave.
San Bernardino, CA 92411

San Diego Astronomy Association
P.O. Box 23215
San Diego, CA 92123
(619) 587-0172

San Francisco Amateur Astronomers
114 Museum Way
San Francisco, CA 94114
(415) 752-9420

San Jose Astronomical Association
3509 Calico Ave.
San Jose, CA 95124
(408) 371-1307

Sonoma County Astronomical Society
P.O. Box 183
Santa Rosa, CA 95404
(707) 528-1034

Stockton Astronomical Society
P.O. Box 243
Stockton, CA 95201
(209) 473-8234

Tulare Astronomical Association, Inc.
P.O. Box 515
Tulare, CA 93275
(209) 685-0585

Ventura County Astronomical Society, Inc.
P.O. Box 982
Simi Valley, CA 93063

COLORADO

Denver Astronomical Society
P.O. Box 10814
Denver, CO 80210

Rocky Mountain Astrophysical
Group
P.O. Box 25233
Colorado Springs, CO 80936
(719) 550-9804

CONNECTICUT

Fairfield County Astronomical
Society
Stamford Museum Observatory
39 Scofieldtown Road
Stamford, CT 06903
(203) 322-1648

Mattatuck Astronomical Society
Mattatuck Community College
Math-Science Division
750 Chase Parkway
Waterbury, CT 06708

Westport Astronomical Society
P.O. Box 5118
Westport, CT 06880

DELAWARE

Delaware Astronomical Society
P.O. Box 652
Wilmington, DE 19899
(215) 444-2966

FLORIDA

Central Florida Astronomical
Society
810 E. Rollins Street
Orlando, FL 32803
(305) 323-8890

Escambia Amateur Astronomers
Association
6235 Omie Circle
Pensacola, FL 32504
(904) 484-1154

Local Group of Deep Sky
Observers
2311 23rd Ave. W.
Bradenton, FL 34205
(813) 747-8334

St. Petersburg Astronomical Club,
Inc.
594 59th Street S.
St. Petersburg, FL 33707
(813) 343-1594

GEORGIA

Atlanta Astronomy Club
5198 Avanti Ct.
Stone Mountain, GA 30088
(404) 498-1240

Oglethorpe Astronomical
Association
Savannah Science Museum
4405 Paulsen Street
Savannah, GA 31405
(912) 355-6705

HAWAII

Hawaiian Astronomical Society
P.O. Box 17671
Honolulu, HI 96817

Maui Astronomy Club
325 Olokani Street
Makawao, Maui, HI 96768
(808) 572-1939

Mauna Kea Astronomical Society
R. R. #1, Box 525
Captain Cook, HI 96704
(808) 328-9201

IDAHO

Boise Astronomical Society
10879 Ashburton Dr.
Boise, ID 83709
(208) 377-5220

ILLINOIS

Astronomical Society at University
of Illinois
349 Astronomy Bldg.
1011 W. Springfield
Urbana, IL 61801
(217) 351-7898

Chicago Astronomical Society
P.O. Box 48504
Chicago, IL 60648
(312) 966-6214

Naperville Astronomical
Association
205 N. Mill Street
Naperville, IL 60540
(312) 355-5357

Northwest Suburban Astronomers
4960 Chambers Dr.
Barrington, IL 60010

Peoria Astronomical Society, Inc.
1125 W. Lake Ave.
Peoria, IL 61604
(309) 347-7285

Twin City Amateur Astronomers
P.O. Box 755
Normal, IL 61761
(309) 454-4164

INDIANA

Evansville Astronomical Society,
Inc.
P.O. Box 3474
Evansville, IN 47733
(812) 922-5681

Fort Wayne Astronomical Society
P.O. Box 6004
Fort Wayne, IN 46896
(219) 747-0774

Indiana Astronomical Society
2 Wilson Dr.
Mooresville, IN 46158
(317) 831-8387

IOWA

Des Moines Astronomical Society, Inc.
2307 49th Street
Des Moines, IA 50310
(515) 274-1873

Quad Cities Astronomical Society
P.O. Box 3706
Davenport, IA 52808
(319) 324-4661

KANSAS

Kansas Astronomical Observers
220 S. Main
Wichita, KS 67202
(316) 264-3174

KENTUCKY

Blue Grass Amateur Astronomy Society
1490 N. Forbes Road
Lexington, KY 40511
(606) 252-6143

LOUISIANA

Pontchartrain Astronomy Society, Inc.
1441 Avenue A
Marrero, LA 70072
(504) 340-0256

Shreveport Astronomical Society, Inc.
1426 Alma Street
Shreveport, LA 71108
(318) 865-2433

MARYLAND

Baltimore Astronomical Society
601 Light Street
Baltimore, MD 21230
(301) 766-6605

Harford County Astronomical Society
P.O. Box 906
Bel Air, MD 21014
(301) 457-5597

Westminster Astronomical Society
3481 Salem Bottom Road
Westminster, MD 21157
(301) 848-6384

MASSACHUSETTS

Amateur Telescope Makers of Boston, Inc.
8 Pond Street
Dover, MA 02030
(508) 785-0352

South Shore Astronomical Society
P.O. Box 429
Jacobs Lane
Norwell, MA 02061
(617) 588-0673

MICHIGAN

Detroit Astronomical Society
14298 Lauder
Detroit, MI 48227
(313) 981-4096

Grand Rapids Amateur Astronomical Association
4 Alten N.E.
Grand Rapids, MI 49503
(616) 454-7645

Warren Astronomical Society
P.O. Box 474
East Detroit, MI 48021
(313) 355-5844

MINNESOTA

Minnesota Astronomical Society
30 E. 10th Street
St. Paul, MN 55101
(612) 451-7680

3M Astronomical Society
14601 55th Street S.
Afton, MN 55001
(612) 733-2690

MISSISSIPPI

Jackson Astronomical Association
6207 Winthrop Circle
Jackson, MS 39206
(601) 982-2317

MISSOURI

Astronomical Society of Kansas City
P.O. Box 400
Blue Springs, MO 64015
(816) 228-4238

St. Louis Astronomical Society, Inc.
4562 Clearbrook Dr.
St. Charles, MO 63303

MONTANA

Astronomical Institute of the Rockies
6351 Canyon Ferry Road
Helena, MT 59601
(406) 442-2208

NEBRASKA

Omaha Astronomical Society
5025 S. 163 Street
Omaha, NE 68135
(402) 896-4417

Prairie Astronomy Club
P.O. Box 80553
Lincoln, NE 68501
(402) 467-4222

NEVADA

Astronomical Society of Nevada
825 Wilkinson Ave.
Reno, NV 89502
(702) 329-9946

Las Vegas Astronomical Society
Clark County Community College Planetarium
3200 E. Cheyenne Ave
Las Vegas, NV 89030
(702) 459-8401

NEW HAMPSHIRE

New Hampshire Astronomical
Society
22 Center Street
Penacook, NH 03303
(603) 753-9225

NEW JERSEY

Amateur Astronomers Association
of Princeton, Inc.
P.O. Box 2017
Princeton, NJ 08540
(609) 396-3630

Amateur Astronomers, Inc.
W. M. Spezzy Observatory
1033 Springfield Ave.
Cranford, NJ 07016
(201) 549-0615

New Jersey Astronomical
Association
Voorhees State Park
P.O. Box 214
High Bridge, NJ 08829
(215) 253-7294

Small Scope Observers Association
4 Kingfisher Place
Audubon Park, NJ 08106
(609) 547-9487

NEW MEXICO

Albuquerque Astronomical Society
P.O. Box 54072
Albuquerque, NM 87153
(505) 299-0891

Astronomical Society of Las
Cruces
P.O. Box 921
Las Cruces, NM 88004
(505) 526-2968

NEW YORK

Amateur Astronomers Association
of New York, Inc.
1010 Park Ave.
New York, NY 10028
(212) 535-2922

Astronomical Society of Long
Island, Inc.
1011 Howells Road
Bay Shore, Long Island, NY
11706
(516) 586-1760

Broome County Astronomical
Society
Roberson-Kopernik Observatory
Underwood Road
Vestal, NY 13850
(607) 748-3685

Buffalo Astronomical Association,
Inc.
Buffalo Museum of Science
Humbolt Parkway
Buffalo, NY 14211

Rockland Astronomy Club
110 Pascack Road
Pearl River, NY 10965
(914) 735-4163

Syracuse Astronomical Society,
Inc.
1115 E. Colvin Street
Syracuse, NY 13210
(315) 458-1454

Westchester Astronomy Club
511 Warburton Ave.
Yonkers, NY 10701
(914) 963-4550

NORTH CAROLINA

Forsyth Astronomical Society
504 Gayron Drive
Winston-Salem, NC 27105
(919) 744-7141

Raleigh Astronomy Club
P.O. Box 10643
Raleigh, NC 27605
(919) 832-NOVA

NORTH DAKOTA

Dakota Astronomical Society
P.O. Box 2539
Bismarck, ND 58502
(701) 256-3620

OHIO

Astronomy Club of Akron, Inc.
5070 Manchester Road
Akron, OH 44319
(216) 644-0230

Cincinnati Astronomical Society
5274 Zion Road
Cleves, OH 45002
(513) 661-3252

Columbus Astronomical Society,
Inc.
P.O. Box 16209
Columbus, OH 43216
(614) 262-9713

Miami Valley Astronomical Society
Dayton Museum of Natural
History
2629 Ridge Ave.
Dayton, OH 45414
(513) 275-7431

Ohio Turnpike Astronomers
Association
1494 Lakeland Ave.
Lakewood, OH 44107
(216) 521-5115

OKLAHOMA

Astronomy Club of Tulsa
P.O. Box 470611
Tulsa, OK 74147
(918) 742-7577

Oklahoma City Astronomy Club
2100 N.E. 52nd Street
Oklahoma City, OK 73111
(405) 424-5545

OREGON

Eugene Astronomical Society
Lane E. S. D. Planetarium
P.O. Box 2680
Eugene, OR 97402
(503) 741-0501

Portland Astronomical Society
2626 S.W. Luradel Street
Portland, OR 97219
(503) 245-6251

PENNSYLVANIA

Amateur Astronomers Association
of Pittsburgh, Inc.
Wagman Observatory
P.O. Box 314
Glenshaw, PA 15116
(412) 224-2510

Astronomical Society of
Harrisburg
1915 Enfield Street
Camp Hill, PA 17011
(717) 975-9799

Delaware Valley Amateur
Astronomers
6233 Castor Ave.
Philadelphia, PA 19149
(215) 831-0485

Lackawanna Astronomical Society
1112 Fairview Road
Clarks Summit, PA 18411
(717) 586-0789

Lehigh Valley Amateur
Astronomical Society, Inc.
620 E. Rock Road
Allentown, PA 18103
(215) 398-7295

RHODE ISLAND

Skyscrapers, Inc.
47 Peeptoad Road
North Scituate, RI 02857
(401) 942-7893

SOUTH CAROLINA

Carolina Skygazers Astronomy
Club
Museum of York County
4621 Mount Gallant Road
Rock Hill, SC 29730
(803) 329-2121

SOUTH DAKOTA

Black Hills Astronomy Society
3719 Locust
Rapid City, SD 57701

TENNESSEE

Barnard Astronomical Society
P.O. Box 90042
Chattanooga, TN 37412
(615) 629-6094

Smoky Mountain Astronomical
Society
P.O. Box 6204
Knoxville, TN 37914
(615) 637-1121

TEXAS

Austin Astronomical Society
P.O. Box 12831
Austin, TX 78711

Fort Worth Astronomical Society,
Inc.
P.O. Box 161715
Fort Worth, TX 76161
(817) 860-6858

JSC Astronomical Society
3702 Townes Forest
Friendswood, TX 77546
(713) 482-3909

San Antonio Astronomical
Association
6427 Thoreau's Way
San Antonio, TX 78239
(512) 654-9784

Texas Astronomical Society
P.O. Box 25162
Dallas, TX 75225
(214) 368-6982

UTAH

Salt Lake Astronomical Society
15 S. State Street
Salt Lake City, UT 84111
(801) 538-2104

VIRGINIA

Northern Virginia Astronomy
Club
6028 Ticonderoga Ct.
Burke, VA 22015
(703) 866-4985

Richmond Astronomical Society
709 Timken Dr.
Richmond, VA 23239
(804) 741-3689

WASHINGTON

Seattle Astronomical Society
852 N.W. 67th Street
Seattle, WA 98115
(205) 523-2787

Spokane Astronomical Society,
Inc.
4140 Cook Street
Spokane, WA 99223
(509) 448-9694

Tacoma Astronomical Society
7101 Topaz Dr. S.W.
Tacoma, WA 98498
(206) 588-9504

WISCONSIN

Milwaukee Astronomical Society
W248 S7040 Sugar Maple Dr.
Waukesha, WI 53186
(414) 662-2987

WYOMING

Cheyenne Astronomical Society
3409 Frontier Street
Cheyenne, WY

GLOSSARY

absolute humidity See *humidity*.

absolute zero The zero point on the Kelvin scale of absolute temperature. No substance can be cooled below zero kelvin, or 0 K. The equivalent temperature on the Celsius scale is −273.15°C.

absorption spectrum See *spectrum*.

acceleration The change in velocity of an object per unit time. Acceleration is a vector quantity. A body is accelerating when it is speeding up, slowing down, or changing direction.

acceleration due to gravity The acceleration of a freely falling body under the influence of the Earth's gravitation at sea level. This term is symbolized by the letter g and is equal to approximately 9.8 m/s^2.

acid A substance that can supply hydrogen ions (protons) to another substance, known as a base.

adiabatic Referring to a process that occurs without loss or gain of heat.

adiabatic cooling A cooling process that occurs without loss or gain of heat. Under certain conditions, a rising air mass may cool adiabatically. If no heat is transferred, a rising air mass will expand and, as it expands, it performs work and therefore cools.

adsorption The process by which molecules from a liquid or gaseous phase become concentrated on the surface of a solid.

air mass A large body of air that has approximately the same temperature and humidity throughout.

air pollution The deterioration of the quality of air that results from the addition of impurities.

albedo A measure of the reflectivity of a surface, measured as the ratio of light reflected to light received. A mirror or bright snowy surface has a high albedo, whereas a rough flat road surface has a low albedo.

allotropes Different forms of the same element.

alpha particle The nucleus of a helium atom.

alternating current (ac) An electric current that oscillates in a wire. When an alternating current is established in a wire, electrical energy is transmitted, but the electrons themselves do not move in a concerted direction. See also *direct current*.

ampere A measure of electrical current equal to the movement of one coulomb of charge past a given point in a wire in one second.

amplitude (of a wave) The magnitude or height of a wave, measured as the distance between the zero point of the wave to the point of maximum displacement. The amplitude is one half of the vertical distance between the crest and the trough.

angle of incidence When a light ray is beamed onto a surface, the angle of incidence is defined as the angle between the light ray and a line drawn perpendicular to the surface.

angle of reflection The angle between a reflected light beam and a line drawn perpendicularly to the surface.

anode The positive terminal in a vacuum tube. The electrons emitted by the cathode are collected at the anode. In an electrochemical cell, the anode is the electrode where oxidation (loss of electrons) occurs.

aquifer An underground layer of rock that is porous and permeable enough to store significant quantities of water.

asteroid One of the small planetary bodies that orbits the Sun. Asteroids range in size from less than 1 km to 1000 km in diameter.

asthenosphere The plastic part of the Earth's mantle just below the lithosphere.

atmosphere The predominantly gaseous envelope that surrounds the Earth.

atmosphere (standard of pressure) The pressure exerted at sea level by a column of mercury 76 cm high. This corresponds to the normal pressure exerted by the Earth's atmosphere at sea level.

atom The fundamental unit of the element.

atomic nucleus The small positive central portion of the atom that contains its protons and neutrons.

atomic number The number of protons in an atomic nucleus.

atomic orbital See *orbital*.

aurora A luminous atmospheric display appearing mainly in the high latitudes that is created by interactions between the particles of the solar wind and the magnetic field of the Earth.

Avogadro's law Equal volumes of all gases (at the same temperature and pressure) contain the same number of molecules.

background radiation The level of radiation on Earth from natural sources.

barometer A device used to measure atmospheric pressure.

barometric pressure The pressure (force/area) exerted by the atmosphere.

base Chemistry: A substance that accepts protons from an acid. The reaction is said to neutralize the acid.

beta particle An electron emitted from an atomic nucleus.

Big Bang An event that is thought to mark the beginning of our Universe. The theory assumes that, some 10 to 20 billion years ago, all matter that was to form the Universe exploded into space from an infinitely compressed state. See also *oscillating universe cosmology*.

black hole A small region of space that contains matter packed so densely that an intense gravitational field is created, from which light cannot escape.

Boyle's law The volume of a gas (at constant temperature) is inversely proportional to its pressure.

branching chain reaction A chain reaction in which each step produces more than one succeeding step.

breeder reactor A nuclear reactor that produces more fissionable material than it consumes.

bit A single unit of binary information such as is stored electronically in a computer.

calorie A unit used to express quantities of thermal energy. When "calorie" is spelled with a small c, it refers to the quantity of heat required to heat 1 g of water 1°C. When "Calorie" is spelled with a capital C, it means 1000 small calories, or one kilocalorie, the quantity of heat required to heat 1000 g (1kg) of water 1°C.

capillary action The movement of water upward against the force of gravity through the action of electrical attractions to the surfaces of small openings.

catalyst A substance that influences (usually speeds up) the rate of a chemical reaction and that is not consumed in the reaction.

cathode The negative terminal in a vacuum tube that emits electrons. These electrons then travel across free space toward the anode. In an electrochemical cell, the cathode is the electrode where reduction (gain of electrons) occurs.

cathode ray A beam of electrons emerging from the cathode in a cathode ray tube.

Celsius scale The temperature scale used in the metric system. On this scale, water freezes at 0°C and boils at 100°C at sea level.

centrifugal force An outward force observed by a person from within an accelerating frame of reference.

centripetal force The inward force that is necessary to keep a body in circular motion.

chain reaction A reaction that proceeds in a series of steps, each step being made possible by the preceding one. See also *branching chain reaction*.

Charles' law The volume of a gas (at constant pressure) is directly proportional to its absolute temperature.

chemical bond The force that holds atoms together to form molecules. See also *covalent bond* and *ionic bond*.

chemical change A transformation that results from making or breaking of chemical bonds.

chemical energy The energy that is absorbed when chemical bonds are broken or that is released when chemical bonds are formed. A substance that can release energy by undergoing chemical reactions is said to have chemical energy.

chemical formula A combination of symbols of elements that shows the composition of a molecule or a substance.

China syndrome A facetious expression referring to a nuclear meltdown in which the hot radioactive mass melts its way into the ground toward China.

chip A tiny circuit board containing many different electronic switching and/or amplifying devices.

chromosphere A turbulent diffuse gaseous layer of the Sun that lies above the photosphere.

climate The composite pattern of weather conditions that can be expected in a given region. Climate refers to yearly cycles of temperature, wind, rainfall, etc., and not to daily variations. See also *weather*.

closed system A system that is isolated so that neither mass nor energy can enter or leave.

cogeneration A tandem operation in which thermal energy that is normally wasted from one industrial process, such as the generation of electricity, is used in another process, such as oil refining. In general, such a system uses fuel more efficiently than would two facilities operating separately.

coherent light A beam of light in which all the component waves are traveling in phase, at the same frequency, and in exactly the same direction. Coherent light is produced by a laser.

comet A celestial body moving about the Sun, usually in a highly elliptical path or orbit. Comets appear to have a fairly dense core surrounded by a "fuzzy," "fiery" halo, but in actuality comets are quite cold. When a comet approaches the Sun, the force of the solar wind blows matter outward from the comet, forming a long "tail."

compound (compound substance) A substance that consists of a fixed composition of elements and has a fixed set of properties.

condensation Conversion of vapor to liquid.

conduction (of thermal energy) The process by which thermal energy is transmitted directly through materials. Conduction occurs because energetic atoms or molecules move rapidly and collide with neighboring atoms or molecules. Kinetic energy is transferred during the collision process, and the neighboring molecules accelerate and become energetic, or "hot."

conductor (electrical) A material that offers little resistance to the movement of electric current.

continental drift The theory stating that the continent-sized masses of the Earth's crust are moving slowly relative to one another.

control rod A neutron-absorbing medium that controls the reaction rate in a nuclear reactor.

convection The process by which thermal energy is transmitted through gases and liquids by the action of currents that circulate in the fluid.

core (of the Earth) The central portion of the Earth, believed to be composed mainly of iron and nickel.

Coriolis effect The deflection of air or water flow caused by the rotation of the Earth.

corona The luminous irregular envelope of highly ionized gas outside the chromosphere of the Sun.

cosmic ray A form of high-energy radiation consisting mainly of high-speed atomic nuclei and other atomic particles that move through space and frequently strike the Earth's atmosphere.

cosmology The study of the origin and the end of the Universe.

coulomb A unit quantity of electricity transported in one second by a current of 1 ampere.

Coulomb's law A relationship that states that the force of attraction (or repulsion) between two charges is directly proportional to the product of the charges and inversely proportional to the square of the distance between them.

covalent bond A chemical bond between atoms that is characterized by shared electrons.

crest The highest point in a wave.

critical condition A condition under which a chain reaction continues at a steady rate, neither accelerating nor slowing down.

critical mass (in a nuclear reaction) The quantity of fissionable material just sufficient to maintain a nuclear chain reaction.

crust (of the Earth) The solid outer layer of the Earth; the portion on which we live.

crystalline solid See *solid*.

decibel (dB) A unit of sound intensity. The decibel scale is a logarithmic scale used in measuring sound intensities relative to the intensity of the faintest audible sound.

degeneracy pressure The strength of the atomic particles that holds a white dwarf star from further collapse.

density Mass per unit volume.

deuterium Isotope of hydrogen with mass number of 2. Also called "heavy hydrogen."

dew Moisture condensed from the atmosphere, usually during the night, when the ground and leaf surfaces become significantly cooler than the surrounding air.

diffraction The ability of a wave to pass around an obstacle or to bend past an opening.

direct current (dc) An electric current moving in one direction only. When a direct current moves through a wire, electrons travel progressively through the wire. See also *alternating current*.

distillation A process in which a liquid is vaporized and the vapor is condensed to a liquid again.

doldrums A region of the Earth near the Equator in which hot, humid air is moving vertically upward, forming a vast low-pressure region. Local squalls and rainstorms are common, and steady winds are rare.

domain (magnetic) Microscopic bundles of magnetic atoms that are held together by electrical forces. The atoms in a domain are aligned so that the entire bundle produces a net magnetic field.

Doppler effect The observed change in frequency of light or sound that occurs when the source of the wave is moving relative to the observer.

dust An airborne substance that consists of solid particles typically having diameters greater than about 1 micrometer.

earthquake A sudden traumatic movement of part of the Earth's crust.

eclipse A phenomenon that occurs when a heavenly body is shadowed by another and therefore rendered invisible. When the Moon lies directly between the Earth and the Sun, we observe a *solar eclipse;* when the Earth lies directly between the Sun and the Moon, we observe a *lunar eclipse.*

electric charge The net quantity of electricity or the electric energy possessed by a substance. It is a measure of the excess or deficiency of electrons.

electric circuit A complete path of conducting materials that allows electric current to flow.

electric current A concerted and continuous movement of charged particles in response to a potential gradient.

electric field A region of space in which electric forces can be detected.

electric force The force that results from the interaction of charged bodies. Electrical force is repulsive if the bodies carry like charges (++ or −−) and is attractive if the bodies are oppositely charged (+− or −+).

electric potential See *volt.*

electrode A terminal in an electric circuit where electrons enter or leave a gas, a vacuum, or an ionic liquid. See also *anode* and *cathode.*

electromagnet A device consisting of an iron core wrapped with wire that is magnetized when a direct current is passed through the wire.

electromagnetic field The combined electric and magnetic field produced by an oscillating charged particle or particles.

electromagnetic induction The induction of an electric current in a wire when a magnetic field changes near the wire or when the wire moves across a magnetic field.

electromagnetic spectrum The entire range of electromagnetic radiation.

electromagnetic wave A periodically oscillating electromagnetic field.

electromagnetism The force generated by magnets or by charged objects.

electron The fundamental atomic unit of negative electricity.

electron shell An energy level in an atom that can accommodate a specific number of electrons and that is designated by a principal quantum number.

element A substance all of whose atoms have the same atomic number.

emission spectrum See *spectrum.*

energy The capacity to perform work. See also specific types of energy such as *chemical, kinetic, nuclear,* and *potential.*

engine A device that converts heat to work.

entropy A thermodynamic measure of disorder. It has been observed that the entropy of an isolated system always increases during any spontaneous process; that is, the degree of disorder always increases.

equinox Either of two times during a year when the Sun shines directly overhead at the Equator. During the equinox, every portion of the Earth receives 12 hours of daylight and 12 hours of darkness.

erosion The process by which parts of the Earth's surface are transported to new locations by water, wind, waves, ice, or other natural agents.

eutrophication The pollution of a body of water by enrichment with nutrients, with a consequent increase in the growth of organisms and a resultant depletion of dissolved oxygen.

evaporation Conversion of liquid to vapor.

excited electron An electron in an atom or molecule that has been promoted to a higher energy level and therefore exists in an excited state.

excited state A state of an atom or molecule in which one or more electrons have absorbed energy and exist in higher energy levels. See also *ground state.*

fault A crack or weak point in the Earth's surface; a potential site of earthquake activity.

fault creep The gradual non-traumatic slippage of land surfaces past each other.

fiber optics The use of glass fibers to transmit information by means of a pulsed laser beam.

field of force See *electric field.*

First law of thermodynamics See *thermodynamics.*

fission (of atomic nuclei) The splitting of atomic nuclei into approximately equal fragments.

fog A low cloud formation usually formed when warm, moisture-laden air is cooled on contact with land or water.

force   Any influence that can cause a body to accelerate. Force is commonly measured in newtons in the SI and pounds in the British system.

frequency   The number of wave disturbances (can be measured as the number of crests) that pass a given point in a specific amount of time. Frequency is usually expressed in cycles/s, or hertz. 1 Hz = 1 cycle/s.

friction   A type of force that opposes the motion of one body past another when the two are in contact.

frontal weather system   A weather system that develops when air masses collide.

fulcrum   The support or point of rest of a lever.

fundamental frequency   The lowest frequency of a musical sound.

fusion (of atomic nuclei)   The combination of nuclei of light elements to form heavier nuclei.

galaxy   A large volume of space containing many billions of stars, all held together by mutual gravitation.

gas   A state of matter that consists of molecules that are moving independently of each other in random patterns.

generator   A device that produces electrical energy when a coil of wire is rotated in a magnetic field. Generators must be driven by some external source of energy.

glacier   A large mass of flowing ice. A glacier usually takes the form of a river of ice.

glass (or glassy solid)   A rigid state of matter in which the atoms or molecules are randomly arranged.

graphite   A soft, black form of the element carbon that consists of crystalline layers, which can slide past one another.

gravitation   A universal force of mutual attraction between all bodies.

gravitation, acceleration due to   See *acceleration due to gravity.*

Great Red Spot   A large atmospheric storm on the surface of Jupiter that was first observed over 300 years ago.

greenhouse effect   The effect produced by certain gases, such as carbon dioxide or water vapor, that causes a warming of the atmosphere by absorption of infrared radiation.

ground state   A state of an atom or molecule in which all the electrons are in their lowest allowed energy levels. See also *excited state.*

gyre   A curved or circular ocean current formed when currents traveling northwards or southwards are deflected by the spin of the Earth and the shape of the continents.

half-life (of a radioactive substance)   The time required for half of a sample of radioactive matter to decompose.

heat   The energy that is transferred from one system to another when two systems at different temperatures are in contact.

heat engine   A mechanical device that converts heat to work.

heat of fusion   The heat required to melt 1 g of a solid at constant temperature.

heat of vaporization   The heat required to vaporize 1 g of a liquid at constant temperature.

heavy oil   Petroleum deposits that are too viscous to be extracted and pumped using conventional oil well technology.

heliocentric theory   The theory, now known to be true, that the Sun, and not the Earth, is the center of the Solar System.

hertz (Hz)   A unit that expresses the frequency of a wave form. When one crest of a wave passes a given point every second, that wave is said to have a frequency of 1 hertz. One hertz is therefore one cycle per second.

heterogeneous   Referring to a nonuniform substance that has different properties and compositions in different regions.

hologram   A three-dimensional photograph produced using laser light.

homogeneous   Referring to a uniform substance having the same properties throughout the sample.

horse latitudes   A region of the Earth, lying at about 30 degrees north and south latitudes, in which air is moving vertically downward, forming a vast high-pressure region. Generally dry conditions prevail, and steady winds are rare.

Hubble's law   A law that relates the red shift of an object outside our galaxy to its distance from Earth. See *red shift.*

humidity   A measure of the amount of moisture in the air. *Absolute humidity* is defined as the amount of water vapor in a given volume of air. *Relative humidity* is defined as the ratio of the amount of moisture in a given volume of air divided by the amount of moisture that can be held by that volume at a given temperature when the air is saturated.

humus   The complex mixture of decayed organic matter that is an essential part of healthy natural soil.

hydrocarbon   A compound of hydrogen and carbon.

hydrologic cycle (water cycle)   The cycling of water, in all its forms, on the Earth.

igneous rock   Rock formed directly from cooling magma.

inclined plane   A ramp or tilted surface used as a machine element.

inertia   That property of a body that compels it to remain at rest or at constant velocity unless it is forced to change.

insulator A material that offers substantial resistance to the movement of an electric current.

interference A process whereby two or more waves combine when they reach a single point in space at the same time. The new wave is formed by the addition of the wave components.

ion An electrically charged atom or group of atoms.

ionic bond A chemical bond formed by attraction between oppositely charged ions.

isoelectronic structures Atoms or ions that have the same electronic composition, such as F^-, Ne, and Na^+.

isomers Different substances that have the same molecular formula, such as ethyl alcohol and dimethyl ether, both C_2H_6O.

isostasy A principle that states that the Earth's crust is floating on denser, fluid layers beneath it.

isotopes Atoms of the same element that have different mass numbers.

joule The fundamental unit of work in the SI. One joule equals 1 newton-meter.

kelvin The SI unit of temperature. One kelvin is the same as a difference in temperature of 1°C. The Kelvin temperature scale starts at 0 K, which equals −273.15°C.

kinetic energy The energy possessed by a moving object, equal to $1/2$ mv^2.

kinetic theory A theory that accounts for the nature of gases by assuming that they consist of independently moving molecules.

laser Acronym for **l**ight **a**mplification by **s**timulated **e**mission of **r**adiation. A device that produces a short, intense flash of coherent light in which all the component waves are traveling in phase, at the same frequency, and in exactly the same direction.

laser fusion A nuclear fusion reaction triggered by a laser beam.

lava The material produced when magma pours rapidly onto the surface of the Earth through fissures in the crust. A site where lava appears is called a volcano.

lens A curved piece of transparent material that refracts light in such a way that the apparent sizes of objects are altered.

lever A rigid bar positioned over a fulcrum that is used as a machine element.

light-year The distance traveled by an electromagnetic wave in one year, approximately $9.5 \cdot 10^{15}$ m.

lighthouse theory The theory that states that a rapidly rotating neutron star may be the source of pulsar signals.

liquid A non-rigid state of matter in which the molecules are arranged rather randomly and adhere to each other just strongly enough to form a well-defined boundary. Liquids therefore flow readily and take the shape of their containers.

liquid crystal A state of matter of substances that are crystalline when undisturbed but that can easily be made to flow like liquids.

liquid metallic hydrogen Hydrogen that is exposed to such extremes of high pressure that it becomes liquid and exhibits metallic properties.

lithosphere The outer shell of the Earth, including the crust and the uppermost portion of the mantle.

longitudinal wave See *wave*.

mach number A unit of speed related to the speed of sound. The mach number is the speed of the object divided by the speed of sound.

machine A type of tool that alters the magnitude or direction of an applied force.

magma A fluid material, lying in the upper layers of the Earth's asthenosphere, consisting of melted rock mixed with various gases such as steam and hydrogen sulfide.

magnetohydrodynamic generator (MHD) A type of electrical generator that operates by passing ions through a magnetic field. MHD systems are more efficient than conventional mechanical generators.

magnetosphere A region of magnetic field around a planet.

mantle The solid but partly semiplastic portion of the Earth that surrounds the central core and lies under the crust.

maria Smooth, flat plains on the surface of the Moon. (Originally these were erroneously thought to be seas—hence the name "maria.")

mass The quantity of matter in an object. Mass is fundamentally defined in terms of the inertia of a body, its resistance to a change in velocity, by the equation: $F = ma$, or $m = F/a$, where m is mass, F is force, and a is acceleration.

mass number The sum of the number of protons and neutrons in an atomic nucleus.

mesosphere The layer of air that lies above the stratosphere and extends from about 33 km upward to 80 km above the surface of the Earth.

metal An element characterized by a great ability to conduct heat and electricity, to reflect light, and to maintain its crystal structure even when its shape is deformed.

metamorphic rock Rock formed when sedimentary or igneous material is altered by heat and pressure within the Earth's crust.

meteor A meteoroid that glows in the Earth's atmosphere when it is accelerated downward by gravity. (Called also *shooting star*.)

meteorite A meteoroid that has fallen to the surface of the Earth.

meteoroid A small bit of solid matter traveling through space.

metric system See *Système International d'Unités.*

MHD generator See *magnetohydrodynamic generator.*

Mid-Atlantic Ridge A large ridge running under the surface of the Atlantic Ocean that is split in the middle by a sharp rift or valley. Igneous material rises out of the Mid-Atlantic Ridge, forming new crustal rock as the tectonic plates move apart.

Milky Way The large spiral galaxy that contains our Solar System.

moderator A medium used in a nuclear reactor to slow down neutrons.

mole The amount of substance that contains as many elementary particles (atoms or molecules) as there are carbon atoms in exactly 12 g of the carbon-12 isotope. Also, the amount of substance that contains $6.02205 \cdot 10^{23}$ elementary particles. Also, the atomic weight of an element or the molecular weight of a compound expressed in grams.

molecule The fundamental particle that characterizes a compound. It consists of a group of atoms held together by chemical bonds.

momentum The product of the mass of a body times its velocity.

monsoon A continental wind system caused by uneven heating of land and ocean surfaces. Monsoons generally blow from the sea to the land in the summer, when the continents are warmer than the ocean, and bring predictable rainstorms. In winter, when the ocean is warmer than the land surfaces, the winds reverse.

net force The vector sum of all the forces acting on a body.

neutralization The reaction of an acid with a base to produce a salt and water.

neutrino A subatomic particle that bears no charge and has very little, if any, mass.

neutron A subatomic particle that is electrically neutral and has a mass approximately equal to that of a proton.

neutron star A small (by stellar standards) dense core of neutrons that remains intact after a supernova explosion. See also *pulsar.*

newton The SI unit of force. One newton is equal to the force needed to give a mass of 1 kg an acceleration of 1 m/s^2.

noble (or "inert") gases The elements of Group 8 of the periodic system, starting with helium.

nodes (of an orbital system) The points where the planes of two orbits intersect. An eclipse of the Sun or the Moon can occur only when the Moon passes through the nodes of the planes of the Earth-Sun and Earth-Moon orbits.

noise Unwanted sound or an unwanted signal.

nuclear energy The energy transferred in nuclear reactions. This energy can be released by natural radioactivity or in the transformations in nuclear reactors or atomic bombs.

nuclear reactor A device that utilizes nuclear reactions to produce useful energy.

nucleus See *atomic nucleus.*

ohm A measure of the electrical resistance of a material.

Ohm's law A relationship that states that the voltage between two points of an electric circuit is equal to the current flowing through the circuit multiplied by the resistance between the two points.

orbital The shape that defines the space occupied by an electron at a given energy level.

ore A rock mixture that contains enough valuable minerals to be mined profitably with currently available technology.

oscillating universe cosmology A version of the Big Bang theory that assumes that the Universe expands and contracts in an endless sequence.

overtone An acoustical frequency that is an even multiple greater than the fundamental frequency.

oxidation The chemical reaction of a substance with oxygen. More generally, oxidation is a loss of electrons.

parallax The apparent displacement of an object caused by the movement of the observer.

parallel circuit An electric circuit with two or more resistors arranged so that any one resistor completes the circuit independently of the others.

pascal The SI unit of pressure. One pascal is one newton per square meter.

periodic table (of the elements) An arrangement of the symbols of the elements that shows that their properties are periodic functions of their atomic numbers.

pH A measure of acidity. pH = −log (hydrogen ion concentration).

photochemical reaction A chemical reaction initiated by a photon.

photoelectric effect A process whereby a ray of light can produce an electric current. The photoelectric effect can be realized if high-frequency light is shined on a cathode constructed of a metal such as potassium or cesium.

photon The smallest burst, or packet, of electromagnetic energy.

photosphere The surface of the Sun visible to us here on Earth.

photosynthesis The process by which chlorophyll-bearing plants use energy from the Sun to convert carbon dioxide and water to sugars.

Planck's constant The constant that relates the frequency of a photon to its energy. $h = 6.63 \times 10^{-34}$ J $\cdot$ 5 s.

plasma A gas at such a high temperature that the electrons have been stripped from their atoms, resulting in a mixture of nuclei surrounded by rapidly moving electrons.

plate tectonics The theory stating that tectonic plates move about and collide with one another. See *continental drift*.

polar ice cap The permanent layer of ice that covers parts of the Arctic Ocean.

pollution The impairment of the quality of some portion of the environment by the addition of harmful substances.

positron A positive electron.

potential energy The energy posessed by an object that can be released sometime in the future. Gravitational potential energy is available when an object at some height has the potential to fall down to a lower level. Other forms of potential energy include chemical and nuclear energy.

power The amount of energy delivered in a given time interval.

ppm Abbreviation for parts per million.

precipitation (of water) All forms in which atmospheric moisture descends to Earth.

pressure Force per unit area.

principal quantum number A whole number, 1, 2, 3, . . . that characterizes an electron shell.

proton A fundamental particle of the atom that bears a unit of positive charge.

protoplanets The planets in their earliest, incipient stage of formation.

protosun The Sun in its earliest, incipient stage of formation. The protosun was a cold condensing agglomeration of dust and gas.

pulsar A neutron star that emits a pulsating radio signal.

quantum A small, discrete quantity. Specifically, a discrete quantity of energy; a photon.

quark A fundamental subatomic particle that, in various combinations, makes up other particles such as protons and neutrons.

quasar A region of space, less than one light-year in diameter and very distant from Earth, that emits extremely large quantities of energy.

radiation The process by which energy is emitted and transmitted as electromagnetic waves.

radio galaxy A galactic-sized collection of stars that emit large quantities of radio-frequency energy.

radioactivity The spontaneous emission of radiation by atomic nuclei.

radioactive dating Measuring the ages of objects by calculations from rates of radioactive decay.

radioisotope A radioactive isotope.

red giant A stage in the life of a star when the core is composed of helium that is not undergoing fusion. A hot shell of hydrogen around this core is fusing at a rapid rate, producing enough energy to cause the star to expand greatly and glow brightly.

red shift The frequency shift of light waves observed in the spectrum of an object traveling away from an observer. This shift is caused by the Doppler effect, which states that the observed frequency of a wave will decrease if the object is traveling away from an observer and increase if the object is approaching the observer.

reduction A chemical change in which oxygen is removed from a substance. More generally, reduction is a gain of electrons.

redundancy In the context of safety systems, redundancy refers to the provision of a series of devices that duplicate each other's functions and that are programmed to go into operation in sequence if a preceding device in the series fails.

reflection The phenomenon that occurs when waves bounce back from an object in their path.

refraction The change in direction of a wave motion as it moves from one transparent medium to another and strikes the second medium at an angle.

relative humidity See *humidity*.

relativistic speeds Speeds close to the speed of light.

residence time (for water) The average time that a water molecule spends in a particular region such as a lake or underground aquifer.

resonance A process whereby a periodic disturbance acts upon a second medium, thereby causing it to vibrate with the same frequency as the original disturbance.

respiration The process by which plants and animals combine oxygen with sugar and other organic matter to produce energy and maintain body functions. Carbon dioxide and water are released as by-products.

revolve To orbit a central point. A satellite revolves around the Earth. See also *rotate*.

rift A deep and very narrow split or crack in the Earth's crust.

rotate To turn or spin on an axis. A top rotates. See also *revolve*.

rotational motion The movement of an object around an axis of rotation.

runoff The flow of water toward the ocean through surface and underground pathways.

salt water intrusion The movement of salt water

from the ocean to terrestrial groundwater supplies that occurs when the water table in coastal areas is reduced.

saturation point   In meteorology, the maximum concentration of water vapor that can ordinarily exist in air at a given temperature.

sea breeze   A local wind caused by uneven heating of land and ocean surfaces.

Second law of thermodynamics   See *thermodynamics*.

sedimentary rock   Rock formed from compressed sediment.

sediments   Small particles of mineral and organic matter deposited by erosion.

seismograph   A device used to detect earthquakes.

seismology   The science of measuring and recording the shock waves of earthquakes.

semiconductor   A material that is a moderately effective conductor of electricity but whose conductivity can be sensitively controlled by regulating various factors. Semiconductors are used in the construction of transistors.

semimetal   A substance that combines metallic and nonmetallic properties.

series circuit   An electrical circuit with two or more resistors arranged so that the electric current travels through each one of them in turn.

shooting star   See *meteor*.

short circuit   A phenomenon that occurs when a circuit is completed with materials of low resistance only. Because the resistance is low, a great deal of current is allowed to flow, and large quantities of heat are generated.

SI   See *Systeme International d'Unites*.

sine wave   A smoothly oscillating symmetrical wave form. Mathematically it is defined as a wave described by the following equation: $y = $ sine x.

sinusoidal   Having the character of a sine wave.

solar cell   A semiconductor device that converts sunlight directly into electricity.

solar collector   A device used to collect solar energy and concentrate it for useful purposes such as space or water heating.

solar design (active)   A solar heating system in which a working substance (usually water or air) is actively pumped from a solar collector to some type of radiator within the building.

solar design (passive)   A series of design features used in building construction to capture solar heat without the use of mechanical collection or pumping.

solar eclipse   See *eclipse*.

solar wind   A stream of atomic particles shot out into space by violent storms occurring in the outer regions of the Sun's atmosphere.

solid (crystalline)   A rigid state of matter in which the atoms or molecules are arranged in an orderly pattern.

solstice   Either of two times per year when the Sun shines directly overhead farthest from the Equator. One solstice occurs on or about June 21 and marks the longest day of the year in the Northern Hemisphere and the shortest day in the Southern Hemisphere; the other solstice occurs on or about December 22, marking the longest day in the Southern Hemisphere and the shortest day in the Northern Hemisphere.

sonic boom   The sharp disturbance of air pressure caused by the reinforcing waves that trail an object moving at supersonic speed.

sonic speed   The speed of sound.

specific heat   The quantity of energy required to raise the temperature of 1 g of a substance 1°C.

spectrum (electromagnetic)   A pattern of wavelengths into which a beam of light or other electromagnetic radiation is separated. The spectrum is seen as colors, or is photographed, or is detected by an electronic device. An *emission spectrum* is obtained from radiation emitted from a source. An *absorption spectrum* is obtained after radiation from a source has passed through a substance that absorbs some of the wavelengths.

speed   The distance traveled by an object in a given time interval.

speed of light   The speed traveled by an electromagnetic wave in a vacuum. The speed of light is a universal constant: $2.998 \cdot 10^8$ m/s.

standard atmosphere   See *atmosphere (standard pressure)*.

standing wave   A wave formed by constructive interference of the original wave whose properties and position do not change with time.

steady state   A condition in which the inflow of material or energy is equal to the outflow.

stimulated emission   Emission of a photon caused by an electron transition that is stimulated by another photon of the proper frequency. Stimulated emission is the fundamental process used to create a laser beam.

stratosphere   A layer of air of fairly constant temperature that lies just above the troposphere.

strong force   The force that holds atomic nuclei together.

structural formula   A chemical formula that shows the sequences of atomic linkages.

subduction   A process whereby one continental plate is forced downward during a collision with another.

sublimation   Direct conversion of solid to vapor.

subsidence The settling of the surface of the ground as an ore, oil, or deep groundwater is removed.

subsonic speed Less than the speed of sound.

sunspot A cool region of the Sun formed by an intense magnetic disturbance. Sunspots are observed as dark blotches on the surface of the Sun.

superconductor A material that offers almost zero resistance to the flow of electric current. Superconductivity is shown by certain materials when they are cooled to temperatures close to absolute zero.

supernova A star that has collapsed under intense gravitation and then exploded, hurling matter into space and sometimes emitting as much energy as an entire galaxy.

supersonic speed Greater than the speed of sound.

synfuels An abbreviation for synthetic fuels. Any fuel that is manufactured by a chemical conversion from one type of fuel to another. The gasoline produced by conversion of coal or extraction of oil shale is a synfuel.

Système International d'Unités (SI) An outgrowth of the metric system of measurement used in all scientific circles and by lay people in most nations of the world. The base units in the SI are: length—meter; mass—kilogram; time—second; electric current—ampere; temperature—kelvin; luminous intensity—candela; and amount of substance—mole.

tectonic plate A large, continent-sized piece of the Earth's crust that may move and collide with other tectonic plates.

temperature A measure of the warmth or coldness of an object with reference to some standard. Temperature should not be confused with thermal energy. Thermal energy is the quantity of energy possessed by a body; the temperature is just a measure of how hot or cold it is.

terrestrial planets Mercury, Venus, Earth, and Mars; the four innermost planets of our Solar System that are all relatively small, dense, and rocky.

thermal energy The combined energy of motion of all the particles in a sample.

thermal pollution A change in the quality of an environment (usually an aquatic environment) caused by raising its temperature.

thermodynamics The science concerned with thermal energy and work and the relationships between them.

First law of thermodynamics Energy cannot be created or destroyed.

Second law of thermodynamics It is impossible to derive mechanical work from any portion of matter by cooling it below the temperature of the coldest surrounding object.

thermonuclear reaction A nuclear reaction, specifically fusion, initiated by a very high temperature.

thermosphere An extremely high and diffuse region of the atmosphere lying above the mesosphere.

tides The cyclic rise and fall of ocean water caused by the gravitational force of the Moon and, to a lesser extent, by the gravitational force of the Sun.

trade winds The winds that blow steadily from the northeast in the Northern Hemisphere and southeast in the Southern Hemisphere between 5 and 30 degrees north and south latitudes.

transformer A device that changes the magnitude of the voltage and current of an electric signal but does not by itself produce electric power.

transistor A semiconductor device used in most modern amplifier and switching circuits. The transistor acts as the heart of a circuit that amplifies a signal without changing its form.

translational motion The movement of an object such that the entire object travels from one place to another.

transpiration The vaporization of water through the tissues of plants, especially through leaf surfaces.

transmutation (of elements) The conversion of one element to another.

transverse wave See *wave*.

tritium Radioactive isotope of hydrogen with mass number of 3.

troposphere The layer of air that lies closest to the surface of the Earth and extends upward to about 12 km.

trough The lowest point in a wave.

uncertainty principle The theory that tells us that we cannot know both the position and the velocity of a particle with infinite accuracy.

valence The chemical combining capacity of an element.

valence shell The highest-energy electron shell of an atom, which houses the electrons usually involved in chemical changes.

vector quantity A quantity that has both magnitude and direction. Velocity, force, and acceleration are all vector quantites.

vein (of a rock) A thin layer of one type of rock embedded in a dominant rock formation.

velocity A description of the speed of a body and its direction of motion. Velocity is a vector quantity.

vibrational motion The movement of an object, or a portion of the object, back and forth without any permanent displacement away from a fixed position. The motion of a struck tuning fork is vibrational.

volcano A fissure in the Earth's crust through which lava, steam, and other substances are expelled.

volt A measure of the electric potential energy per unit charge. Voltage is a potential and must always be measured with respect to some other point.

wake A high-energy wave produced when an object such as a boat or an airplane moves through some medium such as water or air at a rate faster than the speed of the wave.

water table The upper level of water in the zone of saturated subsurface soil and rock.

watt The SI unit of power. A watt is a joule per second.

wave A periodic disturbance in some medium. A wave carries energy from one point to another, but there is no net movement of materials. A *longitudinal* or *elastic wave* is a wave that manifests itself as a series of compressions and expansions of an elastic medium. Sound waves are elastic waves. A *transverse wave* moves at right angles to the motion of the medium along which it travels. *Electromagnetic waves* are qualitatively different from all other waves. They can be propagated in a vacuum.

wavelength The distance between successive disturbances of the same type in a wave, such as between neighboring crests.

weather The temperature, wind, and precipitation conditions that prevail in a given region on a particular day. See also *climate*.

weathering The sum of processes that fracture and decompose surface rock.

weight The force of gravity acting on a body.

white dwarf A stage in the life of a star when fusion has halted and the star glows solely from the residual heat produced during past eras. White dwarfs are very small stars.

work The energy expended when something is forced to move. Work is defined as the force exerted on an object multiplied by the distance that the object is forced to travel.

zone of saturation The subsurface zone of soil and rock that is completely saturated with water.

INDEX